Erratum notice

With this notice we would like to draw to your attention an error that was made in the production of **Faraday Discussions volume: 168 Astrochemistry of Dust, Ice and Gas.**

Unfortunately, an error occurred during the production process which meant that the four discussion sections appear out of order in the online and print issues and therefore do not appear directly after the papers which they discuss.

The discussion section that starts on page 129 is correct and follows the papers it discusses in section 1 (pages 9-128).

The discussion that starts on page 287 refers to the papers that follow in section 3 (pages 313 – 422).

The discussion section that starts on page 423 refers to the papers that follow in section 4 (pages 449-570.)

The discussion that starts on page 571 refers to the papers in section 2 (pages 151-286)

We would like to apologise to our authors and readers for this error and for any inconvenience or confusion this may have caused.

Faraday
Discussions

Astrochemistry of Dust, Ice and Gas

Leiden University, The Netherlands
7–9th April 2014

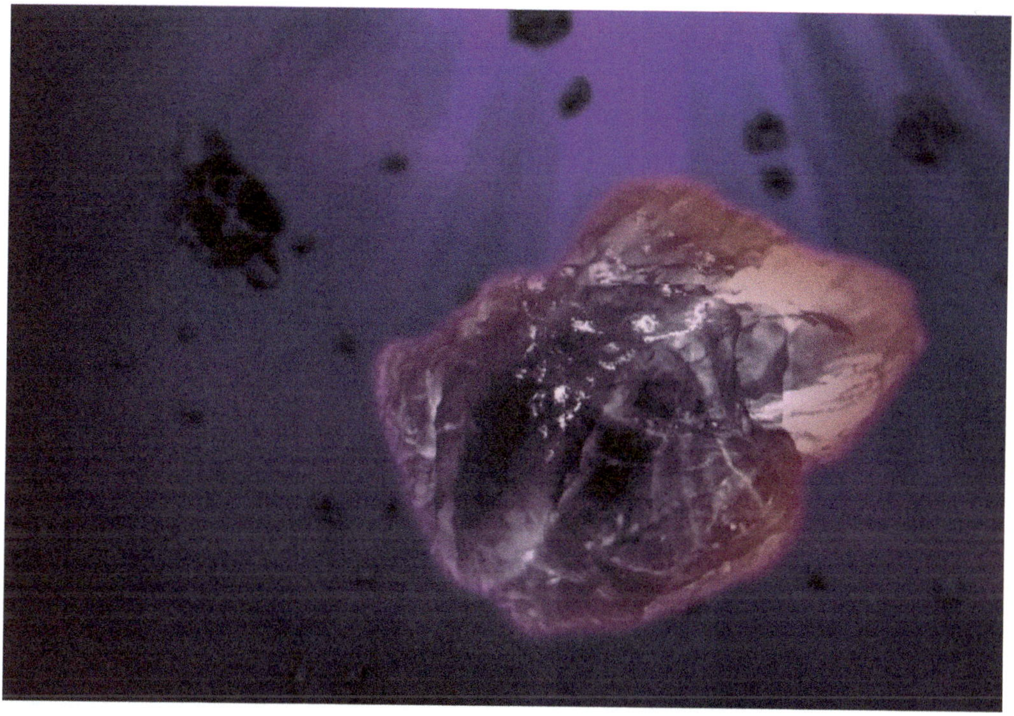

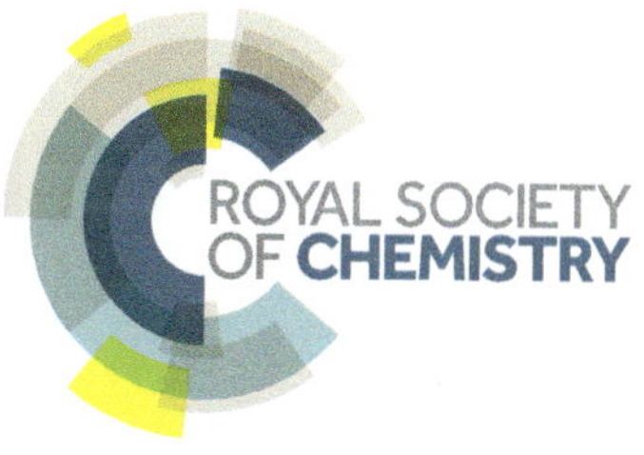

FARADAY DISCUSSIONS

Volume 168, 2014

ROYAL SOCIETY OF CHEMISTRY

The Faraday Division of the Royal Society of Chemistry, previously the Faraday Society, founded in 1903 to promote the study of sciences lying between Chemistry, Physics and Biology.

Editorial staff

Executive Editor
Robert Eagling

Deputy editor
Heather Montgomery

Development editor
Alessia Millemaggi

Editorial Production Manager
Philippa Ross

Publishing editors
William Bergius, Catherine Brown

Publishing assistants
Victoria Bache, Bethany Johnson, Kate McCallum, Ruba Miah

Faraday Discussions (Print ISSN 1359-6640, Electronic ISSN 1364-5498) is published 8 times a year by the Royal Society of Chemistry, Thomas Graham House, Science Park, Milton Road, Cambridge, UK CB4 0WF.

Volume 168 ISBN-13: 978-1-78262-128-7

2014 annual subscription price: print+electronic £827, US $1542; electronic only £785, US $1465. Customers in Canada will be subject to a surcharge to cover GST. Customers in the EU subscribing to the electronic version only will be charged VAT. All orders, with cheques made payable to the Royal Society of Chemistry, should be sent to RSC Order Department, Royal Society of Chemistry, Thomas Graham House, Science Park, Milton Road, Cambridge, CB4 0WF, UK. Tel +44 (0) 1223 432398; E-mail **www.orders@rsc.org**

If you take an institutional subscription to any RSC journal you are entitled to free, site-wide web access to that journal. You can arrange access via Internet Protocol (IP) address at **www.rsc.org/ip**. Customers should make payments by cheque in sterling payable on a UK clearing bank or in US dollars payable on a US clearing bank.

Printed in the UK

Faraday Discussions documents a long-established series of *Faraday Discussion* meetings which provide a unique international forum for the exchange of views and newly acquired results in developing areas of physical chemistry, biophysical chemistry and chemical physics.

Scientific committee
volume 168

Chair

Professor Martin McCoustra (Heriot-Watt University, UK)

Professor Wendy Brown (University College London, UK)
Dr Robin Garrod (Cornell University, USA)
Professor Harold Linnartz (Leiden University, the Netherlands)
Dr Anthony Meijer (University of Sheffield, UK)
Professor Tom Millar (Queens University Belfast, UK)

Faraday standing committe on conferences

Chair

A Mount (Edinburgh, UK)

Secretary
A Becker (RSC, UK)

I Hamley (Reading, UK)
F Manby (Bristol, UK)
G Hutchings (Cardiff, UK)
K Reid (Nottingham, UK)
C Percival (Manchester, UK)

Astrochemistry of Dust, Ice and Gas

Faraday Discussions

www.rsc.org/faraday_d

A General Discussion on the Astrochemistry of Dust, Ice and Gas was held in Leiden, the Netherlands on the 7th, 8th and 9th April 2014.

RSC Publishing is a not-for-profit publisher and a division of the Royal Society of Chemistry. Any surplus made is used to support charitable activities aimed at advancing the chemical sciences. Full details are available from www.rsc.org

CONTENTS

ISSN 1359-6640;   ISBN 978-1-78262-128-7

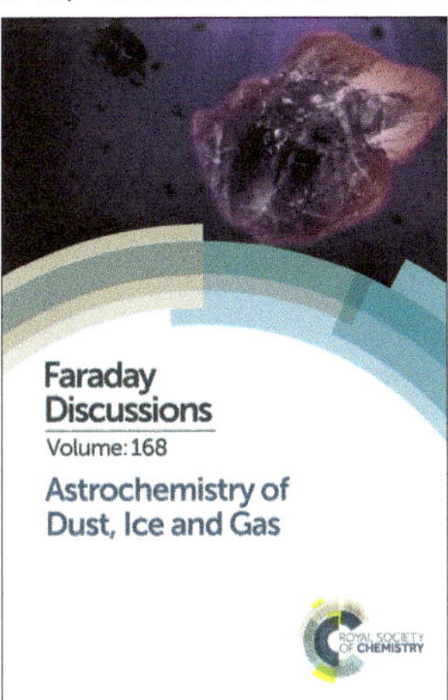

Cover
Image courtesy of the Sackler Laboratory for Astrophysics, University of Leiden, the Netherlands.

INTRODUCTORY LECTURE

PAPERS AND DISCUSSIONS

CONCLUDING REMARKS

ADDITIONAL INFORMATION

PAPER

Astrochemistry of dust, ice and gas: introduction and overview

Ewine F. van Dishoeck[ab]

Received 28th June 2014, Accepted 30th June 2014
DOI: 10.1039/c4fd00140k

A brief introduction and overview of the astrochemistry of dust, ice and gas and their interplay is presented. The importance of basic chemical physics studies of critical reactions is illustrated through a number of recent examples. Such studies have also triggered new insight into chemistry, illustrating how astronomy and chemistry can enhance each other. Much of the chemistry in star- and planet-forming regions is now thought to be driven by gas–grain chemistry rather than pure gas-phase chemistry, and a critical discussion of the state of such models is given. Recent developments in studies of diffuse clouds and PDRs, cold dense clouds, hot cores, protoplanetary disks and exoplanetary atmospheres are summarized, both for simple and more complex molecules, with links to papers presented in this volume. In spite of many lingering uncertainties, the future of astrochemistry is bright: new observational facilities promise major advances in our understanding of the journey of gas, ice and dust from clouds to planets.

1 Introduction

The space between the stars is not empty, but is filled with a very dilute gas: the interstellar medium (ISM). The ISM is far from homogeneous and contains gas with temperatures ranging from more than 10^6 K down to 10 K, and densities from as low as 10^{-4} particles cm^{-3} to more than 10^8 cm^{-3}. The colder and denser concentrations of the gas are called interstellar clouds, and this is where molecules, dust and ice are detected. Even at the upper range of densities, interstellar clouds are still more tenuous than a typical ultra-high vacuum laboratory experiment on Earth. Thus, interstellar space provides a unique environment in which chemistry can be studied under extreme conditions. From an astronomical perspective, dense clouds are also important because they are the nurseries of new generations of stars like our Sun and planets like Jupiter or Earth. This combination makes astrochemistry such a fascinating research field, for both chemists and astronomers alike.

[a]Leiden Observatory, Leiden University, P.O. Box 9513, 2300 RA Leiden, the Netherlands. E-mail: ewine@strw.leidenuniv.nl

[b]Max-Planck-Institute für Extraterrestrische Physik, Giessenbachstrasse 1, Garching, 85748, Germany

Traditional chemistry would predict that virtually no molecules are formed at typical densities of 10^4 cm^{-3} and temperatures of 10 K of dense molecular clouds, with 1000 times more hydrogen than any other chemically interesting element. The detection of nearly 180 different species over the past 45 years (not counting isotopologs) demonstrates the opposite: there is a very rich chemistry in space. Molecules like CO and H_2O have even been detected in distant galaxies out to high redshifts, when the Universe was less than 1 Gyear old.[1] Large Polycyclic Aromatic Hydrocarbon molecules (PAHs) are also detected across the Universe.[2] Closer to home, an increasing variety of molecules (no longer just the simplest ones) are found in luminous galaxies undergoing bursts of star formation at a rate up to 100 times higher than that in our own Milky Way.[3,4] In star-forming clouds within our Galaxy, a myriad of species has been found, some of which are prebiotic, *i.e.*, molecules that are thought to be involved in the processes leading to living organisms. Fullerenes such as C_{60} have also been identified.[5] Finally, simple molecules including H_2O are being detected in the atmospheres of giant exo-planets.[6,7]

Astrochemistry, also known as molecular astrophysics, is 'the study of the formation, destruction and excitation of molecules in astronomical environments and their influence on the structure, dynamics and evolution of astronomical objects' as stated by Dalgarno.[8] This definition includes not only the chemical aspects of the field, but also the fact that molecules are excellent diagnostics of the physical conditions and processes in the regions where they reside. This sensitivity stems from the fact that both the excitation and abundances of molecules are determined by collisions, which in turn are sensitive to gas temperature and density as well as to the radiation from nearby stars.

The main questions in the field of astrochemistry therefore include: how, when and where are these molecules produced and excited? How far does this chemical complexity go? How are they cycled through the various phases of stellar evolution, from birth to death? And, most far-reaching, can they become part of new planetary systems and form the building blocks for life elsewhere in the Universe?

The topic of this symposium — dust, ice and gas — is at the heart of these questions. This brief overview is aimed at providing background information for non specialists on the conditions and ingredients in interstellar space (Section 2), the tools that are used in astrochemistry (telescopes, laboratory experiments, computer simulations) (Section 3), the main characters of this symposium (dust, ice and gas) (Section 4), the 'play' (chemical processes) (Section 5), and the 'plot' (evolution from clouds to planets) (Sections 6–11), followed by some concluding remarks (Section 12). There has been no shortage of excellent and much more detailed (and partly overlapping) reviews of the field over the last few years, see ref. 9–14 for further information.

Throughout this talk the interdisciplinary aspect of this research will be emphasized as a two-way street: astrochemistry needs basic data on molecular spectroscopy and chemical processes, but also inspires new chemical physics through studies of different classes of molecules and reactions that are normally not considered on Earth. Indeed, astrochemistry is a 'blending of astronomy and chemistry in which each area enriches the other in a mutually stimulating interaction'.[8]

Fig. 1 Blow-up of the Hubble space telescope optical image of the Carina nebula, showing dark molecular clouds as well as more diffuse ionized gas between the stars. The clouds are dark at visible wavelengths due to extinction by dust grains; young stars are embedded within them. In the colder regions, the dust particles are covered by ice mantles. Gas-phase molecules are present throughout the dark clouds. The distance to the cloud is 2.3 kpc, and the image covers a region of 0.94 pc in size. Colors: forbidden transitions of [O III] (blue), [N II] (green), and [S II] (red), together with hydrogen Hα (green). Credit: NASA/ESA/M. Livio.

2 The setting: interstellar clouds

2.1 Images of clouds

Fig. 1 presents an Hubble space telescope image of a small part of the Carina star-forming cloud. The colorful nebulae are due to ionized gas at 10^4 K that emits brightly at optical wavelengths and consists of atomic hydrogen recombination lines (primarily Hα) and forbidden electronic transitions of atoms such as [O III], [N II] and [S II]. The dark areas are the dense molecular clouds which contain $\sim$0.1 μm-sized particles of dust composed of silicates and carbonaceous compounds which absorb and scatter the light. They also shield molecules from the dissociating ultraviolet radiation emitted by nearby stars. The dense cloud complexes can be quite large (tens of parsecs† across) and massive (up to 10^5 solar masses‡) but the process of star formation is quite inefficient and only a few % of the mass of the cloud is turned into stars.

The UV radiation absorbed by dust grains is converted to heat, raising their temperatures from $\sim$10 K in the most shielded regions to $\sim$50 K in more exposed

---

† 1 parsec (pc) $= 3.1 \times 10^{18}$ cm $= 3.26$ light year.

‡ 1 $M = 2.0 \times 10^{33}$ g.

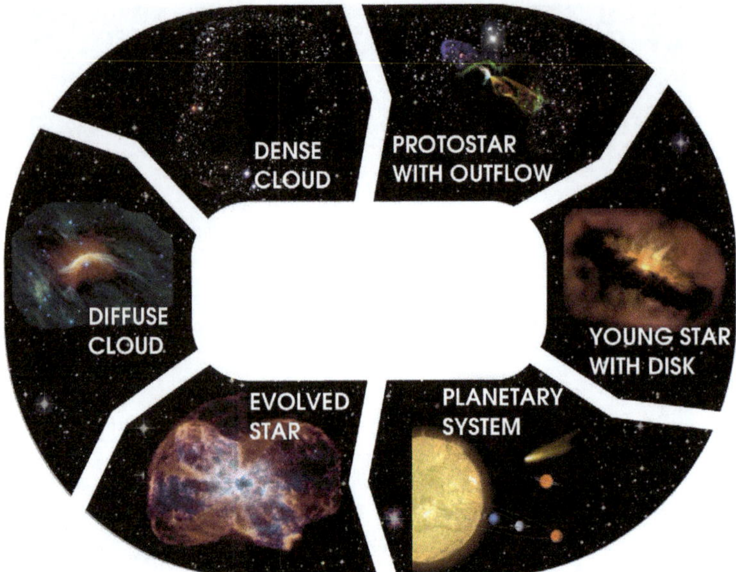

Fig. 2   Various stages in the lifecycle of gas, dust and ice in interstellar and circumstellar clouds. Credit inserts: NASA/ESA/ESO/NRAO.

gas. This heat is then radiated away through thermal emission in the far-infrared part of the spectrum as observed for example with the *Herschel Space Observatory*.

2.2   Birth and death of stars

Fig. 2 illustrates the cycle of material from clouds to stars and planets, and ultimately back to the interstellar medium. Diffuse clouds are low-density clouds ($\sim$100 cm^{-3}) in which UV radiation can penetrate and destroy molecules. They are usually transient structures in the interstellar medium. Dense molecular clouds can be stable for millions of years due to support by turbulence and magnetic fields, but eventually gravity takes over and the densest part of the cloud collapses to form a new star. In the standard scenario, the collapse occurs inside out so the protostar at the center of the cloud continues to grow as material from the envelope accretes onto the star. Because of angular momentum conservation, the material that falls in at later times ends up in a disk around the star where the gas is in Keplerian rotation. Further accretion of material takes place in the innermost part of the disk in magneto-hydrodynamically (MHD) mediated funnel flows onto the star. The MHD processes near the disk-star boundary also result in jets and winds which can escape in a direction perpendicular to the disk. When they interact with the surrounding envelope and cloud, they create shocks and entrain material in bipolar outflows.

With time ($\sim$1 Myear $= 10^6$ years), the opening angle of the wind increases and the envelope is gradually dispersed, revealing a young pre-main sequence star surrounded by a so-called protoplanetary disk. These disks are about the same size as the solar system ($\sim$100 AU§) and contain a few Jupiter masses of gas and

---

§ 1 AU $=$ distance Sun–Earth $= 1.5 \times 10^{13}$ cm.

dust ($1\ M_{\text{Jup}} = 1.9 \times 10^{30}$ g or about 0.1% of the mass of the Sun). Here densities can be as high as 10^{13} cm^{-3} in the inner regions of disks. The increased collision rates lead to gradual coagulation of dust grains to form pebbles, rocks and planetesimals, although the precise mechanisms are not yet understood.[15] The large particles settle to the midplane, where they can form kilometer-sized objects that interact gravitationally to form (proto)planets and eventually a full planetary system (up to 100 Myears). Comets and asteroids are remnant planetesimals that did not end up in one of the planets and were scattered and preserved in the cold outer regions of our Solar system. If they have not been heated during their lifetime, their composition reflects the conditions during the solar system formation.

The young star eventually becomes hot enough for nuclear fusion of hydrogen to ignite, at which stage it enters the stable main-sequence phase for several Gyears. Photospheric temperatures of low-mass stars are typically $T_* = 3000\text{--}5000$ K so they emit most of their radiation at visible (as our Sun) and near-infrared wavelengths. High-mass stars have T_* up to 40 000 K so their Planck spectra peak at far-ultraviolet wavelengths, which are much more effective at destroying molecules.

At the end of the stellar life cycle, the nuclear fuel is exhausted, causing the low-mass stars to become larger and lose part of their mass (Fig. 2). These evolved Asymptotic Giant Branch (AGB) stars are surrounded by circumstellar envelopes, *i.e.*, dense shells of molecular material driven by radiation pressure. Temperatures are high, 2000–3000 K near the stellar photosphere, dropping with radius down to 10 K at the outer edge. Because of the dense and warm conditions, they are a rich source of molecules. Such conditions are also favorable for the condensation of dust grains. Once the envelopes have been dissipated, the central star becomes very hot and emits copious UV radiation, illuminating the surrounding remnant gas producing a so-called 'planetary nebula' (not to be confused with a (proto)planetary disk). Massive stars explode after a much shorter lifetime ($\sim$10 Myears) as supernovae with their cores ending up as black holes or neutron stars.

2.3　UV radiation and cosmic rays

The UV radiation impinging on an interstellar cloud is often approximated by a scaling factor called G_0 or I_{UV} with respect to the average radiation produced by all stars in the solar neighborhood from all directions. This interstellar radiation field (ISRF) I_0 has been estimated by Habing[16] and Draine[17] to have an intensity of about 10^8 photons cm^{-2} s^{-1}, with a relatively flat spectrum between the threshold at 912 Å and 2000 Å. Any photon with energies greater than 13.6 eV (wavelength <912 Å) is absorbed by atomic hydrogen outside the cloud and therefore does not affect the chemistry.

Dust particles absorb UV radiation and thereby shield molecules deeper into the cloud from the harshest radiation. The extinction at visual wavelengths, A_{V}, is defined as $1.086 \times \tau_{\text{d}}$ at 5500 Å, where τ_{d} is the optical depth of the dust. The intensity decreases as $I_{5500} = I_0 10^{-0.4 A_{\text{V}}}$ with depth into a cloud. At UV wavelengths, the decline is much steeper but depends on the grain properties such as size, composition, shape and scattering characteristics.[18]

Table 1 Solar elemental abundances

Element	Abundance	Element	Abundance
H	1.00	Mg	4.0×10^{-5}
He	0.085	Al	2.8×10^{-6}
C	2.7×10^{-4}	Si	3.2×10^{-5}
N	6.8×10^{-5}	S	1.3×10^{-5}
O	4.9×10^{-4}	P	2.6×10^{-7}
Na	1.7×10^{-6}	Fe	3.2×10^{-5}

Cosmic ray particles are another important ingredient of interstellar clouds. These are highly energetic atomic nuclei with >MeV energies. Cosmic rays penetrate much deeper into clouds than UV radiation and ionize a small fraction of atomic and molecular hydrogen needed to kick-start the gas-phase chemistry.

Cosmic rays also maintain a low level of UV radiation deep inside dense clouds:[19] the ionization of H and H_2 produces energetic secondary electrons which excite H_2 into the $B^1\Sigma_u^+$ and $C^1\Pi_u$ electronic states, which subsequently decay through spontaneous emission in the Lyman and Werner bands. The resulting UV spectrum consists of discrete lines and a weak continuum in the 900–1700 Å range.[20] The flux of internally generated UV photons is typically 10^4 photons cm^{-2} s^{-1} but depends on the energy distribution of the cosmic rays (see Fig. 4 of Shen et al.[21]) and the grain properties.

Interstellar clouds are largely neutral, since hydrogen-ionizing photons have been absorbed in the more diffuse surrounding gas. Of the major elements, only carbon can be ionized by the ISRF because its first ionization potential is less than 13.6 eV. Since the abundance of gas-phase carbon with respect to hydrogen is about 10^{-4} (see Table 1), this sets the maximum electron fraction in the cloud to about 10^{-4}. With depth into the cloud, the ionized carbon is converted into neutral atomic and molecular form. Around $A_V = 5$ mag, cosmic rays take over as the main ionizing agent at a rate denoted by ζ in s^{-1}. The resulting ionization fraction depends on the detailed chemistry and grain physics but is typically 10^{-7} or lower, and scales as $(n/\zeta)^{-1/2}$.

3   The tools: telescopes, laboratory

3.1   Importance of multi-wavelength observations

The energy levels of a molecule are quantized into electronic, vibration and rotation states, with decreasing energy difference between two neighboring levels of the same type. Electronic transitions typically occur at optical and UV wavelengths, whereas those between two vibrational levels within the same electronic state take place at infrared wavelengths. Rotational transitions within a given electronic and vibrational state are found at (sub)millimeter and far-infrared wavelengths.

The advantages of optical and UV spectroscopy are that a number of key species, most notably H_2 and atoms, can be observed directly. The oscillator strengths of the transitions involved are large, so even minor species can be detected. The drawback is that only diffuse clouds and translucent clouds with less than a few mag of extinction can be observed along the line of sight toward a background source, since short wavelength photons do not penetrate thicker clouds.

The main advantage of infrared spectroscopy is that not just gases but also solids can be observed, both in absorption and emission. For the simplest case of a harmonic oscillator, the energy levels are given by $\omega_e(v + 1/2)$ where ω_e is the vibrational frequency and v the vibrational quantum number. Note the zero-point vibrational energy for $v = 0$, which has important chemical consequences in cold dark clouds. The strongest vibrational bands of ices, silicates, oxides and PAHs occur at mid- and far-infrared wavelengths.[22] Some important gas-phase molecules without a permanent dipole moment, such as H_3^+, CH_4, C_2H_2 and CO_2, are also only observed through their vibrational bands. Disadvantages are relatively low spectral resolving power (up to $R = \lambda/\Delta\lambda = 10^5$) and moderate oscillator strengths, which means that in practice only a handful of molecules are detected at infrared wavelengths.

The bulk of the interstellar molecules have been discovered at millimeter (mm) wavelengths. For the simplest case of a linear molecule, the rotational energy levels scale as $BJ(J + 1)$ where J is the rotational quantum number and B the rotational constant, which is inversely proportional to the reduced mass of the molecule. Thus, the transitions of light molecules such as hydrides occur at higher frequencies (THz) than those of heavy molecules. The lowest levels have energies E_u/k_B (with $k_B =$ Boltzmann constant) that range from a few to tens of K, and are thus readily excited at typical dense cloud conditions. The advantage of mm observations is high sensitivity to low abundance molecules (down to 10^{-11} with respect to hydrogen) and the fact that the emission can be mapped so that one is not limited to a single line of sight.

3.2   Telescopes

Progress in astronomy is very much driven by large telescopes equipped with highly sensitive detectors that allow investigation of clouds at a variety of wavelengths across the electromagnetic spectrum, each of which tells a different part of the story. In the last decade astrochemistry has been very fortunate to have had access to a number of new powerful telescopes with both imaging and high resolution spectroscopic instruments, the latter crucial for observing molecules. In particular, the *Herschel Space Observatory* was the largest astronomical telescope in space (3.5 m diameter) operative from mid-2009 to mid-2013. Specifically, the HIFI instrument provided very high resolution (R up to 10^7) heterodyne spectroscopy covering the 490–1250 GHz (600–240 μm) and 1410–1910 GHz (210–157 μm) bands for a single pixel on the sky.[23] At mid-infrared wavelengths, the *Spitzer Space Telescope* had a low-resolution ($R = 50$–600) spectrometer at 5–40 μm from 2003–2009, which was particularly powerful to study dust features and ices in large numbers of sources due to its high sensitivity, following on the pioneering results from the *Infrared Space Observatory* (ISO).[22]

On the ground there have also been significant advances. The European Southern Observatory (ESO) Very Large Telescope (VLT) and the Keck and Subaru Observatories provide the most powerful collection of 8–10 m optical-infrared telescopes on Earth. At millimeter wavelengths the single-dish 10 m Caltech Submillimeter Observatory, 15 m James Clerk Maxwell Telescope, the 45 m Nobeyama dish, the IRAM 30 m telescope in Spain and the 12 m Atacama

Pathfinder Experiment (APEX) have opened up the (sub)millimeter regime and have been workhorses for astrochemistry over the past 30 years. To obtain higher spatial resolution, pioneering millimeter interferometers such as the IRAM Plateau de Bure and the CARMA array have provided a first glimpse. Astronomers usually express spatial resolution and scales in arcsec ($''$).¶ The *Atacama Large Millimeter/submillimeter Array* (ALMA), located in north Chile on the Chajnantor plateau at an altitude of 5000 meters, became operational in 2011 and will be transformational for the field. It consists of 54×12 m and 12×7 m antennas which together can form images with a sharpness ranging from $\sim 0.01''$ to a few $''$ over the frequency range from 84 to 900 GHz, and with a sensitivity up to two orders of magnitude higher than any previous instrument.

3.3 Laboratory experiments

The most basic information that astronomers need from laboratory experiments is spectroscopy from UV to millimeter wavelengths, to identify the sharp lines and broad bands observed toward astronomical sources. Techniques range from classical absorption set-ups to the use of cavity ringdown spectroscopy to increase the sensitivity by orders of magnitude.[24] Transition frequencies and strengths of (sub)mm transitions are summarized in the Jet Propulsion Laboratory catalog (JPL)[25] and the Cologne Database for Molecular Spectroscopy (CDMS).[26,27]‖ Databases for vibrational transitions at infrared wavelengths include the HITRAN database[28] and the EXOMOL line lists.**

The spectroscopy of large samples of PAHs has been determined with matrix-isolation studies, with selected PAHs also measured in the gas phase.[29] Spectroscopic databases of solids continue to grow and include the Heidelberg–Jena–St. Petersburg database of optical constants for silicates,[30] various databases for ices[31-34] and carbonaceous material.[35,36] THz spectroscopy of ices is just starting, as reported by Ioppolo *et al.* (DOI: 10.1039/c3fd00154g) in this volume.

The next step is to determine the rates for the various reactions that are expected to form and destroy molecules under space conditions. Laboratory experiments for gas-phase processes have recently been summarized by Smith.[37] Developments include measurements and theory of gaseous neutral–neutral rate coefficients at low temperatures using techniques such as CRESU[38] or crossed molecular beams,[39] branching ratios for dissociative recombination,[40] and rates for photodissociation of molecules exposed to different radiation fields.[41]

For ices, the combination of observations and associated modeling have stimulated the new field of solid-state laboratory astrochemistry, in which modern surface science techniques at ultra-high vacuum conditions are used to quantitatively study various chemical processes. A recent summary of activities across the world is given by Allodi *et al.*[42] Temperature programmed desorption experiments now routinely probe thermal desorption of both pure and mixed ices, providing binding energies that can be used in models.[43-45] Surface chemistry experiments have demonstrated that solid CH_3OH (ref. 46–48) and H_2O (ref.

¶ One arcsec ($''$) is 1/3600 of a degree or 1/206 265 of a radian ($\pi/180/3600$). At a distance of $n \times 100$ pc, 1 arcsec corresponds to $n \times 100$ AU diameter, or $n \times 100\times$ the Sun–Earth distance.

‖ http://spec.jpl.nasa.gov and http://www.astro.uni-koeln.de/cdms/catalog.

** http://www.cfa.harvard.edu/hitran and http://www.exomol.com.

49–53) can indeed form at temperatures as low as 10 K, using beams of atomic H to bombard solid CO, O, O_2 and O_3, thereby confirming routes that were postulated more than 30 years ago by Tielens and Hagen.[54] The process of photodesorption has been measured quantitatively and found to be much more efficient than previously thought,[55,56] and to depend strongly on the wavelength of the incident radiation field.[57,58] Details of the formation of complex organic molecules by UV irradiation and by high-energy particle bombardment[59-64] continue to be elucidated by a variety of experiments. Finally, the techniques to analyze meteoritic and cometary material in the laboratory have improved enormously in the last decade, and now allow studies of samples on submicrometer scale using techniques such as ultra-L^2MS, nano-SIMS and NMR.[65]

3.4   Theory and computation

Theoretical studies can be as important as laboratory experiments in providing information on astrochemically relevant questions. The most important example is that of collisional rate coefficients, needed to determine the excitation of interstellar molecules. Although laboratory experiments can provide some selected data, only theory can provide the thousands of state-to-state cross sections as functions of temperature that are needed in astronomy. The process consists of two steps. First, *ab initio* quantum chemical methods are used to compute the multi-dimensional potential energy surface of the molecule and its collider (usually H_2). This surface is then fitted to a convenient functional or numerical form for use in the second step; the dynamics of the nuclei. Inelastic scattering calculations need to be computed over a large range of collision energies. Full quantum calculations are used at the lowest energies whereas quasi-classical methods are often employed at the highest energies. A large number of systems have been studied over the past two decades, as summarized in a number of reviews.[66-68]

Theoretical studies are also important for computing cross sections and rates for other processes. A prime example is photodissociation rates of small radicals and ions through calculation of the excited electronic states and their transition dipole moments.[69-71] Other examples include radiative association reactions,[72] rates and barriers for neutral–neutral reactions,[73,74] state-selective reactions,[75] and the structure[76] and infrared spectroscopy[77] of large molecules.

Molecular dynamics techniques can also be extended to solid-state processes, as demonstrated by the detailed study of the photodissociation and photodesorption of water ice and its isotopologs.[78-80] Another recent example is the attempt to model the formation of solid CO_2 ice.[81]

4   The main players: dust, ice and gas

4.1   The ingredients: elemental abundances

The astronomer's view of the periodic table is fairly restricted. The local Universe consists primarily of 90% hydrogen with about 8% helium by number. All other elements are called 'metals' by astronomers, even if they are obviously not a metal in a chemical sense. The next most abundant elements are oxygen, carbon, and nitrogen at abundances of only about 4.9, 2.7 and 0.7×10^{-4} that of hydrogen. The abundances summarized in Table 1 are those derived for the solar photosphere by Asplund *et al.*[82] These abundances are thought to apply to the interstellar medium as well, although small variations are possible.[83]

4.2   Interstellar dust

The size, composition, shape and other properties of interstellar dust have been studied through a wide variety of observational techniques, such as extinction, reflection and emission measurements, including polarization techniques, as well as spectroscopy (see reviews and books by Draine[84,85] and Tielens[86]). It is clear that the bulk of the dust grains consist of a mix of amorphous silicates and carbonaceous material, which locks up nearly 100% of the Si, Mg and Fe, $\sim$30% of the oxygen and about 70% of the available carbon. These materials are called 'refractory' since they do not vaporize until temperatures well above 1200 K. They are therefore not available for the 'volatile' gas or ice chemistry. Other types of solids such as iron oxides, carbides, sulfides or metallic iron are also present, but are at most a minor component.

Interstellar grains have a distribution of sizes a following roughly $a^{-3.5}$. This means that most of the surface area for chemistry is in grains that are smaller than the typical 0.1 μm-sized grains that dominate the mass. These smaller grains, down to 0.001 μm or less, dictate the absorption and scattering of UV radiation. In dense cores and circumstellar disks, grain growth up to sizes of a few μm to a few cm has been found.[87,88] At low temperatures, grains in dense cores are covered with a thick ice mantle consisting primarily of H_2O ice. If the bulk of the 'volatile' oxygen is locked up in ice, a typical 0.1 μm-sized silicate grain core is surrounded by about 100 monolayers of water ice.

Polarization observations show that interstellar grains are not spherical but elongated. Whether or not they have a highly irregular and/or porous structure, such as seen in the larger Interplanetary Dust Particles (see Wikipedia for images), is still under debate. The bulk of the surface sites are suited for physisorption of atoms and molecules from the gas, but there are likely also some chemisorption sites available. Indeed, Jones *et al.*[89] (see also this volume) has suggested that grains in diffuse clouds are coated by a layer of amorphous hydrocarbon material, a-C(:H), which will change the surface site properties.

More generally, interstellar grains are not thought to have an active role as catalysts in the chemical sense, in which surface molecules participate in promoting the reaction. Rather, they provide a reservoir where atoms and molecules from the gas can be stored and brought together for a much longer period than possible in the gas. They thus enable reactions with activation barriers that are too slow in the gas such as the hydrogenation of atomic O, C and N. Also, they act as a third body that absorbs the binding energy of a newly formed molecule, thereby stabilizing it before it can dissociate again.

4.3   Interstellar ices

Ices are primarily observed in absorption against bright mid-infrared sources, either embedded in the cloud or behind it (Fig. 3). The Kuiper Airborne Observatory and especially ISO opened up the field of infrared spectroscopy unhindered by the Earth's atmosphere, allowing the first full inventory of interstellar ices. The dominant ice species are simple molecules, H_2O, CO, CO_2, CH_4, NH_3 and CH_3OH, whose presence has been firmly identified thanks to comparison with laboratory

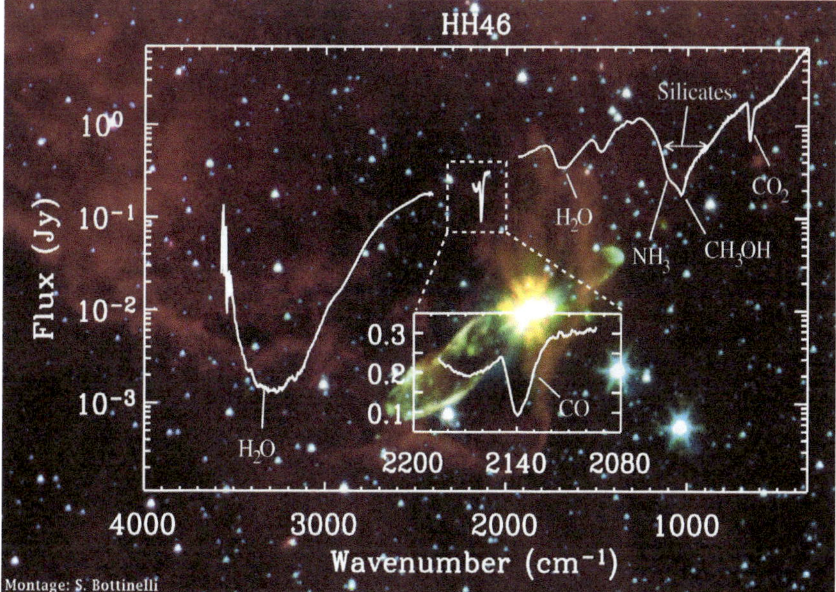

Fig. 3 Detection of ices toward the low-mass protostar HH46 IRS using *Spitzer* data at 5–20 μm and VLT-ISAAC data at 2–5 μm.[90,91] The strong solid CO_2 stretching band at 4.3 μm is missing since it cannot be observed from the ground. The insert shows a blow-up of the strong solid CO band; the weaker feature at 2167 cm⁻¹ is due to OCN⁻. Background: *Spitzer* composite 3 (blue, stars), 4.5 (green, shocked H_2), and 8 μm (red, PAHs) image, showing the embedded protostar with its outflow. Credit NASA/ESA/A. Noriego-Crespo.

spectroscopy.[31,32] These molecules are precisely the species predicted to be produced by hydrogenation and oxidation of the dominant atoms (O, C and N) and molecules (CO) arriving from the gas on the grains at low temperatures.[54] One ion, OCN⁻, has also been convincingly detected along many lines of sight.[92,93] More recent *Spitzer* surveys have extended such studies to large samples of low-mass YSOs (Fig. 3). An important conclusion is that these sources have a remarkably similar overall composition as their high-mass counterparts within factors of two.[94] This implies that the formation mechanism of these 'zeroth generation' ice species is robust and universal across the Galaxy.

Interstellar ices are known to have a layered structure surrounding the silicate or carbonaceous cores: a water-rich and a water-poor layer (also called 'polar' and 'apolar' or 'non-polar' ices) (see Fig. 4). The observational evidence for different layers comes from the shapes of various ice bands, in particular the high-quality line profiles of solid CO:[95,96] a CO molecule surrounded by an H_2O molecule has a slightly different vibrational constant than a CO molecule embedded in CO itself, and these differences can be readily distinguished. The water-rich layer is thought to form early in the evolution of a cloud by hydrogenation of atomic O, once the extinction is a few mag.[97] The bulk of solid CH_4 and NH_3 is likely also formed at this stage. In contrast, the water-poor, CO-rich layer is observed to form in much denser gas (typically $>10^5$ cm^{-3}) when the freeze-out time, $\tau_{fo} \approx 2 \times 10^9/n_H$ years, has become so short compared with the lifetime of the core that the bulk of the heavy elements are removed from the gas 'catastrophically'.[98–100]

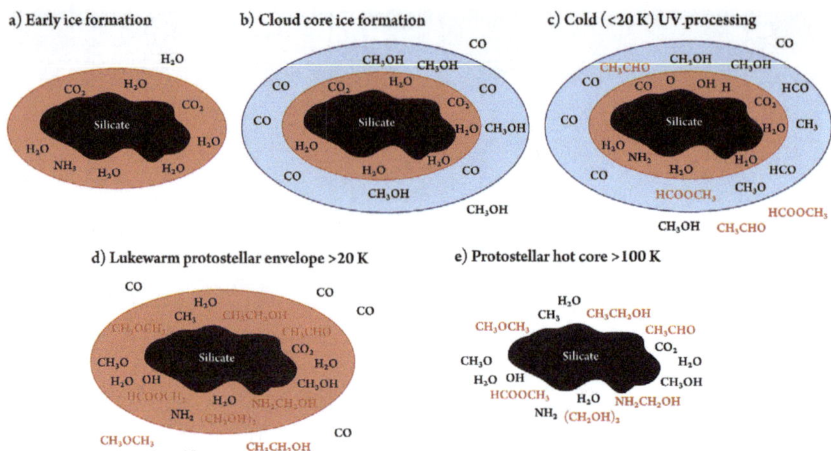

Fig. 4 Proposed evolution of ices during star formation and formation of complex molecules. Pink indicates an H_2O-dominated ('polar') ice and blue a CO-dominated ('non-polar') ice. Early during cloud formation (a) an H_2O-rich ice forms, with minor amounts of CH_4 and NH_3 through hydrogenation of O, C and N. Once a critical density is reached CO freezes out catastrophically (b), providing reactants for CH_3OH ice formation through hydrogenation of CO (0th generation complex organic molecules). In the cold outer envelope (c), photoprocessing of the CO-rich ice results in the production of, *e.g.*, $HCOOCH_3$ (1st generation). Closer to the protostar (d), following sublimation of CO, other complex molecules become abundant. Finally, all ices desorb thermally close to the protostar >100 K (e) (2nd generation). Figure from ref. 101, copyright AAS, reproduced with permission.

4.4   Interstellar gas

About 180 different molecules have been detected in interstellar space, not counting isotopologs.†† The bulk of the molecules are organic, *i.e.*, they contain at least one carbon atom together with (usually) at least one hydrogen atom. Several molecules were first found in space before they were ever synthesized in a laboratory on Earth. This includes ions such as HCO^+ and N_2H^+, and radicals such as the long carbon chains HC_9N and C_6H. Only a few ring molecules have so far been found, with $c\text{-}C_3H_2$ the best known case. This does not necessarily imply that ring molecules are less abundant than linear chains: it may simply be an observational effect since the energy level structure and hence the partition function are less favorable for cyclic molecules. Several of the organic molecules are isomers, *i.e.*, molecules with the same atomic constituents but different structures. Well known examples are HCN and HNC, methyl formate ($HCOOCH_3$), acetic acid (CH_3COOH)[102] and glycolaldehyde ($HCOCH_2OH$), and most recently HNCO *versus* HCNO, HOCN and HONC.[103]

The saturated organic molecules are found primarily in hot cores associated with star formation. In contrast, the unsaturated long carbon-chain molecules are found predominantly in cold dark clouds prior to star formation. Molecules involving metals such as MgCN, AlOH, and FeCN and phospor-containing molecules like HCP are primarily identified in the envelopes around evolved stars.

†† See http://www.astro.uni-koeln.de/cdms/molecules or http://www.astrochemystry.org for summaries.

The rate of detections is still on average 3 new molecules per year. Recent highlights include the detection of new hydrides, such as H_2O^+,[104,105] HCl^+,[106] H_2Cl^+,[107] SH^+ (ref. 108) and SH.[109] Particularly exciting is the recent discovery of the first interstellar noble gas molecule, $^{36}ArH^+$, in the gas associated with the Crab nebula,[110] the remnant of a supernova that exploded in 1054 AD. This identification was facilitated by the realization that in space, ^{36}Ar is more abundant than the ^{40}Ar isotope commonly found on Earth. $^{36}ArH^+$ is now also seen in diffuse clouds along the line of sight to the Galactic Center and other distant sources.[111]

At the opposite end of the size range of interstellar molecules, C_{60} and C_{70} have recently been identified in a young planetary nebula through infrared spectroscopy.[5] Other large molecules such as PAHs are also inferred to be present based on the ubiquitous set of mid-infrared bands that are observed throughout the Universe and which can be ascribed to aromatic C=C and C–H bonds.[2] However, in contrast with all other molecules mentioned above, no individual PAH molecule has yet been identified; likely a complex mix of PAH species makes up the observed features.[112]

5 Interstellar chemistry

5.1 Basic types of processes

Table 2 summarizes the basic types of reactions in space. Because of the low densities, only two-body processes are thought to be important. The rate of a reaction between species X and Y is given by $kn(X)n(Y)$ in cm^{-3} s^{-1}, where k is the reaction rate coefficient in cm^3 s^{-1} and n is the concentration in cm^{-3}. Three-body reactions only become significant at densities above 10^{13} cm^{-3} such as encountered in the atmospheres of AGB stars, the inner midplanes of protoplanetary disks, and in the formation of the first stars in the Universe.

5.1.1 Formation. There are two basic processes by which molecular bonds can be formed. The first one is radiative association of atoms or molecules, in which the binding energy of the new molecule is carried away through the emission of photons. The second process involves formation on the surfaces of

Table 2 Types of Molecular Processes

Bond formation processes		Typical rate coefficient (cm^3 s^{-1})
Radiative association	$X + Y \rightarrow XY + h\nu$	10^{-17} to 10^{-14}
Grain surface formation	$X + Y{:}g \rightarrow XY + g$	$\sim 10^{-17}$
Associative detachment	$X^- + Y \rightarrow XY + e$	$\sim 10^{-9}$
Bond destruction processes		
Photodissociation	$XY + h\nu \rightarrow X + Y$	10^{-10} to 10^{-8} s^{-1}
Dissociative recombination	$XY^+ + e \rightarrow X + Y$	10^{-7} to 10^{-6}
Collisional dissociation	$XY + M \rightarrow X + Y + M$	$\sim 10^{-26}$ cm^6 s^{-1}
Bond rearrangement processes		
Ion-molecule exchange	$X^+ + YZ \rightarrow XY^+ + Z$	10^{-9} to 10^{-8}
Charge-transfer	$X^+ + YZ \rightarrow X + YZ^+$	10^{-9}
Neutral–neutral	$X + YZ \rightarrow XY + Z$	10^{-11} to 10^{-9}

grains, in which the dust particle accomodates the released energy (see Section 5.2). Both of these processes are intrinsically slow. For radiative association, typical timescales for infrared emission are 10^{-3} s^{-1}, about 10 orders of magnitude slower than collision timescales of 10^{-13} s^{-1}, making the process intrinsically very inefficient. Only for specific cases, such as $C^+ + H_2 \rightarrow CH_2^+ + h\nu$, or for large molecules, are the rate coefficients increased by a few orders of magnitude. Radiative association is so slow that it is very difficult to measure properly in a laboratory on Earth, where three-body processes often dominate.[113]

A third process, called associative detachment, is fast but requires the presence of negative ions, the formation of which is the rate limiting step. It plays a minor role in dense cloud chemistry but is important, for example, in the chemistry of the early Universe.

5.1.2 Destruction. Diffuse clouds are permeated by intense ultraviolet radiation, which destroys molecular bonds through the process of photodissociation.[114,115] If the molecule has dissociative excited electronic states below 13.6 eV, photodissociation is very rapid, with typical molecular lifetimes of only ~100–1000 years in the standard ISRF. Photodissociation can also take place by absorption into excited states that are initially bound, but which are subsequently either pre-dissociated or decay through emission of photons into the vibrational continuum of the ground state. The photodissociation of three of the most important interstellar molecules, H_2,[116] CO (ref. 117) and N_2,[118,119] is controlled by these indirect processes. Because the UV absorption lines by which the photodissociation is initiated can become optically thick, these molecules can shield molecules lying deeper into the cloud through 'self-' or 'mutual' shielding.[120]

Molecular ions are efficiently destroyed by the process of dissociative recombination with electrons, which is very rapid at low temperatures. Rate coefficients in 10–30 K gas are typically 10^{-7} to 10^{-6} cm^3 s^{-1}. The dissociative recombination of H_3^+, a key species in the chemistry, has been subject to considerable discussion in the last three decades, but experiments appear to converge on a rapid value of ~10^{-6} cm^3 s^{-1} at low temperatures.[121]

In contrast with the absolute rate coefficients, the branching ratios to the various products are a major uncertainty. Experiments and theory have produced different results. Experiments indicate that three-body product channels (*e.g.*, $H_3O^+ + e \rightarrow OH + H + H$) have a much larger probability than previously thought. The fraction of dissociations leading to the largest stable molecule is often small. For example, the probability of forming H_2O in the dissociative recombination of H_3O^+ is only 17% (ref. 122) and that of CH_3OH in the recombination of $CH_3OH_2^+$ only ~6%.[123]

In dense clouds, destruction of neutral molecules can also occur through chemical reactions (see below). The He$^+$ ion, formed by the cosmic-ray ionization of He, is particularly effective in breaking bonds since 24 eV of energy is liberated in its neutralization (*e.g.*, $He^+ + N_2 \rightarrow He + N^+ + N$ or $He^+ + CO \rightarrow He + C^+ + O$). Collisional dissociation of molecules is only important in regions of very high temperature (>3000 K) and density such as shocks in the vicinity of young stars.

5.1.3 Rearrangement. Once molecular bonds have been formed, they can be rearranged by chemical reactions leading to more complex species. Initially mostly ion-molecule reactions were considered in the models, primarily because the vast majority of them are very rapid down to temperatures of 10 K.[124] If the reaction is exothermic, the simple Langevin theory states that the rate coefficient is independent of temperature, and depends only on the polarizability of the neutral molecule and the reduced mass of the system, leading to typical values of $\sim 10^{-9}$ cm^{-3} s^{-1}. It was subsequently realized that reactions between ions and molecules with a permanent dipole (*e.g.* $C^+ + H_2O$) may be factors of 10–100 larger at low temperatures, because of the enhanced long-range attraction. The ions are produced either by photoionization (C^+) or by cosmic rays producing H^+ or H_2^+. H_2^+ reacts quickly with H_2 to form H_3^+, whereas H^+ can transfer its charge to species like O. These ions subsequently react rapidly with neutral molecules down to very low temperatures, as long as the reactions are exothermic and have no activation barrier.

Over the last two decades, experimental work has demonstrated that radical–radical (*e.g.* $CN + O_2$) and radical–unsaturated molecule (*e.g.* $CN + C_2H_2$) reactions have rate coefficients that are only a factor of ~ 5 lower than those of ion-molecule reactions at low temperatures. Even some radical–saturated molecule reactions can occur in cold clouds (*e.g.* $CN + C_2H_6$). Also, reactions between radicals and atoms with a non-zero angular momentum (*e.g.* $O(^3P_2) + OH$) are fast at low temperatures.[125]

In deciding which reactions are most important in the formation of a certain species, a few simple facts should be kept in mind. First, the abundances of the elements play an important role (see Table 1). Because hydrogen is so much more abundant than any other element, reactions with H and H_2 dominate the networks if they are exothermic. This is only the case for small ions. Most reactions of neutrals and large ions with H or H_2 have substantial energy barriers or are endothermic, and therefore do not proceed at low temperatures. Reactions with the next most abundant species then become important, especially with ions because of their large rate coefficients. The ions are produced either by photoionization by the ISRF at the edges of clouds (*e.g.*, C^+) or by cosmic rays deep inside clouds (*e.g.*, H_3^+). Because the ionization potentials of O and N are larger than 13.6 eV, these elements cannot be photoionized by the ISRF and are therefore mostly neutral in interstellar clouds.

Compilations of reaction rate coefficients together with codes that solve the coupled differential equations include the UMIST 2013 database[126] and the KIDA database.[127]‡‡ The latter website includes the Nahoon (formally known as Ohio State) gas-grain chemistry code by Herbst and co-workers.

5.2 Gas–grain chemistry models

5.2.1 Surface chemistry.
The overall efficiency of surface reactions depends on the probability that the atoms or molecules stick to the grains upon collision, their mobility on the surface, the probability that molecule formation occurs, and finally the probability that the molecule is released back into the gas phase.

The bulk of the surface reactions are assumed to proceed by diffusion of at least one of the two reactants on the surface to find the other partner. This requires a description of the diffusion of the reactant to hop from one surface site to another. Usually, this is described by a standard reaction rate coefficient $K_{hop} = \nu \exp(-E_{hop}/kT_s)$ where ν is the vibrational frequency of the reactant to the surface, E_{hop} is the energy barrier to hop from one site to another and T_s is the surface

---

‡‡ http://www.adfa.net and http://kida.obs.u-bordeaux1.fr.

temperature. In most models, this barrier is taken to be a constant fraction of the binding energy, $E_{hop} = c^{st} \times E_{bind}$ with c^{st} varying from 0.3 to 0.7 in different models.[128] However, this description does not take into account that the surface is rough and that the hopping barriers change from site to site[129] (see Fig. 5). Also, the importance of tunneling at the lowest temperatures is still debated and the formulation for the competition between diffusion and reaction is not clear.

The competing Ely–Rideal mechanism, in which an atom or molecule from the gas lands directly on top of the reactant on the surface can also play a role under some conditions, as do 'hot atom' reactions in which the atom lands on the surface with excess energy that can be used to overcome barriers. Regardless of the precise process, it is clear that surface temperature plays a critical role in the ability of species to react. Light species such as H, H_2, C, N and O can likely hop over the surface to find a reaction partner even at $T_D \approx 10$ K, whereas heavier species are immobile at low dust temperatures. Reactions of these atoms with CO can produce a number of 'cold' complex molecules (see Section 8).[130]

Chemistry occurs not only on the surface of the grains but also deep inside the ice. UV photons can penetrate at least 50 monolayers and dissociate molecules both on top of and deep inside the ice. The resulting atoms and radicals are initially highly mobile because of the excess energy from the dissociation process, but quickly lose this energy on pico-second timescales and become trapped.[78] A second route to producing complex organic molecules invokes that these radicals become mobile and find each other once the ice temperature increases from ∼10 K to 20–40 K.[131,132]

Alternatively, cosmic rays can penetrate ices and create a wealth of chemical complexity, and there is a rich literature on laboratory experiments bombarding ices with high energy particles.[59,62,64] The relative importance of UV *vs.* cosmic ray processing in creating complex organic molecules is still under discussion. Shen *et al.*[21] have argued that UV radiation is more important since it deposits 10× more energy per molecule in the ice than cosmic rays for typical cosmic ray fluxes (see their Table 3 in ref. 21). On the other hand, McCoustra argues in the discussion section of this volume that each cosmic ray triggers multiple events compensating for the lower energy deposition rate. Overall, the chemical consequences of UV and high-energy particle processing of ices may be rather similar,[34] although the very strong CO and N_2 bonds are usually not broken in the UV irradiation case. A new aspect discussed at this conference in papers by Mason *et al.* (DOI: 10.1039/c3fd00004h), Boamah *et al.* (DOI: 10.1039/c3fd00158j), Siemer *et al.* (DOI: 10.1039/c3fd00116d) and Maity *et al.* (DOI: 10.1039/c3fd00121k), is to

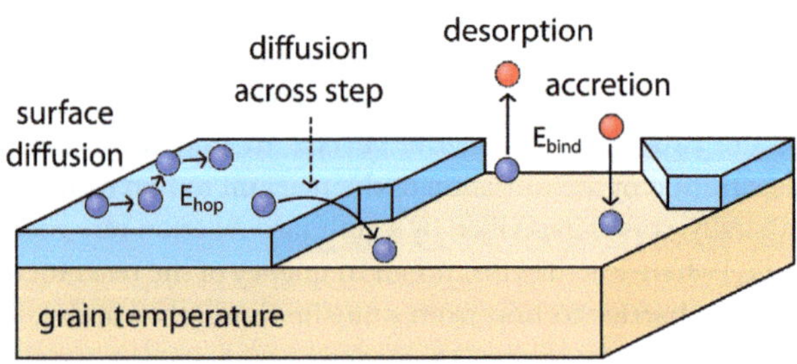

Fig. 5   Illustration of different binding sites on a rough surface.

what extent electron bombardment and charged species in ices can affect the production of complex molecules.

While there is a growing set of laboratory experiments studying reactions in ices of the types described above, it is not yet clear how to translate the laboratory data to astronomical model parameters. Timescales are also an issue: individual processes occur on picosecond timescales but lab experiments usually measure changes of bulk ice as function of fluence over a period of hours. Astronomical applications involve timescales of $>10^5$ years, and further modeling is needed to translate laboratory results to these different regimes.[133,134]

5.2.2 Gas–grain interactions. The standard treatment of the interplay between gas phase and grain surface chemistry is through rate equations,[135] the basics of which are described in more detail elsewhere.[10,11] The standard rate equation approach is known to be inadequate under some conditions, however, especially for models with very small grains and only a few species per grain. Many alternative approaches are being considered such as the modified rate equations, Monte Carlo, Master equation, and hybrid methods, each with their advantages and drawbacks.

The terms coupling the two regimes are accretion of gas-phase species onto the grains and desorption of molecules from the grains back into the gas. The accretion basically involves the rate of collisions of the gas-phase species with the grain and the probability that the species then sticks on the grain. This term is in principle straight-forward to implement and depends on the geometrical surface area of the grains as well as the sticking coefficient S as function of temperature. Usually, S is taken to be 1 for heavy species at low temperatures, as also suggested by laboratory experiments.[44] For the important case of atomic H sticking to ice, the temperature dependence of $S(T)$ has been computed[136] and is now also being measured.[137] Older models have often adopted a value of S less than unity, say 0.8 or 0.9, as a 'fudge factor' to implicitly prevent all species from freezing out onto the grains within 10^5 years, rather than treating the desorption processes explicitly.

In terms of desorption, there is a growing list of mechanisms that can return molecules from the ice into the gas. These include (i) thermal sublimation;[43,138] (ii) UV photodesorption;[56] (iii) cosmic-ray induced spot heating;[139] (iv) cosmic-ray whole grain heating;[139] (v) exothermicity from chemical reactions;[140,141] and (vi) ice mantle explosions.[142] Process (i) is well measured in the laboratory for many species but is clearly not operative in cold dark clouds where dust temperatures are only 10 K. Process (iii) may be effective for weakly bound molecules like CO but not for strongly bound species like H_2O, whereas process (iv) is generally negligible except for on the smallest grains. Processes (v) and (vi) may contribute but are still poorly characterized experimentally.

This leaves process (ii), UV photodesorption, as a prime non-thermal mechanism for getting molecules off the grains in cold dense clouds. The UV radiation is provided not only by the external ISRF but also has contributions from the internally produced UV photons by the interaction of cosmic rays with H_2 (Section 1). Thanks to a series of molecular dynamics simulations[78,79] and laboratory experiments,[56,57] the photodesorption yields are now being quantified for the main ice species like CO and H_2O (see also Fillion *et al.* (DOI: 10.1039/c3fd00129f) in this volume). The process is induced by absorption into excited electronic states (either directly dissociative or indirectly through an exciton state), so the photodissociation cross section or yield is wavelength dependent just as for gas-phase molecules. It is also possible that UV excitation of one molecule kicks out a neighboring molecule, as demonstrated experimentally for the case of N_2 by CO.[55,58]

Basic input parameters for any gas–grain model are the overall elemental abundances, the initial molecular abundances at $t = 0$ in the case of time-dependent models, the primary cosmic ray ionization rate ζ_H and the grain size distribution. The models then provide number densities or concentrations n (in cm^{-3}) of a certain molecule as a function of time and/or position within a cloud. The fractional abundance of the molecule AB with respect to H_2 is given by $n(AB)/n(H_2)$. Observers measure column densities in cm^{-2}, *i.e.*, the number density n in cm^{-3} of a species integrated along a path, $N = \int n dz$. An empirical relation between extinction and the column density of hydrogen nuclei has been found[143,144] $N_H = N(H) + 2N(H_2) = 1.8 \times 10^{21} A_V$ cm^{-2}, which is often used to present model results as a function of A_V rather than pathlength z. For diffuse clouds or PDRs proper integration along the line of sight is done, but for dark clouds local model concentration ratios are usually compared with observed column density ratios.

 5.2.3 Water as an example. An illustrative example of how different routes contribute to the formation of a particular molecule under different conditions is provided by the networks leading to interstellar water[10] (Fig. 6). At low temperatures and densities, the ion-molecule reaction route dominates and produces a low fractional abundance of water around 10^{-7}. At high temperatures such as encountered in shocks, reaction barriers of O and OH with H_2 can be overcome and H_2O is rapidly formed by neutral–neutral reactions. Finally, in cold dense clouds, formation of water ice is very efficient and locks up the bulk of the oxygen not contained in CO. Water can be brought from the ice into the gas phase by photodesorption in cold clouds and by thermal desorption at high temperatures.

5.3 Model results: managing expectations and 'back to basics'

The reliability of the results of any chemical model depend on the accuracy of the rate coefficients of the hundreds or thousands of reactions contained in the databases. The rate coefficients, in turn, are provided by chemical physics experts carrying out the relevant experiments or calculations. Ideally, each rate coefficient in a database should be motivated by a critical evaluation of the chemistry literature on that particular reaction by an independent expert: the latest measurement is not necessarily the best or the most appropriate value for astrochemical applications. Also, often extrapolations beyond measured temperature regimes have to be made and motivated. This is a time consuming process and one that is not highly valued by funding agencies nor by universities in terms of career paths.

 It is therefore important to recognize that astrochemical databases are put together and updated on a 'best effort' basis. Following the good example of atmospheric chemistry, where a two decade long effort has resulted in a set of critically evaluated and motivated rate coefficients,[125] the KIDA database has an (estimated) uncertainty associated with each rate coefficient which can then be propagated in the network. Wakelam *et al.*[145] show examples of such sensitivity analyses for a pure gas-phase chemistry network, demonstrating that even the abundances of simple molecules like H_2O, SO or CH, have an uncertainty of a

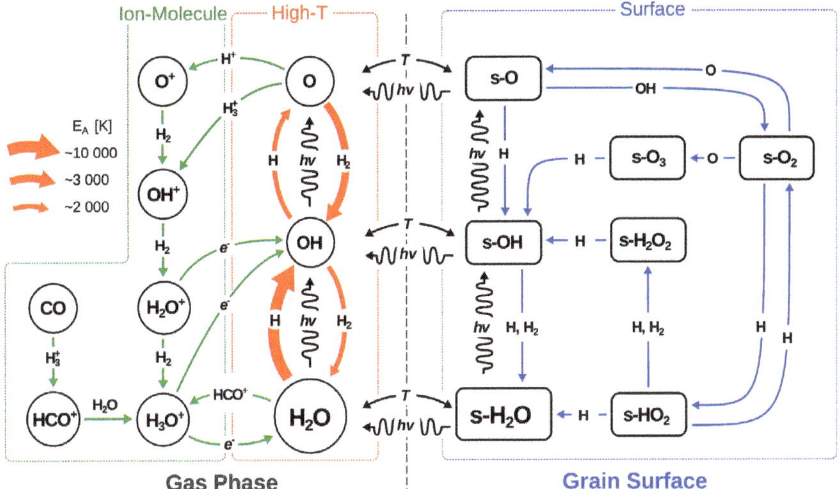

Fig. 6 Simplified water chemistry illustrating the different routes to water through (left) low temperature gas-phase ion-molecule chemistry; (middle) high temperature gas-phase chemistry; and (right) surface chemistry. Reproduced with permission from van Dishoeck et al.[10]

factor of ~3 just from the uncertainties in individual rate coefficients. For larger molecules, the cumulative effect of uncertainties in many more rates easily results in an order of magnitude overall uncertainty. Such differences are comparable to those found when two independent networks are run for the same physical model.

So what constitutes good agreement between models and observations? For diffuse clouds, the column densities of simple molecules like CH, C_2, CN, OH, H_2O and HCl can be reproduced within a factor of 2 or better if the physical conditions of the cloud are independently constrained.[146–148] Even for such relatively simple clouds with well constrained temperatures and densities, there are well-known exceptions like CH^+ that require different physical processes to be added to the model such as turbulence (see Section 6.1). Nevertheless, diffuse clouds and PDRs are still the best laboratories for 'precision astrochemistry'.

For dense clouds, the agreement between models and observations is generally much worse. The chemically-rich dark cloud TMC-1 has served as one of the main testbeds for decades, and here agreement within an order of magnitude for 80% of the observed species is considered good.[149] In spite of considerable laboratory effort, there has not been much progress in this comparison over the past few decades. Should we aim to do better as a community? Or is this uncertainty the inevitable consequence of the cumulative effect of uncertainties in many individual rate coefficients with little hope for improvement? Or do these dicrepancies point to other ingredients that are missing in the models, such as the physical and dynamical evolution of the source or small scale unresolved structure?

For this reason, several groups have started to go 'back to basics' and to ask more specialized questions that can be addressed with more limited networks in order to isolate the critical chemical processes at work. For example, detailed models of well characterized dark cores like B68 or L1544, for which the physical structure is well determined independently, demonstrate that impressive agreement with observations can be achieved for selected species, even on a linear scale.[100,150–152] Similarly, the water abundance profile in low-mass pre- and protostellar cores is well described by a simple network.[153–156]

This leads to two different approaches for modeling observational data.[157] On the one side is the 'forward model' in which a full-blown chemical model is applied to a physical structure of the source and the output is then compared with observations. In such a model, many knobs can be turned to improve comparison with the data and the question is what one has learned if good agreement is finally obtained. The alternative approach is the 'backward' or 'retrieval' model in which a trial abundance of a molecule is taken within a physical model of the source, and this abundance is then varied to obtain best agreement with observations. A more sophisticated case of 'retrieval' is where the trial abundance is inspired by the full chemical models (*e.g.*, by applying abundance jumps at certain locations or adopting a functional form). The hybrid 'back to basics' method described above uses a simple or minimal chemical network of the particular molecule under study to isolate the principal chemical routes. As deeper and higher angular resolution data are becoming available with ALMA, this approach of addressing more focused questions may be key to progress in the field.

In the following, major developments in observations and models of specific regions in the lifecycle from clouds to planets are discussed.

6   Diffuse clouds and dense PDRs

6.1   Diffuse and translucent clouds

The study of diffuse and translucent clouds with visual extinctions of a few mag has obtained a large boost with the recent *Herschel*-HIFI data.[158] Rotational lines of molecules are seen in absorption in diffuse clouds throughout the Galaxy along the lines of sight toward distant far-infrared sources. Usually only the lowest $J = 1$–0 line is observed because higher levels do not have sufficient population under these tenuous conditions. These far-infrared data complement continued studies of other molecules like H_2, C_2, and C_3 which can only be observed by traditional optical and UV absorption lines toward bright stars.[144,159] Only a few molecules, most notably CH and CH^+, can be observed by both techniques, although not for the same lines of sight.

Because H_2 cannot be observed directly toward far-infrared sources, the use of other molecules as tracers of H_2 have been investigated. CH is a good candidate,[160] but one of the best options has proven to be the HF molecule.[161,162] Even though the elemental abundance of fluorine is only 4×10^{-8}, the fluorine chemistry is particularly simple because the $F + H_2 \rightarrow HF + H$ reaction is exothermic and has been very well studied by the chemical physics community. The reaction of HF with C^+ produces CF^+ which has also been detected, further confirming its chemistry.

A summary of the state of the observations and models prior to *Herschel* has been given in a number of reviews.[146,163] The recent detections of OH^+ and H_2O^+ point toward the existence of a new type of diffuse clouds with only a small molecular fraction.[158] The H_2/H ratio must be low, of the order of 10%, because both ions react rapidly with H_2 to eventually form H_3O^+. This mixed H/H_2 phase must be ubiquitous in the interstellar medium since it is seen in absorption along nearly every line of sight in the Galaxy, and even in emission throughout entire galaxies in the local and distant Universe.[1,164] For the same reason, ArH^+ probes nearly pure atomic hydrogen gas, with an even lower H_2/H ratio of 10^{-4}.

The long-standing puzzle of the high abundance of CH^+ in diffuse clouds, first detected nearly 80 years ago,[165] is still not fully solved, although models of the chemistry in turbulent clouds, where the dissipation of turbulent energy provides the heat to overcome the $C^+ + H_2$ endothermicity of 4640 K, is a plausible explanation.[166,167] The physics of interstellar turbulence is not yet understood from first principles, however, and non-Maxwellian motions between ions and neutrals need to be taken into account in the presence of magnetic fields. Similarly, the formation of SH^+, detected both with ground-based telescopes and with *Herschel*-HIFI, requires extra energy for the $S^+ + H_2$ reaction (+9860 K) to proceed.[108,168] Both species have somewhat larger line widths than those of CH and CN, $\sim$4 *vs.* $\sim$2 km s^{-1}, further justifying that additional physical processes need to be invoked to drive the reactions.

6.2   Photon-dominated regions

Photon-dominated or photodissociation regions (PDRs) are dense clouds exposed to intense UV radiation, which controls both the chemistry and heating of the gas. They are the high I_{UV} ($>10^3$), high n_H ($>10^4$ cm^{-3}) versions of the diffuse clouds discussed above. At the edge of the cloud, the gas temperature becomes so high ($\sim$1000 K) that endothermic processes and reactions with energy barriers can proceed, most notably the $C^+ + H_2 \rightarrow CH^+ + H$ and the $O + H_2 \rightarrow OH + H$ reactions.[169] Subsequent reactions with C^+ lead to other characteristic PDR tracers such as CO^+.

The Orion Bar PDR and the Horsehead nebula are prototype examples of dense PDRs which have been studied in great detail observationally. The Orion Bar is the clearest case exhibiting the layered chemical structure: C^+ and PAH emission (pumped by UV) is seen close to the star, followed by warm H_2 emission and then cool CO.[170] This structure is also reflected in the chemistry of other species, with radicals such as C_2H, C_4H and C_3H_2, whose formation benefits from the presence of free atomic carbon, peaking ahead of molecules like $C^{18}O$.[171-174] Deep searches are now starting to reveal more complex organic molecules in PDRs, mostly those belonging to the 'cold' complex type (see Sections 5.2.1 and 8) (Guzmán *et al.* (DOI: 10.1039/c3fd00114h) in this volume).

Because of exposure to intense radiation and high temperatures, excited ro-vibrational levels of molecules can be pumped. The resulting populations are large enough that *state-to-state* processes become significant in dense PDRs. The most prominent example is that of CH^+ formation through reactions of C^+ with $H_2(v,J)$, where the H_2 ro-vibrational levels are pumped by UV radiation. These state-specific reactions were included in PDR models developed in the 1980s and 1990s but have recently been revived in the context of dense PDRs and the surface layers of protoplanetary disks.[175-177]

A second example is *formation pumping*, in which the reaction produces a molecule in an excited ro-vibrational level which then radiates before it collides with H_2. This process must dominate the excitation of those species that react on every collision.[178] Examples are $C^+ + H_2 \rightarrow CH^+(v,J) + H$,[177] $C^+ + OH \rightarrow CO^+(J) + H$ (ref. 179) and the processes leading to the formation of $H_3O^+(J)$.[180]

Finally, excitation by electrons rather than H_2 or H can become significant in PDRs. It explains why the HF line can be seen in emission in the Orion Bar rather than absorption.[181]

7   Cold molecular cores

7.1 H_2D^+ and extreme deuteration

Cold molecular clouds have long been known to harbor high abundances of deuterated molecules such as DCO^+ and DCN,[182] with DCO^+/HCO^+ and DCN/HCN ratios at least three orders of magnitude higher than the overall [D]/[H] ratio of $\sim 2 \times 10^{-5}$. More recently, even doubly- and triply-deuterated molecules such as D_2CO and ND_3 have been detected.[183] The observed ND_3/NH_3 ratio is about 10^{-3}, indicating an extreme deuterium enhancement of a factor of $\sim 10^{12}$.

This huge fractionation has its origin in two factors. First, the zero-point vibrational energy of deuterated molecules is lower than that of their normal counterparts because of their higher reduced mass. This makes their production reactions exothermic. In cold cores, most of the fractionation is initiated by the $H_3^+ + HD \rightarrow H_2D^+ + H_2$ reaction which is exothermic by about 230 K. H_2D^+ then transfers a deuteron to CO or N_2 or another species. Proof of this mechanism comes from direct observations of H_2D^+ in cold clouds.[184,185] Even D_2H^+ has been observed toward these sources.[186,187] Correct calculation of their abundances requires explicit treatment of the nuclear spin states (ortho and para) of all the species involved in the reactions.[188–191]

The extreme fractionations observed in some clouds require an additional explanation beyond just gas-phase reactions. In the centers of cold cores, there is now convincing observational evidence that most of the heavy elements, including CO, are frozen out onto the grains (see Section 4.3). Since CO is the main destroyer of both H_3^+ and H_2D^+, their abundances are even further enhanced when CO is removed from the gas. Indeed, HD then becomes the main reaction partner of H_3^+ and the chemistry rapidly proceeds to a highly deuterated state in which even D_3^+ can become comparable to H_3^+ in abundance.[192] At slightly elevated temperatures, the CH_2D^+ ion, formed by reaction of $CH_3^+ + HD$, becomes more effective in controlling the DCN/HCN ratio, consistent with observations of somewhat warmer PDRs and protoplanetary disks.[193,194]

H_3^+ and HCO^+ are among the most abundant ions in dark clouds and thus set the level of ionization in the cloud. This, in turn, controls the coupling of the gas to the magnetic field which slows down the collapse of the cloud. Since H_3^+ cannot be observed directly in these clouds, the DCO^+/HCO^+ abundance ratio is commonly used to infer the ionization fraction of cold cores. Typical values are 10^{-9} to 10^{-8} in the densest regions.[195]

7.2 H_2O, O_2 and the importance of solid-state chemistry

In the shielded regions of cold cores, the bulk of the heavy elements are in ice mantles (see Section 4.3). The laboratory confirmation of the low temperature production of H_2O ice through hydrogenation of O, O_2 and O_3 forms a highlight of solid-state astrochemistry.[49–52,196,197] In low density regions the route through atomic O dominates, but at higher densities the O_2 route becomes more important, with O_2 formed on the grains rather than being frozen out from the gas. This simple chemistry reproduces well the *Herschel*-HIFI observations of water in pre-stellar cores[153] and protostellar envelopes,[154] provided that cosmic-ray induced UV photodesorption is included in the models to get water off the grains in the central part of the core.

The O_2 route forms solid HO_2 and H_2O_2 as intermediates. The success story of combined laboratory and observational studies became complete with the detection of gas-phase HO_2 (ref. 198) and H_2O_2 (ref. 199) with abundances consistent with their grain surface formation.[200] However, both species have so far been detected in only one cloud, ρ Oph A, with searches in other clouds unsuccessful (Parise *et al.* (DOI: 10.1039/c3fd00115f) in this volume).

Interestingly, ρ Oph A is also one of only two positions where interstellar O_2 gas has been firmly detected through multi-line observations with the Odin and *Herschel* satellites.[201,202] Deep limits in other cold cores confirm the scenario in which most of the O and O_2 must be converted into H_2O ice on the grains before the protostar starts to heat its surroundings.[203,204] Perhaps the grain temperature in ρ Oph A, at around 30 K, is just optimal to prevent atomic oxygen from freezing out and being converted into water ice. A critical parameter in these models is the binding and diffusion energy of atomic O to silicates and ices. The papers by Congiu *et al.* (DOI: 10.1039/c3fd00002a), Lee *et al.* (DOI: 10.1039/c3fd00160a) and He *et al.* (DOI: 10.1039/c3fd00113j) discuss recent laboratory experiments which may suggest somewhat higher values than used before.[205] They also highlight the difficulties in providing a proper formulation for diffusion and desorption of atoms at very low temperatures.

8   Protostars and hot cores: complex molecules

Saturated complex organic molecules such as CH_3OCH_3, $HCOOCH_3$ and CH_3CN are prominently observed toward many (but not all) high-mass protostars as well as toward some low-mass sources[12,14] (Fig. 7). Three generations of complex molecules are distinguished:[14] zeroth generation ice formation in cold dense cores prior to star formation, which leads primarily to CH_3OH through hydrogenation of CO. First generation organics are then formed during the subsequent protostellar warm-up phase when radicals produced by photodissociation of species like CH_3OH become mobile in and on the ices (first generation, *e.g.*, $HCOOCH_3$). Second generation organics result from high temperature gas-phase reactions involving evaporated molecules (Fig. 4).

As the protostar heats the envelope, molecules sublimate from the ice mantles, starting with the most volatile molecules like CO and N_2 in the outer envelope, and followed by species with larger binding energies like CO_2. Once the grain temperature reaches 100 K close to the protostar, even the most strongly bound molecules like H_2O and CH_3OH sublimate from the grains, including any minor species that are trapped in their matrices. These evaporated species then drive a high temperature gas-phase chemistry producing second generation complex molecules for a period of $\sim 10^5$ years after sublimation.[207] Recent determinations of reaction rate coefficients and branching ratios for dissociative recombination show that this gas-phase route is less important than previously thought, however (see Section 5.1).

This scenario can soon be tested in great detail with high angular ALMA observations which can spatially resolve the hot cores and map the different

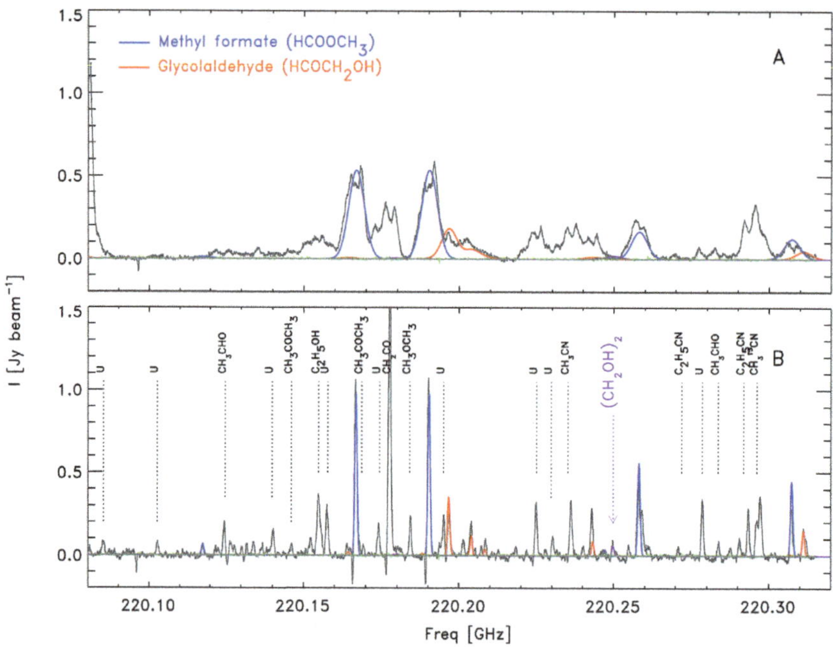

Fig. 7 ALMA data toward the low-mass binary protostar IRAS16293A (upper) and IRAS16293B (lower) obtained with 16 antennas. Fits from LTE models of the methyl formate (blue) and glycolaldehyde (red) emission are overplotted. The purple line indicates the model fit to the possible ethylene glycol transition. Note the high density of lines, including several unidentified features, even in these early ALMA science data. Reproduced from ref. 206, copyright AAS, reproduced with permission.

molecules. A 'sweet taste' of what to expect is the recent detection of the simplest sugar, glycolaldehyde ($HCOCH_2OH$), in the low-mass protostar IRAS 16293-2422B (Fig. 7). The emission comes from a region of only 25 AU in radius, $i.e.$, comparable to the orbit of Uranus in our own solar system.[206] Its abundance with respect to related complex molecules like methyl formate and ethylene glycol is consistent with laboratory experiments of mild photoprocessing of methanol-rich ice mantles,[61] although other reaction routes such as discussed by Boamah et $al.$ (DOI: 10.1039/c3fd00158j) in this volume, are not excluded.

The above scenario requires that the grains spend some time at elevated temperatures so that the radicals become mobile (see Section 5.2). The recent detection of some complex organic molecules in very cold sources that have never been heated much above 10 K therefore came as a surprise.[101,208–210] Most of these species are the same as those identified as 'cold' complex molecules in the survey of hot cores by Bisschop et $al.$,[211] $i.e.$, molecules with excitation temperatures below 100 K originating in the colder outer envelope rather than the hot core. The papers by Öberg et $al.$ (DOI: 10.1039/c3fd00146f) and Guzmán et $al.$ (DOI: 10.1039/ c3fd00114h) image these molecules in protostellar envelopes[212] and PDRs, respectively. Possible explanations that have been put forward include enhanced radiative association rate coefficients for these species or using the exothermicity of the reactions.[213] Alternatively, low temperature surface reactions of atomic C and H with CO can naturally lead to these species, if the binding energy of C to ice is low enough (see Guzmán *et al.* (DOI: 10.1039/c3fd00114h) in this volume). Similarly, reactions with N and H with CO could lead to nitrogen-bearing species like HNCO and even NH_2CHO.

8.1 Protoplanetary disks

Disks around young stars are the birthplaces of planets and are therefore particularly important targets for astrochemistry. However, disks are at least a factor of 1000 smaller than the clouds in which they are formed and they contain only $\sim$1% of the mass. Thus, their emission is readily overwhelmed by that of any cloud and they can only now be properly studied with the new generation of high angular resolution and high sensitivity instruments. Disks are heated by the radiation of their parent star so they have a radial and vertical temperature gradient in both the gas and dust. No single instrument or wavelength probes the entire disk reservoir: a combination of near-, mid-, far-infrared spectroscopy combined with spatially resolved ALMA data is needed (Fig. 8). As a result of this physical structure, disks consist of different chemical layers:[214] at the surface, molecules are dissociated into atoms by the strong UV radiation. Deeper in the disk, the grains are still warm enough to prevent freeze-out and molecules are shielded enough from the UV to survive. Deep in the cold midplane most molecules except H_2, H_3^+ and their isotopologs are frozen out onto the grains.

There has been enormous observational and model activity in this field in recent years – see recent reviews.[195,215–218] Highlights include the *Spitzer* and ground-based detections of hot ($\sim$300–800 K) C_2H_2, HCN, H_2O and CO_2 originating in the inner $\sim$1 AU of disks.[219–225] CH_4 has been found in one disk,[226] but no detection of NH_3 has yet been reported.[227] This rich array of lines of simple molecules is primarily seen in disks around low-mass stars that are cooler than the Sun; they are absent toward higher mass stars with $T_{eff} \approx$ 10 000 K. This points again to the importance of the wavelength dependence of the radiation field and associated photodissociation.[228] The implications for the elemental C and N budget in disks is discussed in the papers by Bergin *et al.* (DOI: 10.1039/c3fd00003j) and Pontoppidan *et al.* (DOI: 10.1039/c3fd00141e) in this volume.

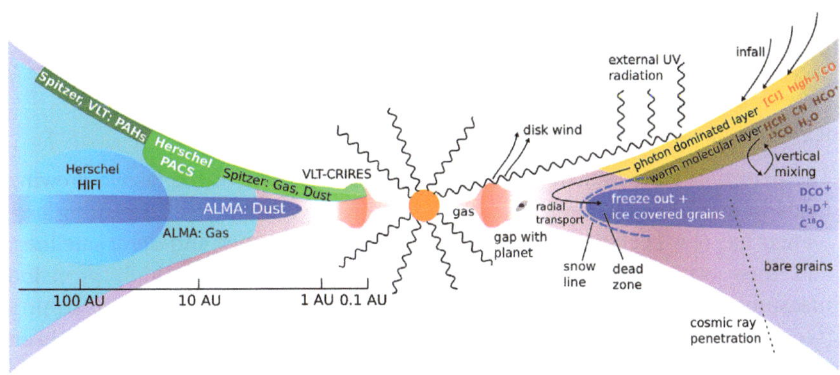

Fig. 8 Cartoon illustration of the various physical and chemical processes taking place in protoplanetary disks (right) that are probed by different observational facilities (left).

Cooler H_2O and OH have been detected somewhat further out and deeper into the disk with *Herschel*-PACS in a few sources.[229,230] The bulk of the cold water reservoir has been revealed by *Herschel*-HIFI in two disks. Because the observed water gas is produced primarily by UV photodesorption of ice, it points to the presence of a large reservoir of underlying water ice.[231]

Another related highlight is the first imaging of 'snow lines' in disks, *i.e.*, the radius where a molecule changes from being primarily in the gas phase to being frozen out as ice. Because of the vertical structure of disks, a 'snow surface' is a more appropriate description. Snow lines are important because they enhance the mass of solids by a factor of a few and thus facilitate planet formation. Also, the coating of grains with water ice enhances the coagulation of grains, the first step in planet formation. Because of their lower binding energies, the snow lines of CO and CO_2 are outside that of H_2O. This selective freeze-out of major ice reservoirs can change the overall elemental [C]/[O] abundance ratio in the gas and thus the composition of the atmospheres of giant planets that are formed there.[94] The CO snowline has been imaged with ALMA through N_2H^+ observations in the nearby TW Hya disk.[232] N_2H^+ is enhanced when its main destroyer, CO, freezes out. Another example is the CO snow line in the HD 163296 disk imaged in CO isotopologs and in the DCO^+ ion, whose abundance peaks when CO freezes out (see Section 7.1).[233,234]

In contrast with these simple species, complex molecules have not yet been detected in disks, not even methanol. The most complex molecules found so far are H_2CO, HC_3N and C_3H_2. One option is that the strong UV radiation in disks prevents the build up of more complex species, although methanol should be detectable,[235] see Walsh *et al.* this volume (DOI: 10.1039/c3fd00135k).

Finally, transitional disks which have a hole or gap in their dust distribution are a hot topic since these gaps are likely caused by planets currently forming in the disk (Fig. 8). ALMA images of several transitional disks show remarkably asymmetric dust structures, pointing to traps induced by the young planets in which the mm-sized grains are collected.[236–238] The survival of molecules like CO and the chemistry of other species in dust-free cavities in transitional disks has been modeled by Bruderer[239,240] and is being tested against ALMA observations for the case of H_2CO.[241] Molecules produced in ices may be observable as well when the disk midplane is exposed to the stellar photons at the outer edge of the cavity and molecules can be desorbed.[242] Altogether, these data form a vivid illustration and warning that gas and dust grains do not necessarily follow each other but that they need to be treated separately in the models.

8.2   Exo-planetary atmospheres

The detection and characterization of exo-planets is one of the fastest growing fields in astronomy – see recent reviews.[6,243] A large number of transiting exoplanets has now been detected, most recently with the *Kepler* and *Corot* satellites, and their atmospheres are starting to be studied through high precision measurements when the planet is in front (primary transit) and behind (secondary eclipse) the star. Detection of molecules like CO and H_2O have been claimed and disputed, but both are now firmly identified in a few targets.[7,244] One of the main science goals of future facilities is to study the composition of exo-planetary atmospheres, not only of Jupiter-like planets but down to the (super-)

Earth regime. Chemical models of such atmospheres are actively being pursued using techniques from astrochemistry,[245,246] coupled with expertise from other fields such as aeronomy.

Liquid water is likely a prerequisite for the emergence of life and much of the water in our oceans on Earth plausibly comes from the impact of asteroids and comets containing ice. The water molecules themselves are mostly formed on the surfaces of grains in the cold interstellar clouds prior to collapse as in Section 7.2. Following the water trail from dense cloud cores through collapsing envelopes to planet-forming disks and exoplanets is a major goal of modern astrochemistry and has recently been summarized.[230]

9 Concluding remarks

"Is astrochemistry useful"? The answer to this question, asked first by Dalgarno in 1986, is a convincing "yes", with astrochemistry now firmly integrated into astronomy and at the same time stimulating chemical physics. Indeed, the interaction between dust, ice and gas has stimulated the development of solid-state astrochemistry and provided new insights into basic chemistry. Continued close interaction on well-defined questions that can be addressed by the combination of laboratory-observations-models is needed.

There has been a clear shift from pure gas-phase chemistry to a gas–dust–ice chemistry over the past decades, making the topic of this Faraday discussion very timely. The solid-state routes to water and methanol, two key components of ices, have been firmly established through laboratory experiments and observations, including detection of the intermediates in the network. Gas–grain chemistry now has predictive power and is likely the basis of many of the complex organic molecules seen in star-forming regions. However, the precise routes are not yet understood, and the presence of the 'cold' complex molecules challenges the conventional theory that the grain temperature needs to be elevated above 10 K to drive the reactions. Other puzzles to be solved include why some sources are more line-rich than others, why oxygen- and nitrogen-containing complex molecules are sometimes spatially separated, and why protoplanetary disks are poor in complex molecules.

Because interstellar clouds and protostars are dynamic entities, there is increased activity to couple the full gas–grain chemical models with hydrodynamical models of cloud formation and collapse. Computing power is now available to do so and impressive simulations are being performed. An alternative approach is to couple chemistry with semi-analytical models of protostar and disk formation.[247] Other factors such as the detailed 3D geometry of the source, dust evolution and gas–dust separation now also need to be taken into account. While these avenues clearly need to be pursued, this paper has also advocated the 'back to basics' approach to elucidate the main chemical and physical processes at play.

This review has focused primarily on small molecules and moderate size complex organic molecules built through reactions in and on ices, highlighting the gas–grain interactions. A different route to molecular complexity starts with the much bigger PAH molecules and carbonaceous material, which are known to exist throughout the Universe. Through a series of UV and X-ray photolysis, radical reactions and combustion-type chemistry, these big molecules can also be transformed into smaller species and organic material that can be incorporated in future solar systems.[9] Also, the carriers of the omni-present Diffuse Interstellar Bands, discovered nearly a century ago in diffuse clouds, are still a mystery – see recent review by Cami and Cox.[248] This alternative chemistry cycle could be an appropriate topic for a future Faraday discussion.

On the observational side, ALMA is *the* astrochemistry machine of the future and is expected to bring advances in several areas.[249] It can search for new molecules at least a factor of 100 deeper than before, including prebiotic molecules such as glycine. Equally importantly, it will survey many more sources than just the 'classical' Orion and SgrB2 regions used so far to hunt for new molecules. ALMA has the spatial resolution to image the relevant physical–chemical scales for the first time, such as thin PDR and shock layers, and it can resolve the hot cores and outflow cavities close to protostars. With its high sensitivity and broad frequency range, it can also probe hotter gas and reveal the importance of state-selective processes. Finally, astrochemical studies of high-redshift galaxies are enabled by ALMA, at the level that was used to study galactic clouds 30 years ago.

Besides ALMA, astrochemistry has other exciting new missions on the horizon. The *Rosetta* spacecraft will rendez-vous with the comet 67P/Churyumov-Gerasimenko in summer 2014 and deploy a lander to investigate its surface composition in late 2014. This mission will give a big boost to studies of the connection between solar system and interstellar material. The 6 m *James Webb Space Telescope* (JWST) will be launched in late 2018 and will have near- and mid-infrared spectroscopic instruments to survey ices and gases with unprecedented sensitivity. Finally, the next generation of Extremely Large Telescopes (ELTs) is starting to be built and will provide the ultimate spatial and spectral resolution at optical- and near-infrared wavelengths. Overall, there will be plenty of material for future exciting new 'plots' and 'plays' in astrochemistry.

Acknowledgements

I am indebted to Magnus Persson (Fig. 2, 5 and 6), Sandrine Bottinelli (Fig. 3) and Simon Bruderer (Fig. 8) for providing figures. Astrochemistry in Leiden is supported by the Netherlands Organization for Scientific Research (NWO), The Netherlands Research School for Astronomy (NOVA), by a Royal Netherlands Academy of Arts and Sciences (KNAW) professor prize, and by European Union A-ERC grant 291141 CHEMPLAN.

References

1 A. Weiß, C. De Breuck, D. P. Marrone, *et al.*, *Astrophys. J.*, 2013, **767**, 88.

2 A. G. G. M. Tielens, *Annu. Rev. Astron. Astrophys.*, 2008, **46**, 289–337.

3 Y. Gao, C. L. Carilli, P. M. Solomon and P. A. Vanden Bout, *Astrophys. J.*, 2007, **660**, L93–L96.

4 S. Martín, in *The Molecular Universe*, ed. J. Cernicharo and R. Bachiller, Cambridge University Press, 2011, vol. IAU Symposium 280, pp. 351–360.

5 J. Cami, J. Bernard-Salas, E. Peeters and S. E. Malek, *Science*, 2010, **329**, 1180.

6 S. Seager and D. Deming, *Annu. Rev. Astron. Astrophys.*, 2010, **48**, 631–672.

7 J. L. Birkby, R. J. de Kok, M. Brogi, E. J. W. de Mooij, H. Schwarz, S. Albrecht and I. A. G. Snellen, *Mon. Not. R. Astron. Soc.: Lett.*, 2013, **436**, L35–L39.

8 A. Dalgarno, *Annu. Rev. Astron. Astrophys.*, 2008, **46**, 1–20.

9 A. G. G. M. Tielens, *Rev. Mod. Phys.*, 2013, **85**, 1021–1081.

10 E. F. van Dishoeck, E. Herbst and D. A. Neufeld, *Chem. Rev.*, 2013, **113**, 9043–9085.

11 E. Herbst, *Phys. Chem. Chem. Phys.*, 2014, **16**, 3344–3359.

12 P. Caselli and C. Ceccarelli, *Astron. Astrophys. Rev.*, 2012, **20**, 56.

13 E. A. Bergin, L. I. Cleeves, U. Gorti, K. Zhang, G. A. Blake, J. D. Green, S. M. Andrews, N. J. Evans, II, T. Henning, K. Öberg, K. Pontoppidan, C. Qi, C. Salyk and E. F. van Dishoeck, *Nature*, 2013, **493**, 644–646.

14 E. Herbst and E. F. van Dishoeck, *Annu. Rev. Astron. Astrophys.*, 2009, **47**, 427–480.

15 J. Blum and G. Wurm, *Annu. Rev. Astron. Astrophys.*, 2008, **46**, 21–56.

16 H. J. Habing, *Bull. Astr. Inst. Netherlands*, 1968, **19**, 421.

17 T. Draine, *Astrophys. J. Suppl.*, 1978, **36**, 595–619.

18 W. G. Roberge, D. Jones, S. Lepp and A. Dalgarno, *Astrophys. J. Suppl.*, 1991, **77**, 287–297.

19 S. S. Prasad and S. P. Tarafdar, *Astrophys. J.*, 1983, **267**, 603–609.

20 R. Gredel, S. Lepp, A. Dalgarno and E. Herbst, *Astrophys. J.*, 1989, **347**, 289–293.

21 C. J. Shen, J. M. Greenberg, W. A. Schutte and E. F. van Dishoeck, *Astron. Astrophys.*, 2004, **415**, 203–215.

22 E. F. van Dishoeck, *Annu. Rev. Astron. Astrophys.*, 2004, **42**, 119–167.

23 T. de Graauw, F. P. Helmich, T. G. Phillips, *et al.*, *Astron. Astrophys.*, 2010, **518**, L6.

24 H. Linnartz, N. Wehres, H. van Winckel, G. A. H. Walker, D. A. Bohlender, A. G. G. M. Tielens, T. Motylewski and J. P. Maier, *Astron. Astrophys.*, 2010, **511**, L3.

25 H. M. Pickett, I. R. L. Poynter, E. A. Cohen, M. L. Delitsky, J. C. Pearson and H. S. P. Muller, *J. Quant. Spectrosc. Radiat. Transfer*, 1998, **60**, 883–890.

26 H. S. P. Müller, S. Thorwirth, D. A. Roth and G. Winnewisser, *Astron. Astrophys.*, 2001, **370**, L49–L52.

27 H. S. P. Müller, F. Schlöder, J. Stutzki and G. Winnewisser, *J. Mol. Struct.*, 2005, **742**, 215–227.

28 L. S. Rothman, I. E. Gordon, A. Barbe, *et al.*, *J. Quant. Spectrosc. Radiat. Transfer*, 2009, **110**, 533–572.

29 C. Boersma, C. W. Bauschlicher, Jr, A. Ricca, A. L. Mattioda, J. Cami, E. Peeters, F. Sánchez de Armas, G. Puerta Saborido, D. M. Hudgins and L. J. Allamandola, *Astrophys. J. Suppl.*, 2014, **211**, 8.

30 T. Henning, *Annu. Rev. Astron. Astrophys.*, 2010, **48**, 21–46.

31 D. M. Hudgins, S. A. Sandford, L. J. Allamandola and A. G. G. M. Tielens, *Astrophys. J. Suppl.*, 1993, **86**, 713–870.

32 H. Linnartz, J.-B. Bossa, J. Bouwman, H. M. Cuppen, S. H. Cuylle, E. F. van Dishoeck, E. C. Fayolle, G. Fedoseev, G. W. Fuchs, S. Ioppolo, K. Isokoski, T. Lamberts, K. I. Öberg, C. Romanzin, E. Tenenbaum and J. Zhen, in *The Molecular Universe*, ed. J. Cernicharo and R. Bachiller, Cambridge University Press, 2011, vol. IAU Symposium 280, pp. 390–404.

33 R. L. Hudson, R. F. Ferrante and M. H. Moore, *Icarus*, 2014, **228**, 276–287.

34 G. M. Munoz Caro, E. Dartois, P. Boduch, H. Rothard, A. Domaracka and A. Jimenez-Escobar, *Astron. Astrophys.*, 2014, **566**, A93.

35 V. Mennella, *Astrophys. J.*, 2010, **718**, 867–875.

36 G. M. Muñoz Caro, G. Matrajt, E. Dartois, M. Nuevo, L. D'Hendecourt, D. Deboffle, G. Montagnac, N. Chauvin, C. Boukari and D. Le Du, *Astron. Astrophys.*, 2006, **459**, 147–159.

37 W. M. Smith, *Annu. Rev. Astron. Astrophys.*, 2011, **49**, 29–66.

38 D. Chastaing, S. D. Le Picard, I. R. Sims and I. W. M. Smith, *Astron. Astrophys.*, 2001, **365**, 241–247.

39 S. B. Morales, C. J. Bennett, S. D. Le Picard, A. Canosa, I. R. Sims, B. J. Sun, P. H. Chen, A. H. H. Chang, V. V. Kislov, A. M. Mebel, X. Gu, F. Zhang, P. Maksyutenko and R. I. Kaiser, *Astrophys. J.*, 2011, **742**, 26.

40 W. D. Geppert and M. Larsson, *Chem. Rev.*, 2013, **113**, 8872–8905.

41 *MPI-Mainz UV/VIS Spectral Atlas of Gaseous Molecules*, ed. H. Keller-Rudek, G. Moortgat, R. Sander and R. Sörensen, 2013, http://www.uv-vis-spectral-atlas-mainz.org.

42 M. A. Allodi, R. A. Baragiola, G. A. Baratta, M. A. Barucci, G. A. Blake, P. Boduch, J. R. Brucato, C. Contreras, S. H. Cuylle, D. Fulvio, M. S. Gudipati, S. Ioppolo, Z. Kaňuchová, A. Lignell, H. Linnartz, M. E. Palumbo, U. Raut, H. Rothard, F. Salama, E. V. Savchenko, E. Sciamma-O'Brien and G. Strazzulla, *Space Sci. Rev.*, 2013, **180**, 101–175.

43 M. P. Collings, M. A. Anderson, R. Chen, J. W. Dever, S. Viti, D. A. Williams and M. R. S. McCoustra, *Mon. Not. R. Astron. Soc.*, 2004, **354**, 1133–1140.

44 S. E. Bisschop, H. J. Fraser, K. I. Öberg, E. F. van Dishoeck and S. Schlemmer, *Astron. Astrophys.*, 2006, **449**, 1297–1309.

45 R. Martín-Doménech, G. M. Muñoz Caro, J. Bueno and F. Goesmann, *Astron. Astrophys.*, 2014, **564**, A8.

46 N. Watanabe and A. Kouchi, *Astrophys. J.*, 2002, **571**, L173–L176.

47 H. Hidaka, N. Watanabe, T. Shiraki, A. Nagaoka and A. Kouchi, *Astrophys. J.*, 2004, **614**, 1124–1131.

48 G. W. Fuchs, H. M. Cuppen, S. Ioppolo, C. Romanzin, S. E. Bisschop, S. Andersson, E. F. van Dishoeck and H. Linnartz, *Astron. Astrophys.*, 2009, **505**, 629–639.

49 S. Ioppolo, H. M. Cuppen, C. Romanzin, E. F. van Dishoeck and H. Linnartz, *Astrophys. J.*, 2008, **686**, 1474–1479.

50 N. Miyauchi, H. Hidaka, T. Chigai, A. Nagaoka, N. Watanabe and A. Kouchi, *Chem. Phys. Lett.*, 2008, **456**, 27–30.

51 F. Dulieu, L. Amiaud, E. Congiu, J. Fillion, E. Matar, A. Momeni, V. Pirronello and J. L. Lemaire, *Astron. Astrophys.*, 2010, **512**, A30.

52 H. M. Cuppen, S. Ioppolo, C. Romanzin and H. Linnartz, *Phys. Chem. Chem. Phys.*, 2010, **12**, 12077–12088.

53 M. Accolla, E. Congiu, G. Manico, F. Dulieu, H. Chaabouni, J. L. Lemaire and V. Pirronello, *Mon. Not. R. Astron. Soc.*, 2013, **429**, 3200–3206.

54 A. G. G. M. Tielens and W. Hagen, *Astron. Astrophys.*, 1982, **114**, 245–260.

55 K. I. Öberg, G. W. Fuchs, Z. Awad, H. J. Fraser, S. Schlemmer, E. F. van Dishoeck and H. Linnartz, *Astrophys. J.*, 2007, **662**, L23–L26.

56 I. Öberg, H. Linnartz, R. Visser and E. F. van Dishoeck, *Astrophys. J.*, 2009, **693**, 1209–1218.

57 E. C. Fayolle, M. Bertin, C. Romanzin, X. Michaut, K. I. Öberg, H. Linnartz and J.-H. Fillion, *Astrophys. J.*, 2011, **739**, L36.

58 M. Bertin, E. C. Fayolle, C. Romanzin, H. A. M. Poderoso, X. Michaut, L. Philippe, P. Jeseck, K. I. Öberg, H. Linnartz and J.-H. Fillion, *Astrophys. J.*, 2013, **779**, 120.

59 M. H. Moore, R. L. Hudson and P. A. Gerakines, *Spectrochim. Acta, Part A*, 2001, **57**, 843–858.

60 J. E. Elsila, J. P. Dworkin, M. P. Bernstein, M. P. Martin and S. A. Sandford, *Astrophys. J.*, 2007, **660**, 911–918.

61 K. I. Öberg, R. T. Garrod, E. F. van Dishoeck and H. Linnartz, *Astron. Astrophys.*, 2009, **504**, 891–913.

62 Y. S. Kim and R. I. Kaiser, *Astrophys. J.*, 2011, **729**, 68.

63 P. A. Gerakines, R. L. Hudson, M. H. Moore and J.-L. Bell, *Icarus*, 2012, **220**, 647–659.

64 F. Islam, G. A. Baratta and M. E. Palumbo, *Astron. Astrophys.*, 2014, **561**, A73.

65 G. D. Cody, E. Heying, C. M. O. Alexander, L. R. Nittler, A. L. D. Kilcoyne, S. A. Sandford and R. M. Stroud, *Proc. Natl. Acad. Sci. U. S. A.*, 2011, **108**, 19171–19176.

66 F. L. Schöier, F. F. S. van der Tak, E. F. van Dishoeck and J. H. Black, *Astron. Astrophys.*, 2005, **432**, 369–379.

67 F. van der Tak, in *The Molecular Universe*, ed. J. Cernicharo and R. Bachiller, Cambridge University Press, 2011, vol. IAU Symposium 280, pp. 449–460.

68 M.-L. Dubernet, M. H. Alexander, Y. A. Ba, *et al.*, *Astron. Astrophys.*, 2013, **553**, A50.

69 E. F. van Dishoeck, B. Jonkheid and M. C. van Hemert, *Faraday Discuss.*, 2006, **133**, 231.

70 M. C. van Hemert and E. F. van Dishoeck, *Chem. Phys.*, 2008, **343**, 292–302.

71 W. H. el-Qadi and P. C. Stancil, *Astrophys. J.*, 2013, **779**, 97.

72 Z. Lin, D. Talbi, E. Roueff, E. Herbst, N. Wehres, C. A. Cole, Z. Yang, T. P. Snow and V. M. Bierbaum, *Astrophys. J.*, 2013, **765**, 80.

73 D. E. Woon and E. Herbst, *Astrophys. J.*, 1997, **477**, 204–208.

74 E. Herbst, R. Terzieva and D. Talbi, *Mon. Not. R. Astron. Soc.*, 2000, **311**, 869–876.

75 X. Li, C. Arasa, M. C. van Hemert and E. F. van Dishoeck, *J. Phys. Chem. A*, 2013, **117**, 12889–12896.

76 D. E. Woon and E. Herbst, *Astrophys. J. Suppl.*, 2009, **185**, 273–288.

77 A. Ricca, C. W. Bauschlicher, Jr and L. J. Allamandola, *Astrophys. J.*, 2013, **776**, 31.

78 S. Andersson, A. Al-Halabi, G.-J. Kroes and E. F. van Dishoeck, *J. Chem. Phys.*, 2006, **124**, 064715.

79 C. Arasa, S. Andersson, H. M. Cuppen, E. F. van Dishoeck and G.-J. Kroes, *J. Chem. Phys.*, 2010, **132**, 184510.

80 J. Koning, G. J. Kroes and C. Arasa, *J. Chem. Phys.*, 2013, **138**, 104701.

81 C. Arasa, M. C. van Hemert, E. F. van Dishoeck and G. J. Kroes, *J. Phys. Chem. A*, 2013, **117**, 7064–7074.

82 M. Asplund, N. Grevesse, A. J. Sauval and P. Scott, *Annu. Rev. Astron. Astrophys.*, 2009, **47**, 481–522.

83 N. Przybilla, M.-F. Nieva and K. Butler, *Astrophys. J.*, 2008, **688**, L103–L106.

84 B. T. Draine, *Annu. Rev. Astron. Astrophys.*, 2003, **41**, 241–289.

85 B. T. Draine, *Physics of the Interstellar and Intergalactic Medium*, Princeton University Press, 2011.

86 A. G. G. M. Tielens, *The Physics and Chemistry of the Interstellar Medium*, Cambridge University Press, Cambridge, 2005.

87 I. Oliveira, J. Olofsson, K. M. Pontoppidan, E. F. van Dishoeck, J.-C. Augereau and B. Merín, *Astrophys. J.*, 2011, **734**, 51.

88 L. Testi, T. Birnstiel, L. Ricci, S. Andrews, J. Blum, J. Carpenter, C. Dominik, A. Isella, A. Natta, J. Williams and D. Wilner, in *Protostars & Planets VI*, ed. H. Beuther, R. Klessen, K. Dullemond, Th. Henning, Univ. Arizona Press, Tucson, 2014, in press.

89 A. P. Jones, L. Fanciullo, M. Köhler, L. Verstraete, V. Guillet, M. Bocchio and N. Ysard, *Astron. Astrophys.*, 2013, **558**, A62.

90 A. C. A. Boogert, K. M. Pontoppidan, F. Lahuis, *et al.*, *Astrophys. J. Suppl.*, 2004, **154**, 359–362.

91 A. C. A. Boogert, K. M. Pontoppidan, C. Knez, *et al.*, *Astrophys. J.*, 2008, **678**, 985–1004.

92 J. H. Novozamsky, W. A. Schutte and J. V. Keane, *Astron. Astrophys.*, 2001, **379**, 588–591.

93 F. A. van Broekhuizen, K. M. Pontoppidan, H. J. Fraser and E. F. van Dishoeck, *Astron. Astrophys.*, 2005, **441**, 249–260.

94 K. I. Öberg, A. C. A. Boogert, K. M. Pontoppidan, S. van den Broek, E. F. van Dishoeck, S. Bottinelli, G. A. Blake and N. J. Evans, II, *Astrophys. J.*, 2011, **740**, 109.

95 A. G. G. M. Tielens, A. T. Tokunaga, T. R. Geballe and F. Baas, *Astrophys. J.*, 1991, **381**, 181–199.

96 K. M. Pontoppidan, H. J. Fraser, E. Dartois, W. Thi, E. F. van Dishoeck, A. C. A. Boogert, L. d'Hendecourt, A. G. G. M. Tielens and S. E. Bisschop, *Astron. Astrophys.*, 2003, **408**, 981–1007.

97 D. C. B. Whittet, C. A. Poteet, J. E. Chiar, L. Pagani, V. M. Bajaj, D. Horne, S. S. Shenoy and A. J. Adamson, *Astrophys. J.*, 2013, **774**, 102.

98 P. Caselli, C. M. Walmsley, M. Tafalla, L. Dore and P. C. Myers, *Astrophys. J.*, 1999, **523**, L165–L169.

99 K. M. Pontoppidan, *Astron. Astrophys.*, 2006, **453**, L47–L50.

100 E. A. Bergin, J. Alves, T. Huard and C. J. Lada, *Astrophys. J.*, 2002, **570**, L101–L104.

101 K. I. Öberg, S. Bottinelli, J. K. Jørgensen and E. F. van Dishoeck, *Astrophys. J.*, 2010, **716**, 825–834.

102 A. Remijan, L. E. Snyder, D. N. Friedel, S.-Y. Liu and R. Y. Shah, *Astrophys. J.*, 2003, **590**, 314–332.

103 D. Quan, E. Herbst, Y. Osamura and E. Roueff, *Astrophys. J.*, 2010, **725**, 2101–2109.

104 V. Ossenkopf, H. S. P. Müller, D. C. Lis, *et al.*, *Astron. Astrophys.*, 2010, **518**, L111.

105 A. O. Benz, S. Bruderer, E. F. van Dishoeck, *et al.*, *Astron. Astrophys.*, 2010, **521**, L35.

106 M. De Luca, H. Gupta, D. Neufeld, M. Gerin, D. Teyssier, B. J. Drouin, J. C. Pearson, D. C. Lis, R. Monje, T. G. Phillips, J. R. Goicoechea, B. Godard, E. Falgarone, A. Coutens and T. A. Bell, *Astrophys. J.*, 2012, **751**, L37.

107 D. C. Lis, J. C. Pearson, D. A. Neufeld, *et al.*, *Astron. Astrophys.*, 2010, **521**, L9.

108 K. M. Menten, F. Wyrowski, A. Belloche, R. Güsten, L. Dedes and H. S. P. Müller, *Astron. Astrophys.*, 2011, **525**, A77.

109 D. A. Neufeld, E. Falgarone, M. Gerin, B. Godard, E. Herbst, G. Pineau des Forêts, A. I. Vasyunin, R. Güsten, H. Wiesemeyer and O. Ricken, *Astron. Astrophys.*, 2012, **542**, L6.

110 M. J. Barlow, B. M. Swinyard, P. J. Owen, J. Cernicharo, H. L. Gomez, R. J. Ivison, O. Krause, T. L. Lim, M. Matsuura, S. Miller, G. Olofsson and E. T. Polehampton, *Science*, 2013, **342**, 1343–1345.

111 P. Schilke, D. A. Neufeld, H. S. P. Müller, C. Comito, E. A. Bergin, D. C. Lis, M. Gerin, J. H. Black, M. Wolfire, N. Indriolo, J. C. Pearson, K. M. Menten, B. Winkel, Á. Sánchez-Monge, T. Möller, B. Godard and E. Falgarone, *Astron. Astrophys.*, 2014, **566**, A29.

112 M. J. F. Rosenberg, O. Berné and C. Boersma, *Astron. Astrophys.*, 2014, **566**, L4.

113 D. Gerlich, *J. Chem. Soc., Faraday Trans.*, 1993, **89**, 2199–2208.

114 E. F. van Dishoeck, in *Rate Coefficients in Astrochemistry*, ed. T. J. Millar and D. A. Williams, Kluwer, Dordrecht, 1988, vol. 146, pp. 49–72.

115 E. F. van Dishoeck and R. Visser, in *Modern Concepts in Laboratory Astrochemistry*, ed. H. Schlemmer, S. Mutschke and T. Giesen, Wiley, 2014, in press.

116 H. Abgrall, E. Roueff and I. Drira, *Astron. Astrophys., Suppl. Ser.*, 2000, **141**, 297–300.

117 R. Visser, E. F. van Dishoeck, S. D. Doty and C. P. Dullemond, *Astron. Astrophys.*, 2009, **495**, 881–897.

118 X. Li, A. N. Heays, R. Visser, W. Ubachs, B. R. Lewis, S. T. Gibson and E. F. van Dishoeck, *Astron. Astrophys.*, 2013, **555**, A14.

119 A. N. Heays, R. Visser, R. Gredel, W. Ubachs, B. R. Lewis, S. T. Gibson and E. F. van Dishoeck, *Astron. Astrophys.*, 2014, **562**, A61.

120 E. F. van Dishoeck and J. H. Black, *Astrophys. J.*, 1988, **334**, 771–802.

121 B. J. McCall, A. J. Huneycutt, R. J. Saykally, T. R. Geballe, N. Djuric, G. H. Dunn, J. Semaniak, O. Novotny, A. Al-Khalili, A. Ehlerding, F. Hellberg, S. Kalhori, A. Neau, R. Thomas, F. Österdahl and M. Larsson, *Nature*, 2003, **422**, 500–502.

122 H. Buhr, J. Stuetzel, M. B. Mendes, O. Novotny, D. Schwalm, M. H. Berg, D. Bing, M. Grieser, O. Heber, C. Krantz, S. Menk, S. Novotny, D. A. Orlov, A. Petrignani, M. L. Rappaport, R. Repnow, D. Zajfman and A. Wolf, *Phys. Rev. Lett.*, 2010, **105**, 103202.

123 W. D. Geppert, M. Hamberg, R. D. Thomas, F. Österdahl, F. Hellberg, V. Zhaunerchyk, A. Ehlerding, T. J. Millar, H. Roberts, J. Semaniak, M. A. Ugglas, A. Källberg, A. Simonsson, M. Kaminska and M. Larsson, *Faraday Discuss.*, 2006, **133**, 177.

124 E. Herbst and W. Klemperer, *Astrophys. J.*, 1973, **185**, 505–534.

125 R. Atkinson, D. Baulch, R. Cox, J. Crowley, R. Hampson and R. Hynes, *Atmos. Chem. Phys.*, 2004, **4**, 1461.

126 D. McElroy, C. Walsh, A. J. Markwick, M. A. Cordiner, K. Smith and T. J. Millar, *Astron. Astrophys.*, 2013, **550**, A36.

127 V. Wakelam, E. Herbst, J.-C. Loison, *et al.*, *Astrophys. J. Suppl.*, 2012, **199**, 21.

128 Y. Aikawa and E. Herbst, *Astron. Astrophys.*, 2001, **371**, 1107–1117.

129 M. Cuppen and E. Herbst, *Astrophys. J.*, 2007, **668**, 294–309.

130 A. G. G. M. Tielens and S. B. Charnley, *Origins Life Evol. Biospheres*, 1997, **27**, 23–51.

131 R. T. Garrod and E. Herbst, *Astron. Astrophys.*, 2006, **457**, 927–936.

132 R. T. Garrod, *Astron. Astrophys.*, 2008, **491**, 239–251.

133 H. M. Cuppen, E. F. van Dishoeck, E. Herbst and A. G. G. M. Tielens, *Astron. Astrophys.*, 2009, **508**, 275–287.

134 H. M. Cuppen, L. J. Karssemeijer and T. Lamberts, *Chem. Rev.*, 2013, **113**, 8840–8871.

135 T. I. Hasegawa and E. Herbst, *Mon. Not. R. Astron. Soc.*, 1993, **263**, 589.

136 A. Al-Halabi and E. F. van Dishoeck, *Mon. Not. R. Astron. Soc.*, 2007, **382**, 1648–1656.

137 E. Matar, H. Bergeron, F. Dulieu, H. Chaabouni, M. Accolla and J. L. Lemaire, *J. Chem. Phys.*, 2010, **133**, 104507.

138 J. C. Brown, H. E. Potts, L. J. Porter and G. Le Chat, *Astron. Astrophys.*, 2011, **535**, A71.

139 A. Leger, M. Jura and A. Omont, *Astron. Astrophys.*, 1985, **144**, 147–160.

140 R. T. Garrod, V. Wakelam and E. Herbst, *Astron. Astrophys.*, 2007, **467**, 1103–1115.

141 F. Dulieu, E. Congiu, J. Noble, S. Baouche, H. Chaabouni, A. Moudens, M. Minissale and S. Cazaux, *Nat. Sci. Rep.*, 2013, **3**, 1338.

142 L. B. D'Hendecourt, L. J. Allamandola, F. Baas and J. M. Greenberg, *Astron. Astrophys.*, 1982, **109**, L12–L14.

143 R. C. Bohlin, B. D. Savage and J. F. Drake, *Astrophys. J.*, 1978, **224**, 132–142.

144 B. L. Rachford, T. P. Snow, J. D. Destree, T. L. Ross, R. Ferlet, S. D. Friedman, C. Gry, E. B. Jenkins, D. C. Morton, B. D. Savage, J. M. Shull, P. Sonnentrucker, J. Tumlinson, A. Vidal-Madjar, D. E. Welty and D. G. York, *Astrophys. J. Suppl.*, 2009, **180**, 125–137.

145 V. Wakelam, E. Herbst, J. Le Bourlot, F. Hersant, F. Selsis and S. Guilloteau, *Astron. Astrophys.*, 2010, **517**, A21.

146 E. F. van Dishoeck, in *The Molecular Astrophysics of Stars and Galaxies*, Clarendon Press, Oxford, 1998, vol. 4, p. 53.

147 D. Hollenbach, M. J. Kaufman, D. Neufeld, M. Wolfire and J. R. Goicoechea, *Astrophys. J.*, 2012, **754**, 105.

148 N. Flagey, P. F. Goldsmith, D. C. Lis, M. Gerin, D. Neufeld, P. Sonnentrucker, M. De Luca, B. Godard, J. R. Goicoechea, R. Monje and T. G. Phillips, *Astrophys. J.*, 2013, **762**, 11.

149 R. Terzieva and E. Herbst, *Astrophys. J.*, 1998, **501**, 207–220.

150 M. Tafalla, P. C. Myers, P. Caselli, C. M. Walmsley and C. Comito, *Astrophys. J.*, 2002, **569**, 815–835.

151 E. Keto and P. Caselli, *Mon. Not. R. Astron. Soc.*, 2010, **402**, 1625–1634.

152 M. Padovani, C. M. Walmsley, M. Tafalla, P. Hily-Blant and G. Pineau Des Forêts, *Astron. Astrophys.*, 2011, **534**, A77.

153 P. Caselli, E. Keto, E. A. Bergin, M. Tafalla, Y. Aikawa, T. Douglas, L. Pagani, U. A. Yíldíz, F. F. S. van der Tak, C. M. Walmsley, C. Codella, B. Nisini, L. E. Kristensen and E. F. van Dishoeck, *Astrophys. J.*, 2012, **759**, L37.

154 J. C. Mottram, E. F. van Dishoeck, M. Schmalzl, L. E. Kristensen, R. Visser, M. R. Hogerheijde and S. Bruderer, *Astron. Astrophys.*, 2013, **558**, A126.

155 E. Keto, J. Rawlings and P. Caselli, *Mon. Not. R. Astron. Soc.*, 2014, **440**, 2616–2624.

156 M. Schmalzl, *et al.*, *Astron. Astrophys.*, 2014, submitted.

157 S. D. Doty, F. L. Schöier and E. F. van Dishoeck, *Astron. Astrophys.*, 2004, **418**, 1021–1034.

158 M. Gerin, M. de Luca, J. Black, *et al.*, *Astron. Astrophys.*, 2010, **518**, L110.

159 M. R. Schmidt, J. Krełowski, G. A. Galazutdinov, D. Zhao, M. A. Haddad, W. Ubachs and H. Linnartz, *Mon. Not. R. Astron. Soc.*, 2014, **441**, 1134–1146.

160 R. Gredel, E. F. van Dishoeck and J. H. Black, *Astron. Astrophys.*, 1993, **269**, 477–495.

161 D. A. Neufeld, P. Sonnentrucker, T. G. Phillips, *et al.*, *Astron. Astrophys.*, 2010, **518**, L108.

162 P. Sonnentrucker, D. A. Neufeld, T. G. Phillips, *et al.*, *Astron. Astrophys.*, 2010, **521**, L12.

163 T. P. Snow and B. J. McCall, *Annu. Rev. Astron. Astrophys.*, 2006, **44**, 367–414.

164 P. P. van der Werf, K. G. Isaak, R. Meijerink, *et al.*, *Astron. Astrophys.*, 2010, **518**, L42.

165 A. E. Douglas and G. Herzberg, *Astrophys. J.*, 1941, **94**, 381.

166 B. Godard, E. Falgarone and G. Pineau Des Forêts, *Astron. Astrophys.*, 2009, **495**, 847–867.

167 E. Falgarone, B. Godard, J. Cernicharo, *et al.*, *Astron. Astrophys.*, 2010, **521**, L15.

168 B. Godard, E. Falgarone, M. Gerin, D. C. Lis, M. De Luca, J. H. Black, J. R. Goicoechea, J. Cernicharo, D. A. Neufeld, K. M. Menten and M. Emprechtinger, *Astron. Astrophys.*, 2012, **540**, A87.

169 A. Sternberg and A. Dalgarno, *Astrophys. J. Suppl.*, 1995, **99**, 565.

170 A. G. G. M. Tielens, M. M. Meixner, P. P. van der Werf, J. Bregman, J. A. Tauber, J. Stutzki and D. Rank, *Science*, 1993, **262**, 86–89.

171 M. R. Hogerheijde, D. J. Jansen and E. F. van Dishoeck, *Astron. Astrophys.*, 1995, **294**, 792–810.

172 M. H. D. van der Wiel, F. F. S. van der Tak, V. Ossenkopf, M. Spaans, H. Roberts, G. A. Fuller and R. Plume, *Astron. Astrophys.*, 2009, **498**, 161–165.

173 J. Pety, D. Teyssier, D. Fossé, M. Gerin, E. Roueff, A. Abergel, E. Habart and J. Cernicharo, *Astron. Astrophys.*, 2005, **435**, 885–899.

174 M. Gerin, J. R. Goicoechea, J. Pety and P. Hily-Blant, *Astron. Astrophys.*, 2009, **494**, 977–985.

175 M. Agúndez, J. Cernicharo and J. R. Goicoechea, *Astron. Astrophys.*, 2008, **483**, 831–837.

176 W.-F. Thi, F. Ménard, G. Meeus, C. Martin-Zaïdi, P. Woitke, E. Tatulli, M. Benisty, I. Kamp, I. Pascucci, C. Pinte, C. A. Grady, S. Brittain, G. J. White, C. D. Howard, G. Sandell and C. Eiroa, *Astron. Astrophys.*, 2011, **530**, L2.

177 B. Godard and J. Cernicharo, *Astron. Astrophys.*, 2013, **550**, A8.

178 J. H. Black, *Faraday Discuss.*, 1998, **109**, 257.

179 P. Stäuber and S. Bruderer, *Astron. Astrophys.*, 2009, **505**, 195–203.

180 D. C. Lis, P. Schilke, E. A. Bergin, M. Gerin, J. H. Black, C. Comito, M. De Luca, B. Godard, R. Higgins, F. Le Petit, J. C. Pearson, E. W. Pellegrini, T. G. Phillips and S. Yu, *Astrophys. J.*, 2014, **785**, 135.

181 F. F. S. van der Tak, V. Ossenkopf, Z. Nagy, A. Faure, M. Röllig and E. A. Bergin, *Astron. Astrophys.*, 2012, **537**, L10.

182 A. Wootten, in *Astrochemistry*, ed. M. S. Vardya and S. P. Tarafdar, Kluwer, Dordrecht, 1987, vol. IAU Symposium 120, pp. 311–318.

183 C. Ceccarelli, P. Caselli, D. Bockelée-Morvan, O. Mousis, F. Pizzarello, F. Robert and D. Semenov, in *Protostars & Planets VI*, ed. H. Beuther, R. Klessen, K. Dullemond, Th. Henning, Univ. Arizona Press, Tucson, 2014, in press.

184 R. Stark, F. F. S. van der Tak and E. F. van Dishoeck, *Astrophys. J.*, 1999, **521**, L67–L70.

185 P. Caselli, F. F. S. van der Tak, C. Ceccarelli and A. Bacmann, *Astron. Astrophys.*, 2003, **403**, L37–L41.

186 C. Vastel, T. G. Phillips and H. Yoshida, *Astrophys. J.*, 2004, **606**, L127–L130.

187 B. Parise, A. Belloche, F. Du, R. Güsten and K. M. Menten, *Astron. Astrophys.*, 2011, **526**, A31.

188 L. Pagani, M. Salez and P. G. Wannier, *Astron. Astrophys.*, 1992, **258**, 479–488.

189 L. Pagani, P. Lesaffre, M. Jorfi, P. Honvault, T. González-Lezana and A. Faure, *Astron. Astrophys.*, 2013, **551**, A38.

190 C. M. Walmsley, D. R. Flower and G. Pineau des Forêts, *Astron. Astrophys.*, 2004, **418**, 1035–1043.

191 O. Sipilä, E. Hugo, J. Harju, O. Asvany, M. Juvela and S. Schlemmer, *Astron. Astrophys.*, 2010, **509**, A98.

192 H. Roberts, E. Herbst and T. J. Millar, *Astrophys. J.*, 2003, **591**, L41–L44.

193 B. Parise, S. Leurini, P. Schilke, E. Roueff, S. Thorwirth and D. C. Lis, *Astron. Astrophys.*, 2009, **508**, 737–749.

194 K. I. Öberg, C. Qi, D. J. Wilner and M. R. Hogerheijde, *Astrophys. J.*, 2012, **749**, 162.

195 E. A. Bergin and M. Tafalla, *Annu. Rev. Astron. Astrophys.*, 2007, **45**, 339–396.

196 Y. Oba, N. Watanabe, T. Hama, K. Kuwahata, H. Hidaka and A. Kouchi, *Astrophys. J.*, 2012, **749**, 67.

197 H. Chaabouni, M. Minissale, G. Manic, E. Congiu, J. A. Noble, S. Baouche, M. Accolla, J. L. Lemaire, V. Pironello and F. Dulieu, *J. Chem. Phys.*, 2012, **137**, 234706.

198 P. Bergman, B. Parise, R. Liseau, B. Larsson, H. Olofsson, K. M. Menten and R. Güsten, *Astron. Astrophys.*, 2011, **531**, L8.

199 B. Parise, P. Bergman and F. Du, *Astron. Astrophys.*, 2012, **541**, L11.

200 F. Du and B. Parise, *Astron. Astrophys.*, 2011, **530**, A131.

201 B. Larsson, R. Liseau, L. Pagani, *et al.*, *Astron. Astrophys.*, 2007, **466**, 999–1003.

202 R. Liseau, P. F. Goldsmith, B. Larsson, *et al.*, *Astron. Astrophys.*, 2012, **541**, A73.

203 E. A. Bergin, M. R. Hogerheijde, C. Brinch, *et al.*, *Astron. Astrophys.*, 2010, **521**, L33.

204 U. A. Yıldız, K. Acharyya, P. F. Goldsmith, E. F. van Dishoeck, G. Melnick, R. Snell, R. Liseau, J.-H. Chen, L. Pagani, E. Bergin, P. Caselli, E. Herbst, L. E. Kristensen, R. Visser, D. C. Lis and M. Gerin, *Astron. Astrophys.*, 2013, **558**, A58.

205 D. Jing, J. He, J. R. Brucato, G. Vidali, L. Tozzetti and A. De Sio, *Astrophys. J.*, 2012, **756**, 98.

206 J. K. Jørgensen, C. Favre, S. E. Bisschop, T. L. Bourke, E. F. van Dishoeck and M. Schmalzl, *Astrophys. J.*, 2012, **757**, L4.

207 S. B. Charnley, A. G. G. M. Tielens and T. J. Millar, *Astrophys. J.*, 1992, **399**, L71–L74.

208 H. G. Arce, J. Santiago-García, J. K. Jørgensen, M. Tafalla and R. Bachiller, *Astrophys. J.*, 2008, **681**, L21–L24.

209 K. I. Öberg, N. van der Marel, L. E. Kristensen and E. F. van Dishoeck, *Astrophys. J.*, 2011, **740**, 14.

210 A. Bacmann, V. Taquet, A. Faure, C. Kahane and C. Ceccarelli, *Astron. Astrophys.*, 2012, **541**, L12.

211 S. E. Bisschop, J. K. Jørgensen, E. F. van Dishoeck and E. B. M. de Wachter, *Astron. Astrophys.*, 2007, **465**, 913–929.

212 K. I. Öberg, M. D. Boamah, E. C. Fayolle, R. T. Garrod, C. J. Cyganowski and F. van der Tak, *Astrophys. J.*, 2013, **771**, 95.

213 A. I. Vasyunin and E. Herbst, *Astrophys. J.*, 2013, **769**, 34.

214 Y. Aikawa, G. J. van Zadelhoff, E. F. van Dishoeck and E. Herbst, *Astron. Astrophys.*, 2002, **386**, 622–632.

215 E. A. Bergin, in *XVII Special Courses at the National Observatory of Rio de Janeiro*, AIP Conference Proceedings, 2014, in press.

216 T. Henning and D. Semenov, *Chem. Rev.*, 2013, **113**, 9016–9042.

217 K. M. Pontoppidan, C. Salyk, E. A. Bergin, S. Brittain, B. Marty, O. Mousis and K. L. Oberg, in *Protostars & Planets VI*, ed. H. Beuther, R. Klessen, K. Dullemond and Th. Henning, Univ. Arizona Press, Tucson, 2014, in press.

218 A. Dutrey, D. Semenov, E. Chapillon, U. Gorti, S. Guilloteau, F. Hersant, M. Hogerheijde, M. Hughes, G. Meeus, H. Nomura, V. Piétu, C. Qi and V. Wakelam, in *Protostars & Planets VI*, ed. H. Beuther, R. Klessen, K. Dullemond and Th. Henning, Univ. Arizona Press, Tucson, 2014, in press.

219 F. Lahuis, E. F. van Dishoeck, A. C. A. Boogert, K. M. Pontoppidan, G. A. Blake, C. P. Dullemond, N. J. Evans, II, M. R. Hogerheijde, J. K. Jørgensen, J. E. Kessler-Silacci and C. Knez, *Astrophys. J.*, 2006, **636**, L145–L148.

220 J. S. Carr and J. R. Najita, *Science*, 2008, **319**, 1504.

221 C. Salyk, K. M. Pontoppidan, G. A. Blake, F. Lahuis, E. F. van Dishoeck and N. J. Evans, II, *Astrophys. J.*, 2008, **676**, L49–L52.

222 I. Pascucci, D. Apai, K. Luhman, T. Henning, J. Bouwman, M. R. Meyer, F. Lahuis and A. Natta, *Astrophys. J.*, 2009, **696**, 143–159.

223 C. Salyk, K. M. Pontoppidan, G. A. Blake, J. R. Najita and J. S. Carr, *Astrophys. J.*, 2011, **731**, 130.

224 K. M. Pontoppidan, C. Salyk, G. A. Blake, R. Meijerink, J. S. Carr and J. Najita, *Astrophys. J.*, 2010, **720**, 887–903.

225 J. R. Najita, J. S. Carr, K. M. Pontoppidan, C. Salyk, E. F. van Dishoeck and G. A. Blake, *Astrophys. J.*, 2013, **766**, 134.

226 E. L. Gibb and D. Horne, *Astrophys. J.*, 2013, **776**, L28.

227 A. M. Mandell, J. Bast, E. F. van Dishoeck, G. A. Blake, C. Salyk, M. J. Mumma and G. Villanueva, *Astrophys. J.*, 2012, **747**, 92.

228 E. F. van Dishoeck, *Proc. Natl. Acad. Sci. U. S. A.*, 2006, **103**, 12249–12256.

229 D. Fedele, S. Bruderer, E. F. van Dishoeck, J. Carr, G. J. Herczeg, C. Salyk, N. J. Evans, J. Bouwman, G. Meeus, T. Henning, J. Green, J. R. Najita and M. Güdel, *Astron. Astrophys.*, 2013, **559**, A77.

230 E. F. van Dishoeck, E. A. Bergin, D. C. Lis and J. I. Lunine, in *Protostars & Planets VI*, ed. H. Beuther, R. Klessen, K. Dullemond and Th. Henning, Univ. Arizona Press, Tucson, 2014, in press.

231 M. R. Hogerheijde, E. A. Bergin, C. Brinch, L. I. Cleeves, J. K. J. Fogel, G. A. Blake, C. Dominik, D. C. Lis, G. Melnick, D. Neufeld, O. Panić, J. C. Pearson, L. Kristensen, U. A. Yıldız and E. F. van Dishoeck, *Science*, 2011, **334**, 338.

232 C. Qi, K. I. Öberg, D. J. Wilner, P. D'Alessio, E. Bergin, S. M. Andrews, G. A. Blake, M. R. Hogerheijde and E. F. van Dishoeck, *Science*, 2013, **341**, 630–632.

233 G. S. Mathews, P. D. Klaassen, A. Juhász, D. Harsono, E. Chapillon, E. F. van Dishoeck, D. Espada, I. de Gregorio-Monsalvo, A. Hales, M. R. Hogerheijde, J. C. Mottram, M. G. Rawlings, S. Takahashi and L. Testi, *Astron. Astrophys.*, 2013, **557**, A132.

234 C. Qi, P. D'Alessio, K. I. Öberg, D. J. Wilner, A. M. Hughes, S. M. Andrews and S. Ayala, *Astrophys. J.*, 2011, **740**, 84.

235 C. Walsh, T. J. Millar, H. Nomura, E. Herbst, S. Widicus Weaver, Y. Aikawa, J. C. Laas and A. I. Vasyunin, *Astron. Astrophys.*, 2014, **563**, A33.

236 N. van der Marel, E. F. van Dishoeck, S. Bruderer, T. Birnstiel, P. Pinilla, C. P. Dullemond, T. A. van Kempen, M. Schmalzl, J. M. Brown, G. J. Herczeg, G. S. Mathews and V. Geers, *Science*, 2013, **340**, 1199–1202.

237 S. Casassus, G. van der Plas, M. Sebastian Perez, W. R. F. Dent, E. Fomalont, J. Hagelberg, A. Hales, A. Jordán, D. Mawet, F. Ménard, A. Wootten, D. Wilner, A. M. Hughes, M. R. Schreiber, J. H. Girard, B. Ercolano, H. Canovas, P. E. Román and V. Salinas, *Nature*, 2013, **493**, 191–194.

238 L. M. Pérez, A. Isella, J. M. Carpenter and C. J. Chandler, *Astrophys. J.*, 2014, **783**, L13.

239 S. Bruderer, *Astron. Astrophys.*, 2013, **559**, A46.

240 S. Bruderer, N. van der Marel, E. F. van Dishoeck and T. A. van Kempen, *Astron. Astrophys.*, 2014, **562**, A26.

241 N. van der Marel, E. F. van Dishoeck, S. Bruderer and T. A. van Kempen, *Astron. Astrophys.*, 2014, **563**, A113.

242 L. I. Cleeves, E. A. Bergin, T. J. Bethell, N. Calvet, J. K. J. Fogel, J. Sauter and S. Wolf, *Astrophys. J.*, 2011, **743**, L2.

243 N. Madhusudhan, H. Knutson, J. Fortney and T. Barman, in *Protostars & Planets VI*, ed. H. Beuther, R. Klessen, K. Dullemond and Th. Henning, Univ. Arizona Press, Tucson, 2014, in press.

244 D. Deming, A. Wilkins, P. McCullough, A. Burrows, J. J. Fortney, E. Agol, I. Dobbs-Dixon, N. Madhusudhan, N. Crouzet, J.-M. Desert, R. L. Gilliland, K. Haynes, H. A. Knutson, M. Line, Z. Magic, A. M. Mandell, S. Ranjan, D. Charbonneau, M. Clampin, S. Seager and A. P. Showman, *Astrophys. J.*, 2013, **774**, 95.

245 C. Helling, P. Woitke, P. B. Rimmer, I. Kamp, W.-F. Thi and R. Meijerink, *Life*, 2014, **4**, 142–173.

246 M. Agúndez, V. Parmentier, O. Venot, F. Hersant and F. Selsis, *Astron. Astrophys.*, 2014, **564**, A73.

247 R. Visser, S. D. Doty and E. F. van Dishoeck, *Astron. Astrophys.*, 2011, **534**, A132.

248 *The Diffuse Interstellar Bands*, ed. J. Cami and N. L. J. Cox, Cambridge University Press, 2014, vol. 297.
249 E. Herbst and T. Millar, in *Low Temperatures and Cold Molecules*, ed. I. W. M. Smith, Imperial College, London, 2008, p. 1.

Faraday Discussions

PAPER

The chemistry of planet-forming regions is not interstellar

Klaus M. Pontoppidan[*a] and Sandra M. Blevins[ab]

Received 15th December 2013, Accepted 4th February 2014

DOI: 10.1039/c3fd00141e

Advances in infrared and submillimeter technology have allowed for detailed observations of the molecular content of the planet-forming regions of protoplanetary disks. In particular, disks around solar-type stars now have growing molecular inventories that can be directly compared with both prestellar chemistry and that inferred for the early solar nebula. The data directly address the old question of whether the chemistry of planet-forming matter is similar or different and unique relative to the chemistry of dense clouds and protostellar envelopes. The answer to this question may have profound consequences for the structure and composition of planetary systems. The practical challenge is that observations of emission lines from disks do not easily translate into chemical concentrations. Here, we present a two-dimensional radiative transfer model of RNO 90, a classical protoplanetary disk around a solar-mass star, and retrieve the concentrations of dominant molecular carriers of carbon, oxygen and nitrogen in the terrestrial region around 1 AU. We compare our results to the chemical inventory of dense clouds and protostellar envelopes, and argue that inner disk chemistry is, as expected, fundamentally different from prestellar chemistry. We find that the clearest discriminant may be the concentration of CO_2, which is extremely low in disks, but one of the most abundant constituents of dense clouds and protostellar envelopes.

1 Introduction

Planets are ultimately built from matter that is inherited from diffuse gas in the interstellar medium. The chemical state of this material continuously evolves as it travels a complex path through a dense molecular cloud, protostellar envelope and protoplanetary disk. The elements in the material are differentially partitioned into gases and solids, and apart from a small contribution from nuclear reactions with stellar and cosmic radiation, their total abundances remain largely unchanged. Conversely, their molecular carriers, including those carrying the bulk of some elements, may change dramatically along this path, and depending

[a]Space Telescope Science Institute, 3700 San Martin Drive, MD 21218, Baltimore, USA. E-mail: pontoppi@stsci.edu; Tel: +1 410 338 4744

[b]Catholic University of America, Physics Department, Washington, DC 20064, USA

on their volatility the local elemental abundances may be altered by hydrodynamic transport processes.[1,2] This rich history of pre-planetary matter is validated by strong differences in the chemical content of primitive chondrites from the 3 AU region of the solar nebula,[3] comets that formed beyond tens of AU,[4] and protostellar envelopes.[2]

If planet-forming chemistry, in bulk, is an active, local process, this means that exoplanetary systems, particularly terrestrial planets, are formed from material with chemical properties that are unlike those of their natal envelopes and, likely, cometary systems. This scenario would not be surprising, since, in the solar system, it has long been known that almost all of the matter within a few AU was subjected to temperatures sufficiently high to fundamentally alter its chemistry.[5] It is therefore of great interest to determine whether strong chemical evolution is occurring in typical protoplanetary disks, not only in the case of refractory material[6] but also for ice-forming molecules such as water, CO_2 and volatile organics. A consequence of strong evolution is that prestellar-like chemistry cannot be used directly for input into planet formation models; their input must come from direct observations of protoplanetary disks in cooperation with thermochemical modeling applied to physical conditions relevant for inner disks.

In this paper, we consider the recent evidence that the inner planet-forming regions of protoplanetary disks (<10 AU) have a radically different chemistry than that of the dense interstellar medium and the typical protostellar envelope. That the molecular emission from inner disks is indicative of an active chemistry has already been suggested, based on simple zero-dimensional calculations,[7] albeit with significant degeneracies.[8] As one step toward eliminating such retrieval degeneracies, we are developing more sophisticated models, along with additional constraints. We present preliminary results using two-dimensional radiative transfer models to retrieve molecular abundances from planet-forming regions with greater accuracy than has previously been possible. Specifically, we will compare the molecular inventory obtained from dense clouds to that of the protoplanetary disk around RNO 90 – a classical T Tauri star of roughly 1 solar mass.

1.1   The bulk chemistry of dense clouds in protostellar envelopes

Prior to the formation of the star and disk, nearly all of the CNO carriers are bound in the solid state, either as ices or in refractory silicates and carbon compounds.[9,10] CO, and possibly N_2, persists the longest in the gas-phase, but even these freeze out at the lowest temperatures. Using infrared absorption spectroscopy on background or young stars, a very accurate, and nearly complete, estimate of the molecular inventory in dense clouds has been compiled.[11] Interstellar ices are dominated by H_2O and CO but with large contributions from CO_2 (~25% relative to water), and occasionally CH_3OH (up to 25%, but more typically 5%). The most abundant ice species are characterized by remarkably universal chemical signatures, and the concentrations of e.g., CO_2, CH_4 and various volatile organics do not vary by more than a factor of 2 within nearby quiescent dense molecular clouds.[2] There are some occasional exceptions to this characteristic "ice signature", including CH_3OH,[9] but, generally speaking, dense cloud bulk ice chemistry is quite unique. As we shall see, perhaps CO_2, one of the most stable cloud species, offers the strongest evidence that planet-forming chemistry is far from interstellar.

1.2 Observations of water and organics in disks with Spitzer

To discuss the similarities and differences in bulk chemistries among distinct stages in protostellar and planetary evolution, it is important to have a relatively complete observational inventory to quantify the majority of the elemental constituents. A significant fraction of missing CNO could be an indication of either uncertain measurements, or that it is sequestered in a chemically and radiatively inactive reservoir. An example of the latter case could be where a gas-phase carrier is observed in a region where a significant part of the element is depleted in ices. It is well known that gas-phase observations of molecules in cold, dense clouds may provide little information on the bulk chemistry, since most species are frozen out as ices. This is particularly a problem if geometric circumstances prevent the use of infrared absorption spectroscopy. For instance, the flattened structure of disks typically do not allow for a line of sight to line up with an infrared background source. However, at 1–2 AU in typical protoplanetary disks, ice condensation is much less of a problem, since gas and dust temperatures are hundreds of K or more, effectively preventing any ice formation.

The Spitzer Space Telescope obtained sensitive mid-infrared spectroscopy of hundreds of protoplanetary disks around young stars across the stellar mass range.[12-14] Many of the higher-quality spectra revealed a multitude of lines from many bulk carriers of the volatile elements carbon, oxygen and nitrogen.[7,15] The species that have so far been seen include H_2O, CO, HCN, C_2H_2, CO_2 and OH. The lines come from rovibrational states, or high rotational states in the case of water, with upper level energies ranging from a few hundred K to a few thousand K.[16] Furthermore, the wavelength coverage was broad and provided strict upper limits on most molecules potentially carrying a significant fraction of the total elemental abundance of C, N and O. With these data sets it became possible to begin constructing a complete chemical inventory of planet-forming regions.[17]

1.3 Previously derived inner disk concentrations

One challenge is that the Spitzer spectra do not resolve the individual lines, leading to strong line blending. In the analysis of the Spitzer spectra, sophisticated models were often considered over-powered, and relative molecular concentrations were therefore calculated using highly simplified "slab models", in which the disk is modeled as a dustless slab of gas at rest with a single kinetic temperature and level populations in thermodynamic equilibrium (TE). While these assumptions are unlikely to hold for protoplanetary disks, with the possible exception of TE, such models are usually able to reproduce the observed spectra of organics and water, at least over a limited wavelength range. Indeed, most slab models are in fact degenerate, resulting in a wide range of allowed concentrations.

Carr and Najita[12] used a slab model in which they allowed all of the parameters, a single temperature (T), column density (N), and surface emitting area (A), to be free. They found CO : H_2O of $\sim$1, and HCN : H_2O of a few percent. Significantly, this required the emitting area of HCN to be much smaller than that of H_2O, although the temperatures were about the same ($\sim$600 K). CO_2 was also found to have a smaller emitting area than that of H_2O, but at a much lower temperature of $\sim$350 K and a wide range of allowed concentrations, up to 15% but as low as 0.2%. Salyk *et al.*[8] did not allow all parameters to be free, but fixed the emitting areas to that of water, essentially assuming that the chemistry remains constant throughout the inner few AU of the disks, resulting in similar $CO : H_2O$ ratios of ~ 1, but much lower $HCN : H_2O$ ratios of 0.1% or less. Further, for CO_2 the concentration was as low as 0.1%, but was also found to be highly degenerate.

While arriving at different results, the two studies are actually in general and quantitative agreement in that smaller emitting areas lead to higher column densities and vice versa. One problem is that in both cases there is no guarantee that the derived combination of temperatures and emitting areas are consistent with a disk temperature structure. This is the aspect that we now attempt to address; if a realistic disk structure is imposed will this eliminate some of the degenerate parameter space, and allow us to measure inner disk chemistries with precision high enough to determine whether they are different from those measured in protostellar envelopes?

2 Concentration retrieval using two-dimensional radiative transfer models

One possible way to limit the degeneracies of slab models is to require that all observed lines from all species are reproduced by a single two-dimensional disk structure in Keplerian rotation, with a self-consistent temperature structure. To this end, we use a two-dimensional radiative transfer model to retrieve the molecular concentrations from infrared spectroscopic observations to higher levels of accuracy than previously achieved with single-slab models. In particular, the higher-order model yields estimates for the absolute local concentration in the disk photospheres $(n(X)/n(H))$. The modeling procedure is based on the commonly used continuum radiative transfer code RADMC,[18] along with the line raytracer, RADLite.[19]

There are still a number of simplifying assumptions in order to keep the problem contained. For instance, we set the gas temperature to that of the dust, which is a good approximation for the hydrogen column density of the molecular layer at mid-infrared wavelengths $(>10^{22})$.[20] For atomic species and highly optically thick transitions probing higher layers, this approximation would likely be less valid. Similarly, we assume that all molecular level populations, except for CO, are in thermodynamic equilibrium. This is generally justified by the high densities of the gas that form the lines $(n \sim 10^8\text{--}10^{12} \text{ cm}^{-3})$, although future studies should include a general non-LTE treatment to verify this assumption. The non-LTE calculations carried out for CO are based on the escape probability methodology of Woitke *et al.*,[21] and are described in greater detail by Lockwood *et al.* We also assume a constant gas-to-dust ratio that was increased from the canonical value of 100 to 500 to include some dust settling. Since the disk structure is fixed by dust emission (Blevins *et al.*, in preparation), a different choice of gas-to-dust ratio will simply scale the total disk mass as $\sim g2d/500$, and the absolute chemical concentrations as $\sim 500/g2d$. Relative concentrations are not affected.

2.1 Modeling procedure

For a given disk, we begin by fitting the two-dimensional dust distribution to the observed spectral energy distribution (SED), after correcting for any interstellar extinction. The free parameters include the disk mass, the shape of the disk surface (the flaring index), and the height and radius of the disk. Optimization of these parameters results in a dust temperature and mean intensity spectrum at every location in the disk. Given the dust temperature model and a molecular concentration at every (R,z) point in the disk, line spectra are rendered with the raytracer, RADLite. In other investigations, the local molecular concentrations are often generated using a chemical model,[21] but since we are interested in determining what the concentrations are, independent of any model, we parameterize the spatial concentration distribution using an inner and outer abundance, along with a transition radius. The transition radius could be synonymous with a snow line – the radius at which water or another molecule freezes out – or it could be a purely chemical boundary. The same methods were used to model the water vapor emission for TW Hya.[22]

Since we only fit two parameters for each molecule (the outer abundance can typically be set to 0 in this context), we can find the best solution by generating a grid of models on those two parameters and minimizing the difference between the model spectra and the data after matching their continua using an additive straight line to correct for any minor differences ($\sim$10% in this case). Note that RADLite does carry out a self-consistent calculation of the dust continuum, but matching the model continuum to that of the broad band data to much better than 10% requires a great deal of fine tuning with little effect on the conclusions. Nevertheless, the immediate advantage of using a two-dimensional disk model, constrained by continuum observations, is that the temperature of the gas forming molecular lines is no longer a free parameter, as it is when using an otherwise unconstrained slab model. Further, the spectra due to each separate molecule are linked because they ultimately have to be produced by the same disk model. Using a slab model there are no such links, and each molecule is completely independent of another.

Altogether, we simultaneously model H_2O, CO, HCN, C_2H_2 and CO_2, accounting for about 2000 individually rendered lines and line-images in the covered spectral ranges.

2.2 Example disk: RNO 90

Protoplanetary disks with strong molecular emission are common, and since slab models produce very similar results for different sources,[8] we can choose any one as an example. In the future, more elaborate studies may include larger disk samples. For this paper we chose RNO 90, a young solar-mass star (G5) with strong water vapor emission in the mid-infrared. RNO 90 is located near (but not inside) the LDN 43 dark cloud, roughly 10 degrees from the well-known ρ Ophiuchus young stellar cluster, at a distance generally assumed to be that of the cluster of 125 pc.[23,24] The disk is optically thick in the infrared at all radii, and there have been no observations of structural "complications" as seen in more evolved disks, such as TW Hya,[25] although RNO 90 is also not as thoroughly observed. The geometric parameters of the RNO 90 disk are summarized in Table 1.

2.3 Observations

RNO 90 has one of the highest quality mid-infrared molecular spectra available,[15] and has an increasing amount of geometric and structural information available, including submillimeter imaging,[26] Herschel spectroscopy,[27] as well as high

Table 1 Two-dimensional model parameters for RNO 90

Structural parameters

Stellar mass	M_*	$1.5 M_\odot$
Disk mass	M_{disk}	$1.5 \times 10^{-2} M_\odot$
Gas-to-dust ratio	M_{gas}/M_{dust}	500
Outer radius	R_{out}	150 AU
Flaring index	α	0.06
Dust sublimation temperature	T_{sub}	1800 K
H_2O inner concentration	$n(H_2O)/n(H)$	2.5×10^{-3}
H_2O transition radius	$R(H_2O)$	4 AU
H_2O outer concentration	$n(H_2O)/n(H)$	0
CO inner concentration	$n(CO)/n(H)$	7.5×10^{-5}
CO transition radius	$R(CO)$	4 AU
CO outer concentration	$n(CO)/n(H)$	0
C_2H_2 inner concentration	$n(C_2H_2)/n(H)$	10^{-5}
C_2H_2 transition radius	$R(C_2H_2)$	0.2 AU
C_2H_2 outer concentration	$n(C_2H_2)/n(H)$	0
HCN inner concentration	$n(HCN)/n(H)$	7.5×10^{-6}
HCN transition radius	$R(HCN)$	0.5 AU
HCN outer concentration	$n(HCN)/n(H)$	6.5×10^{-7}
CO_2 inner concentration	$n(CO_2)/n(H)$	2.5×10^{-7}
CO_2 transition radius	$R(CO_2)$	4 AU
CO_2 outer concentration	$n(CO_2)/n(H)$	0

resolution CO spectroscopy and spectro-astrometry.[28] Specifically, we use the 10–40 μm Spitzer spectrum[15] (with a resolving power of 500 km s^{-1}) in combination with the high resolution spectrum of the 4.7 μm CO rovibrational band obtained with CRIRES (resolving power of $\sim$3 km s^{-1}) on the Very Large Telescope.[28] The CO spectrum is highly complementary to the Spitzer spectrum, since the former fully resolved the line profiles and velocity structure, while the latter fails to resolve most individual lines. The details of the data reduction can be found in the references given.

3 The observed inner disk chemistry of RNO 90

The infrared spectrum of RNO 90 is typical in that it shows strong emission lines from, at least, CO, H_2O, OH, C_2H_2, HCN and CO_2. We first fit a water model to the many water lines in the 10–16 μm region. There are sufficient water lines to fit a more detailed radial structure, but to limit the scope of this paper, we use a simple step function with a sharp phase transition at the snow line. For RNO 90, this occurs at 2 AU in the mid-plane, and at roughly 4 AU nearer to the surface. Note that most of the lines considered have high enough upper level energies that they predominantly trace the region well inside 4 AU. Thus, the water model is not sensitive to the exact choice of boundary. Once the water model parameters are fixed, we fit the other molecular species, one at a time, to their major spectral bands as indicated in Fig. 1 and 2. The absolute concentrations are listed in Table 1, and the relative (to water) concentrations are shown in Table 2, where they are compared to the column density values from different slab models.

3.1 CO

The CO rovibrational lines, which are fully velocity-resolved by CRIRES, are fitted without degeneracies. Their profiles and strengths are matched well by the structural model of RNO 90, which supports our choice of inner disk structure (see Fig. 2). One departure is that the $\nu = 2 - 1$ line profiles are more double-peaked than the data. This may likely be corrected by adopting a smooth transition of the inner disk edge, rather than the sharp inner edge used by RADMC. The inner disk CO/H model concentration is 7.5×10^{-5}, which is very close to the canonical value supported by thermochemical models.[29] This also matches the abundance of dense clouds in the interstellar medium.[30] That the CO concentrations are so close to the canonical values supports the absolute values derived for other chemical species formed in the same gas.

3.2 HCN

In the case of HCN, which has a spectrally resolved Q branch band structure, the model fit is definite. Conversely, the slab models are degenerate between optically thick, high column density parameters and optically thin, high temperature parameters.[8] Since the temperature distribution is fixed in our 2D model, we find that the HCN emission must be optically thick. To reproduce the total band strength we are driven toward a model in which the HCN emission is much closer to the star than the water emission. Further, the HCN bands (near 14.0 and 14.28 μm) cannot be fitted simultaneously with a constant abundance model. Thus, in the case of HCN, we impose a boundary around 0.5 AU with a high concentration on the inside of the boundary, and a slightly lower concentration outside; both regions contribute significantly to the combined HCN emission spectrum.

3.3 C_2H_2

C_2H_2 behaves like HCN, only with a greater difference between the inner and outer abundances. The band shape seems to requires a very high concentration in

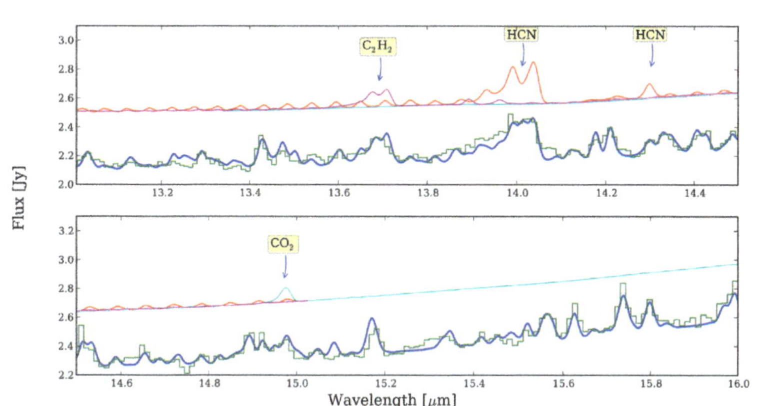

Fig. 1 Best-fitting two-dimensional model of the Spitzer mid-infrared spectrum of the RNO 90 protoplanetary disk. The continuum of the model has been additively adjusted by ~10% to match the observations exactly. The total spectrum is overplotted on the data, while model spectra of the individual species (HCN, C_2H_2 and CO_2) are offset for clarity. Unmarked model features are due to water.

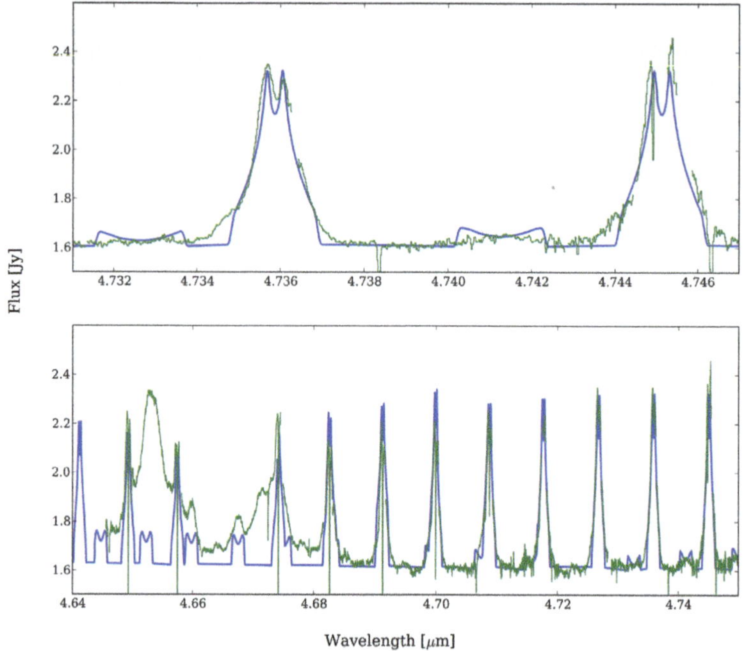

Fig. 2 As in Fig. 1, but for the rovibrational CO fundamental band, as observed with CRIRES. The double-peaked line structure is due to the Keplerian rotation of the disk. The top panel shows the line profiles, while the lower panel shows a wider view of the CO fundamental spectrum. The broad lines at 4.655 and 4.673 μm are due to atomic hydrogen.

Table 2 Comparison of observed relative molecular concentrations using different retrieval methods

		Carr and Najita, 2008, 2011	Salyk *et al.*, 2011	RADLite
Model type	Band	Slab	Slab with fixed area	2D dust + gas
$n(CO)/n(H_2O)$	ν_1	0.63 ± 0.2	0.79	0.03
$n(C_2H_2)/n(H_2O)$	ν_5	$(1.0 \pm 0.5) \times 10^{-2}$	2.5×10^{-4}	4×10^{-3}
$n(HCN)/n(H_2O)$	ν_2	$(8 \pm 5) \times 10^{-2}$	1.6×10^{-3}	3×10^{-3}
$n(CO_2)/n(H_2O)$	ν_2	$(8 \pm 7) \times 10^{-2}$	8×10^{-4}	1×10^{-4}

the innermost region of the disk (the current model uses 0.2 AU). The band is not as well resolved or as strong as the HCN band in RNO 90, so the C_2H_2 parameters are still uncertain, even with the two-dimensional model.

3.4 CO_2

CO_2 is a particularly interesting case. The CO_2 Q branch is not spectrally resolved, but it can be fitted sufficiently with a constant concentration defined throughout the disk (out to the water transition radius of 4 AU). The best fitting concentration has a very low value of 10^{-4} relative to water. Analogous to the slab models, the maximum concentration can be increased by introducing a radial chemical boundary, beyond which the concentration drops to 0 (or a very low value). However, even if the transition radius is set very low, there is still a large area of the disk where the $n(CO_2)/n(H_2O)$ concentration is several orders of magnitude below the dense cloud/protostellar value of 0.25. It is difficult to explain this if the inner disk volatiles are directly derived, without chemical alteration, from evaporating icy bodies that were originally formed at low temperatures in the outer disk or in a protostellar envelope.

4   Discussion

It is clear from the derived molecular concentrations that they do not resemble those of prestellar ices (Fig. 3). In particular, CO_2 has much lower concentrations in disks. This is in particularly stark contrast to the nearly universal (within a factor of 2) concentration of 25% CO_2 relative to water in low-mass star forming regions.[31]

Conversely, thermochemical models predict relatively low CO_2 concentrations[20,32] in the planet-forming regions of protoplanetary disks. Indeed, the high concentration of CO_2 (and other species) in interstellar ices has long been seen as strong evidence for the action of surface chemistry.[33] The model concentrations of HCN also tend to be low, with most of the nitrogen carried by N_2 in the inner disk. As can be seen from the absolute concentrations (relative to H) listed in Table 3, some of the observed values are consistent with gas-phase thermochemical models. The exceptions are water, which is more abundant by as much as an order of magnitude, and C_2H_2, which is observed to be several orders of magnitude more abundant.

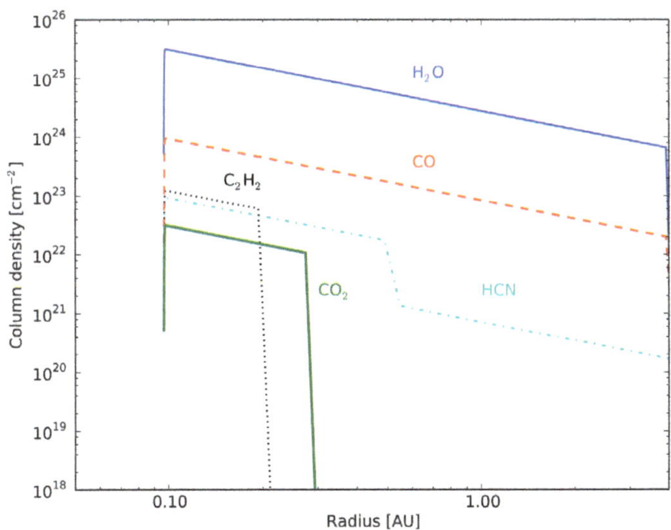

Fig. 3   Vertically integrated molecular column densities as a function of disk radius. The observations trace only a small fraction (~0.1%) of the total column included in the disk, so the values of the figure should be considered an extrapolation.

Table 3 Comparison of the observed absolute molecular concentrations with thermo-chemical disk models at 1 AU

	Willacy and Woods, 2009[32]	Najita *et al.*, 2011[20]	RADLite (this paper)
$n(H_2O)$	1.4×10^{-4}	10^{-6}–10^{-4}	5×10^{-3}
$n(CO)$	2.2×10^{-5}	10^{-4}	7.5×10^{-5}
$n(C_2H_2)$	7.2×10^{-7}	10^{-10}–10^{-8}	1×10^{-5a}
$n(HCN)$	5.1×10^{-6}	10^{-7}	5×10^{-6}
$n(CO_2)$	1.4×10^{-6}	10^{-10}–10^{-7}	2.5×10^{-7}

[a] The C_2H_2 concentration is valid inside 0.2 AU. Outside of this, it is at least an order of magnitude lower.

One potential way of explaining the increase of the water concentration (beyond what is possible with the solar abundance of oxygen – $\sim$4 $\times$ 10^{-4} from ref. 34) is if the water has been preferentially enriched by advection (inward migration of icy bodies against the disk pressure gradient).[1] Such bodies might be preferentially water-rich if they formed outside the water snow line, but inside the CO_2, HCN and CO lines. This was proposed by Banzatti *et al.*[35] to explain the high water column densities suggested by some slab models. Alternately, we may have underestimated the gas-to-dust ratio. Increasing this will place higher column densities into the line formation region, and therefore decrease the absolute concentrations, bringing water into the canonical range, but suppressing CO_2 and HCN below the thermochemical models.

It is arguably less surprising that the relative concentrations indicate a very different chemistry from those of ices, even if the material ultimately originated in evaporating icy bodies. This could be related to the extremely short chemical time scales at the relevant temperatures and densities.[36] Note that the time scale for gas-phase chemistry in the molecular layer of the inner disk is extremely fast (<1 h inside 10 AU). The same region is somewhat shielded from direct irradiation by ultra-violet photons from the central star and accretion shock, such that photochemistry time scales are likely somewhat longer (>1 year).

Another aspect that we have not addressed by modeling a single source is the known strong dependence of the chemistry on stellar type. Hot young stars (Herbig Ae) have disks with weak or absent molecular infrared emission,[15] while disks around very low-mass stars have weak water, but strong organic lines.[37] A common two-dimensional model, such as that presented here, is needed to derive comparable chemical concentrations, and get a complete picture of protoplane-tary chemistry across the stellar mass range.

It seems that whichever way we look at the data, the chemistry of inner disks does not appear to be reflective of a protostellar origin. We would not expect to see strong water emission at the lower canonical water abundance, but would have expected to see a very strong CO_2 feature. Other species would also show strong features at prestellar abundances, including CH_3OH and NH_3, but are yet to be seen.[38] However, it is very likely that new and interesting chemistry will appear in the mid-infrared once we have space-based spectroscopy at higher spectral resolution (such as will be available with the James Webb Space Telescope). With such data we will be in a better position to understand what indeed drives planet-forming chemistry, knowing that inner disks are highly active chemical factories.

Acknowledgements

This work is based in part on observations made with the Spitzer Space Telescope, which is operated by the Jet Propulsion Laboratory, California Institute of Technology under a contract with NASA. Support for this work was provided by NASA through an award issued by JPL/Caltech. The work is based in part on observations made with ESO Telescopes at the La Silla Paranal Observatory under programme ID 179.C-0151.

References

1 F. J. Ciesla and J. N. Cuzzi, *Icarus*, 2006, **181**, 178–204.
2 K. I. Öberg, A. C. A. Boogert, K. M. Pontoppidan, S. van den Broek, E. F. van Dishoeck, S. Bottinelli, G. A. Blake and N. J. Evans, II, *Astrophys. J.*, 2011, **740**, 109.
3 E. R. D. Scott, *Annu. Rev. Earth Planet. Sci.*, 2007, **35**, 577–620.
4 M. J. Mumma and S. B. Charnley, *Annu. Rev. Astron. Astrophys.*, 2011, **49**, 471–524.
5 L. Grossman, *Geochim. Cosmochim. Acta*, 1972, **36**, 597–619.
6 R. van Boekel, M. Min, C. Leinert, L. B. F. M. Waters, A. Richichi, O. Chesneau, C. Dominik, W. Jaffe, A. Dutrey, U. Graser, T. Henning, J. de Jong, R. Köhler, A. de Koter, B. Lopez, F. Malbet, S. Morel, F. Paresce, G. Perrin, T. Preibisch, F. Przygodda, M. Schöller and M. Wittkowski, *Nature*, 2004, **432**, 479–482.
7 J. S. Carr and J. R. Najita, *Astrophys. J.*, 2011, **733**, 102.
8 C. Salyk, K. M. Pontoppidan, G. A. Blake, J. R. Najita and J. S. Carr, *Astrophys. J.*, 2011, **731**, 130.
9 K. M. Pontoppidan, E. F. van Dishoeck and E. Dartois, *Astron. Astrophys.*, 2004, **426**, 925–940.
10 D. C. B. Whittet, *Astrophys. J.*, 2010, **710**, 1009–1016.
11 A. C. A. Boogert, K. M. Pontoppidan, C. Knez, F. Lahuis, J. Kessler-Silacci, E. F. van Dishoeck, G. A. Blake, J.-C. Augereau, S. E. Bisschop, S. Bottinelli, T. Y. Brooke, J. Brown, A. Crapsi, N. J. Evans, II, H. J. Fraser, V. Geers, T. L. Huard, J. K. Jørgensen, K. I. Öberg, L. E. Allen, P. M. Harvey, D. W. Koerner, L. G. Mundy, D. L. Padgett, A. I. Sargent and K. R. Stapelfeldt, *Astrophys. J.*, 2008, **678**, 985–1004.
12 J. S. Carr and J. R. Najita, *Science*, 2008, **319**, 1504.
13 C. Salyk, K. M. Pontoppidan, G. A. Blake, F. Lahuis, E. F. van Dishoeck and N. J. Evans, II, *Astrophys. J.*, 2008, **676**, L49–L52.
14 I. Pascucci, D. Apai, K. Luhman, T. Henning, J. Bouwman, M. R. Meyer, F. Lahuis and A. Natta, *Astrophys. J.*, 2009, **696**, 143–159.
15 K. M. Pontoppidan, C. Salyk, G. A. Blake, R. Meijerink, J. S. Carr and J. Najita, *Astrophys. J.*, 2010, **720**, 887–903.
16 R. Meijerink, K. M. Pontoppidan, G. A. Blake, D. R. Poelman and C. P. Dullemond, *Astrophys. J.*, 2009, **704**, 1471–1481.
17 J. E. Bast, F. Lahuis, E. F. van Dishoeck and A. G. G. M. Tielens, *Astron. Astrophys.*, 2013, **551**, A118.
18 C. P. Dullemond and C. Dominik, *Astron. Astrophys.*, 2004, **417**, 159–168.
19 K. M. Pontoppidan, R. Meijerink, C. P. Dullemond and G. A. Blake, *Astrophys. J.*, 2009, **704**, 1482–1494.

20 J. R. Najita, M. Ádámkovics and A. E. Glassgold, *Astrophys. J.*, 2011, **743**, 147.

21 P. Woitke, I. Kamp and W.-F. Thi, *Astron. Astrophys.*, 2009, **501**, 383–406.

22 K. Zhang, K. M. Pontoppidan, C. Salyk and G. A. Blake, *Astrophys. J.*, 2013, **766**, 82.

23 L. Loinard, R. M. Torres, A. J. Mioduszewski and L. F. Rodríguez, *Astrophys. J.*, 2008, **675**, L29–L32.

24 E. E. Mamajek, *Astron. Nachr.*, 2008, **329**, 10.

25 R. L. Akeson, R. Millan-Gabet, D. R. Ciardi, A. F. Boden, A. I. Sargent, J. D. Monnier, H. McAlister, T. ten Brummelaar, J. Sturmann, L. Sturmann and N. Turner, *Astrophys. J.*, 2011, **728**, 96.

26 S. M. Andrews and J. P. Williams, *Astrophys. J.*, 2007, **671**, 1800–1812.

27 L. Podio, I. Kamp, C. Codella, S. Cabrit, B. Nisini, C. Dougados, G. Sandell, J. P. Williams, L. Testi, W.-F. Thi, P. Woitke, R. Meijerink, M. Spaans, G. Aresu, F. Ménard and C. Pinte, *Astrophys. J.*, 2013, **766**, L5.

28 K. M. Pontoppidan, G. A. Blake and A. Smette, *Astrophys. J.*, 2011, **733**, 84.

29 R. P. Hein Bertelsen, I. Kamp, M. Goto, G. van der Plas, W.-F. Thi, L. B. F. M. Waters, M. E. van den Ancker and P. Woitke, *Astron. Astrophys.*, 2014, **561**, A102.

30 R. L. Dickman, *Astrophys. J. Suppl.*, 1978, **37**, 407–427.

31 K. M. Pontoppidan, G. A. Blake, E. F. van Dishoeck, A. Smette, M. J. Ireland and J. Brown, *Astrophys. J.*, 2008, **684**, 1323–1329.

32 K. Willacy and P. M. Woods, *Astrophys. J.*, 2009, **703**, 479–499.

33 L. B. D'Hendecourt, L. J. Allamandola and J. M. Greenberg, *Astron. Astrophys.*, 1985, **152**, 130–150.

34 N. Grevesse, M. Asplund, A. J. Sauval and P. Scott, *Astrophys. Space Sci.*, 2010, **328**, 179–183.

35 A. Banzatti, M. Meyer, K. Pontoppidan and S. Bruderer, *Protostars and Planets VI*, Heidelberg, July 15–20, 2013. Poster #2S034, 2013, p. 34.

36 D. Semenov and D. Wiebe, *Astrophys. J. Suppl.*, 2011, **196**, 25.

37 I. Pascucci, G. Herczeg, J. S. Carr and S. Bruderer, *Astrophys. J.*, 2013, **779**, 178.

38 A. M. Mandell, M. J. Mumma, G. A. Blake, B. P. Bonev, G. L. Villanueva and C. Salyk, *Astrophys. J.*, 2008, **681**, L25–L28.

PAPER

Exploring the origins of carbon in terrestrial worlds[†]

Edwin Bergin,[a] L. Ilsedore Cleeves,[a] Nathan Crockett[b] and Geoffrey Blake[b]

Received 16th January 2014, Accepted 12th February 2014

DOI: 10.1039/c4fd00003j

Given the central role of carbon in the chemistry of life, it is a fundamental question as to how carbon is supplied to the Earth, in what form and when. We provide an accounting of carbon found in solar system bodies, and in particular a comparison between the organic content of meteorites and that in identified organics in the dense interstellar medium (ISM). Based on this accounting, identified organics created by the chemistry of star formation could contain at most ~15% of the organic carbon content in primitive meteorites and significantly less for cometary organics, which represent the putative contributors to starting materials for the Earth. In the ISM ~30% of the elemental carbon exists as CO, either in gaseous form or in ices, with a typical abundance of $\sim 10^{-4}$ (relative to H_2). Recent observations of the TW Hya disk find that the gas phase abundance of CO is reduced by an order of magnitude compared to this value. We explore an explanation for this observation whereby the volatile CO is destroyed *via* gas phase processes, providing an additional source of carbon for organic material to be incorporated into planetesimals and cometesimals. This chemical processing mechanism requires warm grains (>20 K), partially ionized gas, and sufficiently small (a_{grain} < 10μm) grains, *i.e.* a larger *total* grain surface area, such that freeze-out is efficient. Under these conditions, static (non-turbulent) chemical models predict that a large fraction of the carbon nominally sequestered in CO can be the source of carbon for a wide variety of organics that are present as ice coatings on the surfaces of warm pre-planetesimal dust grains.

1   Introduction

The birth of terrestrial worlds involves the accumulation of material over a potentially wide range of distances in the young pre-planetary solar nebula disk (*e.g.* Morbidelli *et al.*[1] and references therein). Our general expectation is thus that Earth formed from material both interior to and beyond its location at

[a]*University of Michigan, Department of Astronomy, 500 Church St., Ann Arbor, MI 48109, USA. E-mail: ebergin@umich.edu; Fax: +1 734 763 6317; Tel: +1 734 615 8720*

[b]*California Institute of Technology, Division of Geological & Planetary Sciences, MS 150-21, Pasadena, CA 91125, USA*

1 Astronomical Unit from the Sun. This is important as prevailing theories argue that, due to the hot temperatures of pre-planetary materials, the Earth formed dry with little water. For that reason, the Earth is thought to have received water from hydrated rocks (asteroids) and perhaps icy planetesimals that exist beyond the so-called solar nebula disk† snow line; although alternative theories exist (see summary presented by van Dishoeck *et al.*[2]).

We aim to explore some of the chemical/physical mechanisms associated with a potentially related topic: how the Earth received its carbon. Here we will draw upon previous works[3,4] to discuss the "carbon problem" that outlines the likelihood that the Earth (using meteoritic material as a proxy) is carbon-poor relative to the carbon available at its birth. Two additional key facets are notable:

1. Most carbon in the solar nebular disk was in volatile form. In the interstellar medium carbon monoxide accounts for $\sim$30% of elemental carbon. Thus it is reasonable to assume that a large fraction of the volatile carbon in a disk resides in gaseous or solid state CO.

2. Most meteoritic and much of cometary carbon is in organic form as opposed to volatile CO, CO_2, or even CH_3OH. In this regard, meteorites represent a likely source of carbon, and other volatiles such as water, to the young Earth.[1,5,6]

In this contribution we explore potential connections between volatile CO and organic/molecular ices that might contribute to pre-planetary rocks. This exploration is motivated by a recent indirect measurement of the CO abundance in a nearby disk,[7] showing that CO is not a significant reservoir (<10%) of elemental carbon in layers where CO is expected to be in the gaseous phase (*i.e.* not a result of CO freeze-out). Thus carbon is somehow processed from volatile CO and potentially into the form of pre-meteoritic organics. Favre *et al.*[7] hypothesized that one mechanism for this to occur is *via* the X-ray ionization of He in the surface layers of the disk. The resulting He^+ atoms can react with gaseous CO and gradually extract the carbon, which can then be processed *via* disk chemistry into more complex, and less volatile forms, that can locally freeze onto the surface of cold dust grains. This process was first noted in an earlier Faraday discussion by Aikawa *et al.*[8] Here we present new models that directly illustrate this effect and its key dependencies on sources of disk ionization and the surface area of small grains.

In §2 we will outline the carbon problem in the context of the composition of bodies in the solar system and interstellar medium. We also explore the question of how much of the organic carbon in solar system bodies might have originated in the interstellar medium. We will also summarize and expand upon the result of Favre *et al.*[7] to support the inference that chemical processing of volatile CO appears to be active. In §3 we will present new results from a chemical model that explores the time dependence of CO chemical processing in the context of a realistic disk model, and in §4 we will summarize the implications for the overall chemical evolution of carbon within pre-planetary materials.

---

† In this contribution we will refer to the disk of material out of which our planetary system was formed as the solar nebula disk and other extra-solar disk systems as protoplanetary disks.

	

2   Carbon in the solar system, ISM, and protoplanetary disks

2.1   Carbon in the solar system and the interstellar medium

Fig. 1 illustrates the dramatic differences in the amount of elemental carbon relative to silicon incorporated into planets, comets, and asteroids represented by meteorites. Exploring the relative amounts of carbon, we see that icy bodies (*i.e.* comets) contain an abundance of carbon in the form of ices and dust (perhaps amorphous carbon) comparable to that seen in the Sun. In contrast, the Earth's mantle has many orders of magnitude less carbon (relative to Si) than was available during formation as represented by the Sun or the diffuse ISM. The Earth's mantle is predominantly comprised of silicate minerals with a total carbon content of $\sim10^{23}$ g.[9] This is a tiny fraction of the Earth's mass ($\sim0.002\%$), but we lack knowledge regarding the carbon content in the Earth's core. Thus while the mantle is clearly carbon-poor relative to the elemental carbon available at birth, we cannot make definitive statements regarding the bulk Earth. Meteorites, in particular undifferentiated carbonaceous chondrites (class CI), have long been posited as tracers of the starting materials of the Earth and these rocks have an order of magnitude depletion of carbon relative to the amount available at formation.

The severe depletions of C suggest that carbon must have been present in fairly volatile forms in the solar nebula disk. One suggestion is that Earth's carbon was supplied from meteoritic organics.[5,6] We present a new perspective on this issue through the analysis of the most complete spectral observations (in terms of

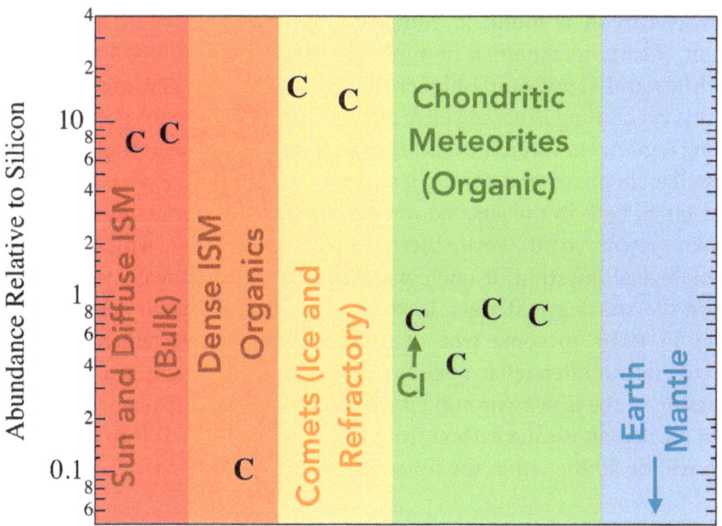

Fig. 1   Plot of total elemental carbon abundance in various solar system bodies (Sun; comets Hale-Bopp and Halley; 4 chondritic meteorites, classes CI, CV, CM, CO; Earth mantle); and the interstellar medium relative to elemental silicon. Figure adapted from Lee *et al.*[3] and Pontoppidan *et al.*[4] and references for abundance estimates are given in those publications. The abundances of dense ISM organics is taken from Crockett *et al.*[10] and compiled for the first time here.

spectral coverage) of the star-forming molecular gas inventory in the Herschel spectrum of Orion KL.[10] This region is the nearest example of massive star formation (an environment analogous to that in which the Sun was thought to have formed).[11] The unique Orion spectrum is the first example in which we can examine a near complete chemical census (to observational sensitivity limits) in the highest temperature gas in which we know the ices have been fully evaporated. Since organics are theorized to form on the surfaces of dust as ices, we are therefore tracing the composition of simple organics provided (as ice) to the formation of stellar and planetary systems.

For this comparison we have summed up the carbon abundances in detected organics to provide a total organic carbon abundance relative to hydrogen, and normalized these to the silicon abundance from present day cosmic abundance estimates.[12] Orion KL abundances were calculated from the column densities estimated by matching emission from numerous transitions and normalization to the total gas (H_2) column using over 7 transitions of $C^{18}O$ and also $C^{17}O$ (taken from Plume *et al.*[13]); thus these abundances are very robust. The errors are dominated by the beam couplings between the detected organic and the Herschel beam (which we derived from the ALMA science verification data of Orion) and are likely below $\sim$25%. For this comparison we have summed up the carbon abundances in detected organics to provide a total organic carbon abundance relative to hydrogen and normalized these to the silicon abundance from present day cosmic abundance estimates,[12] assuming all silicon resides in dust grains.

Exploring the most abundant, detected, and identified organics in space, we find that the chemistry associated with stellar birth might have created up to $\sim$15% of the organics that could be placed into meteorites. Carbon in comets is found primarily in the dust with a ratio of 2 : 1 between dust : ice,[14] at least in comet Halley, which has *in situ* measurements. Furthermore, about 50% of the refractory carbon is found in complex organic form,[14] which means that the amount of carbon contained in organic (refractory) form is $\gtrsim$5 times that of CI chondrites and significantly higher than the identified ISM organic molecules. Now this accounting is based solely on identified organic molecules and one star-forming region. However, it is also the most complete accounting of the extent of interstellar chemistry to date, and it suggests that the chemistry occurring prior to stellar birth, both in the gas and ice, cannot account for the organic material in the solar system record. We are therefore "missing" carbon, which is a significant astrobiological question, if one considers that astronomically we wish to characterize the starting materials. Instead the carbon found in solar system rocks resides in PAHs to some type of carbonaceous grains, which may originate primarily in the interstellar medium. However, there might also be subsequent processing in the disk environment which could rearrange major carbon carriers and in particular, extract carbon from volatiles and place it into different forms, which will be addressed in the following section.

2.2 Volatile carbon abundance in the TW Hya protoplanetary disk

In general, the determination of chemical abundances in protoplanetary disks is inherently uncertain. This is mainly due to the difficulty of measuring the molecular hydrogen content of the disk, *i.e.* the total gas mass. Because H_2 does not emit at the temperatures characteristic of much of the disk, we are forced to use calibrated proxies, which often have large uncertainties; see discussion in Bergin *et al.*[15] and Williams and Cieza.[16] For TW Hya, the closest planet-forming disk at 51 pc,[17] the determination of chemical abundances is more directly enabled *via* the detection of the fundamental transition of HD at $\sim$112 μm by Bergin *et al.*[15] In the case of HD, the HD/H$_2$ ratio is well characterized to be $3.0 \pm 0.2 \times 10^{-5}$ by Linsky.[18] With rotational energy spacings better matched to the gas temperature and a weak dipole, the emission of HD $J = 1$–0 is in local thermodynamic equilibrium (LTE) and can readily be converted to a gas mass. However, while HD co-exists with H$_2$, the HD $J = 1$–0 transition has a $\Delta E_{1-0} = 128.5$ K, and as a direct result, this transition does not emit appreciably at temperatures below 20 K. The total estimation of the H$_2$ disk-mass thus requires knowledge of the gas thermal structure.[15,19] However, for the determination of the volatile CO abundance this weakness becomes a strength, as gaseous CO is believed to freeze onto grains at temperatures below $\sim$20 K, and thus the emission from both species directly traces the same warm gas.

Thus Favre *et al.*[7] used the ratio of optically thin C^{18}O emission to that of HD in TW Hya (with isotopic ratios) to provide a measure of the CO abundance in the so-called warm molecular layer.[20] The estimated CO abundance is $\leq 10^{-5}$, an order of magnitude below the expected value in the ISM; a level that has been confirmed by an independent analysis.[21] Favre *et al.*[7] discussed a variety of potential pitfalls and theories to explain this low abundance. We explore two of these here: differences in the self-shielding between HD and C^{18}O, and chemical processing.

In the former case HD and C^{18}O are both capable of self-shielding from the destructive effects of ultraviolet photons (including the effects of mutual shielding *via* H$_2$ and CO). If HD reaches full shielding at a layer significantly above that of C^{18}O then the abundance measurement would be a lower limit as HD would trace more mass than C^{18}O. To explore this question, in Fig. 2 we show the self-shielding factor f for each molecule, as a function of the total H$_2$ column, using

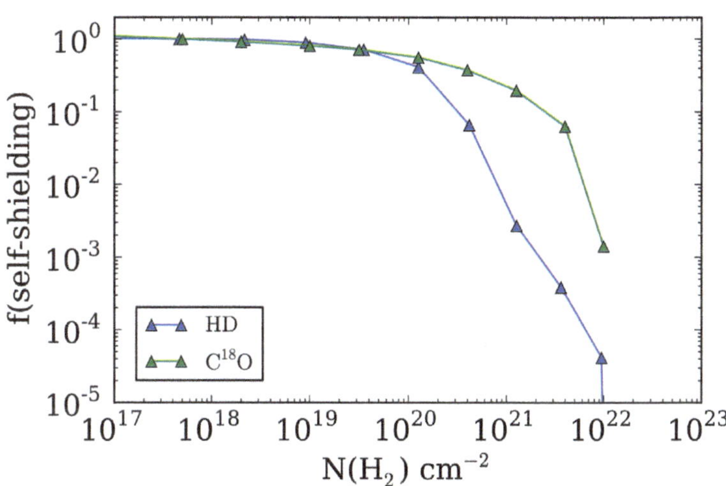

Fig. 2 Self-shielding factors for HD and C^{18}O as a function of the molecular hydrogen column density. Density structure taken from a cut in the disk physical model, presented in §3.2, at 20 AU.

the self-shielding expressions provided by Wolcott-Green and Haiman[22] for HD and Visser *et al.*[23] for $C^{18}O$, and a disk vertical density profile at $R = 20$ AU (see model details below). This demonstrates that HD does indeed self-shield prior to $C^{18}O$ (at $N(H_2) \sim 10^{21}$ cm^{-2} and $\sim 10^{22}$ cm^{-2}, respectively). However, while this difference in self-shielding column exists, the total column density of gas at $R = 20$ AU is $N(H_2) = 7 \times 10^{24}$ cm^{-2} and is above 20 K at all heights. Thus the mass contained in this layer where the differential self shielding occurs constitutes a negligible fraction of the mass (<1%) and cannot account for the order of magnitude deficit in CO abundance relative to interstellar.

The age of TW Hya has some uncertainty, but is believed to be in the range of 3–10 Myr,[24,25] which is older than typical T Tauri disk systems.[16] Within this time frame it is possible that there might be disk chemical processing that extracts carbon from volatile CO and places it into molecular material of lower volatility that may contribute to organics incorporated into cometesimals and planetesimals. In the following sections we will outline a potential theoretical mechanism that will operate in any disk system that is exposed to a source of radiation capable of ionizing helium, in this instance, either X-rays or cosmic rays.

3 Evolution of the volatile carbon reservoir

3.1 Potential CO reprocessing mechanism

A simple mechanism to extract carbon from CO relies on the fact that in dense $(n > 10^5$ cm$^{-3})$ and cold $(T < 50$ K) gas the timescales for molecules to collide with dust grains are short, $t < 10^5$ yrs.[26–28] In this case the chemistry becomes dominated by gas phase freeze-out as molecules are adsorbed on grain surfaces. This effect is well documented in disk systems. In the solar nebula disk, the presence of volatile CO in solar system comets clearly shows that pre-cometary ices contained highly volatile species.[29] In extra-solar disks the formation of pre-planetesimal ices is inferred due to the low abundances of various molecules measured in disks compared to their interstellar medium abundances[30–32] and in some cases *via* direct observations.[33,34]

The disk surface is directly exposed to stellar UV and X-ray radiation producing sharp radial and vertical thermal gradients.[35,36] This leads to vertical chemical stratification into roughly three domains: (1) the photon-dominated surface where molecules are photodissociated, (2) a "warm molecular layer" where dust temperatures are warm enough $(T_d > 20$ K) for CO and other species with similar volatility to be present in UV-shielded gas, (3) the dense $(n \gg 10^5$cm$^{-3})$ cold $(T \leq 20$ K) midplane where molecules are frozen as ices.[20,37] The "warm molecular layer"[20] should be distinguished from the snow line,[38] which typically refers to the radial abundance gradient in the midplane with the snow line located at the sublimation front.

For CO the warm molecular layer exists when the dust temperature is >20 K; however there are a host of more complex carbon-bearing molecules that remain frozen on grain surfaces.[39,40] In this layer, if CO_2 remains frozen, the elemental C/O ratio will be close to unity because water remains as ice for temperatures below $\sim$100–150 K (depending upon pressure)[41] and the carbon and oxygen resides primarily in CO. If there is a source of radiation capable of ionizing helium atoms then the following sequence of reactions will extract small amounts of carbon in ionized atomic form (below ζ is the ionization rate in s^{-1}):

$$He + \zeta \rightarrow He^+ + e \tag{1}$$

$$He^+ + CO \rightarrow C^+ + O + He. \tag{2}$$

Much of this free carbon reforms as CO in the gas, but a fraction can follow a variety of gas-phase or surface pathways to make less volatile ices such as CO_2, hydrocarbons, or CH_3OH. In this fashion the grains operate as a sink to the carbon chemistry as they trap any volatile that has a sublimation temperature below the temperature of the dust grains. Over time the CO abundance might become eroded, a facet noted and discussed by Aikawa *et al.*[42] and elucidated further below.

3.2   Chemical model

To explore this theory and its dependencies more directly we have used the chemical models of Fogel *et al.*[43] and Cleeves *et al.*[44] Briefly these models adopt realistic disk physical structures (gas density and temperature, dust temperature, dust properties) motivated by resolved disk observations and the overall spectral energy distribution.[16] We adopt the observed UV field of TW Hya[45] and X-ray[46] radiation fields for T Tauri systems and use a Monte-Carlo radiation transfer code to solve for the position and wavelength dependent UV and X-ray radiation field within the disk. The model X-ray flux has a total luminosity of $L_{XR} = 10^{29.5}$ erg s^{-1} between 0.1–10 keV with absorption/scattering cross-sections taken from Bethell and Bergin.[47] This code directly includes the propagation of Ly α photons *via* H-atom resonant scattering and dust absorption/scattering;[48] Ly α radiation dominates the stellar UV emission generated by magnetospheric accretion.[45,49] More directly, the adopted physical model is the same as used by Cleeves *et al.*[50] and the reader is referred to that publication for greater detail.

With the physical structure defined we then use a detailed chemical network[51] including gas grain interactions (freeze-out, sublimation, cosmic ray induced desorption, photodesorption) to solve for the time dependence of predicted chemical abundances. A limited amount of grain surface reactions were included that create H_2, H_2O, OH, CH_4, CO_2, H_2CO, and CH_3OH ice. Thus we have (roughly) included pathways for the hydrogenation of O, C, and CO on grain surfaces. This surface chemistry is quite limited compared to the wide variety of potential pathways,[52,53] but allows for an exploration of what might happen to the carbon chemistry. We note that in this regard the end point of the carbon surface chemistry is CH_3OH and its ice abundance is thus an upper limit. The detailed physical structure is given in Fig. 3.

The ionization energy of He is 24.59 eV, which requires either X-ray or cosmic ray ionization. X-ray ionization dominates the deep disk surface below the UV photon penetration depth and also significantly contributes to the ionization of the midplane in the inner tens of AU. Cosmic rays are required to ionize the midplane in the outer ($R > 50$ AU) disk.[50,54] Cleeves *et al.*[50] explored the question of cosmic ray attenuation in the T Tauri system as cosmic ray penetration can be limited by the stellar magnetic field, stellar winds, and disk winds. They predict that the cosmic ray ionization rate within disk systems is likely severely reduced compared to the rate in the general interstellar medium. In this light, since we are using the same model with identical radiation fields, we have adopted their

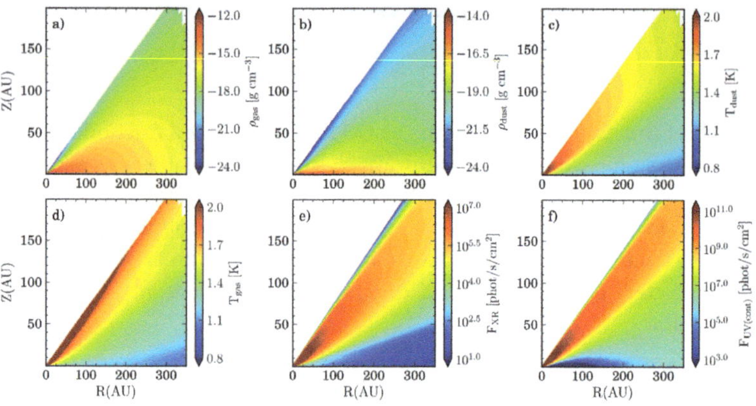

Fig. 3 Physical parameters of the adopted disk model (and Figure) taken from Cleeves *et al.*[50] In panel (a) we provide the gas density, (b) the dust density, (c) dust temperature, (d) gas temperature. All parameters are shown as a function of disk radial and vertical position. In panels (e) and (f) we show the derived UV and X-ray photon flux from our simulations.

ionization structure and explored two models assuming the disk is exposed to the interstellar cosmic ray radiation field and an attenuated model commensurate with cosmic ray modulation comparable to the Sun at solar maximum. An additional factor is the time scale for molecular freeze-out onto grain surfaces as the gas-phase processing of CO only occurs when the grains act as sinks to molecules created in the gas. The freeze-out timescale is $\tau_{fo} = n_{gr}\sigma_{gr}v$, where n_{gr} is the space density of grains, σ_{gr} is the cross-section, and v is the velocity of the gaseous atoms. $n_{gr}\sigma_{gr}$ is set by the distribution of grain sizes. The standard grain size distribution in the interstellar medium follows a power-law of $dn(a)/da \propto a^{-3.5}$, where a is the grain radius.[55] This is the starting point for disk simulations. Assuming a lower cutoff of 20 Å, $n_{gr}\sigma_{gr} \simeq 2 \times 10^{-21}n$ cm^{-1}, as listed by Hollenbach *et al.*,[28] the freeze-out timescale, τ_{fo}, is:

$$\tau_{fo} = 5 \times 10^3 \,\text{yrs} \left(\frac{10^5 \,\text{cm}^{-3}}{n}\right) \left(\frac{20 \,\text{K}}{T}\right)^{\frac{1}{2}} \sqrt{A} \tag{3}$$

where A is the molecule mass in atomic mass units. In our simulations we will adopt this standard value and a second model that assumes grain growth from 0.1 μm to 10 μm which increases the freeze-out timescale by a factor of 1000.

3.3 Results

3.3.1 He ionization rate.

In Fig. 4 we show the predicted ionization rate of He atoms using the standard physical model provided in Fig. 3. The two panels show the ionization rate of He assuming the disk is exposed to the standard interstellar cosmic ray radiation field and one that is attenuated due to magnetized stellar winds. In both cases the disk surface is dominated by X-ray radiation, but the X-ray dominated regime is significantly larger in the case of attenuated cosmic rays. The major difference lies in the midplane. In the model with cosmic rays attenuated by a magnetized stellar wind, radionuclides dominate the ionization of the midplane providing a floor to the ionization rate, which is $\sim 8 \times 10^{-19}$ s^{-1} at

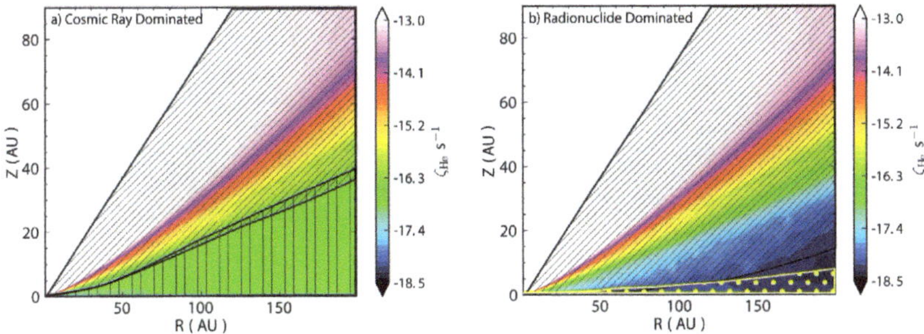

Fig. 4   Ionization rate for He atoms as a function of position in the standard disk model. Two models are shown. Both models include X-ray ionization and radionuclides as described in § 3.2. In panel (a) the model is exposed to the standard interstellar cosmic ray field and in panel (b) the cosmic ray field is reduced as consistent with the attenuation seen in our solar system at the maximum sunspot cycle. In this case, the "floor" to the ionization is provided by radionuclides at the initial abundances determined for the solar nebula disk. In each plot diagonal lines denote where the ionization is dominated by X-rays, vertical lines by cosmic rays, and yellow dots dominated by radionuclide decay.

the midplane at $R = 50$ AU. We have assumed solar nebula disk abundances for the radioactive agents, which include ^{26}Al, ^{36}Cl and ^{60}Fe decay.[56,57] For a discussion regarding the likelihood that disk systems contain radionuclide ionization see Cleeves $et\ al.$[57] and Jura $et\ al.$[58]

A key facet is that as a noble gas, helium is expected to remain in the gas-phase, and, as long as a source of ionization exists, then He^+ will be present in the system. In equilibrium, and in the warm molecular layer where water is frozen as ice, the He^+ abundance can be determined by the balance between its ionization rate and destruction with CO and H_2. The neutralizing reaction with H_2 has a rate of $k_1 = 7.2 \times 10^{-15}$ cm^3 s^{-1} (H_2^+ product channel) and $k_2 = 3.7 \times 10^{-14} e^{(-35\ K/T)}$ cm^3 s^{-1} ($H^+ + H$ product channel),[59] and $k_{CO} = 1.6 \times 10^{-9}$ cm^3 s^{-1} is the rate for the reaction between He^+ and CO.[60] In the following we will sum both channels for the reaction between H^+ with H_2 ($k_{H_2} = k_1 + k_2$) and assume a temperature of 30 K. Thus in equilibrium,

$$n_{He^+}(x_{CO} n_{H_2} k_{CO} + n_{H_2} k_{H_2}) = \zeta_{He} x_{He} n_{H_2}$$

where x_{CO} and $x_{He} = 0.28$ are the CO and helium abundances respectively, both relative to H_2. Solving for n_{He^+} we find,

$$n_{He^+} = 0.28 \zeta_{He}/(x_{CO} k_{CO} + k_{H_2}).$$

The timescale for gas phase CO destruction is then $\tau_{CO} = 1/n_{He^+} k_{CO}$. A typical ionization rate in the deep surface, where X-ray ionization dominates and the dust temperature is $\sim$30 K, is $\sim 10^{-15}$ s^{-1} to 10^{-16} s^{-1} (see Fig. 4). Thus the timescale for CO destruction via reaction with He^+ is $\sim 10^4$–10^6 yrs, depending on ζ_{He^+} and on whether the CO abundance is 10^{-4} (relative to H_2) or significantly lower, at which point neutralization with H_2 dominates.

3.3.2   Evolution of CO abundance. In Fig. 5 we provide the time dependence of the CO abundance structure in our standard disk model exposed to an unattenuated interstellar cosmic ray radiation field. The top panel shows a disk where

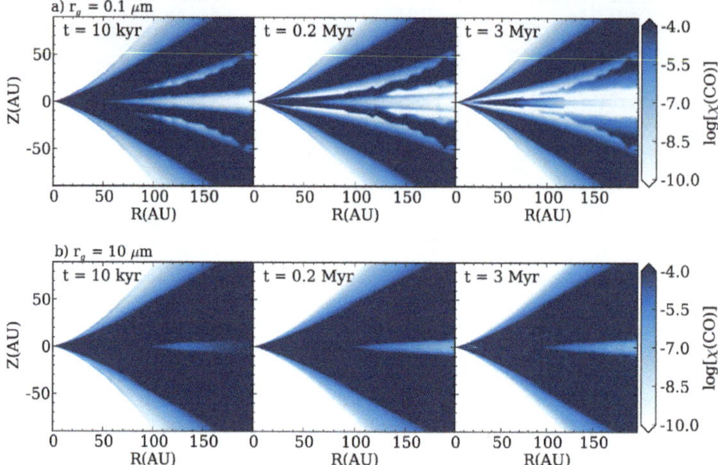

Fig. 5   Chemical abundance of CO (relative to H_2) as a function of position in the disk at selected times. Top panels are for the standard model with interstellar (0.1μm) grains and the bottom panels assume grain growth to 10μm size.

the grain size distribution is similar to the interstellar medium and the bottom panel, a disk that has undergone significant grain evolution. Focusing first at 10 kyr, the warm molecular layer is seen as the large radial band of "normal" CO abundance (10^{-4}) on the surface. In the outer disk $(R > 50\ AU)$ significant abundance structure is predicted with two areas of CO depletion. First is the midplane $(Z = 0)$ where CO is frozen on grains with a snow line, or sublimation front, near 50 AU in this model. The second zone of CO depletion is the radial band seen in the middle of the warm molecular layer. This region is where the reaction with He^+ and subsequent gas phase chemical processing is active. This exists in the middle of the warm molecular layer and is due to the rapid He^+ reaction timescales as the result of X-ray ionization. At later evolutionary steps this zone expands both radially and vertically leading to significant CO depletion in the original warm molecular layer. Note that the CO gas phase depletion seen near the midplane also expands at later times. This is not a result of greater freeze-out but rather it is gas-phase processing of CO; the lower ionization rates provided by scattered X-rays or cosmic rays lead to longer He^+ reaction timescales.

Protoplanetary disks undergo significant grain evolution.[16,61] Thus the bottom panels illustrate how this mechanism depends on the evolution of grain surface area with grain growth. For larger grain sizes (10 μm) the chemical processing mechanism becomes inefficient in destroying CO, with only minor differences seen at 3 Myr compared to earlier times in the same model. In both cases, the mechanism at work begins with carbon extraction from CO, initially giving C^+. C^+ is presumed to remain in the gaseous phase and not freeze-out due to the expectation that grains are negatively charged, which will result in recombination and gas phase ejection of the neutralized product.[62] The C^+ ion is reactive and can gradually find its way into a variety of carbon bearing molecular ions. The molecular ions will more rapidly dissociatively recombine with free electrons or on negatively charged grain surfaces to make a neutral molecule. The carbon-bearing neutral molecule can deplete onto grain surfaces, provided the grain temperature is below its respective sublimation temperature. However, if grain collision timescales are long, the gas-phase chemistry has sufficient time to approach equilibrium which ultimately places the carbon back into CO.

Two points are notable here. It is possible that CO chemical processing is only active at earlier stages prior to significant grain growth, and as grains grow this mechanism becomes less effective. Along the same lines, the layer near the midplane where CO ice is frozen on grains is substantially reduced in the grain growth model, due to the decrease in the total grain surface area and a subsequent increase in the time-scale for freeze out. The ice-dominated layer will be underestimated in this model, which does not include full time-dependent grain evolution, because grains will tend to trap ices as they grow. In essence, whenever the smaller ice coated grains coagulate and grow to larger sizes it is expected that they maintain the pre-existing ice coatings. An additional effect is the expected stratification in the grain size distribution with grains growing to larger sizes in the midplane and smaller grains lofted to greater heights due to coupling to gas motions.[63] For instance, it is well known that small grains do exist on the surfaces of protoplanetary disks as scattering requires the presence of grains with smaller sizes, while the presence of larger grains is inferred *via* detection of centimeter-wave continuum excess emission.[64] Thus the layers where CO chemical processing is active may vary in effectiveness depending on the available surface area of dust grains for gas phase depletion.

3.3.3 **Possible carbon reservoirs in disk surface.** In the preceding section we illustrated the diminishment of the CO gas phase abundance in gas exposed to ionization and a high total grain surface area. The carbon extracted from CO *via* this mechanism is placed on grains in a variety of forms depending on the local physical environment. In Fig. 6 we illustrate the effects of this process in a vertical cut of abundances at 10 AU, in the giant planet-forming zone taken at a timescale of 1 Myr. In this plot we refer to abundances as percentages of volatile carbon. Thus even an abundance of 10^{-2} is high ($\sim 10^{-6}$, relative to H_2) by astronomical detection standards for gaseous molecules which are as low as 10^{-11} in the dense ISM and perhaps somewhat higher in smaller disk systems. Fig. 6a presents the model with the disk exposed to an unattenuated cosmic ray radiation field with $0.1 \mu m$ grains. In this panel the vertical surface layer where this process is active is clearly seen as a drop in the CO abundance near $Z = 1.1$ AU. Deeper in the disk, in the midplane, a similar depletion of gas phase CO is predicted. At both points ($Z = 1.1$ AU and the midplane) there is a rise in the abundance of CO_2, CH_3OH, HC_3N, C_3, and C_4H_2 ice.

The production of these species proceeds through a strict gas phase process. The destruction of CO *via* He^+ atoms creates C^+ ions. It has been known for some time that chemical models, starting with carbon in C^+ within H_2 dominated gas, predict that, prior to the reformation of CO, the chemistry produces a host of complex molecular species.[65,66] If the dust temperature is below the sublimation temperature for a given complex species then it will freeze onto the grain. In this instance the disk densities are sufficiently high ($n > 10^8$ cm^{-3}) that the radiative association reaction between C^+ and H_2, creating CH_2^+, dominates along with a contribution of recombination on negatively charged grains. One potential competitor to this process is the reaction between C^+ and H_2O, which provides an indirect channel to CO *via* HCO^+. However, in these layers the dust grain

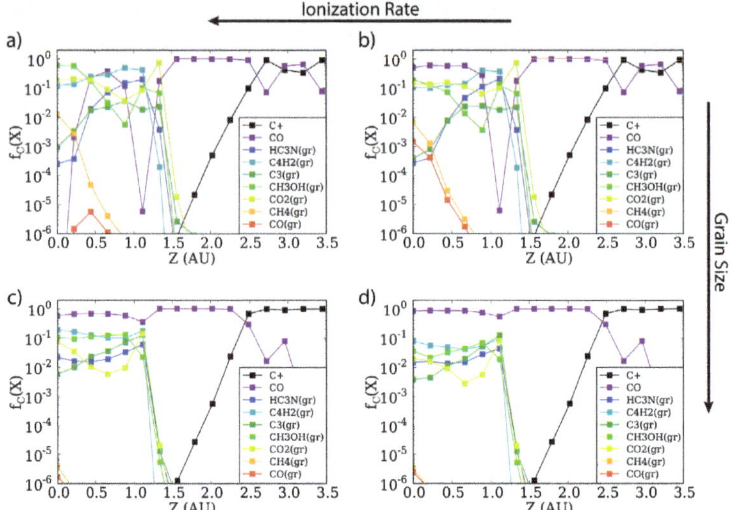

Fig. 6 Plot illustrating dominant volatile carbon carriers in either the gas phase or on the grain surface. Plot is taken from the standard model at ~1 Myr at a radius of 10 AU. Abundances relative to the total amount of volatile carbon (*i.e.* the initial CO abundance) are shown as a function of disk vertical height. All the colors are the same across the panels. Panel (a) unattenuated cosmic ray radiation field and 0.1 μm grains; (b) model with wind-attenuated cosmic rays and 0.1 μm grains; (c) unattenuated cosmic ray radiation field and 10 μm grains; (d) model with wind attenuated cosmic rays and 10 μm grains. Arrows indicate trends.

temperature is well below that required for water sublimation (>100–150 K, depending on pressure).[41] Instead a tenuous layer of water vapor is predicted to exist deeper inside the exposed disk surface induced by UV photodesorption, *e.g.* Dominik *et al.*[67] and Hollenbach *et al.*[28] In our model shown in Fig. 6a, C^+ destruction with H_2 is two orders of magnitude more effective than the reaction with photodesorbed water. Here we have assumed a photodesorption yield of 10^{-3}, motivated through the laboratory work by Öberg *et al.*[68] A significantly higher yield would be required for this reaction to become important, which is inconsistent with existing water observations.[32,69] However, inside the water sublimation front this mechanism must therefore become inefficient.

One beneficiary of this gas phase process is CH_3OH which we use as an example of the end products. When CO is destroyed, gas phase formation of H_2CO leads to a production of H_3CO^+ *via* a reaction with abundant HCO^+. H_3CO^+ then reacts with CH_4 to form $CH_3OCH_4^+$, which recombines with electrons in the gas or on grains to create gas phase CH_3OH, which freezes onto grains. Similar sets of ion-molecule reactions lead to the other abundant ices.

In Fig. 6b model results are presented assuming a reduced ionization rate by cosmic rays and a large surface area of grains. At the surface the presence of strong X-ray ionization powers CO destruction and the creation of a similar suite of carbon-bearing molecules. In the midplane the reduced level of ionization, which is now powered by scattered high energy X-rays, leads to less CO destruction and factors of a few less complex ice production. However, we note that the process is still active and longer timescales will lead to activation. Exploring models with reduced grain surface area (Fig. 6c, d) we see that the effectiveness of this mechanism is significantly reduced, at least when viewed from the perspective of the CO abundance. The reduction in the effectiveness of gas phase CO processing is due to longer freeze-out timescales allowing for significant CO re-formation to occur. Regardless, processing still occurs at reduced levels providing astronomically significant amounts of a variety of complex species.

In Fig. 7 the primary *volatile* carbon carriers are shown at 1 Myr for each of our four models as a function of disk radial and vertical position. In the model with the highest ionization and largest grain surface area (Fig. 7a), the general chemical structure on the surface of the disk for carbon species is seen for the outer disk: $C^+ \rightarrow CO$. In the midplane CO is frozen on grains but above the midplane CO chemical processing leads primarily to the creation of CO_2 ice. This can occur *via* two pathways. In the gas a portion of the oxygen freed by CO destruction can react with H_3^+ ultimately leading ($^2/_3$ of the time) to the production of OH. Existing gas phase CO will react with the hydroxyl producing CO_2, which can freeze onto a grain surface. An alternate route exists on grain surfaces in layers where CO is allowed to freeze onto grains in moderate amounts, at which point free OH on grain surfaces reacts with a small (80 K) energy barrier, directly making CO_2 ice. These are very localized on top of the regions where CO

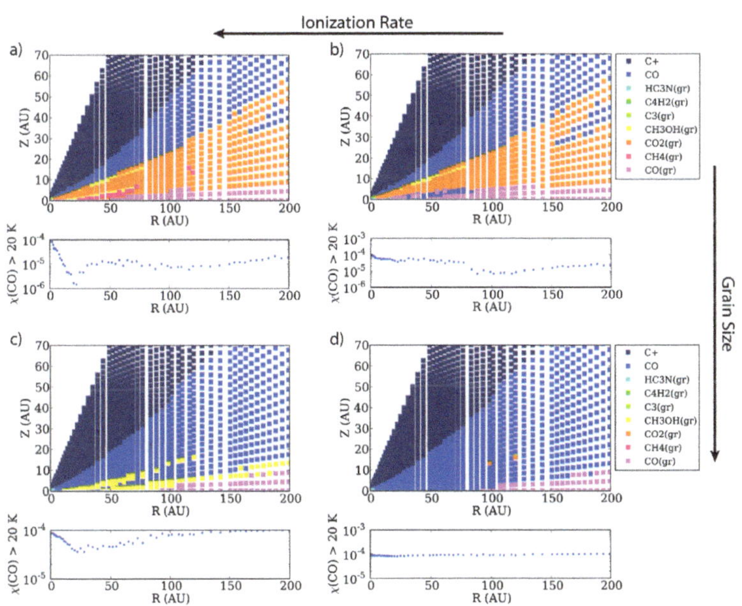

Fig. 7 Plot illustrating dominant volatile carbon carrier at each *R*, *Z* position in the gridded disk model. In this instance the dominant volatile carbon carrier contains over 50% of the carbon that initially resided in CO (and sometimes is CO). All the colors are the same across the panels. All panels present the standard model at 1 Myr. Panel (a) unattenuated cosmic ray radiation field and 0.1 µm grains; (b) model with wind-attenuated cosmic rays and 0.1 µm grains; (c) unattenuated cosmic ray radiation field and 10 µm grains; (d) model with wind attenuated cosmic rays and 10 µm grains. Arrows indicate trends. Underneath each panel the vertically averaged gas phase CO abundance above 20 K is shown as a function of radius for each respective model.

can be frozen onto grains and contributes at some level to the abundance of CH_3OH as well.

Similar results are seen for the model with a reduced cosmic ray ionization rate, but interstellar grain surface area (panel b). In this model the active radio-nuclides provide a floor for the ionization rate still allowing for CO chemical processing throughout the disk. One notable difference is seen in the midplane inside of the nominal CO sublimation front (near 70 AU) where CO sublimation and the presence of ionization leads to CO destruction and creation of CO_2 ice. In the model with reduced ionization this effect is not observed, at least within the first 1 Myr. The volatile CO abundance in the warm layers above the nominal freeze-out temperature of 20 K is substantially reduced in models with a high ionization rate. If the ionization rate is reduced, the CO processing mechanism is more effective in the outer disk than the inner disk. This is generally due to a lower equilibrium ion abundance at higher densities (*i.e.* more frequent recombination), thus leading to the ineffectiveness of midplane CO destruction in the inner 70 AU.

In models with lower grain surface area (Fig. 7c, d) the effectiveness of CO chemical processing is significantly reduced. Beyond the midplane snow line the disk has the anticipated chemical structure for volatile carbon $C^+ \rightarrow CO \rightarrow CO_{gr}$. If the disk is directly exposed to interstellar cosmic rays there is some processing, which is more effective in the inner disk where freeze-out is relatively more efficient due to higher gas densities. This is most directly seen in the vertically averaged CO abundance in panel (c), where the CO abundance decreases to a minimum at 30 AU and then increases as the rising dust temperature reduces the number of available sinks on grain surface (*i.e.* complex molecules begin to evaporate). Thus the second part of the mechanism for the sequestration of carbon on grain surfaces is hampered. Finally in the model with low disk ionization and grain surface area, little change is seen in the CO abundance.

In all models the dust temperature is what sets the specific carbon sink, as the chemistry tends to move towards the most abundant molecule the gas-phase chemistry can produce that will locally freeze-out. Thus in the inner tens of AU a diversity of carbon carriers are predicted, *e.g.* HC_3N, C_4H_2, HC_5N, and C_6H_2. We stress that these predictions regarding the specific carbon carrier are highly uncertain as our model includes an incomplete surface chemistry. Moreover there exist large uncertainties in the gas–grain interaction, as different binding energies for the various carbon carriers would change model predictions. One intriguing facet is the general tendency of theoretical binding energies to be successively higher for larger molecules,[52] a facet that is broadly reproduced in lab experiments.[40,70] This is consistent with the changing nature of the model predictions for the species that are the carbon-endpoints of this process. In summary, when provided a source of ionization, across a range of dust temperatures, a variety of potential carbon-sinks for chemistry can readily be created.

4   Implications for carbon incorporation into planetesimals and planets

In this contribution we have explored the potential for chemistry in the planet-forming disk to alter chemical abundances inherited from the interstellar medium. In this case we refer to carbon monoxide, which forms in the gas long before stellar birth and is provided by collapse to the young protoplanetary disk. The chemical processing mechanism requires grains to be warm (>20 K), partially ionized, and a total grain surface area per hydrogen above that of 1/1000 of the value in the interstellar medium (*i.e.* $n_{gr}\sigma_{gr}/n \simeq 2 \times 10^{-24}$ cm^{-1}). If these conditions are met then the timescales for CO chemical destruction are short enough and there is sufficient ionization in the gas for ion-molecule reactions to slowly create a host of gas phase molecules with less volatility than CO that freeze onto grain surfaces. For the most part this mechanism can extract carbon from CO within the disk surface layers, but can also be active in the disk midplane, inside the CO snow line. Beyond the CO snow line any carbon locked as CO ice will remain unaltered, thus cometary CO is preserved. A true disk is significantly more complex than explored in the static models presented here, as both the solids and the gaseous components can undergo systemic (advection) and semi-random (turbulent) motions. In the case of the small grains with sizes $\sim$0.1μm, they are coupled to the gas and *via* turbulence can migrate radially and verti-cally,[71] while intermediate size grains (cm to m-sizes) will undergo inward radial drift due to differential rotation speeds compared to the gaseous disk.[72] Thus there is substantial dynamical evolution which we have not explored. This will influence the effectiveness of gas phase chemical processing of CO in both directions. Finally there is the likelihood of the significant radial redistribution of materials as giant planets form, interact with the gaseous disk, leading to inward and outward movement depending on the timescales of gas dissipation. Thus the ultimate composition of planetary materials may be comprised of material that originated at a variety of distances, *e.g.* Walsh *et al.*[73]

Observationally there is now evidence of below ISM CO abundance in the warm ($T > 20$ K) gas in one system, a result that we have shown is not due to differences in the self-shielding of HD and C^{18}O. Similar effects need to be corroborated in other systems before concluding that this is a generic result of disk chemical evolution. However, molecular ions, and stellar X-rays, are readily detected in disk systems confirming the presence of sources of ionization.[37,74] In the disk, helium atoms will be in the vapor state; thus He^{+} ions are present to react with CO on short timescales given the high densities. The only question is how effective grains are at providing a sink. In this light there are a number of additional chemical effects that are unexplored in our models that could lead to increased chemical complexity of the molecules species created from volatile carbon. Beyond CH$_3$OH, we have not fully explored the potential for grain surface chemistry itself to lead to the creation of more complex organics with stronger bonds to the grain surface. In warmer regions with $T \sim 20$–30 K, grain surfaces are more conducive to a larger number of grain surface reaction pathways as heavier radicals become free to move and scan the surface.[75] Moreover, laboratory experiments of interstellar-analog molecular ices (*e.g.* H$_2$O, CH$_3$OH, CO) show that exposure to high levels of UV flux fosters the formation of a wide array of very complex organics similar to those found in meteorites.[76,77] In addition Ciesla and Sandford[71] demonstrated that the dynamical motions of 1.0 μm grains leads to the likelihood that the surface ices are subject to repeated exposure to UV photons at levels similar to the laboratory experiments. Thus the volatile carbon nominally locked in CO could readily be a source of the organic matter seen in meteoritic bodies.

It is important to note that the question of carbon incorporated into terrestrial worlds is complex as it depends on the source terms (carbon-rich rocks or icy comets) and on whether the supplied carbon remains near the surface of the forming planet or is sequestered deeper into its core. Our exploration of the fate of volatile carbon and its possible relation to the beginnings of organics in meteorites or comets is only one potential piece of this puzzle. We have not discussed the fate of solid state or refractory carbon, such as amorphous carbon grains or PAHs which is another important aspect to consider.[3,78,79] Looking forward, with the Atacama Large Millimeter Array in operation, there will be more data to compare to models to directly explore the fate of molecules formed in the dense interstellar medium. Thus, the results presented here are the beginnings of our more direct exploration of the origins of material to be supplied to forming planets.

Acknowledgements

This work was supported by funding from the National Science Foundation grant AST-1008800 and AST-1344133 (INSPIRE).

References

1 A. Morbidelli, J. I. Lunine, D. P. O'Brien, S. N. Raymond and K. J. Walsh, *Annu. Rev. Earth Planet. Sci.*, 2012, **40**, 251–275.

2 E. van Dishoeck, E. A. Bergin, D. C. Lis and J. I. Lunine, *Protostars and Planets VI*, 2013, in press.

3 J.-E. Lee, E. A. Bergin and H. Nomura, *Astrophys. J.*, 2010, **710**, L21–L25.

4 K. Pontoppidan, C. Salyk, E. A. Bergin, S. Brittain, B. Marty, O. Mousis and K. I. Öberg, *Protostars and Planets VI*, 2013, in press.

5 B. Marty, *Earth Planet. Sci. Lett.*, 2012, **313**, 56–66.

6 F. Albarede, C. Ballhaus, J. Blichert-Toft, C.-T. Lee, B. Marty, F. Moynier and Q.-Z. Yin, *Icarus*, 2013, **222**, 44–52.

7 C. Favre, L. I. Cleeves, E. A. Bergin, C. Qi and G. A. Blake, *Astrophys. J.*, 2013, **776**, L38.

8 Y. Aikawa, T. Umebayashi, T. Nakano and S. Miyama, *Faraday Discuss.*, 1998, **109**, 281.

9 R. Dasgupta and M. M. Hirschmann, *Earth Planet. Sci. Lett.*, 2010, **298**, 1–13.

10 N. R. Crockett, E. A. Bergin, J. L. Neill, C. Favre, *et al.*, *Astrophys. J.*, 2014, in press.

11 F. C. Adams, *Annu. Rev. Astron. Astrophys.*, 2010, **48**, 47–85.

12 M. F. Nieva and N. Przybilla, *Astron. Astrophys.*, 2012, **539**, A143.

13 R. Plume, E. A. Bergin, T. G. Phillips, D. C. Lis, S. Wang, N. R. Crockett, E. Caux, C. Comito, P. F. Goldsmith and P. Schilke, *Astrophys. J.*, 2012, **744**, 28.

14 M. N. Fomenkova, *Space Sci. Rev.*, 1999, **90**, 109–114.

15 E. A. Bergin, L. I. Cleeves, U. Gorti, K. Zhang, G. A. Blake, J. D. Green, S. M. Andrews, N. J. Evans, II, T. Henning, K. Öberg, K. Pontoppidan, C. Qi, C. Salyk and E. F. van Dishoeck, *Nature*, 2013, **493**, 644–646.

16 J. P. Williams and L. A. Cieza, *Annu. Rev. Astron. Astrophys.*, 2011, **49**, 67–117.

17 E. E. Mamajek, *Astrophys. J.*, 2005, **634**, 1385–1394.

18 J. L. Linsky, *Space Sci. Rev.*, 1998, **84**, 285–296.

19 U. Gorti, D. Hollenbach, J. Najita and I. Pascucci, *Astrophys. J.*, 2011, **735**, 90.
20 Y. Aikawa, G. J. van Zadelhoff, E. F. van Dishoeck and E. Herbst, *Astron. Astrophys.*, 2002, **386**, 622–632.
21 J. P. Williams and W. M. J. Best, *ApJ*, 2013, in press.
22 J. Wolcott-Green and Z. Haiman, *Mon. Not. R. Astron. Soc.*, 2011, **412**, 2603–2616.
23 R. Visser, E. F. van Dishoeck and J. H. Black, *Astron. Astrophys.*, 2009, **503**, 323–343.
24 W. Hoff, T. Henning and W. Pfau, *Astron. & Astrophys.*, 1998, **336**, 242–250.
25 W. D. Vacca and G. Sandell, *Astrophys. J.*, 2011, **732**, 8.
26 E. F. van Dishoeck, G. A. Blake, B. T. Draine and J. I. Lunine, *Protostars and Planets III*, 1993, pp. 163–241.
27 E. A. Bergin and M. Tafalla, *Annu. Rev. Astron. Astrophys.*, 2007, **45**, 339–396.
28 D. Hollenbach, M. J. Kaufman, E. A. Bergin and G. J. Melnick, *Astrophys. J.*, 2009, **690**, 1497–1521.
29 M. J. Mumma and S. B. Charnley, *Annu. Rev. Astron. Astrophys.*, 2011, **49**, 471–524.
30 A. Dutrey, S. Guilloteau and M. Guelin, *Astron. & Astrophys.*, 1997, **317**, L55–L58.
31 J. H. Kastner, B. Zuckerman, D. A. Weintraub and T. Forveille, *Science*, 1997, **277**, 67–71.
32 M. R. Hogerheijde, E. A. Bergin, C. Brinch, *et al.*, *Science*, 2011, **334**, 338–340.
33 K. M. Pontoppidan, C. P. Dullemond, E. F. van Dishoeck, G. A. Blake, A. C. A. Boogert, N. J. Evans, II, J. E. Kessler-Silacci and F. Lahuis, *Astrophys. J.*, 2005, **622**, 463–481.
34 M. Honda, A. K. Inoue, M. Fukagawa, A. Oka, T. Nakamoto, M. Ishii, H. Terada, N. Takato, H. Kawakita, Y. K. Okamoto, H. Shibai, M. Tamura, T. Kudo and Y. Itoh, *Astrophys. J.*, 2009, **690**, L110–L113.
35 I. Kamp and C. P. Dullemond, *Astrophys. J.*, 2004, **615**, 991–999.
36 H. Nomura, Y. Aikawa, M. Tsujimoto, Y. Nakagawa and T. J. Millar, *Astrophys. J.*, 2007, **661**, 334–353.
37 E. A. Bergin, Y. Aikawa, G. A. Blake and E. F. van Dishoeck, *Protostars and Planets V*, 2007, p. 751.
38 C. Hayashi, *Prog. Theor. Phys., Suppl.*, 1981, **70**, 35–53.
39 M. P. Collings, J. W. Dever, H. J. Fraser and M. R. S. McCoustra, *Astrophysics and Space Science*, 2003, **285**, 633–659.
40 M. P. Collings, M. A. Anderson, R. Chen, J. W. Dever, S. Viti, D. A. Williams and M. R. S. McCoustra, *Mon. Not. R. Astron. Soc.*, 2004, **354**, 1133–1140.
41 H. J. Fraser, M. P. Collings, M. R. S. McCoustra and D. A. Williams, *Mon. Not. R. Astron. Soc.*, 2001, **327**, 1165–1172.
42 Y. Aikawa, T. Umebayashi, T. Nakano and S. M. Miyama, *Astrophys. J.*, 1997, **486**, L51.
43 J. K. J. Fogel, T. J. Bethell, E. A. Bergin, N. Calvet and D. Semenov, *Astrophys. J.*, 2011, **726**, 29.
44 L. I. Cleeves, E. A. Bergin, T. J. Bethell, N. Calvet, J. K. J. Fogel, J. Sauter and S. Wolf, *Astrophys. J.*, 2011, **743**, L2.
45 G. J. Herczeg, J. L. Linsky, J. A. Valenti, C. M. Johns-Krull and B. E. Wood, *Astrophys. J.*, 2002, **572**, 310–325.

46 T. Preibisch, Y.-C. Kim, F. Favata, E. D. Feigelson, E. Flaccomio, K. Getman, G. Micela, S. Sciortino, K. Stassun, B. Stelzer and H. Zinnecker, *Astrophys. J. Suppl.*, 2005, **160**, 401–422.

47 T. J. Bethell and E. A. Bergin, *Astrophys. J.*, 2011, **740**, 7.

48 T. J. Bethell and E. A. Bergin, *Astrophys. J.*, 2011, **739**, 78.

49 E. Schindhelm, K. France, G. J. Herczeg, E. Bergin, H. Yang, A. Brown, J. M. Brown, J. L. Linsky and J. Valenti, *Astrophys. J.*, 2012, **756**, L23.

50 L. I. Cleeves, F. C. Adams and E. A. Bergin, *Astrophys. J.*, 2013, **772**, 5.

51 I. W. M. Smith, E. Herbst and Q. Chang, *Mon. Not. R. Astron. Soc.*, 2004, **350**, 323–330.

52 T. I. Hasegawa, E. Herbst and C. M. Leung, *Astrophys. J. Suppl.*, 1992, **82**, 167–195.

53 S. B. Charnley and S. B. Rodgers, *Astronomical Society of the Pacific Conference Series*, 2009, pp. 29–+.

54 J. Igea and A. E. Glassgold, *Astrophys. J.*, 1999, **518**, 848–858.

55 J. S. Mathis, W. Rumpl and K. H. Nordsieck, *Astrophys. J.*, 1977, **217**, 425–433.

56 T. Umebayashi and T. Nakano, *Astrophys. J.*, 2009, **690**, 69–81.

57 L. I. Cleeves, F. C. Adams, E. A. Bergin and R. Visser, *Astrophys. J.*, 2013, **777**, 28.

58 M. Jura, S. Xu and E. D. Young, *Astrophys. J.*, 2013, **775**, L41.

59 S. E. Barlow, PhD thesis, University of Colorado at Boulder, 1984.

60 J. B. Laudenslager, W. T. Huntress, Jr. and M. T. Bowers, *J. Chem. Phys.*, 1974, **61**, 4600–4617.

61 A. Natta, L. Testi, N. Calvet, T. Henning, R. Waters and D. Wilner, Protostars and Planets V, (University of Arizona Press; Tucson, Az), ed. B. Reipurth, D. Jewitt and K. Keil, 2007, 767–781.

62 Y. Aikawa, E. Herbst and F. N. Dzegilenko, *Astrophys. J.*, 1999, **527**, 262–265.

63 C. P. Dullemond and C. Dominik, *Astron. Astrophys.*, 2004, **421**, 1075–1086.

64 D. J. Wilner, P. T. P. Ho, J. H. Kastner and L. F. Rodríguez, *Astrophys. J.*, 2000, **534**, L101–L104.

65 T. E. Graedel, W. D. Langer and M. A. Frerking, *Astrophys. J. Suppl.*, 1982, **48**, 321–368.

66 C. M. Leung, E. Herbst and W. F. Huebner, *Astrophys. J. Suppl.*, 1984, **56**, 231–256.

67 C. Dominik, C. Ceccarelli, D. Hollenbach and M. Kaufman, *Astrophys. J.*, 2005, **635**, L85–L88.

68 K. I. Öberg, H. Linnartz, R. Visser and E. F. van Dishoeck, *Astrophys. J.*, 2009, **693**, 1209–1218.

69 E. A. Bergin, M. R. Hogerheijde, C. Brinch, *et al.*, *Astron. Astrophys.*, 2010, **521**, L33.

70 K. I. Öberg, R. T. Garrod, E. F. van Dishoeck and H. Linnartz, *Astron. Astrophys.*, 2009, **504**, 891–913.

71 F. J. Ciesla and S. A. Sandford, *Science*, 2012, **336**, 452.

72 S. J. Weidenschilling and J. N. Cuzzi, *Protostars and Planets III*, 1993, pp. 1031–1060.

73 K. J. Walsh, A. Morbidelli, S. N. Raymond, D. P. O'Brien and A. M. Mandell, *Nature*, 2011, **475**, 206–209.

74 T. Henning and D. Semenov, *Chemical Reviews*, 2013, **113**, 9016.

75 R. T. Garrod, S. L. W. Weaver and E. Herbst, *Astrophys. J.*, 2008, **682**, 283–302.

76 M. P. Bernstein, S. A. Sandford, L. J. Allamandola, S. Chang and M. A. Scharberg, *Astrophys. J.*, 1995, **454**, 327.
77 M. Nuevo, S. N. Milam, S. A. Sandford, B. T. De Gregorio, G. D. Cody and A. L. D. Kilcoyne, *Adv. Space Res.*, 2011, **48**, 1126–1135.
78 M. E. Kress, A. G. G. M. Tielens and M. Frenklach, *Adv. Space Res.*, 2010, **46**, 44–49.
79 J. C. Bond, D. S. Lauretta and D. P. O'Brien, *Icarus*, 2010, **205**, 321–337.

PAPER

Complex molecule formation around massive young stellar objects

Karin I. Öberg,[*a] Edith C. Fayolle,[a] John B. Reiter[b]
and Claudia Cyganowski[c]

Received 18th December 2013, Accepted 6th February 2014
DOI: 10.1039/c3fd00146f

Interstellar complex organic molecules were first identified in the hot inner regions of massive young stellar objects (MYSOs), but have more recently been found in many colder sources, indicating that complex molecules can form at a range of temperatures. However, individually these observations provide limited constraints on how complex molecules form, and whether the same formation pathways dominate in cold, warm and hot environments. To address these questions, we use spatially resolved observations from the Submillimeter Array of three MYSOs together with mostly unresolved literature data to explore how molecular ratios depend on environmental parameters, especially temperature. Towards the three MYSOs, we find multiple complex organic emission peaks characterized by different molecular compositions and temperatures. In particular, CH_3CCH and CH_3CN seem to always trace a lukewarm ($T \approx$ 60 K) and a hot ($T > 100$ K) complex chemistry, respectively. These spatial trends are consistent with abundance–temperature correlations of four representative complex organics – CH_3CCH, CH_3CN, CH_3OCH_3 and CH_3CHO – in a large sample of complex molecule hosts mined from the literature. Together, these results indicate a general chemical evolution with temperature, *i.e.* that new complex molecule formation pathways are activated as a MYSO heats up. This is qualitatively consistent with model predictions. Furthermore, these results suggest that ratios of complex molecules may be developed into a powerful probe of the evolutionary stage of a MYSO, and may provide information about its formation history.

1   Introduction

Complex Organic Molecules (COMs)† were first detected towards massive young stellar objects, in so-called hot cores. These hot cores are characterized by intense

[a] Harvard-Smithsonian Center for Astrophysics, 60 Garden St, MS 16, Cambridge, MA, 02138, USA. E-mail: koberg@cfa.harvard.edu
[b] University of Virginia, Charlottesville, VA, USA
[c] St. Andrews University, UK

† For the purpose of this paper, COMs are defined as hydrogen-rich organics with three or more heavy elements.

and crowded millimeter and sub-millimeter spectra, where most of the lines are attributed to COMs, and gas temperatures exceed 100 K.[1,2] In the past couple of decades, COMs have also been detected towards low-mass protostars,[3–5] often referred to as hot corinos, protostellar envelopes,[6] outflows,[7] and, most recently, two pre-stellar cores.[8–10] Together, these observations suggest that COM chemistry can take place over a much larger range of temperatures than previously assumed, and that if properly understood, specific COM compositions and ratios could be used as precise probes of different aspects of low- and high-mass star formation.

This existence of low-temperature formation pathways of COMs does not exclude, however, the possibility that additional formation pathways of COMs are activated as the source temperature is increased. Indeed, there is evidence for significantly different COM compositions both between different kinds of sources (Fig. 1)[11] and across a single massive young stellar object (MYSO) where the COM abundance profiles were resolved across a range of temperatures.[12] In this paper we aim to combine spatially resolved observations of a small sample of MYSOs with statistics from single-point observations of a diverse sample of COM hosts to explore whether there are clear trends in the COM chemistry between different classes of sources, and how such trends relate to source characteristics, especially temperature.

The focus on source temperature as a potential COM chemistry regulator is motivated by the current most popular astrochemical scenario, where COM formation takes place in three distinct stages or generations as material heats up during star formation.[2,13,14] The precursors of COMs, or the zeroth generation, form in cold molecular clouds, where low temperatures ($\sim$10 K) and moderately high densities ($\sim$10^5 cm^{-3}) result in the efficient freeze-out of all elements and molecules heavier than H and He onto interstellar dust grains. While H atoms do not form permanent ice layers, they can still reside long enough on grain and ice surfaces to be an important reaction partner at low temperatures. At 10 K, H is

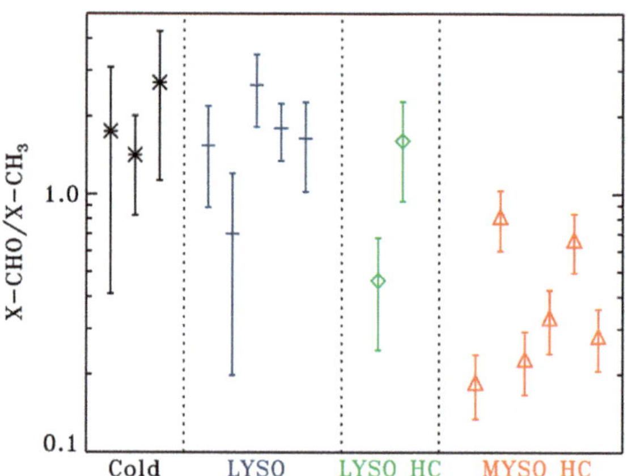

Fig. 1 The ratio of (CH$_3$CHO + HCOOCH$_3$) : (CH$_3$OCH$_3$ + CH$_3$CH$_2$OH) abundances toward cold sources (pre-stellar cores and an outflow), low-mass protostellar envelopes, low-mass hot cores, and traditional high-mass hot cores.[11] Based on laboratory experiments, the relative abundance of CHO-bearing molecules should trace the relative importance of cold (CO-ice rich) and warm ice COM chemistry.[18] Literature column densities without uncertainties were assigned a 20% uncertainty.

also orders of magnitude more mobile than heavier elements and molecules, such as C and CO, and this stage is therefore characterized by hydrogenation reactions on grain surfaces to form, for example, CH_3OH from CO. Some heavier atom additions must also occur to form *e.g.* CO_2, and possibly HNCO and other small COMs.[11,15,16] In the gas phase this stage is characterized by ion-neutral reactions, which may be responsible for some COMs, especially hydrocarbons.

COM ice formation is expected to become efficient in the following phase when the cloud core collapses to form a protostar, resulting in heating due to dissipation of gravitational energy. At these elevated temperatures (>30 K) diffusion of larger molecules and radicals in the ice is possible. Radicals are continuously produced in the ice through dissociative absorption of high-energy photons and electrons. When mobile, these radicals can combine to form complex molecules, so called first generation COMs.[2] The final stage of COM chemistry happens when the gas temperature exceeds 100 K, the hot core stage, releasing all the ice into the gas phase and also enabling further gas-phase processing into even more complex species (second generation COMs[2,17]). This model has been very successful in explaining the kinds of molecules that appear in hot cores and can also explain the different excitation temperatures extracted for different COMs in these sources.[14]

Comprehensive tests of this formation scenario, have been scarce, however, which limits our basic understanding of COM formation. In this study we combine submillimeter observations of three MYSOs with statistics drawn from a large number of previous observational studies to explore how the COM chemistry depends on temperature across and between sources. The observations are briefly described in Section 2. In Section 3 we present emission maps of key COM lines toward the three MYSOs, and present new strategies to quantify the COM distributions. The extracted COM column density ratios are then used together with literature values to evaluate which, if any, aspects of the COM chemistry are regulated by source temperature. The implications of the results are discussed in Section 4, followed by a brief summary of the results of this paper and open questions that remain on COM formation.

2 Observations and source characteristics

The three MYSOs, W3 IRS5, NGC 7538 IRS 9 and NGC 7538 IRS 1 (Table 1) are all nearby (d < 3 kpc), and have comparable luminosities and virial envelope mass estimates. All sources are associated with multiple molecular outflows, and thus multiple YSOs within a larger structure. NGC7538 IRS9 and NGC7538 IRS1 are located in Perseus, and NGC7538 IRS1 is known to house a bright hot core[19] and is situated in the middle of a cluster of continuum peaks and molecular outflows.

Table 1 Source characteristics[28]

Source	RA	Dec	d kpc	L $10^4 L_\odot$	M_V $10^4 M_\odot$
W3 IRS5	02 21 53.1	+61 52 20	2.0	17	2.0
NGC7538 IRS9	23 11 52.8	+61 10 59	2.7	3.5	1.1
NGC7538 IRS1	23 11 36.7	+61 11 51	2.7	13	1.0

W3 IRS5 is associated with at least five young stellar objects, two of which are massive,[20-24] and is known to present strong S-bearing molecular lines.[25] In summary, all three sources are known to have complicated structures with many different potential origins of complex molecular emission. This is a nuisance when interpreting spatially unresolved observations, but in the age of interferometry this increases the potential value of each MYSO data set: resolving several different environments with a single observation enables an efficient exploration of COM chemistry environmental dependencies.

All three MYSOs are known to host complex organics (Fayolle *et al.*),[12] and to present a different molecular composition on large and small scales. In this study, we focus on resolving the chemical differentiation on the individual core scale using observations from the Submillimeter Array‡ (SMA). W3 IRS5, NGC 7538 IRS 9 and NGC 7538 IRS 1 were observed on July 29th and August 15th (extended configuration) and October 15th (compact configuration) 2011 in good to excellent weather: τ_{225GHz} was 0.09 on the 29th of July, 0.1 on the 15th of August, and 0.07 on the 15th of October 2011.

The combined range of baselines was 16–226 m. The SMA correlator was set up to obtain a spectral resolution of $\sim$1 km s^{-1} using 128 channels for each of the 46 chunks, covering 227–231 GHz in the lower sideband and 239–243 GHz in the upper sideband. Absolute flux calibration was done with Callisto. The quasars 1924–292 and 3c84 were used as bandpass calibrators for the compact observations, and 3c454.3 and 3c279 were used to calibrate the 29th of July and 15th of July observations respectively. The quasars 0014+612 and 0102+584 were used as gain calibrators for NGC7538 IRS9 and IRS1, and 0244+624, 0359+509 and 0102+584 were used for W3 IRS5.

Routine calibration tasks were performed using the CASA software package,[26] including phase and amplitude self-calibration. The continuum was subtracted separately for the upper and lower sideband for each observational data set in CASA, using line-free channels. The continuum-subtracted compact and extended data were combined for each source for imaging and CLEANing in CASA using robust weighting, which resulted in synthesized beam sizes of 2.0$''$ × 1.7$''$ for NGC 7538 IRS 9 and W3 IRS5, and 1.5$''$ × 1.7$''$ for NGC7538 IRS 1. The primary beam of the SMA at these wavelengths is $\sim$50$''$. Considering the baseline coverage, all emission at scales larger than 18$''$ is completely filtered out, and angular structures smaller than 7$''$ are required to filter out less than 50% of the emission.[27] These observations are thus not sensitive to large-scale envelope COM chemistry, as explored in our previous papers using the IRAM 30m (Fayolle *et al.*, submitted).[12]

3   Results

3.1 Molecular image analysis

Fig. 2 shows the continuum and line images toward W3 IRS5, NGC 7538 IRS 9 and NGC 7538 IRS 1. Each source displays multiple continuum peaks as expected from previous studies. These peaks are marked and referred to throughout the

‡ The Submillimeter Array is a joint project between the Smithsonian Astrophysical Observatory and the Academia Sinica Institute of Astronomy and Astrophysics. It is funded by the Smithsonian Institute and the Academia Sinica.

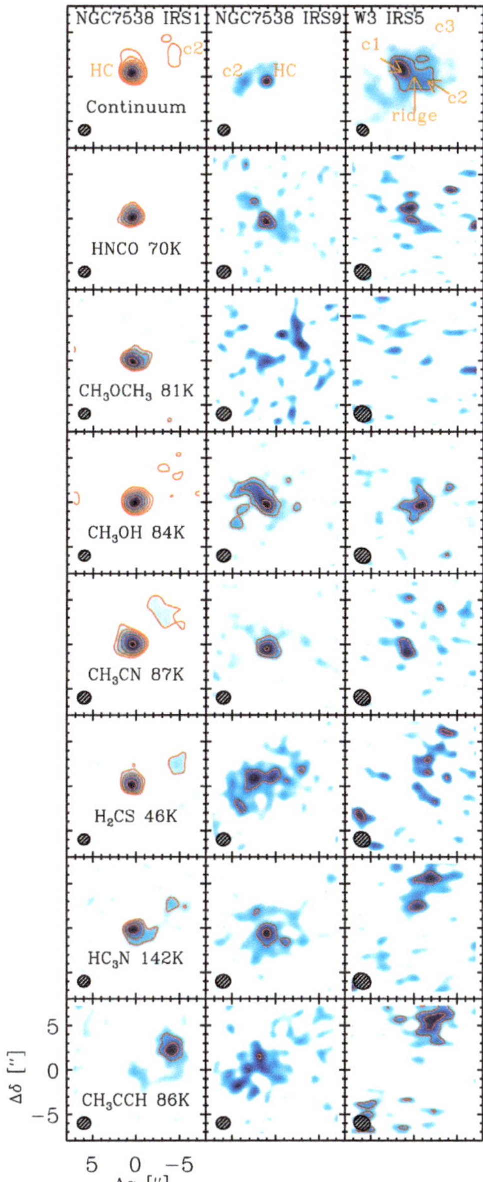

Fig. 2 Continuum and molecular line emission maps. The continuum maps in the top panels show the locations of the brightest continuum emission and the CH$_3$CCH core in the NW of W3 IRS 5. The panel labels indicate the species and upper energy level in K (other line characteristics are listed in Table 2). The red contour marks (0.05, 0.1, 0.2, 0.4, 0.8, 1.6) Jy beam^{-1} in the continuum panels, and (0.5, 1, 2, 4, 8) Jy km s^{-1} beam^{-1} in the molecular line panels. The synthesized beam is displayed in the bottom left corner of each panel. The RA and Dec offsets are with respect to the positions listed in Table 1.

text when discussing the chemical differentiation within these sources. The molecular images display the spectrally integrated (across the line FWHM), spatially resolved emission profiles of 7 organic emission lines, including lines from the COMs HNCO, CH_3CN, and CH_3CCH detected in all sources. Multiple lines were detected for each molecule. The lines in Fig. 2 were selected to have similar upper level energies (Table 2), which minimizes the differences in emission profiles due to differences in excitation characteristics. This allows us to use these line emission patterns to visualize the spatial extents of different molecules, *i.e.* the different emission profiles should be mainly driven by differences in molecular abundance profiles.

In all three sources there is a clear difference in emission profiles between different molecular lines. Some line emission, *e.g.* HNCO and CH_3CN, mostly originates from a single unresolved component, which can be associated with a continuum peak (the hot core, in the case of NGC 7538 IRS1). Other lines, especially those belonging to CH_3CCH, are distributed more diffusely, and the CH_3CCH line emission peaks are not associated with any detected continuum. Three molecules, CH_3OH, H_2CS and HC_3N, display different kinds of profiles across the small sample: in NGC 7538 IRS1 most of their emission can be attributed to the unresolved hot core, while the emission is clearly resolved in the other two sources. The differences in the spatial distributions of these different groups of COMs are suggestive of multiple COM formation pathways, which may be associated with the theoretically predicted low-temperature gas and ice chemistry, luke-warm ice chemistry, and ice evaporation in hot cores.

Despite our careful selection of lines in Fig. 2, we cannot completely neglect the fact that emission from individual lines always depends on a combination of molecular abundance and excitation. To establish a firm relationship between COM line emission and abundance distributions thus requires the imaging of multiple lines. This is especially true in cases where the continuum and line emission distributions are complex, such as W3 IRS5. Even in these moderately rich sources, images of all detected lines (50+) would be difficult to absorb, and therefore not very informative. Instead Fig. 3 shows, in a single figure, the line emission peaks of all clearly detected COM and S-molecule lines toward W3 IRS5 together with the continuum. The line emission peaks were determined by fitting 2D Gaussians in the image plane, integrated over the FWHM of each line, in CASA. The uncertainties of these fits (typically $\sim$0.5$''$) are also shown.

Most SO_2 lines are clustered toward the 'c1' continuum peak from Fig. 2, while CH_3OH line emission seems to define the second continuum peak 'c2'. CH_3CN and HNCO are not obviously associated with either peak, but rather reside in a ridge connecting the two clumps. CH_3CCH lines form another cluster in the

Table 2   Molecular line data for imaged lines in Fig. 2[a]

Species	Frequency (GHz)	$\log (A_{ij})$	E_u (K)	d_u[b]
HNCO	241.774	−3.71	69	23
CH_3OCH_3	241.946	−3.78	81	378
CH_3OH	241.833	−4.41	84	22
CH_3CN	239.133	−2.93	87	54
H_2CS	240.267	−3.69	46	15
HC_3N	227.419	−3.03	141	51
CH_3CCH	239.252	−4.84	86	58

[a] From CDMS and JPL spectral databases.[29,30] [b] Degeneracy in the upper level.

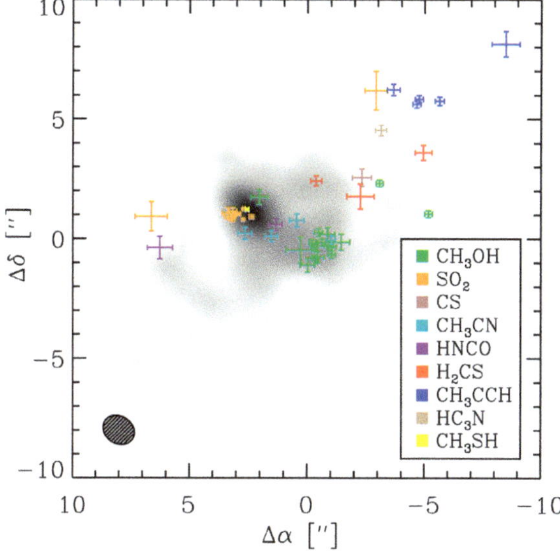

Fig. 3 The continuum emission (1 mm) toward W3 IRS5 (gray scale), and the spatially resolved line emission peaks of all detected COMs and S-bearing molecules toward this source. The error bars mark the uncertainties in the peak positions based on Gaussian fits. Most lines are heavily clustered by species. A few exceptions are CH_3CN, which is spread out across the ridge connecting the SO_2 and CH_3OH cores, and H_2CS. The two continuum peaks seem to be chemically defined by SO_2 and CH_3OH line emission.

northwest, and the space in between this cluster and the continuum peaks is associated with H_2CS, HC_3N and some CH_3OH. There is also some diffuse emission of HNCO and SO_2 to the east. These molecular line clusters partly coincide with the chemical regions identified by Wang *et al.*,[24] *i.e.* both studies found 'c1' and 'c2' to be defined by SO_2 and HNCO, and by CH_3OH, respectively. Wang *et al.* classified the 'c1' peak as a pole-on outflow and the 'c2' peak as a budding hot core. Both studies also found CS and H_2CS in the NW region, but the discovery of CH_3CCH is new, and so is the realization that the ridge connecting 'c1' and 'c2' seems to be chemically distinct, and the source of N-bearing COMs.

The sizes of the different line emission regions are displayed in Fig. 4, based on the Gaussian fitting parameters (major and minor axes, center, and position angle). The line emission peaking on the 'c1' core is compact. The other line emission centers display a range of emission sizes. Large ratios between the major and minor axes are mainly associated with very weak lines, suggesting that the 'true' line emission profiles are typically close to circular or unresolved.

Based on the previous two figures, SO_2, CH_3OH and CH_3CN have multiple emission centers, which may be indicative of different lines tracing different excitation conditions. To explore this, Fig. 5 shows the distribution of emission centers for these lines as a function of upper energy levels. In all cases, there are line-energy dependent differences in the emission centers. For CH_3OH, the line emission that peaks away from 'c2' comes from low energy lines. Furthermore, within the 'c2' line emission cluster the higher energy lines (E_{up} > 100 K) peak closer to the western side of 'c2'. There are only 4 detected CH_3CN lines and they are distributed along the ridge, with the lowest level line in the NE and the highest

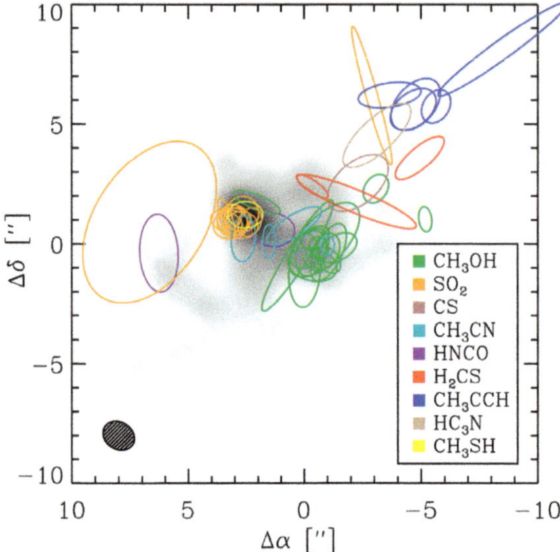

Fig. 4 The continuum emission (1 mm) toward W3 IRS5 (gray scale) and the spatially resolved line emission peaks of all detected COMs and S-bearing molecules toward this source. The ellipses mark the best-fit 2D Gaussians ($\times 0.5$ for visibility). It is clear that for each species the best fit ellipses vary in size, but some of this is due to differences in emission line strength rather than actual emission region differences. The emission regions centered on the c1 core do seem significantly more compact compared to all other emissions, however.

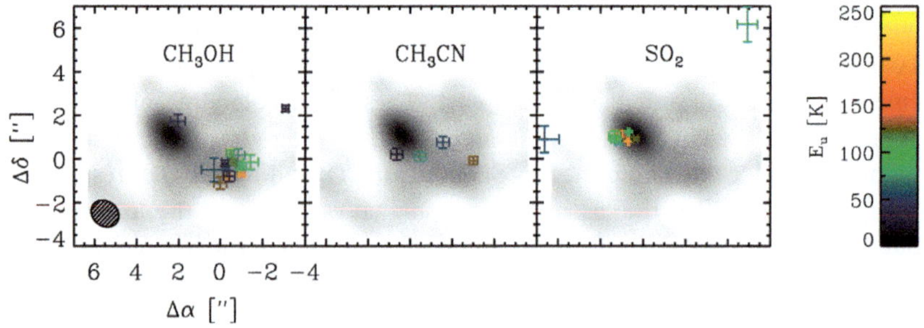

Fig. 5 The continuum emission (1 mm) toward the central cores in W3 IRS5 (gray scale) and the spatially resolved line emission peaks of CH_3OH, CH_3CN and SO_2 lines, color-coded by their upper energy levels. In each line there seems to be a temperature gradient. CH_3OH and SO_2 high temperature lines seem to trace the very center of their respective cores, while the CH_3CN line emission centers suggest an increasing temperature from east to west across the ridge.

level line in the SW, indicative of a temperature gradient in the same direction as seen for CH_3OH. Finally, the SO_2 line emission distribution presents a similar trend to CH_3OH, with lower lying lines more offset from the 'c1' core.

3.2 Excitation temperatures and column densities

The line emission maps presented in Fig. 2 could theoretically be converted into column density maps, using the emission from all detected lines of each species at each spatial resolution element. If local thermal equilibrium (LTE) in each pixel and optically thin lines can be assumed, this would be rather straightforward using *e.g.* rotational diagrams.[31] Such an approach should be very fruitful with ALMA, whose orders of magnitude greater sensitivity and densely populated antenna array will result in very high SNR and image fidelity.

The current observations have too low SNR to allow for this kind of detailed analysis, however. Instead, Fig. 6 shows the spectra extracted from all 8 continuum and molecular emission peak centers marked in Fig. 2 using a 2''-diameter circular mask. As expected from the image analysis in the previous section, the spectra appear very different toward the different sources, and toward the different peaks within each source. In particular, the maximum intensity of the CH_3OH 5–4 ladder is orders of magnitude lower toward W3 IRS5 and NGC 7538 IRS9 compared to NGC 7538 IRS1. SO_2 is also relatively more important in the latter two sources. There are also some trends across the sample: strong CH_3OH lines appear to coincide with strong CH_3CN lines, while strong CH_3CCH lines are associated with no or very weak emission of any other molecule, including CH_3OH.

We use these spectra to extract excitation temperatures at each position using CH_3OH, CH_3CCH or CH_3CN lines, dependent on the availability of lines, and the rotational diagram method. We could not identify a sufficient number of lines of any of these species toward the 'c1' (S-core) or 'c2' core in W3 IRS5 or the 'c2' core in NGC7538 IRS9, and exclude these positions from further analysis. The derived temperatures in the remaining cores range between 60 and 200 K and are shown in Table 3. The rotational temperature toward the W3 IRS5 ridge is, however, highly uncertain because of the small number and low SNR of the CH_3CN lines

The listed excitation temperatures are used to determine column density (limits) of CH_3OH and the four COMs, CH_3CN, CH_3OCH_3, CH_3CCH and CH_3CHO, which are expected to be products of a diverse set of complex formation pathways.[13] For each molecule, we use between one and five detected lines to derive a column density, or the spectral rms to determine a 3σ upper limit. This approach results in substantial uncertainties, especially since it is *a priori* not clear that all species in the same line of sight have the same excitation temperature, but provides a first quantitative constraint on how the complex chemistry varies at these scales.

To check the accuracy of the derived column densities we compared with previous studies where possible. Two of the sources, NGC 7538 IRS1 HC and W3 IRS5 c2, have been observed previously by others.[24,32] In the case of NGC 7538 IRS1 HC, the derived CH_3OH, CH_3CN and CH_3OCH_3 columns are consistent with previous observations, while the CH_3CHO and CH_3CCH columns are at least an order of magnitude higher in our study. Previous studies were single dish however, and assumed a beam dilution factor commensurate with the expected 100 K hot core boundary for most molecules, except for designated cold molecules CH_3CCH and CH_3CHO, which presented low excitation temperatures. The discrepancy is therefore readily resolved if CH_3CHO and CH_3CCH abundances are enhanced in the hot core compared to the envelope, even if most of the single-dish emission does indeed originate in the envelope, resulting in a low average excitation temperature. In the case of W3 IRS5, the derived CH_3OH column densities are also consistent between our and previous observations.

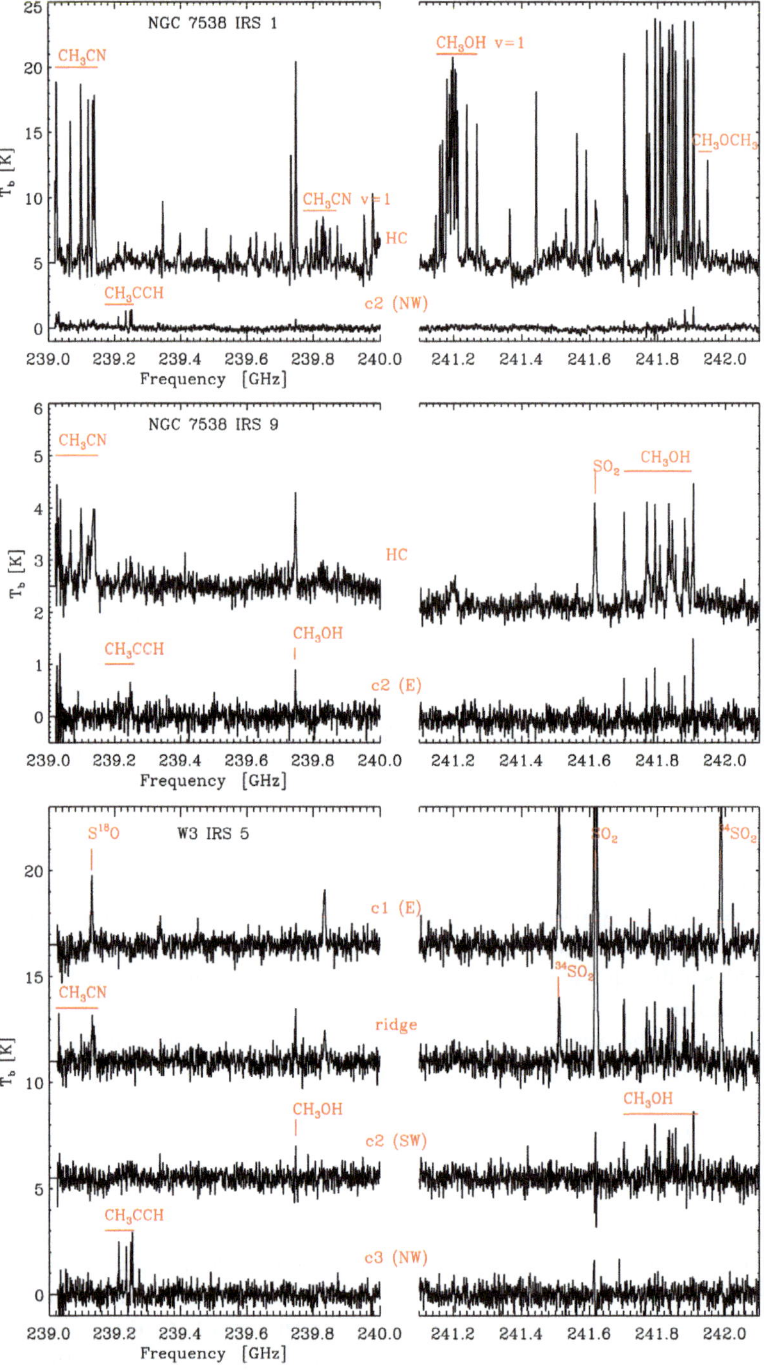

Fig. 6 Spectra extracted from the positions marked in Fig. 2, using 2″-diameter circular masks. Key molecular ladders and lines are marked.

Table 3 Excitation temperatures and COM column densities in MYSO cores

Core	X	$T_{\mathrm{rot}}(X)$ [K]	$N_{\mathrm{CH_3OH}}$ [cm^{-2}]	$N_{\mathrm{CH_3CN}}$ [cm^{-2}]	$N_{\mathrm{CH_3OCH_3}}$ [cm^{-2}]	$N_{\mathrm{CH_3CCH}}$ [cm^{-2}]	$N_{\mathrm{CH_3CHO}}$ [cm^{-2}]
W3 IRS5 ridge	CH$_3$OH	$\sim$130	$4[2] \times 10^{15}$	$6[6] \times 10^{13}$	$<6 \times 10^{14}$	$<2 \times 10^{15}$	$<3 \times 10^{14}$
W3 IRS5 c3	CH$_3$CCH	78 [30]	$3[4] \times 10^{14}$	$<1 \times 10^{13}$	$<3 \times 10^{14}$	$4[2] \times 10^{15}$	$<1 \times 10^{14}$
NGC 7538 IRS9 HC	CH$_3$OH	119 [29]	$4.3[0.5] \times 10^{15}$	$4[1] \times 10^{14}$	$<2 \times 10^{14}$	$4[1] \times 10^{15}$	$<2 \times 10^{14}$
NGC 7538 IRS1 HC	CH$_3$CN	196 [15]	$1.3[0.1] \times 10^{17}$	$8.4[0.4] \times 10^{14}$	$8[2] \times 10^{15}$	$4[3] \times 10^{15}$	$6[1] \times 10^{15}$
NGC 7538 IRS1 c2	CH$_3$CCH	61 [15]	$9[3] \times 10^{14}$	$2[2] \times 10^{13}$	$<2 \times 10^{14}$	$2[1] \times 10^{15}$	$<2 \times 10^{14}$

The resulting CH$_3$OH column densities span 10^{14} – 10^{17} cm^{-2}, with higher column densities toward the warmer sources. CH$_3$CN column densities vary over two orders of magnitude and seem correlated with temperature as well. In contrast, the CH$_3$CCH column densities are almost constant across this small sample. CH$_3$OCH$_3$ and CH$_3$CHO are only detected toward the NGC7538 IRS1 hot core, but are known to exist on larger scales toward the other two MYSOs (Fayolle *et al.*, submitted).

3.3   COM compositional dependencies

To explore whether trends observed on small scales with respect to temperature are also present across samples of COM hosts, we have gathered COM data from the literature based on three criteria: (1) detection of CH$_3$OH, (2) determination of a CH$_3$OH, CH$_3$CN or CH$_3$CCH excitation temperature in the same line of sight, and (3) detection or limits on at least one of the four COMS, CH$_3$CN, CH$_3$OCH$_3$, CH$_3$CCH and CH$_3$CHO. The resulting sample is listed in Table 4. It comprises 42 observations in 35 lines of sight (for seven targets there are measurements of both the extended and compact COM composition). The sample contains massive hot cores, MYSOs dominated by either cold or warm chemistry, low mass hot cores and envelopes, Galactic Center sources, IRDCs, an outflow and a cold pre-stellar core. It is thus representative of the full range of observed COM sources in space. The source diversity is also apparent when inspecting the reported excitation temperatures: 14 sources have excitation temperatures below 30 K (the traditional 0th generation regime), 10 sources have temperatures between 30 and 100 K (the temperature range when 1st generation COM chemistry should be active), and 18 sources have excitation temperatures above 100 K, typical for hot cores. The observed CH$_3$OH column densities span five orders of magnitude, between 10^{13} and 10^{18} cm^{-2}. For all parameters we use the reported uncertainties where available, and otherwise assume a typical 20% error.

Fig. 7 shows that there is a strong trend between excitation temperature and CH$_3$OH column density, consistent with the observed temperature dependence within our MYSO sample. Below 100 K, the average $\log_{10}(N_{\mathrm{CH_3OH}}) = 14.9$ for the whole sample with a standard deviation of 0.6. At or above 100 K, $\log_{10}(N_{\mathrm{CH_3OH}}) = 17.5$ with a standard deviation of 0.9. Within these two sub-

Table 4 Previous observations of representative complex organic molecules

Source	Type	T_{rot} [K]	N_{CH_3OH} [cm^{-2}]	N_{CH_3CN} [cm^{-2}]	$N_{CH_3OCH_3}$ [cm^{-2}]	N_{CH_3CCH} [cm^{-2}]	N_{CH_3CHO} [cm^{-2}]
IRAS20126 + 4104[33]	HC	300	$1.1[0.1] \times 10^{17}$	1.3×10^{15}	1×10^{16}	—	—
IRAS20126 + 4104[33]	MYSO env	14[1]	$2.2[0.9] \times 10^{15}$	—	—	3.5×10^{14}	$< 5 \times 10^{13}$
IRAS18089-1732[33]	HC	300	$2.0[0.2] \times 10^{17}$	$> 3.5 \times 10^{15}$	$< 1 \times 10^{17}$	—	—
IRAS18089-1732[33]	MYSO env	15[2]	$2.4[1.4] \times 10^{15}$	—	—	9.4×10^{14}	$< 5 \times 10^{13}$
G31.41 + 0.31[33]	HC	200	$1.0[0.2] \times 10^{18}$	$> 2.0 \times 10^{16}$	5×10^{17}	—	—
G31.41 + 0.31[33]	MYSO env	14[2]	$1.2[1.4] \times 10^{16}$	—	—	1.4×10^{15}	5×10^{14}
AFGL 2591[32]	HC	147[11]	4.7×10^{16}	$< 3.5 \times 10^{15}$	$< 7.7 \times 10^{15}$	7.9×10^{14}	3.1×10^{12}
G24.78[32]	HC	211[13]	2.8×10^{17}	5.9×10^{16}	1.2×10^{17}	2.3×10^{15}	1.3×10^{13}
G75.78[32]	HC	113[7]	1.1×10^{17}	1.8×10^{15}	2.3×10^{16}	8.6×10^{14}	2.1×10^{13}
NGC 6334 IRS1[32]	HC	178[10]	9.7×10^{17}	2.9×10^{16}	5.8×10^{17}	5.2×10^{15}	1.2×10^{14}
NGC 7538 IRS1[32]	HC	156[10]	1.2×10^{17}	8.2×10^{15}	1.6×10^{16}	8.4×10^{14}	2.8×10^{13}
W 3(H2O)[32]	HC	139[8]	1.0×10^{18}	7.0×10^{15}	1.5×10^{17}	1.5×10^{15}	3.5×10^{13}
W 33A[32]	HC	259[16]	2.0×10^{17}	2.7×10^{16}	2.7×10^{16}	1.3×10^{15}	3.0×10^{13}
NGC7538 IRS9[34]	MYSO	25[2]	$9[1] \times 10^{14}$	$1[0.3] \times 10^{13}$	$5[2] \times 10^{13}$	$1.2[0.3] \times 10^{15}$	$3.1[0.4] \times 10^{13}$
W3 IRS5[34]	MYSO	64[6]	$3.2[0.4] \times 10^{14}$	$4[1] \times 10^{12}$	$1.5[0.6] \times 10^{13}$	$7[2] \times 10^{14}$	$< 1.1 \times 10^{13}$
AFGL 490[34]	MYSO	25[2]	$2.4[0.4] \times 10^{14}$	$6[2] \times 10^{12}$	$8[3] \times 10^{13}$	$4[2] \times 10^{14}$	$< 1.3 \times 10^{13}$
B1-b[10,35]	CC	10[5]	4×10^{14}	—	3×10^{12}	—	$5[1] \times 10^{12}$
SMM4-W[11]	LYSO	11[1]	$2.2[0.7] \times 10^{15}$	—	2.4×10^{13}	—	1.3×10^{14}
SMM1[11]	LYSO	16[1]	$2.5[0.3] \times 10^{14}$	—	1.3×10^{13}	—	1.7×10^{13}
SMM4[11]	LYSO	13[1]	$1.1[0.1] \times 10^{15}$	—	8.4×10^{12}	—	2.3×10^{13}
L1157[7]	outflow	12[2]	1.5×10^{15}	$1[0.5] \times 10^{12}$	—	—	—
NGC1333 IRAS 4a[2,5,36]	LYSO	24 [2]	5.1×10^{14}	—	1.1×10^{14}	—	—
NGC1333 IRAS 4b[2,5,36]	LYSO	34[4]	3.5×10^{14}	$9[1] \times 10^{11}$	$< 6.7 \times 10^{13}$	—	3.5×10^{13}
NGC1333 IRAS 2a[5,36]	LYSO	101[16]	3.4×10^{14}	$4[1] \times 10^{12}$	$< 2 \times 10^{14}$	—	3.5×10^{13}
IRAS16293[2-4]	LYSO	85	8.8×10^{14}	—	$[1.8] \times 10^{14}$	—	3.5×10^{13}
IRAS16293 A[37,38]	LYSO HC	100	1.1×10^{18}	3×10^{15}	—	—	$< 1 \times 10^{14}$
IRAS16293 B[37,38]	LYSO HC	100	5×10^{17}	5×10^{15}	—	—	1.5×10^{15}

Table 4 (Contd.)

Source	Type	T_{rot} [K]	N_{CH_3OH} [cm^{-2}]	N_{CH_3CN} [cm^{-2}]	$N_{CH_3OCH_3}$ [cm^{-2}]	N_{CH_3CCH} [cm^{-2}]	N_{CH_3CHO} [cm^{-2}]
Sgr B2(N)[39]	GC halo	45	1×10^{16}	—	—	—	$4.0[0.6] \times 10^{14}$
Sgr B2(N)[39]	GC HC	238	5×10^{18}	$6[2] \times 10^{17}$	$8[0.8] \times 10^{15}$	—	—
Sgr B2(M)[39]	GC halo	40	1×10^{16}	—	—	—	$2.7[0.5] \times 10^{14}$
Sgr B2(M)[39]	GC HC	150	8×10^{18}	$3[2] \times 10^{17}$	—	—	—
G19.61−0.23[40]	HC	151[6]	$5.2[6] \times 10^{17}$	$4.1[0.8] \times 10^{16}$	$1.4[0.1] \times 10^{16}$	—	$9.7[0.8] \times 10^{15}$
G34.26 + 0.15 NE[41,42]	HC	150	3.4×10^{17}	$[1.3] \times 10^{16}$	5.7×10^{16}	—	—
G34.26 + 0.15 SE[41]	HC	150	2.6×10^{17}	—	3.4×10^{16}	—	—
IRDC316.76-1[43]	IRDC	38	$4.6[1.1] \times 10^{14}$	—	—	6.2×10^{14}	—
IRDC316.76-2[43]	IRDC	53	2.5×10^{13}	—	—	$3[2] \times 10^{14}$	—
IRDC317.71-2[43]	IRDC	12	1.6×10^{14}	—	—	612×10^{14}	—
IRDC019.30-1[43]	IRDC	25	$1.5[0.1] \times 10^{15}$	—	$< 7.6 \times 10^{13}$	$1.1[0.2] \times 10^{14}$	$1.5[1.0] \times 10^{14}$
IRDC028.34-6[43]	IRDC	43	$1.2[0.1] \times 10^{15}$	—	$< 1.6 \times 10^{14}$	$2.8[0.2] \times 10^{14}$	$9.1[0.8] \times 10^{13}$
IRDC011.11-4[43]	IRDC	28	$2.3[0.8] \times 10^{14}$	—	—	$1.9[0.4] \times 10^{14}$	$7[5] \times 10^{14}$
IRDC028.34-3[43]	IRDC	32	$4.3[1.7] \times 10^{15}$	—	$< 9.9 \times 10^{13}$	$9.7[1.5] \times 10^{13}$	$1.7[1.1] \times 10^{14}$
NGC7129 FIRS2[44]	IMC	80	2×10^{14}	3.6×10^{12}	—	—	—

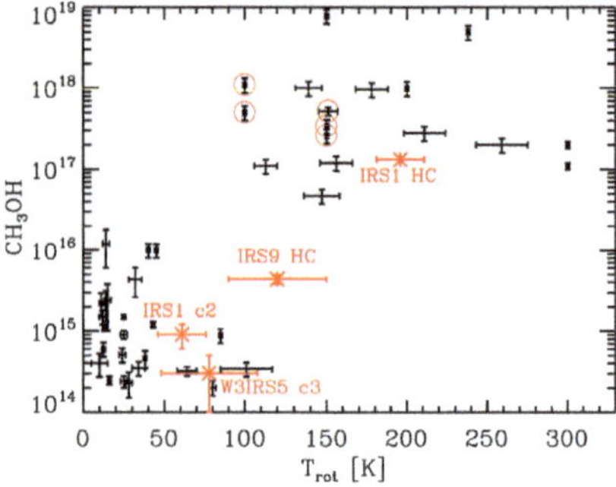

Fig. 7 CH₃OH column density *vs.* excitation temperature for the sample in Table 4 (black, and black with red circles for resolved observations) and our spatially resolved sample in Table 3 (red stars). Together these two parameters define two clearly distinguishable groups in the literature sample: cold ($T < 100$ K) COM sources and hot cores. In contrast, our small spatially resolved sample suggests a more continuous distribution.

groups there are no additional correlations visible. Furthermore, there seems to be a clear break in column densities at 100 K. This result may be somewhat biased, however, by the fact that most of the plotted values were acquired with single-dish telescopes, using a beam-dilution estimate. A common assumption is that if the excitation temperature is high, the emission area is small, *e.g.* a hot core, resulting in a potentially inflated column density estimate. We therefore also looked only at resolved observations, and find a similar trend as in the unresolved

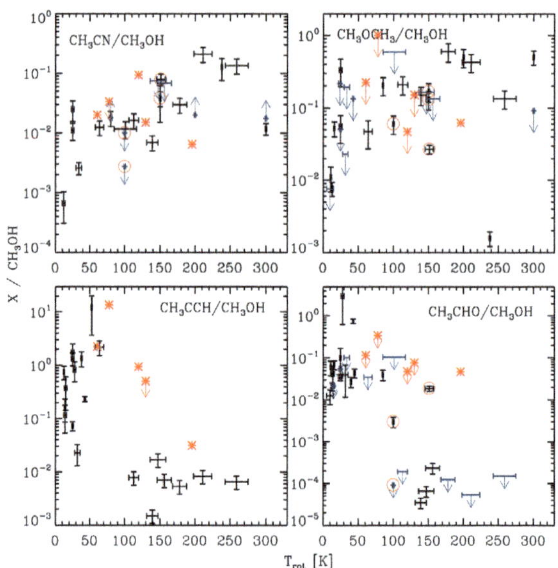

Fig. 8 The column density ratios of CH₃CN, CH₃OCH₃, CH₃CCH, CH₃CHO, and CH₃OH plotted as a function of excitation temperatures. Sources from the literature are shown in black, and our MYSOs in red (error bars are not shown here for sake of visibility – they are plotted in Fig. 9, however, for the same molecular ratios). Spatially resolved observations from the literature are marked with a circle.

sample, except that NGC 7538 IRS9 HC seems to fall somewhere in between the two groupings, perhaps suggestive of a rare transitional object between a MYSO and a luminous hot core. In general, there thus seems to be a well-defined CH_3OH evaporation front toward both low- and high-mass star forming regions.

Fig. 8 shows the relationship between ratios of CH_3CN, CH_3OCH_3, CH_3CCH, CH_3CHO and CH_3OH and the observed excitation temperatures. In the sample as a whole, $CH_3CN : CH_3OH$ correlates with temperature. This correlation is significant at the 99.9% level, based on the computation of the Spearman's rank correlation coefficient. No correlation is seen in our spatially resolved sample, but this may be a question of sample size, since the scatter around the general trend is considerable. CH_3OCH_3 shows no sign of correlation. Both $CH_3CCH : CH_3OH$ and $CH_3CHO : CH_3OH$ appear to depend inversely on the temperature when comparing the $T < 100$ K and $T > 100$ K sources, *i.e.* these two ratios seem to distinguish between the hot core and colder sources quite well. Furthermore, the $CH_3CCH : CH_3OH$ ratio increases with temperature between 10 and 50 K, indicative of formation at these temperatures.

We specifically checked whether there is any difference between previous unresolved and spatially resolved observations, and find none for CH_3CN or CH_3OCH_3, except that NGC 7538 IRS1 has an unusually small amount of CH_3CN compared to other sources with similar excitation temperatures. CH_3CN may be underestimated in this source, however, since there are clear line asymmetries that indicate that some self absorption is present. For $CH_3CHO : CH_3OH$, the spatially resolved hot cores appear to have a higher $CH_3CHO : CH_3OH$ ratio compared to those inferred from single-dish observations, suggestive of there being two CH_3CHO distributions and formation pathways, a cold one that

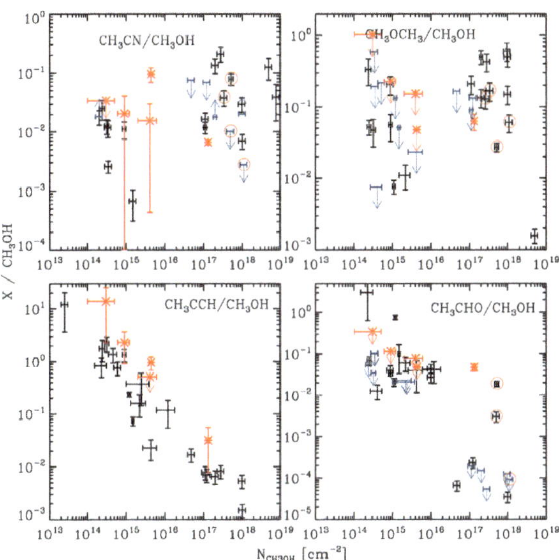

Fig. 9   The column density ratios of CH_3CN, CH_3OCH_3, CH_3CCH, CH_3CHO, and CH_3OH plotted as a function of CH_3OH column densities. Sources from the literature are in black, and our MYSOs in red. Spatially resolved observations from the literature are marked with a circle.

dominates single-dish observations, and a warm component that is picked up with spatially resolved observations. Our MYSO core measurements also indicate a more continuous fall-off with temperature of $CH_3CCH : CH_3OH$ across the 100 K line, indicative of a less sharp distinction between sources colder and warmer than 100 K, similar to what was seen for CH_3OH column densities in Fig. 7.

Fig. 7 suggests that the total CH_3OH column density is an equally good measure of hot core activity as excitation temperature, and we therefore also investigated its relationship with different $COM : CH_3OH$ ratios. Perhaps surprisingly, Fig. 9 shows that $CH_3CN : CH_3OH$ is not correlated with CH_3OH column density, nor is $CH_3OCH_3 : CH_3OH$. $CH_3CCH : CH_3OH$ is very strongly inversely correlated with the CH_3OH column, but this is related to the relatively small spread in CH_3CCH column densities across the sample, as reported above. CH_3CHO shows a slight decrease with CH_3OH column densities up to 10^{16} cm^{-2}. Beyond 10^{16} cm^{-2}, $CH_3CHO : CH_3OH$ abundances vary by three orders of magnitude, which supports the analysis of the temperature trend in Fig. 8, *i.e.* CH_3CHO is common at low temperatures, but in some sources it is also present in excess in hot cores.

4 Discussion

The observed dependencies (or lack thereof) of complex organic molecule abundances on temperature can be used to put empirical constraints on COM chemistry. Most complex molecules have many potential formation pathways with different expected dependencies on temperature. This has been explored extensively in chemical models of hot cores. For example, CH_3CN can form on grains *via* $CH_3 + CN$ at $\sim$30 K. It can also form from HCN in the gas phase, when HCN evaporates at $\sim$40 K, through $CH_3^+ + HCN$ association followed by a recombination reaction.[14] This gas-phase CH_3CN re-accretes onto the grains, and can then thermally evaporate at around 90 K.

COM abundances are often compared with CH_3OH abundances, because of its ubiquity and its proposed early formation at 10 K in ices. The presence of CH_3OH in the gas phase implies efficient desorption at all temperatures, but clearly a much more efficient desorption pathway is activated at 100 K. This is consistent with the theoretical expectation of low levels of non-thermal desorption at all temperatures (resulting in the release of a fraction of a percent of the ice into the gas phase) and rapid thermal desorption above 100 K. The scattering in CH_3OH abundances below 100 K is probably due to a combination of different initial CH_3OH ice abundances and different non-thermal desorption efficiencies. Above 100 K, different initial CH_3OH ice abundances and destruction pathways are the likely causes of the observed order of magnitude column density variation. Most COMs are expected to desorb thermally and non-thermally, similarly to CH_3OH, though this has yet to be quantified. Therefore differences in $COM : CH_3OH$ abundance ratios across MYSOs and samples of sources should be mainly due to differences in the chemical evolution or initial conditions.

One of the clearest trends in this study is the increase of CH_3CN column density, and further the increase in $CH_3CN : CH_3OH$ with temperature. This increase appears to be smooth, and cannot be due to a sudden onset of CH_3CN thermal evaporation at a single temperature. Rather it implies an increasingly efficient CH_3CN formation in the ice or gas as the temperature increases between

10 and 100 K, and potentially the onset of a second hot gas-phase chemistry above 100 K to explain the increase in $CH_3CN : CH_3OH$ in this temperature range. The continuous increase in $CH_3CN : CH_3OH$ with temperature means that with ALMA, this ratio could become a powerful diagnostic of the evolutionary stage of MYSOs.

CH_3OCH_3 has an almost constant ratio of 14% with respect to CH_3OH (the lower and upper quartiles are 5 and 33% respectively), indicative of co-formation or a constant conversion factor of CH_3OH into CH_3OCH_3 at all temperatures. This is most readily explained if CH_3OCH_3 forms from CH_3OH ice dissociation chemistry and then co-desorbs both thermally and non-thermally. $CH_3OCH_3 : CH_3OH$ ratios may thus serve as a signpost of the overall efficiency of conversion of simple ices into more complex ones.

$CH_3CCH : CH_3OH$ displays two clear trends. Firstly, the CH_3CCH column densities vary significantly less than any other complex molecule, which results in a steep inverse relationship between $CH_3CCH : CH_3OH$ and CH_3OH column density. This is indicative of early CH_3CCH formation and no hot core chemistry contribution to its abundance. Secondly, the $CH_3CCH : CH_3OH$ ratio peaks at around 50 K. Thus at low temperatures, CH_3CCH production increases compared to CH_3OH non-thermal desorption, indicative of either a thermal desorption pathway below 50 K, or cold gas-phase chemistry, or a combination of the two, $e.g.$ thermal evaporation of CH_4 at around 25 K followed by gas-phase formation of CH_3CCH. In either case $CH_3CCH : CH_3OH$ has the potential to constrain the relative importance of hot core and non-hot core driven complex chemistry in unresolved observations encompassing both envelopes and protostellar cores.

The $CH_3CHO : CH_3OH$ abundances also display a negative dependence on temperature, but the trend is complicated by a few outliers that suggest that in some sources there exists a high-temperature formation pathway of CH_3CHO. The fact that these outliers are all spatially resolved observations is notable, and as more spatially resolved observations appear we may find that an increase in the CH_3CHO abundance is a common feature of hot core chemistry. This second formation pathway could be either due to warm ice chemistry followed by thermal desorption with CH_3OH, or to hot gas-phase chemistry. Quantifying the CH_3CHO abundances at all scales will be important to understanding differences between cold and warm COM formation pathways and how these pathways together shape the final complex organic composition.

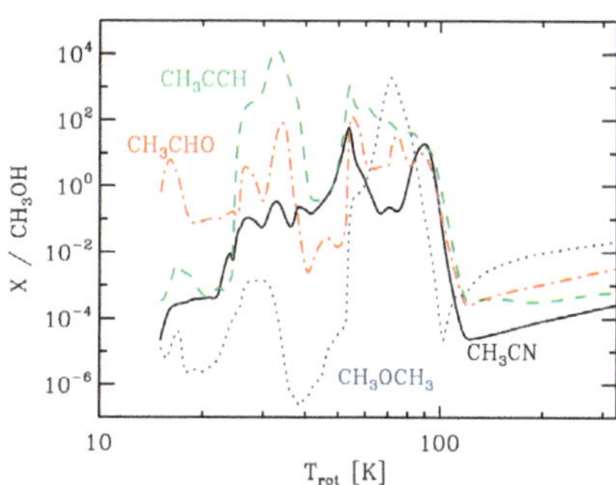

Fig. 10   Model predictions of the COM : CH_3OH ratios as a function of temperature during massive star formation, adapted from Garrod (2013).[14]

This set of empirical constraints can also be used to directly evaluate models of the complex chemistry during star formation, including the predicted onset of hot core activity, the temperature dependence of different COM ratios, and the sensitivity of different ratios to environmental factors other than the current temperature.

Regardless of the warm-up time scale, recent models predict that CH_3OH and most other COMs should thermally evaporate at $\sim$100 K, resulting in rapid increases in CH_3OH and COM column densities, often by many orders of magnitude. This compares well with the bimodal distribution of CH_3OH column densities in our sample, defined by a sharp transition at 100 K from CH_3OH columns of $\sim$$10^{15}$ cm^{-2} to $\sim$$10^{17}$ cm^{-2}. This confirms (1) the model predictions of a very sharp transition between CH_3OH ice and gas dominated regimes, *i.e.* a well-defined snow line, and (2) that the current estimates of the temperature location of this snow line are approximately right.

To evaluate the agreement between models and observations for specific COM abundances, Fig. 10 shows model predictions for COM : CH_3OH ratios, using a recent model that includes a detailed ice surface and bulk chemistry treatment, a suite of desorption mechanisms, and a full gas-phase chemistry network, and assuming a medium warm-up time scale (*i.e.* 2×10^5 years for warm-up to 200 K).[14] For CH_3CN : CH_3OH, the theoretical abundance ratio increase of two orders of magnitude with temperature up to 100 K agrees well with observations. The model drop-off at 100 K is not seen in observations, however. CH_3OCH_3 : CH_3OH displays a more complex dependence on temperature, consistent with the lack of a clear temperature dependence in the observational sample. CH_3CCH : CH_3OH is predicted to increase by eight orders of magnitude between 15 and 30 K, followed by a slow drop-off at 30–100 K, and a steep decline at 100 K. Qualitatively, the observations tell a similar story, with an increasing ratio between 10 and 50 K, followed by a flattening or decline, and then a steep drop at around 100 K. Quantitatively, there is little agreement, however, and the current network seems to under-produce CH_3CCH below 20 K, and over-produce it at 25–40 K. Finally, the predicted flat CH_3CHO : CH_3OH trend followed by a steep drop at 100 K is in excellent agreement with most observations, but does not account for the hot CH_3CHO component seen in some sources. In summary, current state-of-the-art astrochemical models capture many of the observed trends in this data set, but there seem to be some hot formation pathways missing.

Finally, the scattering in COM : CH_3OH temperature relations should probe the sensitivity of the COM chemistry to environmental conditions other than temperature. For example, observationally, CH_3CCH columns are remarkably constant, and therefore models predicting a strong dependence on the environment for this molecule would be problematic. In contrast, CH_3CN : CH_3OH varies by more than an order of magnitude in each temperature bin, and model predictions indicating that this ratio should be sensitive to *e.g.* collapse timescale are perfectly consistent with this observed scatter.[13]

5   Conclusions

Based on the combined analysis of the spatial distribution of COMs toward three MYSOs and, mostly spatially unresolved, COM observations in the literature, we find that:

1. MYSOs display a systematic chemical differentiation including spatially separated CH_3CN, CH_3OH and CH_3CCH emission peaks, where the CH_3CN and

CH_3OH peaks trace hot material and CH_3CCH colder material. Toward the MYSO W3 IRS5 there are at least four distinguishable chemical peaks: SO_2 + CH_3SH, CH_3CN + HNCO, CH_3OH and HC_3N + H_2CS + CH_3CCH.

2. CH_3OH column densities are observed to strongly depend on whether the derived excitation temperatures are above or below 100 K, but do not vary significantly with temperature within these two temperature bins.

3. The CH_3CCH : CH_3OH ratio is correlated with temperature at 10–50 K and anti-correlated with temperature above 50 K. The CH_3CN : CH_3OH ratio increases with temperature, and, with a few notable exceptions, CH_3CHO : CH_3OH decreases sharply around 100 K. These observations together suggest that CH_3CCH only forms at low temperatures, that CH_3CHO forms efficiently at low temperatures, but has a second high-temperature formation pathway that sometimes becomes activated, and that the relative formation efficiency of CH_3CN with respect to CH_3OH increases with temperature. This is generally consistent with model predictions.

4. The formation of some COMs is already efficient at low temperatures, but the COM composition strongly depends on the source temperature, and the combination of CH_3CN, CH_3OH and CH_3CCH observations may therefore be used to define the evolutionary stage of an object in terms of its COM chemistry.

In general, spatially resolved observations are key to deconvolving the contributions from cores at different chemical stages, which will typically be confused by single-dish observations, and thus constrain how the complex chemistry depends on its environment and the evolutionary stage of the core. This approach becomes particularly powerful when multiple emission lines of each molecule are imaged, enabling constraints on both the chemical and temperature structures of a MYSO region. Temperature is not the only regulator of chemistry, however, and to explain, for example, differences between N and S cores requires either differences in initial chemical conditions (ice compositions) or the dominance of energy sources other than passive heating from the central star.

This leads to the question of the relative importance of nature and nurture for COM chemistry, and how to observationally test it. Can we for example identify ratios that exclusively trace the initial conditions, *i.e.* ratios of COMs that evolve similarly with temperature, but depend strongly on the initial ice composition or collapse history? Addressing this question observationally requires spatially resolved observations that can follow the chemistry from envelope to core. In general high-spatial resolution is key to unravelling the complex chemical structures of MYSOs, but it is still an open question how MYSO spectral data cubes can be best mined for information, and how this information is best displayed and quantified.

Acknowledgements

The authors are grateful to Robin T. Garrod for sharing his complex chemistry model results.

References

1 G. A. Blake, E. C. Sutton, C. R. Masson and T. G. Phillips, *Astrophys. J.*, 1987, **315**, 621–645.

2 E. Herbst and E. F. van Dishoeck, *Annu. Rev. Astron. Astrophys.*, 2009, **47**, 427–480.

3 E. F. van Dishoeck, G. A. Blake, D. J. Jansen and T. D. Groesbeck, *Astrophys. J.*, 1995, **447**, 760.

4 S. Cazaux, A. G. G. M. Tielens, C. Ceccarelli, A. Castets, V. Wakelam, E. Caux, B. Parise and D. Teyssier, *Astrophys. J.*, 2003, **593**, L51–L55.

5 S. Bottinelli, C. Ceccarelli, J. P. Williams and B. Lefloch, *Astron. Astrophys.*, 2007, **463**, 601–610.

6 K. I. Öberg, N. van der Marel, L. E. Kristensen and E. F. van Dishoeck, *Astrophys. J.*, 2011, **740**, 14.

7 H. G. Arce, J. Santiago-García, J. K. Jørgensen, M. Tafalla and R. Bachiller, *Astrophys. J.*, 2008, **681**, L21–L24.

8 K. I. Öberg, S. Bottinelli, J. K. Jørgensen and E. F. van Dishoeck, *Astrophys. J.*, 2010, **716**, 825–834.

9 A. Bacmann, V. Taquet, A. Faure, C. Kahane and C. Ceccarelli, *Astron. Astrophys.*, 2012, **541**, L12.

10 J. Cernicharo, N. Marcelino, E. Roueff, M. Gerin, A. Jiménez-Escobar and G. M. Muñoz Caro, *Astrophys. J.*, 2012, **759**, L43.

11 K. I. Öberg, A. C. A. Boogert, K. M. Pontoppidan, S. van den Broek, E. F. van Dishoeck, S. Bottinelli, G. A. Blake and N. J. Evans, II, *Astrophys. J.*, 2011, **740**, 109.

12 K. I. Öberg, M. D. Boamah, E. C. Fayolle, R. T. Garrod, C. J. Cyganowski and F. van der Tak, *Astrophys. J.*, 2013, **771**, 95.

13 R. T. Garrod, S. L. W. Weaver and E. Herbst, *Astrophys. J.*, 2008, **682**, 283–302.

14 R. T. Garrod, *Astrophys. J.*, 2013, **765**, 60.

15 A. G. G. M. Tielens and W. Hagen, *Astron. Astrophys.*, 1982, **114**, 245–260.

16 R. T. Garrod and T. Pauly, *Astrophys. J.*, 2011, **735**, 15.

17 S. B. Charnley, A. G. G. M. Tielens and T. J. Millar, *Astrophys. J.*, 1992, **399**, L71–L74.

18 K. I. Öberg, R. T. Garrod, E. F. van Dishoeck and H. Linnartz, *Astron. Astrophys.*, 2009, **504**, 891–913.

19 F. F. S. van der Tak, E. F. van Dishoeck and P. Caselli, *Astron. Astrophys.*, 2000, **361**, 327–339.

20 F. F. S. van der Tak, P. G. Tuthill and W. C. Danchi, *Astron. Astrophys.*, 2005, **431**, 993–1005.

21 S. T. Megeath, T. L. Wilson and M. R. Corbin, *Astrophys. J.*, 2005, **622**, L141–L144.

22 J. A. Rodón, H. Beuther, S. T. Megeath and F. F. S. van der Tak, *Astron. Astrophys.*, 2008, **490**, 213–222.

23 L. Chavarría, F. Herpin, T. Jacq, J. Braine, S. Bontemps, A. Baudry, M. Marseille, F. van der Tak, B. Pietropaoli, F. Wyrowski, R. Shipman, W. Frieswijk, E. F. van Dishoeck, J. Cernicharo, R. Bachiller, M. Benedettini, A. O. Benz, E. Bergin, P. Bjerkeli, G. A. Blake, S. Bruderer, P. Caselli, C. Codella, F. Daniel, A. M. di Giorgio, C. Dominik, S. D. Doty, P. Encrenaz, M. Fich, A. Fuente, T. Giannini, J. R. Goicoechea, T. de Graauw, P. Hartogh, F. Helmich, G. J. Herczeg, M. R. Hogerheijde, D. Johnstone, J. K. Jørgensen, L. E. Kristensen, B. Larsson, D. Lis, R. Liseau, C. McCoey, G. Melnick, B. Nisini, M. Olberg, B. Parise, J. C. Pearson, R. Plume, C. Risacher, J. Santiago-García, P. Saraceno, J. Stutzki, R. Szczerba, M. Tafalla, A. Tielens,

T. A. van Kempen, R. Visser, S. F. Wampfler, J. Willem and U. A. Yıldız, *Astron. Astrophys.*, 2010, **521**, L37.

24 K.-S. Wang, T. L. Bourke, M. R. Hogerheijde, F. F. S. van der Tak, A. O. Benz, S. T. Megeath and T. L. Wilson, *Astron. Astrophys.*, 2013, **558**, A69.

25 F. P. Helmich, D. J. Jansen, T. de Graauw, T. D. Groesbeck and E. F. van Dishoeck, *Astron. Astrophys.*, 1994, **283**, 626–634.

26 J. P. McMullin, B. Waters, D. Schiebel, W. Young and K. Golap, *Astronomical Data Analysis Software and Systems XVI*, 2007, p. 127.

27 D. J. Wilner and W. J. Welch, *Astrophys. J.*, 1994, **427**, 898–913.

28 F. F. S. van der Tak, E. F. van Dishoeck, N. J. Evans, II and G. A. Blake, *Astrophys. J.*, 2000, **537**, 283–303.

29 H. M. Pickett, R. L. Poynter, E. A. Cohen, M. L. Delitsky, J. C. Pearson and H. S. P. Müller, *J. Quant. Spectrosc. Radiat. Transfer*, 1998, **60**, 883–890.

30 H. S. P. Müller, S. Thorwirth, D. A. Roth and G. Winnewisser, *Astron. Astrophys.*, 2001, **370**, L49–L52.

31 P. F. Goldsmith and W. D. Langer, *Astrophys. J.*, 1999, **517**, 209–225.

32 S. E. Bisschop, J. K. Jørgensen, E. F. van Dishoeck and E. B. M. de Wachter, *Astron. Astrophys.*, 2007, **465**, 913–929.

33 K. Isokoski, S. Bottinelli and E. F. van Dishoeck, *Astron. Astrophys.*, 2013, **554**, A100.

34 E. C. Fayolle, K. I. Öberg, R. Garrod, S. Bisschop and E. van Dishoeck, *Astron. Astrophys.*, 2014, submitted.

35 K. I. Öberg, S. Bottinelli and E. F. van Dishoeck, *Astron. Astrophys.*, 2009, **494**, L13–L16.

36 S. Maret, C. Ceccarelli, A. G. G. M. Tielens, E. Caux, B. Lefloch, A. Faure, A. Castets and D. R. Flower, *Astron. Astrophys.*, 2005, **442**, 527–538.

37 Y. Kuan, H. Huang, S. B. Charnley, N. Hirano, S. Takakuwa, D. J. Wilner, S. Liu, N. Ohashi, T. L. Bourke, C. Qi and Q. Zhang, *Astrophys. J.*, 2004, **616**, L27–L30.

38 S. E. Bisschop, J. K. Jørgensen, T. L. Bourke, S. Bottinelli and E. F. van Dishoeck, *Astron. Astrophys.*, 2008, **488**, 959–968.

39 A. Nummelin, P. Bergman, Å. Hjalmarson, P. Friberg, W. M. Irvine, T. J. Millar, M. Ohishi and S. Saito, *Astrophys. J. Suppl.*, 2000, **128**, 213–243.

40 S.-L. Qin, Y. Wu, M. Huang, G. Zhao, D. Li, J.-J. Wang and S. Chen, *Astrophys. J.*, 2010, **711**, 399–416.

41 B. Mookerjea, E. Casper, L. G. Mundy and L. W. Looney, *Astrophys. J.*, 2007, **659**, 447–458.

42 S. Watt and L. G. Mundy, *Astrophys. J. Suppl.*, 1999, **125**, 143–160.

43 T. Vasyunina, A. I. Vasyunin, E. Herbst, H. Linz, M. Voronkov, T. Britton, I. Zinchenko and F. Schuller, *Astrophysical J.*, 2014, **780**(1), 19.

44 A. Fuente, R. Neri and P. Caselli, *Astron. Astrophys.*, 2005, **444**, 481–493.

PAPER

Chemical complexity in the Horsehead photodissociation region

Viviana V. Guzmán,[a] Jérôme Pety,[ab] Pierre Gratier,[ab] Javier R. Goicoechea,[c] Maryvonne Gerin,[b] Evelyne Roueff,[d] Franck Le Petit[d] and Jacques Le Bourlot[d]

Received 29th November 2013, Accepted 23rd January 2014
DOI: 10.1039/c3fd00114h

The interstellar medium is known to be chemically complex. Organic molecules with up to 11 atoms have been detected in the interstellar medium, and are believed to be formed on the ices around dust grains. The ices can be released into the gas-phase either through thermal desorption, when a newly formed star heats the medium around it and completely evaporates the ices; or through non-thermal desorption mechanisms, such as photodesorption, when a single far-UV photon releases only a few molecules from the ices. The first mechanism dominates in hot cores, hot corinos and strongly UV-illuminated PDRs, while the second dominates in colder regions, such as low UV-field PDRs. This is the case of the Horsehead were dust temperatures are $\simeq 20-30$ K, and therefore offers a clean environment to investigate the role of photodesorption. We have carried out an unbiased spectral line survey at 3, 2 and 1mm with the IRAM-30m telescope in the Horsehead nebula, with an unprecedented combination of bandwidth, high spectral resolution and sensitivity. Two positions were observed: the warm PDR and a cold condensation shielded from the UV field (dense core), located just behind the PDR edge. We summarize our recently published results from this survey and present the first detection of the complex organic molecules HCOOH, CH_2CO, CH_3CHO and CH_3CCH in a PDR. These species together with CH_3CN present enhanced abundances in the PDR compared to the dense core. This suggests that photodesorption is an efficient mechanism to release complex molecules into the gas-phase in far-UV illuminated regions.

1 Introduction

Molecular lines are used to trace the structure of the interstellar medium (ISM) and the physical conditions of the gas in different environments, from high-z

[a]IRAM, 300 rue de la Piscine, 38406 Saint Martin d'Hères, France. E-mail: guzman@iram.fr

[b]LERMA-LRA, UMR 8112, Observatoire de Paris and École normale Supérieure, 24 rue Lhomond, 75231 Paris, France

[c]Centro de Astrobiología, CSIC-INTA, Carretera de Ajalvir, Km 4, Torrejón de Ardoz, 28850 Madrid, Spain

[d]LUTH UMR 8102, CNRS and Observatoire de Paris, Place J. Janssen, 92195 Meudon Cedex, France

galaxies to proto-planetary disks. However, the interpretation of molecular observations for most of these objects is hampered by the complex source geometries, and the small angular sizes in the sky compared with the angular resolution of current instrumentation, that prevent us from resolving the different gas components, and hence to know which specific region each molecule actually traces. Therefore, in order to fully benefit from the diagnostic power of the molecular lines, the formation and destruction paths of the molecules must be quantitatively understood. This challenging task requires the contribution of theoretical models, laboratory experiments and observations. Well-defined sets of observations of simple *template* sources are key to benchmark the predictions of theoretical models. In this respect, the Horsehead nebula has proven to be a good template source of low-UV field irradiated environments because it is close-by ($\sim$400 pc), it has a simple geometry (edge-on) and its gas density is well constrained. Moreover, in contrast to other Galactic photo-dissociation regions (PDRs), like the Orion Bar and Mon R2 which present large radiation fields ($\chi \simeq 10^{4}$–10^{5}), the Horsehead is illuminated by a weaker radiation field ($\chi \sim 60$) and thus better resembles the majority of the far-UV illuminated neutral gas in the Galaxy. Furthermore, the dust grains in the Horsehead have temperatures of $\simeq$20–30 K, which is not enough to thermally desorb most of the ices. The Horsehead therefore offers a clean environment in which to isolate the role of photo-desorption of ices on dust grains.

Observations by the Infrared Space Observatory (ISO) and Spitzer have shown that dust grains are covered by ice mantles in the cold envelopes surrounding high-mass protostars,[1,2] low-mass protostars[3–6] and in isolated dense cores.[7] These studies revealed that the ice mantles consist mostly of H_2O, CO_2 and CO, with smaller amounts of CH_3OH, CH_4, NH_3 and H_2CO. More complex prebiotic molecules, such as glycine (NH_2CH_2COOH) could also form on the ices around dust grains. Although their exact formation is unclear, it is believed that the simplest prebiotic molecules have an interstellar origin.[8,9] Indeed, numerous amino acids, which are the building blocks of proteins, have been found in meteorites.[10] In addition, glycine, which is the simplest amino acid, has been detected in samples returned by NASA's Stardust spacecraft from comet Wild 2.[11] Despite controversial detection claims,[12–14] glycine or other more complex amino acids have not been detected in the interstellar medium yet. The most complex molecules detected in the interstellar medium so far, are glycolaldehyde[15] ($CH_2(OH)CHO$), acetamide[16] (CH_3CONH_2), aminoacetonitrile[17] (NH_2CH_2CN), and ethyl formate[18] (C_2H_5OCHO). This shows the high degree of chemical complexity that can be reached in the interstellar medium.

Simpler, but still complex organic molecules, such as methanol (CH_3OH), ketene (CH_2CO), acetaldehyde (CH_3CHO), formic acid (HCOOH), formamide (NH_2CHO), propyne (CH_3CCH), methyl formate ($HCOOCH_3$), and dimethyl ether (CH_3OCH_3), are widely observed in hot cores of high-mass protostars,[25–27] and also in hot-corinos of low-mass protostars.[28,29] The complex molecules observed in protostars have been classified in three different generations by Herbst and van Dishoeck,[30] depending on their formation mechanism. The zeroth generation species form through grain surface processes in the cold (<20 K) pre-stellar stage (*e.g.*, H_2CO and CH_3OH). First generation species form from surface reactions between photodissociated products of the zeroth generation species in the warm-up (20–100 K) period. Finally, second generation species form in the hot (>100 K) gas from the evaporated zeroth and first generation species in the so called hot-core phase. Although it is clear that grain surface processes play an important role in the formation of complex molecules, the exact formation mechanism of most complex molecules is still debated.

Bisschop *et al.*[27] observed several complex molecules toward seven high-mass protostars, and classified them as cold (T < 100 K) and hot (T > 100 K) molecules based on their rotational temperatures. The hot molecules include H_2CO, CH_3OH, HNCO, CH_3CN, $HCOOCH_3$ and CH_3OCH_3, while the cold molecules include HCOOH, CH_2CO, CH_3CHO, and CH_3CCH. The cold molecules are expected to be present in the colder envelope around the hot-core. Öberg *et al.*[38] studied the spatial distribution of complex molecules around a high-mass protostar and found that CH_2CO, CH_3CHO and CH_3CCH are indeed abundant in the cold envelope. They classified them as zeroth order molecules because their formation must require very little heat.

Complex organic molecules may trace other environments than hot cores and hot corinos. They are also present in the cold UV-shielded gas. Bacmann *et al.*[39] detected CH_3OCH_3, CH_3OCHO, CH_2CO and CH_3CHO in a cold ($T_{kin} \sim 10$ K) prestellar core. These observations challenged the current formation scenario of complex molecules on dust grains, because the diffusion reactions that lead to the formation of species are not efficient on dust grains with temperatures of $\sim$10 K. CH_3CHO and CH_2CO have also been detected in the dark cloud TMC-1.[40,41] CH_2CO and CH_3CHO have also been detected in a $z = 0.89$ spiral galaxy located in front of the quasar PKS1830-211.[42]

In this paper, we present the results of an unbiased line survey performed with the IRAM-30m telescope in a classic star forming region, the Horsehead nebula. We describe the observations in section 2. In section 3 we present a summary of the recently published results of the line survey. In section 4 we present new unpublished results about the first detection of complex molecules in a PDR. We discuss these observations in section 5 and conclude in section 6.

2 Observations

2.1 Deep pointed integrations: the Horsehead WHISPER line survey

With the purpose of providing a benchmark to chemical models we have performed a complete and unbiased line survey: the Horsehead WHISPER (Wideband High-resolution Iram-30m Surveys at two Positions with Emir Receivers, PI: J. Pety). Two positions were observed: 1) the HCO peak, which is characteristic of the photo-dissociation region at the surface of the Horsehead nebula,[21] and 2) the DCO^+ peak, which belongs to a cold condensation located less than 40$''$ away from the PDR edge, where HCO^+ and other species are highly deuterated.[23] Hereafter we refer to these two positions as the PDR and dense core, respectively. The combination of the new EMIR receivers at the IRAM-30m telescope and the Fourier transform spectrometers (FTS) yields a spectral survey with unprecedented combination of bandwidth (36 GHz at 3mm, 34 GHz at 2mm and 73 GHz at 1mm), spectral resolution (49 kHz at 3 and 2mm; and 195 kHz at 1mm), and sensitivity (median noise 8.1 mK, 18.5 mK and 8.6 mK at 3, 2 and 1mm respectively). A detailed presentation of the observing strategy and data reduction process will be given in a forthcoming paper. In short, all frequencies were observed with two different frequency tunings and the Horsehead PDR and dense core positions were alternatively observed every 15 min in position-switching mode with a common fixed off-position. This observing strategy allows us to remove potential ghost lines that are incompletely rejected from a strong line in the image sideband (the typical rejection of the EMIR sideband-separating mixers is only 13dB).

The line density at 3 mm is, on average, 5 and 4 lines/GHz in the PDR and dense core, respectively. At 2 and 1 mm, the line density is 1 line/GHz in both the PDR and dense core. The contribution of molecular lines to the total flux at 1.2 mm is estimated to be 14% at the PDR and 16% at the dense core. Approximately 30 species (plus their isotopologues) are detected with up to 7 atoms in the PDR and the dense core.

2.2 IRAM-30m and PdBI maps

Fig. 1 displays the integrated emission of the C_2H, CH_3CHO, HCO, CF^+, DCO^+, p–H_2CO, and CH_3OH–A lines as well as the 1.2mm continuum emission. The observation parameters are summarized in Table 1. A reference is given where a detailed description of the observations and data reduction can be found for each map.

The A- and E-type CH_3CHO 5_{15}–4_{14} lines at 93.581 GHz and 93.595 GHz were observed simultaneously with the A- and E-type CH_3CHO 6_{16}–5_{15} lines at 112.249 GHz and 112.254 GHz during $\sim$17 hours of average summer weather in August and September 2013. We used the two polarizations of the EMIR receivers and the FTS backends at 49 kHz spectral resolution. We used the position-switching, on-the-fly observing mode. The off-position offsets were $(\delta RA, \delta Dec) = (100'',0'')$, that

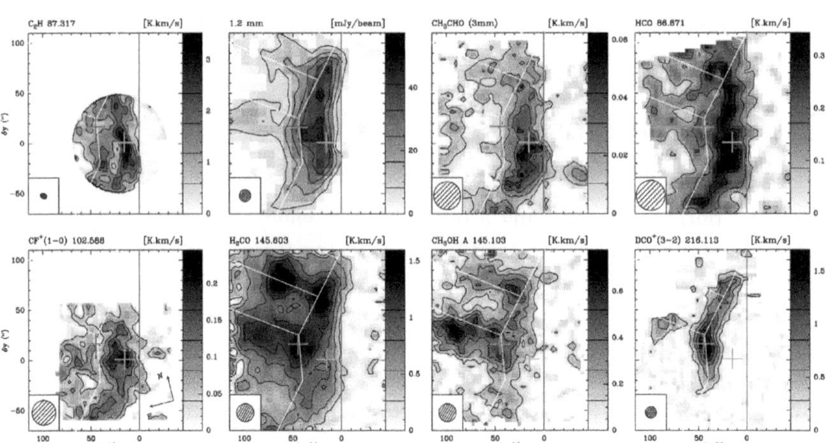

Fig. 1 IRAM-30m and PdBI maps of the Horsehead edge. Maps were rotated by 14° counter-clockwise around the projection center, located at $(\delta x,\delta y) = (20'',0'')$, to bring the exciting star direction in the horizontal direction. The horizontal zero, marked by the red vertical line, delineates the PDR edge. The crosses show the positions of the PDR (green) and the dense core (blue), where deep integrations were performed in the Horsehead WHISPER line survey (PI: J.Pety). The white lines delineate the arc-like structure of the DCO^+ emission. The spatial resolution is plotted in the bottom-left corner. Values of contour levels are shown in the respective image lookup table. The emission of all lines is integrated between 10.1 and 11.1 km s^{-1}.

Table 1 Observation parameters for the maps shown in Fig. 1. The projection center of the maps is $\alpha_{2000} = 05^h40^m54.27^s$, $\delta_{2000} = -02°28'00''$

Molecule	Transition	Frequency GHz	Instrument	Beam arcsec	PA (°)	Int. Time hours	T_{sys} K(T)	Noise K(T_{mb})	Ref.
Continuum at 1.2 mm		250.000000	30m/MAMBO	11.7	—	—	—	—	Hily-Blant et al.[19]
CCH	$1,3/2(2)-0,1/2(1)$	87.316898	PdBI/C&D	7.2×5.0	54				Pety et al.[20]
CH$_3$CHO	$5_{15}-4_{14},6_{16}-5_{15}$	93.5,112.2	30m/EMIR	30.0	0				This work
HCO	$1_{01}3/2,2-0_{00}1/2,1$	86.670760	30m/AB100	29.9	0	$2.6/5.0^a$	133	63	Gerin et al.[21]
CF$^+$	$1-0$	102.587533	30m/EMIR	25.4	0	2.5	88	0.13	Guzmán et al.[22]
DCO$^+$	$3-2$	216.112582	30m/HERA	11.4	0	$1.5/2.0^a$	230	0.10	Pety et al.[23]
p-H$_2$CO	$2_{02}-1_{01}$	145.602949	30m/EMIR	17.8	0	$7.4/12.9^a$	208	0.17	Guzman et al.[24]
CH$_3$OH–A	3_0-2_0	145.103152	30m/EMIR	17.9	0	$7.4/12.9^a$	208	0.095	Guzman et al.[24]

^{a} Two values are given for the integration time: the on-source time and the telescope time.

is, the H II region ionized by σOri and free of molecular emission. We observed along and perpendicular to the direction of the exciting star in zigzags (*i.e.*, $\pm$ the lambda and beta scanning direction). From our knowledge of the IRAM-30m telescope, we estimate the absolute position accuracy to be $3''$.

The IRAM-30m data were processed with the GILDAS/CLASS software. The data were first calibrated to the T_A^* scale using the chopper-wheel method.[43] The data were converted to main-beam temperatures (T_{mb}) using the forward and main-beam efficiencies (F_{eff} and B_{eff}). The resulting amplitude accuracy is 10%. We then computed the experimental noise by subtracting a zeroth-order baseline from every spectra. A systematic comparison of this noise value with the theoretical noise computed from the system temperature, the integration time, and the channel width allowed us to filter out outlier spectra. The spectra where then gridded to a data cube through a convolution with a Gaussian kernel. In order to increase the signal-to-noise ratio, we smoothed the four A- and E-type CH_3CHO lines to the largest angular resolution and then averaged all the data. The averaged map, which has a final resolution of $30''$, is shown in Fig. 1.

3　Recent results from the Horsehead WHISPER line survey

3.1　CF^+: a tracer of C^+ and a measure of the fluorine abundance

CF^+, which was only detected in the Orion Bar before,[44] was detected toward the illuminated edge of the Horsehead nebula by Guzmán *et al.*[22] The CF^+ ion, which is formed by reactions of HF and C^+, is expected to be the second most important fluorine reservoir, after HF, in regions where C^+ is abundant.[45] Indeed, the CF^+ emission is concentrated toward the edge of the Horsehead, delineating the western edge of the DCO^+ emission (see Fig. 1). Theoretical models predict that there is a significant overlap between CF^+ and C^+ at the edges of molecular clouds. Therefore, we propose that CF^+ can be used as a proxy of C^+, but that can be observed from ground-based telescopes, unlike C^+ for which we need to go to space. This can be a powerful tool, because the [C II] 157.8 μm line is the main cooling mechanism of the diffuse gas, and the cooling of the medium, which allows the gas to compress, is a crucial step in the formation of new stars. Moreover, given the simple chemistry of fluorine and assuming that the CF^+ destruction is dominated by dissociative recombination with electrons with little contribution from photodissociation (which is true only in low-UV PDRs), one obtains that the CF^+ column density is proportional to the column density of HF. Then, assuming that in molecular clouds all fluorine is in its molecular form, the elemental abundance of fluorine can be derived directly from CF^+ observations. We infer $F/H = (0.6–1.5) \times 10^{-8}$ in good agreement with the one found in diffuse molecular clouds,[46] and somewhat lower than the solar value[47] and the one found in the diffuse atomic gas.[48] Finally, because the Horsehead shows narrow emission lines, in contrast to other PDRs like the Orion Bar, Guzmán *et al.*[49] were able to resolve the two hyperfine components in the CF^+ $J = 1$–0 line and to compare with *ab initio* computations of the CF^+ spin rotation constant. The derived theoretical value of $C_I = 229.2$ kHz agrees well with the observations. The Horsehead is thus a good laboratory for precise spectroscopic studies of species present in far-UV illuminated environments.

3.2   Detection of a new molecule in space, tentatively attributed to $l\text{-}C_3H^+$

Thanks to the sensitive observations and large bandwidth covered by the Horsehead WHISPER line survey, a consistent set of 8 lines were detected toward the PDR position, that could not be associated to any transition listed in the public line catalogs. The observed lines can be well fitted with a linear rotor model, implying a closed-shell molecule. The deduced rotational constant value is close to that of C_3H. In addition, the spatial distribution of the species integrated emission has a shape similar to radical species such as HCO, and small hydrocarbons such as C_2H (see Fig. 1). Therefore, Pety et al.[35] attributed the detected lines to the small hydrocarbon cation $l\text{-}C_3H^+$.

In the family of small hydrocarbons, Pety et al.[20] found that C_3H and C_3H_2 are about 1 order of magnitude more abundant in PDRs than current pure gas-phase models predict. An additional formation mechanism is therefore needed. One possibility to explain the observed high abundance of hydrocarbons in PDRs is the so called *Top-Down* model. In this scenario polycyclic aromatic hydrocarbons (PAHs) are fragmented into small hydrocarbons in PDRs due to the strong UV fields.[20,31,50] In the same way, PAHs are formed by photo-evaporation of very small grains.[51,52] The discovery of C_3H^+, which is an intermediate species in the gas-phase formation scenario, brings further constraints to the formation pathways of the small hydrocarbons. Indeed, we find a $l\text{-}C_3H^+$ abundance which is too low to explain the observed abundance of the other related small hydrocarbons by means of pure gas-phase chemical reactions.

The lines detected in the Horsehead have been detected in other environments, like the Orion Bar (Cuadrado et al. in prep) and Sgr B2,[53] which confirms the presence of the carrier in the ISM. But the attribution of the unidentified lines to $l\text{-}C_3H^+$ has been questioned by Huang et al.,[54] because their theoretical calculations of the spectroscopic constants of C_3H^+ differ from the ones inferred from our observations. Fortenberry et al.[55] proposed that a more plausible candidate is the hydrocarbon anion C_3H^-. However, if the unknown species is the anion, it would be the first anion detected in the Horsehead, and the ratio of C_3H^- to neutral C_3H would be $\sim$57%, which is higher than any anion to neutral ratio detected in the ISM so far. In addition, the lines were not detected in the dark cloud TMC 1, where other anions have been already detected. Moreover, because C_3H^- is an asymmetric rotor, the lines detected in the Horsehead would correspond to the $K_a = 0$ ladder and the lines from the $K_a = 1$ ladder should also be detected. We find no evidence of the $K_a = 1$ lines of C_3H in the observations of the Horsehead PDR.[56] For all these reasons it would be unexpected that the carrier of the unidentified lines is the anion, C_3H^-. The observations favor the assignment of the unidentified species to the hydrocarbon cation, C_3H^+, as the most likely candidate. However, a direct measurement in the laboratory is necessary to provide a definitive answer and close the controversy created by these observations in the Horsehead. Ongoing high-angular PdBI observations of this species in the Horsehead PDR will allow us to better constrain the chemistry of small hydrocarbons in the near future.

3.3   Photo-desorption of dust grain ice mantles: H_2CO and CH_3OH

Relatively simple organic molecules, like H_2CO and CH_3OH, are key species in the synthesis of more complex molecules in the ISM,[57–59] that could eventually end up

Table 2 Summary of abundances with respect to total hydrogen nuclei ($N_H = N(X)/(N(H) + 2N(H_2))$)) toward the PDR and dense core. The column densities of the total hydrogen nuclei are $N_H = 3.8 \times 10^{22}$ cm^{-2} (PDR) and $N_H = 6.4 \times 10^{22}$ cm^{-2} (dense core)

Species	Beam ('')	PDR Abundance	Offsets	Core Abundance	Offsets	Ref.
$C^{18}O$	6.5 × 4.3	1.9×10^{-7}	(−6,4)	—	—	Pety et al.[20]
C_2H	7.2 × 5.0	1.4×10^{-8}	(−6,4)	—	—	Pety et al.[20]
c-C_3H	28	2.7×10^{-10}	(−10,0)	—	—	Teyssier et al.[31]
l-C_3H	28	1.4×10^{-10}	(−10,0)	—	—	Teyssier et al.[31]
c-C_3H_2	6.1 × 4.7	1.1×10^{-9}	(−6,4)	—	—	Pety et al.[20]
l-C_3H_2	27	$<4.6 \times 10^{-11}$	(−10,0)	—	—	Teyssier et al.[31]
C_4H	6.1 × 4.7	1.0×10^{-9}	(−6,−4)	—	—	Pety et al.[20]
C_6H	28	2.2×10^{-11}	(−6,4)	—	—	Agúndez et al.[32]
CS	10	2.0×10^{-9}	(4,0)	2.9×10^{-9}	(21,15)	Goicoechea et al.[33]
$C^{34}S$	16	9.2×10^{-11}	(4,0)	9.1×10^{-11}	(21,15)	Goicoechea et al.[33]
HCS^+	29	1.7×10^{-11}	(4,0)	1.2×10^{-11}	(21,15)	Goicoechea et al.[33]
HCO	6.7 × 4.4	8.4×10^{-10}	(−5,0)	$<8.0 \times 10^{-11}$	(20,22)	Gerin et al.[21]
HCO^+	28	9.0×10^{-10}	(−5,0)	3.9×10^{-9}	(20,22)	Goicoechea et al.[34]
$H^{13}CO^+$	6.8 × 4.7	1.5×10^{-11}	(−5,0)	6.5×10^{-11}	(20,22)	Goicoechea et al.[34]
HOC^+	28	4.0×10^{-12}	(−5,0)	—	(20,22)	Goicoechea et al.[34]
CO^+	10	$<5.0 \times 10^{-13}$	(−5,0)	—	(20,22)	Goicoechea et al.[34]
DCO^+	12	—	(−5,0)	8.0×10^{-11}	(20,22)	Pety et al.[23]
CF^+	25	5.7×10^{-10}	(−5,0)	$<6.9 \times 10^{-11}$	(20,22)	Guzmán et al.[22]
C_3H^+	27	3.1×10^{-11}	(−5,0)	—	(20,22)	Pety et al.[35]
o-H_2CO	6.1 × 5.6	1.9×10^{-10}	(−5,0)	1.5×10^{-10}	(20,22)	Guzmán et al.[36]
p-H_2CO	6.1 × 5.6	9.5×10^{-11}	(−5,0)	5.0×10^{-11}	(20,22)	Guzmán et al.[36]
HDCO	18	—	(−5,0)	2.5×10^{-11}	(20,22)	Guzmán et al.[36]
D_2CO	24	—	(−5,0)	1.6×10^{-11}	(20,22)	Guzmán et al.[36]
CH_3OH–E	6.1 × 5.6	7.0×10^{-11}	(−5,0)	1.0×10^{-10}	(20,22)	Guzman et al.[24]
CH_3OH–A	6.1 × 5.6	5.3×10^{-11}	(−5,0)	1.3×10^{-10}	(20,22)	Guzman et al.[24]
CH_3CN	27	2.5×10^{-10}	(−5,0)	7.9×10^{-12}	(20,22)	Gratier et al.[37]
CH_3NC	25	4.1×10^{-11}	(−5,0)	$<7.8 \times 10^{-12}$	(20,22)	Gratier et al.[37]
HC_3N	30	6.3×10^{-12}	(−5,0)	7.9×10^{-12}	(20,22)	Gratier et al.[37]
t-HCOOH	29	5.2×10^{-11}	(−5,0)	1.4×10^{-11}	(20,22)	This work
o-CH_2CO	30	1.3×10^{-10}	(−5,0)	4.2×10^{-11}	(20,22)	This work
p-CH_2CO	26	1.8×10^{-11}	(−5,0)	7.3×10^{-12}	(20,22)	This work
CH_3CHO–E	27	1.4×10^{-11}	(−5,0)	3.9×10^{-12}	(20,22)	This work
CH_3CHO–A	27	5.4×10^{-11}	(−5,0)	2.0×10^{-11}	(20,22)	This work
CH_3CCH	29	4.4×10^{-11}	(−5,0)	3.0×10^{-10}	(20,22)	This work

in proto-planetary disks, and hence in new planetary systems. They are also used to probe the temperature and density of the gas in different astrophysical sources.[60–62] Both H_2CO and CH_3OH have been detected in a wide range of interstellar environments such as dark clouds, proto-stellar cores and comets, with high abundances (10^{-6}–10^{-9}) with respect to total hydrogen. Unlike H_2CO, which can be formed efficiently in both the gas-phase and on the surfaces of dust grains, CH_3OH is thought to be formed mostly on the ices, through the successive additions of hydrogen atoms to adsorbed CO molecules.

Guzmán et al.[36] and Guzman et al.[24] observed several millimeter lines of H_2CO and CH_3OH toward the PDR and dense core positions in the Horsehead. The inferred abundances from the observations (see Table 2) were compared to PDR models that include either pure gas-phase chemistry or both gas-phase and grain surface chemistry. Pure gas-phase models cannot reproduce the observed abundances of either H_2CO or CH_3OH at the PDR position. Both species are therefore mostly formed on the surface of dust grains, probably through the successive hydrogenation of CO ices and are subsequently released into the gas-phase through photodesorption. At the dense core, on the other hand, photodesorption of ices is needed to explain the observed abundance of CH_3OH, while a pure gas-phase model can reproduce the observed H_2CO abundance. The different formation routes for H_2CO at the PDR and dense core suggested by the models is strengthened by the different *ortho*-to-*para* ratios derived from the observations (~3 at the dense core, ~2 at the PDR).

In addition to the lines detected in the WHISPER survey, we obtained high-angular resolution (6″) maps of H_2CO and CH_3OH with the IRAM-PdBI. Fig. 1 shows the H_2CO and CH_3OH single-dish 30m maps that were used for the short-spacing of the PdBI observations. The H_2CO emission map presents a peak at the dense core position, while CH_3OH presents a dip in its emission at the same position. The observations thus suggest that CH_3OH is depleted in the dense core. This way, gas-phase CH_3OH is present in an envelope around the dense core, while H_2CO is present in both the envelope and the dense core itself. Indeed, we expect photodesorption to be more efficient at the PDR than at the far-UV shielded dense core. We thus conclude that photo-desorption of ices is an efficient mechanism to release species into the gas-phase in far-UV illuminated regions.

3.4 Nitrile molecules: CH_3CN, CH_3NC and HC_3N

Moderately complex nitriles like CH_3CN and HC_3N are easily detected in (massive) star forming regions. In particular, the CH_3CN emission has been found to be enhanced in star forming regions containing an ultracompact H II region.[63] CH_3CN is thought to be a good tracer of the physical conditions in warm and dense regions. Gratier *et al.*[37] detected several lines of CH_3CN and HC_3N in the PDR and dense core. CH_3NC and C_3N are also detected toward the PDR. The observations show that the chemistry of HC_3N and CH_3CN is quite different. Indeed, we find that CH_3CN is 30 times more abundant in the far-UV illuminated gas than in the far-UV shielded core, while HC_3N has a similar abundance in both positions. The high abundance of CH_3CN inferred in the PDR is surprising

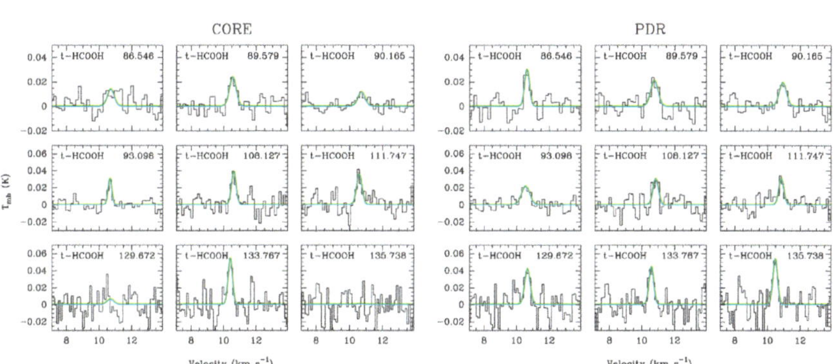

Fig. 2 HCOOH lines detected toward the dense core (*left*) and PDR (*right*). For each line, the same scale is used at both positions for ease of comparison.

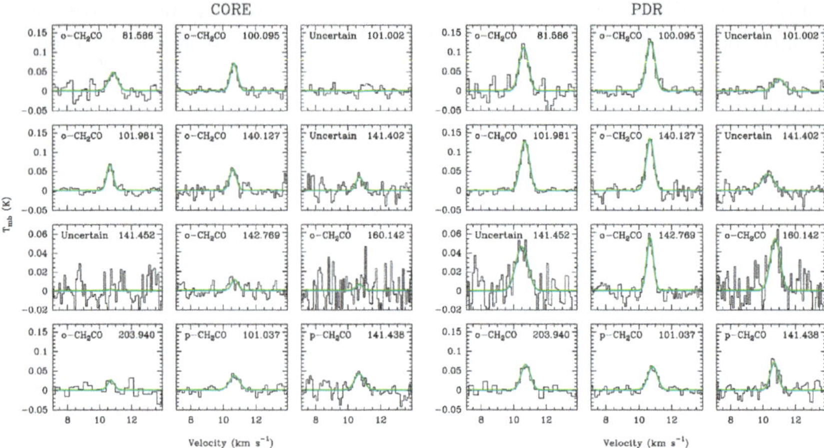

Fig. 3 CH_2CO lines detected toward the dense core (*left*) and PDR (*right*). For each line, the same scale is used at both positions for ease of comparison.

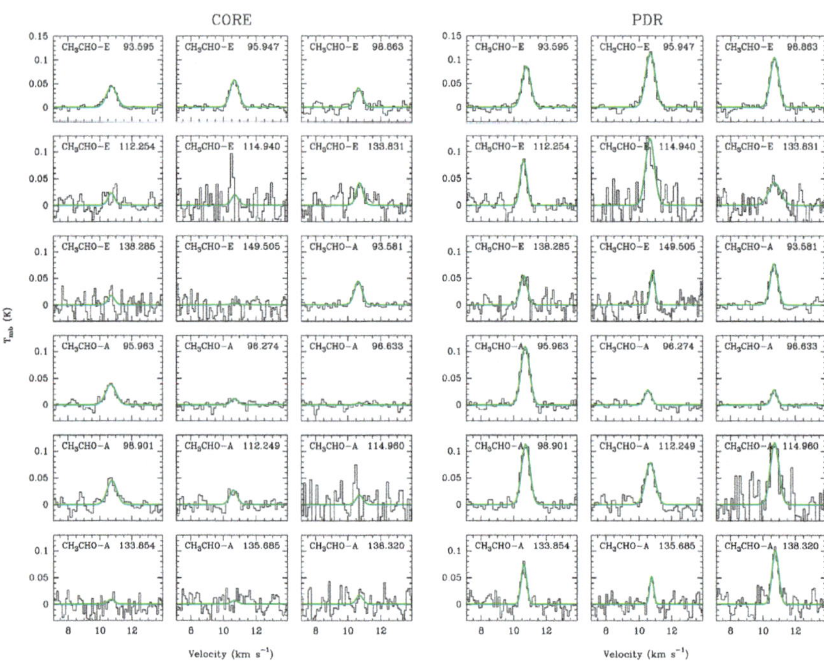

Fig. 4 CH_3CHO lines detected toward the dense core (*left*) and PDR (*right*). For each line, the same scale is used at both positions for ease of comparison.

because the photodissociation of this complex molecule is expected to be efficient in far-UV illuminated regions. The observed abundance in the PDR cannot be reproduced by current pure gas-phase chemical models. We have shown that photodesorption is an efficient mechanism to release H_2CO and CH_3OH in the PDR, but the case of CH_3CN is even more extreme as it is 30 times more abundant

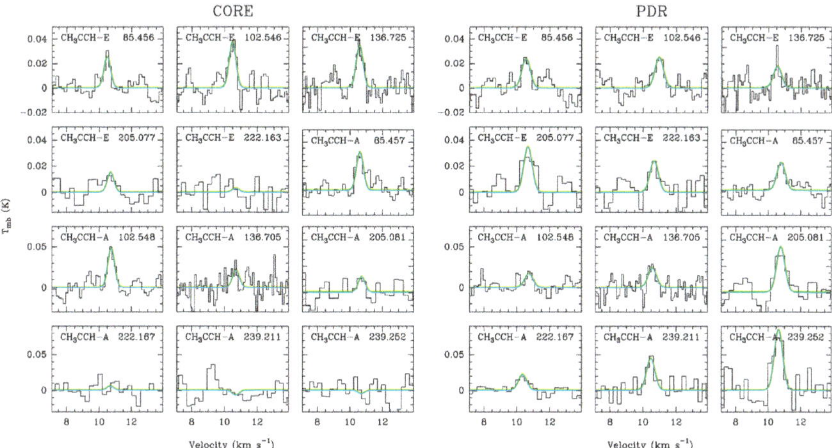

Fig. 5 CH₃CCH lines detected toward the dense core (*left*) and PDR (*right*). For each line, the same scale is used at both positions for ease of comparison.

there than in the dense core, while H_2CO presents similar abundances in both positions. This shows that there is something specific in the chemistry of CH_3CN in FUV-illuminated regions. CH_3CN could be produced on the ices by the photo-processing of N bearing species followed by photodesorption, but it could also be produced in the gas-phase if the abundances of its gas-phase precursors, HCN and CH, are enhanced. The detection of CH_3NC at the PDR, which results in a CH_3NC/CH_3CN isomeric ratio of 0.15, suggests that CH_3NC could also form on the surfaces of dust grains through far-UV irradiation of CH_3CN ices leading to isomerization.

4   Other complex molecules in PDRs

Within the WHISPER line survey, several lines of HCOOH, CH_2CO, CH_3CHO and CH_3CCH are detected. These are presented in Fig. 2–5. The spectroscopic parameters and Gaussian fit results of the detected lines are listed in Appendix A. The spectroscopic parameters are taken from the CDMS[64] and JPL[65] databases. Several lines of ketene and acetaldehyde are clearly detected ($S/N > 5\sigma$). The formic acid and propyne present several but fainter (2σ–5σ) lines toward the PDR and dense core positions. In order to confirm the correct identification of these molecules, we have modeled the spectrum of each species assuming LTE and optically thin emission, and checked that there are no predicted lines missing in the line survey. Both *ortho* and *para* forms of CH_2CO are detected, as well as both E and A forms of CH_3CHO and CH_3CCH. All the lines detected of the formic acid correspond to the *trans* isomer. CH_2CO and CH_3CHO lines are brighter toward the PDR position than toward the dense core, while HCOOH and CH_3CCH lines have similar brightness in both positions.

The beam-averaged column density of each molecule was estimated using rotational diagrams because no collisional coefficients are available. The detected lines cover a sufficiently large energy range to derive a rotational temperature. The resulting rotational diagrams are shown in Fig. 6. The inferred abundances with

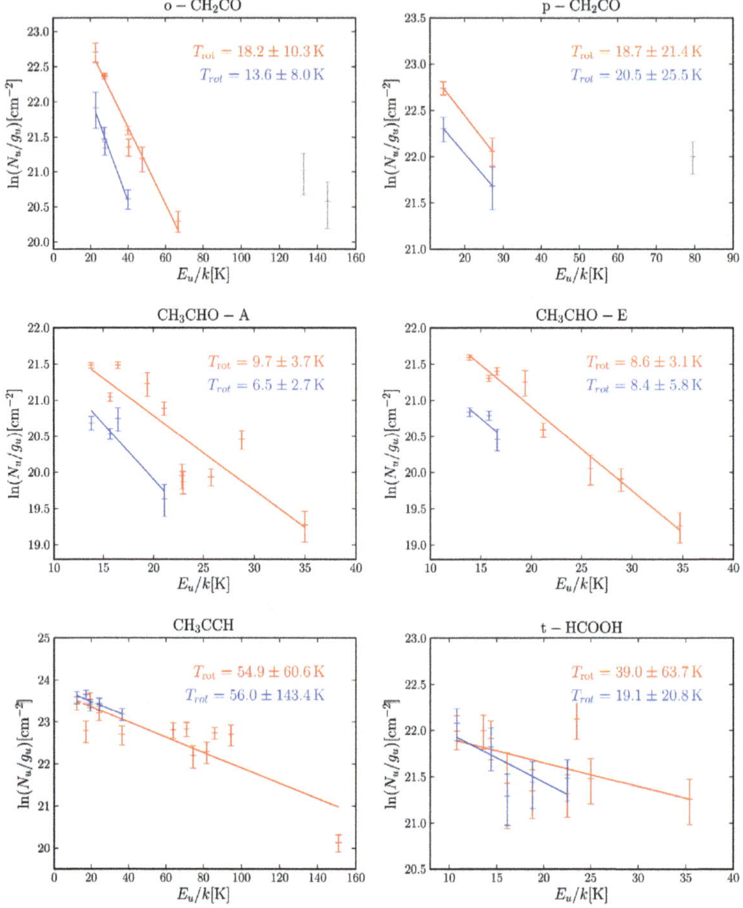

Fig. 6 Rotational diagrams for the PDR (red) and dense core (blue). The gray points correspond to the three lines detected at the PDR whose identification is uncertain. They are not considered in the fit.

respect to H nuclei are summarized in Table 2. The partition function was computed independently for *ortho* and *para* nuclear spin forms (for ketene), and for E and A symmetry forms (for CH_3CHO), by direct summation over the energy levels.

Formic acid

The data in the rotational diagram of HCOOH present a large scatter because all the detected lines are weak and therefore have a larger uncertainty than the lines of the other complex molecules. The rotational temperature is poorly constrained. The inferred abundances of $\sim 5 \times 10^{-11}$ (PDR) and $\sim 1 \times 10^{-11}$ (core) are thus also uncertain and should be considered as an order of magnitude estimate. Deeper integration times are needed to better constrain the HCOOH abundance.

Ketene

The *ortho* and *para* symmetries of CH_2CO were treated as different species. When including all the o-CH_2CO and p-CH_2CO lines detected in the PDR, the fit results in rotational temperatures of 154 K and 120 K for o-CH$_2$CO and p-CH$_2$CO, respectively. These temperatures are much larger than the kinetic temperature at the PDR ($\sim$60 K). When the three lines of CH$_2$CO with energies above 80 K that are detected at the PDR (gray points in Fig. 6) are not considered in the rotational diagrams, the derived rotational temperatures decrease to 18 K for both $ortho$ and $para$ species. This temperature agrees much better with the expected sub-thermal excitation in the Horsehead, and also with the derived rotational temperatures at the dense core. The enhanced emission of the three lines with E_u > 80 K could be the result of an excitation effect. However, these lines are broader (0.8 km s^{-1}) than the other CH$_2$CO lines and than those of other species detected in the Horsehead PDR (the typical linewidth is 0.6 km s^{-1}). In addition, the velocity of these three lines differs by 0.2 km s^{-1} from the systemic velocity of 10.7 km s^{-1} found for most species in the Horsehead. Cummins $et\ al.$[25] detected several o-CH$_2$CO lines toward Sgr B2, including the 5_{33}–4_{32} at 101.002 GHz, which is one of the three lines detected in the Horsehead with E_u > 80 K. They also obtained a large rotational temperature, which led them to remove this line from the fit and consider the identification as uncertain. At the dense core position in the Horsehead, the derived rotational temperature for o-CH$_2$CO is $\sim$14 K. Only two lines of p-CH$_2$CO are detected at the dense core, resulting in a rotational temperature of $\sim$20 K. Ketene is $\sim$3 times more abundant in the PDR than in the dense core, with abundances of 1.5×10^{-10} (PDR) and 4.9×10^{-11} (core). The $ortho$-to-$para$ ratio is poorly constrained, resulting in $o/p = 7.1 \pm 4.2$ (PDR) and $o/p = 5.7 \pm 3.9$ (dense core).

Acetaldehyde

The E and A symmetries of CH$_3$CHO were also treated as different species. Acetaldehyde has a large dipole moment (2.7 Debye, see Table 3) compared to methanol (1.7 Debye). Since the critical density is proportional to μ^2, sub-thermal effects are important for CH$_3$CHO.[27] Indeed, the derived rotational temperatures for CH$_3$CHO (6–10 K) are the lowest ones of all the molecules discussed here. The derived column density of CH$_3$CHO is $\sim$3 times larger in the PDR than in the dense core. The inferred E/A ratio is $\sim$0.3 $\pm$ 0.1 (PDR) and $\sim$0.2 $\pm$ 0.2 (core), $i.e.$, much lower than unity in both positions.

Fig. 1 displays the averaged E and A CH$_3$CHO lines at 93.6 GHz. The CH$_3$CHO emission clearly peaks at the PDR position, delineating the edge of the Horsehead nebula. The CH$_3$CHO emission resembles the HCO emission at the 30m telescope angular resolution of 30$''$, which is concentrated in a narrow structure peaking at the PDR. Higher-angular resolution (6$''$) observations by Gerin $et\ al.$[21] showed that the HCO emission traces a filament of $\sim$12$''$ width. The similarities between the

Table 3   Dipole moments of complex molecules

Species	Dipole moment (Debye)	Reference
HCOOH	1.4	Kim $et\ al.$[66]
CH$_2$CO	1.45	Hannay and Smyth[67]
CH$_3$CHO	2.7	Kleiner $et\ al.$[68]
CH$_3$CCH	0.78	Burrell $et\ al.$[69]

emission of HCO and CH_3CHO thus suggest that CH_3CHO also arises from a narrow filament that peaks at the PDR position. Assuming a filament of $12''$ centered at the PDR, we estimate that $\sim$10% of the CH_3CHO emission detected at the dense core corresponds to beam pick-up from the PDR due to the large beam at 93 GHz $(27'')$. The remaining emission towards the core line of sight could arise from the cloud surface, as was found for CS[33] and HCO.[21]

Propyne

If $CH_3CCH–E$ and $CH_3CCH–A$ are treated as different species, the derived rotational temperatures at the PDR are $\sim$70 K and $\sim$53 K for E and A symmetries, respectively. At the dense core, a rotational temperature of $\sim$70 K is inferred for the E symmetry. A rotational temperature cannot be inferred for $CH_3CCH–A$ at the dense core because the two lines that are detected are faint. The two symmetries are therefore considered as the same species, resulting in a rotational temperature of $\sim$55 K and a total column density for the E- and A-type CH_3CCH of $\sim$2 $\times$ 10^{13} cm^{-2} at both the PDR and dense core positions. The inferred column density at the PDR position does not change when the E and A symmetries are separated.

5   Discussion

Ketene and acetaldehyde are thought to form on the surface of dust grains. Indeed, CH_3CHO has been proposed as a candidate for the 7.41 μm absorption feature observed toward high-mass protostars. Ketene and acetaldehyde are thought to form together on the ices through C and H atom additions to CO.[30] The expected sequence is

$$CO \xrightarrow{\text{H}} HCO \xrightarrow{\text{C}} HCCO \xrightarrow{\text{H}} CH_2CO \xrightarrow{\text{2H}} CH_3CHO. \tag{1}$$

Ketene can also be formed from reactions between C_2H_2 and O in irradiated H_2O-rich and CO_2-rich ices, as shown by recent laboratory experiments.[70] Laboratory experiments also show that reactions between C_2H_4 and O can produce acetaldehyde and its isomer, ethylene oxide (CH_2OCH_2).[71] CH_3CO and CH_3CHO could also be formed in the gas-phase, through ion–molecule and neutral–neutral reactions. Indeed, gas-phase models predict abundances that are comparable to those measured in some of the high-mass protostars observed by Bisschop *et al.*[27]

The exact formation path of HCOOH on ices is unclear, though HCOOH ices have been observed in star-forming regions.[72] Several formation paths on grain surfaces have been proposed in the past. It could form from the addition of H and O atoms to CO or to CO_2.[73] Garrod *et al.*[74] proposed that HCOOH could form through reactions between HCO and OH. More recently, Ioppolo *et al.*[75] have studied the hydrogenation of the HO–CO complex in the laboratory and showed it is an efficient formation route to HCOOH.

The abundances derived in the Horsehead PDR for HCOOH, CH_2CO, and CH_3CHO, are 3–4 times larger toward the PDR than toward the dense core. The case of CH_3CN is even more extreme as it is $\sim$30 times more abundant in the PDR than in the dense core.[37] CH_3CCH is only 1.5 times more abundant in the PDR than in the dense core. In contrast, methanol is $\sim$2 times *less* abundant in the PDR than in the dense core. When comparing the abundances of the different molecules, we found that CH_3CCH is one order of magnitude more abundant (3–4 $\times 10^{-10}$) than CH_2CO and CH_3CHO (2–7 $\times 10^{-11}$). Contrary to the other complex molecules, which present [X]/[CH_3OH] ratios of lower than 1, CH_3CCH is ~4 times more abundant than methanol in the PDR, and ~1.3 times more abundant than methanol in the dense core. Öberg *et al.*[38] also found large CH_3CCH abundances (~1 relative to methanol) toward the high-mass protostar NGC 7538 IRS9. They found the CH_3CCH/CH_3OH abundance ratio to be significantly different to what models including grain surface processes predict, which suggests that an important cold formation pathway is missing for CH_3CCH.

Fig. 7 shows a comparison between the abundances derived in the Horsehead and those derived toward the hot corino sources from Bisschop *et al.*[27] and toward the prestellar core L1689B.[39] CH_3OH is several orders of magnitude more abundant toward the hot core sources than in the Horsehead. The abundances of the other complex molecules with respect to H_2 vary over ~1 order of magnitude between the different hot core sources, and are comparable to the abundances derived in the Horsehead. The abundances with respect to CH_3OH are also shown in the right panel of Fig. 7. In this case, the abundances of complex molecules are ~3 orders of magnitude larger in the Horsehead than in the hot cores. However, this could be a consequence of methanol and the other complex molecules tracing different regions in the hot core sources. Methanol probably traces the hot ($T_{kin} > 100$ K) gas where species have evaporated from the grains, while the other complex molecules trace the colder envelope around the protostars, where the ices have not completely evaporated but can be photodesorbed.

The fact that we only detect cold molecules (HCOOH, CH_2CO, CH_3CHO and CH_3CCH) and none of the hot molecules (*e.g.*, CH_3OCH_3 and $HCOOCH_3$), is in agreement with the idea that the cold molecules are zeroth or first generation species formed on the cold grain surfaces and trace the warm/cold envelope around protostars, because their formation probably requires little energy. In the Horsehead, the enhanced abundances toward the PDR compared to the dense core (for HCOOH, CH_2CO, and CH_3CHO), suggests that their formation is more efficient in the presence of far-UV photons. The similarities between the HCO and CH_3CHO emission maps also suggests that the CH_3CHO abundance at the PDR

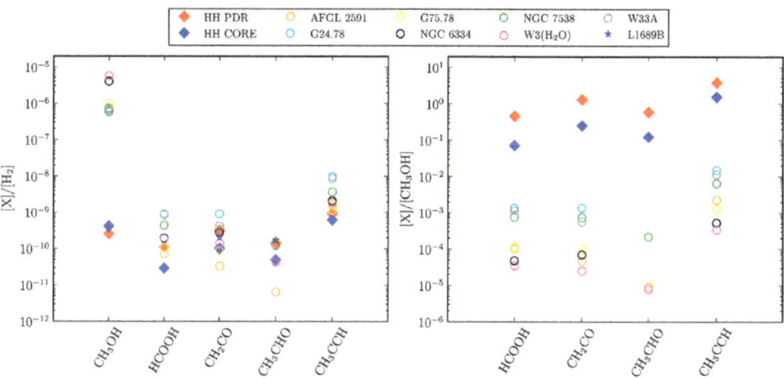

Fig. 7   Abundances with respect to H_2 (*left*) and with respect to CH_3OH (*right*) toward the hot core sources from Bisschop *et al.*[27] (*open circles*), the cold prestellar core from Bacmann *et al.*[39] (*blue star*) and the Horsehead PDR and dense core (*red and blue diamonds*).

could be even higher than estimated here, if the emission arises from a narrow filament like HCO. This could be the result of an efficient photodesorption in the PDR, due to the larger radiation field compared to the dense core, which is consistent with the much lower than unity E/A ratio (statistical value) we infer from the observations. But it could also indicate that the formation on the grains itself is more efficient in the PDR, due to a better mobility of the molecules in ice mantles. Indeed, recent laboratory experiments have shown that the diffusion of molecules is active at $T_{dust} \gtrsim 30$ K, allowing reactions to proceed faster when the ices are warmed by far-UV photons.[76,77] Dust temperatures in the Horsehead PDR range from $T_{dust} \sim 30$ K in the PDR to $T_{dust} \sim 20$ K in the dense core.[78]

6   Conclusions

We have carried out an unbiased spectral line survey at 3, 2 and 1mm with the IRAM-30m telescope in the Horsehead nebula, with an unprecedented combination of bandwidth, high spectral resolution and sensitivity. Two positions were observed: the warm photodissociation region (PDR) and a cold condensation shielded from the UV field, located less than $40''$ away from the PDR edge. The results of this survey include 1) the detection of CF^+, which can be used as a new diagnostic of UV illuminated gas and a potential proxy of the C^+ emission associated to molecular gas; 2) the detection of a new species in the ISM, the small hydrocarbon C_3H^+, which confirms the top-down scenario of formation of the small hydrocarbons from PAHs and photo-erosion; 3) the detection of H_2CO, CH_3OH and CH_3CN, which reveals that photo-desorption of ices is an efficient mechanism to release molecules into the gas phase; and 4) the first detection of the complex organic molecules, HCOOH, CH_2CO, CH_3CHO and CH_3CCH in a PDR, which reveals the degree of chemical complexity reached in the UV illuminated neutral gas. Complex molecules are usually considered as hot-core tracers. The detection of these molecules in PDRs shows that they can survive in the presence of far-UV radiation, and their formation could even be enhanced due to the radiation. This opens the possibility of detecting complex molecules in other far-UV illuminated regions, such as protoplanetary disks, in the future. From this work we conclude that grain surface chemistry and non-thermal desorption are crucial processes in the ISM and therefore must be incorporated into photo-chemical models to interpret the observations.

A Observational tables

Acknowledgements

V.G. thanks support from the Chilean Government through the Becas Chile scholarship program. This work was also funded by grant ANR-09-BLAN-0231-01 from the French Agence Nationale de la Recherche as part of the SCHISM project. J.R.G. thanks the Spanish MICINN for funding support through grants AYA2009-07304 and CSD2009-00038. J.R.G. is supported by a Ramón y Cajal research contract from the Spanish MICINN and co-financed by the European Social Fund.

Table 4 Observation parameters of the deep integrations of the HCOOH lines detected toward the PDR and dense core[a]

Molecule	Transition	ν GHz	E_u K	A_{ul} s^{-1}	g_u	Line area mK km s^{-1}	Velocity km s^{-1}	FWHM km s^{-1}	T_{peak} mK	RMS mK	Peak S/N
PDR											
t-HCOOH	4_{14}–3_{13}	86.546	13.6	6.0×10^{-6}	9	14.3 ± 2.5	10.63	0.41	32.6	6.0	5
t-HCOOH	4_{04}–3_{03}	89.579	10.8	7.0×10^{-6}	9	13.9 ± 2.6	10.74	0.58	22.6	5.8	4
t-HCOOH	4_{22}–3_{21}	90.165	23.5	6.0×10^{-6}	9	10.2 ± 2.2	10.91	0.51	18.7	5.4	3
t-HCOOH	4_{13}–3_{12}	93.098	14.4	8.0×10^{-6}	9	10.8 ± 2.4	10.54	0.51	19.7	6.0	3
t-HCOOH	5_{15}–4_{14}	108.127	18.8	1.3×10^{-5}	11	11.3 ± 2.9	10.83	0.38	27.7	9.8	3
t-HCOOH	5_{05}–4_{04}	111.747	16.1	1.4×10^{-5}	11	13.2 ± 5.1	10.82	0.41	30.0	9.3	3
t-HCOOH	6_{16}–5_{15}	129.672	25.0	2.2×10^{-5}	13	17.8 ± 4.1	10.66	0.40	42.1	13.2	3
t-HCOOH	6_{06}–5_{05}	133.767	22.5	2.5×10^{-5}	13	17.1 ± 3.9	10.57	0.36	45.2	14.1	3
t-HCOOH	6_{24}–5_{23}	135.738	35.4	2.3×10^{-5}	13	16.1 ± 3.5	10.46	0.30	49.9	13.9	4
CORE											
t-HCOOH	4_{04}–3_{03}	89.579	10.8	7.0×10^{-6}	9	17.2 ± 2.8	10.56	0.59	27.4	6.2	4
t-HCOOH	4_{13}–3_{12}	93.098	14.4	8.0×10^{-6}	9	11.4 ± 2.6	10.64	0.34	31.4	6.5	5
t-HCOOH	5_{15}–4_{14}	108.127	18.8	1.3×10^{-5}	11	16.3 ± 3.5	10.61	0.38	40.1	9.9	4
t-HCOOH	5_{05}–4_{04}	111.747	16.1	1.4×10^{-5}	11	11.5 ± 3.1	10.54	0.30	36.3	9.0	4
t-HCOOH	6_{06}–5_{05}	133.767	22.5	2.5×10^{-5}	13	26.2 ± 5.6	10.39	0.44	55.9	13.5	4

[a] Note: All temperatures are given in the main beam temperature scale.

Table 5 Observation parameters of the deep integrations of the CH_2CO lines detected toward the PDR and dense core[a]

Molecule	Transition	ν GHz	E_u K	A_{ul} s^{-1}	g_u	Line area mK km s^{-1}	Velocity km s^{-1}	FWHM km s^{-1}	T_{peak} mK	RMS mK	Peak S/N
PDR											
o-CH_2CO	4_{13}–3_{12}	81.586	22.9	5.0×10^{-6}	27	83.1 ± 10.7	10.62	0.69	113.4	19.8	6
o-CH_2CO	5_{15}–4_{14}	100.095	27.5	1.0×10^{-5}	33	87.9 ± 3.6	10.70	0.64	128.7	7.4	17
o-CH_2CO	5_{14}–4_{13}	101.981	27.8	1.1×10^{-5}	33	89.0 ± 3.2	10.67	0.65	127.7	7.1	18
o-CH_2CO	7_{17}–6_{16}	140.127	40.0	2.9×10^{-5}	45	77.9 ± 4.7	10.66	0.55	133.0	12.4	11
o-CH_2CO	7_{16}–6_{15}	142.769	40.5	3.1×10^{-5}	45	68.4 ± 8.1	10.66	0.47	135.3	25.6	5
o-CH_2CO	8_{18}–7_{17}	160.142	47.6	4.5×10^{-5}	51	102.7 ± 13.5	10.74	0.72	134.2	35.7	4
o-CH_2CO	10_{19}–9_{18}	203.940	66.9	9.3×10^{-5}	63	45.9 ± 7.0	10.72	0.67	64.8	10.6	6
p-CH_2CO	5_{05}–4_{04}	101.037	14.5	1.1×10^{-5}	11	45.3 ± 3.1	10.81	0.70	60.8	6.8	9
p-CH_2CO	7_{07}–6_{06}	141.438	27.2	3.1×10^{-5}	15	42.9 ± 6.7	10.67	0.56	71.6	17.9	4
o-CH_2CO[d]	5_{33}–4_{32}[b]	101.002	132.8	7.0×10^{-6}	33	31.0 ± 4.5	10.98	0.87	33.6	8.5	4
o-CH_2CO[d]	7_{35}–6_{34}[c]	141.402	145.4	2.5×10^{-5}	45	50.2 ± 8.0	10.28	1.03	45.9	16.9	3
o-CH_2CO[d]	7_{25}–6_{24}	141.452	79.5	2.8×10^{-5}	15	39.1 ± 6.9	10.48	0.83	44.4	15.7	3
CORE											
o-CH_2CO	4_{13}–3_{12}	81.586	22.9	5.0×10^{-6}	27	34.3 ± 8.6	10.86	0.69	47.0	17.8	3
o-CH_2CO	5_{15}–4_{14}	100.095	27.5	1.0×10^{-5}	33	38.9 ± 3.3	10.67	0.52	70.7	7.8	9
o-CH_2CO	5_{14}–4_{13}	101.981	27.8	1.1×10^{-5}	33	33.3 ± 3.2	10.65	0.47	66.6	7.8	9
o-CH_2CO	7_{17}–6_{16}	140.127	40.0	2.9×10^{-5}	45	30.6 ± 4.2	10.60	0.51	56.5	11.8	5
p-CH_2CO	5_{05}–4_{04}	101.037	14.5	1.1×10^{-5}	11	29.5 ± 3.9	10.68	0.73	37.7	8.0	5
p-CH_2CO	7_{07}–6_{06}	141.438	27.2	3.1×10^{-5}	15	31.1 ± 7.0	10.68	0.69	42.5	17.0	3

[a] Note: All temperatures are given in the main beam temperature scale. [b] Blended with the 5_{32}–4_{31} line. [c] Blended with the 7_{34}–6_{33} line. [d] The line identification is uncertain.

Table 6 Observation parameters of the deep integrations of the CH_3CHO lines detected toward the PDR and dense core[a]

Molecule	Transition	ν GHz	E_u K	A_{ul} s^{-1}	g_u	Line area mK km s^{-1}	Velocity km s^{-1}	FWHM km s^{-1}	T_{peak} mK	RMS mK	Peak S/N
PDR											
CH_3CHO-E	$5_{15}-4_{14}$	93.595	15.8	2.5×10^{-5}	22	57.5 ± 2.5	10.75	0.63	86.0	5.2	16
CH_3CHO-E	$5_{05}-4_{04}$	95.947	13.9	2.8×10^{-5}	22	81.6 ± 3.0	10.67	0.68	113.5	6.1	19
CH_3CHO-E	$5_{14}-4_{13}$	98.863	16.6	3.0×10^{-5}	22	68.2 ± 3.5	10.68	0.61	104.7	8.0	13
CH_3CHO-E	$6_{16}-5_{15}$	112.254	21.2	4.5×10^{-5}	26	41.8 ± 4.1	10.59	0.47	83.4	11.1	8
CH_3CHO-E	$6_{06}-5_{05}$	114.940	19.4	5.2×10^{-5}	26	88.2 ± 15.3	10.65	0.67	124.3	32.0	4
CH_3CHO-E	$7_{07}-6_{06}$	133.831	25.9	8.2×10^{-5}	30	36.2 ± 7.4	10.70	0.83	41.0	16.6	2
CH_3CHO-E	$7_{16}-6_{15}$	138.285	28.9	8.6×10^{-5}	30	30.6 ± 4.7	10.60	0.52	54.9	14.5	4
CH_3CHO-E	$8_{18}-7_{17}$	149.505	34.7	1.1×10^{-4}	34	20.5 ± 4.2	10.81	0.33	59.1	16.8	4
CH_3CHO-A	$5_{15}-4_{14}$	93.581	15.7	2.5×10^{-5}	22	44.3 ± 2.6	10.66	0.54	76.9	6.0	13
CH_3CHO-A	$5_{05}-4_{04}$	95.963	13.8	2.8×10^{-5}	22	73.1 ± 2.8	10.70	0.63	108.5	5.8	19
CH_3CHO-A	$5_{24}-4_{23}$	96.274	22.9	2.4×10^{-5}	22	13.5 ± 2.3	10.55	0.48	26.7	5.6	5
CH_3CHO-A	$5_{23}-4_{22}$	96.633	23.0	2.4×10^{-5}	22	12.3 ± 1.9	10.68	0.41	28.5	5.1	6
CH_3CHO-A	$5_{14}-4_{13}$	98.901	16.5	3.0×10^{-5}	22	73.8 ± 3.4	10.73	0.62	112.7	7.7	15
CH_3CHO-A	$6_{16}-5_{15}$	112.249	21.1	4.5×10^{-5}	26	56.0 ± 5.1	10.68	0.69	76.3	11.4	7
CH_3CHO-A	$6_{06}-5_{05}$	114.960	19.4	5.0×10^{-5}	26	83.6 ± 13.6	10.71	0.67	117.5	28.8	4
CH_3CHO-A	$7_{07}-6_{06}$	133.854	25.8	7.9×10^{-5}	30	30.8 ± 3.6	10.61	0.39	73.9	11.9	6
CH_3CHO-A	$7_{25}-6_{24}$	135.685	35.0	7.9×10^{-5}	30	15.4 ± 3.2	10.76	0.29	50.5	13.2	4
CH_3CHO-A	$7_{16}-6_{15}$	138.320	28.8	8.6×10^{-5}	30	52.9 ± 6.8	10.74	0.49	100.6	17.0	6
CORE											
CH_3CHO-E	$5_{15}-4_{14}$	93.595	15.8	2.5×10^{-5}	22	34.4 ± 2.2	10.71	0.74	43.5	4.2	10
CH_3CHO-E	$5_{05}-4_{04}$	95.947	13.9	2.8×10^{-5}	22	38.5 ± 2.4	10.66	0.63	57.1	5.4	10

Table 6 (*Contd.*)

Molecule	Transition	ν GHz	E_u K	A_{ul} s^{-1}	g_u	Line area mK km s^{-1}	Velocity km s^{-1}	FWHM km s^{-1}	T_{peak} mK	RMS mK	Peak S/N
CH$_3$CHO–E	5_{14}–4_{13}	98.863	16.6	3.0×10^{-5}	22	26.6 ± 4.0	10.63	0.60	41.5	8.8	5
CH$_3$CHO–A	5_{15}–4_{14}	93.581	15.7	2.5×10^{-5}	22	26.7 ± 2.0	10.62	0.57	44.1	4.6	10
CH$_3$CHO–A	5_{05}–4_{04}	95.963	13.8	2.8×10^{-5}	22	33.1 ± 3.1	10.67	0.80	38.9	5.7	7
CH$_3$CHO–A	5_{14}–4_{13}	98.901	16.5	3.0×10^{-5}	22	35.5 ± 5.7	10.74	0.74	44.8	9.5	5
CH$_3$CHO–A	6_{16}–5_{15}	112.249	21.1	4.5×10^{-5}	26	16.0 ± 3.5	10.62	0.55	27.2	9.9	3

[a] Note: All temperatures are given in the main beam temperature scale.

Faraday Discussions

Table 7 Observation parameters of the deep integrations of the CH_3CCH lines detected toward the PDR and dense core[a]

Molecule	Transition	ν GHz	E_u K	A_{ul} s^{-1}	g_u	Line area mK km s^{-1}	Velocity km s^{-1}	FWHM km s^{-1}[b]	T_{peak} mK	RMS mK	Peak S/N
PDR											
CH_3CCH-E	5_1-4_1	85.456	19.5	5.93×10^{-7}	22	15.2 ± 2.7	10.54	0.60	23.9	6.8	3
CH_3CCH-E	6_1-5_1	102.546	24.4	1.05×10^{-6}	26	16.0 ± 2.7	10.95	0.60	25.0	7.6	3
CH_3CCH-E	8_1-7_1	136.725	36.7	2.50×10^{-6}	34	16.8 ± 3.7	10.53	0.60	26.3	12.2	2
CH_3CCH-E	12_1-44_1	205.077	71.2	8.95×10^{-6}	50	22.3 ± 3.7	10.67	0.60	34.9	7.6	4
CH_3CCH-E	13_1-12_1	222.163	81.9	1.14×10^{-5}	27	15.3 ± 3.8	10.66	0.60	23.9	7.9	3
CH_3CCH-A	5_0-4_0	85.457	12.3	6.18×10^{-7}	22	14.6 ± 2.3	10.77	0.60	22.8	5.7	4
CH_3CCH-A	6_0-5_0	102.548	17.2	1.08×10^{-6}	26	10.8 ± 2.7	10.72	0.60	17.0	7.1	2
CH_3CCH-A	8_3-7_3	136.705	94.6	2.25×10^{-6}	34	15.2 ± 3.7	10.55	0.60	23.8	11.8	2
CH_3CCH-A	12_0-11_0	205.081	64.0	9.02×10^{-6}	50	22.0 ± 4.0	10.76	0.60	34.4	7.9	4
CH_3CCH-A	13_0-12_0	222.167	74.6	1.15×10^{-5}	27	14.1 ± 3.7	10.35	0.60	22.1	7.6	2
CH_3CCH-A	14_3-13_3	239.211	151.1	9.96×10^{-5}	58	28.6 ± 5.9	10.46	0.60	44.8	12.8	3
CH_3CCH-A	14_0-13_0	239.252	86.1	1.40×10^{-5}	58	54.2 ± 7.5	10.63	0.60	84.9	16.1	5
CORE											
CH_3CCH-E	5_1-4_1	85.456	19.5	5.93×10^{-7}	22	13.5 ± 1.9	10.48	0.50	25.4	5.5	4
CH_3CCH-E	6_1-5_1	102.546	24.4	1.05×10^{-6}	26	20.0 ± 2.9	10.48	0.50	37.6	8.9	4
CH_3CCH-E	8_1-7_1	136.725	36.7	2.50×10^{-6}	34	27.4 ± 4.0	10.56	0.50	51.6	14.2	3
CH_3CCH-A	5_0-4_0	85.457	12.3	6.18×10^{-7}	22	16.8 ± 2.4	10.59	0.50	31.5	6.3	4
CH_3CCH-A	6_0-5_0	102.548	17.2	1.08×10^{-6}	26	25.7 ± 2.8	10.74	0.50	48.2	8.5	5

[a] Note: All temperatures are given in the main beam temperature scale. [b] The line width was fixed to guide the Gaussian fit at low signal-to-noise ratio.

References

1 E. L. Gibb, D. C. B. Whittet, W. A. Schutte, A. C. A. Boogert, J. E. Chiar, P. Ehrenfreund, P. A. Gerakines, J. V. Keane, A. G. G. M. Tielens, E. F. van Dishoeck and O. Kerkhof, *Astrophys. J.*, 2000, **536**, 347–356.

2 E. L. Gibb, D. C. B. Whittet, A. C. A. Boogert and A. G. G. M. Tielens, *Astrophys. J. Suppl.*, 2004, **151**, 35–73.

3 A. C. A. Boogert, K. M. Pontoppidan, C. Knez, F. Lahuis, J. Kessler-Silacci, E. F. van Dishoeck, G. A. Blake, J.-C. Augereau, S. E. Bisschop, S. Bottinelli, T. Y. Brooke, J. Brown, A. Crapsi, N. J. Evans, II, H. J. Fraser, V. Geers, T. L. Huard, J. K. Jørgensen, K. I. Öberg, L. E. Allen, P. M. Harvey, D. W. Koerner, L. G. Mundy, D. L. Padgett, A. I. Sargent and K. R. Stapelfeldt, *Astrophys. J.*, 2008, **678**, 985–1004.

4 K. M. Pontoppidan, A. C. A. Boogert, H. J. Fraser, E. F. van Dishoeck, G. A. Blake, F. Lahuis, K. I. Öberg, N. J. Evans, II and C. Salyk, *Astrophys. J.*, 2008, **678**, 1005–1031.

5 K. I. Öberg, A. C. A. Boogert, K. M. Pontoppidan, G. A. Blake, N. J. Evans, F. Lahuis and E. F. van Dishoeck, *Astrophys. J.*, 2008, **678**, 1032–1041.

6 S. Bottinelli, A. C. A. Boogert, J. Bouwman, M. Beckwith, E. F. van Dishoeck, K. I. Öberg, K. M. Pontoppidan, H. Linnartz, G. A. Blake, N. J. Evans, II and F. Lahuis, *Astrophys. J.*, 2010, **718**, 1100–1117.

7 A. C. A. Boogert, T. L. Huard, A. M. Cook, J. E. Chiar, C. Knez, L. Decin, G. A. Blake, A. G. G. M. Tielens and E. F. van Dishoeck, *Astrophys. J.*, 2011, **729**, 92.

8 R. T. Garrod, *Astrophys. J.*, 2013, **765**, 60.

9 J. E. Elsila, J. P. Dworkin, M. P. Bernstein, M. P. Martin and S. A. Sandford, *Astrophys. J.*, 2007, **660**, 911–918.

10 D. P. Glavin, A. S. Burton, J. E. Elsila, J. P. Dworkin, Q.-Z. Yin and P. Jenniskens, *Lunar and Planetary Institute Science Conference Abstracts*, 2013, p. 1189.

11 J. E. Elsila, D. P. Glavin and J. P. Dworkin, *Meteorit. Planet. Sci.*, 2009, **44**, 1323–1330.

12 L. E. Snyder, F. J. Lovas, J. M. Hollis, D. N. Friedel, P. R. Jewell, A. Remijan, V. V. Ilyushin, E. A. Alekseev and S. F. Dyubko, *Astrophys. J.*, 2005, **619**, 914–930.

13 P. A. Jones, M. R. Cunningham, P. D. Godfrey and D. M. Cragg, *Mon. Not. R. Astron. Soc.*, 2007, **374**, 579–589.

14 M. R. Cunningham, P. A. Jones, P. D. Godfrey, D. M. Cragg, I. Bains, M. G. Burton, P. Calisse, N. H. M. Crighton, S. J. Curran, T. M. Davis, J. T. Dempsey, B. Fulton, M. G. Hidas, T. Hill, L. Kedziora-Chudczer, V. Minier, M. B. Pracy, C. Purcell, J. Shobbrook and T. Travouillon, *Mon. Not. R. Astron. Soc.*, 2007, **376**, 1201–1210.

15 J. M. Hollis, F. J. Lovas and P. R. Jewell, *Astrophys. J.*, 2000, **540**, L107–L110.

16 J. M. Hollis, F. J. Lovas, A. J. Remijan, P. R. Jewell, V. V. Ilyushin and I. Kleiner, *Astrophys. J.*, 2006, **643**, L25–L28.

17 A. Belloche, K. M. Menten, C. Comito, H. S. P. Müller, P. Schilke, J. Ott, S. Thorwirth and C. Hieret, *Astron. Astrophys.*, 2008, **482**, 179–196.

18 A. Belloche, R. T. Garrod, H. S. P. Müller, K. M. Menten, C. Comito and P. Schilke, *Astron. Astrophys.*, 2009, **499**, 215–232.

19 P. Hily-Blant, D. Teyssier, S. Philipp and R. Güsten, *Astron. Astrophys.*, 2005, **440**, 909–919.

20 J. Pety, D. Teyssier, D. Fossé, M. Gerin, E. Roueff, A. Abergel, E. Habart and J. Cernicharo, *Astron. Astrophys.*, 2005, **435**, 885–899.

21 M. Gerin, J. R. Goicoechea, J. Pety and P. Hily-Blant, *Astron. Astrophys.*, 2009, **494**, 977–985.

22 V. Guzmán, J. Pety, P. Gratier, J. R. Goicoechea, M. Gerin, E. Roueff and D. Teyssier, *Astron. Astrophys.*, 2012, **543**, L1.

23 J. Pety, J. R. Goicoechea, P. Hily-Blant, M. Gerin and D. Teyssier, *Astron. Astrophys.*, 2007, **464**, L41–L44.

24 V. V. Guzman, J. R. Goicoechea, J. Pety, P. Gratier, M. Gerin, E. Roueff, F. Le Petit, J. Le Bourlot and A. Faure, *ArXiv e-prints*, 2013.

25 S. E. Cummins, R. A. Linke and P. Thaddeus, *Astrophys. J. Suppl.*, 1986, **60**, 819–878.

26 G. A. Blake, E. C. Sutton, C. R. Masson and T. G. Phillips, *Astrophys. J.*, 1987, **315**, 621–645.

27 S. E. Bisschop, J. K. Jørgensen, E. F. van Dishoeck and E. B. M. de Wachter, *Astron. Astrophys.*, 2007, **465**, 913–929.

28 E. F. van Dishoeck, G. A. Blake, D. J. Jansen and T. D. Groesbeck, *Astrophys. J.*, 1995, **447**, 760.

29 S. Cazaux, A. G. G. M. Tielens, C. Ceccarelli, A. Castets, V. Wakelam, E. Caux, B. Parise and D. Teyssier, *Astrophys. J.*, 2003, **593**, L51–L55.

30 E. Herbst and E. F. van Dishoeck, *Annu. Rev. Astron. Astrophys.*, 2009, **47**, 427–480.

31 D. Teyssier, D. Fossé, M. Gerin, J. Pety, A. Abergel and E. Roueff, *Astron. Astrophys.*, 2004, **417**, 135–149.

32 M. Agúndez, J. Cernicharo, M. Guélin, M. Gerin, M. C. McCarthy and P. Thaddeus, *Astron. Astrophys.*, 2008, **478**, L19–L22.

33 J. R. Goicoechea, J. Pety, M. Gerin, D. Teyssier, E. Roueff, P. Hily-Blant and S. Baek, *Astron. Astrophys.*, 2006, **456**, 565–580.

34 J. R. Goicoechea, J. Pety, M. Gerin, P. Hily-Blant and J. Le Bourlot, *Astron. Astrophys.*, 2009, **498**, 771–783.

35 J. Pety, P. Gratier, V. Guzmán, E. Roueff, M. Gerin, J. R. Goicoechea, S. Bardeau, A. Sievers, F. Le Petit, J. Le Bourlot, A. Belloche and D. Talbi, *Astron. Astrophys.*, 2012, **548**, A68.

36 V. Guzmán, J. Pety, J. R. Goicoechea, M. Gerin and E. Roueff, *Astron. Astrophys.*, 2011, **534**, A49.

37 P. Gratier, J. Pety, V. Guzmán, M. Gerin, J. R. Goicoechea, E. Roueff and A. Faure, *Astron. Astrophys.*, 2013, **557**, A101.

38 K. I. Öberg, M. D. Boamah, E. C. Fayolle, R. T. Garrod, C. J. Cyganowski and F. van der Tak, *Astrophys. J.*, 2013, **771**, 95.

39 A. Bacmann, V. Taquet, A. Faure, C. Kahane and C. Ceccarelli, *Astron. Astrophys.*, 2012, **541**, L12.

40 H. E. Matthews, P. Friberg and W. M. Irvine, *Astrophys. J.*, 1985, **290**, 609–614.

41 W. M. Irvine, P. Friberg, N. Kaifu, K. Kawaguchi, Y. Kitamura, H. E. Matthews, Y. Minh, S. Saito, N. Ukita and S. Yamamoto, *Astrophys. J.*, 1989, **342**, 871–875.

42 S. Muller, A. Beelen, M. Guélin, S. Aalto, J. H. Black, F. Combes, S. J. Curran, P. Theule and S. N. Longmore, *Astron. Astrophys.*, 2011, **535**, A103.

43 A. A. Penzias and C. A. Burrus, *Annu. Rev. Astron. Astrophys.*, 1973, **11**, 51.

44 D. A. Neufeld, P. Schilke, K. M. Menten, M. G. Wolfire, J. H. Black, F. Schuller, H. S. P. Müller, S. Thorwirth, R. Güsten and S. Philipp, *Astron. Astrophys.*, 2006, **454**, L37–L40.

45 D. A. Neufeld, M. G. Wolfire and P. Schilke, *Astrophys. J.*, 2005, **628**, 260–274.

46 P. Sonnentrucker, D. A. Neufeld, T. G. Phillips, M. Gerin, D. C. Lis, M. de Luca, J. R. Goicoechea, J. H. Black, T. A. Bell, F. Boulanger, J. Cernicharo, A. Coutens, E. Dartois, M. Kaźmierczak, P. Encrenaz, E. Falgarone, T. R. Geballe, T. Giesen, B. Godard, P. F. Goldsmith, C. Gry, H. Gupta, P. Hennebelle, E. Herbst, P. Hily-Blant, C. Joblin, R. Kołos, J. Krełowski, J. Martín-Pintado, K. M. Menten, R. Monje, B. Mookerjea, J. Pearson, M. Perault, C. M. Persson, R. Plume, M. Salez, S. Schlemmer, M. Schmidt, J. Stutzki, D. Teyssier, C. Vastel, S. Yu, E. Caux, R. Güsten, W. A. Hatch, T. Klein, I. Mehdi, P. Morris and J. S. Ward, *Astron. Astrophys.*, 2010, **521**, L12.

47 M. Asplund, N. Grevesse, A. J. Sauval and P. Scott, *Annu. Rev. Astron. Astrophys.*, 2009, **47**, 481–522.

48 T. P. Snow, J. D. Destree and A. G. Jensen, *Astrophys. J.*, 2007, **655**, 285–298.

49 V. Guzmán, E. Roueff, J. Gauss, J. Pety, P. Gratier, J. R. Goicoechea, M. Gerin and D. Teyssier, *Astron. Astrophys.*, 2012, **548**, A94.

50 A. Fuente, A. Rodrıguez-Franco, S. Garcıa-Burillo, J. Martın-Pintado and J. H. Black, *Astron. Astrophys.*, 2003, **406**, 899–913.

51 O. Berné, C. Joblin, Y. Deville, J. D. Smith, M. Rapacioli, J. P. Bernard, J. Thomas, W. Reach and A. Abergel, *Astron. Astrophys.*, 2007, **469**, 575–586.

52 P. Pilleri, J. Montillaud, O. Berné and C. Joblin, *Astron. Astrophys.*, 2012, **542**, A69.

53 B. A. McGuire, P. B. Carroll, R. A. Loomis, G. A. Blake, J. M. Hollis, F. J. Lovas, P. R. Jewell and A. J. Remijan, *ArXiv e-prints*, 2013.

54 X. Huang, R. C. Fortenberry and T. J. Lee, *Astrophys. J.*, 2013, **768**, L25.

55 R. C. Fortenberry, X. Huang, T. D. Crawford and T. J. Lee, *Astrophys. J.*, 2013, **772**, 39.

56 B. A. McGuire, P. B. Carroll, P. Gratier, V. Guzmán, J. Pety, E. Roueff, M. Gerin, G. A. Blake and A. J. Remijan, *ApJ*, 2014, **783**, 36.

57 M. P. Bernstein, J. P. Dworkin, S. A. Sandford, G. W. Cooper and L. J. Allamandola, *Nature*, 2002, **416**, 401–403.

58 G. M. Muñoz Caro, U. J. Meierhenrich, W. A. Schutte, B. Barbier, A. Arcones Segovia, H. Rosenbauer, W. H.-P. Thiemann, A. Brack and J. M. Greenberg, *Nature*, 2002, **416**, 403–406.

59 R. T. Garrod, S. L. W. Weaver and E. Herbst, *Astrophys. J.*, 2008, **682**, 283–302.

60 J. G. Mangum and A. Wootten, *Astrophys. J. Suppl.*, 1993, **89**, 123–153.

61 S. Leurini, P. Schilke, K. M. Menten, D. R. Flower, J. T. Pottage and L.-H. Xu, *Astron. Astrophys.*, 2004, **422**, 573–585.

62 J. G. Mangum, J. Darling, C. Henkel and K. M. Menten, *Astrophys. J.*, 2013, **766**, 108.

63 C. R. Purcell, R. Balasubramanyam, M. G. Burton, A. J. Walsh, V. Minier, M. R. Hunt-Cunningham, L. L. Kedziora-Chudczer, S. N. Longmore, T. Hill, I. Bains, P. J. Barnes, A. L. Busfield, P. Calisse, N. H. M. Crighton, S. J. Curran, T. M. Davis, J. T. Dempsey, G. Derragopian, B. Fulton, M. G. Hidas, M. G. Hoare, J.-K. Lee, E. F. Ladd, S. L. Lumsden, T. J. T. Moore, M. T. Murphy, R. D. Oudmaijer, M. B. Pracy, J. Rathborne, S. Robertson, A. S. B. Schultz, J. Shobbrook, P. A. Sparks, J. Storey and T. Travouillion, *Mon. Not. R. Astron. Soc.*, 2006, **367**, 553–576.

64 H. S. P. Müller, S. Thorwirth, D. A. Roth and G. Winnewisser, *Astron. Astrophys.*, 2001, **370**, L49–L52.

65 H. M. Pickett, R. L. Poynter, E. A. Cohen, M. L. Delitsky, J. C. Pearson and H. S. P. Müller, *J. Quant. Spectrosc. Radiat. Transfer*, 1998, **60**, 883–890.

66 H. Kim, R. Keller and W. D. Gwinn, *J. Chem. Phys.*, 1962, **37**, 2748–2750.

67 N. B. Hannay and C. P. Smyth, *J. Am. Chem. Soc.*, 1946, **68**, 2408–2409.

68 I. Kleiner, F. J. Lovas and M. Godefroid, *J. Phys. Chem. Ref. Data*, 1996, **25**, 1113–1210.

69 P. M. Burrell, E. Bjarnov and R. H. Schwendeman, *J. Mol. Spectrosc.*, 1980, **82**, 193–201.

70 R. L. Hudson and M. J. Loeffler, *Astrophys. J.*, 2013, **773**, 109.

71 M. D. Ward and S. D. Price, *Astrophys. J.*, 2011, **741**, 121.

72 J. V. Keane, A. G. G. M. Tielens, A. C. A. Boogert, W. A. Schutte and D. C. B. Whittet, *Astron. Astrophys.*, 2001, **376**, 254–270.

73 A. G. G. M. Tielens and W. Hagen, *A&A*, 1982, **114**, 245–260.

74 R. Garrod, I. H. Park, P. Caselli and E. Herbst, *Faraday Discuss.*, 2006, **133**, 51.

75 S. Ioppolo, H. M. Cuppen, E. F. van Dishoeck and H. Linnartz, *Mon. Not. R. Astron. Soc.*, 2011, **410**, 1089–1095.

76 V. Vinogradoff, F. Duvernay, G. Danger, P. Theulé, F. Borget and T. Chiavassa, *Astron. Astrophys.*, 2013, **549**, A40.

77 F. Mispelaer, P. Theulé, H. Aouididi, J. Noble, F. Duvernay, G. Danger, P. Roubin, O. Morata, T. Hasegawa and T. Chiavassa, *Astron. Astrophys.*, 2013, **555**, A13.

78 J. R. Goicoechea, M. Compiègne and E. Habart, *Astrophys. J.*, 2009, **699**, L165–L168.

DISCUSSIONS

General discussion

DOI: 10.1039/c4fd90001d

Professor Dulieu opened the discussion of the paper by Ewine van Dishoeck: I would like to thank you for emphasising the interplay between laboratory astrophysics, models and observations. As an example, you showed that the case study of water has benefited of recent efforts and progresses from the three approaches. I would like to know, if you have some indications about what topics of laboratory astrophysics we should focus on in the coming years. Shall we choose another "target" molecule like methanol or a larger COM? Shall we focus on carbon cycle and reactivity? Is there any precise measurement needed to be done?

Professor van Dishoeck responded: The comprehensive set of laboratory experiments carried out by independent groups across the world on the formation of water and methanol ice have indeed been highlights in solid-state astrochemistry over the past decade. A next step could be to go systematically through the series of reactions postulated by Tielens and Charnley.[1] They start with solid CO and add sequentially H, C, N or O to make more complex organic molecules. Some of these routes have been tested, for example by Bisschop et al.,[2] but many have never been studied in the laboratory. Reactions with atomic carbon have not yet been studied but may indeed well be key to explaining several of the COMs.

1 A. G. G. M. Tielens and S. B. Charnley, *Origins Life Evol. Biospheres*, 1997, **27**, 23.
2 S. E. Bisschop, G. W. Fuchs, A. C. A. Boogert, E. F. van Dishoeck and H. Linnartz, *Astron. Astrophys.*, 2007, **470**, 749.

Professor Meuwly commented: In returning to Prof. van Dishoeck's note that more "physics-based" approaches would benefit astrophysical modeling, I would like to make the following remarks.

Computational approaches in reaction kinetics have made significant progress over the past few years. This is mainly related to the possibility to carry out large numbers (thousands of single point calculations) of high-quality (MP2 or CCSD(T)) single point calculations to characterize the intermolecular interactions.[1–3] Furthermore, with the development of reactive molecular dynamics (MD) simulations, classical MD simulations can be used to carry out a statistically significant number of simulations to compute thermal rates.[4–6] In the following, a few recent studies along these lines are briefly mentioned:

The charge transfer reaction $N_2^+ + N_2 \rightarrow N_2 + N_2^+$ with the initial N_2^+ prepared in its lowest quantum state was investigated on a fully-dimensional PES.[2] This work established that the experimentally observed angular momentum in the N_2^+ product originates from the bending dynamics in the N_4^+ intermediate.

Validation of the computational model consisted of comparing the rate constant with earlier experiments.[2,7]

Current work on the $NO(^2\pi) + O(^3P)$ reaction using similar computational techniques establish that classical MD simulations are an ideal means to compute thermal rates.[5]

To conclude: the point to make here is that with the advent of efficient computational methods to carry out fully-dimensional studies using a statistically significant number of MD trajectories, we are now in a position to more completely characterize chemical reactions in "simple systems".[1,3,8]

1 J. Yosa and M. Meuwly, *J. Phys. Chem. B*, 2011, **115**, 14350.
2 X. Tong, T. Nagy, J. Yosa, M. Germann, M. Meuwly and S. Willitsch, *Chem. Phys. Lett.*, 2012, **547**, 1.
3 J. Y. Reyes, T. Nagy and M. Meuwly, submitted.
4 T. Nagy, J. Y. Reyes and M. Meuwly, *J. Chem. Theory Comput.*, 2014, in print.
5 J. C. Castro Palacio, T. Nagy and M. Meuwly, to be submitted.
6 M. Soloviov and M. Meuwly, *J. Chem. Phys.*, 2014, in print.
7 M. J. Frost, S. Kato, V. M. Bierbaum and S. R. Leone, *J. Chem. Phys.*, 1994, **100**, 6359.
8 S. Lutz and M. Meuwly, *ChemPhysChem*, 2012, **13**, 305.

Professor Kaiser said: You outlined that from an astronomical viewpoint, an agreement between observed and modeled fractional abundances of distinct molecules of a factor of 10 is a "pretty good agreement". However, from the viewpoint of a physical chemist, this is not satisfactory. It is clear that for the next decade, astronomical data such as that from ALMA will be cutting edge, so the problem is certainly not with the observations, but with the models. Just considering the formation of COMs on interstellar grains, multiple pathways exist such as (i) the interaction of ionization radiation with "bulk" ices and formation of COMs at 10 K *via* suprathermal (non-equilibrium, non-Arrhenius) chemistry, (ii) the formation of COMs *via* radical–radical recombination within the "bulk" of the ices upon warming up (thermal chemistry), and (iii) the formation of COMs *via* radical–radical recombination on the surfaces of the ices upon warming up (thermal chemistry). But current models mainly account for item (ii) and ignore some 90+% of the ongoing chemistry in the "bulk" of the ices. Can you comment on your prospects of improving these models so that they become "predictive" models in the near future?

Professor van Dishoeck answered: I fully agree that the factor 10 is not satisfactory and that the community should work hard to bring this down to a factor of a few. This can only come through a combined laboratory/modeling effort. First there is a continued need for well defined quantitative laboratory measurements of the three routes to complex organic molecules that you describe. Second, the difficult step is then to translate these experiments into numbers that the theorists can put into their astrochemical models; here there still is a disconnect that needs to be bridged. Finally, the modelers then need to consider multi-layer ice grain in their models and be able to switch back and forth between the microscopic and the macroscopic picture. There is still a long way to go! In parallel, some qualitative or relative predictions directly from experiments on which molecules do or do not form, and in which ratios, will already be very helpful.

Professor Kamp remarked: From my viewpoint, protoplanetary disk models exist with a high level of physical and/or chemical complexity taken into account.

Here observations are lagging behind due to sensitivity and the lack of spatial resolution. We need comprehensive multi-wavelengths observations of dust and gas, because lots of physical quantities need to be determined before even starting detailed chemical modeling.

Professor van Dishoeck replied: I agree that models are ahead of observation for the case of protoplanetary disks, although that will change soon with the advent of ALMA, with new VLT(I) instruments, and with JWST at the end of the decade. However, this does not remove my concern that models in general (not just for disks) have grown very complex and that one often learns more by going 'back to basics' and just consider a few key reactions or a parametrized approach inspired by more complex models in order to nail down the critical processes. Another key point is to formulate more sharply 'what is the question that one tries to address with what data set?' Often the answer does not require data across all wavelengths.

Dr Semenov opened the discussion of the paper by Klaus M. Pontoppidan: Why are the observed CO_2 abundances in the inner regions of protoplanetary disks (at radii ~ 1 AU) so low? In contemporary gas–grain disk chemical models, the production of CO_2 is very efficient in these warm disk regions (T > $\sim$40–80 K) thanks to the slightly endothermic neutral–neutral reaction of CO with OH. Could it be due to the fact that H_2O self-shields itself from the dissociating stellar UV radiation, and thus abundances of OH are low, and the CO + OH reaction is, in fact, not that effective?

Dr Pontoppidan responded: It is not clear how the CO_2 abundance is balanced in the disk photosphere. One aspect to be cognizant of is the vertical distribution of CO_2. Unless vertical mixing is very efficient, most models predict a very stratified CO_2 abundance structure, and it is possible that we happen to be tracing a layer that is relatively deficient, perhaps due to the shielding process that you are describing.[1,2] However, I think this discussion underscores my suggestion that CO_2 is a particularly interesting tracer of inner disk chemistry, and one that would be very interesting to observe in the future, for instance with the JWST.

1 K. Willacy and P. M. Woods, *Astrophys. J.*, 2009, **703**, 479.
2 D. Semonov and D. Wiebe, *Astrophys. J., Suppl. Ser.*, 2011, **196**, 25.

Professor Millar asked: Could you say something about how this discussion on CO_2 in disks relates to the detection (and non-detection) of CO_2 ice and gas in interstellar clouds?

Dr Pontoppidan replied: In clouds, CO_2 is ubiquitous in very high abundances at densities above 10^{-4} per H or so, but is usually seen in the solid form at the low temperatures of the bulk cloud material. In some cases, CO_2 gas is seen in absorption toward young stars (both massive[1] and low-mass[2]), albeit at somewhat lower abundances. The relevant comparison, I think, is between the ice abundance in clouds and the gas abundance in inner disks (where temperatures ensure that all CO_2 is found in the gas-phase).

1 A. M. S. Boonman, E. F. van Dishoeck, F. Lahuis and S. Doty, *Astron. Astrophys.*, 2003, **399**, 1063.
2 F. Lahuis, E. F. van Dishoeck, A. C. A. Boogert, K. M. Pontoppidan, G. A. Blake, C. P. Dullemond, N. J. Evans II, M. R. Hogerheijde, J. K. Jorgensen, J. E. Kessler-Silacci and C. Knez, *Astrophys. J.*, 2006, **636**, 145.

Professor van der Tak opened the discussion of Edwin Bergin's paper: For the CO abundance in the disk, you quote a limit of 2×10^{-5}, but since you are detecting both species, why is this a limit?

Professor Bergin answered: CO and HD have different excitation characteristics for the $J = 1$–0 transition. HD is significantly more emissive for T > 50 K. Both are unemissive below 20 K (C18O because of freezeout and HD because the $J = 1$ level is unpopulated), so that helps. Thus we explored a range of potential temperatures with HD emitting at 20, 40, 60 K, and C18O at 20 K. This is what led to the limit – different assumptions regarding the differential excitation leads to differing limits. I provided the upper end of the bound.

Professor Kamp continued the discussion of the paper by Klaus M. Pontoppidan: How important is the choice of dust opacity for deriving the various molecular abundances? Is the continuum optical depth not limiting how much of the molecular reservoir is visible to the observer?

Dr Pontoppidan responded: The dust opacity will limit the observable column of molecular gas, especially if the disk is well-mixed and most of the solids are in the form of small grains (which may or may not be the case). However, the advantage is that all of the infrared molecular tracers are found in a relatively narrow wavelength region between 5 and 20 microns, where the dust opacity function is almost flat. Furthermore, the upper level energies of the various species are similar, or at least have large energy overlaps. This means they can be trusted to trace very similar volumes of gas. This is generally confirmed by high resolution spectroscopy, showing that, *e.g.*, CO and H_2O have very similar line shapes.

Professor van Dishoeck said: Can you comment on the importance of IR pumping in the excitation of the levels from which you observe the infrared emission?

Dr Pontoppidan replied: The disk photospheres coincide with warm or hot dust that is close to being optically thick with respect to its own radiation. This means that the molecular gas is bathed in a $\sim$500 K radiation field with almost a 4π coverage in solid angle. Therefore, infrared pumping has the potential to be a very efficient excitation mechanism, given the right channel. Especially for transitions with high critical densities, such as rovibrational CO ($v = 1$–0) and the high-lying H_2O lines in the mid-infrared, the two-dimensional radiative transfer models that we are working on indicate that infrared pumping will be the dominant excitation mechanism at densities below $\sim 10^{-10}$ per H. Conversely, UV pumping is important a bit higher in the disk and for very high-lying levels ($v \geq 2$).

Professor Herbst commented on Edwin Bergin's paper: The chemistry you discussed reminded me of a paper by Garrod *et al.*[1] in which cold core gas–grain chemistry was studied over an unphysically long time scale of 10^8 years. One of the results is that most of the carbon goes into the form of heavy hydrocarbons.

1 R. T. Garrod, V. Wakelam and E. Herbst, *Astron. Astrophys.*, 2007, **467**, 1103.

Professor Bergin noted: This is the analog to what we propose. However in the disk the ionization rates are higher and the densities are also significantly higher which leads this mechanism to be active at shorter timescales.

Dr Jones returned to the discussion of the paper by Klaus M. Pontoppidan: I would like to play the devil's advocate here and throw a bit of a spanner into the works, because it seems as though our current understanding of dust evolution in the transition from the diffuse to the dense ISM neglects an important accretion process. In the low extinction outer regions of molecular clouds ($0.3 < A_V < 1.5$ mag.) it now appears necessary to include the accretion of a hydrogen-rich hydrogenated amorphous carbon (a-C:H) mantle onto grains[1] before ice mantles accrete deeper within the clouds ($A_V > \sim 1.5$ mag.). The accretion of C and H atoms (and perhaps also O, N and S as hetero-atom dopants in a-C:H) from the gas phase leaves all grain surfaces coated with macro-molecular H-rich a-C:H material (H atom fraction, $X_H \sim 0.5$ or H/C ~ 1), which is rich in CH_n aliphatic groups ($n = 1, 2, 3$). Thus, the grain surfaces onto which the ice mantles eventually accrete is a rather chemically-active substrate that could perhaps contribute to some interesting surface chemistry.[2,3]

1 M. Koehler, A. P. Jones and N. Ysard, 2014, in preparation.
2 A. P. Jones, *Astron. Astrophys.*, 2012, **555**, A98.
3 A. P. Jones, *Planet. Space Sci.*, 2014, in press.

Dr Kama asked a question relating to Edwin Bergin's paper: You are taking Orion KL as an example, because it is a high-mass and very luminous source and you can be certain that all ices are evaporated, so you can trace the full budget of carbon-bearing organics in the gas. Given the different timescale and thermal or UV processing history in low-mass protostars, such as IRAS 16293-2422, would you expect a different amount of carbon locked up in complex organics in such sources, which perhaps would be a more straightforward analogue of the proto-solar environment?

Professor Bergin answered: This is a fair point and I have stressed that I only have one source. It certainly is the case that things could be different amongst sources. The question is what processing (chemical) happens after evaporation *vs.* what happened in the initial state. From my perspective, as long as I am seeing the "carbon" that was placed on the grains during the cold phase, it is fine if there is reshuffling at later stages. So that suggests that low mass and high mass might have some similarity at this level. However, clearly this must be proven *via* observation and analysis.

Dr Garrod enquired: Figure 6 in your paper shows a range of carbon-bearing species to be formed abundantly in the inner disk through gas-phase chemistry, including methanol (CH_3OH). In interstellar environments we typically think of methanol as a grain surface-produced molecule, due to the inefficiency of the gas-phase formation mechanisms. Is gas-phase production of methanol actually the dominant process in the environment that you have modeled, and are we seeing an unusual gas-phase mechanism at work?

Professor Bergin replied: The answer to this is yes – this is a different gas phase mechanism. As outlined in the contribution "when CO is destroyed, gas phase formation of H_2CO leads to a production of H_3CO^+ *via* a reaction with abundant HCO^+. H_3CO^+ then reacts with CH_4 to form $CH_3OCH_4^+$, which recombines with electrons in the gas or on grains to create gas phase CH_3OH".

Dr Pontoppidan asked: In Figure 6, most grain species show a sharp drop in abundance at a height of 1.5 AU. Is this primarily due to sublimation?

Professor Bergin responded: This is primarily due to photodesorption along with sublimation for the most volatile species.

Professor Herbst queried: The gas-phase mechanism for the production of methanol starts with the radiative association between CH_3^+ and H_2O. Such reactions are normally most efficient under 100 K. What is the temperature range of the protoplanetary regime you studied?

Professor Bergin answered: This is the key reaction and for the most part the chemistry for CH_3OH production occurs at T > 40 K.

Professor Millar said: A recent paper by Sipila and co-workers[1] argues that in dark clouds HD is destroyed through interaction with grains. He argues that the cosmic-ray ionisation of HD eventually leads to the accretion of D atoms on to cold grains and the conversion of these atoms to HDO, rather than HD, through surface chemistry. He finds that the amount of HD can be reduced significantly, perhaps by as much as 15%. Do you think that such a process is at work in disks and, if so, how would it affect your analysis?

1. O. Sipila, P. Caselli and J. Harju, *Astron. Astrophys.*, 2013, **554**, A92.

Professor Bergin remarked: I agree with this idea. In fact a colleague of mine, Jeong-Eun Lee and I explored this issue ourselves in a paper that was submitted to ApJ and presented to the IAU symposium on Astrochemistry a couple of years ago.[1] This issue is brought up with respect to the use of HD as a mass tracer and we need to know the abundance to derive the mass of hydrogen gas. If this is working in the disk system then the masses derived from HD are lower limits. In this particular case, for TW Hya, the mass derived from assuming normal abundances is already quite high and significantly higher masses straddle the realm of gravitational stability. Moreover, the HD abundance in Jupiter's atmosphere is roughly consistent with the interstellar value.

1 J.-E. Lee and E. A. Bergin, *Astrophys. J.*, submitted.

Professor van Dishoeck asked: Do we have deep enough observational limits on various carbon-bearing molecules in disks to test your model? If not, how much deeper do we need to go and what would be the best species to target?

Professor Bergin responded: The answer is no we do not have deep enough limits. The first place to start is with CO isotopologues. Then searches for H_2CO, CH_3OH would clearly be potential end products. Simple hydrocarbons are also produced so perhaps exploring the ones that can be detected (C_2H, C_3H_2) might shed some light. Beyond that it gets into a guessing game as to what might be detectable – HC_3N was predicted in one case. If this mechanism is active then perhaps we might infer a decay of CO and a rise in one of the detectable tracers. I would imagine after a few more years of ALMA operations we will have a clearer idea of what might be detected.

Professor Linnartz commented: In the past mainly bottom-up scenarios for molecule formation have been considered, *i.e.*, the formation of new species starting from smaller ones. More recently, top-down formation schemes (also resulting in the formation of smaller species) have been proposed, *e.g.*, the case where PAHs photofragment. In your FD168 paper, it is explicitly stated that PAHs are not considered, though these may be important. This is actually confirmed in a number of papers. In one of the most recent ones ("Quadrupole ion trap / time-of-flight photo-fragmentation spectrometry of the hexa-peri-hexabenzocoronene (HBC) cation"),[1] the authors show that a large PAH (HBC – $C_{42}H_{18}$), once put in the spotlight, performs a molecular striptease, resulting both in small (molecule) fragments, that become available for bottom-up reaction schemes, as well as larger species, most noticeably graphene or fullerene like species.

This is illustrated in Fig. 1, taken from the paper by Zhen *et al.*[1] The figure shows ion-trap time-of-flight mass spectra of HBC cations irradiated with laser light. The systematic fragmentation illustrates that a top-down scenario has the potential to contribute substantially to new molecules as well.

Can you comment on the question to what extent such top-down scenarios should be considered in models as well?

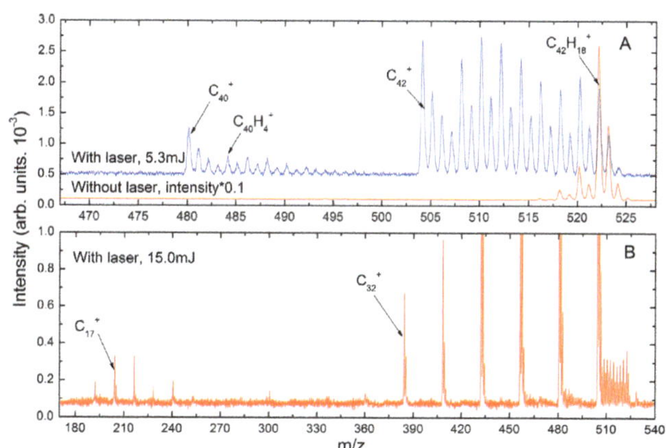

Fig. 1 Time-of-flight mass spectrum of HBC radical cations upon irradiation with 532 nm laser light (5.3 mJ (A) and 15 mJ (B)) compared to a non-irradiated sample. The data show that once put in the spotlight, the PAH starts a molecular striptease, reflected by *e.g.* sequential H-atom and C_2-unit losses. Figure taken from Fig. 3 in **ref. 1**.

1 J. Zhen, D. M. Paardekooper, A. Candian, H. Linnartz and A. G. G. M. Tielens, *Chem. Phys. Lett.*, 2014, **592**, 211.

Professor Bergin replied: Nearly 50% of the carbon resides in some solid state form that is provided to disk material. As we noted in previous work,[1] this carbon needs to be destroyed prior to forming solar system rocks. The main method of destruction we proposed was oxidation, where we explored the timescales of destruction for pure carbon grains (amorphous carbon, graphite). Recently we have also explored PAH chemistry in the surface of the disk, building on work presented by Kress *et al.*[2] In our work[3] we have placed PAH chemistry into our 2D disk model and explored the timescales of PAH destruction as well. In the current chemistry (starting with one PAH, pyrene) the primary recipient of top-down chemistry is CO. But of course this is one PAH and one process (oxidation), and photochemistry could also contribute. This will be a fruitful field of exploration at the cross-currents of astronomy and chemistry for some time to come.

1 J. E. Lee, E. A. Bergin and H. Nomura, *Astrophys. J., Lett.*, 2010, **710**, L21.
2 M. E. Kress, A. G. G. M. Tielens and M. Frenklach, *Adv. Space Res.*, 2010, **46**, 44.
3 D. E. Anderson *et al.* 2014, in prep.

Professor van Dishoeck said: In addition to destroying PAHs by UV radiation in the inner disk, there is also the possibility that PAHs are frozen out in the ices and thereby add to the chemical complexity of carbonaceous material in ices. This freeze-out probably happens already in the pre- and protostellar phases. Remarkably, no PAH features have been detected from low mass protostars with Spitzer-IRS or VLT (sample of more than 50 sources), suggesting that the abundance of PAHs in prostellar envelopes is at least an order of magnitude lower than in the general ISM.[1]

1 V. C. Geers, E. F. van Dishoeck, K. M. Pontoppidan, F. Lahuis, A. Crapsi, C. P. Dullemond and G. A. Blake, *Astron. Astrophys.*, 2009, **495**, 837.

Dr Jones remarked: In circumstellar disc regions, photons energetic enough to ionise He will also destroy hydrocarbon nanoparticles (a-C(:H), PAHs), which will undergo dissociative destruction by C_2 ejection. This dissociative destruction process has been evaluated for energy input from ion and electron collisions and cosmic rays.[1–3] Elisabetta Micelotta is currently evaluating the effects of hydrocarbon nanoparticle photodissociation in HII regions using the same formalism and this could eventually help, at least in part, to explain carbon grain processing in the outer disc regions.

1 E. R. Micelotta, A. P. Jones and A. G. G. M. Tielens, *Astron. Astrophys.*, 2010, **510**, A36.
2 E. R. Micelotta, A. P. Jones and A. G. G. M. Tielens, *Astron. Astrophys.*, 2010, **510**, A37.
3 E. R. Micelotta, A. P. Jones and A. G. G. M. Tielens, *Astron. Astrophys.*, 2011, **526**, A52.

Dr Walsh asked: What effect do you expect vertical mixing within the disk to have on the destruction of gas-phase CO *via* reaction with He$^+$? Could CO observations then help determine the degree of turbulent mixing present within a protoplanetary disk?

Professor Bergin replied: This is important. I would think that it would depend on the timescale of the vertical mixing. If the mixing is fast it could remove CO from the X-ray ionized layer where the He^+ reaction timescales will be short. However if mixing is slow it would bring fresh CO from lower layers into the X-ray dominated regime where the CO chemical destruction would be enhanced. This is an area where some new simulations would help.

Dr Heays asked a question in relation to the paper by Klaus M. Pontoppidan: The simple physical parameterisation you have used for a protoplanetary disk works very well for reproducing the observed molecular lines. No doubt future observations of other spectral regions will not be so well reproduced by this model. Would you consider it beneficial to make your model more complex and with more adjustable parameters in that event?

Dr Pontoppidan answered: As in any exercise modeling empirical data, the number of parameters is always a balance between addressing simply stated high-level questions while keeping an eye on underlying, and potentially biasing, complexity. I think this model is no different. Currently, however, many models introduce many parameters that are not directly justified by the data and there is no clear method to quantitatively assess their relative importance, especially compared to parameters that are not yet included. One example is complex chemical networks (currently included by multiple groups) *versus* bulk elemental transport through advection (not yet included as far as I know). In short, parameters should be added, but judiciously, and in order of their importance relative to high level understanding.

Dr Hornekaer asked a question relating to the paper by Edwin Bergin: During the presentation you showed a model for destruction of PAHs by reactions with oxygen. What reaction rates was this based on and were only small PAHs included, which may be more easily destroyed than larger PAHs?

Professor Bergin replied: Of course there are a range of progenitor PAHs so I cannot fully answer the fate of all potential progenitors. The models we used were based on the work of Kress *et al.*[1]

1 M. E. Kress, A. G. G. M. Tielens and M. Frenklach, *Adv. Space Res.*, 2010, **46**, 44.

Professor Kamp returned to the discussion of the paper by Klaus M. Pontoppidan: You used a radial abundance profile in the models to derive the molecular abundances. How important would be a vertical parametrization of the molecule concentrations, *e.g.* OH sits on top of water, CO on top of water?

Dr Pontoppidan responded: This is a very interesting point and I agree that this is worthy of attention in the future. Of course, the model presented here does not actually treat OH, specifically, because it has strong non-thermal excitation properties as well as the layering you mention. The other species may well be co-spatial, at least, if we are to believe self-shielding models such as those of Bethell and Bergin.[1] Furthermore, given how close the layers are to each other, it should be coupled with a discussion or a model of vertical mixing. There may be chemical tracers of the amount of turbulence in the upper disk layers, and given how much attention that physical parameter has received lately, such models could lead to important physical insights.

1 T. Bethell and E. Bergin, *Science*, 2009, **326**, 1675.

Professor Heard said: In figure 2 of the paper, most easily seen in the upper panel, but also present in some of the other main lines in the lower panel, it seems that the fitting to the line-shapes consistently misses the wider part of the peak at lower flux intensity. This I think is a consequence of setting $n(CO)/n(H) = 0$ for any distance greater than 4 AU. The greater values of AU correspond to the lowest densities and hence the broadest part of the peaks in figure 2. This suggests that for a radius > 4 AU that the value of $n(CO)/n(H)$ approaches zero more gradually than the step function at 4 AU. I recognise that the model deliberately has a simple parameterisation (and that it is acknowledged that the smaller peaks in figure 2 (upper) do not have the correct fitted shape at all), but it does seem that there is additional information in the very high quality spectra which is not being extracted, namely fitting of the wider parts of the spectral features to obtain more information on the functional form of the CO density at larger disk radii. Is it planned to extend the model parameters in order to obtain such information?

Dr Pontoppidan answered: In general, you are right that there is more information in the detailed line shapes than what is being modeled here. The part of the disk structure that is not well-modeled is actually the inner-most radii at $\sim$0.1 AU. These regions have the highest velocities and temperatures and are therefore represented in the line wings, as well as the high-lying ($v = 2$–1) lines. The main reason that they are not important for the purpose of this paper is that the main $v = 1$–0 CO lines, as well as those of water, CO_2 and organics lines are formed at somewhat larger radii ($\sim$1 AU). That said, a future study would include modeling the inner rim of the disk to match the CO line wings and compare those to, for instance, the higher-energy ro-vibrational water lines often seen at 3 μm.[1]

1 C. Salyk, K. M. Pontoppidan, G. A. Blake, F. Lahuis, E. F. van Dishoeck and N. J. Evans, *Astrophys. J., Lett.*, 2008, **676**, L49.

Professor Schram returned to the discussion of the paper by Edwin Bergin: My question is: are there also maps with detailed information on electron density and

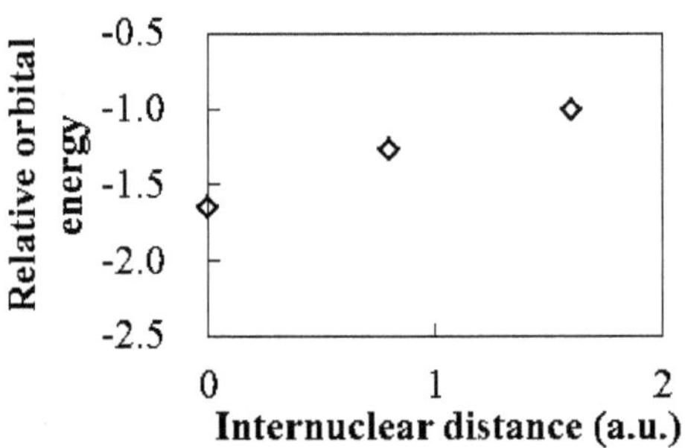

Fig. 2   Orbital energy for H–H nuclear fusion.

electron temperature, like the maps becoming available for the various (*e.g.* carbon containing) molecules? The question is more general than on this subject. The question is related to possible influence of charged particles (also of electrons) on the chemistry and on possible influence of plasma dynamic effects on structure. Can model results on electron densities and temperatures be supported by measured data? The electron temperature can have effect on the charge of clusters, PAHs *etc.*; what is assumed in the models on charge of particulate matter?

Professor Bergin replied: We have predictions of this from chemical models. But the constraints on the models are not high enough (at present) for us to claim definitive numbers. There are a number of modeling groups that have this information that could be contacted to help understand the import of electrons (*e.g.* Cleeves/Bergin, Woitke/Kamp (ProDiMo), Semenov, Walsh/Nomura, Aikawa).

Dr Fukushima commented: I would like to comment on the electronic state analysis of nuclear fusion. Occasionally, carbon atoms are made through nuclear fusions represented step by step as $H + H \rightarrow He$, $He + He \rightarrow Be$ and $Be + He \rightarrow C$. The linear combination of atomic orbitals (LCAO) method employs basis atomic functions, with sizes roughly of an atom's radius. Meanwhile, the size of a pyramid-shaped basis function for the finite element method (FEM) is about 1/50 of an atom's radius and more suitable for the fusion description compared with atomic orbitals.[1] Fig. 2 shows the relative orbital energy with the lowest energy of a hydrogen molecule as a function of the internuclear distance. The orbital energy decreases with the interatomic distance. FEM is thus convenient for the analysis of systems under abnormal conditions. Nuclear fusion is an interdisciplinary field between chemistry and physics (nuclear reactions at short internuclear distances lie within the regime of physics, while chemical reactions at long internuclear distances are the concern of chemistry). Numerical calculations were performed on the TSUBAME2.5 supercomputer in the Tokyo Institute of Technology, supported by the MEXT Open Advanced Research Facilities Initiative (in Japan).

1 K. Fukushima and H. Sato, *J. Comput. Chem., Jpn.*, 2013, **12**, 197.

Professor McCoustra communicated regarding the paper by Edwin Bergin: In your model of the proto-planetary disk you have added PAH oxidation chemistry. Were you aware that under some conditions, hydrocarbon oxidation chemistry can result in PAH, and ultimately soot, formation?[1a-d] What impact might this have on your models?

1 (a) K.-H. Homann, *Angew. Chem. Int. Ed.*, 1998, **37**, 2434; (b) C. S. McEnally, L. D. Pfefferle, B. Atakan and K. Kohse-Höinghaus, *Prog. Energy Combust. Sci.*, 2006, **32**, 247; (c) D. R. Tree and K. I. Svensson, *Prog. Energy Combust. Sci.*, 2007, **33**, 272; (d) H. Wang, *Proc. Combust. Inst.*, 2011, **33**, 41.

Professor Bergin communicated in reply: We modeled carbon grain destruction (PAH, graphite) with the end product in the chemistry seemingly leading to CO, for the most part. So the short answer is at present we have not accounted for the possibility of soot formation. However, as you note there is a vast literature on combustion that might be relevant and we will explore whether the physical conditions in the disk would lead to soot production.

Professor Price continued the discussion of the paper by Edwin Bergin by communicating: Is the double ionization of CO included in the potential CO destruction pathways in your modeling? Dissociative double ionization of CO creates highly translationally energetic [several eV] C^+ (and O^+) ions. The branching ratio between the population of dissociative and "long-lived" states of CO^{++}, by photon and electron ionization of CO, is skewed quite strongly in favour of the dissociative process which forms C^+ and O^+. The highly "hyper-thermal" C^+ ions from dissociative double ionization can participate in nominally endo-thermic reaction pathways in collisions with other species. Would such pathways provide a further way of destroying CO?

Professor Bergin communicated in response: This is not included in our model, but there are several explorations in the literature that have included such chemistry. It is an important effect to consider but other pathways will likely dominate the chemistry.

Dr Hickson returned to the discussion of the paper by Ewine van Dishoeck by communicating: In a recent study,[1] we investigated the gas-phase reactivity of the methylidyne (CH) radical with water:

$$CH + H_2O \rightarrow H + H_2CO \tag{1a}$$

$$CH + H_2O \rightarrow H_2 + HCO \tag{1b}$$

below room temperature (50 K – 296 K) using a supersonic flow (Laval nozzle) reactor. This reaction, which is predicted to lead to $H + H_2CO$ as the major products, presents a negative temperature dependence; that is, its rate increases as the temperature falls, reaching values as large as a few 10^{-10} cm^3 molecule^{-1} s^{-1} at 50 K. This gas-phase reaction is interesting to note as being the only bar-rierless neutral–neutral reaction involving water so far to have been studied at low temperature, and it could play a role in certain warm astrochemical environments such as in the envelopes surrounding protostars where water abundances are high due to the sublimation of water ice mantles surrounding interstellar dust grains. In dense interstellar clouds however, its relevance will be somewhat limited, given the low abundances of both gas-phase water and the CH radical (due to its high reactivity).

It occurred to me, however, that this process could be particularly important on the surface of interstellar grains in dense clouds, where water ice makes up a large fraction of the grain mantle. Atomic carbon, which is predicted to be present in the gas-phase at abundances as high as 10^{-4} with respect to total hydrogen, is expected to react to form CO in the gas-phase, but also to deplete onto interstellar grains where hydrogenation will occur to form (at least initially) the CH radical in sites likely to be adjacent to water molecules.

Do you think that CH radicals formed in this way could react efficiently with water ice either present in the bulk or on the surface of interstellar ices? If so, could this process be an efficient way to form methanol on interstellar ices? The currently recognized methanol formation mechanism relies on the successive hydrogenation of CO on interstellar grains,

$$H + CO \rightarrow HCO \tag{2}$$

$$H + HCO \rightarrow H_2CO \tag{3}$$

$$H + H_2CO \rightarrow H_3CO \,/\, H_2COH \tag{4}$$

$$H_3CO \,/\, H_2COH + H \rightarrow CH_3OH \tag{5}$$

for which reactions (2) and (4) present substantial barriers and are therefore rate limiting steps. Reaction (1) could be an alternative route for methanol formation through the barrierless formation of H_2CO (thereby bypassing the first rate limiting step). Alternatively, direct formation of $H_3CO \,/\, H_2COH$ could occur (thereby bypassing both rate limiting steps) as these are intermediates in the CH + H_2O reaction mechanism, and are likely to be more efficiently stabilized on surfaces than for the equivalent gas-phase process.

1 K. M. Hickson, P. Caubet and J.-C. Loison, *J. Phys. Chem. Lett.*, 2013, **4**, 2843.

Professor van Dishoeck communicated in reply: Thanks for pointing out this reaction of CH + H_2O and your nice experiments. I fully agree that it is worth investigating to what extent this reaction could affect the formation of methanol in the solid state at very low temperatures. Note that the existing laboratory experiments, as summarized by Fuchs *et al.*,[1] have shown that the hydrogenation of CO to form HCO and H_2CO is rapid down to 12 K, in spite of the barrier. The temperature range over which this alternative route may be important may therefore be limited.

1 G. W. Fuchs, H. M. Cuppen, S. Ioppolo, C. Romanzin, S. E. Bisschop, S. Andersson, E. F. van Dishoeck and H. Linnartz, *Astron. Astrophys.*, 2009, **505**, 629.

Mr Ligterink opened the discussion of the paper by Viviana Guzmán: How do you decide which lines to include and exclude in the rotational diagram? For example in Figure 3 of the paper, second column, a line has been excluded, which in my opinion should be included if I compare it with other identifications. Can you explain why?

Dr Guzmán Veloso answered: There are two main reasons to exclude those three ketene lines in the PDR (and one line in the dense core). The first one is their linewidth; they all present larger linewidths (0.8 km s^{-1}) than the other lines (0.6 km s^{-1}). The second one is their systemic velocity; they are all shifted by 0.2 km s^{-1} from the systemic velocity of the Horsehead PDR ($\sim$10.6 km s^{-1}). However, we cannot rule out that an excitation effect could produce the higher than expected intensity of these lines.

Dr Semenov opened the discussion of the paper by Karin Öberg: I do not think it is easy to interpret plots like the one you have shown, that compare observed molecular ratios of complex organic molecules *vs.* source temperature. The problem here is that, if we believe in a current paradigm of star formation, which goes from a warm dilute cloud to a dense cold cloud and then to a hot core, then warm sources should also be older and thus more chemically evolved compared to colder ones. That is, one deals with a rather complicated situation where higher temperatures on the X-axis also imply older ages.

Dr Öberg replied: It is absolutely correct that changes in density, temperature and timescales are intertwined. A potential path to disentangling these different source characteristics is to compare temperature trends toward low- and high-mass young stellar objects; low and high-mass YSOs are expected to both have different density profiles and collapse time scales. This would require a significant observational effort, however, since the number of detections toward low-mass YSOs is still quite low.

Dr Guzman-Ramirez asked: In the models you presented, a single source model was used. In high mass stars it is likely there will be a binary system – have you thought of using a binary system?

Dr Öberg responded: The current models actually include no information on source geometry since they are in effect 0D single-point models. When comparing such models to spatially resolved observations we map the model results at different temperatures to the source radii corresponding to the temperature in question based on source-specific protostellar envelope models, *i.e.* we effectively use the 0D model results to construct a 1D model. The same technique could in principle be used to construct a more complicated 2D or 3D chemical structure if a more complex geometry is necessitated by *e.g.* protostellar multiplicity.

Dr Pontoppidan queried: In terms of the overall budget, what is the fraction of carbon in the organics?

Dr Öberg answered: This question cannot be directly answered considering only the information in the presented observations. However, it can be worked out considering that these observations show that the ratio of specific COMs to CH_3OH is typically on the order of 10% at all temperatures and evolutionary stages. Assuming this is true for the 10 most common COMs and that each COM has on average 2 carbons, this gives a total COM : CH_3OH ratio of 2 : 1. From ISO and Spitzer we further know that the typical CH_3OH/H_2O abundance is ~ 0.05, though 0.25 is observed toward a few sources,[1] and that the H_2O abundance with respect to H_2 is $\sim 10^{-4}$. Assuming that these high abundances of CH_3OH are maintained, *i.e.* they are not the main source of the observed COMs, the total amount of carbon bound up in complex organics would typically be $\sim 10^{-5}$ with respect to H_2. This is a significant proportion of the total carbon budget ($\sim 10\%$ of the non-refractory carbon budget), but far from a dominant one in the average source.

1 K. I. Öberg, A. C. A. Boogert, K. M. Pontoppidan, S. van den Broek, E. F. van Dishoeck, S. Bottinelli, G. A. Blake and N. J. Evans II, *Astrophys. J.*, 2011, **740**, 109.

Professor van Dishoeck commented on the paper by Viviana Guzmán: Nice work; it is very interesting to see that the complex molecules that you detect in the

Horsehead PDR at precisely those which Bisschop *et al.* have classified as the 'cold' complex molecules found in their survey of massive YSOs, *i.e.*, molecules that have rotational temperatures less than 100 K and are thus thought to exist primarily in the colder extended envelope rather than just the compact hot core with temperatures of a few hundred K. My question is the reverse: do you have deep limits on the traditional hot core molecules such as dimethyl ether and methyl formate which have high rotational temperatures of a few hundred K toward massive YSOs? If so, what are their abundance limits relative to methanol?

1 S. E. Bisschop, J. K. Jorgensen, E. F. van Dishoeck, E. B. M. de Wachter, *Astron. Astrophys.*, 2007, **465**, 913.

Dr Guzmán Veloso answered: We derive upper limits of $<10^{-11}$ for the abundances of dimethyl ether and methyl formate with respect to total hydrogen nuclei ($NH + 2N(H_2)$). These correspond to abundances of <0.2 with respect to methanol. This was computed assuming an excitation temperature of 10K and considering the low-energy lines covered in the 3 mm band which has a rms of $\sim$8 mK. This upper limit is comparable to the abundances found in massive YSOs with respect to methanol ($\sim$0.1).

Professor van Dishoeck posed the same question to Karin Öberg: Would you like to comment on this?

Dr Öberg replied: In the literature sample assembled for our paper, more than half of the sources have excitation temperatures below 100 K, the traditional hot core temperature boundary. On average these sources have similar CH_3OCH_3/CH_3OH abundance ratios compared to their warmer counterparts, but an order of magnitude lower CH_3CN/CH_3OH ratios. It thus seems like some classical hot core tracers are widespread with respect to CH_3OH also at earlier and colder stages, while others are really much more abundant once you get to the warmer and more evolved stages.

Professor Herbst returned to the discussion of the paper by Viviana Guzmán: Do you have an estimate for the size of the radiation field in the so-called "edge" region? It would be interesting to use interferometry to measure the abundances of complex molecules *vs.* radiation field more closely.

Dr Guzmán Veloso responded: The impinging radiation field is 60 times the standard interstellar radiation field.[1] The HCO peak emission is located at AV$\sim$1.5, so the attenuated radiation field at this position is about 3 times the standard interstellar radiation field. Interferometric observations would indeed be interesting and we will try with ALMA in cycle 3.

1 B. T. Draine and A Lazarian, *Astrophys. J.*, 1998, **494**, L19.

Dr Ellinger said: As you said: "Complex organic molecules are believed to be formed on the ices around dust grains. The ices can be released in the gas phase either through thermal ... or through non-thermal desorption mechanisms, such as photo-desorption when a single far-UV photon releases only a few molecules from the ices?" When the species given as examples (HCOOH, CH_2O, CH_3CHO ...)

are formed, an energy in the range of 5–7 eV is released into the local environment. Knowing that the average adsorption energies of the COMs are in the order of 0.3 to 1.5 eV and that the cohesion energy of the ices is less than 0.7 eV, could you comment on why you need an extra energetic input from photo-desorption?

Dr Guzmán Veloso answered: Indeed, the energy released upon formation of these molecules can desorb them into the gas-phase, but we do not know how efficient this process is compared to direct photodesorption. We invoke the need of photodesorption in the Horsehead because the observed complex molecules are 2–4 times more abundant in the UV exposed PDR than in the UV-shielded dense core. This strongly suggests that FUV photons are playing a role in the formation of these complex molecules. Perhaps we should start talking about photo-processing instead of photodesorption to include other mechanisms such as chemical desorption following the photo-dissociation of molecules in the ices.

Dr Jones asked: In our previous work,[1] it was postulated that the UV photo-fragmentation processing of small hydrogenated amorphous carbon grains, a-C(:H), and their reaction with O (combustion), would be a viable source of CO and hydrocarbon molecular "fragments" in PDRs, *i.e.*, of the observed CCH, c-C_3H_2, C_4H, C_3H^+.[2,3] This scenario appears to be consistent with the observations. Do you have any thoughts on this?

1 A. P. Jones, *Astron. Astrophys.*, 2012, **540**, A2.
2 J. Pety, P. Gratier, V. Guzman *et al.*, *Astron. Astrophys.*, 2012, **548**, A68.
3 J. Pety, D. Teyssier, D. Fosse *et al.*, *Astron. Astrophys.*, 2005, **435**, 885.

Dr Guzmán Veloso replied: It is very likely that the UV photo-fragmentation of small carbon grains could lead to the enhanced abundances of the small hydrocarbons that are observed in PDRs. This is indeed consistent with the observations in the Horsehead, where the gas phase abundance of C_3H^+, gas-phase precursor of CCH and C_3H_2, is consistent with gas-phase models in sharp contrast with the abundance of CCH and C_3H_2 that gas-phase models underestimate by orders of magnitude.

Moreover, CH_3CN presents an abundance 30 times higher towards the Horsehead PDR than towards the UV-shielded dense core. We interpret this as the need for a new chemical route to form CH_3CN in UV illuminated gas. We propose that the high abundance of CH_3CN could result from the photo-fragmentation of small grains or very complex molecules containing CN. Indeed, there is an anti-correlation between CN and DIBs absorption measurements in the optical/UV[1] and you have proposed during the Cosmic Dust VI meeting in Kobe that DIBs carriers are (sub)-mm-sized products of the UV photo-fragmentation of hydrogenated amorphous carbon grains, a-C(:H), and their heteroatom-doped variants, a-C:H:X (where X may be O, N, Mg, Si, Fe, S, Ni, P).

1 T. Weselak, G. A. Galazutdinov, F. A. Musaev and J. Krelowski, *Astron. Astrophys.*, 2008, **484**, 381.

Professor Bergin returned to the discussion of the paper by Karin Öberg: Your work shows that there are some similarities between models of organic formation and observations, but the match is often quite limited (in terms of molecular species). How do we do better?

Dr Öberg responded: We definitely need larger samples both in terms of sources and species, which in many cases will require ALMA, both to disentangle emission from different nearby sources and to detect a large number of species in each source. Equally important, we need to begin analyzing such data using different statistical methods to actually test whether perceived similarities or differences between models and observations are significant or not. For this approach to work we also need more quantitative predictions on organic molecule abundances in different environments, so that there is something to actually compare observations with. It will probably be especially fruitful to be able to quantitatively compare ratios of different species within molecular 'families' as well as how the importance of different 'families' (*e.g.* alcohols) differ between different sources with well-determined temperature, density and radiation properties.

Dr Garrod asked: Ted, you raised the question of uncertainties and/or inaccuracies in the chemical models that are used to reproduce observations such as those presented by Karin Öberg. In fact, the chemical models produce good agreement with observations in cases where the observed physical conditions and spatial structure are well defined and can be explicitly included within the model treatment. However, the most chemically-advanced models have so far only been applied to generic physical conditions (such as gas density, visual extinction, CR ionization rate), during a time-dependent warm-up phase. The results of these generic models may be compared with observations of specific sources on an *ad hoc* basis, by mapping temperature-dependent fractional chemical abundances onto observed source-temperature profiles, as is done here by Öberg. (This also allows molecular spectroscopic simulations of the source to be conducted).[1] Because the observable gas-phase chemistry is strongly dependent on thermal desorption of molecules from the dust grains, which is highly sensitive to the dust temperature, such a treatment can be used to understand the spatial arrangement of molecules in the source. The treatment would, however, be improved by the use of an explicit density profile during the actual simulation of the chemistry, rather than simply mapping the (fixed gas-density) computed fractional abundances onto the observed density/temperature structure.

Naturally, the ability of chemical models to reproduce observed chemical abundances and structure within specific sources is dependent on our ability to determine the underlying physical structure (and time-dependence) of those sources. The best estimates of the structure of specific hot-cores typically consist of spherically-symmetric physical profiles, although it is to be expected that ALMA observations in particular will highlight substructures that may strongly affect the chemical profiles.

In short, successful reproduction of observed molecular emission data toward MYSOs depends not only on accurate chemical models, but on well-defined source physical conditions, as well as effective methods to translate chemical kinetics simulations into data that may be directly compared with observations, *i.e.* spectral emission predictions.

1 R. T. Garrod, *Astrophys. J.*, 2013, **765**, 60.

Professor Bergin replied: As a practitioner I fully appreciate your statements and indeed this is something that I have been advocating for years. But we also have to recognize there are some (severe) limits to our knowledge and we make some assumptions that may or may not be correct. In a sense the detailed models need strong observational/laboratory guidance, but we need (as a field) to focus on connected systems to explore the actual mechanisms that might be active. This is particularly the case for models of grain surface chemistry.

Professor van der Tak asked a question regarding the paper by Viviana Guzmán: Your rotation temperatures for different species range from 8 K to 50 K which is well beyond the uncertainties (Fig. 6 in paper). What is your interpretation of this; critical density differences or chemical differentiation?

Dr Guzmán Veloso answered: The rotational temperatures found for propyne and the formic acid are not well constrained, and given their large uncertainties they could be as low as 6–18 K as was found for ketene and acetaldehyde. The dipole moment of acetaldehyde is $\sim$2 times larger than that of the other molecules, and it thus has a larger critical density. The gas temperature in the PDR is 60 K, so both CH_2CO and CH_3CHO are sub-thermally excited in this case. Hence, the difference between acetaldehyde and ketene is probably a critical density effect.

Professor Millar commented on the paper by Karin Öberg: Could you comment on the possibility of observing isotopologues to infer routes to molecule formation on ices. In particular, it is known that the enhancement of ^{13}C in CO leads to a deficiency in ^{13}C atoms. Thus, in principle, one has a means of tracing molecules thought to be formed from CO, for example H_2CO, and those which are not. One might even hope to use this technique to probe the alternative formation routes suggested for CO_2. How likely is it that such observations might be made in the near future?

Dr Öberg responded: ^{13}C isotopologues of common COMs should be quite easy to observe with ALMA toward sources that are rich in COMs, *i.e.* hot cores, in terms of pure signal-to-noise. A combination of incomplete spectral databases, line confusion, and optically thick lines for the main isotopologues pose potential problems for identifying and interpreting such lines, however. If these issues are addressed, including very careful radiative transfer modeling to extract ratios between specific ^{13}C and ^{12}C isotopologues, I expect that this line of inquiry will be possible in the immediate future for the least complex COMs, *e.g.* CH_3CN and CH_3OCH_3. A variation of this idea is to compare the ratios of different ^{13}C isotopologues, *i.e.* $^{13}CH_3CH_2OH$ and $CH_3^{13}C_2OH$ to test whether different parts of COMs have different chemical origins or if all form from CH_3OH and thus ultimately from CO ice. This would simplify the radiative transfer calculations, but require more complete spectral databases for isotopologues.

Professor Millar returned to the discussion of the paper by Viviana Guzmán: You mentioned that CH_3CN is enhanced by a factor of about 30 in the PDR. How does this fit in with your suggestion that enhancements are the result of the photodesorption of ices? As far as I am aware, there are no models (or experiments) that suggest the formation of this molecule in ices. Do you know of any irradiation experiments on mixed ices that show its formation?

Dr Guzmán Veloso answered: We are not aware of any ice experiments that lead to the formation of CH_3CN. There are however models that form it through $CH_3 + CN$ reactions on ices.[1] However, that fact that this species is 30 times more abundant in the PDR than in the dense core while the other complex molecules are only 2–4 times more abundant in the PDR suggests that something different is happening for CH_3CN. A gas-phase formation route from species with enhanced abundances in the PDR or photo-fragmentation of small grains or very complex molecules containing CN are also possible scenarios (see question by Anthony Jones).

1 R. T. Garrod, private communication.

Dr Walsh asked: What is the estimated strength of the radiation field at the position in the PDR where the complex molecules, *e.g.*, CH_3CHO, are observed? Have the authors considered running some simple models to check if the rate of formation on (or within) ice mantles on dust grains followed by photodesorption competes with the rate of destruction *via* photodissociation?

Dr Guzmán Veloso responded: The attenuated radiation field at the PDR position is about 3 times the standard interstellar radiation field (see earlier question by Eric Herbst). We have not run any models for the most complex molecules. We did run models for H_2CO and CH_3OH and found that the formation on grains followed by photodesorption is very efficient compared to photodissociation.

Dr Cuppen enquired: Many photodesorption processes occur through photodissociation where the radical products are ejected into the gas phase. Here you observe the photodesorption of methanol that is still intact. Can you comment on this?

Dr Guzmán Veloso answered: Indeed the laboratory experiments have not yet succeeded in the observation of photodesorption of methanol, but rather the dissociated fragments (HCO, H_2CO, *etc.*). One possibility is that some methanol gets kicked out of the ices when a photon hits a neighbour species.

Dr Fayolle returned to Dr. Cuppen's comment on CH_3OH photodesorption and brought some experimental information to the discussion: We investigated the VUV photodesorption of pure CH_3OH ice using the same wavelength dependent technique as successfully done for CO, N_2 and CO_2 (see paper 24 by Fillion *et al.*[1]). We do identify H_2, CO, CO_2, and CH_4 photoproducts desorbing from the surface at 15 K during VUV irradiation, but we do not observe intact CH_3OH, for which we can only estimate a photodesorption rate upper limit of about $\sim 10^{-4}$ molecules per incoming photon in the pure ice case.

1 J.-H. Fillion, E. C. Fayolle, X. Michaut, M. Doronin, L. Philippe, J. Rakovsky, C. Romanzin, N. Champion, K. I. Öberg, H. Linnartz and M. Bertin, *Faraday Discuss.*, 2014, **168**, DOI: 10.1039/c3fd00129f.

Professor van Dishoeck said: A small comment on the process of photo-desorption of ices. Even though the process is often initiated by absorption into a dissociative excited electronic state leading to fragmentation of the molecule, one can still get intact molecules from the ice by at least two processes. First, the fragments have excess energy and are thus mobile through the ice so that they can find each other again to recombine. Because the recombination releases bond energy, the molecule then has enough energy to desorb. Second, one of the energetic fragments can 'kick out' a neighboring intact molecule.

These processes happen in the top few layers for most ices (CO being one of the main exceptions). For H_2O, HDO and D_2O these processes have been described in detail in a series of papers based on molecular dynamics simulations.[1–4]

1 S. Andersson, A. Al-Halabi, G. J. Kroes and E. F, van Dishoeck, *J. Chem. Phys.*, 2006, **124**, 064715.
2 S. Andersson and E. F. van Dishoeck, *Astron. Astrophys.*, 2008, **491**, 907.
3 C. Arasa, S. Andersson, H. M. Cuppen, E. F. van Dishoeck and G. J. Kroes, *J. Chem. Phys.*, 2011, **134**, 164503.
4 J. Koning, G. J. Kroes and C. Arasa, *J. Chem. Phys.* 2013, **138**, 104701.

Professor Fillion remarked: In addition to the 'kick-out' mechanism and exothermic recombination process recalled by E. van Dishoeck, an indirect photodesorption mechanism arising from a desorption induced by electronic transitions in CO seems to be a very interesting route to explain the non-thermal desorption of other intact organic molecules in the gas phase from CO-rich ices. In this mechanism, the energy gained by the electronic relaxation of CO is transferred to neighbouring surface molecules that can desorb. This is clearly shown in the paper by Fillion *et al.*[1] in the case of CO_2. As stated by E. Fayolle, our experiment indicates that irradiation of pure CH_3OH leads mainly to photo-products and daughter species, but rarely to the production of intact methanol into the gas phase.

1 J.-H. Fillion, E. C. Fayolle, X. Michaut, M. Doronin, L. Philippe, J. Rakovsky, C. Romanzin, N. Champion, K. I. Öberg, H. Linnartz and M. Bertin, *Faraday Discuss.*, 2014, **168**, DOI: 10.1039/c3fd00129f.

Professor Herbst asked in relation to the paper by Karin Öberg: This is a rather flippant question, but why did you change the earlier definition of complex organic molecules used by Herbst and van Dishoeck?[1] Was it to eliminate carbon-chain species? Secondly, if we do change the definition, don't you think that we should use the words "partially saturated" rather than "saturated" species?

1 E. Herbst and E. F. van Dishoeck, *Annu. Rev. Astron. Astrophys.*, 2009, **47**, 427.

Dr Öberg replied: In the paper, complex molecules are defined as hydrogen-rich molecules with three or more heavy elements. "Hydrogen-rich" is of course a rather vague definition, but the aim with this definition was to eliminate hydrogen-poor carbon chains formed through ion–molecule gas-phase chemistry, but not their more, but not completely, saturated cousins such as CH_3CCH.

Professor Rawlings returned to the discussion of the introductory paper by Ewine van Dishoeck: The issue of what happens to the enthalpy of formation for molecular species formed on dust grain surfaces has largely been ignored in models of surface chemistry. On the one hand H_2 formation is believed to result in 100% desorption of the nascent molecule, whilst on the other hand, if even a small percentage of the water that is formed on grains is desorbed then there appears to be a serious mismatch between chemical models and observations. Yet, the formation of these species is typically exothermic by several eV, as compared to surface binding energies of 0.1eV. It seems inconceivable, and at odds with laboratory experiments, that all of this energy is efficiently transmitted to the grain substrate so that no desorption occurs.

Professor van Dishoeck replied: I agree that this point needs to be investigated further even though it is very difficult, both from a computational and a laboratory point of view. On the other hand, the binding energy of H_2 to water ice should be much less than that of H_2O to water ice, so I do not see a conflict once the first few layers of the water ice mantle have formed.

PAPER

Efficient diffusive mechanisms of O atoms at very low temperatures on surfaces of astrophysical interest

Emanuele Congiu,*[a] Marco Minissale,[ac] Saoud Baouche,[a] Henda Chaabouni,[a] Audrey Moudens,[a] Stephanie Cazaux,[b] Giulio Manicò,[c] Valerio Pironello[c] and François Dulieu[a]

Received 12th January 2014, Accepted 20th January 2014

DOI: 10.1039/c4fd00002a

At the low temperatures of interstellar dust grains, it is well established that surface chemistry proceeds *via* diffusive mechanisms of H atoms weakly bound (physisorbed) to the surface. Until recently, however, it was unknown whether atoms heavier than hydrogen could diffuse rapidly enough on interstellar grains to react with other accreted species. In addition, models still require simple reduction as well as oxidation reactions to occur on grains to explain the abundances of various molecules. In this paper we investigate O-atom diffusion and reactivity on a variety of astrophysically relevant surfaces (water ice of three different morphologies, silicate, and graphite) in the 6.5–25 K temperature range. Experimental values were used to derive a diffusion law that emphasizes that O atoms diffuse by quantum mechanical tunnelling at temperatures as low as 6.5 K. The rates of diffusion on each surface, based on modelling results, were calculated and an empirical law is given as a function of the surface temperature. The relative diffusion rates are $k_{H2Oice} > k_{sil} > k_{graph} \gg k_{expected}$. The implications of efficient O-atom diffusion over astrophysically relevant time-scales are discussed. Our findings show that O atoms can scan any available reaction partners (*e.g.*, either another H atom, if available, or a surface radical like O or OH) at a faster rate than that of accretion. Also, as dense clouds mature, H_2 becomes far more abundant than H and the O : H ratio grows, and the reactivity of O atoms on grains is such that O becomes one of the dominant reactive partners together with H.

Introduction

In the cold regions of the Universe, where temperatures are lower than 10 K and densities are far weaker than those attainable on earth, a rich chemistry is

---

[a]LERMA Université de Cergy-Pontoise and Observatoire de Paris, ENS, UPMC, UMR 8112 du CNRS, 5, mail Gay Lussac, 95000 Cergy Pontoise cedex, France. E-mail: emanuele.congiu@u-cergy.fr

[b]Kapteyn Astronomical Institute, PO Box 800, 9700 AV Groningen, The Netherlands

[c]Dipartimento di Fisica e Astronomia, Università di Catania, via Santa Sofia 64, 95123 Catania, Sicily, Italy

initiated on the surfaces of minuscule dust particles.[1-4] The species weakly bound to the surface play a central role in the evolution of the pristine chemistry governed by the diffusion of reactive species.[5] In space, thermal atom-addition induced chemistry occurs mostly at low temperatures ($\sim$10 K), $i.e.$, in the innermost parts of the clouds where newly formed species are protected from radiation to a great extent by dust particles. These regions are parts of collapsing envelopes that feed young stellar objects and that provide the original material from which comets and ultimately planets are made.[6] Hydrogenation of interstellar ices can induce the formation of new species in the solid phase and, therefore, this has been the topic of recent laboratory-based studies.[7-10] The efficient surface formation of the bulk of interstellar ices, $i.e.$, water, methanol, formaldehyde, and formic acid, has been demonstrated through H-atom additions of CO and/or O_2 ices under relevant interstellar conditions.[11-13] In particular, the formation of the most important and abundant ice of all, amorphous solid water (ASW), was proved to be the principal product of all possible chemical pathways involving O and H atoms/molecules or the hydroxyl radical (O_3, O_2, O, OH, H and H_2).[14-20] So far, diffusion has only been explored experimentally for H atoms.[21,22] Nevertheless, if we assume that only H atoms are mobile and can diffuse over the dust grain surface, it becomes difficult to explain the observed evidence for CO_2 – the second most abundant condensed species in grain mantles – as well as the rich molecular diversity of the general interstellar medium. CO_2 is believed to form in the solid phase via several energetic[23-26] and non-energetic[23-32] mechanisms, $i.e.$, CO + OH and CO + O reactions at 10 K. The reaction CO + OH leading to CO_2, however, occurs in competition with H_2O formation via the H + OH (barrierless) pathway,[30] which is a much faster route whenever H is the only mobile species. For this reason, the CO + OH reaction alone cannot account for the CO_2 ice observed in quiescent regions.[33,34]

Furthermore, it seems that a sort of depth ($i.e.$, age) segregation exists between the three most abundant ices (H_2O, CO_2, CO):[35,36] water tends to be concentrated in the layers forming the inner (and older) part of the mantles, while CO_2 and CO abundance increases in the outer (and more recent) layers. CO accretes onto icy grain surfaces at a rate proportional to the gas density,[37,38] almost certainly because dust grains are not cold enough in low density regions ($e.g.$, diffuse clouds), so that CO residence time on the surface is extremely short. Conversely, the reason why H_2O (together with other fully hydrogenated species like CH_3OH and NH_3)[35] is mostly found in the inner part (near to the silicate/carbonaceous core), and CO_2 is found in the outer layers, is not fully understood. This could actually be explained by assuming the presence of efficient diffusive processes of O atoms on the icy grains, as recent laboratory works by our group demonstrated on realistic dust grain analogues, such as water ice and silicate at very low temperatures.[39,40] Also, this scenario is consistent with the CO_2 formation timeline proposed by Noble $et\ al.$,[41] which was tested through concurrent observations of CO, CO_2, and H_2O ices.

Here we present new results of diffusion constants of O atoms calculated on graphite, and compare them to the diffusion constants previously calculated on amorphous silicate and three different surfaces of water ice: porous, compact amorphous, and crystalline water ice. In this comparative study, the measured diffusion constants on each surface are given as a function of surface temperature. These data are modelled to determine a simple analytical expression that accurately reproduces the diffusion constants of O atoms with surface temperature. Also, by changing the model to incorporate both a classical (Arrhenius-type) law and a quantum tunnelling description, we are able to address some key physical questions, namely what diffusive process is at play in the 6–25 K temperature range.

Experimental

The FORmation of MOLecules in the InterStellar Medium (FORMOLISM) experimental set-up has been developed with the purpose of studying the reaction and interaction of atoms and molecules on surfaces simulating dust grains under interstellar conditions (the relevance of substrate, low density, and very low temperatures of $\sim$10 K).[42] FORMOLISM is composed of an ultrahigh vacuum chamber with a base pressure of $\sim$$10^{-10}$ mbar, a rotatable quadrupole mass spectrometer (QMS) and an oxygen-free high-conductivity copper sample holder. The sample holder is attached to the cold finger of a closed-cycle He cryostat and can be cooled to 6 K. The temperature is measured with a calibrated silicon diode clamped on the sample holder and controlled by a Lakeshore 334 controller to $\pm$0.2 K with an accuracy of $\pm$1 K in the 8–400 K range. The apparatus is also equipped with a reflection–absorption infrared spectroscopy (RAIRS) facility, used to probe the deposited or produced species *in situ*.[7] Reactants are introduced into the vacuum chamber *via* two separate triply differentially pumped beam lines aimed at the cold surface. Each beam line, in its first stage, consists of an air-cooled quartz tube surrounded by a microwave cavity for dissociating select species (*e.g.*, H_2, O_2, N_2). In this study, only one beam line was used to dissociate and deposit O atoms. Typical values of the dissociation rate are $\sim$70%, which means that the O–O_2 mixture sent onto the sample is in the ratio 14 : 3 (*i.e.*, 0.7 2O : 0.3 O_2). Atoms are cooled and thermalised to 300–400 K upon impact with the surfaces of the quartz tube. We also determined that O atoms and O_2 molecules exiting the source are in their ground states, ^3P and $X^3\Sigma_g^-$, respectively. The beam flux was calibrated by using temperature-programmed desorption (TPD) to determine the O_2 exposure time required to saturate the O_2 monolayer.

Five surfaces were investigated in this study: porous ASW ($H_2O_{(p)}$), non-porous ASW ($H_2O_{(np)}$), crystalline ice ($H_2O_{(c)}$), amorphous olivine-type silicate (SiO_x), and an oxidised slab of highly oriented pyrolytic graphite (HOPG). The SiO_x, HOPG, and other carbon-based surfaces mimic bare dust grains in molecular clouds and have been previously used in investigations of heterogeneous chemistry on dust grain analogues. The silicate sample is amorphous in nature, as evidenced by infrared spectroscopic studies,[43] while TPD experiments reveal the surface to be non-porous on the molecular scale.[44] The HOPG surface used in the experiments is a ZYA-grade HOPG sample, which had been previously exposed to an O-atom beam (oxidised) to avoid surface changes during the experimental sequences. For the water substrates, ice films were grown on top of the silicate surface by spraying water vapour from a microchannel array doser located 2 cm in front of the surface. The water vapour was obtained from deionised water which had been purified by several freeze–pump–thaw cycles, carried out under vacuum. $H_2O_{(p)}$ and $H_2O_{(np)}$ mimic the ASW, which makes up the bulk of interstellar ice, and $H_2O_{(c)}$ mimics the crystalline ice seen in some star-forming regions. To produce $H_2O_{(np)}$, water was dosed while the surface was held at a constant temperature of 110 K. $H_2O_{(p)}$

was grown at 10 K on top of $H_2O_{(np)}$, and then the composite film was annealed to 90 K to stabilize the surface morphology before subsequent heating–cooling runs between 6.5 and 90 K. The sub-layer of $H_2O_{(np)}$ has a thickness of ~50 ML (1 ML = 10^{15} molecules cm^{-2}), and its purpose is to isolate the ensuing $H_2O_{(p)}$ films from the SiO_x substrate.[45] To form $H_2O_{(c)}$, the surface was held at 120 K during the deposition, then flash heated at 50 K min^{-1} to 145 K. For each type of ice surface, the temperature was held constant until the background pressure in the chamber stabilised, before cooling it back down to the range 6–25 K, at which temperature O atoms were dosed onto the respective surfaces.

Results

In Fig. 1 we show the type of raw experimental data we obtain: O_2 and O_3 mass spectra (TPD performed at a heating rate of 10 K min^{-1}) are recorded after deposition of a $O + O_2$ dose on graphite at a given surface temperature (Fig. 1, left panel). O_2 desorbs between 27 and 50 K, while ozone desorption is observed between 55 and 75 K (directly at mass 48 amu, or *via* the O_2^+ fragments at mass 32 amu). O (16 amu) desorption was never observed. The area of the peaks (proportional to the amount of species formed on the surface) changes depending on the coverage ($O + O_2$ dose) and on the surface temperature. The right panel of Fig. 1 shows the RAIR spectrum of the v_3-ozone band at 1043.5 cm^{-1} recorded at 6.5 K after deposition of 0.3 ML of $O + O_2$ on graphite at the same temperature. O_3 and O_2 yields are calculated after depositions of $O + O_2$ performed by (a) varying the dose (coverage) in the range 0.05–1 ML and keeping the surface temperature (T_s) fixed, and (b) dosing with a fixed amount and varying the deposition temperature between 6 and 30 K. The acquisition of two sets of data is motivated by the nature of the mechanisms through which surface reactions leading to O_2 and O_3 proceed. The two main mechanisms involved in surface catalysis are the Langmuir–Hinshelwood (LH) mechanism, in which O_2 and O_3 form in an O-atom diffusion process, and the Eley–Rideal (ER) mechanism, or 'prompt' mechanism,

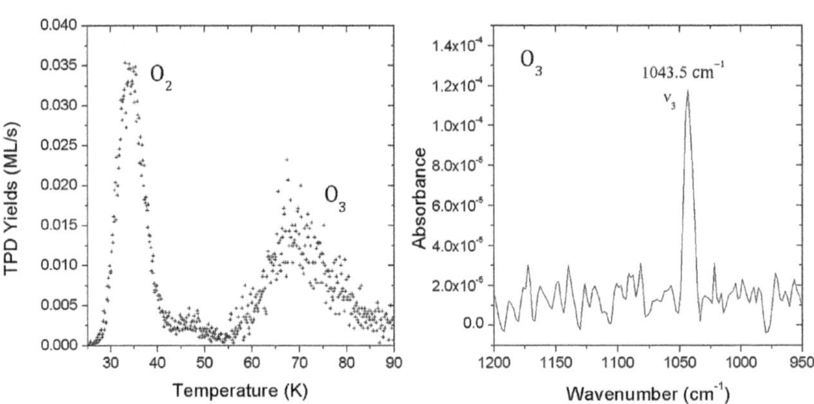

Fig. 1 Left panel: O_2 and O_3 TPD traces obtained after deposition of 0.3 ML of $O + O_2$ on oxidised HOPG held at 6.5 K. Right panel: RAIR spectrum recorded after deposition of 0.3 ML of $O + O_2$ on oxidised HOPG at 6.5 K; the absorption band at 1043.5 cm^{-1} is due to the v_3 asymmetric stretching mode of O_3.

in which an impinging O-atom reacts directly with an adsorbed O or O_2. The LH mechanism is highly dependent on the surface temperature as it affects the mobility of species on the surface. This justifies deposition of O_2 and O_3 at different surface temperatures. The ER mechanism, on the other hand, is independent of T_s and becomes more efficient with the increase in surface coverage, and therefore we also investigated the O_2 and O_3 formation at several O + O_2 initial doses. In two previous papers[39,40] dealing with the diffusion of oxygen atoms on water ice and amorphous silicate, we showed the complete series of TPDs performed at several O + O_2 coverages and at various surface temperatures. From the resulting O_3 and O_2 yields, we inferred that the $O_3 : O_2$ ratio increases with initial coverage, as an incoming O-atom is more likely to find O_2 molecules at higher coverages, and with surface temperature because at higher temperatures the mobility of O atoms is favoured and ozone formation is more efficient.

If we focus on the dependency of diffusion on T_s, and at coverages less than 0.5 ML – more suitable to the astrophysical context – the LH mechanism can be fairly considered to be the main process governing surface reactions involving oxygen atoms and molecules. One may argue, however, that our experiments – based on linear thermal ramps starting from deposition temperature T_s – can affect the diffusion of O atoms and the actual O_2 and O_3 yields. To rule this possibility out, we planned an experiment in which TPD and RAIRS results could be compared in order to confirm or discard the possible role of the heating ramp in O diffusion. In Fig. 2, we present the amount of ozone formed after depositing a given dose of O + O_2 on graphite. The experiment was performed using two different procedures. First, after 0.3 ML of O + O_2 was dosed at various temperatures, a RAIR spectrum was recorded and then a TPD was started for each

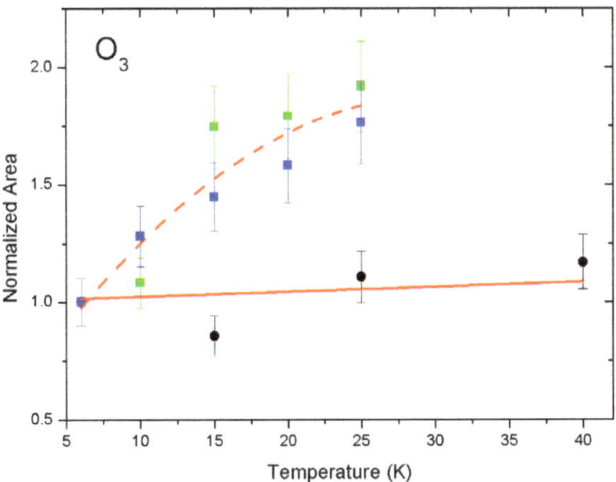

Fig. 2   Ozone yields derived from TPD peak areas (blue squares) and RAIRS ozone ν_3 band area (green squares) after deposition of 0.3 ML of O + O_2 on oxidised HOPG at 6.5, 10, 15, 20, and 25 K. The black circles represent the RAIRS ozone ν_3 band areas obtained by depositing a single 0.3 ML dose of O + O_2 on oxidised HOPG held at 6.5 K and then recording a RAIR spectrum at different surface temperatures (6.5, 15, 25, and 40 K). The dashed and the solid red lines are the fits of the experimental values and serve as a guide to the eye. All ozone yields were normalised to the O_3 yield obtained from oxidised HOPG after deposition of 0.3 ML O + O_2 at 6.5 K.

deposition. The blue squares in Fig. 2 indicate the TPD ozone yields after the deposition of 0.3 ML of O + O$_2$ on oxidised HOPG kept at 6.5, 10, 15, 20 and 25 K. The green squares were obtained by integration of the band area of ozone at 1043.5 cm^{-1}, recorded after each deposition. Secondly, after depositing 0.3 ML of O + O$_2$ onto graphite at 6.5 K, we monitored the evolution of the ozone IR band intensity with temperature. By this second method we could estimate the formation of ozone during the thermal ramp. The results of this experiment are shown by the black circles in Fig. 2. They give the ozone IR band area as a function of temperature after a single 0.3 ML dose of O + O$_2$ at 6.5 K. For temperatures greater than 40 K, the O$_3$ band does not increase because O$_2$ desorbs and the O + O$_2$ reaction can no longer take place. The data presented in Fig. 2, in summary, show the yield of ozone formed following a given dose of O + O$_2$ deposited at different T_s between 6 and 25 K (T_s greater than 25 K would make O$_2$ mobile as well, and hence add another degree of complexity to this study). It is clear that ozone formation efficiency increases quickly with the deposition temperature (see the squares in Fig. 2), while the contribution to ozone provided by the thermal ramp – which is likely to induce diffusion of the residual O atoms – is very small (see black circles in Fig. 2). In addition, if we consider the IR data at 15 K, the O$_3$ yield obtained after O + O$_2$ deposition at 15 K (green square) is much higher than the ozone yield after deposition at 6.5 K and then heated to 15 K (black circle). This fact confirms that all the chemistry has occurred at the deposition temperature; and if this is true at $T_s = 6.5$ K, then it is also the case at any other temperature higher than 6.5 K.

Experimental data are then inserted into a model composed of a series of rate equations used to simulate the O$_2$ and O$_3$ formation yields according to coverage and surface temperature. The model includes both LH and ER mechanisms, and it allows reactions to occur during the deposition phase, as well as during the heating phase (TPD). A complete account of our model is given in Minissale *et al.*[40] Here we will focus on the diffusion rates k and the different methods by which they are calculated. We already alluded to the fact that reactions mostly occur during the exposure phase. The diffusion of atoms during the heating phase is small because no more than a few percent of the deposited O atoms remain available on the surface in the low coverage regime. The effect of possible diffusion during the TPD lies within the error bars of the experimental data, and can be neglected. For this reason, in the following, we will address only the diffusion constants at a fixed temperature for each of the substrates investigated.

The diffusion coefficients k include two components due to quantum tunnelling and thermal motion:[46]

$$k = k_{qt} + k_{tm}$$

In our model, k can be treated as a free numerical parameter during the deposition phase at constant temperature, owing to the fact that the evolution of the coverage with time is known and provides a strong constraint. Therefore, the resulting k values are a set of constants giving the diffusion rate at given temperatures, although no information can be inferred about the nature of the diffusive process. In Fig. 3, the diffusion constants k that we obtained for various substrate compositions are plotted as a function of temperature. An important finding of this comparative study is that diffusion coefficients on water ices

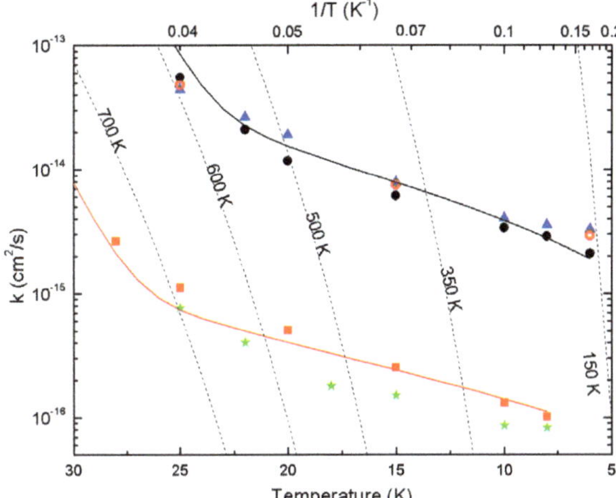

Fig. 3   Diffusion constants k of O atoms obtained on $H_2O_{(p)}$ (open pink circles), on $H_2O_{(np)}$ (black circles), $H_2O_{(c)}$ (blue triangles), amorphous silicate (red squares), and oxidised HOPG (green stars), plotted as a function of surface temperature. The dashed lines represent a series of Arrhenius-type laws generated using five values of E_{diff} (from 150 to 700 K). The two solid lines are the best fits of the experimental values obtained through the quantum tunneling diffusion law for O atoms on $H_2O_{(np)}$ (black solid line) and amorphous silicate (red solid line); see Table 1 for the best fit values of a (barrier width) and E_a (barrier height).

(regardless of morphology) are about one order of magnitude greater than those on silicate and graphite.

Diffusion coefficients k *vs.* T_s can be displayed in several ways according to the law used to describe them, namely, k may have (a) an empirical law built for fitting the experimental values, (b) an Arrhenius-law form, with the activation energy E_{diff} free to vary, or (c) a quantum-tunnelling form with a barrier width and height. A detailed analytical or numerical solution of the dependence of k on T_s can help give some insight into the physical nature of the diffusion process at play.

In case (a), the empirical law we use for fitting the diffusion efficients as a function of surface temperature has the form:

$$k_{emp}(T) = k_0 + \alpha(T/10)^\beta. \tag{1}$$

The diffusion coefficients given by eqn (1) and eqn (2) provide the diffusion probability of exploring a fraction of 1 ML per unit time, and can be converted into the usual units $cm^2\ s^{-1}$ (Fig. 3, Fig. 4, and Fig. 6) by assuming that there are 10^{15} sites cm^{-2}.

Fig. 4 displays a fit of the experimental values obtained on amorphous silicate according to the empirical law given in eqn (1). $k_0\ (s^{-1})$ can be considered to be the minimum value of k, or the value k tends towards near $T = 0$ K. α is a free parameter with values between 0 and 1 $K^{-1}\ s^{-1}$, and it accounts for diffusion efficiency differences between the various substrates. α is 1 for water ice, and about 0.1 for graphite and silicate. The dependency on the surface temperature is governed by the factor $(T/10)^\beta$; the exponent β can have a value between 3 and 4, with variations due to the surface nature, although the best fits give typical values of $\sim$3.5.

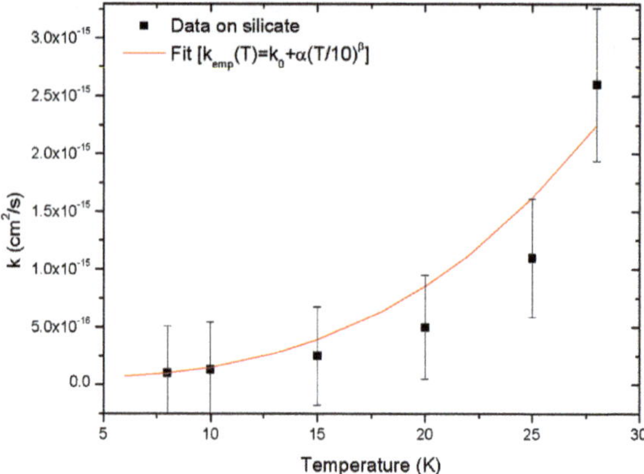

Fig. 4   Black squares represent diffusion constants of O atoms on amorphous silicate as a function of temperature. The red solid line is a best fit of diffusion constant *vs.* temperature obtained using the empirical law given in eqn (1); see Table 1 for the best fit values of k_0, α, and β.

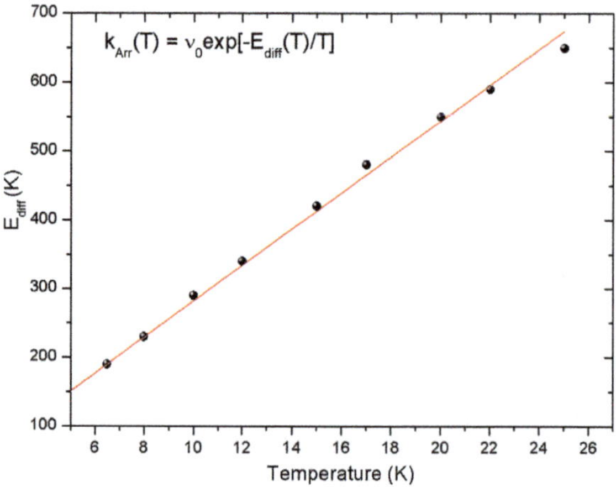

Fig. 5   The energy barrier for diffusion on $H_2O_{(np)}$ as a funcion of surface temperature in the case where diffusion constants are derived from the Arrhenius-type law given in eqn (2). The red solid line represents a linear fit of $E_{diff}(T)$. A single value of E_{diff} cannot satisfy the whole set of diffusion constants observed in the 6–25 K temperature range; see Table 1 for the interval of E_{diff} values needed to obtain k_{Arr} between 6 and 25 K.

In case (b), the classical Arrhenius law used to model the diffusion coefficients k is:

$$k_{Arr}(T) = \nu_0 \exp[-E_{diff}(T)/T]. \tag{2}$$

E_{diff} is the diffusion barrier expressed in kelvin (eV/k_B) and the pre-exponential factor $\nu_0 \, (= 10^{12} \, s^{-1})$ can be viewed as a trial frequency for attempting a new event. In Fig. 5 we present a fit of the diffusion coefficients k on non-porous ASW obtained using the Arrhenius law. Fig. 5 actually displays the activation energies for diffusion (E_{diff}) as a function of temperature. In fact, according to eqn (2), a suitable set of E_{diff} can be used to derive one diffusion coefficient for each temperature. It is thus possible to link each of these diffusion coefficients to an Arrhenius behaviour, and find one energy barrier at each temperature, as shown in Fig. 5 (see also the dashed lines in Fig. 3). It should be noted, however, that an Arrenius-law form in which E_{diff} is fixed (independent of T), or where the distribution of E_{diff} is given, is not able to fit the data. This is why we discarded the Arrhenius-type behaviour of k as it made no physical sense to us. In fact, a systematic increase of the Arrhenius barrier with temperature seemed to us an *ad hoc* solution. Also, it implies that at low temperatures ($\sim$6 K) diffusion occurs through low diffusion barriers (*e.g.*, $E_{diff} = 170$ K). If such low barriers actually exist, they represent fast connections between adsorption sites. Why then would these low energy barriers vanish at higher temperatures? To put it in other terms, why and how would atoms diffuse through slow pathways (high diffusion barriers) at high temperatures ($\sim$20 K) if faster pathways exist? We consider this scenario unlikely and not physically reasonable.

In Fig. 6 we show a comparison between the classical behaviour (described by an Arrhenius-type law) of the diffusion coefficients as reported by Karssmeijer *et al.*[47] for CO molecules on hexagonal water ice, and the trend that we observe experimentally for O atoms on amorphous silicate and crystalline water ice. It is clear that our experimental values do not follow an Arrhenius behaviour, suggesting that a classical description is insufficient to explain the experimental data (squares and triangles in Fig. 6). In fact, in a pure thermal diffusion the slope is very different, and if we fit the data using a classical Arrhenius law, we get values

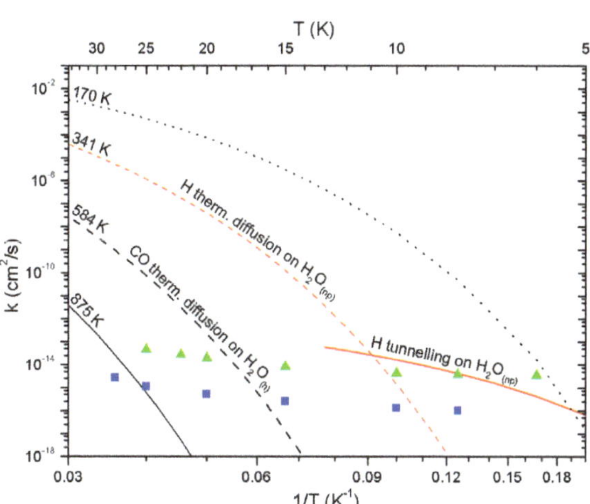

Fig. 6 Comparison between diffusion behaviours of CO molecules, H atoms, and O atoms. Blue squares and green triangle represent O-atom diffusion constants as a function of surface temperature on amorphous silicate and crystalline water ice, respectively (this work). The black dashed line represents the thermal diffusion (Arrhenius behaviour) of CO molecules on hexagonal water ice found in Karssemeijer *et al.*[47] The red solid line and the red dashed line display the H-atom tunnelling (6–13 K) and the H thermal diffusion ($T_s > 13$ K), respectively, obtained by Senevirathne *et al.*[48] on compact amorphous water ice. The difference between the slopes of CO and O behaviors, and the similarity between the slopes of H tunnelling and O data, corroborate the conclusion that O atoms diffuse *via* quantum tunnelling in the surface temperature domain between 6 and 22 K.

of ν_0 and E_{diff} that are not physically acceptable. Therefore, a quantum mechanical approach ought to be used to account for the deviations from the classical trend. Our results on oxygen atoms are consistent with a tunneling-dominated diffusion observed for H atoms on $H_2O_{(np)}$ by Senevirathne et al.[48] in the 6–13 K temperature range (the slopes of H- and O-diffusion constant behaviours are similar). They also found that the diffusion of H atoms is enhanced around 13 K, as occurs to O atoms at around 22 K, just where classical thermal motion begins to predominate over quantum processes.[39,48] Hama et al.[49] found that the temperature border between quantum and classical diffusion of H atoms is likely to be at $T_s < 10$ K. We found that at very low temperatures the diffusion of O atoms is better simulated by quantum tunneling through a square barrier.[39,50] The physical parameters we use to describe such a quantum jump are the width a and the height E_a of the barrier. The choice of a square barrier, the simplest shape of potential, was made purposefully to show that the right trend is obtained if one uses quantum-tunneling diffusion, not because we believed that a square barrier was the right one. We believe that any other more realistic potential shape we could use would not fundamentally change the results, and would still be unrealistic given the complexity of the distribution of diffusion barriers. We did not try to obtain the best fit of our data, but tried to show that the right trend is obtained if one uses quantum-tunneling diffusion (see solid lines in Fig. 3). Hence, we chose to model the quantum diffusion with two parameters which have a simple physical meaning, although they correspond to macroscopic values that come from the interplay of many different microscopic situations.

The values of k and of all the parameters used to fit the diffusion coefficients on each substrate, using the three methods, are listed in Table 1.

The diffusion coefficients of O atoms calculated on water ices are one order of magnitude greater than those found on amorphous silicate and oxidised HOPG, i.e. the O diffusive mechanism is more efficient on icy grains. Also, as opposed to the case of H atoms, there is no difference between the efficiency of O mobility on the three types of water ices investigated ($H_2O_{(p)}$, $H_2O_{(np)}$, and $H_2O_{(c)}$). In the light of our experimental results, we can only observe and simply report this finding. In fact, dealing with atoms makes it very difficult to derive key parameters such as the energy barrier for diffusion, or even the energy barrier for desorption, hence no pertinent assumption can be made to explain these findings from a physicochemical point of view. However, to give a physical explanation of our results is beyond the scope of this paper, since we believe that quantum calculations and simulations will be necessary to thoroughly describe O diffusion mechanisms at low temperatures.

Table 1 Best fit parameters of the three methods used to model the diffusion constants for O diffusion on five different grain surface analogues

	Quantum tunnelling		Arrhenius law	Empirical law		
	a (Å)	E_a (K)	$E_{\text{diff}[6<T_s<25]}$ (K)	k_0 (10^{-15})	α	β
Porous ASW	0.69 ± 0.10	530 ± 70	$170 < E_{\text{diff}} < 600$	1.30	1	3
Non-porous ASW	0.70 ± 0.05	520 ± 60	$170 < E_{\text{diff}} < 600$	1.21	1	3
Crystalline water ice	0.69 ± 0.05	500 ± 50	$170 < E_{\text{diff}} < 600$	1.42	1	3
Amorphous silicate	0.67 ± 0.10	720 ± 70	$290 < E_{\text{diff}} < 740$	0.15	0.1	4
Oxidised HOPG	0.67 ± 0.10	740 ± 60	$290 < E_{\text{diff}} < 740$	0.1	0.1	4

Astrophysical implications

As far as the diffusion of O atoms is concerned, it turns out that, whenever a diffusive process exists, this has an impact on the chemistry occurring at the surface of dust grains. In fact, either the formation of some species may be enhanced, or at least the relative abundances of the final products affected, if O diffusion is efficient. An important example of how O-atom mobility can modulate the abundances of key species of ices in the ISM is the case of the $H_2O : CO_2$ ratio.

In dense quiescent molecular clouds, hydrogen atoms have always been thought to be the only mobile species on the surface of icy grains. Most of the molecular variety observed in interstellar ices has long been considered the outcome of H-atom addition reactions involving O, O_2, O_3, CO, N, and NO. Water, for example, is the final and most stable species of all the chemical networks between H and O, O_2 and O_3, which justifies its role as the most abundant ice in the Universe. If the reactive partner of H is CO, then CH_3OH is obtained *via* a series of successive hydrogenations. On the other hand, if H is the only mobile species able to scan the entire surface of the grain,[21] it may be difficult to explain the abundance of CO_2, the second most abundant condensed species. CO_2 can also be formed *via* energetic processes by irradiating ice mixtures of H_2O and CO with UV photons or ions. In the dense core of molecular clouds, however, these processes may not apply, and CO_2 can only be formed *via* non-energetic mechanisms, *i.e.*, the reactions CO + OH and CO + O. If these chemical routes leading to CO_2 involved only species which are not mobile at 10 K, then CO_2 formation would be greatly hindered by the rate of accretion and the high mobility of H atoms, which are able to reach CO, OH,

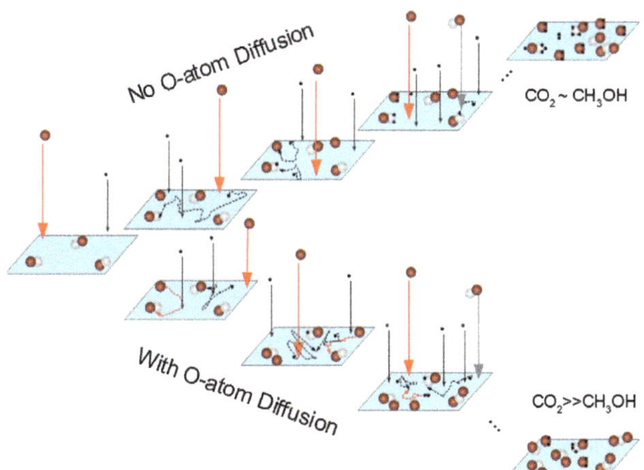

Fig. 7   A schematic view of the two possible scenarios concerning the H_2O/CO_2 balance on icy grains in dense molecular clouds. If H atoms (black dots) are the only mobile species, H-addition mechanisms are dominant and the formation of water (H + OH/O/O_2/O_3 → H_2O) and other H-saturated species (*e.g.*, CO + 3H → CH_3OH) is favoured. On the other hand, if O atoms (red circles) can also diffuse at very low temperatures, the formation of CO_2 in dense clouds may proceed *via* non-energetic reactions (CO + O and CO + OH) as well, making it possible that CO_2 is the second most abundant ice in the ISM.

and O long before these species can meet to form carbon dioxide. Our present and previous works introduce strong arguments to suggest that O atoms too are mobile at very low temperatures. This implies that the formation rate of CO_2 in dense clouds is governed by a balance between the accretion rate of H atoms and the diffusion rate of O atoms on the surface of dust grains. The cartoon in Fig. 7 shows that when the accretion rate of H atoms is dominant, H_2O and CH_3OH are for the most part the final products; when the diffusion rate of O atoms prevails, formation of CO_2 (and O_3) is favoured.

With this in mind, we carried out some calculations to show the evolution of the relative abundances of H atoms and O atoms on the surfaces of dust grains and – assuming that both species are mobile at low temperaures – how this balance can affect the chemistry within interstellar clouds of various densities. In fact, different environments are characterized by different densities; the abundances of species in the gas phase change and this entails a change in the accretion time-scales of particles on dust grains. In diffuse clouds, hydrogen is mainly present in its atomic form and is by far the most abundant atomic species. In dark clouds, hydrogen is mainly present in its molecular form, so H atoms become a rather rare reactant, with $[H] : [H_2] \approx 10^{-3}$ (see, $e.g.$, Li and Goldsmith[51]). The number density of H atoms is mostly governed by the destruction of H_2 due to cosmic rays. This value, almost independent of the density of the cloud, is of the order 1 H cm^{-3}. On the other hand, the $[O] : [H_2]$ ratio remains approximately constant (10^{-4}), thus the number of atomic O, unlike H, is proportional to the density of the cloud (see, for example, Table 1 in Caselli et $al.$[52]). For a cloud with number density of 10^4 cm^{-3}, the $[H] : [O]$ ratio is ~1 : 0.75, while for a denser cloud with a density of 10^5 cm^{-3}, the $[H] : [O]$ ratio is ~1/7. Therefore, for very dense clouds, O is the most abundant species in atomic form, can accrete on grains and, provided the O atoms are mobile, can subsequently react with other species before these become fully saturated by H-additions. The accretion rates of H atoms and the diffusion coefficients of O are then the key factors to be compared in order to determine

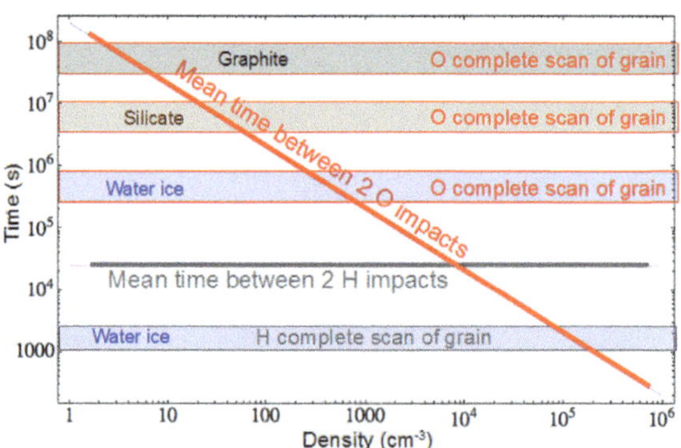

Fig. 8   Time intervals between two impacts of H and O, and the times employed to scan a whole grain (shaded horizontal bands) on various surfaces of interest are plotted as a function of the density of the cloud. The time interval between two arrivals of H is constant, as the density of H atoms remains rather constant regardless of the density of the medium, while the abundance of O atoms increases with cloud density.

at what density of the medium oxidation reactions become comparable to H-atom additions.

In Fig. 8 we show the time interval between two impacts of particles of the same species (H or O) on a single dust grain, as a function of the density n of the cloud. The time intervals between the two arrivals are derived from the actual particle flux of a given species. The interstellar flux of a species accreting on dust grains can be calculated as follows:

$$\Phi_x = 1/4 n_x v_x \qquad (3)$$

where n_x is the density of species x in the gas phase and $v_x = (8k_BT/\pi m_x)^{0.5}$ is its mean velocity. Φ_x is thus expressed in particles cm^{-2} s^{-1}. For our calculation, we can approximate the dust grains to spheres with a typical radius $r = 0.1$ μm, with accessible surface area $A = 4\pi r^2$. The time interval between the impacts of two particles is then:

$$t = (n_x v_x A/4)^{-1}. \qquad (4)$$

In Fig. 8, the grey solid line represents the time interval between the impact of two hydrogen atoms, calculated by assuming a constant density of H atoms, $n_H = 2.3$ cm^{-3} (from Li and Goldsmith[51]). The density of O atoms is proportional to the density of the clouds, n, namely, $n_O = 5 \times 10^{-4} n$. The time interval between the arrival of two O atoms is displayed as a red solid line, which clearly shows that the arrival of oxygen atoms becomes more frequent (shorter time between the two impacts) with the increasing density of the cloud. The grey and red lines cross at a density n of around 10^4 cm^{-3}. This suggests that for cloud densities of $\sim 10^4$ cm^{-3} the accretion rates of H and O are comparable, and, given that both species can diffuse, oxidation reactions on the grains may play a role, although H-atom additions are still dominant due to the higher mobility of H. In Fig. 8 we also indicate the mean time O atoms need to complete a scan of all the adorption sites on the surface of one typical grain as used above, with radius $= 0.1$ μm and 10^6 absorption sites (10^{15} sites cm^{-2}). The mean times needed for a complete scan of the grain surface were calculated for a surface temperature of 10 K by using the diffusion constants k of O atoms on each substrate presented in this work, taking into account that $k = 10^{-15}$ cm^2 s^{-1} corresponds to one jump per unit time. For H atoms, the mean time for scanning the entire surface of water ice was derived by the energy barrier for diffusion of 255 K (at 10 K) given by Matar et al.[21] Again, in Fig. 8, it is interesting to observe the intersection occurring at $n \sim 10^5$ cm^{-3} between the red line and the band giving the mean time H atoms employ to scan the whole surface of the grain. This implies that at cloud densities of $\sim 10^5$ cm^{-3} or greater, the diffusion and accretion rates of H atoms are smaller than the accretion rate of O atoms. Therefore, in very dense clouds, oxygen atoms may become the dominant reaction partner, able to react with CO and produce CO_2, as well as reacting with H to produce OH. Since H atoms are rare in this environment, OH will not be readily transformed into water via hydrogenation, and the hydroxyl radical is likely to react with the abundant CO molecules to form CO_2.

Acknowledgements

The LERMA-LAMAp team in Cergy acknowledges the support of the national PCMI programme founded by CNRS, and the Conseil Regional d'Ile de France through SESAME programmes (contract I-07597R). MM acknowledges financial support by LASSIE, a European FP7 ITN Communities Seventh Framework Programme under grant agreement no. 238258. MM also thanks Prof. Tomellini (Università di Roma Tor Vergata) for fruitful discussions.

References

1 A. G. G. M. Tielens, *Rev. Mod. Phys.*, 2013, **85**, 1021.

2 A. G. G. M. Tielens and W. Hagen, *Astron. Astrophys.*, 1982, **114**, 245.

3 L. Pagani, A. Bacmann, S. Cabrit and C. Vastel, *Astron. Astrophys.*, 2007, **467**, 179.

4 E. Herbst and E. F. van Dishoeck, *Annu. Rev. Astron. Astrophys.*, 2009, **47**, 427.

5 H. M. Cuppen and E. Herbst, *Astrophys. J.*, 2007, **668**, 294.

6 S. A. Sandford, J. A. Aléon, C. M. O'D. Alexander, *et al.*, *Science*, 2006, **314**, 1720.

7 E. Congiu, H. Chaabouni, C. Laffon, P. Parent, S. Baouche and F. Dulieu, *J. Chem. Phys.*, 2012, **137**, 054713.

8 E. Congiu, G. Fedoseev, S. Ioppolo, F. Dulieu, H. Chaabouni, S. Baouche, J. L. Lemaire, C. Laffon, P. Parent, T. Lamberts, II. M. Cuppen and H. Linnartz, *Astrophys. J. Lett.*, 2012, **750**, L12.

9 G. Fedoseev, S. Ioppolo, T. Lamberts, J. F. Zhen, H. M. Cuppen and H. Linnartz, *J. Chem. Phys.*, 2012, **137**, 054714.

10 P. Theule, F. Borget, F. Mispelaer, G. Danger, F. Duvernay, J. C. Guillemin and T. Chiavassa, *Astron. Astrophys.*, 2011, **534**, A64.

11 N. Watanabe, A. Nagaoka, H. Hidaka, *et al.*, *Planet. Space Sci.*, 2006, **54**, 1107.

12 G. W. Fuchs, H. M. Cuppen, S. Ioppolo, *et al.*, *Astron. Astrophys.*, 2009, **505**, 629.

13 H. M. Cuppen, S. Ioppolo, C. Romanzin and H. Linnartz, *Phys. Chem. Chem. Phys.*, 2010, **12**, 12077.

14 N. Miyauchi, H. Hidaka, T. Chigai, A. Nagaoka, N. Watanabe and A. Kouchi, *Chem. Phys. Lett.*, 2008, **456**, 27.

15 S. Ioppolo, H. M. Cuppen, C. Romanzin, E. F. van Dishoeck and H. Linnartz, *Astrophys. J.*, 2008, **686**, 1474.

16 F. Dulieu, L. Amiaud, E. Congiu, J.-H. Fillion, E. Matar, A. Momeni, V. Pirronello and J. L. Lemaire, *Astron. Astrophys.*, 2010, **512**, A30.

17 D. Jing, J. He, J. Brucato, A. De Sio, L. Tozzetti and G. Vidali, *Astrophys. J.*, 2011, **741**, L9.

18 C. Romanzin, S. Ioppolo, H. M. Cuppen, E. F. van Dishoeck and H. Linnartz, *J. Chem. Phys.*, 2011, **134**, 084504.

19 Y. Oba, N. Watanabe, T. Hama, K. Kuwahata, H. Hidaka and A. Kouchi, *Astrophys. J.*, 2012, **749**, 67.

20 H. Chaabouni, M. Minissale, G. Manicò, E. Congiu, J. A. Noble, S. Baouche, M. Accolla, J. L. Lemaire, V. Pirronello and F. Dulieu, *J. Chem. Phys.*, 2012, **137**, 234706.

21 E. Matar, E. Congiu, F. Dulieu, A. Momeni and J. L. Lemaire, *Astron. Astrophys.*, 2008, **492**, L17.

22 N. Watanabe, Y. Kimura, A. Kouchi, T. Chigai, T. Hama and V. Pirronello, *Astrophys. J. Lett.*, 2010, **714**, L233.

23 P. A. Gerakines, W. A. Schutte and P. Ehrenfreund, *Astron. Astrophys.*, 1996, **312**, 289.

24 M. E. Palumbo, G. A. Baratta, J. R. Brucato, *et al.*, *Astron. Astrophys.*, 1998, **334**, 247.

25 C. S. Jamieson, A. M. Mebel and R. I. Kaiser, *Astrophys. J. Suppl.*, 2006, **163**, 184.

26 V. Mennella, M. E. Palumbo and G. A. Baratta, *Astrophys. J.*, 2004, **615**, 1073.

27 T. P. M. Goumans, M. A. Uppal and W. A. Brown, *Mon. Not. R. Astron. Soc.*, 2008, **384**, 1158.

28 Y. Oba, N. Watanabe, A. Kouchi, T. Hama and V. Pirronello, *Astrophys. J. Lett.*, 2010, **712**, L174.

29 S. Ioppolo, Y. van Boheemen, H. M. Cuppen, E. F. van Dishoeck and H. Linnartz, *Mon. Not. R. Astron. Soc.*, 2011, **413**, 2281.

30 J. A. Noble, F. Dulieu, E. Congiu and H. J. Fraser, *Astrophys. J.*, 2011, **735**, 121.

31 U. Raut and R. Baragiola, *Astrophys. J. Lett.*, 2011, **737**, L14.

32 M. Minissale, E. Congiu, G. Manicò, V. Pirronello and F. Dulieu, *Astron. Astrophys.*, 2013, **559**, A49.

33 A. Nummelin, D. C. B. Whittet, E. L. Gibb, P. A. Gerakines and J. E. Chiar, *Astrophys. J.*, 2001, **558**, 185.

34 K. M. Pontoppidan, *Astron. Astrophys.*, 2006, **453**, L47.

35 E. L. Gibb, D. C. B. Whittet, A. C. A. Boogert and A. G. G. M. Tielens, *Astrophys. J. Suppl.*, 2004, **151**, 35.

36 K. I. Oberg, A. C. A. Boogert, K. M. Pontoppidan, *et al.*, *Astrophys. J.*, 2011, **740**, 109.

37 Y. Aikawa, N. Ohashi, S.-i. Inutsuka, E. Herbst and S. Takakuwa, *Astrophys. J.*, 2001, **552**, 639.

38 K. M. Pontoppidan, H. J. Fraser, E. Dartois, *et al.*, *Astron. Astrophys.*, 2003, **408**, 981.

39 M. Minissale, E. Congiu, S. Baouche, H. Chaabouni, A. Moudens, F. Dulieu, M. Accolla, S. Cazaux, G. Manicò and V. Pirronello, *Phys. Rev. Lett.*, 2013, **111**, 053201.

40 M. Minissale, E. Congiu and F. Dulieu, *J. Chem. Phys.*, 2014, **140**, 074705.

41 J. A. Noble, H. J. Fraser, Y. Aikawa, K. M. Pontoppidan and I. Sakon, *Astrophys. J.*, 2013, **775**, 85.

42 L. Amiaud, J. H. Fillion, S. Baouche, F. Dulieu, A. Momeni and J. L. Lemaire, *J. Chem. Phys.*, 2006, **124**, 094702.

43 Z. Djouadi, L. D'Hendecourt, H. Leroux, A. P. Jones, J. Borg, D. Deboffle and N. Chauvin, *Astron. Astrophys.*, 2005, **440**, 179.

44 A. Noble, E. Congiu, F. Dulieu and H. J. Fraser, *Mon. Not. R. Astron. Soc.*, 2012, **421**, 768.

45 I. Engquist, I. Lundstroem and B. Liedberg, *J. Phys. Chem.*, 1995, **99**, 1225.

46 S. Cazaux and A. Tielens, *Astrophys. J.*, 2004, **604**, 222.

47 L. J. Karssemeijer, A. Pedersen, H. Jònsson and H. M. Cuppen, *Phys. Chem. Chem. Phys.*, 2012, **14**, 10844.

48 B. Senevirathne, S. Andersson, F. Dulieu and G. Nyman, *Phys. Chem. Chem. Phys.*, submitted manuscript available at: https://dl.dropboxusercontent.com/u/9601281/Senevirathne_etal2014.pdf).

49 T. Hama, K. Kuwahata, N. Watanabe, A. Kouchi, Y. Kimura, T. Chigai and V. Pirronello, *Astrophys. J.*, 2012, **757**, 185.

50 A. Messiah, *Quantum Mechanics*, Wiley, Amsterdam, 1973.

51 D. Li and P. F. Goldsmith, *Astrophys. J.*, 2003, **585**, 823.

52 P. Caselli, T. Stantcheva, O. M. Shalabiea, V. I. Shematovich and E. Herbst, *Planet. Space Sci.*, 2002, **50**, 1257.

PAPER

Single and double addition of oxygen atoms to propyne on surfaces at low temperatures

Helen J. Kimber, Courtney P. Ennis† and Stephen D. Price*

Received 11th December 2013, Accepted 14th January 2014

DOI: 10.1039/c3fd00130j

Experiments designed to simulate the low temperature surface chemistry occurring in interstellar clouds provide clear evidence of a reaction between oxygen atoms and propyne ice. The reactants are dosed onto a surface held at a fixed temperature between 14 and 100 K. After the dosing period, temperature programmed desorption (TPD), coupled with time-of-flight mass spectrometry, are used to identify two reaction products with molecular formulae C_3H_4O and $C_3H_4O_2$. These products result from the addition of a single oxygen atom, or two oxygen atoms, to a propyne reactant. A simple model has been used to extract kinetic data from the measured yield of the single-addition (C_3H_4O) product at surface temperatures from 30–100 K. This modelling reveals that the barrier of the solid-state reaction between propyne and a single oxygen atom (160 ± 10 K) is an order of magnitude less than that reported for the gas-phase reaction. In addition, estimates for the desorption energy of propyne and reaction rate coefficient, as a function of temperature, are determined for the single addition process from the modelling. The yield of the single addition product falls as the surface temperature decreases from 50 K to 30K, but rises again as the surface temperature falls below 30 K. This increase in the rate of reaction at low surface temperatures is indicative of an alternative, perhaps barrierless, pathway to the single addition product which is only important at low surface temperatures. The kinetic model has been further developed to characterize the double addition reaction, which appears to involve the addition of a second oxygen atom to C_3H_4O. This modelling indicates that this second addition is a barrierless process. The kinetic parameters we extract from our experiments indicate that the reaction between atomic oxygen and propyne could occur under on interstellar dust grains on an astrophysical time scale.

1   Introduction

The elemental composition of the known universe comprises almost exclusively light atoms (~99.9% hydrogen and helium). However, to date, over 160 different

Chemistry Dept., UCL, 20 Gordon Street, London, UK, WC1H 0AJ. E-mail: s.d.price@ucl.ac.uk
† Current address: Australian Synchrotron, 800 Blackburn Road, Clayton, Vic. 3168, Australia

molecules have been detected in the interstellar medium.[1] The vast majority of these interstellar molecules contain hydrogen. The distribution of these molecules is far from uniform across the interstellar medium (ISM), where dense interstellar clouds are observed to harbour relatively large densities and varieties of complex molecules that have been proposed as precursors to biologically relevant species.[2,3] In these interstellar clouds, molecular lifetimes are extended with respect to other more diffuse parts of the ISM. Specifically, the relative opacity of these dense interstellar clouds shields the cloud's centre from high energy photons, allowing relatively fragile molecules to survive for extended periods.

In dense interstellar clouds, molecules are detected in abundances that cannot be completely accounted for by known gas-phase kinetics.[4] Hence, it is now widely accepted that there is a contribution to these molecular abundances from reactions on and within the molecular ices on the surfaces of interstellar dust. This interstellar dust comprises predominantly silicate or carbonaceous particles, with a typical diameter of 100 nm,[5] making up typically 1% of the mass of an interstellar cloud.[6]

In general, bimolecular reactions on surfaces are thought to follow one of two prototypical reaction pathways. The first is the Langmuir-Hinshelwood (LH) mechanism where both reactive species are initially adsorbed and thermalised on the surface.[7] Diffusive processes then allow the reactants to encounter one another and react. The second mechanism, the Eley-Rideal (ER) pathway, involves a thermalised surface molecule undergoing direct reaction with an incident, and potentially energetic, gas-phase species.[7] To correctly model the contribution of these heterogeneous reactions to the abundances of interstellar molecules, kinetic data is required. The measurement of such kinetic data is the objective of our experimental efforts.[8,9]

One relatively abundant interstellar molecule, that is considered to play an important role in astrochemical processes, is the small unsaturated hydrocarbon, propyne (CH_3CCH).[10,11] Interstellar propyne was first identified in the Milky Way in the giant molecular cloud Sagittarius B2.[11,12] Propyne has also been observed in cold cloud cores and lukewarm corinos in the Milky Way.[13] Propyne has also been detected across the disc of the distant Messier 82 (M82) galaxy.[14] In M82, the propyne-to-methanol ratio ($[CH_3CCH]/[CH_3OH] > 8$) is higher than for comparable starburst galaxies such as NGC 253 ($[CH_3CCH]/[CH_3OH] \approx 1$).[15] The propyne component of a prototypical starburst galaxy such M82 and NGC 253 resides primarily in interstellar clouds, where thermal dust grain chemistry is potentially important. Here we distinguish *thermal* surface chemistry, where reactions occur on the surface without the input of additional energy, with *activated* chemistry, where reactions are initiated and molecules subsequently processed, by external agents such as cosmic rays.[16]

To our knowledge, despite propyne being repeatedly detected in the ISM, the reactivity of propyne on cold surfaces, such as those of dust grains, has not yet been investigated. In contrast, the chemistry of methanol on interstellar surfaces has been extensively studied.[17-20] Given their comparable abundances, the surface reactivity of propyne appears to be overdue an investigation. Such an investigation is the target of this study.

In addition to its detection in interstellar clouds, propyne has also been observed in planetary environments in our Solar System. Specifically, propyne has been detected in stratospheres of Jupiter and Saturn, displaying column densities of $1.5 \pm 4\times10^{14}$ molecules cm^{-2} and 2×10^{15} molecules cm^{-2} respectively.[21,22] Interstellar ice composition is considered to be important when discussing the abundance of propyne in the Saturnian and Jovian atmospheres. Here, volatile molecules, such as propyne, are thought to originate from molecules trapped as clathrates in the water ice that aggregated during the formation of the planetisimals that preceded these planets.[23–28] Propyne has also recently been identified in the atmosphere of Uranus[29] and the atmosphere of Saturn's moon Titan, where it appears enriched at northern latitudes.[30,31] In all cases propyne is identified by its infrared absorption band (ν_9) centred at 15.8 μm.

Oxygen is the third most abundant element in the ISM, after hydrogen and helium.[32] There has been considerable recent interest in accounting for the relatively low abundance (the so-called depletion) of oxygen in the gas-phase in the interstellar medium.[33–35] It has been proposed that interstellar grains could act as a sink for oxygen atoms, but it appears that the necessary depletion cannot be generated simply by the incorporation of oxygen atoms into the structures of the dust grains themselves.[34] One possible additional sink for oxygen atoms is perhaps their reaction with the organic component of the molecular ice mantles that coat the dust grains in some interstellar clouds. The reaction we investigate in this paper, the addition of oxygen atoms to propyne on a cold surface, could be considered representative of oxygen atom depletion by reaction with organic ices.

Oxygen atom diffusion has recently been studied on an amorphous water ice surface.[36] In this work Minissale *et al*[36] report that oxygen atom diffusion exhibits a quantum-classical transition at a surface temperature of 20 K. That is, above 20 K a classical motion involving barrier 'hopping' between adsorption sites is the dominant O atom migration mechanism, but at surface temperatures below 20 K quantum tunnelling of the O atoms becomes dominant.

The gas-phase reaction of propyne with O atoms has been studied both experimentally and computationally.[37–44] In the gas phase the reaction produces CO and the CH_3CH diradical which form from a vibrationally excited methyl ketene.[45] At high pressures, or in a condensed medium where intermediates can be collisionally stabilized, the adduct that results from O atom addition is expected to simply relax to a more stable isomer such as methyl ketene or propenal.[38,46]

In this paper we present a study of the reaction of propyne with atomic oxygen on a cold surface. These experiments generate kinetic data for this reaction under astrophysically relevant conditions. These studies build on our previous investigations of the reaction of oxygen atoms with alkenes[9] and sulphur containing molecules.[8]

2   Experimental procedure

The experimental apparatus employed in this study, which has been described in detail before,[8,9] is designed to probe the reactions of molecules with atomic species on molecular ices deposited on a graphite surface held at a specific temperature. The surface temperatures investigated are pertinent to heterogeneous reactivity on dust surfaces in the ISM. The products of any surface reactions are detected by a temperature-programmed desorption (TPD) methodology. In the TPD experiment, either an electron beam or a laser is used to ionize the molecules

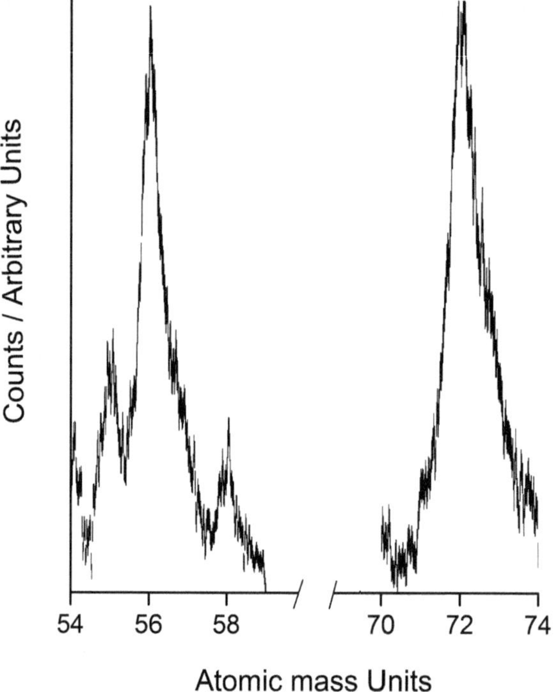

Fig. 1 Sections of representative mass spectra recorded during the TPD phase showing peaks for $m/z = 56$ and $m/z = 72$, corresponding to the single and double addition products. See text for details.

formed on the surface when they are desorbed as the surface is heated. The ions resulting from this ionization of the desorbed molecules are detected and identified by a time-of-flight mass spectrometer (TOFMS).[47]

The experimental apparatus consists of two vacuum chambers. The 'source' chamber has a base pressure of approximately 10^{-7} Torr. This chamber houses a microwave discharge cell used to generate O atoms from O_2. Previous experiments have determined an O_2 dissociation efficiency of approximately 20% from this source,[9] so O_2 is still the majority species in our "O atom beam".

In investigating the low-temperature reactions induced on surfaces by our O atom source one must be aware, as discussed before,[9] that ozone can be formed by reactions of O atoms with O_2 on the surface. However, the oxygen atom addition reactions we observe proceed efficiently at surface temperatures well above 30 K, where ozone is not present on the surface.[9] This observation indicates that the oxygenated products we observe are not the result of the reaction between propyne and ozone. Supporting this analysis is the observation that ozone only reacts with unsaturated hydrocarbons at low temperatures following irradiation.[48] We also note that no reaction is observed if no microwaves are applied to the O atoms source and simply O_2 is dosed onto the surface.

Gas from the O atom source, undergoes significant differential pumping in the source chamber before being piped into the 'target' chamber. A second deposition line attached to the source chamber allows the dosing of stable molecules, in this case propyne, onto the cold surface in addition to the oxygen atoms. Again,

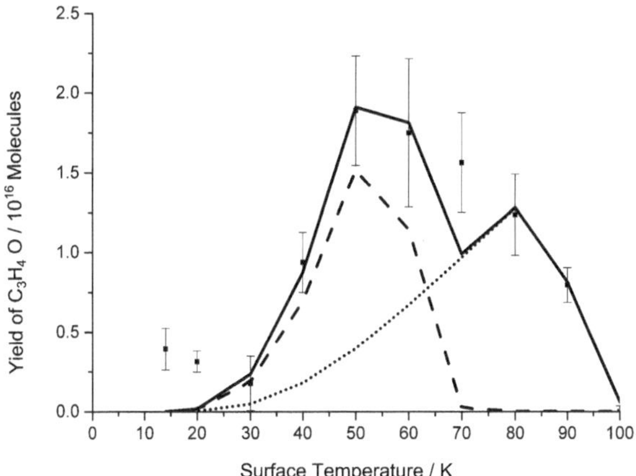

Fig. 2   Summed yield of the single and double addition product, formed following the co-deposition of propyne (C_3H_4) and O atoms, as a function of surface temperature. Squares: experimental data; solid line: model; dash LH mechanism; dot ER mechanism. The error bars associated with the experimental results represent two standard deviations from four repeats at each surface temperature. The kinetic model used to derive the fit shown employs the paramenters listed in Table 1 and an O atom desorption energy of 15.3 kJ mol^{-1}.

the propyne dosing line undergoes significant differential pumping in the source chamber before being directed into the target chamber. This differential pumping arrangement serves to allow sufficient pressure in the microwave source for stable operation, whilst dosing the target surface at an acceptably low rate. The UHV target chamber has a base pressure of approximately 10^{-10} Torr when the dosing gases are switched off, but has a pressure of approximately 10^{-8} Torr during the dosing process.

The PTFE tubes which direct the reactants from the source chamber into the target chamber terminate close above a highly oriented pyrolytic graphite (HOPG) substrate. This substrate can be cooled to close to 10 K and heated to above 500 K. Propyne and $O_2/O^{\cdot}$ are co-deposited on the HOPG substrate at a fixed surface temperature (in the range from 14 K to 100 K). The fluxes of the propyne and O atoms are 1.0×10^{15} cm^{-2} s^{-1} and 1.3×10^{14} cm^{-2} s^{-1} respectively.[9] After one hour, the delivery lines are evacuated and the substrate is allowed to cool from the fixed deposition temperature to 14.0 ± 0.5 K. A current is then passed through a tantalum strip heater to slowly raise the surface temperature to $\sim$200 K. As the substrate temperature increases, the molecules on the surface desorb at their specific sublimation temperatures and enter the source region of the TOFMS which is located in front of the sample. It should be noted that the dosing fluxes employed in our experiments rapidly result in a multilayer propyne/O_2/O ice. Thus, formally the reactivity we observe should be considered representative of this surface. However, since the propyne molecules are physisorbed at the surface, we can assume that their electronic structure is largely unperturbed upon adsorption, and thus the reactivity we observe should be broadly representative of physisorbed propyne, irrespective of the precise nature of the surface on which the molecule is adsorbed.

Molecules desorbed from the surface are ionized in the source region of the TOFMS by a pulsed beam of 200 eV electrons running with a duty cycle of 32 μs.

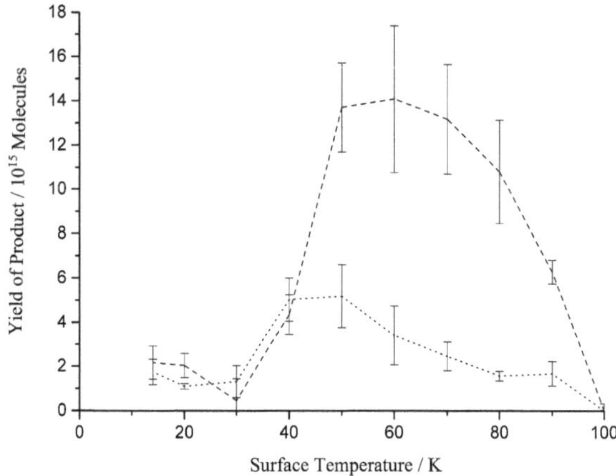

Fig. 3 Experimental yield of the single (dashed line) and double (dotted line) addition products following the co-deposition of propyne (C_3H_4) and O atoms. The error bars associated with the experimental results represent two standard deviations from four experiments at each surface temperature. The lines linking the points serve only to guide the eye.

The electron pulses have a duration of approximately 1 μs. After each pulse of electrons crosses the source, a voltage is applied to the repeller plate of the TOFMS to accelerate any ions formed in the spectrometer's source region towards the detector for analysis. The detector is a pair of multichannel plates (MCPs), the output from which is amplified and discriminated before being passed to a time-to-digital convertor (TDC). The TDC is triggered by the same pulse generator that controls the electron gun and the repeller plate voltage. Times from the TDC are recorded as a histogram of time-of-flight *versus* counts, *i.e.* a mass spectrum. During the TPD phase, as the temperature of the surface is slowly increased, a new mass spectrum is generated every second. At the end of each TPD experiment, these mass spectra are combined to generate a two-dimensional data set of ion intensity at each mass-to-charge (m/z) ratio against the substrate temperature during the heating process. This dataset can then be used to generate, for example, a mass spectrum for a particular temperature range or to see how the intensity of a particular ion (at a given m/z) varies as the surface temperature increases. In this work we recorded these TPD datasets after dosing propyne and O atoms onto the surface at a range of surface temperatures between 14 K and 100 K, performing four separate dosing/TPD measurements at each surface temperature investigated.

3.   Results

Whenever propyne and O atoms are allowed to react at surface temperatures below 100 K, we observe signals at mass-to-charge ratios (m/z) of 56 and 72 in the TOF mass spectra recorded during the TPD phase (Figure 1). These two signals possess different TPD profiles, with the signal at $m/z = 56$ appearing at lower desorption temperatures. Thus, it seems evident that these two mass spectral signals correspond to two different products. These two signals in the mass spectrum are consistent with the empirical formulae C_3H_4O and $C_3H_4O_2$ and so can clearly be identified with single and double addition of O atoms to propyne.

To determine the yield of each of these products, the total integrated mass spectral signals at $m/z = 56$ and $m/z = 72$ in each TPD measurement is determined. This procedure involves selecting the temperature range over which each product desorbs and summing the mass spectra in this temperature range for one TPD measurement to form an integrated mass spectrum. A background mass spectrum, recorded in an identical experiment carried out with the microwave discharge off, is then subtracted from this integrated mass spectrum. This background correction procedure confirms the product signals are a result of the oxygen atoms generated in the discharge. The intensities of the resulting mass spectral peaks are determined from this "corrected" integrated spectrum by peak fitting. This integration procedure is repeated for each separate experiment, and the results averaged for each surface temperature at which the dosing was carried out. The above integration procedure gives the relative ion signals at $m/z = 56$ and $m/z = 72$ in terms of ion counts at the MCP detector. To compare with the kinetic model, with which we interpret our results (see below), we require the product yields in terms of the number of molecules formed on the surface. To transform our mass spectral ion counts to these absolute units requires calibration of the mass spectral intensities. To carry out this transformation, as reported before,[8,9] we measure the mass spectrum resulting from a TPD dataset recorded following the adsorption of a known dose of propyne when the surface is held at 12 K. The integrated propyne mass spectral intensity in such a TPD dataset allows us to calculate the proportionality constant, the detection efficiency, between the dose of molecules and the mass spectral signal, assuming a sticking probability of 1 in the calibration experiment. Such a sticking coefficient is likely to be an excellent characterization of the interaction of propyne with the 12 K surface. The detection efficiency we measure for propyne must now be adjusted to represent the detection of the two product molecules. Since the ionization cross sections at 200 eV are not known for propyne and the two products (C_3H_4O and $C_3H_4O_2$), we must assume they are the same for all three species. Since we integrate just the parent ion signal in the mass spectrum, we must also adjust the detection efficiency to reflect the parent to fragment ion ratio in the mass spectrum of propyne and in the mass spectra of the product species. For the single addition product we used the parent to fragment ion ratio in a propanal mass spectrum measured in our apparatus. For the double addition product we estimated the fragment to parent ratio in the standard mass spectrum of methyl glyoxal from the NIST reference database.[49] The above procedure allows us to estimate the yield of the product ions on an absolute scale from our TPD spectra, and report the yield as a function of the dosing temperature.

We initially report the *total* product yield, the summed yields of the single and double addition products, as a function of dosing surface temperature (Figure 2). These experimental results show that below 30 K the yield of C_3H_4O gently increases with decreasing temperature. Above a surface temperature of 30 K the product yield increases to a maximum at 50 K and then decreases to zero at a surface temperature of 100 K. We also report the individual yields for the single addition product and the double addition product (Figure 3). Figure 3 shows that the same general trend with surface temperature is exhibited by the yields of both of the individual addition products as was described for the total product yield

(Figure 2). We also note the yields for the single and double addition products are very similar below a surface temperature of 40 K. Above a surface temperature of 40K the yield of the single addition product is much larger than that of the double addition product (Figure 3).

4   Data reduction

To extract estimates of kinetic parameters from our experimental product yields, we fit the experimental data with a simple kinetic model. This kinetic model has been described before in the literature,[8,9] but has also been extended here to model sequential addition of two oxygen atoms to propyne.

The kinetic model involves the two prototypical mechanisms by which the reaction of O atoms with propyne can proceed on the surface. Firstly, the LH mechanism where both reactants are adsorbed and thermalised on the substrate:

$$C_3H_4(ads) + O(ads) \rightarrow C_3H_4O(ads) \tag{1}$$

Secondly, the ER mechanism, where an adsorbed species undergoes reaction with a gas-phase partner:

$$C_3H_4(ads) + O(g) \rightarrow C_3H_4O(ads) \tag{2}$$

In our experiments, as we have described before,[8,9] the molecular species should have a significantly larger adsorption energy than the O atom. Thus, to simplify our model we only consider the form of ER reactivity where the O atom is the gas-phase reactant and the propyne molecule is adsorbed on the surface. This is because when both reactants are adsorbed on the surface, the rate for LH reactivity should dominate. Given the arguments above, the rates for the LH and ER mechanisms are therefore:

$$r_{LH} = k_{LH} \, [C_3H_4(ads)] \, [O(ads)] \tag{3}$$

$$r_{ER} = k_{ER} \, [C_3H_4(ads)] \, F_O \tag{4}$$

where r_i is the rate of reaction and k_i is the rate coefficient. In these equations i denotes the LH or ER mechanism, $[C_3H_4(ads)]$ and $[O(ads)]$ are the surface concentrations (in molecules cm^{-2}) of the reactants absorbed on the surface and F_O is the deposition fluence of oxygen atoms.

At the low surface temperatures employed in this study, and given the inertness of the surface, it is safe to assume that the reactants are physisorbed and, as physisorption is dominant, there is no limit on the number of accessible adsorption sites. That is, once the first layer of molecular ice, comprising reactants and products, is deposited a second layer will then readily form on top of the first layer, eventually building up a multi-layer ice. In our model, we assume only the uppermost monolayer is accessible to the incident reactants, that is, we assume the reactants radicals cannot penetrate into the ice. Thus, the maximum surface coverage of each species is constrained to 10^{15} molecules cm^{-2} in our model,[8,9] the standard value for an accessible monolayer.

The temperature variation of the rate coefficients for the two surface reactions can be described by an Arrhenius expression:

$$k_i = A_i \exp(-E_i/RT) \tag{5}$$

where i again denotes the LH or ER mechanism. In equation 5, A_i is a pre-exponential factor and E_i is the activation energy of the pathway. To simplify the modelling and reduce the number of free parameters, the activation energy is assumed to be the same for both the LH and ER mechanisms.

Since both LH and ER reactions can occur simultaneously, the overall rate r of formation of the single addition product is given by:

$$r = k_{LH} \, [C_3H_4(ads)] \, [O(ads)] + k_{ER} \, [C_3H_4(ads)] \, F_o \tag{6}$$

As we see in Eq. 6, the reaction rate is dependent upon the surface concentration of propyne and oxygen atoms. For propyne the three factors that control the surface concentration are given in Equation 7:

$$\frac{d[C_3H_4]}{dt} = F_{propyne} - r - r_{Des,propyne} \tag{7}$$

In equation 7, $F_{propyne}$ is the flux of propyne onto the substrate and $r_{Des,propyne}$ is the rate of propyne desorption from the surface. The surface concentration of propyne can be found by integrating Equation 7 with respect to deposition time t. As described before, the propyne flux, $F_{propyne}$, is estimated experimentally from the propyne flow rate and the pressure/pumping characteristics of the vacuum chamber.[33,35] The depletion of propyne due to the reaction r is calculated using equation 6. Finally, the desorption rate of propyne, $r_{Des,propyne}$, can be evaluated using a second Arrhenius equation at each surface temperature:

$$r_{Des,propyne} = A_{Des,propyne} \exp(-E_{Des,propyne}/RT) \, [C_3H_4(ads)] \tag{9}$$

where $A_{Des,propyne}$ is the pre-exponential factor for the desorption of propyne and $E_{Des,propyne}$ is the desorption energy for propyne. As discussed before, since the desorption of the multi-layer ice in our experiment just reveals another layer of the ice for reactions, only desorption in the monolayer regime will affect the surface concentrations and hence first-order desorption kinetics are used in Equation (9).[33,35]

The concentration of oxygen atoms is evaluated analogously to the concentration of propyne, allowing for the dissociation efficiency of the source.[33,35] Desorption of the products of the reactions are not considered in our model. Such a simplification is justifiable as the products from the surface reaction between propyne and oxygen are heavier and more polar than the reactants and so should have larger desorption energies. This conclusion is confirmed by the TPD profiles we observe for the products.

As discussed above, the different TPD profiles of the two products, $m/z = 56$ and $m/z = 72$, confirm that the single addition signal, $m/z = 56$, is from a separate product to that responsible for the double addition signal at $m/z = 72$. That is, the signal at $m/z = 56$ is clearly not a mass spectral fragment of the species at $m/z = 72$. Given that the reaction conditions rule out a concerted three-body reaction, it seems clear that the reaction that forms the double addition product occurs in two steps. Firstly, a single oxygen atom adds to propyne to form a C_3H_4O species. Then, in the second step, a second oxygen atom adds to the C_3H_4O to form the double addition product ($C_3H_4O_2$). Hence, all the double addition product we detect was at some point during the reaction a single addition product. Thus, to extract the kinetic parameters for the initial single addition step, we fit our kinetic model to the summed experimental yields of the single and double addition products.

Some preliminary considerations need to be addressed before fitting the kinetic model to the experimental data (Figure 2) in order to extract the kinetic parameters. The experimental yield curves we determined in previous investigations of the reactions of O displayed unambiguous evidence of the parallel operation of both the LH and ER mechanisms.[9] Specifically, a clear peak in the yield was observed at surface temperatures just below the desorption temperature of the O atoms. At higher surface temperatures another peak in the yield was observed due to efficient operation of the ER pathway. The experimental yield profile for O + propyne (Figure 2) does not show this distinct double peaked structure. However it is apparent that the LH mechanism is still a significant contributor to the yield of C_3H_4O, since we see a sharp drop in the yield at surface temperatures where the O atoms no longer stick efficiently to the surface, above approximately 50K. Our kinetic model does not give a good fit to the experimental data using the LH mechanism alone since above 50 K, where the O atoms no longer stick efficiently, there is still a significant yield of C_3H_4O. Therefore, although there is no observed double peaked structure in the experimental yield, both the LH and ER processes appear to be contributing to the product yield. Thus we have used both the LH and ER reactions in our kinetic model to fit the experimental data.

To model the yield of C_3H_4O at a given surface temperature, we numerically integrate Equation 6 for a time period equal to the experimental dosing period. The total number of product molecules formed can then be compared with the experimental data. To achieve a fit with the experimental data, we can vary the pre-exponential factors and the activation energy for the chemical reaction and also the desorption energies of the reactants. Again to constrain the number of free parameters, the pre-exponential factors for the desorption of the two reactants from the surface are kept fixed. Usually such pre-exponential factors ($A_{\text{Des,O}}$ and $A_{\text{Des,propyne}}$) are taken as the vibrational frequency of the adsorbate–surface bond. In previous work we employed a value for $A_{\text{Des,O}}$ of 3.10×10^{12} s^{-1} and we employ this value again here.[9,50] To the best of our knowledge there are no available surface vibrational frequencies for propyne, so for $A_{\text{Des,propyne}}$ we employ the value of 2.33×10^{12} s^{-1}, which is that calculated for an acetylene-graphite bond. [9,51]

We first attempted to fit the kinetic model to the experimental data, constraining the pre-exponential factors for the LH and ER mechanisms to be equal ($A_{\text{LH}} = A_{\text{ER}}$) and also constraining the activation energies ($E_{\text{LH}} = E_{\text{ER}}$) for both mechanisms to be equal. Under these constraints the fit to the experimental data is poor. From this poor fit of the highly constrained model we concluded that we needed a larger rate coefficient for the ER mechanism than for the LH mechanism. If the activation energy for surface diffusion of the O atoms is not significant, we would expect the activation energies for the LH and ER mechanisms to be equal. Thus, to allow a larger ER rate coefficient in the model we permit A_{ER} to take a different value to A_{LH}. This additional degree of freedom allows us to achieve a much better, although not perfect, fit to the experimental data (Figure 2) as discussed below. In attempting this fitting it is clear that the barrier for the LH

Table 1 Kinetic parameters characterizing the single addition of O atoms to propyne, to form the single addition product C_3H_4O, as a function of surface temperature. As discussed in the text, these parameters have been extracted by fitting a kinetic model to the experimental data we record for this surface reaction

(E_i/R) /K i = LH or ER	$10^{16}A_{LH}$/ cm^2molecule^{-1}s^{-1}	$10^{16}A_{ER}$/ cm^2molecule^{-1} s^{-1}	$E_{Des,propyne}$/ kJ mol^{-1}	$E_{Des,O}$/ kJ mol^{-1}
160 ± 10	0.95 ± 0.2	2.45 ± 0.4	20.8 ± 0.3	14 ± 2

mechanism is well constrained by the form of the data to 160 ± 10 K. Similarly, the fitting shows the propyne binding energy is also well constrained by the experimental data; we comment further on the O atom binding energy below. The desorption energies for propyne and oxygen are found to be 20.8 ± 0.3 kJ mol^{-1} and 14.0 ± 1.2 kJ mol^{-1} respectively.

The parameters for this satisfactory fit are reported in Table 1 and the fit is shown graphically in Figure 2. As is clear in Figure 2, the reported fit still does not satisfactorily reproduce experimental yield at $T = 70$ K. Possible physical explanations for this poor fit at a surface temperature of 70 K include the presence of more than one oxygen atom binding site, as discussed further below.

As mentioned earlier, we have also extended our model to attempt to extract kinetic parameters for the double addition reaction. This extension involves the addition of two further rate equations for the formation of the double addition product from the single addition product:

$$r'_{LH} = k'_{LH} [C_3H_4O(ads)] [O(ads)] \tag{10}$$

$$r'_{ER} = k'_{ER} [C_3H_4O (ads)] F_O \tag{11}$$

Again, we integrate these equations to derive the yield of double addition product at the different surface temperatures, allowing the single addition product to form with the kinetic parameters listed in Table 1. Under these constraints we find that to fit the yield of the double addition product (Figure 3) requires a negative activation energy (-60 ± 20 K).

5   Discussion

The oxygen atom desorption energy which gives the best fit to the experimental data in our kinetic model is 15.3 kJ mol^{-1}. The interaction energy of an oxygen atom and pyrene (representative of a bridge site in graphite) is calculated to be 11.6 kJ mol^{-1}.[50] As discussed above, in our experiments the graphite substrate is saturated after a matter of seconds during dosing. This means that although our substrate is graphite, the majority of the oxygen atoms are interacting with a propyne–oxygen matrix. Previous experimental work has shown that the desorption energy of an oxygen atom from an ethene–oxygen matrix and from a propene–oxygen matrix is, similarly to the oxygen–graphite system, about 12 kJ mol^{-1}.[9] The oxygen atom desorption energy which best fits our data is markedly larger than these previous values for the binding of oxygen atoms to small organic molecules. If we use a value of 12.9 kJ mol^{-1} in our simulation (Figure 4), we

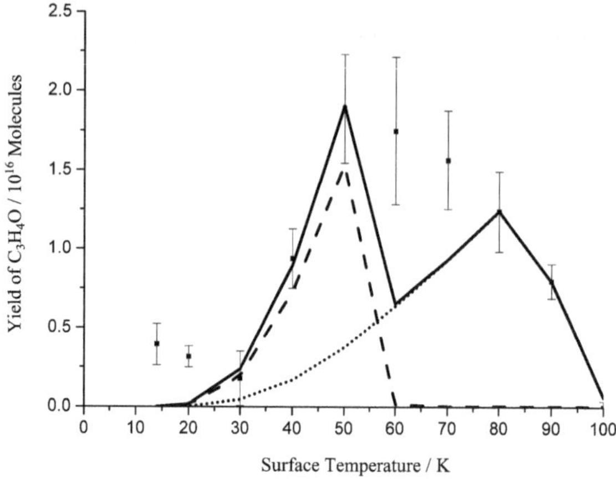

Fig. 4 Summed yield of the single and double addition product, formed following the co-deposition of propyne (C_3H_4) and O atoms, as a function of surface temperature. Squares: experimental data; solid line: model; dash LH mechanism; dot ER mechanism. The error bars associated with the experimental results represent two standard deviations from four repeats at each surface temperature. The kinetic model used to derive the fits shown employs the paramenters listed in Table 1 and an O atom desorption energy of 12.9 kJ mol^{-1}.

satisfactorily reproduce the rise in the yield of the product over the surface temperature range from 30 K to 50 K, but a larger desorption energy, or a component of the oxygen atoms possessing a larger desorption energy, is required to fit the data point at a surface temperature of 60 K (Figure 2). From this analysis it appears that the experimental data might be best represented with at least two desorption energies for the oxygen atoms. Given the consequent increase in the number of free parameters, we are reluctant to use such a desorption energy distribution for the oxygen atoms as we feel we risk over-interpreting our experimental data. We simply report that the oxygen atom desorption energy is not that well defined by the form of our experimental data, and is best represented by a value of 14 ± 2 kJ mol^{-1}, this value perhaps representing a distribution of desorption energies.

The propyne desorption energy extracted using our kinetic model is 20.8 ± 0.3 kJ mol^{-1}. To our knowledge there are no literature values for the desorption energy of propyne from interstellar ice analogues or a graphite surface. The value of this desorption energy agrees nicely with the observed trend that an alkyne's desorption energy is smaller than that of its alkene equivalent.[51] Specifically, the work of Rubes et al.[51] shows that ethene's desorption energy from a graphite (0001) surface is 17.36 ± 0.0361 kJ mol^{-1} whilst acetylene's desorption energy from the same surface is 14.67 ± 0.0273 kJ mol^{-1}. In agreement with this trend, previous work has shown that the desorption energy of propene from propene ice is 21.4 ± 0.3 kJ mol^{-1},[9] a value that is slightly larger than the equivalent value we extract here for the desorption of propyne.

Our modelling yields a value for the activation energy of single O atom addition to propyne of 160 ± 10 K, a value larger than the comparable reaction of O atoms with propene (145 ± 10 K).[9] Such an increase in the activation energy is in accord with chemical intuition, as the reactive C–C bond in propyne is stronger than that in propene. Such a trend is also supported by the fact we observe no reaction between O atoms and acetylene at a range of surface temperatures between 14 K and 100 K. Previous work has shown that the activation energy (for the LH mechanism) for the reaction between ethene and O atoms is 190 ± 45 K.[9] The absence of any signals in our experiments for the addition of O atoms to acetylene indicates the barrier for this process is significantly above 200 K. Such a conclusion is in accord with the trend of a rise in the activation energy, for the reaction with O atoms, on changing the reactant from the alkene to the corresponding alkyne.

More generally, our results show that, under our surface conditions, the methyl substituted organic reactant (propene and propyne) is more reactive with oxygen atoms that the corresponding smaller molecule (ethene and acetylene). A similar trend is observed in combustion chemistry[43] where the activation energy for the reaction of O atoms with propyne is smaller than that for acetylene.

The pre-exponential factors we extract from the fit of the kinetic model to the experimental data are $9.8 \pm 0.2 \times 10^{-17}$ cm^1 molecule^{-1} s^{-1} and $2.45 \pm 0.4 \times 10^{-16}$ cm^1 molecule^{-1} s^{-1} for the LH and ER reactions respectively. These pre-exponential factors are approximately a factor of 10 smaller than we report for the reaction between oxygen atoms and alkenes.[9] This trend in pre-exponential factors is also present in the gas phase, where the pre-exponential factors for the reaction of O atoms with alkynes are smaller than those for the reaction with alkenes.[38,52] As encompassed by these pre-exponential factors, and discussed above, to fit our experimental data (Figure 2), the rate coefficient for the ER reaction must be larger than the rate coefficient for the LH process. One reason for this difference might be the fact that in the ER process the O atoms are not thermalized with the surface. Indeed, in our experiment the O atoms will possess a kinetic energy distribution representative of 300 K. One way to test such a hypothesis experimentally would be to cool the incident O atom beam, but the methodology for such cooling is not well established experimentally. A greater rate constant for the ER mechanism, in comparison with the LH mechanism, has also been reported for the reactions between O atoms and O_2 on cold surfaces and again assigned to non-thermal O atoms arriving at the surface.[36]

As discussed above, at surface temperatures below 30 K the yield of the single and double addition products increases as the surface temperature decreases.

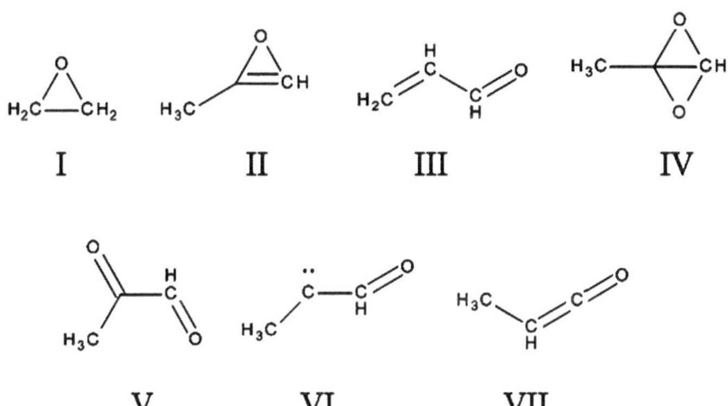

Fig. 5 Organic structures relevant to the discussion of the reaction of O atoms with propyne.

Our simple kinetic model does not account for the increase in the product yield at temperatures below 30 K. A similar reactive pathway, with an increasing yield at low temperatures, has been observed before for reactions of O atoms with CS_2.[8] Reactivity at such surface temperatures is particularly pertinent to the ISM. Since the rate of this low temperature channel rises with falling surface temperature, it could be that this low temperature process is effectively barrierless. Since this low temperature reactivity is only present between surface temperatures of 14 K and 30 K, performing any sort of kinetic fit for our data is not statistically meaningful. Hence we simply report that the rate coefficient at 14 K for the formation of the single addition product is $(6 \pm 2) \times 10^{-19}$ cm^1 molecule^{-1} s^{-1}. We note that this low temperature rate is at least comparable to that observed for the reaction of O atoms with CS_2,[8] and that tunneling processes have been recently reported for O atoms at low surface temperatures.[36]

Our key finding for the addition of a second oxygen atom to the C_3H_4O single addition product is that this is a "barrierless" process. Such a conclusion is not unexpected since the initial addition of an oxygen atom to propyne is likely to initially result in a radical species. The subsequent reaction of this radical with a further oxygen atom might be expected to exhibit a negative activation energy.[53]

As shown in Figure 3, the yields of the single (C_3H_4O) and double ($C_3H_4O_2$) addition products are similar at surface temperatures below 40 K. Above 40 K there is a dramatic increase in the yield of the single addition product with respect to the double addition product. The observed increase in the yield of the single addition product at surface temperatures above 40 K perhaps implies that the reactive intermediate resulting from single addition isomerizes above 40 K to yield a more stable isomer of C_3H_4O, an isomer that is less susceptible to attack by a second oxygen atom.

Our previous work has shown that the surface reaction between ethene (C_2H_4) and O atoms results in the formation of ethylene oxide (Structure I, Figure 5).[9] Similarly it is reasonable to propose that the first intermediate formed by the reaction of O atoms with propyne is an oxirene (II, Figure 5). We would expect this oxirene to be reactive, but potentially to be stabilized on a low temperature surface. Such an interpretation agrees nicely with our deduction above that at surface temperatures below 40 K we have a reactive single addition product on the surface, but above a surface temperature of 40 K this reactive single addition product isomerizes to a less reactive species. There are many (10) possible isomerisation products for the reactive oxirene, but several of these would also be highly reactive. However, one stable molecule which can be formed from the isomerisation of the oxirene is propenal (III, Figure 5), the isomerisation occurring *via* a pathway involving a carbene.[54] It has been shown that the rearrangement of an oxirene to a ketocarbene is barrierless.[55] Indeed, the first investigation of the gas-phase reactivity of O atoms with propyne concluded that, if energy could be efficiently lost from the initial addition product, the ketocarbene (VI Figure 5) formed from the oxirene will relax to methyl ketene (VII Figure 5) or to propenal (III Figure 5).[38]

The structure of the double addition product is more speculative. An obvious candidate for the initial product of the second addition is the bicyclic structure IV (Figure 5). This strained intermediate might be expected to be very reactive and will probably rapidly rearrange. There are a number of possible rearranged structures for this primary bicyclic adduct, with one stable rearrangement product being methyl glyoxal (V Figure 5).

6　Astrophysical implications

Our experiments show that on interstellar dust grain analogues, under UHV conditions and at low temperature, oxygen atoms can readily add to the carbon–carbon triple bond of a propyne molecule. The reaction is detectable at surface temperatures of 14 K and above, and is most efficient at 50 K. We estimate our experiments to be the equivalent of approximately 10^5–10^6 years of exposure to O atoms in an interstellar cloud.[9,33,35] It is therefore possible that this pathway is active in interstellar clouds. In addition, given the observed kinetics, the yield from the reaction of propyne with oxygen atoms may also be significant when the cloud begins to warm.[56]

Our experiments reveal a marked difference in the heterogeneous reactivity of acetylene and propyne on low temperature surfaces, a difference in reactivity that may be pertinent in the ISM. Indeed, we note that the reactions of molecular ices with incident oxygen atoms are not well represented in recent descriptions of gas-grain reaction networks.[57]

As discussed in the Introduction, propyne has been observed in a variety of environments in the ISM, and in these environments the surface chemistry we report in this article may be relevant. We also previously discussed the need for an O atom sink in interstellar dust clouds, to account for the observed gas-phase depletion. We postulate that the stable molecule resulting from the single addition of O atoms to propyne on a cold surface is propenal (III Figure 5). This reaction, along with other O atom reactions, may contribute to that O atom sink. Propenal (C_3H_4O) has been observed in the ISM towards the star forming region Sagittarius B2(N).[58] Hollis *et al.* postulate that the formation of interstellar propenal is due to hydrogen addition to propynal.[58] However, the work reported in this paper suggests that, in addition, the surface chemistry of propyne may also be a source of propenal. Clearly modelling work is required to support this suggestion. Encouragingly, propyne has also been detected in Sagittarius B2[59] in accord with the idea that propenal may be formed from the heterogeneous oxidation of propyne by oxygen atoms.

To the best of our knowledge no molecule with the empirical formula of the double addition product ($C_3H_4O_2$) has been observed in the interstellar medium. However, given the observed facility of the double addition process, our experimental data would suggest that at least one isomer of $C_3H_4O_2$, perhaps methylglyoxal, could be present during the warm-up phase of an interstellar cloud where propyne ice has been thermally processed by oxygen atoms.

7　Conclusions

This paper reports the first laboratory investigation of the heterogeneous reaction of propyne with oxygen atoms at temperatures relevant to the ISM. Our data shows this reaction, which initially forms an adduct due to single addition (C_3H_4O) proceeds efficiently at surface temperatures below 100 K. As the surface temperature decreases from 100 K the yield of the reaction increases to reach a maximum at 50 K and then falls to a minimum at 30K. Modelling the yield of the reaction at surface temperatures above 30 K allows the extraction of an activation energy for the reaction of 160 $\pm$ 10 K. We also detect a product corresponding to the addition of two oxygen atoms to propyne. This double-addition product appears to form in a sequential reaction of the single addition product with another oxygen atom. The addition of this second oxygen atom to the single addition product appears to be barrierless. We propose structures for the intermediates and products of the O + propyne reaction which are consistent with our observations and the known behaviour of these organic compounds. We postulate that the addition of oxygen atoms to propyne to form propenal could act as a source of this molecule in the ISM.

Acknowledgements

We would like to thank University College London and the Max-Planck Institute for Astronomy for support of HJK. We also acknowledge support from the European Community's Seventh Framework Programme (FP7/2007–2013) under the LASSIE training network. We also gratefully acknowledge very helpful discussions with Eric Herbst, Serena Viti, Wendy Brown, Michael Ward and Angela Occhiogrosso.

8 References

1 S. Kwok and Y. Zhang, *Nature*, 2011, **479**, 80–83.

2 Y. J. Kuan, S. B. Charnley, H. C. Huang, W. L. Tseng and Z. Kisiel, *Astrophys. J.*, 2003, **593**, 848–867.

3 A. G. G. M. Tielens, *The Physics and Chemistry of the Interstellar Medium*, Cambridge Univ. Press, Cambridge, 1st ed edn., 2005.

4 D. J. Burke and W. A. Brown, *Phys. Chem. Chem. Phys.*, 2010, **12**, 5947–5969.

5 L. R. Nittler and C. M. O. D. Alexander, *Highlights of Astronomy*, 2006, **14**, 17–20.

6 D. A. Williams and E. Herbst, *Surf. Sci.*, 2002, **500**, 823–837.

7 H. J. Fraser, M. R. S. McCoustra and D. A. Williams, *Astron. Geophys.*, 2002, **43**, 10–18.

8 M. D. Ward, I. A. Hogg and S. D. Price, *Mon. Not. R. Astron. Soc.*, 2012, **425**, 1264–1269.

9 M. D. Ward and S. D. Price, *Astrophysical Journal*, 2011, 741.

10 B. E. Turner, E. Herbst and R. Terzieva, *Astrophys. J. Suppl.*, 2000, **126**, 427–460.

11 S. Martin, R. Mauersberger, J. Martin-Pintado, C. Henkel and S. Garcia-Burillo, *Astrophys. J. Suppl.*, 2006, **164**, 450–476.

12 E. Churchwell and J. M. Hollis, *Astrophys. J.*, 1983, **272**, 591–608.

13 E. Herbst and E. F. van Dishoeck, in *Annual Review of Astronomy and Astrophysics, Vol 47*, eds. R. Blandford, J.Kormendy and E. VanDishoeck, Annual Reviews, Palo Alto, 2009, vol. 47, pp. 427–480.

14 A. Fuente, S. Garcia-Burillo, M. Gerin, D. Teyssier, A. Usero, J. R. Rizzo and P. de Vicente, *Astrophys. J.*, 2005, **619**, L155–L158.

15 R. Mauersberger, C. Henkel, C. M. Walmsley, L. J. Sage and T. Wiklind, *Astronomy & Astrophysics*, 1991, **247**, 307–314.

16 S. Maity and R. I. Kaiser, *Astrophysical Journal*, 2013, 773.

17 K. Hiraoka, N. Mochizuki and A. Wada, in *Astrochemistry: From Laboratory Studies to Astronomical Observations*, eds. R. I. Kaiser, P. Bernath, Y. Osamura, S. Petrie and A. M. Mebel, 2006, vol. 855, pp. 86–99.

18 T. Kasamatsu, T. Kaneko, T. Saito and K. Kobayashi, *Bull. Chem. Soc. Jpn.*, 1997, **70**, 1021–1026.

19 Y. Dede and I. Ozkan, *Phys. Chem. Chem. Phys.*, 2012, **14**, 2326–2332.

20 Y. Takano, H. Masuda, T. Kaneko and K. Kobayashi, *Chem. Lett.*, 2002, 986–987, DOI: 10.1246/cl.2002.986.

21 T. Fouchet, E. Lellouch, B. Bezard, H. Feuchtgruber, P. Drossart and T. Encrenaz, *Astronomy and Astrophysics*, 2000, **355**, L13–L17.

22 T. deGraauw, H. Feuchtgruber, B. Bezard, P. Drossart, T. Encrenaz, D. A. Beintema, M. Griffin, A. Heras, M. Kessler, K. Leech, E. Lellouch, P. Morris, P. R. Roelfsema, M. RoosSerote, A. Salama, B. Vandenbussche, E. A. Valentijn, G. R. Davis and D. A. Naylor, *Astronomy and Astrophysics*, 1997, **321**, L13–L16.

23 Y. Alibert, O. Mousis and W. Benz, *Astrophys. J.*, 2005, **622**, L145–L148.

24 Y. Alibert, O. Mousis, C. Mordasini and W. Benz, *Astrophys. J.*, 2005, **626**, L57–L60.

25 D. Gautier, F. Hersant, O. Mousis and J. I. Lunine, *Astrophys. J.*, 2001, **559**, L183–L183.

26 D. Gautier, F. Hersant, O. Mousis and J. I. Lunine, *Astrophys. J.*, 2001, **550**, L227–L230.

27 F. Hersant, D. Gautier and J. I. Lunine, *Planet. Space Sci.*, 2004, **52**, 623–641.

28 O. Mousis, Y. Alibert and W. Benz, *Astron. Astrophys.*, 2006, **449**, 411–415.

29 M. Burgdorf, G. Orton, J. van Cleve, V. Meadows and J. Houck, *Icarus*, 2006, **184**, 634–637.

30 A. Coustenis, B. Bezard, D. Gautier, A. Marten and R. Samuelson, *Icarus*, 1991, **89**, 152–167.

31 N. A. Teanby, P. G. J. Irwin, R. de Kok, A. Jolly, B. Bezard, C. A. Nixon and S. B. Calcutt, *Icarus*, 2009, **202**, 620–631.

32 S. I. B. Cartledge, J. T. Lauroesch, D. M. Meyer and U. J. Sofia, *Astrophys. J.*, 2004, **613**, 1037–1048.

33 U. Hincelin, V. Wakelam, F. Hersant, S. Guilloteau, J. C. Loison, P. Honvault and J. Troe, *Astronomy & Astrophysics*, 2011, 530.

34 D. C. B. Whittet, *Astrophys. J.*, 2010, **710**, 1009–1016.

35 E. B. Jenkins, *Astrophys. J.*, 2009, **700**, 1299–1348.

36 M. Minissale, E. Congiu, S. Baouche, H. Chaabouni, A. Moudens, F. Dulieu, M. Accolla, S. Cazaux, G. Manico and V. Pirronello, *Physical Review Letters*, 2013, 111.

37 S. L. Zhao, W. Q. Wu, H. M. Zhao, H. Wang, C. F. Yang, K. H. Liu and H. M. Su, *J. Phys. Chem. A*, 2009, **113**, 23–34.

38 C. A. Arrington and D. J. Cox, *J. Phys. Chem.*, 1975, **79**, 2584–2586.

39 G. Y. Adusei, A. S. Blue and A. Fontijn, *J. Phys. Chem.*, 1996, **100**, 16921–16924.

40 E. N. Aleksandrov, I. V. Dubrovina and S. N. Kozlov, *Kinet. Catal.*, 1981, **22**, 394–396.

41 J. M. Brown and B. A. Thrush, *Trans. Faraday Soc.*, 1967, **63**, 630.

42 K. H. Homann and C. Wellmann, *Ber. Bunsen-Ges. Phys. Chem.*, 1983, **87**, 527–532.

43 J. R. Kanofsky, D. Lucas, F. Pruss and D. Gutman, *J. Phys. Chem.*, 1974, **78**, 311–316.

44 G. Q. Xing, X. Huang, X. B. Wang and R. Bersohn, *J. Chem. Phys.*, 1996, **105**, 488–495.

45 M. E. Umstead, R. G. Shortridge and M. C. Lin, *Chem. Phys.*, 1977, **20**, 271–276.

46 H. E. Avery and S. J. Heath, *J. Chem. Soc., Faraday Trans. 1*, 1972, **68**, 512.

47 H. p. Schreibe and A. g. Mackinno, *Can. J. Chem.*, 1968, **46**, 1033.

48 M. Hawkins and L. Andrews, *J. Am. Chem. Soc.*, 1983, **105**, 2523–2530.

49 N. M. S. D. Center and D. S. E. Stein, in *NIST Chemistry WebBook, NIST Standard Reference Database Number 69, Eds. P.J. Linstrom and W.G. Mallard*, National Institute of Standards and Technology, Gaithersburg MD, 20899, http://webbook.nist.gov, (retrieved December 4, 2013).

50 H. Bergeron, N. Rougeau, V. Sidis, M. Sizun, D. Teillet-Billy and F. Aguillon, *J. Phys. Chem. A*, 2008, **112**, 11921–11930.

51 M. Rubes, J. Kysilka, P. Nachtigall and O. Bludsky, *Phys. Chem. Chem. Phys.*, 2010, **12**, 6438–6444.

52 R. Atkinson and R. j. Cvetanov, *J. Chem. Phys.*, 1972, **56**, 432–437.

53 I. R. Sims and I. W. M. Smith, *Annu. Rev. Phys. Chem.*, 1995, **46**, 109–137.

54 T. S. S. Rao and S. Awasthi, *THEOCHEM*, 2008, **848**, 98–113.

55 J. P. Toscano, M. S. Platz and V. Nikolaev, *J. Am. Chem. Soc.*, 1995, **117**, 4712–4713.

56 A. Occhiogrosso, S. Viti, M. D. Ward and S. D. Price, *Mon. Not. R. Astron. Soc.*, 2012, **427**, 2450–2456.

57 P. Theule, F. Duvernay, G. Danger, F. Borget, J. B. Bossa, V. Vinogradoff, F. Mispelaer and T. Chiavassa, *Adv. Space Res.*, 2013, **52**, 1567–1579.

58 J. M. Hollis, P. R. Jewell, F. J. Lovas, A. Remijan and H. Mollendal, *Astrophys. J.*, 2004, **610**, L21–L24.

59 S. L. E. D. Buhl, *Molecules in the Galactic Environment*, Wiley-Interscience, New York, 1973.

PAPER

Reaction kinetics and isotope effect of water formation by the surface reaction of solid H_2O_2 with H atoms at low temperatures

Yasuhiro Oba,* Kazuya Osaka, Naoki Watanabe, Takeshi Chigai and Akira Kouchi

Received 26th November 2013, Accepted 17th January 2014

DOI: 10.1039/c3fd00112a

We performed laboratory experiments on the formation of water and its isotopologues by surface reactions of hydrogen peroxide (H_2O_2) with hydrogen (H) atoms and their deuterated counterparts (D_2O_2, D) at 10–30 K. High-purity H_2O_2 (>95%) was prepared *in situ* by the codeposition of molecular oxygen and H atoms at relatively high temperatures (45–50 K). We determined that the high-purity H_2O_2 solid reacts with both H and deuterium (D) atoms at 10–30 K despite the large activation barriers (~2000 K). Moreover, the reaction rate for H atoms is approximately 45 times faster than that for D atoms at 15 K. Thus, the observed large isotope effect indicates that these reactions occurred through quantum tunneling. We propose that the observed HDO/H_2O ratio in molecular clouds might be a good tool for the estimation of the atomic D/H ratio in those environments.

Introduction

Water (H_2O) is the predominant solid constituent of icy layers of submicron-sized interstellar grains. Because of the potential importance of H_2O for chemical evolution in molecular clouds (MCs), elucidating the formation mechanism of H_2O in those environments is important. Although H_2O formation is possible by gas phase reactions at low temperatures,[1] the observed large abundance of H_2O cannot be explained only by the gas-phase synthesis.[2] Therefore, it is generally accepted that grain-surface reactions are crucial for producing H_2O in MCs.

Tielens and Hagen[3] proposed that H_2O formation is initiated by hydrogenation of atomic oxygen (O), molecular oxygen (O_2), and ozone (O_3), and is completed by the following reactions:

$$OH + H \rightarrow H_2O, \tag{1}$$

N19W8, Kita-ku, Sapporo, Hokkaido, Japan. E-mail: oba@lowtem.hokudai.ac.jp; Fax: +81 11 706 7142; Tel: +81 11 706 5477

$$OH + H_2 \rightarrow H_2O + H, \tag{2}$$

$$H_2O_2 + H \rightarrow H_2O + OH. \tag{3}$$

Since reaction (1) is a radical–radical reaction, it should proceed immediately once reactants encounter each other on the surface. This reaction has been studied experimentally by several groups.[4–7] In contrast to reaction (1), reactions (2) and (3) have large activation barriers (>2000 K) in the gas phase.[8,9] However, despite such a large barrier, these two reactions were proposed to contribute significantly to H_2O formation in dense MCs.[10] Since reactions having such large barriers do not occur thermally in MCs, reactions (2) and (3) require quantum tunneling. The quantum tunneling rate k_q is expressed by the following equation, assuming a rectangular activation barrier with a height E_a and width a:[11]

$$k_q \approx \nu_0 \exp[-(2a/\hbar)(2mE_a)^{1/2}], \tag{4}$$

where ν_0 and m represent the frequency of harmonic motion and the mass of the reaction, respectively. Since temperature is not included in the equation, k_q does not depend on the reaction temperature. Further details about quantum tunneling reactions have been described elsewhere.[11–14]

We have recently studied reaction (2) experimentally by the codeposition of nonenergetic OH with H_2 and isotopologues such as OD, HD, and D_2 on a substrate and determined that the reactions occur at 10 K.[14] In addition, significant isotope effects were observed, and reactions of OH and OD abstracting a D atom from HD and D_2 were approximately ten times slower than those abstracting an H atom from H_2 and HD. This isotope effect can be explained by the difference in the effective mass of tunneling reactions.[14]

A number of research studies have been conducted on reaction (3). In these previous studies, O_2 was used as an initial reactant rather than hydrogen peroxide (H_2O_2), which was produced by the successive hydrogenation of O_2 as follows:

$$O_2 + H \rightarrow HO_2, \tag{5}$$

$$HO_2 + H \rightarrow H_2O_2. \tag{6}$$

For example, Miyauchi et al.[15] exposed solid O_2 layers to H or D atoms at 10 K and determined that (i) the rate of O_2 hydrogenation (reaction (5)) is equal to that of O_2 deuteration, (ii) the rate of reaction (3) is slower than that of reaction (5), and (iii) the rate of reaction (3) is eight times faster than that of the following isotopically substituted reaction (7):

$$D_2O_2 + D \rightarrow D_2O + OD. \tag{7}$$

Miyauchi et al. considered that the rate difference between reactions (3) and (7) would be due to the isotope effect of quantum tunneling.[15] However, in the O_2 hydrogenation experiments, the formation of both H_2O_2 and H_2O occurs in the sample solid. Furthermore, the parent O_2 molecule is IR-inactive. Thus, multiparameter fittings, which often cause significant errors, are necessary to obtain the rates for reactions (3) and (5). Moreover, recent studies suggested that H_2O may form by another exothermic reaction in typical experimental conditions for O_2 hydrogenation:[16,17]

$$OH + OH \rightarrow H_2O + O. \qquad (8)$$

In addition to reaction (3), OH is expected to form by the following pathway:

$$HO_2 + H \rightarrow H_2O_2{}^* \rightarrow 2OH, \qquad (9)$$

where $H_2O_2{}^*$ is a reaction intermediate. The reaction of two OH also yields H_2O_2 on the surface:

$$OH + OH \rightarrow H_2O_2. \qquad (10)$$

The branching ratio of barrierless reactions (8) to (10) was determined experimentally[17] to be 1 to 4 at 40 K and theoretically[18] to be 1 to 9 on a cold substrate. In any case, reactions (8) and (10) may compete during O_2 hydrogenation experiments to some extent, making it more difficult to obtain reliable kinetic parameters for reaction (3). In the case of CO hydrogenation experiments, a similar problem occurred because both formaldehyde (H_2CO) and methanol (CH_3OH) were produced in a single experiment.[19-23] However, additional experiments using H_2CO as an initial reactant have enabled us to better understand the reaction kinetics and isotope effects of CO and H_2CO hydrogenation.[13,24] Similarly, the use of H_2O_2 as an initial reactant is desirable for studying the kinetics and isotope effects of reaction (3). However, because of difficulty in using pure H_2O_2, to date such an experiment has not been performed.

In the present study, we performed experimental studies on the formation of H_2O *via* reaction (3) and its isotope effect using high-purity (>95%) solid H_2O_2 and D_2O_2.

Experimental

Apparatus and experimental conditions for water formation

All experiments were performed using the Apparatus for SUrface Reaction in Astrophysics (ASURA) system. The ASURA primarily comprises a main chamber and an atomic source. An aluminum (Al) substrate was mounted at the center of the main chamber and all reactions were performed on the substrate at 10–30 K. Hydrogen (H) and deuterium (D) atoms were produced by the dissociation of H_2 and D_2 molecules, respectively, in a microwave discharge plasma, and were cooled by multiple interactions with the inner wall of the aluminum pipe, which was cooled to 100 K. We confirmed that the formed H and D atoms were well thermalized to the pipe temperature.[25] Further details of the ASURA have been described elsewhere.[25,26]

The fluxes of H and D atoms were not directly measured in the present experimental setup; they were estimated by comparing the effective rates of CO hydrogenation and deuteration with those reported by Hidaka *et al.*,[27] which were obtained under the same experimental conditions. Briefly, amorphous H_2O ice (a-H_2O) with a thickness of approximately ten monolayers (ML; $\sim 10^{15}$ molecules cm^{-2}) was produced by vapor deposition on the substrate at 15 K, followed by the deposition of CO with a thickness of ~ 0.8 ML. The column density of H_2O and CO was calculated from the peak area and the previously published band strengths, as described by Hidaka *et al.*[27] The band strengths for the CO stretching of CO and OH stretching of H_2O are 2.0×10^{-16} and 1.1×10^{-17} cm molecules^{-1}, respectively.[28] The obtained effective rates were 1.4 and 0.21 min^{-1} for the hydrogenation and deuteration of CO, respectively (see Results section below for the determination of effective rate constants). These values are a factor of 3.3 and 6.4 larger than the effective rates of CO hydrogenation and deuteration, respectively, compared to those reported by Hidaka *et al.*,[27] whose fluxes of both H and D atoms were 2.6×10^{14} atoms cm^{-2} s^{-1}. Assuming that the surface density of H and D atoms correlates linearly with their fluxes, the fluxes of H and D atoms in the present study were estimated to be 8.7×10^{14} and 1.7×10^{15} atoms cm^{-2} s^{-1}, respectively, which corresponds to a D and H atom flux ratio of $\sim$2. The variations in the H and D fluxes are expected to be less than 10% during and between each experiment.

Approximately 1 ML of solid H_2O_2 or its deuterated counterpart D_2O_2 was produced on the substrate by the procedures shown in the next section. H_2O_2 and D_2O_2 were exposed to H (D) atoms at 10–30 K. Reaction products were monitored *in situ* by a reflection absorption Fourier Transform Infrared (FTIR) spectrometer with a resolution of 4 cm^{-1} in the spectral range from 700 to 4000 cm^{-1}. The column density of H_2O_2 and D_2O_2 was calculated from the peak area and previously published band strengths.[15] The band strengths used were 2.1×10^{-17} and 1.5×10^{-17} cm molecule^{-1} for the OH and OD bending bands at 1385 and 1039 cm^{-1}, respectively.

Experiments were also performed on an amorphous D_2O ice (a-D_2O; $\sim$30 ML) vapor-deposited at 10 K. The band strength used was 1.3×10^{-16} cm molecule^{-1} for the OD stretching band.[15]

Preparation of high-purity solid hydrogen peroxide

In previous studies,[29,30] high-purity H_2O_2 (>97%) has been prepared by distilling commercially available H_2O_2 solution under vacuum. The distilled H_2O_2 was introduced into a reaction substrate through a transfer line made with nonreactive materials such as glass to avoid the catalytic decomposition of H_2O_2 on metal surfaces. However, in a typical apparatus, which comprises a stainless steel chamber and gas lines, it is not easy to obtain pure H_2O_2 using this procedure without significant modification.

Thus, we produced high-purity solid H_2O_2 *in situ* by the codeposition of H atoms with O_2 molecules on a substrate at relatively high temperatures. In our previous study, we determined that H_2O_2-rich ice tends to form by the O_2/H codeposition at high temperatures (>30 K) and with an increasing proportion of O_2 relative to H atoms.[31] For example, when O_2 and H were codeposited with an O_2/H ratio of $\sim$2 $\times$ 10^{-3} at 20 K, the main product was H_2O with a small amount of H_2O_2 (H_2O/H_2O_2 $\sim$ 5). When the same experiment was performed with an O_2/H ratio of $\sim$9 $\times$ 10^{-3} at 40 K, the main product was H_2O_2 with $\sim$15% contamination of H_2O. We further extended this experiment for the production of solid H_2O_2 with higher purity suitable for studying the kinetics and the isotope effect of reaction (3). One of the major advantages of this sample preparation method is that it is possible to study the isotope effect of reaction (3) using different isotopes (*i.e.*, H and D), which was not possible in previous studies.[15,32]

Gaseous O_2 was introduced into the main chamber through a capillary plate. From the pressure inside the main chamber, the O_2 flux was estimated to be 1.0×10^{14} molecules cm^{-2} s^{-1}, which is two to four orders of magnitude larger than that used by Oba *et al.*[31] The H atoms were codeposited with O_2 onto the substrate at 45–50 K. In the present experiments, pure H_2O_2 solid and the solid on a-D_2O were used as reactants. The amount of solid H_2O_2 was ~1 ML. After the formation of solid H_2O_2, the microwave was switched off, the supply of all gases (H_2 and O_2) was stopped, and the substrate temperature was raised to 70 K to remove residual O_2 from the sample solid. H_2O contamination in the solid H_2O_2 was <5%, which was confirmed by the IR spectrum of the product (Figure 1a). We believe that this small amount of contamination does not significantly impact the kinetics of reaction (3). We confirmed this by a temperature-programmed desorption experiment wherein little O_2 remained on the substrate after the sample treatment. The produced H_2O_2 was then cooled to the desired temperatures (10–30 K) for hydrogenation or deuteration experiments. Moreover, when D atoms were used, high-purity D_2O_2 was formed (Figure 1b).

Results

H_2O_2 + H and H_2O_2 + D

Figure 2 shows IR absorption spectra of solid H_2O_2 (top) and H_2O (bottom) for comparison and the difference spectra of 1 ML pure solid H_2O_2 after H atom exposure for up to 10 min at 15 K (middle). In the difference spectra, the peaks below and above the baseline represent decreases of initial reactant and increase of reaction products, respectively. With increasing H atom fluence, the peak intensity for the OH bending of H_2O_2 at 1385 cm^{-1} decreased, and new peaks appeared at 3000–3600, 2850, and 1660 cm^{-1}. The peak at 1660 cm^{-1} is attributable to the OH bending of H_2O formed by reaction (3). Based on the peak position of OH stretching bands of solid H_2O_2 and H_2O (Figure 2), the H_2O band at 3000–3600 cm^{-1} has a substantial overlap with that of H_2O_2, and the peak shape would be attributable to the sum of H_2O_2 decrease and H_2O increase. In addition, H_2O_2 has another strong peak at 2827 cm^{-1}, which is often assigned to the $v_2 + v_6$ combination band.[33] If the amount of H_2O_2 decreases after H atom exposure, the intensity of the peak should also decrease. However, a peak was observed slightly above the baseline at 2850 cm^{-1} after H atom exposure, unlike other peaks at ~3300 and 1385 cm^{-1} (Figure 2). These apparently contradictory observations will be discussed later in the Discussion section. Two small peaks

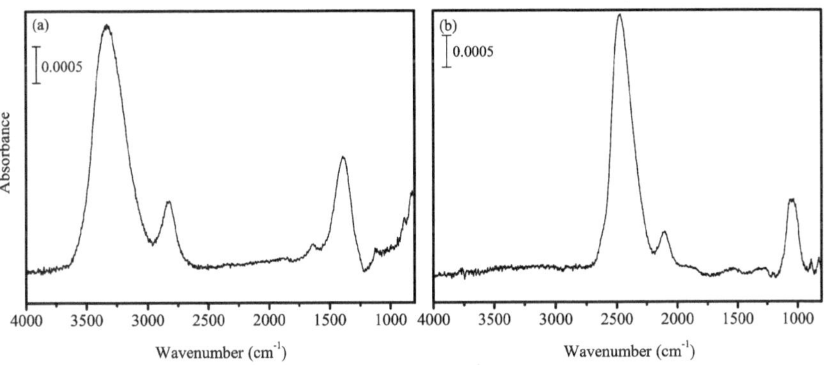

Fig. 1   FTIR spectra of (a) H_2O_2 and (b) D_2O_2 produced at 45 K.

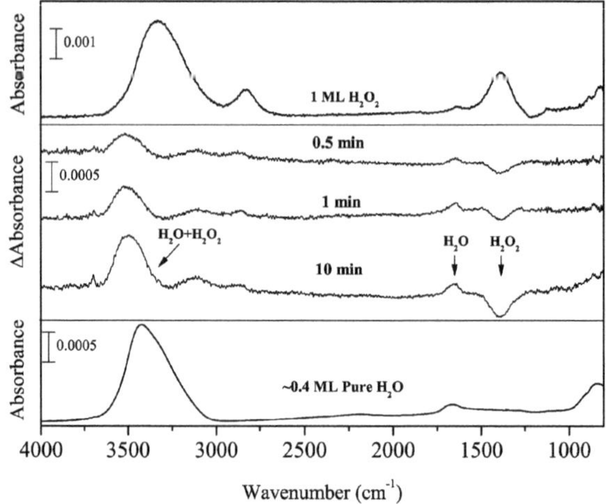

Fig. 2 FTIR spectra of H_2O_2 (top) and pure H_2O (bottom) and the variations in the difference spectra of H_2O_2 after exposure to H atoms for up to 10 min at 15 K (middle).

appeared at 3697 and 3721 cm^{-1} after H atom exposure (Figure 2), which are attributed to the 3- and 2-coordinated water molecules, respectively.[34]

Figure 3 shows the difference spectra of 1 ML H_2O_2 after D atom exposure for up to 120 min at 15 K. If H_2O_2 reacts with D atoms, HDO is expected to form as a main product by the following reaction:

$$H_2O_2 + D \rightarrow HDO + OH. \tag{11}$$

After D atom exposure to H_2O_2, the peak area for OH stretching and bending bands decreased. New peaks appeared at 3490, 2527, and 1492 cm^{-1}, which are

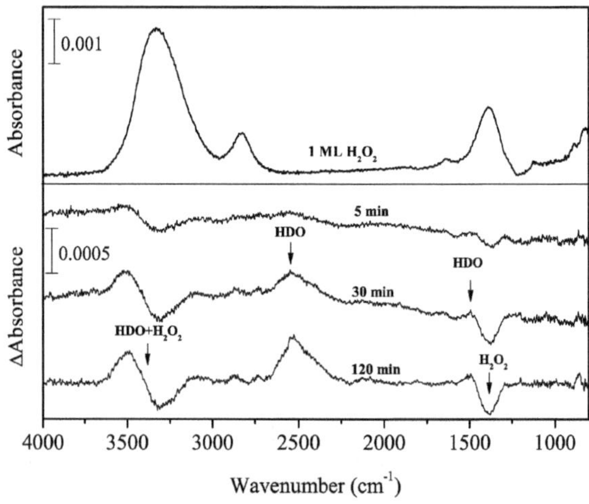

Fig. 3 Variations in the difference spectra of H_2O_2 after exposure to D atoms for up to 120 min at 15 K.

typically observed for solid HDO at low temperatures.[35,36] This result clearly indicates that HDO was formed by reaction (11) at 15 K. Although the peak area for OH stretching and bending bands decreased after D atom exposure, the $\nu_2 + \nu_6$ combination band at 2860 cm^{-1} increased slightly. Very small peaks appeared at 2722 and 3693 cm^{-1} after D atom exposure (Figure 3), the former of which is probably derived from the dangling OD bond of HDO.[37] The latter peak could be the dangling OH bond of HDO; however, the assignment is uncertain because of low S/N.

Figure 4 plots variations in the column density of solid H_2O_2 normalized to the unexposed initial amount after exposure to H or D atoms. We fitted the plots in Figure 4 to the following single-exponential decay function to obtain the kinetic parameters for reactions (3) and (11):

$$\Delta[H_2O_2]_t/[H_2O_2]_0 = A(e^{-k_n[X]t} - 1), \tag{12}$$

where A is a saturation value, t is the H or D atom exposure time, k_n is the rate constant of reaction (n), and [X] is the number density of X atoms (X = H or D) on the surface. Unfortunately, it is difficult to measure [X] in the present experiment; thus, the product $k_n[X]$ is obtained as a fitting parameter for equation (12). Hereafter, $k_n[X]$ is denoted as the effective rate constant k_n' in the present study. We assume that during exposure, [X] is independent of time and is governed mainly by the balance between the flux of X atoms, the sticking coefficient of the impinging atoms, and the loss of atoms by X–X recombination. We obtained $k_3' = 7.2 \times 10^{-1}$ and $k_{11}' = 3.2 \times 10^{-2}$ min^{-1} at 15 K, and the k_3'/k_{11}' ratio was 23.

In the present study, we do not determine the absolute yields of reaction products with the decrease in the column density of reactants, because the band strengths of the products (H_2O or HDO) have been reported only for a transmission method. Using these reported values may cause a large error (<50%)[31] when those band strengths are used in a reflection method. However, it does not

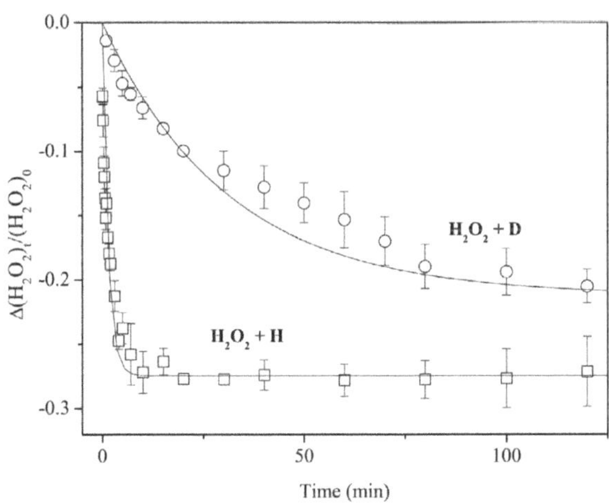

Fig. 4 Variations in the column density of H_2O_2 normalized to the initial amount as a function of exposure time of H or D atoms at 15 K. Solid lines are single-exponential decay fits to the plots.

affect the values of effective rate constants because the term of band strength is not included in equation (12). In addition, a portion of the reaction products could desorb from the substrate upon formation,[38,39] making the interpretation of the yields of products more difficult.

$D_2O_2 + H$ and $D_2O_2 + D$

Figure 5 shows variations in the difference spectra of pure solid D_2O_2 after exposure to H atoms on the Al substrate at 15 K. With increasing D atom fluence, the peak intensity of OD stretching and bending bands at 2467 and 1045 cm^{-1}, respectively, decreased, and new peaks appeared at 3455, 2587, and 1477 cm^{-1}. By comparing the peak positions with those given in the literature,[36] we determined that these new peaks are attributable to HDO, indicating that solid D_2O_2 reacted with H atoms to yield HDO at 15 K:

$$D_2O_2 + H \rightarrow HDO + OD. \tag{13}$$

The peak intensity of the $\nu_2 + \nu_6$ combination band for D_2O_2 (2126 cm^{-1})[33] increased slightly, which is opposite to the behavior that would be observed if D_2O_2 was consumed by reaction (13). The dangling OH band of HDO was observed at 3694 cm^{-1} while the dangling OD band was not confirmed probably due to low S/N (Figure 5).

Figure 6 shows variations in the difference spectra after D atom exposure to D_2O_2 for up to 120 min at 15 K. This experiment was performed to study reaction (7):

$$D_2O_2 + D \rightarrow D_2O + OD. \tag{7}$$

This reaction has been studied in previous O_2 deuteration experiments.[15,32] In the present study, with increasing D atom fluence, new peaks appeared at 2562 and

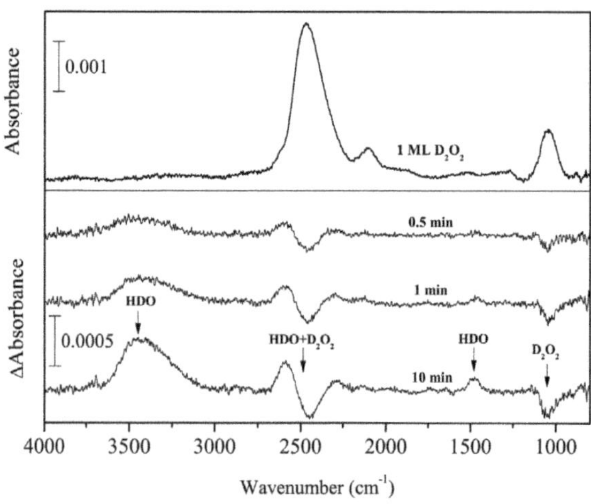

Fig. 5   Variations in the difference spectra of D_2O_2 after exposure to H atoms for up to 10 min at 15 K.

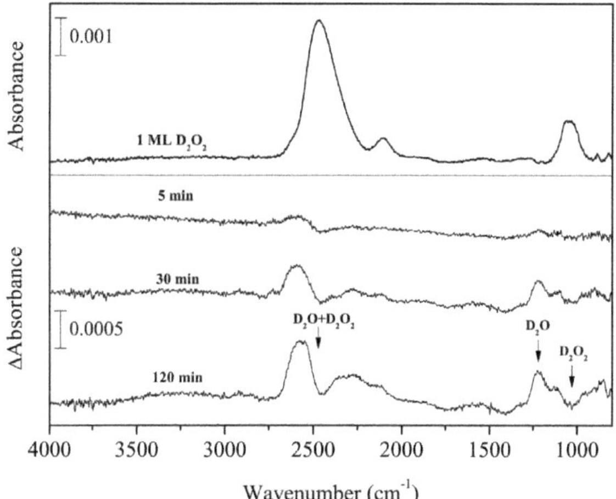

Wavenumber (cm⁻¹)

Fig. 6 Variations in the difference spectra of D_2O_2 after exposure to D atoms for up to 120 min at 15 K.

$1220\ \mathrm{cm}^{-1}$, concurrent with the decrease of D_2O_2 (Figure 6). The new peaks can be attributed to the OD stretching and bending of D_2O, which clearly indicates that reaction (7) occurred to yield D_2O on the surface at 15 K. A small peak was observed after exposure to D atoms at $2727\ \mathrm{cm}^{-1}$, which is assigned to the 3-coordinated dangling OD band of D_2O.[34] A 2-coordinated dangling OD band (at $2748\ \mathrm{cm}^{-1}$)[34] was not firmly identified, probably due to the low S/N of the spectrum. The peak intensity of the $\nu_2 + \nu_6$ combination band for D_2O_2 increased as well (Figure 6).

Figure 7 shows variations in the column density of solid D_2O_2 normalized to the unexposed initial amount after H or D atom exposure at 15 K. The relative abundance of D_2O_2 by reaction (13) reaches a saturation value of approximately

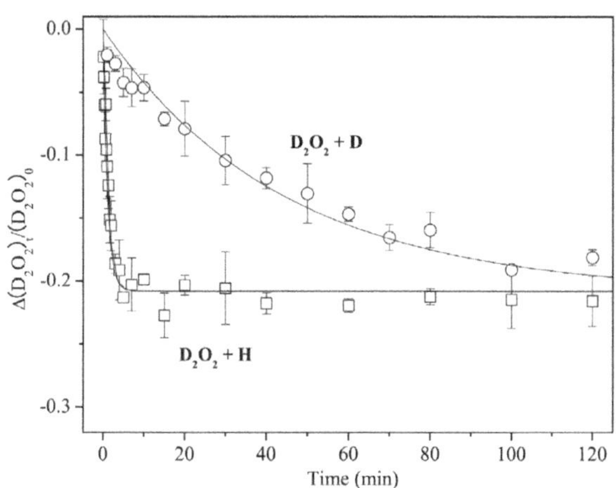

Fig. 7 Variations in the column density of D_2O_2 normalized to the initial amount as a function of H or D atom exposure times at 15 K. Solid lines are single-exponential decay fits to the plots.

−0.2 after a few minutes. In contrast, the decrease of D_2O_2 by reaction (7) was much slower, reaching the almost same saturation value after 150 min. By fitting the plots in Figure 7 into single exponential decay function (12) where $[H_2O_2]$ is replaced with $[D_2O_2]$, we obtained $k_{13}' = 8.9 \times 10^{-1}$ and $k_7' = 2.3 \times 10^{-2}$ min^{-1}, and the k_{13}'/k_7' ratio was 38.

Reactions on amorphous D_2O ice

Hydrogenation and deuteration of solid H_2O_2 and D_2O_2 were also performed on vapor-deposited a-D_2O with a thickness of ~30 ML at 15 K. Figure 8 shows an IR spectrum of 1 ML solid H_2O_2 produced on a-D_2O and the difference spectra after H atom exposure for up to 5 min. The peak area of the $v_2 + v_6$ combination band before H atom exposure was larger by a factor of two than that on the Al substrate, although the peak area of other bands (OH stretching and bending) was almost equal.

With increasing H atom fluence, the intensities of peaks at 3233, 2852, and 1392 cm^{-1} decreased and new peaks appeared at 3422 and 1635 cm^{-1}. These observations clearly indicate that H_2O was formed by reaction (13). Notably, the peak intensity for the $v_2 + v_6$ combination band at 2852 cm^{-1} decreased after H atom exposure. This behavior is straightforward since H_2O_2 was consumed by reaction (13); however, interestingly, this is opposite to the result for the same reactions of pure H_2O_2 on the Al substrate described above. In addition to reaction (13), we confirmed in separate experiments that reactions (11), (13), and (7) occur on a-D_2O at 15 K. Two small peaks appeared at 3697 and 3719 cm^{-1} after exposure to H atoms (Figure 8), which are attributed to the 3- and 2-coordinated dangling OH bands, respectively.[34]

We determined the effective rate constants with statistical errors for each reaction (Table 1) by fitting the attenuation of H_2O_2 and D_2O_2 into a single exponential decay function (12). As a general trend, values of k' are larger for reactions on a-D_2O.

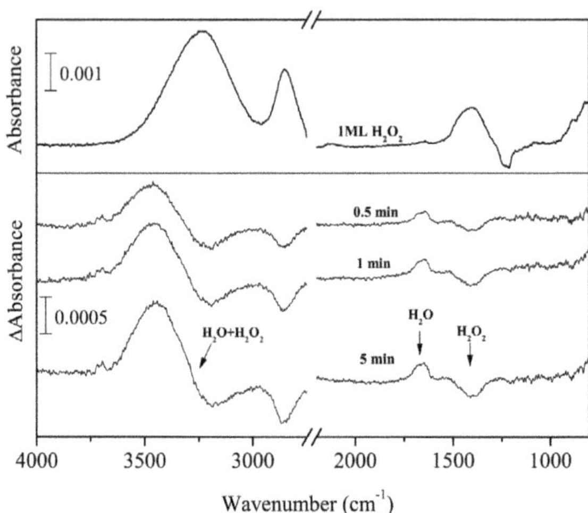

Fig. 8 Variations in the difference spectra of H_2O_2 produced on a-D_2O for up to 5 min at 15 K.

Table 1 Effective rate constants with statistical errors determined in the present study at 15 K

Reaction	Reaction number	Substrate	Effective rate (min^{-1})	E_a (K)[a]	Reduced mass (u)
$H_2O_2 + H$	3	Al	$7.2 \pm 0.6 \times 10^{-1}$	2508	0.97
		a-D_2O	$9.9 \pm 0.6 \times 10^{-1}$		
$H_2O_2 + D$	11	Al	$3.2 \pm 0.1 \times 10^{-2}$	2355	1.89
		a-D_2O	$4.0 \pm 0.6 \times 10^{-2}$		
$D_2O_2 + H$	13	Al	$8.9 \pm 1.1 \times 10^{-1}$	2540	0.97
		a-D_2O	$9.2 \pm 0.6 \times 10^{-1}$		
$D_2O_2 + D$	7	Al	$2.3 \pm 0.2 \times 10^{-2}$	2384	1.89
		a-D_2O	$2.2 \pm 0.8 \times 10^{-2}$		

[a] Taquet *et al.*[40]

The column densities of H_2O formed on the Al and a-D_2O by reaction (3) were calculated using the band strength of the OH-bending at 1635 cm^{-1} (1.2×10^{-17} cm molecule^{-1}).[28] We determined that the H_2O yield on a-D_2O was approximately two orders of magnitude larger than that on the Al at the same H fluence, although the amount of H_2O_2 consumption on H atom exposure was identical for a-D_2O and Al (Figure 9).

Temperature dependence of reaction kinetics

Reaction (3) was studied using pure H_2O_2 at 10, 20, and 30 K as well as at 15 K. Reaction (3) occurred at all temperatures. We obtained kinetic parameters for reaction (3) (k_3') at each temperature following the procedures described above. Figure 10 shows variations in the relative abundance of pure H_2O_2 after H atom exposure at 10–30 K. The temperature dependence shows that the saturation value of H_2O_2 becomes larger with increasing temperature up to 20 K; however, at

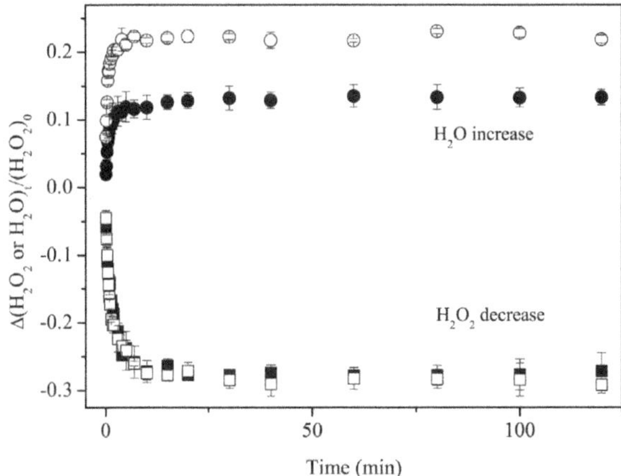

Fig. 9 Variations in the column densities of H_2O (circle) and H_2O_2 (square) normalized to the initial H_2O_2 amount obtained after H atom exposure to H_2O_2 at 15 K. Open and filled symbols represent experimental results on a-D_2O and Al substrate, respectively.

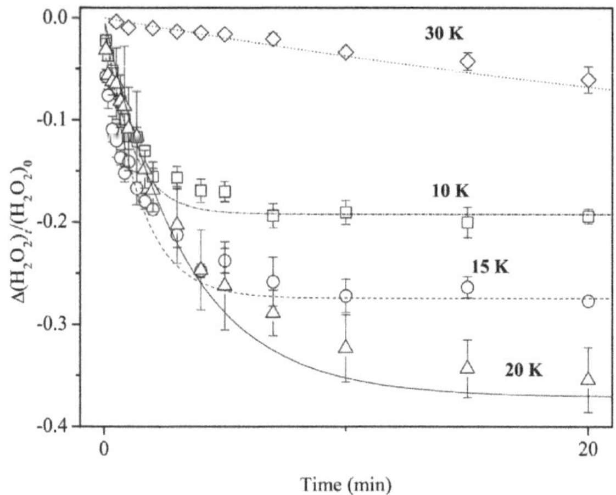

Fig. 10   Variations in the column density of pure H_2O_2 normalized to the initial amount after exposure to H atoms at 10 (square), 15 (circle), 20 (triangle), and 30 (diamond) K. Lines are single-exponential decay fits to the plots.

30 K, the reaction is very slow and the saturation value is much less than that below 20 K. The effective rate constant is the largest at 10 K and decreases with increasing temperature. These features were also obtained for reactions on a-D_2O ice (Figure 11).

Discussion

Quantum tunneling and isotope effect

The reaction of H_2O_2 with H atoms and that of their deuterated counterparts has a large activation barrier (>2000 K) in the gas phase.[9,40,41] Therefore, these reactions are expected to proceed through quantum tunneling at 10–30 K even on the

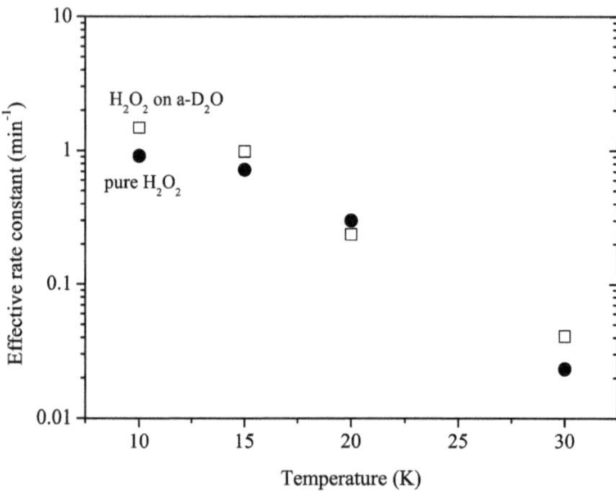

Fig. 11   Temperature dependence of the effective rate constant for reaction (3): filled circle and open square symbols represent the rate for pure H_2O_2 and H_2O_2 on a-D_2O, respectively.

surface, as mentioned earlier. A tunneling reaction strongly depends on the transmission mass of the activation barrier of reaction, as can be seen in equation (4). Theoretical studies proposed that H_2O formation by reaction (3) initiates from the formation of an intermediate by H atom addition to one O atom in H_2O_2, followed by the cleavage of the O–O bond.[9,41] In general, the tunneling mass in the two-body addition reaction is described by the reduced mass μ.[12,13] As shown in Table 1, the mass dependence of the reaction rate is evident; the lighter mass results in a faster reaction rate. Thus, we conclude that the reactions of H_2O_2/D_2O_2 with H/D atoms proceed by quantum tunneling.

The ratios of the effective rate constants between reactions (3) and (11) (k_3'/k_{11}') are 23 and 25 for pure H_2O_2 and H_2O_2 on a-D_2O, respectively. Since the effective rate constant k_n' is expressed as $k_n[X]$ where X = H or D, the ratio of the effective rate constants is not equal to the ratio of the actual reaction rate constants if [H] $\neq$ [D]. Assuming that the value of [X] has a linear correlation with the flux of X atoms in the present experiment, the ratios of k_3 to k_{11} (k_3/k_{11}) become 44 and 48 for pure H_2O_2 and H_2O_2 on a-D_2O, respectively, because the D atom flux is approximately twice that of the H atom flux. Similarly, the k_{13}/k_7 ratio was calculated to be 75 and 81 for pure D_2O_2 and D_2O_2 on a-D_2O, respectively. Taquet et al.[40] calculated a tunneling probability P_n through an Eckart potential barrier for reaction (n). In the present study, if we compare the P_3/P_{11} ratio with the k_3/k_{11} and the P_{13}/P_7 ratio with the k_{13}/k_7, we obtain $k_3/k_{11} = \sim 2 \times P_3/P_{11}$ and $k_{13}/k_7 = \sim 3 \times P_{13}/P_7$. These differences are not surprising because the tunneling probability is significantly affected by the shape of the potential barrier, and it is very difficult to determine the tunneling rate with an accurate barrier for surface reactions. In fact, Taquet et al.[40] noted that, although the Eckart model provides a significant improvement over square barriers, which were typically used for this type of calculation, the tunneling probabilities of some reactions at low temperatures can be underestimated (or overestimated).

We obtained the k_3/k_7 ratios of 61 and 88 for pure H_2O_2/D_2O_2 and H_2O_2/D_2O_2 on a-D_2O, respectively. These values are about one order of magnitude larger than those reported by Miyauchi et al., who determined the k_3'/k_7' ratio to be 8 in their O_2 hydrogenation/deuteration experiments at 10 K. Since the flux of H atoms was the same with that of D atoms in the previous experiment,[15] the k_3'/k_7' ratio is certainly equal to the k_3/k_7 ratio on the assumption that the ratio of surface number densities between H and D atoms is same as that of the fluxes. The large difference may arise for several reasons. First, multi-parameter fittings may cause significant errors of k_3 and k_7 in their work. Second, H_2O and D_2O would form in O_2 hydrogenation/deuteration via separate pathways: OH + OH → H_2O + O and OD + OD → D_2O + O, respectively, which are both barrierless reactions. This would lead to an overestimation of the k_3/k_7 ratio. In the present study, these uncertainties are removed because H_2O_2 or D_2O_2 are used as the initial reactants. Thus, we believe that our calculated value of k_3/k_7 is more reliable than that obtained in O_2 hydrogenation/deuteration experiments.

Next, we compare the rates of reactions (3) and (5). The k_5/k_3 ratio in the O_2 hydrogenation experiment was reported to be 3.3.[15] Based on the fact that reaction (5) is barrierless and reaction (3) has a large barrier (>2000 K),[9] the value obtained by Miyauchi et al. seems to be rather small. Subsequently, we calculated the k_5/k_3 ratio using k_3' obtained in the present study. The value of k_5' reported by Miyauchi et al. (12.8 min^{-1})[15] was not used for this calculation because their conditions such as atom fluxes and sample amounts were significantly different from ours. Instead, we used the value of k_5' (5.2 min^{-1}) obtained in the following experiment: solid O_2 (~3 ML) on an amorphous D_2O ice (30 ML) was exposed to H atoms (flux: 2×10^{14} atoms cm^{-2} s^{-1}) at 10 K.[42] After the correction of the flux difference, we obtained the k_5/k_3 ratio of ~15, which is approximately five times larger than that reported by Miyauchi et al.[15] Although the sample compositions and the fluxes of H atoms in these two experiments do not exactly match, we believe that the present result would better represent the k_5/k_3 ratio compared to previous studies.

Notably, the observed isotope effect in the hydrogen/deuterium addition to H_2O_2 (k_3/k_{11}) is much larger than that for one of the astrochemically-important tunneling surface reactions: the hydrogenation/deuteration of CO. Hidaka et al.[27] reported that the rate of CO hydrogenation (k_H) was larger than CO deuteration (k_D), with a k_H/k_D ratio of 12.5. This isotope effect is approximately four times smaller than that of the hydrogenation/deuteration of H_2O_2 observed in the present study. This large difference is rather surprising because the barrier height for the hydrogenation/deuteration of H_2O_2 is similar to that of CO.[40,43] We believe that the difference might have been due to accumulation of multiple factors such as differences in residence time of H/D atoms on H_2O_2 and CO and the shape of the potential barriers.

As mentioned previously, we assume that the surface density of H atoms is the same as that of D atoms when the flux is equal. However, in a series of experiments under high H and D flux conditions, this may not always be true. Even under the same flux conditions for H and D atoms, the surface density of D atoms ([D]) may become larger than that of H atoms ([H]). Since H atoms can diffuse faster on the surface than D atoms, the recombination probability for H atoms is higher than that for D atoms,[12] resulting in a lower value of [H] than that of [D] even under the same fluxes. In that case, the ratio of [H]/[D] becomes smaller than 0.5, yielding larger values of k_3/k_{11} and k_{13}/k_7 than those reported in the present study. Further studies related to the surface densities of H and D atoms are necessary.

Dependence of reaction efficiency on temperature and type of substrate

The effective rate constants for reaction (3) decrease with increasing substrate temperature (Figure 11). This is opposite to the typical Arrhenius-type behavior where the reaction rate has a positive correlation with temperature. In addition, because reaction (3) occurred through quantum tunneling, the reaction rate should have lesser dependence on the surface temperatures. The decrease of the effective rate can be explained well by the decrease in the number density of H atoms on the surface with increasing temperature because the effective rate k_3' is expressed by $k_3[H]$. This trend was also observed in CO hydrogenation and deuteration, H_2CO hydrogenation, H–D substitution of CH_3OH, and O_2 hydrogenation on low-temperature surfaces.[11,42]

The difference of the effective rate constant between reactions of pure H_2O_2 and H_2O_2 on a-D_2O is the largest at 10 K and decreases with increasing temperature (Figure 11). The large difference observed at lower temperatures (<15 K) may be explained by the difference in the number density of H atoms on the surface because a-D_2O would have a much larger surface area compared to planar Al substrate, as reported for CO hydrogenation by Hidaka *et al.*[44] who determined that the surface area for a-H_2O with thickness equivalent to $\sim$17 ML is approximately nine times larger than that for crystalline H_2O. In contrast, at elevated temperatures where the residence time of H atoms on the surface is very short even on a-D_2O, the difference could little be affected by the density of H atoms, which can lead to similar effective rate constant values under both conditions (Figure 11).

In Figure 9, H_2O yield for the sample of pure H_2O_2 on Al substrate is lower than that for H_2O_2 on a-D_2O, although the decrease of H_2O_2 is approximately the same for both samples. This difference may be partly explained by immediate desorption of H_2O formed as a product at the reaction preferentially occurred on the Al substrate. This behavior is also observed in O_2 + D experiments on amorphous silicates and a-H_2O, where the yield of D_2O is higher for reactions on a-H_2O.[39] The heat of reaction (3) (285 kJ mol^{-1} = 2.9 eV) is partly partitioned to the reaction product H_2O and OH. This energy would be sufficient for H_2O to desorb from the substrate (Al or a-D_2O). For the H_2O_2 on a-D_2O sample, interaction of H_2O_2 and H_2O product with the a-D_2O surface is much stronger than that with the Al surface because of hydrogen bonding. Therefore, the reaction heat can dissipate into a-D_2O more easily, and desorption at the reaction surface may be reduced. Furthermore, since a-D_2O has a large surface area because of pores and cracks, the desorbed H_2O can be retrapped on the surface of a-D_2O. This type of H_2O relaxation and trap has been demonstrated both experimentally[45–47] and theoretically[48,49] to occur during the photodesorption of H_2O from bulk a-H_2O. In addition to H_2O, OH should also form by reaction (3) and can be the source of another H_2O formation. Accordingly, H_2O and OH trapped on the surface of a-D_2O may explain the larger column density of the formed H_2O than that on the Al substrate (Figure 9).

As shown in Figure 1, solid H_2O_2 produced on the Al substrate has representative strong absorptions at three positions: 1385 (v_2), 2827 ($v_2 + v_6$), and 3326 cm^{-1} (v_1).[33] Similarly, for solid H_2O_2 on a-D_2O, these three peaks are slightly shifted on the spectrum at 1392, 2852, and 3233 cm^{-1} (Figure 8). The difference in the peak position for each band is not very large (<25 cm^{-1}). However, the peak area of the $v_2 + v_6$ combination band was larger by a factor of two for the sample on a-D_2O, although the peak area of other bands (v_1 and v_2) was identical on both substrates. A relatively strong peak at $\sim$2850 cm^{-1} for solid H_2O_2 is typically assigned to the $v_2 + v_6$ combination band;[33,50] however, some studies considered that this assignment is disputable.[51,52] Ignatov *et al.*[51] proposed that this peak is attributable to an OH valence band of the H_2O_2 molecule placed in the unusual surface environment. The present results imply that the peak intensity at $\sim$2850 cm^{-1} might correlate with the configuration of H_2O_2 on the substrate. In addition, the fact that H_2O_2 forms hydrogen bonds more on a-D_2O than on Al could result in the different IR features at $\sim$2850 cm^{-1}. In other words, the number of hydrogen bonds between H_2O_2 and surrounding H_2O may constrain the peak intensity at $\sim$2850 cm^{-1}. If this is the case, it could explain the contradictory behavior of the peak intensity after exposure to atoms (*e.g.* Figures 2 and 8). More detailed experimental and theoretical studies are necessary for the precise peak assignment. Thus, we propose that care should be taken when this IR peak is used for quantification of solid H_2O_2 in various environments.

Dangling OH (OD) bands of reaction products

We confirmed that the formed solid H_2O and its isotopologues (HDO and D_2O) show the dangling OH (OD) bands (*e.g.*, Figure 2). The presence of dangling OH bands in the IR spectrum of solid H_2O typically indicates that it is amorphous and has a microporous structure.[34] However, we have reported that dangling OH bands are not observed in solid H_2O formed by codeposition of O_2 and H atoms at 10 K and that the formed ice is amorphous but has a compact structure.[31] These different results suggest that the formed ice structure depends on experimental conditions and that the properties of the reaction products are strikingly different between each study (multilayers with O_2 contaminants *vs.* very thin layers with little contaminants).

Astrophysical implication

Deuterated water (HDO and D_2O) has been identified in the gas phase with HDO/H_2O and D_2O/H_2O ratios up to the order of 10^{-2} and 10^{-3}, respectively, toward protostars.[53–57] In contrast, deuterated water was not clearly identified in the solid phase; only the upper limit of HDO was determined (HDO/H_2O < 0.2%–2%).[58,59] Despite the detection of deuterated water in the gas phase only, it is reasonable to consider that deuterated water is also produced by surface reactions on interstellar grains at very low temperatures.

The following reactions are possible to yield water isotopologues from H_2O_2 and its isotopologues:

$$H_2O_2 + H \rightarrow H_2O + OH, \tag{3}$$

$$H_2O_2 + D \rightarrow HDO + OH, \tag{11}$$

$$D_2O_2 + H \rightarrow HDO + OD, \tag{13}$$

$$D_2O_2 + D \rightarrow D_2O + OD, \tag{7}$$

$$HDO_2 + H \rightarrow HDO + OH, \tag{14a}$$

$$HDO_2 + H \rightarrow H_2O + OD, \tag{14b}$$

$$HDO_2 + D \rightarrow HDO + OD, \tag{15a}$$

$$HDO_2 + D \rightarrow D_2O + OH. \tag{15b}$$

We did not study reactions (14) and (15) because of difficulty in producing high-purity HDO_2. Two types of reaction products are possible for reactions (14) and (15). Since the reduced mass of a reaction to produce the intermediate does not depend on the type of products, we expect that reactions (14a) and (14b) occur to the same extent statistically. In fact, the tunneling probability for reaction (14a) is calculated to be identical with that of reaction (14b).[40] The same is true for reactions (15a) and (15b). Thus, for simplicity, we made a rough assumption based on the obtained results that H atom addition reactions such as reactions (3) and (13) occur faster by a factor of 50 than D atom addition reactions such as reactions (7) and (11). In addition, we assume that the atomic D/H ratio is constant and the surface diffusion rates of H and D atoms are the same. Under these assumptions, we determined that the HDO/H_2O ratio was almost identical to the atomic D/H ratio, while the D_2O formation was negligible ($\sim10^{-6}$ that of H_2O). On the other hand, assuming no isotope effect on the reactions, the HDO/H_2O ratio becomes twice as large as the atomic D/H ratio; thus, quantum tunneling would cause a decrease of the HDO/H_2O ratio in this reaction pathway. In other words, if quantum tunneling corrections are not included in a chemical model, the obtained result should overestimate the value of the HDO/H_2O ratio.

We extend this discussion to other water formation pathways, *i.e.*, reactions (1) $(OH + H \rightarrow H_2O)$ and (2) $(OH + H_2 \rightarrow H_2O + H)$. For reaction (2), we have experimentally determined the relative efficiency of reactions for all possible isotopologues;[14] H atom abstraction reactions are approximately ten times more efficient than D atom abstraction reactions. Assuming that OH and OD are formed by surface reactions O + H and O + D, respectively, HD/H_2 ratio is 10^{-5}, and no D_2 is present, the HDO/H_2O ratio was very much consistent with the atomic D/H ratio.

In contrast to the former two reaction pathways, the reactions of OH or OD with H or D atoms do not have an activation barrier. Thus, the product ratio would strongly depend on the atomic D/H ratio. Since there are two possible reactions to yield HDO ($OH + D \rightarrow HDO$ and $OD + H \rightarrow HDO$), statistically the HDO/H_2O ratio is twice as large as the atomic D/H ratio. With regard to the formation of D_2O, this reaction pathway is the most favorable among the three reaction pathways; the D_2O/H_2O ratio is statistically the square of the value of the D/H ratio. For example, the D_2O/H_2O ratio is 10^{-4} if the atomic D/H is 10^{-2}. This value is two to four orders of magnitude higher than that estimated by other pathways under the same assumptions. Therefore, the OD + D reaction is the only possible pathway to effectively yield D_2O in MCs.

Moreover, it is important to note that, on grain surfaces, the deuterium fractionation of water occurs only during the formation by surface reactions at the typical temperature of MCs (~10 K). This is supported by the fact that H_2O does not react with D atoms at <15 K[60] and thermal H–D exchange does not occur at <100 K,[61] which prevents the D-enrichment by H–D substitution with other deuterium-enriched species after the formation of H_2O on the grains. The reactivity of water with D atoms at low temperatures differs from that of organic species such as CH_3OH,[25,60] H_2CO,[13,27] and CH_3NH_2[62] where H–D exchange occurs for these molecules by reacting with D atoms at temperatures as low as 10 K.

Based on the present and previous experimental results for water formation, we suggest that the HDO/H_2O ratio might be a key parameter to estimate the atomic D/H ratio during the formation of water. Namely, the following relationship is roughly derived from experimental studies: $(D/H)_{atom} \leq (HDO/H_2O) \leq 2(D/H)_{atom}$ or $1/2(HDO/H_2O) \leq (D/H)_{atom} \leq (HDO/H_2O)$, where $(D/H)_{atom}$ and (HDO/H_2O) represent the atomic D/H on grains and the HDO/H_2O ratio formed by surface reactions, respectively. Note that these relationships were derived with reference to experimental studies for surface reactions. In addition to surface reactions, deuterium fractionation of water by gas phase reactions may also be possible.[63] Moreover, energetic processes induced by UV and cosmic rays might cause hydrogen isotopic fractionation of water during its decomposition and interactions with other ice components, both of which may modify the HDO/H_2O ratio. Therefore, we propose that further collaborative theoretical and experimental studies on surface and gas phase chemistries are necessary to construct a complete chemical model regarding the evolution of the water D/H ratio in MCs.

Acknowledgements

The authors thank Drs H. Hidaka and T. Hama for fruitful discussions at the earlier stages of manuscript preparation. We also thank an anonymous referee for providing constructive comments. This work is partly supported by a Grant-in-Aid for Scientific Research from the Japan Society for the Promotion of Science. Y.O. has received funding from the Kurita Water and Environment Foundation.

References

1 E. Herbst and W. Klempere, *Astrophys. J.*, 1973, **185**, 505–533.
2 T. I. Hasegawa, E. Herbst and C. M. Leung, *Astrophys. J. Suppl.*, 1992, **82**, 167–195.
3 A. Tielens and W. Hagen, *Astron. Astrophys.*, 1982, **114**, 245–260.
4 F. Dulieu, L. Amiaud, E. Congiu, J. H. Fillion, E. Matar, A. Momeni, V. Pirronello and J. L. Lemaire, *Astron. Astrophys.*, 2010, **512**, 5.
5 K. Hiraoka, T. Miyagoshi, T. Takayama, K. Yamamoto and Y. Kihara, *Astrophys. J.*, 1998, **498**, 710–715.
6 D. P. Jing, J. He, M. Bonini, J. R. Brucato and G. Vidali, *J. Phys. Chem. A*, 2013, **117**, 3009–3016.
7 D. P. Jing, J. He, J. Brucato, A. De Sio, L. Tozzetti and G. Vidali, *Astrophys. J.*, 2011, **741**, 5.
8 R. Atkinson, D. L. Baulch, R. A. Cox, J. N. Crowley, R. F. Hampson, R. G. Hynes, M. E. Jenkin, M. J. Rossi and J. Troe, *Atmos. Chem. Phys.*, 2004, **4**, 1461–1738.
9 H. Koussa, M. Bahri, N. Jaidane and Z. Ben Lakhdar, *THEOCHEM*, 2006, **770**, 149–156.
10 H. M. Cuppen and E. Herbst, *Astrophys. J.*, 2007, **668**, 294–309.
11 N. Watanabe and A. Kouchi, *Prog. Surf. Sci.*, 2008, **83**, 439–489.
12 T. Hama and N. Watanabe, *Chem. Rev.*, 2013, **113**, 8783–8839.
13 H. Hidaka, M. Watanabe, A. Kouchi and N. Watanabe, *Astrophys. J.*, 2009, **702**, 291–300.
14 Y. Oba, N. Watanabe, T. Hama, K. Kuwahata, H. Hidaka and A. Kouchi, *Astrophys. J.*, 2012, 749.
15 N. Miyauchi, H. Hidaka, T. Chigai, A. Nagaoka, N. Watanabe and A. Kouchi, *Chem. Phys. Lett.*, 2008, **456**, 27–30.
16 H. M. Cuppen, S. Ioppolo, C. Romanzin and H. Linnartz, *Phys. Chem. Chem. Phys.*, 2010, **12**, 12077–12088.
17 Y. Oba, N. Watanabe, A. Kouchi, T. Hama and V. Pirronello, *Phys. Chem. Chem. Phys.*, 2011, **13**, 15792–15797.
18 T. Lamberts, H. M. Cuppen, S. Ioppolo and H. Linnartz, *Phys. Chem. Chem. Phys.*, 2013, **15**, 8287–8302.
19 G. W. Fuchs, H. M. Cuppen, S. Ioppolo, C. Romanzin, S. E. Bisschop, S. Andersson, E. F. van Dishoeck and H. Linnartz, *Astron. Astrophys.*, 2009, **505**, 629–639.
20 C. Pirim and L. Krim, *Chem. Phys.*, 2011, **380**, 67–76.

21 N. Watanabe and A. Kouchi, *Astrophys. J.*, 2002, **571**, L173–L176.

22 N. Watanabe, T. Shiraki and A. Kouchi, *Astrophys. J.*, 2003, **588**, L121–L124.

23 N. Watanabe, A. Nagaoka, T. Shiraki and A. Kouchi, *Astrophys. J.*, 2004, **616**, 638–642.

24 H. Hidaka, N. Watanabe, T. Shiraki, A. Nagaoka and A. Kouchi, *Astrophys. J.*, 2004, **614**, 1124–1131.

25 A. Nagaoka, N. Watanabe and A. Kouchi, *J. Phys. Chem. A*, 2007, **111**, 3016–3028.

26 N. Watanabe, A. Nagaoka, H. Hidaka, T. Shiraki, T. Chigai and A. Kouchi, *Planet. Space Sci.*, 2006, **54**, 1107–1114.

27 H. Hidaka, A. Kouchi and N. Watanabe, *J. Chem. Phys.*, 2007, 126.

28 P. A. Gerakines, W. A. Schutte, J. M. Greenberg and E. F. van Dishoeck, *Astron. Astrophys.*, 1995, **296**, 810–818.

29 M. J. Loeffler and R. A. Baragiola, *J. Phys. Chem. A*, 2011, **115**, 5324–5328.

30 R. G. Smith, S. B. Charnley, Y. J. Pendleton, C. M. Wright, M. M. Maldoni and G. Robinson, *Astrophys. J.*, 2011, **743**, 13.

31 Y. Oba, N. Miyauchi, H. Hidaka, T. Chigai, N. Watanabe and A. Kouchi, *Astrophys. J.*, 2009, **701**, 464–470.

32 S. Ioppolo, H. M. Cuppen, C. Romanzin, E. F. van Dishoeck and H. Linnartz, *Astrophys. J.*, 2008, **686**, 1474–1479.

33 J. A. Lannon, F. D. Verderam and R. W. Anderson, *J. Chem. Phys.*, 1971, **54**, 2212–2223.

34 V. Buch and J. P. Devlin, *J. Chem. Phys.*, 1991, **94**, 4091–4092.

35 J. P. Devlin, *J. Mol. Struct.*, 1990, **224**, 33–43.

36 D. F. Hornig, H. F. White and F. P. Reding, *Spectrochim. Acta*, 1958, **12**, 338–349.

37 J. P. Devlin, *J. Chem. Phys.*, 2000, **112**, 5527–5529.

38 R. T. Garrod, V. Wakelam and E. Herbst, *Astron. Astrophys.*, 2007, **467**, 1103–1115.

39 H. Chaabouni, M. Minissale, G. Manico, E. Congiu, J. A. Noble, S. Baouche, M. Accolla, J. L. Lemaire, V. Pirronello and F. Dulieu, *J. Chem. Phys.*, 2012, **137**, 234706.

40 V. Taquet, P. S. Peters, C. Kahane, C. Ceccarelli, A. Lopez-Sepulcre, C. Toubin, D. Duflot and L. Wiesenfeld, *Astron. Astrophys.*, 2013, **550**, 23.

41 B. A. Ellingson, D. P. Theis, O. Tishchenko, J. Zheng and D. G. Truhlar, *J. Phys. Chem. A*, 2007, **111**, 13554–13566.

42 Y. Oba, N. Miyauchi, T. Chigai, H. Hidaka, N. Watanabe and A. Kouchi, in *Physics and Chemistry of Ice 2010*, Hokkaido University Press, 2011, Y. Furukawa, G. Sazaki, T. Uchida, and N. Watanabe, Eds., 361–368.

43 D. E. Woon, *Astrophys. J.*, 2002, **569**, 541–548.

44 H. Hidaka, N. Miyauchi, A. Kouchi and N. Watanabe, *Chem. Phys. Lett.*, 2008, **456**, 36–40.

45 T. Hama, M. Yokoyama, A. Yabushita, M. Kawasaki, S. Andersson, C. M. Western, M. N. R. Ashfold, R. N. Dixon and N. Watanabe, *J. Chem. Phys.*, 2010, **132**, 8.

46 A. Yabushita, T. Hama, M. Yokoyama, M. Kawasaki, S. Andersson, R. N. Dixon, M. N. R. Ashfold and N. Watanabe, *Astrophys. J.*, 2009, **699**, L80–L83.

47 K. I. Oberg, H. Linnartz, R. Visser and E. F. van Dishoeck, *Astrophys. J.*, 2009, **693**, 1209–1218.

48 S. Andersson and E. F. van Dishoeck, *Astron. Astrophys.*, 2008, **491**, 907–916.

49 C. Arasa, S. Andersson, H. M. Cuppen, E. F. van Dishoeck and G. J. Kroes, *J. Chem. Phys.*, 2010, **132**, 12.

50 A. Engdahl, B. Nelander and G. Karlstrom, *J. Phys. Chem. A*, 2001, **105**, 8393–8398.

51 S. K. Ignatov, A. G. Razuvaev, P. G. Sennikov and O. Schrems, *THEOCHEM*, 2009, **908**, 47–54.

52 P. G. Sennikov, S. K. Ignatov and O. Schrems, *ChemPhysChem*, 2005, **6**, 392–412.

53 H. M. Butner, S. B. Charnley, C. Ceccarelli, S. D. Rodgers, J. R. Pardo, B. Parise, J. Cernicharo and G. R. Davis, *Astrophys. J.*, 2007, **659**, L137–L140.

54 C. Ceccarelli, C. Dominik, E. Caux, B. Lefloch and P. Caselli, *Astrophys. J.*, 2005, **631**, L81–L84.

55 A. Coutens, C. Vastel, E. Caux, C. Ceccarelli, S. Bottinelli, L. Wiesenfeld, A. Faure, Y. Scribano and C. Kahane, *Astron. Astrophys.*, 2012, **539**, 12.

56 B. Parise, E. Caux, A. Castets, C. Ceccarelli, L. Loinard, A. Tielens, A. Bacmann, S. Cazaux, C. Comito, F. Helmich, C. Kahane, P. Schilke, E. van Dishoeck, V. Wakelam and A. Walters, *Astron. Astrophys.*, 2005, **431**, 547–554.

57 C. Vastel, C. Ceccarelli, E. Caux, A. Coutens, J. Cernicharo, S. Bottinelli, K. Demyk, A. Faure, L. Wiesenfeld, Y. Scribano, A. Bacmann, P. Hily-Blant, S. Maret, A. Walters, E. A. Bergin, G. A. Blake, A. Castets, N. Crimier, C. Dominik, P. Encrenaz, M. Gerin, P. Hennebelle, C. Kahane, A. Klotz, G. Melnick, L. Pagani, B. Parise, P. Schilke, V. Wakelam, A. Baudry, T. Bell, M. Benedettini, A. Boogert, S. Cabrit, P. Caselli, C. Codella, C. Comito, E. Falgarone, A. Fuente, P. F. Goldsmith, F. Helmich, T. Henning, E. Herbst, T. Jacq, M. Kama, W. Langer, B. Lefloch, D. Lis, S. Lord, A. Lorenzani, D. Neufeld, B. Nisini, S. Pacheco, J. Pearson, T. Phillips, M. Salez, P. Saraceno, K. Schuster, X. Tielens, F. van der Tak, M. H. D. van der Wiel, S. Viti, F. Wyrowski, H. Yorke, P. Cais, J. M. Krieg, M. Olberg and L. Ravera, *Astron. Astrophys.*, 2010, **521**, 5.

58 E. Dartois, W. F. Thi, T. R. Geballe, D. Deboffle, L. d'Hendecourt and E. van Dishoeck, *Astron. Astrophys.*, 2003, **399**, 1009–1020.

59 B. Parise, T. Simon, E. Caux, E. Dartois, C. Ceccarelli, J. Rayner and A. Tielens, *Astron. Astrophys.*, 2003, **410**, 897–904.

60 A. Nagaoka, N. Watanabe and A. Kouchi, *Astrophys. J.*, 2005, **624**, L29–L32.

61 J. P. Devlin and V. Buch, *J. Chem. Phys.*, 2007, **127**, 4.

62 Y. Oba, T. Chigai, Y. Osamura, N. Watanabe and A. Kouchi, *Meteorit. Planet. Sci.*, 2014, **49**, 117–132.

63 H. Roberts, E. Herbst and T. J. Millar, *Astron. Astrophys.*, 2004, **424**, 905–917.

Faraday Discussions

PAPER

Diffusion of atomic oxygen relevant to water formation in amorphous interstellar ices

Myung Won Lee and Markus Meuwly*

Received 24th December 2013, Accepted 11th February 2014

DOI: 10.1039/c3fd00160a

Molecular dynamics (MD) simulations together with accurate physics-based force fields are employed to determine the mobility of atomic oxygen in amorphous ice at low temperatures, characteristic for conditions in interstellar ices. From the simulations it is found that the mobility of atomic oxygen ranges from 60 to 480 $\text{Å}^2\text{ ns}^{-1}$ in amorphous ice at temperatures between 50 and 200 K. Hence, the simulations establish that atomic oxygen is mobile to a certain degree and a chemical mechanism for water formation involving oxygen mobility is a realistic scenario. This is also confirmed by the computed migration barriers for oxygen diffusion by multiple umbrella sampling simulations, which yield barriers for diffusion in the range of 0.7–1.9 kcal mol^{-1}. The physics-based force field – based on a multipolar expansion of the electrostatic interactions – yields more pronounced energetics for oxygen migration pathways compared to the conventional point-charge models employed in typical simulations. Once formed, the computed solvation free energy suggests that atomic oxygen thermodynamically prefers to be localized inside amorphous ice and is available for chemical reaction, which may be relevant to water formation in and on grains.

1 Introduction

Water is an essential component in astrophysical environments, where it has been detected in various forms. Bulk water itself is present in the form of amorphous solid water (ASW), which is the main component of interstellar ices.[1] The structure – and transitions between different forms – of ASW is typically inferred from spectroscopic measurements,[1,2] although recently an interference-based method has also been employed.[3] ASW is able to support a highly porous structure which is characterized by large scale internal cavities which can potentially retain large quantities of guest species in molecular or atomic form.[4] Under laboratory conditions the water ices seem to be amorphous in nature[5] whereas the morphologies of ices in the interstellar medium are more uncertain.[6]

Department of Chemistry, University of Basel, Klingelbergstrasse 80, 4056 Basel, Switzerland

It is believed that pristine water can be formed and grows on the icy dust grains in interstellar clouds from hydrogen and oxygen atoms, due to the inefficiency of the gas-phase formation of water.[7-10] At present, however, little is known about the mechanisms underlying water formation in and on dust grains, where the amorphous form of water seems to dominate. A Monte Carlo study showed that the prevalence of a particular water formation channel depends on the local environment.[7] In translucent and diffuse clouds (with temperatures $T \lesssim 50$ K)[11] the main route went through the atomic oxygen channel whereas the molecular oxygen (O_2) together with the ozone (O_3) route dominated in dense cold molecular clouds ($T \approx 10$ K).[7] The reactions leading to water formation through atomic and molecular oxygen and ozone have been studied extensively over the past few years.[12-15]

Independent of the particular reaction channel(s) assumed, atomic oxygen plays an essential role in water formation.[16] Several scenarios for the elementary reaction steps involved in water formation under astrophysical and interstellar conditions have been put forward in the past.[13,16-19] As reaction partners within an assumed reaction network are not necessarily generated or present in immediate spatial proximity to each other (independent of whether surface or bulk reactions are considered), the reactants' mobilities strongly affect the formation rates of particular molecular species. Such information is extremely difficult to obtain from experiments, which makes direct simulation an attractive and often the only alternative. Some of the models explicitly assume a certain mobility of the participating reagents which is not guaranteed and incompletely understood at the low temperatures characteristic of such environments. One such relationship for the hopping probability P is $P = \exp(-[E_D/200] \times \Delta x)$, which depends on the desorption energy (E_D) of the species of interest relative to water, and the average distance Δx between adsorption sites.[7] However, since in amorphous ice – which is the form of radiatively processed water ices – Δx is characterized by a distribution, and E_D is often only incompletely known from direct measurements for the species of interest, a considerable amount of uncertainty is involved in employing P values obtained from such estimates. Hence, more direct approaches for estimating the mobility of relevant chemical species are desirable. One such method is atomistic molecular dynamics simulations. However, it is worth mentioning that very recently an experimental characterization of oxygen atom surface diffusion on amorphous water ices has been carried out which reports an activation energy between 0.5 and 1 kcal mol^{-1} for surface diffusion.[20]

Depending on the particular reaction network model used, free atomic oxygen is implicated in several elementary steps in water formation under interstellar conditions. One possibility is the sequential addition of hydrogen atoms to adsorbed oxygen atoms O_{ads}:[14,17,21]

$$O_{ads} + H_{ads} \rightarrow OH_{ads} \tag{1}$$

$$OH_{ads} + H_{ads} \rightarrow H_2O_{ads} \tag{2}$$

This is possible as the number density of atomic oxygen is among the largest of all elements (several 10^{-4} relative to n_H) in interstellar space.[22,23] Also, depending on the astrophysical environment, n_O can exceed n_H and be comparable to n_{CO}.[23] The oxygen atom adsorbed on the amorphous ice may diffuse into the bulk ice and then migrate inside it. In one scenario (the "O_2 route"), atomic oxygen is involved in the formation of O_2 and O_3 from their respective atomic and molecular precursors.[7,17,24] Under these circumstances, oxygen diffusion is relevant. On the other hand, if hydrogen atoms diffuse on the surface (or within the bulk) and react with a localized oxygen atom ("oxygen route")[14] the mobility of the oxygen atoms is probably less relevant, although it may still contribute to enhancing the formation rates of OH, which is the first step for H_2O production.

Other reactions that have been proposed to be involved in water formation, in which atomic oxygen does not directly participate, include the collision of H_2 and OH to form $H + H_2O$, and $H_2O_2 + H \rightarrow H_2O + OH$.[7] Finally, water could also be formed by the addition of hydrogen atoms to molecular oxygen:[19,25]

$$H + O_{2,s} \rightarrow HO_{2,ads} \tag{3}$$

$$H + HO_{2,ads} \rightarrow H_2O_{2,ads} \tag{4}$$

$$H + H_2O_{2,ads} \rightarrow H_2O_{ads} + OH_{ads} \tag{5}$$

where O_2 is formed from the collision of two atomic oxygens. Hence, oxygen mobility is relevant at various levels in water synthesis under interstellar conditions.

As it is difficult to investigate the reactions or movement of the atomic species occurring remotely, laboratory experiments and computer simulations are essential to elucidate elementary processes and understand the process in more detail. In the present work, classical molecular dynamics (MD) simulations are used to estimate the mobility of atomic oxygen in bulk amorphous water ice at low temperature. As evidenced above, the migration of atomic oxygen is essential for water formation for several scenarios, but quantitative data on whether and to what extent it is mobile are lacking. In addition, the free energy barrier for the migration and the solvation free energy of the atomic oxygen in amorphous ice are estimated.

One essential ingredient in atomistic simulations is the quantitative and meaningful description of the intermolecular interactions. In conventional MD simulations, point charges are used to describe the electrostatic interactions in the system. A conventional point-charge (PC) model is not suitable to describe atomic oxygen, as its total charge is zero. However, an oxygen atom in its ground electronic state (^3P) has two unpaired electrons and the electronic density is not spherically symmetric, which gives rise to a nonzero quadrupole moment on the atom. Such nonspherical charge distributions can be conveniently described within an atomic orbital framework which is, for example, provided by a multipolar representation. As has been shown in previous works,[26–29] the inclusion of multipole moments in the simulation systems can give more realistic results in MD simulations. It has been shown recently that a multipolar force field with physically motivated parametrization can describe nuclear dynamics, spectroscopy, and thermodynamics quantitatively.[28,30,31]

In the present work a quadrupolar model for atomic oxygen is employed to realistically describe the dynamics and cavity migration in amorphous ice at low temperatures. First, the computational methods are presented. Next, oxygen migration and its solvation free energy are characterized, and finally conclusions are drawn.

2   Computational methods

2.1   Intermolecular interactions and molecular dynamics

All molecular dynamics (MD) simulations were carried out with the CHARMM program[32] with provisions for multipolar interactions.[26] The system considered includes one oxygen atom, O, inside a cubic box of amorphous ice of edge length $\approx$ 31 Å containing 997 water molecules. The density of the system is close to that of water at standard conditions. A snapshot of the simulation system is shown in Fig. 1.

All simulations were carried out with periodic boundary conditions (PBCs). Nonbonded interactions (electrostatic and van der Waals) for particles whose interatomic distances are larger than 12 Å were ignored. A shift function was used for the electrostatic interactions and the van der Waals interactions were switched between 10 and 12 Å. A Nosé–Hoover thermostat[33,34] was used in the NVT simulations, and the equations of motion were propagated with a time step of $\Delta t = 0.4$ fs to account for the flexible water molecules (see below).

For the water molecules, a flexible model based on the parametrization by Kumagai, Kawamura, and Yokokawa (KKY) was used.[35] The functional form of the KKY potential for the stretching and bending energies is:

$$E_{str} = D_e\{1 - \exp[-\beta(r - r_0)]\}^2 - D_e \tag{6}$$

and

$$E_{bend} = 2f_k\sqrt{k_1 k_2}\,\sin^2(\theta - \theta_0) \tag{7}$$

where $k_i = 1/\{\exp[g_r(r_i - r_m)] + 1\}$, in which r_i is the distance of one O–H bond of the water molecule and g_r and r_m are the force field parameters reported in the work of Kumagai $et\ al.$[35] The parameters used in the present work are identical

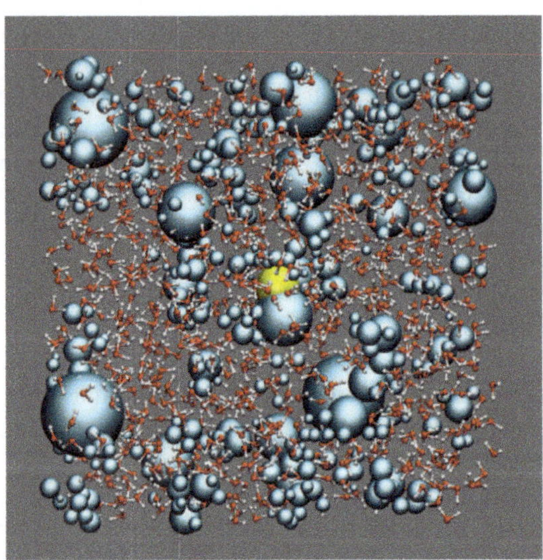

Fig. 1   A snapshot of the system studied. The yellow sphere represents atomic oxygen and the cyan beads delineate the cavities in amorphous ice. Water molecules are shown in ball-and-stick representation.

except for the bending parameter (f_k) which needed to be changed to match the experimental bending frequency. As has been pointed out previously,[36] the value of f_k from the original KKY parametrization seems to be several orders of magnitude too large and overestimates the bending frequency. This is probably related to the fact that the original parameters were fitted to thermodynamic rather than spectroscopic data.[35,37] The present parametrization has also been used in recent work of water mobility at the solid–water interface and found to perform well for such applications.[38]

Electrostatic interactions for water were described by previously determined geometry-independent multipole moments.[27] Point dipole and quadrupole moments were placed on the water oxygen atoms in addition to the point charges. For water hydrogen atoms, only point charges were used. For the atomic oxygen, a point quadrupole moment was placed in order to describe its interactions with the water molecules more accurately (see below). Such a water model has been validated in spectroscopic simulations and correctly reproduces experimentally observed infrared spectra and water diffusion.[38,39]

For the MTP model of the oxygen atom, DFT/B3LYP calculations with the aug-cc-pVQZ basis set were carried out for a single oxygen atom using Gaussian 03 (G03).[40] The level of theory was chosen so as to be consistent with the electrostatics used for the water molecules. Then, the GDMA program[41] was used to obtain multipole parameters up to quadrupole moments. Only one component of the quadrupole tensor, $Q_{20} = -0.948\ ea_0^2$, is nonzero. As a comparison, values obtained from previous calculations are -0.968 (ROHF), -1.021 (CASSCF), and -1.036 (CASPT2) in atomic units. [42] As the quadrupole moment is the only nonvanishing contribution (except for van der Waals interactions), it is of interest to put this into context with other small molecules: the quadrupole moment for atomic oxygen is also comparable to that of molecular NO ($1.14\ ea_0^2$) and somewhat smaller than that of CN^- ($3.67\ ea_0^2$).

When higher multipole moments are employed on interaction centers in MD simulations, local coordinate systems are required. They relate the orientation of the atom-centered MTPs (*i.e.* "orbitals") to the frame of the entire molecule and the laboratory frame.[43] As for a single atomic site, no neighboring atoms rigidly linked to it exist, and the instantaneous reference axis system needs to be set up differently.[44] In the present case, the direction of the axis which gives the lowest energy is determined by scanning the grid points of the solid angle given by HEALPix.[45] In the present work, grid points with $N_{\text{side}} = 2$ were employed, which gives 48 pixels. Thanks to the symmetry of the oxygen atom and the grid points, only half of the pixels are required.

For the Lennard-Jones (LJ) parameters of water, ε and r_{min}, those of the TIP3P model were used in all simulations. For atomic oxygen the values from the universal force field (UFF)[46] were adopted, *i.e.* $\varepsilon = 0.060$ kcal mol^{-1} and $r_{\text{min}}/2 = 1.75$ Å. Overall, the oxygen atoms interact with their environment through charge–quadrupole and van der Waals (O–O$_{\text{water}}$ and O–H$_{\text{water}}$) interactions.

Atomistic simulations for a single oxygen atom in amorphous ice were started by an initial heating and equilibration simulation at 300 K. Then, the temperature of the system was quenched to 50, 100, or 200 K. Upon cooling, the system was equilibrated for $\sim$20 ps of *NPT* simulations. Next, 4 ps of *NVT* equilibration was carried out, followed by production simulations 100 ps in length. Overall, 80 independent trajectories were run, each 100 ps in length, totaling 8 ns of data from which the observables were determined. As ASW has no evident long range structure, it is difficult to directly compare structural determinants. The distribution of pore sizes in the present work is reported in Fig. 4, together with the typical separation between the pores. However, to the best of our knowledge no such information is available from direct experimentation. Nevertheless, it is worth mentioning that an early electron diffraction study at temperatures between 15 K and 188 K found transitions between three amorphous forms of water ice, I_ah (high-density amorphous ice), I_al (low-density amorphous ice), and I_ar (restrained amorphous ice). It was suggested that the persistence of I_ar above 144 K explained the anomalous gas retention and release from water-rich ices at temperatures above 150 K.[2]

For comparison, simulations with a conventional TIP3P water model were also carried out without any electrostatic interactions on the atomic oxygen, *i.e.* a standard force field model. Hence, the oxygen atom only interacted with its environment through van der Waals interactions between the water-oxygen and hydrogen atoms. For this, SHAKE[47,48] was used to constrain the bonds involving hydrogen atoms and the time step was increased to $\Delta t = 1$ fs.

2.2 Analysis and determination of observables

Diffusion coefficient. Traditionally, the diffusion coefficient is determined from the mean square displacement.[49] As in the present case – diffusion of a particle in a constraining and slowly moving environment at low temperature – the process in question is activated (hopping of atomic oxygen from one cavity to another) and slow, such an analysis would require extremely long MD simulations. Hence, a different approach was pursued here, as described in more detail below. Furthermore, simulations with varying interaction strengths between the oxygen atom and the water matrix were also carried out, whereby the nonbonded interactions were scaled by factors of 0.2, 0.4, 0.6, and 0.8, respectively. This will also facilitate diffusion of the oxygen atom and improve the statistics, which can be extrapolated to the full interaction strength.

The number of jumps of the atomic oxygen from one cavity to another in the amorphous ice was determined from the coordinates of the atomic oxygen. First, a running average over 4000 steps was calculated for each coordinate of the atomic oxygen along the x-, y-, and z-directions, and averaged coordinates were taken every 200 steps. The unit vector in the direction of the displacement of the atomic oxygen was constructed for each pair of consecutive coordinate sets. Then the angle γ between two consecutive unit vectors was calculated. If $\gamma \lesssim 78°$, the atomic oxygen was regarded to be moving in the same direction. The value $78°$ was determined from an analysis of the jump distances and directions and upon visual inspection of selected trajectories. If the oxygen atom moves in the same direction for more than 2000 steps (0.8 ps), this sequence of movements has been considered as a jump, as long as the total displacement is larger than 1.5 Å, which will be denoted as Δr_{thr}. The average displacement over all the jumps has also been computed and used as the jump distance Λ.

Alternatively, a phenomenological model based on a jump frequency Γ has been devised, from which a diffusion coefficient D has been estimated according to:[37]

$$D = \alpha \Lambda^2 \Gamma, \tag{8}$$

where α is the number of equivalent paths for a jump and Λ is the jump distance. For two-dimensional diffusion in ice I$_h$, $\alpha = 2$ has been employed,[37] which is also the value used here. However, this value is probably larger in the present situation and could be optimized for the case of three-dimensional diffusion in amorphous ice. As this is only a phenomenological analysis, and more rigorous treatments based on free energy simulations are discussed further below, no attempt was made to refine the value of α. The diffusion coefficients obtained from this analysis are therefore lower bounds. The jump distance Λ was obtained by averaging actual jump distances from the trajectories. The temperature dependence of $D(T)$ is assumed to follow an Arrhenius-type behaviour,[37]

$$D = D_0 \exp\left(-\frac{E_a}{RT}\right), \tag{9}$$

where D_0 is a constant, E_a is the activation energy, and R is the gas constant. From this expression, an activation energy can be estimated based on the number of jumps observed in the simulations.

Free energy simulations. A more rigorous estimate for the transition barrier heights is available from free energy simulations. Such techniques can also be employed to determine solvation free energies and are briefly summarized below.

Umbrella sampling (US) is used to estimate the free energy barrier for the movement of an oxygen atom from one cavity to another. US employs an external biasing potential $U^{(i)}(\delta) = K^{(i)}(\delta - \delta_0^{(i)})^2$ to constrain a system around a value $\delta_0^{(i)}$ of the reaction coordinate, and then determines the probability distribution $P(\delta)$ to find the system at reaction coordinate δ. For each window i a corresponding free energy

$$F^{(i)}(\delta) = -k_B T \ln[P^{(i)}(\delta)] - U^{(i)}(\delta) + C^{(i)} \tag{10}$$

can be obtained, where $C^{(i)}$ is a constant for the i-th window and $P^{(i)}(\delta)$ is the equilibrium distribution in the presence of the biasing potential $U^{(i)}(\delta)$ for the same window. The constant $C^{(i)}$ is in general different for each window, and can be determined using the Weighted Histogram Analysis Method (WHAM).[50,51] In the present work, δ is the distance between the atomic oxygen and the plane at the barrier separating the two cavities which can hold the oxygen atom. The plane is composed of 3 water oxygen atoms and is approximately perpendicular to the path followed by the oxygen atom when it makes a transition from one cavity to another. Overall, 17 windows with δ_0 values ranging from -4.0 to 4.0 Å with a separation of 0.5 Å and a force constant of $K = 3.0$ kcal mol^{-1} Å^{-2} for all intervals were employed. For each window, the system was equilibrated for 2 ps with the umbrella potential, which was followed by 40 ps of simulation during which data collection took place. Six independent snapshots from different jump instances at 100 K with full intermolecular interactions were selected for the umbrella sampling simulations.

Free energy differences between two states A and B can also be determined by alternative methods, including thermodynamic integration (TI),[52,53] free energy perturbation (FEP) simulations,[54] or the Bennett acceptance ratio (BAR) approach.[55] Several methods were used in order to verify consistency between the results obtained. In TI a coupling parameter λ is introduced, which is 0 for A (reactant – or oxygen not solvated in ASW) and 1 for B (product – or oxygen fully solvated in ASW), and the potential energy of the system is written as $U(\mathbf{r}_1,\ldots,\mathbf{r}_N,\lambda) = (1-\lambda)U_A(\mathbf{r}_1,\ldots,\mathbf{r}_N) + \lambda U_B(\mathbf{r}_1,\ldots,\mathbf{r}_N)$. Then, the free energy change for the reaction A $\rightarrow$ B is obtained from

$$\Delta F(A \rightarrow B) = \int_0^1 d\lambda \left\langle \frac{\partial U}{\partial \lambda} \right\rangle_\lambda = \int_0^1 d\lambda \langle U_B - U_A \rangle_\lambda = \int_0^1 d\lambda \langle U_{ow} \rangle_\lambda, \tag{11}$$

where U is the potential energy, $\langle \cdots \rangle_\lambda$ denotes an ensemble average at λ, and U_{ow} are the solute–solvent interactions.[31] The thermodynamic integration is carried out by numerical integration of $\langle U_{ow} \rangle_\lambda$ for λ between 0 and 1, where $\langle \cdots \rangle_\lambda$ is computed by averaging over the distribution $\exp[-\beta U(\mathbf{r}_1,\ldots,\mathbf{r}_N,\lambda)]$ from the simulations performed at a particular value of λ, which include (0.2, 0.4, 0.6, 0.8, and 1.0) in the present work.

In FEP the free energy difference between states A and B can be computed from

$$\Delta F(A \rightarrow B) = -k_B T \ln\left\langle e^{-\beta(U_B - U_A)} \right\rangle_A. \tag{12}$$

While the equation above can be applied if states A and B are very similar, in many cases they can be quite different, such as the system considered in the current work. To reduce the error, the path from A to B is divided into multiple steps, and the free energy difference for the change from A to B is obtained by summing the free energy differences of all consecutive steps. The same λ values as used in the thermodynamic integration are used for FEP in the present work, so as to reuse the trajectories generated for the thermodynamic integration.

The BAR method is an algorithm which uses the fact that any function $f(x)$ satisfying $f(x)/f(-x) = e^{-x}$ and for any value of C, the equation

$$e^{-\beta(\Delta F - C)} = \frac{\langle f(\beta(U_B - U_A - C)) \rangle_A}{\langle f(\beta(U_A - U_B + C)) \rangle_B}, \tag{13}$$

holds. Here, ΔF is the free energy difference and U_A is the potential energy of system A. The optimal choice of $f(x)$ and C has been shown to be $f(x) = 1/(1 + e^x)$ and $C \approx \Delta F$.[55] Similarly to the case of FEP, the path from A to B is divided into multiple steps, and eqn (13) is applied to each consecutive step. The free energy change from A to B is obtained by summing free energy differences of all consecutive steps. The same trajectories are used for TI, FEP, and BAR methods.

3   Results and discussion

3.1   Oxygen migration

The movement of atomic oxygen in bulk amorphous water ice at temperatures of 50, 100, and 200 K was investigated from multiple independent MD simulations each 100 ps in length. The temperature range chosen covers the high side of astrophysically relevant temperatures, which is typically $T \approx 10$ K[17] up to $T \lesssim 50$ K,[11] and goes up to temperatures at which ASW are assumed and have been reported to exist.[2] Such relatively high temperatures in the simulations are called for as the expected mobility of an oxygen atom is expected to be slow and very long simulation times would be required to observe transitions between cavities in unbiased simulations at all. From the temperature dependence of the computed observables, one may extrapolate to estimate the behaviour at lower temperatures.

Fig. 2 shows the temporal variation of the x-, y- and z-coordinates of the atomic oxygen at 50, 100, and 200 K from a representative 100 ps simulation with the KKY water model. Typical jumps, which occur along all three spatial directions in the

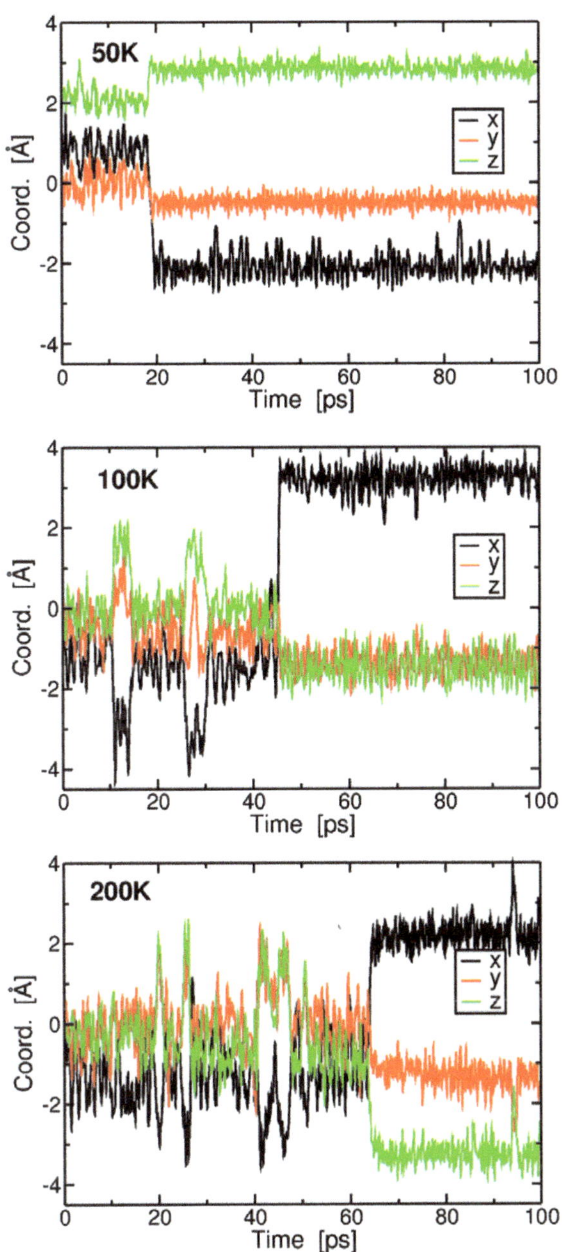

Fig. 2 Temporal variation of the x-, y- and z-coordinates of atomic oxygen in 100-ps simulations at 50, 100, and 200 K with full intermolecular interactions using the KKY water model.

present case, range over 2 to 4 Å, see Fig. 2. The traces also establish that the migration is an activated, jump-like process, and does not resemble conventional diffusion.

Two-dimensional projections of an oxygen translocation at the three simulation temperatures are reported in Fig. 3. On the time scale of the simulation (100 ps) at 100 K (middle panel) the positions of the water molecules are quasi-stationary, as is expected for ice, while the oxygen atom visits at least three

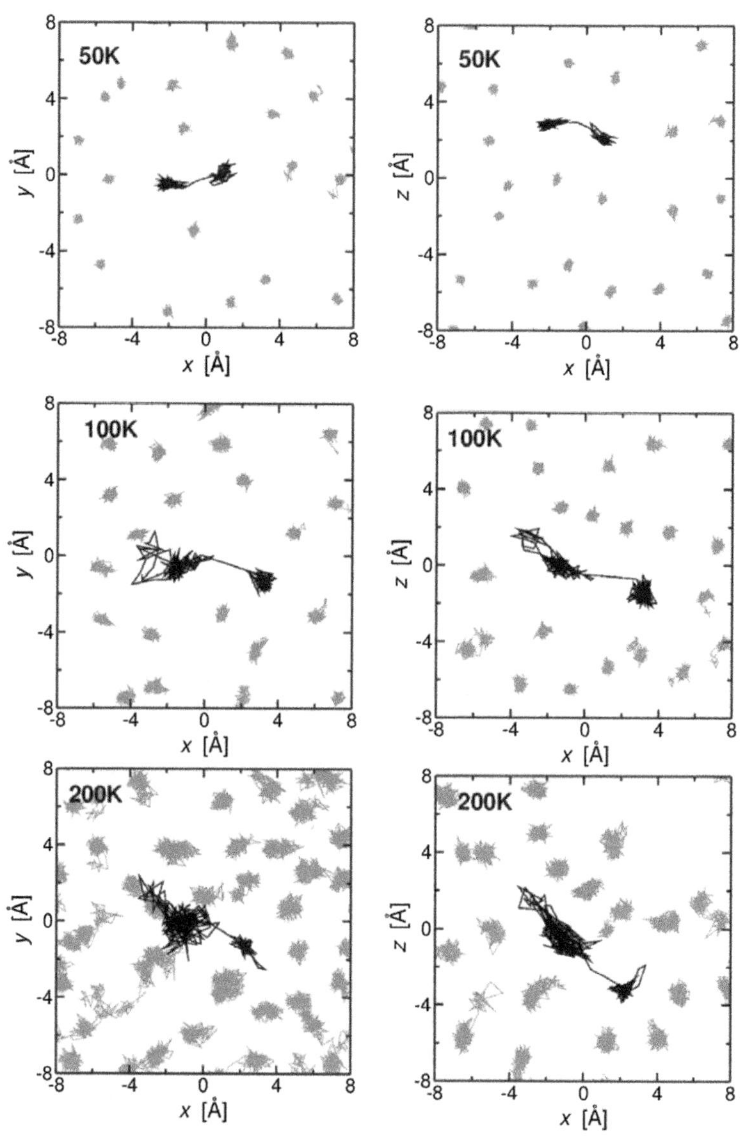

Fig. 3 Trajectories of atomic oxygen projected onto the xy- and xz-planes shown in black in the left and right panels, respectively, from 100-ps simulations at 50, 100, and 200 K. The trajectories of water oxygen atoms located in the range -2.0 Å $< z < 0.5$ Å or -2.0 Å $< y < 0.5$ Å are shown as gray lines.

different sites within the ice matrix. Also, the irregular distribution of the water molecules in both the *xy*- and *xz*-planes is evident. At the temperatures of interest, spontaneous migration of atomic oxygen in amorphous ice between neighboring cavities was observed between 7–33 times on the 1 ns timescale. To increase the number of transitions, simulations were also carried out with decreased interaction strength between the atomic oxygen and water molecules, as the barrier height for the transition is directly influenced by the $O \cdots H_2O$ interactions. This was achieved by simultaneously scaling the electrostatic and van der Waals interactions by a constant factor $\lambda < 1$. Specifically, simulations of 8 ns (80 × 100 ps) were carried out for each value of λ (0.2, 0.4, 0.6 and 0.8).

The migration of an oxygen atom in amorphous ice directly depends on the size and distribution of the cavities. For this, the cavities were analyzed in more detail. The positions and volumes of the cavities were determined with the SURFNET program,[56] which can detect the cavities in a heterogeneous distribution of matter, such as in amorphous ice or in proteins, by placing probe spheres at specific locations which are subsequently grown until they fill space. The volumes of cavities inside amorphous ice were computed for cavities whose volumes are larger than 20 Å^3 from the snapshots at 50, 100, and 200 K, respectively. Only snapshots obtained with full nonbonded interactions between atomic oxygen and water were considered. The distribution of the cavity volumes is shown in the inset of Fig. 4. A useful measure for the probability of finding a cavity within a distance r of another cavity is provided by the radial distribution function $g(r)$ of the cavity centers. From the same trajectories used for the cavity volume analysis, the radial distribution function was determined and is shown in Fig. 4.

3.2 Oxygen diffusion

The number of transitions for atomic oxygen from 80 × 100 ps simulations in amorphous ice is shown in Fig. 5 at various temperatures. As expected, the number of transitions is larger at higher temperature and at decreased intermolecular interaction strength. At 50 K, the number of transitions is generally

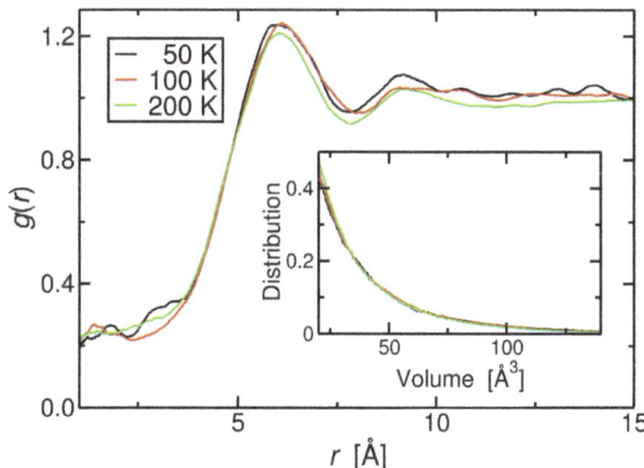

Fig. 4 Radial distribution functions of cavities in amorphous ice at 50, 100, and 200 K. The inset shows the distribution of the volumes of cavities in amorphous ice at 50, 100, and 200 K.

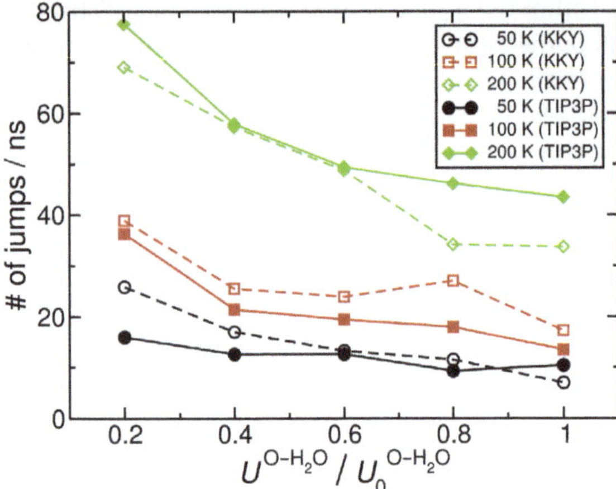

Fig. 5 Average number of jumps per nanosecond of atomic oxygen between cavities in the amorphous ice during the 80 × 100 ps simulations (MTP with KKY, dashed lines) and 80 × 250 ps simulations (standard electrostatics with TIP3P water model, solid lines) at 50, 100, and 200 K, shown in black, red, and green, respectively. The interaction potential between the atomic oxygen and the water molecules, $U^{O\cdots H_2O}$, was scaled by 0.2, 0.4, 0.6, 0.8, and 1.0 from the original value, $U_0^{O\cdots H_2O}$.

small, ~5 transitions per nanosecond, and increases with increasing temperature. Still, due to the small number of transitions, the error in the number of jumps per nanosecond is appreciable. As MTP simulations are quite time consuming, an additional 80 × 250 ps (*i.e.* 20 ns in total) simulations without MTP on atomic oxygen were also carried out using the TIP3P water model with SHAKE. These simulations provide an independent validation and assessment of the importance of refined electrostatics on the oxygen atom. The results are reported in Fig. 5. Due to the increased length of the simulations and the improved statistics, the results are smoother. However, overall the results from MTP and point charge simulations are similar, and suggest that the 10 ns simulations are capable of qualitatively capturing oxygen migration in amorphous ice.

To estimate the error bars of the number of jumps obtained from the simulations, the entire data set was divided into 4 subgroups, and the number of jumps per nanosecond was determined for each subgroup. From this, the standard deviation was determined. The approximate error bar thus obtained for the 80 × 100 ps simulations (MTP and KKY for water) is ~13 ns^{-1}, whereas for the 80 × 250 ps simulations (standard electrostatics and TIP3P) the error is 4 ns^{-1}. The error bars tend to decrease at higher temperature. The positions of water molecules in the amorphous ice for different λ values are similar, as the simulations start from the same initial structures. While separate equilibrations for each λ value were carried out, water molecules do not move away under the conditions used for the simulations.

To quantify the free energy change along the transition paths of atomic oxygen from one cavity to another, umbrella sampling as described in the "Computational methods" section was used. A typical computed free energy profile is shown

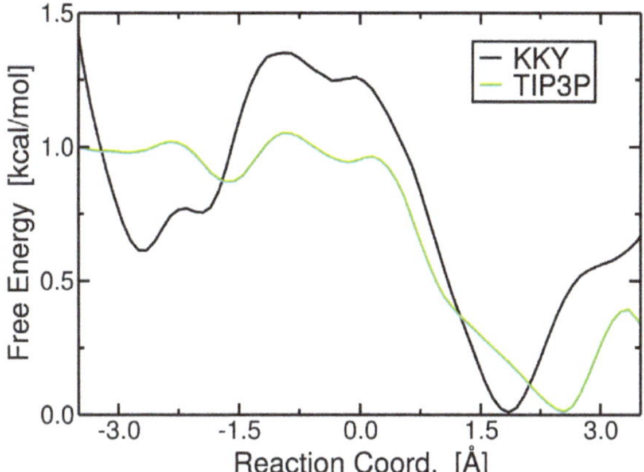

Fig. 6 Free energy profile for a single oxygen migration step from one cavity to the nearest neighbor at 100 K. Umbrella sampling was used for 40 ps for each window with full van der Waals and electrostatic interactions with the KKY water model (black curve). For comparison, the same calculations were carried out with a PC model for oxygen and the TIP3P water model starting from the same initial coordinate file (green curve). The profiles were obtained using 17 windows along the progression coordinate.

in Fig. 6. Umbrella sampling was carried out for both the MTP oxygen with the KKY water model and PC oxygen with the TIP3P water model. For direct comparison, the same initial coordinate set was used for both simulations. The barriers ranged from 0.5 to 1.5 kcal mol^{-1}. It is noted that the barrier height is lower in the case of a conventional force field for the oxygen atom together with the TIP3P water model, which does not involve multipolar interactions. This is also what has been found in simulations of CO in myoglobin.[57] The minima along the migration pathway occur at similar reaction coordinates, but for the MTP charge model, stabilization in the local minima is more pronounced.

For the particular situation encountered here, a local maximum in the free energy profile is found around $\delta \approx -0.9$ Å. At either side of this maximum, local minima of different depths are found. For the umbrella sampling shown in Fig. 6, the estimated barrier height from the right minimum is $\sim$1.4 kcal mol^{-1}, while that from the left is $\sim$0.8 kcal mol^{-1}. The values of the barrier heights from six different sets of umbrella sampling simulations range from 0.7 to 1.9 kcal mol^{-1} (0.7, 0.9, 1.3, 1.4, 1.8 and 1.9 kcal mol^{-1}), hence the situation in Fig. 6 is a typical one.

The diffusion coefficients at different temperatures have also been estimated using the phenomenological relationship in eqn (8), with $\alpha = 2$ and the values of Λ and Γ obtained from the simulations with full intermolecular interactions ($\Lambda \approx$ 2.1, 2.2, and 2.7 Å and $\Gamma \approx$ 7.0, 17.2, and 33.6 ns^{-1} at 50, 100, and 200 K, respectively). This yields $D = 60$, 174, and 480 Å^2 ns^{-1} at 50, 100, and 200 K, respectively. From the diffusion coefficients $D(T)$ at different temperatures (see Fig. 7) an activation energy $E_a \approx 0.27$ kcal mol^{-1} is found.

To better understand the effect of selecting different jumps on the activation energy, additional analysis with $\Delta r_{thr} = 2.0$ Å was carried out. This corresponds to excluding smaller jumps included in the previous analysis with $\Delta r_{thr} = 1.5$ Å. The

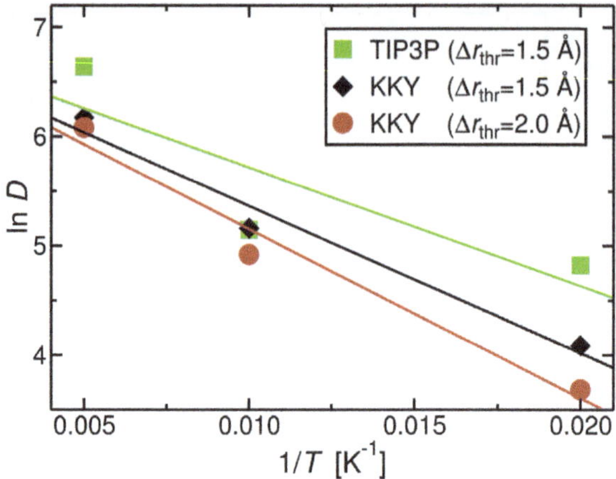

Fig. 7 The plot of ln$D$ *vs.* $1/T$ computed from trajectories generated with the TIP3P water model (green), and those with KKY water model (black and red). The green and black lines were computed with $\Delta r_{thr} = 1.5$ Å (see "Computational methods" section), and the red line with $\Delta r_{thr} = 2.0$ Å. The estimated E_a values from the slopes calculated from the linear regressions are 0.22, 0.27, and 0.31 kcal mol^{-1} (green, black, and red, respectively).

activation energy with $\Delta r_{thr} = 2.0$ Å is 0.31 kcal mol^{-1}, while that with $\Delta r_{thr} = 1.5$ Å is 0.27 kcal mol^{-1}, as displayed in Fig. 7. This shows that selecting only the pronounced jumps increases the activation energy. It is also noted that the activation energy obtained from trajectories generated with the KKY water model is higher than that with the TIP3P model, which is in agreement with the change in barrier height computed from umbrella sampling simulations using KKY and TIP3P models.

When comparing the barrier height from the umbrella sampling with the activation energy $E_a \approx 0.27$ kcal mol^{-1} from an Arrhenius-like expression, it is found that the magnitude of the barrier height is larger than the activation energy. This difference may be caused by the applied umbrella potential which can affect the local structure in the umbrella sampling simulations and thus can lead to changes in the barrier height. However, as no appreciable water restructuring was observed during the umbrella sampling simulations, this is less likely. Rather, the phenomenological character of eqn (8), insufficient sampling of the unbiased MD simulations, and incomplete knowledge of the number of independent paths (α) suggest that typical barrier heights for oxygen migration in amorphous ice are of the order of 1 kcal mol^{-1}. This agrees quite well with recent experiments for oxygen diffusion on water ice at temperatures between 6 and 25 K, which found a best-fit value for the activation energy of 520 K (1.0 kcal mol^{-1}) and a range of acceptable model parameters between 300 and 500 K (0.6 to 1.0 kcal mol^{-1}).[20] Previous values employed in astrophysical models and reported in the literature range from 240 K (0.5 kcal mol^{-1}) (oxygen diffusion in solid xenon),[58] to 400 K (0.8 kcal mol^{-1})[17] and 900 K (1.8 kcal mol^{-1}).[59]

As the linearity in ln$D$ *vs.* $1/T$ shown in Fig. 7 is not fully apparent, errors in the estimated activation energy from the slope cannot be avoided. On the other hand, for the umbrella sampling, more pronounced jumps were selected which are expected to exhibit larger barrier heights. Therefore, the barrier height computed from umbrella sampling is expected to be somewhat larger than the activation energy estimated from the temperature dependence of the diffusion coefficient.

3.3 Solvation free energy

The solvation free energy of atomic oxygen in amorphous ice was computed using TI, FEP, and BAR methods. The results are summarized in Table 1. The computed solvation free energy of atomic oxygen is considerably smaller than that of diatomic anions in water,[31] due to the absence of charge. The differences among the results obtained from different methods are larger in the range of λ between 0.0 and 0.2, while the results from various methods coincide within less than 0.06 kcal mol^{-1} for each interval in the case of intervals with $\lambda \geq 0.2$, as can be seen in the table. Similar behaviour has also been observed in previous work on the solvation free energy of cyanide or hydroxide ion in water,[31] where a much finer grid of λ values was used for $\lambda \leq 0.1$ in order to increase the accuracy in this range. No finer grids were employed in the present calculations as the magnitude of the free energy is small, which makes it difficult to decrease the relative error due to the inherent inaccuracy in the free energy calculation. In the case of TI, it is also possible to decompose the contribution to the solvation free energy into the electrostatic interactions (elec) and van der Waals interactions (vdW). The decomposed values in kcal mol^{-1} are -0.88 (elec) and -0.85 (vdW) at 50 K, -0.84 (elec) and -0.60 (vdW) at 100 K, and -0.89 (elec) and -0.07 (vdW) at 200 K. Without a quadrupole moment on the oxygen atom, the contribution of the electrostatic interactions is zero. It is clear that electrostatic interactions play a significant role, further stabilizing the atom in the environment of the

Table 1 Solvation free energies of atomic oxygen in amorphous ice computed with TI, FEP, and BAR at 50, 100, and 200 K given in kcal mol^{-1}. The values from each λ interval and the sum over all intervals are shown.

Temperature	Range of λ	TI	FEP	BAR
50 K	0.0–0.2	-0.23	0.03	0.03
	0.2–0.4	-0.29	-0.29	-0.29
	0.4–0.6	-0.34	-0.38	-0.34
	0.6–0.8	-0.40	-0.42	-0.39
	0.8–1.0	-0.46	-0.48	-0.44
	Total	-1.72	-1.54	-1.43
100 K	0.0–0.2	-0.15	0.80	0.80
	0.2–0.4	-0.22	-0.23	-0.24
	0.4–0.6	-0.29	-0.32	-0.31
	0.6–0.8	-0.36	-0.36	-0.35
	0.8–1.0	-0.43	-0.42	-0.41
	Total	-1.43	-0.53	-0.51
200 K	0.0–0.2	-0.02	0.84	0.84
	0.2–0.4	-0.10	-0.12	-0.12
	0.4–0.6	-0.19	-0.25	-0.25
	0.6–0.8	-0.28	-0.31	-0.31
	0.8–1.0	-0.37	-0.36	-0.35
	Total	-0.97	-0.19	-0.19

amorphous ice, and this is what should be observed from simulations with and without MTP electrostatics, as shown below.

4   Conclusions

Classical MD simulations of atomic oxygen migration in amorphous ice using physically motivated force fields establish that intersite hopping occurs on the sub-nanosecond time scale at low temperatures ($T \leq 200$ K) with rates ranging from 4.5 to 15.5 ns^{-1}. The barrier heights from the present work are $\sim$1 kcal mol^{-1}, which compares well with estimates of 0.5 kcal mol^{-1} to 1.8 kcal mol^{-1} from previous work, and most favourably with a recent experimental study of oxygen on an amorphous water surface.[20] Together with the present findings (sub-kcal mol^{-1} activation energy and favourable solvation of oxygen in ASW) these insights suggest that oxygen-driven chemistry may play a more important role in interstellar environments than previously assumed.

The fact that conventional PC force fields underestimate barrier heights by up to 50% compared to MTP-based models is consistent with findings for carbon monoxide migration in myoglobin, for which PC-based force fields also system-atically underestimate migration barriers.[57] Free energy barriers computed from rigorous umbrella sampling simulations are larger than the activation energies estimated from a phenomenological model based on the number of intersite jumps and fitting of the Arrhenius-like equation to the temperature-dependence of the diffusion coefficient. This difference may be attributed to insufficient statistics in counting the number of jumps for calculating the diffusion coeffi-cient and the assumed number of pathways α between two neighboring sites.

The results of the present study are essential for rationalizing the formation and abundances of oxygen-containing molecular species in cold interstellar environments. It has generally been assumed that the chemistry in such envi-ronments is primarily driven by hydrogen diffusion. However, given that the number densities of atomic hydrogen and atomic oxygen are comparable[23] and noting that the barrier for oxygen migration on ASW[20] and in ASW (this work) is of the order of 1 kcal mol^{-1} or below, oxygen chemistry will play an important role in both the formation of water (through the O_2 pathway) and the synthesis of larger organic molecules. It will be interesting to extend the present work to oxygen diffusion on ASW surfaces, which has already been done for CO diffusion.[60]

Acknowledgements

The authors gratefully acknowledge financial support from the Swiss National Science Foundation through grant 200021-117810 and to the NCCR-MUST. The authors thank Dr T. Nagy for discussions.

References

1 W. Hagen, A. Tielens and J. Greenberg, *Chem. Phys.*, 1981, **56**, 367–379.

2 P. Jenniskens and D. Blake, *Science*, 1994, **265**, 753–756.

3 J. B. Bossa, K. Isokoski, M. S. de Valois and H. Linnartz, *Astron. Astrophys.*, 2012, **545**, A82.

4 A. Bar-Nun, J. Dror, E. Kochavi and D. Laufer, *Phys. Rev. B*, 1987, **35**, 2427–2435.

5 Y. Oba, N. Miyauchi, H. Hidaka, T. Chigai, N. Watanabe and A. Kouchi, *Astrophys. J.*, 2009, **701**, 464–470.

6 J. Keane, A. Tielens, A. Boogert, W. Schutte and D. Whittet, *Astron. Astrophys.*, 2001, **376**, 254–270.

7 H. M. Cuppen and E. Herbst, *Astrophys. J.*, 2007, **668**, 294–309.

8 A. G. G. M. Tielens, *Rev. Mod. Phys.*, 2013, **85**, 1021–1081.

9 N. Miyauchi, H. Hidaka, T. Chigai, A. Nagaoka, N. Watanabe and A. Kouchi, *Chem. Phys. Lett.*, 2008, **456**, 27–30.

10 R. Papoular, *Mon. Not. R. Astron. Soc.*, 2005, **362**, 489–497.

11 N. Flagey, P. F. Goldsmith, D. C. Lis, M. Gerin, D. Neufeld, P. Sonnentrucker, M. De Luca, B. Godard, J. R. Goicoechea, R. Monje and T. G. Phillips, *Astrophys. J.*, 2013, **762**, 11.

12 K. Hiraoka, T. Miyagoshi, T. Takayama, K. Yamamoto and Y. Kihara, *Astrophys. J.*, 1998, **498**, 710–715.

13 H. Mokrane, H. Chaabouni, M. Accolla, E. Congiu, F. Dulieu, M. Chehrouri and J. L. Lemaire, *Astrophys. J. Lett.*, 2009, **705**, L195–L198.

14 F. Dulieu, L. Amiaud, E. Congiu, J.-H. Fillion, E. Matar, A. Momeni, V. Pironello and J. L. Lemaire, *Astron. Astrophys.*, 2010, **512**, A30.

15 C. Romanzin, S. Ioppolo, H. M. Cuppen, E. F. van Dishoeck and H. Linnartz, *J. Chem. Phys.*, 2011, **134**, 084504.

16 H. M. Cuppen, S. Ioppolo, C. Romanzin and H. Linnartz, *Phys. Chem. Chem. Phys.*, 2010, **12**, 12077–12088.

17 A. G. G. M. Tielens and W. Hagen, *Astron. Astrophys.*, 1982, **114**, 245–260.

18 H. M. Cuppen and E. Herbst, *Astrophys. J.*, 2007, **668**, 294–309.

19 S. Ioppolo, H. M. Cuppen, C. Romanzin, E. F. van Dishoeck and H. Linnartz, *Astrophys. J.*, 2008, **686**, 1474–1479.

20 M. Minissale, E. Congiu, S. Baouche, H. Chaabouni, A. Moudens, F. Dulieu, M. Accolla, S. Cazaux, G. Manicó and V. Pironello, *Phys. Rev. Lett.*, 2013, **111**, 053201.

21 T. P. M. Goumans, C. R. A. Catlow, W. A. Brown, J. Kästner and P. Sherwood, *Phys. Chem. Chem. Phys.*, 2009, **11**, 5431–5436.

22 V. Wakelam and E. Herbst, *Astrophys. J.*, 2008, **680**, 371–383.

23 P. Caselli, T. Stantcheva, O. Shalabiea, V. Shematovich and E. Herbst, *Planet. Space Sci.*, 2002, **50**, 1257–1266.

24 S. Ioppolo, H. M. Cuppen, C. Romanzin, E. F. van Dishoeck and H. Linnartz, *Phys. Chem. Chem. Phys.*, 2010, **12**, 12065–12076.

25 N. Miyauchi, H. Hidaka, T. Chigai, A. Nagaoka, N. Watanabe and A. Kouchi, *Chem. Phys. Lett.*, 2008, **456**, 27–30.

26 N. Plattner and M. Meuwly, *Biophys. J.*, 2008, **94**, 2505–2515.

27 N. Plattner, M. W. Lee and M. Meuwly, *Faraday Discuss.*, 2010, **147**, 217–230.

28 M. W. Lee and M. Meuwly, *J. Phys. Chem. A*, 2011, **115**, 5053–5061.

29 M. W. Lee, N. Plattner and M. Meuwly, *Phys. Chem. Chem. Phys.*, 2012, **14**, 15464–15474.

30 M. W. Lee, J. K. Carr, M. Göllner, P. Hamm and M. Meuwly, *J. Chem. Phys.*, 2013, **139**, 054506.

31 M. W. Lee and M. Meuwly, *Phys. Chem. Chem. Phys.*, 2013, **15**, 20303–20312.

32 B. R. Brooks, C. L. Brooks, III, A. D. Mackerell, Jr, L. Nilsson, R. J. Petrella, B. Roux, Y. Won, G. Archontis, C. Bartels, S. Boresch, A. Caflisch, L. Caves, Q. Cui, A. R. Dinner, M. Feig, S. Fischer, J. Gao, M. Hodoscek, W. Im,

K. Kuczera, T. Lazaridis, J. Ma, V. Ovchinnikov, E. Paci, R. W. Pastor, C. B. Post, J. Z. Pu, M. Schaefer, B. Tidor, R. M. Venable, H. L. Woodcock, X. Wu, W. Yang, D. M. York and M. Karplus, *J. Comput. Chem.*, 2009, **30**, 1545–1614.

33 S. Nosé, *J. Chem. Phys.*, 1984, **81**, 511–519.

34 W. G. Hoover, *Phys. Rev. A: At., Mol., Opt. Phys.*, 1985, **31**, 1695–1697.

35 N. Kumagai, K. Kawamura and T. Yokokawa, *Mol. Simul.*, 1994, **12**, 177–186.

36 C. J. Burnham, J. C. Li and M. Leslie, *J. Phys. Chem. B*, 1997, **101**, 6192–6195.

37 T. Ikeda-Fukazawa, S. Horikawa, T. Hondoh and K. Kawamura, *J. Chem. Phys.*, 2002, **117**, 3886–3896.

38 P. K. Gupta and M. Meuwly, *Faraday Discuss.*, 2013, **167**, 329–346.

39 J. J. Szymczak, F. D. Hofmann and M. Meuwly, *Phys. Chem. Chem. Phys.*, 2013, **15**, 6268–6277.

40 M. J. Frisch, G. W. Trucks, H. B. Schlegel, *et al.*, *Gaussian 03, Revision B.01*, Pittsburgh, PA, 2003.

41 A. J. Stone, *J. Chem. Theory Comput.*, 2005, **1**, 1128–1132.

42 M. Medveď, P. W. Fowler and J. M. Hutson, *Mol. Phys.*, 2000, **98**, 453–463.

43 C. Kramer, P. Gedeck and M. Meuwly, *J. Chem. Theory Comput.*, 2013, **9**, 1499–1511.

44 M. W. Lee and M. Meuwly, *J. Phys. Chem. B*, 2012, **116**, 4154–4162.

45 K. M. Górski, E. Hivon, A. J. Banday, B. D. Wandelt, F. K. Hansen, M. Reinecke and M. Bartelmann, *Astrophys. J.*, 2005, **622**, 759–771.

46 A. K. Rappé, C. J. Casewit, K. S. Colwell, W. A. Goddard, III and W. M. Skiff, *J. Am. Chem. Soc.*, 1992, **114**, 10024–10035.

47 J.-P. Ryckaert, G. Ciccotti and H. J. C. Berendsen, *J. Comput. Phys.*, 1977, **23**, 327–341.

48 M. Yoneya, H. J. C. Berendsen and K. Hirasawa, *Mol. Simul.*, 1994, **13**, 395–405.

49 D. Frenkel and B. Smit, *Understanding Molecular Simulation: From Algorithms to Applications*, Academic Press, New York, 1996.

50 S. Kumar, J. M. Rosenberg, D. Bouzida, R. H. Swendsen and P. A. Kollman, *J. Comput. Chem.*, 1992, **13**, 1011–1021.

51 S. Kumar, J. M. Rosenberg, D. Bouzida, R. H. Swendsen and P. A. Kollman, *J. Comput. Chem.*, 1995, **16**, 1339–1350.

52 J. G. Kirkwood, *J. Chem. Phys.*, 1935, **3**, 300–313.

53 T. P. Straatsma and J. A. McCammon, *J. Chem. Phys.*, 1991, **95**, 1175–1188.

54 R. W. Zwanzig, *J. Chem. Phys.*, 1954, **22**, 1420–1426.

55 C. H. Bennett, *J. Comput. Phys.*, 1976, **22**, 245–268.

56 R. A. Laskowski, *J. Mol. Graphics*, 1995, **13**, 323–330.

57 N. Plattner and M. Meuwly, *Biophys. J.*, 2012, **102**, 333–341.

58 A. Benderskii and C. Wight, *J. Chem. Phys.*, 1996, **104**, 85–94.

59 S. Cazaux, V. Cobut, M. Marseille, M. Spaans and P. Caselli, *Astron. Astrophys.*, 2010, **522**, A74.

60 N. Plattner, J. D. Doll and M. Meuwly, *J. Chem. Phys.*, 2010, **133**, 044506.

Faraday Discussions

PAPER

Polycyclic aromatic hydrocarbons – catalysts for molecular hydrogen formation

A. L. Skov, J. D. Thrower† and L. Hornekær

Received 20th December 2013, Accepted 11th February 2014

DOI: 10.1039/c3fd00151b

Polycyclic aromatic hydrocarbons (PAHs) have been shown to catalyse molecular hydrogen formation. The process occurs *via* atomic hydrogen addition reactions leading to the formation of super-hydrogenated PAH species, followed by molecular hydrogen forming abstraction reactions. Here, we combine quadrupole mass spectrometry data with kinetic simulations to follow the addition of deuterium atoms to the PAH molecule coronene. When exposed to sufficiently large D atom fluences, coronene is observed to be driven towards the completely deuterated state ($C_{24}D_{36}$) with the mass distribution peaking at 358 amu, just below the peak mass of 360 amu. Kinetic models reproduce the experimental observations for an abstraction cross-section of $\sigma_{abs} = 0.01$ Å^2 per excess H/D atom, and addition cross-sections in the range of $\sigma_{add} = 0.55$–2.0 Å^2 for all degrees of hydrogenation. These findings indicate that the cross-section for addition does not scale with the number of sites available for addition on the molecule, but rather has a fairly constant value over a large interval of super-hydrogenation levels.

Introduction

Polycyclic aromatic hydrocarbons (PAHs) are ubiquitous in the interstellar medium (ISM) [1] and are believed to play an important role in determining a range of physical parameters, such as the ionization fraction and heating rates in the ISM,[1] as well as influencing the chemical composition of the ISM through catalysis of chemical reactions.[1–12]

Specifically, observations have suggested that high abundances of PAHs in photo-dissociation/photon-dominated regions (PDRs) are related to high rates of molecular hydrogen formation.[2,3] This has led to the suggestion that PAHs might be involved in H_2 formation in PDRs through the chemisorption of H-atoms on PAHs.[3] The presence of such super-hydrogenated PAH species is supported by

Department of Physics and Astronomy and Intersciplinary Nanoscience Center (iNANO), Aarhus University, Ny Munkegade 120, 8000 Aarhus, Denmark. E-mail: liv@phys.au.dk; Fax: +45 8612 0740; Tel: +45 871 56336
† Current address: Physikalisches Institut, Westfälische Wilhelms-Universität Münster, Germany.

observations which have revealed aliphatic C–H stretch features at 3.4 μm alongside the 3.3 μm aromatic C–H stretch.[4,5] This supports the results of models showing that large PAHs can exist with high degrees of super-hydrogenation in many environments.[6] Although the aliphatic C–H stretch can also arise as a result of the presence of alkane side chains, experimentally obtained IR emission spectra of neutral PAHs with additional peripheral H atoms are more consistent with observations.[7]

Theoretical calculations[10–14] and experimental measurements[15–21] strengthen this hypothesis by demonstrating that PAHs may indeed catalyse molecular hydrogen formation. The catalytic activity of both PAH cations and neutrals have been investigated, since the charge state of PAHs is strongly dependent on the far UV flux.[22] Furthermore, studies of the catalytic activity of both gas phase and condensed phases of PAH molecules have been conducted, mimicking the phase variations between interstellar environments. In diffuse cloud environments PAHs are thought to exist in the gas phase, while in dense clouds IR absorption features attributable to PAHs have been observed toward protostellar objects, indicating the presence of both gas[23] and condensed[24] phase PAHs.

Theoretical studies indicate that both PAH cations and neutrals might act as catalysts for H_2 formation.[10–14] This is supported by experimental investigations demonstrating hydrogen addition to gas phase PAH cations[15–17] and to condensed PAH neutrals on grain surfaces.[18–20]

In a series of measurements employing surface adsorbed coronene ($C_{24}H_{12}$) as a prototypical PAH which possesses three inequivalent binding sites for additional H atoms, the catalytic activity was observed to proceed *via* the formation of super-hydrogenated species, followed by abstraction reactions where excess H-atoms react with incoming H-atoms to form molecular hydrogen.[18,19] Furthermore, it was observed that PAH molecules could abstract H atoms chemisorbed on the underlying graphite surface.[21] Infrared spectroscopy measurements using a 300 K H-atom beam were used to derive an addition cross-section of $\sigma_{add} = 1.1$ Å^2, and an abstraction cross-section of $\sigma_{abs} = 0.06$ Å^2 for each excess H/D atom on the super-hydrogenated coronene molecule,[18] while for higher energy H atom beams, such as the one employed here, a slightly smaller cross-section for the addition of the first H atom to the coronene molecule of $\sigma_{add,first} = 0.55$ Å^2 was found.[19]

Here, we combine quadrupole mass spectrometry data with kinetic simulations to follow the addition of hydrogen atoms to the coronene molecule. We identify addition and abstraction cross-sections for the super-hydrogenated species as a function of the degree of super-hydrogenation, yielding a good match between experimental and simulated data. We show that the measured mass distributions are simulated well by a model employing an abstraction cross-section of $\sigma_{abs} = 0.01$ Å^2 per excess H/D atom, a factor of 6 lower than the measured cross-section for 300 K beams, and an addition cross-section which is initially $\sigma_{add} = 0.55$ Å^2, increasing to a plateau value of *ca.* $\sigma_{add} = 2.0$ Å^2 for intermediate degrees of hydrogenation before falling off to a value around $\sigma_{add} = 1.0$ Å^2 for addition of the final six hydrogen atoms. The findings indicate that the cross-section for addition does not scale with the number of available sites for addition on the molecule, but rather has a fairly constant value over a large interval of super-hydrogenation levels.

Methods

Experimental methods

All measurements were performed in an ultrahigh vacuum (UHV) chamber in which a base pressure of $<1 \times 10^{-10}$ mbar is routinely obtained. The highly oriented pyrolytic graphite (HOPG, SPI grade 1) substrate was supported in a Ta holder fixed to a water-cooled copper mount. This grade of HOPG is characterised by large grain sizes and a low density of step edges and grain boundaries. Sample heating was performed through electron bombardment on the rear of the sample holder. The HOPG was cleaved with adhesive tape prior to mounting in the UHV chamber and annealed to 1200 K under UHV conditions to remove any residual contaminants. The temperature of the HOPG was measured with a type-C thermocouple secured between the front face of the substrate and the Ta holder. During all depositions a substrate temperature of 290 ± 1 K was maintained.

Coronene $(C_{24}H_{12}$; Aldrich, sublimed, 99%) films were grown by thermal evaporation from the bulk sample using a home built temperature controlled Knudsen cell type molecular doser. A cell temperature of 438 K was typically used, giving a deposition rate of *ca.* 0.03 monolayers s^{-1} (ML s^{-1}). The distance between the doser and the substrate was minimised to reduce deposition on the Ta holder. A beam of D atoms was produced using a hot capillary thermal cracker to dissociate D_2 (Air Liquide; N30, >99.9%). A capillary temperature of *ca.* 2300 K was used, yielding a typical D-atom flux of 3×10^{14} cm^{-2} s^{-1} to within a factor of two. This estimate was made by considering the operational parameters (capillary temperature, feed pressure, distance to substrate) of the source.[25,26] D, rather than H atoms were used to enable the detection of abstraction–addition reactions leading to the replacement of the original hydrogen atoms on the coronene molecule with deuterium atoms. Temperature programmed desorption (TPD) measurements were performed by heating the substrate linearly to 1200 K at a rate of 1 K s^{-1} with the aid of a PID controller (Lakeshore Model 340). Desorbing coronene and deuterated coronene species were detected with a quadrupole mass spectrometer (QMS; Extrel CMS LLC) with a cross-beam ionization source. The mass range of this QMS ($m/z = 500$) was sufficient to detect the coronene parent ion $(C_{24}H_{12}{}^{+}$; $m/z = 300$) and those of all deuterated species up to the deuterated analogue of perhydrocoronene $(C_{24}D_{36}{}^{+}$; $m/z = 360$). This mass range was scanned in 1 s, allowing for the detection of all deuterated products during an individual TPD measurement.

Simulated mass distributions

To extract addition and abstraction cross-sections from the experimental data, kinetic simulations of the evolution of the mass distribution as a function of D atom fluence were performed. These calculations started with a molecular distribution where all molecules are in the form of coronene $(C_{24}H_{12}$; $m/z = 300$), and the addition of D atoms with a flux of 3×10^{14} cm^{-2} s^{-1}, as in the experiments, was then considered. The molecular distribution was evaluated through a series of time steps. For each time step, there are three possible reactions that molecules can undergo:

$$C_{24}H_xD_y + D \rightarrow C_{24}H_xD_{y+1} \tag{1}$$

$$C_{24}H_xD_y + D \rightarrow C_{24}H_xD_{y-1} + D_2 \tag{2a}$$

$$C_{24}H_xD_y + D \rightarrow C_{24}H_{x-1}D_y + HD \tag{2b}$$

where (1) is an addition reaction and (2a) and (2b) are reactions in which the incoming D atom abstracts a D or H atom to form D_2 or HD respectively. The rate of an addition reaction involving a super-hydrogenated coronene molecule with n excess H/D atoms was determined by the D atom flux and a super-hydrogenation degree dependent cross-section, $\sigma_{add}(n)$, for addition of D atoms to the molecule. Since the cross-section for addition was treated as being only dependent on the number of excess H/D atoms on the molecule, n, it is in essence an average over additions into all available sites on all possible conformers of super-hydrogenated coronene for a given degree of super-hydrogenation. Hence, the simulation is not site specific and provides only an average addition cross-section for a given super-hydrogenation degree. In the case of $n = 0$, where no excess H/D atoms have been added to the coronene molecule, the cross-section derived from previous measurements on the system,[19] $\sigma_{add}(0) = 0.55$ Å^2 was used. For $n > 0$, $\sigma_{ad}(n)$ was a variable parameter. The rate of abstraction reactions scales linearly with the D atom flux, the number of excess H/D atoms already present on the molecule, n, and the abstraction cross-section, σ_{abs}, for each excess H/D atom. No abstraction reactions were allowed for $n = 0$, *i.e.* the molecules will always have at least 12 H/D atoms. Two different models have been investigated, where the abstraction cross-section has either been taken to be equal to the abstraction cross-section measured for a 300 K H atom beam, $\sigma_{abs} = 0.06$ Å^2 per excess H/D atom[18] (simulation 1), or has been optimized to give a better fit to the data, (simulations 2 and 3). The cross-sections for both hydrogen isotopes were taken to be equal. When the molecular distribution had evolved for a time period matching the experimental conditions, the final mass distribution was calculated by modifying the masses of the molecular distribution to reflect the correct $^{13}C : {}^{12}C$ ratio, arising from the natural abundance of ^{13}C in the coronene sample, and by including possible fragmentation in the mass spectrometer ionization process. Fragmentation patterns measured for coronene were used for all super-hydrogenated species, since no information on fragmentation of the super-hydrogenated species is available. The fragmentation leads to the loss of 1 H/D atom in 7.5%, 2 H/D atoms in 9.7%, 3 H/D atoms in 2.8% and 4–5 H/D atoms in less than 1% of all molecules in the distribution.

Results and discussion

To determine cross-sections for the addition of H atoms to (hydrogenated) coronene as a function of the degree of hydrogenation, we prepared a monolayer of coronene by exposing a HOPG surface to the coronene source for 60 s, which was sufficient to produce a film 2–3 ML thick, followed by annealing to 390 K to desorb the multilayers,[11] The coronene monolayer was then exposed to the desired fluence of atomic deuterium, and a temperature programmed desorption (TPD) measurement for all masses from 285 amu to 365 amu was carried out. The measured TPD spectra were time integrated and normalized to the total ion count to give the total relative yield for each mass component. Each component contained multiple species, since several different conformers exist for each hydrogenation degree and different hydrogenation degrees can yield the same mass depending on the ratio of hydrogen and deuterium atoms in the molecule. By repeating the measurement for several increasing D atom fluences, the evolution of the mass distribution with D atom fluence can be obtained. In the following, the experimental data are compared to a series of simulations employing different strategies for determining addition and/or abstraction cross-sections.

In Fig. 1–3 the measured mass distribution for super-hydrogenated coronene as a function of D atom exposure is displayed as black columns. The initial mass distribution displayed in panels (a) before any exposure to hydrogen is dominated by the parent coronene molecule ($m/z = 300$), with minor peaks at mass 301 amu and 302 amu due to the natural abundance of ^{13}C, as well as minor peaks just below mass 300 amu which we ascribe to fragmentation in the QMS ionization region. Following exposure to atomic deuterium we observe a reduction in the desorption yield obtained for the parent ion signal at $m/z = 300$, and the appearance of peaks corresponding to other higher mass species, which are

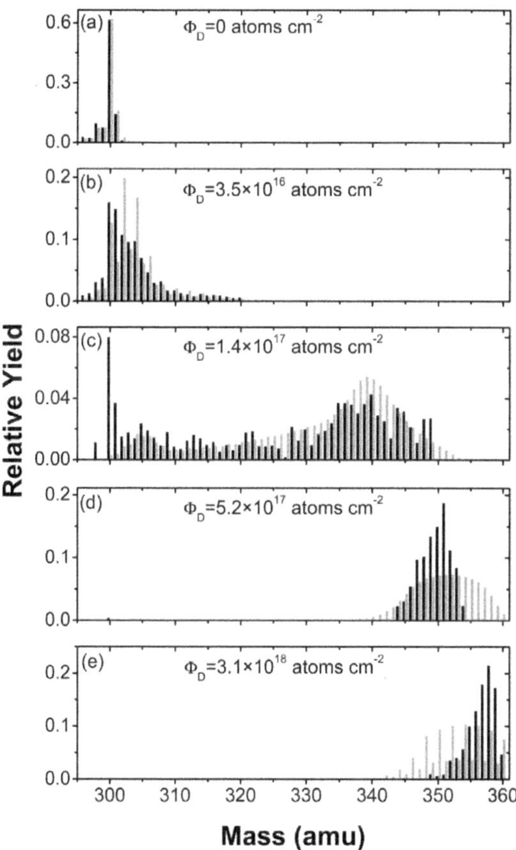

Fig. 1 Experimental (black) and simulated (grey) mass distributions. The cross-sections used for the simulations are given in Table 1 under Simulation 1. Specifically, the abstraction cross-section per excess H/D atom, σ_{abs}, is fixed at the experimentally determined value for 300 K H atom beams: $\sigma_{abs} = 0.06$ Å^2.

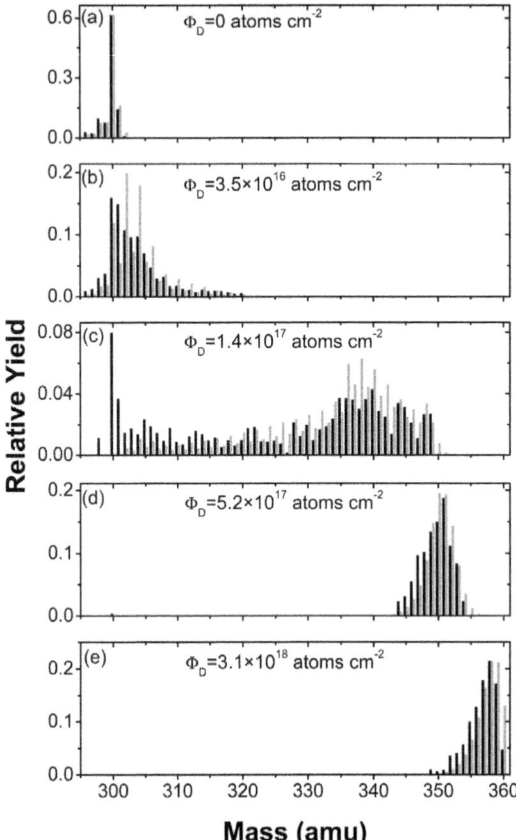

Fig. 2 Experimental (black) and simulated (grey) mass distributions using the cross-sections given in Table 1 under Simulation 2. $\sigma_{abs} = 0.01$ Å^2 has been used, since this value gives the best fit to the mass distributions at long exposures.

ascribed to super-hydrogenated coronene species.[19] For intermediate D atom fluences of $\Phi_D = 1.4 \times 10^{17}$ atoms cm^{-2} (panels (c)), a very broad mass distribution stretching from 300 to 350 amu is observed, similar to previous observations.[19] Here we exposed the sample to even higher D atom fluences of up to $\Phi_D = 3.1 \times 10^{18}$ atoms cm^{-2}. For these highest fluences, the mass distribution is observed to narrow and peak at mass 358 amu, with a detectable signal even at 360 amu. Such high masses indicate close to complete super-hydrogenation of coronene involving the addition of an excess D atom to all carbon atoms in the molecule, including those on the central ring, and the exchange of almost all H atoms on the parent coronene molecule with D atoms *via* abstraction reactions. It should be noted that the deuterated analogue of the fully hydrogenated perhydrocoronene molecule ($C_{24}D_{36}$) has a mass of 360 amu, and is thus the expected upper limit.

The experimental data are compared to the simulated mass distributions, shown by grey columns, obtained by using the cross-sections displayed in Table 1. For the simulations in Fig. 1, the abstraction cross-section per excess H/D atom (each H/D atom beyond the initial 12 H atoms), σ_{abs}, was held fixed at the

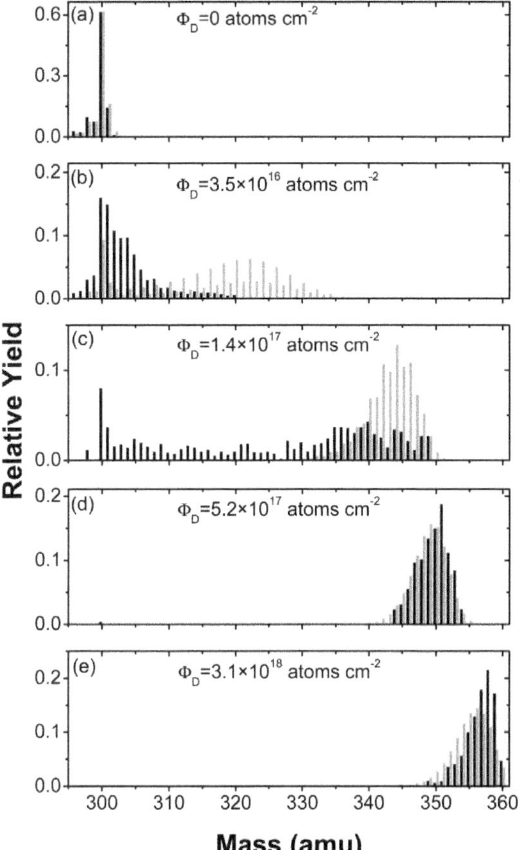

Fig. 3 Experimental (black) and simulated (grey) mass distributions using the cross-sections given in Table 1 under Simulation 3. The addition cross-sections have been chosen to depend linearly on the number of available sites for hydrogenation, $0.15\ \text{Å}^2 \times (24 - n)$, $\sigma_{\text{abs}} = 0.01\ \text{Å}^2$.

experimentally determined value for a 300 K beam of H atoms: $\sigma_{\text{abs}} = 0.06\ \text{Å}^2$. The mass distribution at lower masses is found to be reproduced well by the series of addition cross-sections displayed in Table 1, Simulation 1. The addition cross-sections used in this simulation fall into three groups: (i) for small n, $\sigma_{\text{add}}(n)$ is small and matches the initial addition cross-section, $\sigma_{\text{add}}(0) = 0.55\ \text{Å}^2$, increasing to $\sigma_{\text{add}}(3) = 1.0\ \text{Å}^2$ for the addition of the 4th D atom. (ii) For intermediate n, $\sigma_{\text{add}}(n)$ is at a larger constant value of $\sigma_{\text{add}}(n) = 2.5\ \text{Å}^2$, while (iii) for $n > 17$, $\sigma_{\text{add}}(n)$ drops to a smaller value of $\sigma_{\text{add}}(n) = 1.3\ \text{Å}^2$.

The small value for the addition cross-section at low n reflects the small cross-section for the very first addition reaction, for which an energy barrier of 60 meV is predicted by density functional theory (DFT) calculations.[10] Although these calculations reveal reduced energy barriers for subsequent addition reactions on neighbouring sites, the majority of addition sites on the molecule will still have high barriers, thus resulting in an overall small cross-section for the first few addition reactions. The good correspondence between the measured and

Table 1 Abstraction (σ_{abs}) and addition ($\sigma_{add}(n)$) cross-sections used in the simulations displayed in Fig. 1–3. All values are in units of Å^2

	Simulation 1	Simulation 2	Simulation 3
σ_{abs} per site	0.06	0.01	0.01
$\sigma_{add}(0)$	0.55	0.55	0.55
$\sigma_{add}(1)$	0.55	0.55	3.45
$\sigma_{add}(2)$	0.55	0.55	3.30
$\sigma_{add}(3)$	1.00	1.00	3.15
$\sigma_{add}(4)$	2.50	2.00	3.00
$\sigma_{add}(5)$	2.50	2.00	2.85
$\sigma_{add}(6)$	2.50	2.00	2.70
$\sigma_{add}(7)$	2.50	2.00	2.55
$\sigma_{add}(8)$	2.50	2.00	2.40
$\sigma_{add}(9)$	2.50	2.00	2.25
$\sigma_{add}(10)$	2.50	2.00	2.10
$\sigma_{add}(11)$	2.50	2.00	1.95
$\sigma_{add}(12)$	2.50	2.00	1.80
$\sigma_{add}(13)$	2.50	2.00	1.65
$\sigma_{add}(14)$	2.50	2.00	1.50
$\sigma_{add}(15)$	2.50	2.00	1.35
$\sigma_{add}(16)$	2.50	2.00	1.20
$\sigma_{add}(17)$	2.50	2.00	1.05
$\sigma_{add}(18)$	1.30	1.00	0.90
$\sigma_{add}(19)$	1.30	1.00	0.75
$\sigma_{add}(20)$	1.30	1.00	0.60
$\sigma_{add}(21)$	1.30	1.00	0.45
$\sigma_{add}(22)$	1.30	1.00	0.30
$\sigma_{add}(23)$	1.30	1.00	0.15

simulated data obtained for the intermediate hydrogenation regime using a constant cross-section indicates that the overall cross-section for addition reactions does not scale significantly with the number of sites available for addition. Finally, the reduced addition cross-section at hydrogenation levels above $n = 17$ may be related to the reduced cross-section for addition reactions to the center sites of the molecule once outer edge and edge sites are fully saturated. The agreement between simulated and experimental data at the highest D-atom fluences is not as good, and the simulated mass distribution is clearly broader than that obtained experimentally. In addition, the simulations show a clear odd–even mass oscillation which is not observed in the experimental data. This odd–even oscillation reflects an almost complete abstraction of all the original H atoms on the coronene parent molecule. Hence, both the broader distribution and clear odd–even oscillation in the simulated data indicate that the value for the abstraction cross-section employed in the simulations is too high. This discrepancy likely arises from the employed cross-sections being obtained from measurements using a 300 K H atom beam, while in the present experiment 2300 K D atoms were used.

Based on theoretical calculations, very small or even vanishing energy barriers are expected for the abstraction of excess H/D atoms, with quantum dynamical calculations on a simplified version of the system indicating that the energy dependence of the reaction is fairly complex.[27] Hence, at present a detailed treatment of the energy dependence of the abstraction cross-section is not possible. In order to investigate the effect of a possible lower abstraction cross-section on the evolution of the mass distribution, simulations using reduced abstraction cross-sections were carried out.

In Fig. 2, experimental and simulated mass distributions using the cross-sections given in Table 1 under Simulation 2 are displayed. An abstraction cross-section of $\sigma_{abs} = 0.01$ Å^2 per excess H/D atom was used, since this value reproduces best the mass distributions at long exposures. This abstraction cross-section allows a reasonable match between measured and simulated data over a wide fluence range. Again, the addition cross-sections giving the best simulation of the experimental data fall into three groups: (i) for small n, $\sigma_{add}(n)$ is small and matches the initial addition cross-section, $\sigma_{add}(0) = 0.55$ Å^2, increasing to $\sigma_{add}(3)$ $= 1.0$ Å^2 for the addition of the 4th D atom. (ii) For intermediate n, $\sigma_{add}(n)$ has a larger constant value of $\sigma_{add}(n) = 2.0$ Å^2, while (iii) for $n > 17$, $\sigma_{add}(n)$ drops to a smaller value of $\sigma_{add}(n) = 1.0$ Å^2. This simulation is seen to give an overall good agreement between the experimental and simulated mass distributions over the entire fluence interval. However, the odd–even variation appearing in the simulated data for short and intermediate H atom fluences is not observed in the experimental data. This could be related to differences in fragmentation in the QMS for odd and even mass H/D atom species, which are not incorporated in the present model.

Fig. 3 shows the best fit for long exposure times using a constant cross-section per available addition site (*i.e.* the addition cross-sections were chosen to depend linearly on the number of available sites for hydrogenation: $\sigma_{add}(n) = (24 - n) \times 0.15$ Å^2, see Table 1, Simulation 3). Such models could not reproduce the evolution in the mass distribution over the entire interval of D atom fluences, indicating that the cross-section for H addition does not scale significantly with the number of available sites. Thus, the set of cross-section values used in Simulation 2 yield the best agreement with the experimental mass distributions.

In Fig. 4, the evolution of several specific masses as a function of D atom exposure is displayed and compared to those obtained through the model simulations discussed above. Low masses are seen to initially increase in intensity with D atom fluence as they are formed through addition reactions. With increasing fluence these masses go through a maximum and then decrease as they are further hydrogenated to higher mass species. For the largest fluences, an increase in the intensity of the highest mass species, above 350 amu, demonstrates that the system is, under our experimental conditions, driven towards complete hydrogenation. The best fit over a wide range of masses is observed for the model used in Simulation 2 (see Fig. 2) (red curve) with a low abstraction cross-section of $\sigma_{abs} = 0.01$ Å^2 per excess H/D atom and addition cross-sections ranging from $\sigma_{add}(n) = 0.55–2.0$ Å^2. Fig. 4(f) shows the total ion count, integrated over the entire mass range considered, as a function of D atom fluence. The total ion count undergoes a sharp decrease for small fluence, reaching a limiting value of around 25% of the initial value. While the QMS sensitivity may be species dependent, it is unlikely that this effect can account for the observed decay in total ion count. This suggests that up to 75% of the originally deposited coronene molecules may be lost from the surface prior to detection during the TPD. Several factors may be responsible for such a loss. It is known that the super-hydrogenated coronene species formed through D atom addition are less strongly bound

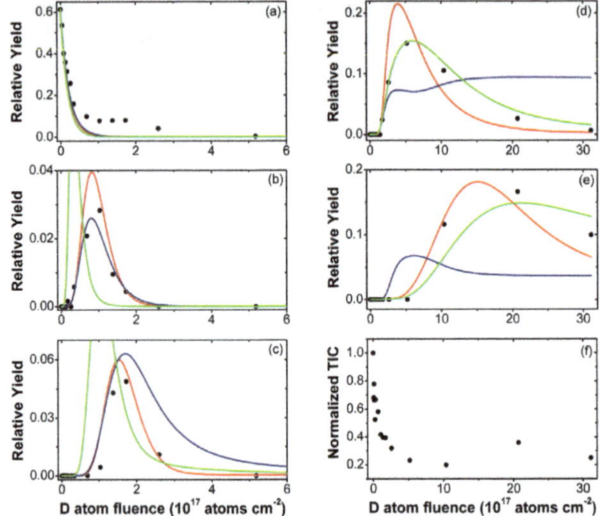

Fig. 4 Evolution in the abundance of (a) mass 300 amu, (b) 320 amu, (c) 340 amu, (d) 350 amu and (e) 355 amu as a function of D atom exposure. Black dots are the experimental data, blue curves are Simulation 1 (see Fig. 1) with a high abstraction cross-section, red curves are Simulation 2 (see Fig. 2) with a lower abstraction cross-section, while green curves are Simulation 3 (see Fig. 3), with a constant cross-section per available addition site. (f) Evolution of the total ion count.

to the HOPG surface.[19,28] Furthermore, according to DFT calculations,[10] many of the addition reactions are exothermic. The energy released may be sufficient to result in desorption of the less strongly bound hydrogenated species. As a result, the reported cross-sections must be considered lower limits.

Conclusions

The experimental measurements presented here demonstrate that when exposed to sufficiently large hot D atom fluences, coronene is driven towards the completely deuterated state ($C_{24}D_{36}$, $m/z = 360$), with the mass distribution peaking at $m/z = 358$, just below the peak mass of 360. Furthermore, a significant loss of molecules from the surface is observed during the hydrogenation process. This loss could potentially be ascribed to chemi-sputtering. The measured mass distributions are simulated well by a model employing an abstraction cross-section of $\sigma_{abs} = 0.01$ Å^2 per excess H/D atom, a factor of 6 lower than the measured cross-section for 300 K beams, and an addition cross-section which is initially $\sigma_{add} = 0.55$ Å^2, increases to a plateau value of $ca.$ $\sigma_{add} = 2.0$ Å^2 for intermediate degrees of hydrogenation, before falling off to a value around $\sigma_{add} = 1.0$ Å^2 for the addition of the final six hydrogen atoms. The addition cross-sections thus vary only within the range $\sigma_{add}(n) = 0.55$–2.0 Å^2 for all degrees of hydrogenation. These findings are in good agreement with the average addition cross-sections measured for low energy D atom beams of $\sigma_{add} = 1.1$ Å^2, and indicate that the cross-section for addition does not scale with the number of available sites for addition on the molecule, but rather has a fairly constant value over a large interval of super-hydrogenation levels. This may indicate that the incoming D atoms are mobile on the coronene molecule and are able to scan the molecule for available addition sites. Similar behaviour has been observed for the hydrogenation of graphite.[29]

The presented results extend previous investigations of the hydrogenation and catalytic activity of coronene[9,18,19] by providing experimental evidence of the formation of even higher degrees of super-hydrogenation of coronene *via* addition reactions. Furthermore, the kinetic simulations have provided quantitative values for the cross-section for addition reactions as a function of the degree of hydrogenation of the coronene molecule. The experimental observation of a coronene species with very high degrees of super-hydrogenation indicates that such species could be formed under interstellar conditions, lending support to the interpretation of observations of IR features attributed to super-hydrogenated PAHs.[4,5] Given the presence of neutral PAH species within PDRs, the present results may have implications for H_2 formation in such regions. In particular, the cross-section for molecular hydrogen formation *via* abstraction reactions obtained from measurements with 300 K H atoms is sufficiently high to explain the observed increased molecular hydrogen formation rates in these regions, provided that the PAHs display very high degrees of super-hydrogenation (the cross-section scales with the number of excess H atoms on the molecule).[18] The data reported here, which suggest that the addition cross-section is largely independent of the degree of hydrogenation, indicate that this may indeed be the case. However, further investigations are needed to determine the average hydrogenation degree of PAHs in these regions. In particular, the H atom energy dependence of the cross-sections for H atom addition and molecular hydrogen forming abstraction reactions, as well as the cross-sections for UV induced hydrogen loss from super-hydrogenated PAHs, must be determined.

Acknowledgements

We acknowledge financial support from the European Research Council under ERC starting grant "HPAH" no. 208344, and the European Commission's 7th Framework Programme through the "LASSIE" ITN under grant agreement number 238258.

Notes and references

1 A. G. G. M. Tielens, *Rev. Mod. Phys.*, 2013, **85**, 1021–1081.

2 E. Habart, F. Boulanger, L. Verstraete, G. P. des Fortes, E. Falgarone and A. Abergel, *Astron. Astrophys.*, 2003, **397**, 623–634.

3 E. Habart, F. Boulanger, L. Verstraete, C. M. Walmsley and G. P. des Forets, *Astron. Astrophys.*, 2004, **414**, 531–544.

4 M. P. Bernstein, S. A. Sandford and L. J. Allamandola, *Astrophys. J.*, 1996, **472**, L127.

5 G. C. Sloan, J. D. Bregman, T. R. Geballe, L. J. Allamandola and E. Woodward, *Astrophys. J.*, 1997, **474**, 735.

6 V. Le Page, T. P. Snow and V. M. Bierbaum, *Astrophys. J.*, 2003, **584**, 316.

7 D. R. Wagner, H. Kim and R. J. Saykally, *Astrophys. J.*, 2000, **545**, 854; C. W. Bauschlicher, *Astrophys. J. Lett.*, 1998, **509**, L125–L127.

8 V. Le Page, T. P. Snow and V. M. Bierbaum, *Astrophys. J.*, 2009, **704**, 274–280.

9 E. Rauls and L. Hornekær, *Astrophys. J.*, 2008, **679**, 531–536.

10 P. Cassam-Chenaï, F. Pauzat and Y. Ellinger, *AIP Conf. Proc.*, 1994.

11 *Molecules and Grains in Space*, ed. I. Nenner, AIP, Melville, NY, p. 543; C. W. Bauschlicher, *Astrophys. J.*, 1998, **509**, L125.

12 M. P. Bernstein, J. E. Elsila, J. P. Dworkin, *et al.*, *Astrophys. J.*, 2002, **576**, 1115.

13 V. Le Page, T. P. Snow and V. M. Bierbaum, *Astrophys. J.*, 2009, **704**, 274.

14 P. M. Goumans, *Mon. Not. R. Astron. Soc.*, 2011, **415**, 3129–3134.

15 T. P. Snow, V. Le Page, Y. Keheyan and V. M. Bierbaum, *Nature*, 1998, **391**, 259–260.

16 L. Boschman, G. Reitsma, S. Cazaux, T. Schlatholter, R. Hoekstra, M. Spaans and O. Gonzalez-Magana, *Astrophys. J. Lett.*, 2012, **761**, L33.

17 B. Klærke, Y. Toker, D. B. Rahbek, L. Hornekær and L. H. Andersen, *Astron. Astrophys.*, 2013, **549**, A84.

18 V. Mennella, L. Hornekær, J. Thrower and M. Accolla, *Astrophys. J. Lett.*, 2012, **745**, L2.

19 J. D. Thrower, B. Jørgensen, E. E. Friis, S. Baouche, V. Mennella, A. C. Luntz, M. Andersen, B. Hammer and L. Hornekær, *Astrophys. J.*, 2012, **752**, 3.

20 J. D. Thrower, E. E. Friis, A. L. Skov, B. Jørgensen and L. Hornekær, *Phys. Chem. Chem. Phys.*, 2014, **16**, 3381.

21 Y. Fu, J. Szczepanski and N. C. Polfer, *Astrophys. J.*, 2012, **744**, 61.

22 E. L. O. Bakes and A. G. G. M. Tielens, *Astrophys. J.*, 1998, **499**, 258.

23 J. D. Bregman and P. Temi, *Astrophys. J.*, 2001, **554**, 126.

24 J. E. Chiar, A. G. G. M. Tielens, D. C. B. Whittet, *et al.*, *Astrophys. J.*, 2000, **537**, 749.

25 K. G. Tschersich and V. von Bonin, *J. Appl. Phys.*, 1998, **84**, 4065–4070.

26 K. G. Tschersich, *J. Appl. Phys.*, 2000, **87**, 2565–2573.

27 M. Bonfanti, S. Casolo, G. F. Tantardini and R. Martinazzo, *Phys. Chem. Chem. Phys.*, 2011, **13**, 16680–16688.

28 J. D. Thrower, E. E. Friis, A. L. Skov, L. Nilsson, M. Andersen, L. Ferrighi, B. Jørgensen, S. Baouche, R. Balog, B. Hammer and L. Hornekær, *J. Phys. Chem. C*, 2013, **117**, 13520–13529.

29 H. Cuppen and L. Hornekær, *J. Chem. Phys.*, 2008, **128**, 174707–13529.

PAPER

Electron induced chemistry: a new frontier in astrochemistry

Nigel J. Mason, Binukumar Nair, Sohan Jheeta and Ewelina Szymańska

Received 31st January 2014, Accepted 11th February 2014
DOI: 10.1039/c4fd00004h

The commissioning of the ALMA array and the next generation of space telescopes heralds the dawn of a new age of Astronomy, in which the role of chemistry in the interstellar medium and in star and planet formation may be quantified. A vital part of these studies will be to determine the molecular complexity in these seemingly hostile regions and explore how molecules are synthesised and survive. The current hypothesis is that many of these species are formed within the ice mantles on interstellar dust grains with irradiation by UV light or cosmic rays stimulating chemical reactions. However, such irradiation releases many secondary electrons which may themselves induce chemistry. In this article we discuss the potential role of such electron induced chemistry and demonstrate, through some simple experiments, the rich molecular synthesis that this may lead to.

1 Introduction

The next decade promises to be the 'decade of astrochemistry', with recent discoveries preluding an impending explosion of chemical data from new telescopes (such as ALMA) that will reveal the 'molecular universe'.[1] The identification of now more than one hundred eighty molecular species in the Interstellar Medium (ISM) (http://www.astro.uni-koeln.de/cdms/molecules, 2013), ranging from the simplest (and most abundant) diatomics H_2 and CO, to the more complex organics (including aldehydes, esters and ketones), provides significant challenges to our understanding of chemistry under the 'extreme conditions' that prevail in the ISM. The ISM is both cold (<10 K) and empty (pressures equivalent to 10^{-13} Torr) such that it may appear that reaction rates will be slow even when two reactants collide in the vastness of space, negating the probability of molecular assembly. Faced with this apparent misnomer a new hypothesis has been formulated that suggests chemical synthesis may occur on the surfaces of small (micron) sized dust grains found in the ISM (with about 1% abundance by mass). These dust grains act as 'chemical factories', accreting simple molecular material over the long lifetime of the ISM dust clouds ready for 'triggering' by an

The Department of Physical Sciences and CEPSAR, The Open University, Walton Hall, Milton Keynes, MK7 6AA, United Kingdom. E-mail: nigel.mason@open.ac.uk; Fax: +441908 6554192; Tel: +44 1908 655253

external energy source to form more complex molecules that are subsequently desorbed from the grain surface to be detected by terrestrial observations based on their spectroscopic signatures (in microwave, THz and IR). Direct reactions may occur when an impinging atom/molecule interacts with the icy surface, *e.g.* the recombination of two H atoms, $H + H \rightarrow H_2$, is believed to be the only method for forming the H_2 background that dominates space. Similarly, hydrogenation of CO on dust grains may lead to the formation of methanol in a four step process:[2]

$$CO + H \rightarrow HCO \rightarrow H_2CO \rightarrow CH_3OH$$

However, such routes are speculative and appear inefficient, with many integrated steps being needed to create the large species such as cyanomethanimine, $NHCHCN$, and ethanamine, $CH_3CH_2NH_2$, recently identified using radio-astronomy.[3] Since cyanomethanimine is believed to be one step in the process that can lead to the nucleobase adenine, while ethanamine is thought to play a role in forming alanine, one of the twenty amino acids in the genetic code, the presence of such complex species in the ISM suggests that prebiotic chemistry seeding life itself may start in the ISM and be an integral part of the star and planet formation process.

Ice covered dust grains in the ISM are embedded in the interstellar radiation field comprising cosmic rays (90% of which are protons), and once dust clouds collapse to form a protostar may be subject to UV radiation. UV induced photochemistry has been the subject of extensive laboratory studies which demonstrate that irradiation of ice analogues may lead to the synthesis of complex (prebiotic) molecules including amino acids.[4-6] These studies have traditionally used lamp sources to mimic the ISM UV spectrum, but more recently[7] have used synchrotron radiation to irradiate ice analogues, demonstrating that UV induced synthesis is both practical and wavelength dependent, but the product yields were quite low.

Deeper in the cloud the UV field is weak, and chemistry may be induced by deeper penetrating X-rays or cosmic rays (protons) which produce similar products to those observed with UV radiation of the same ice, albeit with different (higher) yields.[8] The similarity in the products suggests that there may be a common chemistry occurring in such irradiation phenomena, and it is now postulated that this may be chemistry induced by electrons. Secondary electrons are produced when any ionising radiation interacts with solid ices,[9] indeed one high energy cosmic ray may release 10^4 secondary electrons in its passage through a single ice covered dust grain, and such electrons may themselves trigger chemistry, with the mean kinetic energy of these electrons below 20 eV. Thus if we are to understand the synthesis of complex molecular species in the ice mantles of dust grains in the ISM, it is necessary to understand the chemistry induced by the most abundant species in any ice processing – electrons, and often low energy electrons close to the ionisation potential of absorbed molecular species. In this paper we will explore typical chemistry induced by electrons within thin layers of ice mantle as experienced on dust grains in the ISM.

2   Experimental apparatus

The electron irradiation experiments described in this paper were performed inside an ultra-high vacuum (UHV) chamber at the Molecular Physics Laboratory at the Open University at Milton Keynes, UK. Fig. 1 presents a schematic illustration of the experimental setup employed. The chamber aims to simulate the physical conditions in the ISM by providing a base pressure of below 2×10^{-8} mbar, with the residual gas being mainly molecular hydrogen, as in the ISM. A clean ZnSe substrate cooled by a closed cycle He cryostat system provides a surface on which ISM analogues may be prepared by 'background' deposition. The vapour is introduced at a rate of 1×10^{-6} mbar over 300 s. Typically, ice layers are formed at 14 K and are amorphous in nature. The substrate may be heated to initiate phase changes (from amorphous to crystalline ice) and to initiate Temperature Programmed Desorption (TPD) to analyse the products of electron irradiation. The ice samples may also be analysed *in situ* by Fourier transform infrared (FTIR) spectrometry using a NICOLET FTIR spectrometer operating in the 3500–800 cm^{-1} region with a spectral resolution of 2 cm^{-1}. Infrared spectra of the clean substrates were acquired before each experiment for background correction purposes.

Irradiation of the ice samples was performed using a collimated electron beam generated from a Kimball Physics electron gun. The electron beam current was measured with a Faraday cup mounted at the end of the electron gun. The bombarded area was about 0.28 cm^2, and the beam current at the sample was estimated to be 1.7 µA, $\sim 10^{13}$ electrons s^{-1}. The employed electron flux at the sample was roughly 3.7×10^{13} electrons cm^{-2} s^{-1} (2.3×10^{15} electrons cm^{-2} min^{-1}). This flux is much higher than that encountered in any astrophysical context (Table 1); for example, the electron flux at the Earth's orbit is about 1×10^{7} electrons cm^{-2} s^{-1},[11] such that one minute of irradiation in the laboratory corresponds to roughly 6 years in space outside the Earth's atmosphere.

In the present set of experiments we used 100–5 keV electrons to irradiate the ices. The penetration depth of 2 keV electrons may be estimated using Monte Carlo simulations drawn from the CASINO code.[12] The penetration depth (for 95% of the beam) is around 100 nm, resulting in a LET (Linear Energy Transfer) of roughly 20 keV m^{-1}. For the employed electron flux, and assuming a typical sample density of 1.16 g cm^{-3}, the energy delivered by the beam into the sample

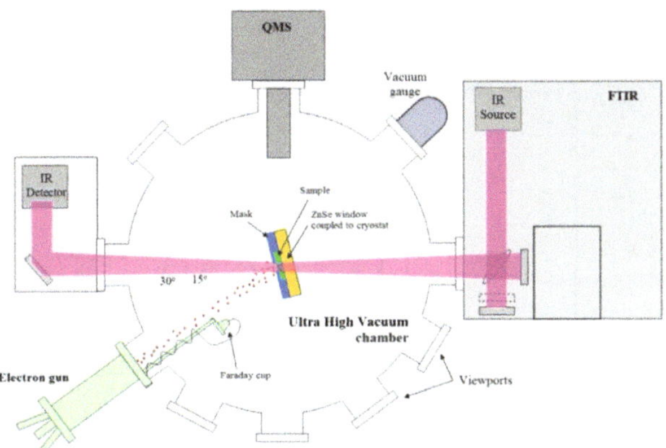

Fig. 1 Schematic diagram of the experimental apparatus used in the electron irradiation experiments. See text for details.

every second (ice layers within the first 100 nm) was about 0.8 eV molecule^{-1}. This value is the same amount of energy delivered by cosmic rays to a typical interstellar ice grain over the few billions (10^9) of years that a dust grain is expected to survive. Thus in the present experiments we may deposit as much energy into the irradiated ice as a typical ice covered dust grain in the ISM may receive in its entire life cycle. Thus we must be aware that in astrochemistry analogue experiments (with photons or electrons) we are greatly enhancing the molecular synthesis rates that occur in these regions. In mitigation it should however be noted that once/as a protostar forms there may be a period of more intense bombardment by such particles.

3.   Electron induced molecular synthesis

3.1   O$_2$ ice films

As a first, simple, example of electron induced synthesis, consider a pure film of molecular oxygen.[13,14] Irradiation of such a film by electrons with energies above ~10 eV leads to the dissociation of molecular oxygen to release reactive O atoms which can react with remaining O$_2$ species. What follows is characteristic of the well-known chemistry in the Earth's stratosphere:

$$O + O_2 + M \rightarrow O_3 + M$$

The formation of ozone requires the presence of a 'third body', M, to stabilise the highly excited nascent ozone. In the Earth's stratosphere, M is commonly N$_2$ or another O$_2$, and thus ozone formation is limited to lower altitudes where the probability of a three body reaction is sufficiently high. In contrast, in a solid film M maybe the surface itself or neighbouring molecules. Hence rates of ozone formation in a solid ice film are high. Further details of the dynamics of such a molecular synthesis may be determined by using a mixture of $^{16}O_2$ and $^{18}O_2$ leading to the formation of many different isotopologues of ozone (Fig. 2).[14]

This simple experiment demonstrates many of the key features of electron ice irradiation experiments. It is efficient (ozone yields are observed as soon as irradiation starts), and depends strongly on the temperature of the ice, which

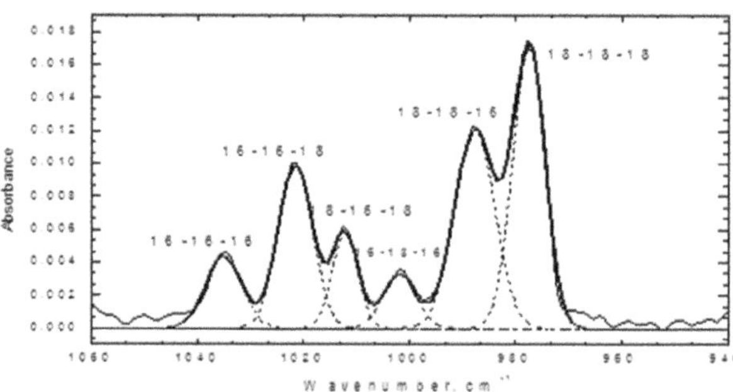

Fig. 2 The formation of six ozone isotopomers and isotopologues, $^{16}O^{16}O^{16}O$, $^{18}O^{18}O^{18}O$, $^{16}O^{16}O^{18}O$, $^{18}O^{18}O^{16}O$, $^{16}O^{18}O^{16}O$, and $^{18}O^{16}O^{18}O$, has been studied in electron-irradiated solid oxygen $^{16}O_2$ and $^{18}O_2$ (1 : 1) ice at 11 K.

Table 1 Estimated UV photon and 1 MeV proton fluxes in different regions of space, and doses experienced by ices. Table adapted from ref. 10

Region	Residence time of tces in region (years)	UV Flux, 10 eV photon (eV cm^{-2} s^{-1})	Energy dose (eV molecule^{-1})	Cosmic ray photons Flux, 1 MeV proton (eV cm^{-2} s^{-1})	Energy dose (eV molecule^{-1})
IS cold dense cloud (0.02 µm ice)	10^6, 10^7	1.4×10^4	0.4	1×10^6	0.3
Oort cloud and KBO Region	4.6×10^9	9.6×10^8	10^8 (top 0.015 µm layer)	Energy dependent flux	150 (0.1 µm), 55–5 (1–5 m)
Jovian satellites, *e.g.* Europa	$<1 \times 10^8$	4×10^{13}	10^{13} (top 0.015 µm layer)	Magnetospheric ions, 7.8×10^{13}	150 (1 cm 10^4 years)
Typical lab experiment	4.6×10^{-4}	8.6×10^{14}	14	7×10^{10}	9

influences the mobility of product reactants. Indeed if the ice is left dormant after a period of irradiation at a low temperature (14 K) (*e.g.* for an hour) and then subsequently warmed ozone yields increase during the warming, since O atoms trapped in the irradiated ice become mobile and react away from the site of their original production.[13] Similarly, in space a radical released by one cosmic ray may remain dormant in the ice for long periods until another reactive species is produced in a second irradiation event or until the grain is warmed (*e.g.* by a shock wave).[15] Ice covered dust grains have therefore been labelled 'radical bombs', with the energy released by such reactions leading to the desorption of products from the surface so they may be observed. Such observations also pose questions as to the validity of simple TPD as a method for determining products of irradiation, since the TPD warming may itself initiate radical reactions, and therefore a combination of *in situ* (FTIR) studies of the irradiated ice with TPD is recommended for such astrochemical studies.

The temperature dependence of the product yields may also be counterintuitive. The ozone monomer column density is highest in ices prepared at the lowest

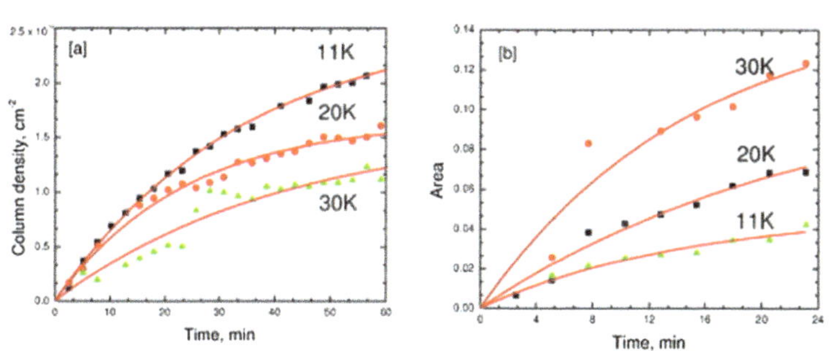

Fig. 3 (a) Temporal growth of the column density of the ozone monomer. (b) Temporal growth of the [O$_3$···O] complex.

temperature of 11 K (Fig. 3), while the column density of the $[O_3 \cdots O]$ complex increases as the temperature of the oxygen matrix is increased. This may be due to the presence of nascent O_2 in the form of weakly bound dimers (O_4) with the O + O_2 reaction occurring within the dimer, a process more commonly called an 'intercluster' reaction, in contrast to the case at higher temperatures, were the dimer is unstable and the ice may be treated as independent O_2 molecules in a solid ice matrix.[13,14]

Having reviewed the possible physico-chemical processes that may occur in the irradiation of ices, we will now illustrate the variety of electron induced chemistry that may occur in some simple ISM ice analogues.

3.2   Irradiation of methanol ice

Methanol (CH_3OH) has been detected through infrared spectroscopy in some low- and high-mass protostars such as W33A and RAFGL 7009, in comets such as Hale–Bopp, and in other solar system bodies such as 5145 Pholus, a centaur planetoid. It is the sixth most abundant molecule after H_2O, CO, CO_2, NH_3 and O_2 in protostars ($\sim$6%) and comets ($\sim$2%), such that it is proposed to be an important precursor of more complex prebiotic molecules, the synthesis of which may be induced by UV or cosmic ray bombardment.

Fig. 4 shows the production of CO and CO_2 from the electron irradiation of methanol. The products and yields show little energy dependence, with 100 eV, 1 keV or 5 keV bombardment producing very similar yields, suggesting that the bulk of the formation is induced by lower energy electrons as the incident electrons thermalize in the ice. Indeed, these results are in good agreement with the low energy electron experimental data.[16,17]

Fig. 4 also shows that ice modification begins to take place almost immediately upon irradiation. The $^{12}CO_2$ column density is lower than that of ^{12}CO, as might be expected if CO liberated from CH_3OH is a precursor to CO_2 formation. Other products readily detected by FTIR spectroscopy of the ice are H_2CO and CH_4, demonstrating the rich chemistry in the ice film upon electron irradiation. The loss of nascent methanol is reflected in the formation of these products by a conservation of total molecular mass (Fig. 5). Irradiation products gradually increase in concentration until they are present with sufficient intensity that they in turn are dissociated by irradiating electrons; this is demonstrated in the

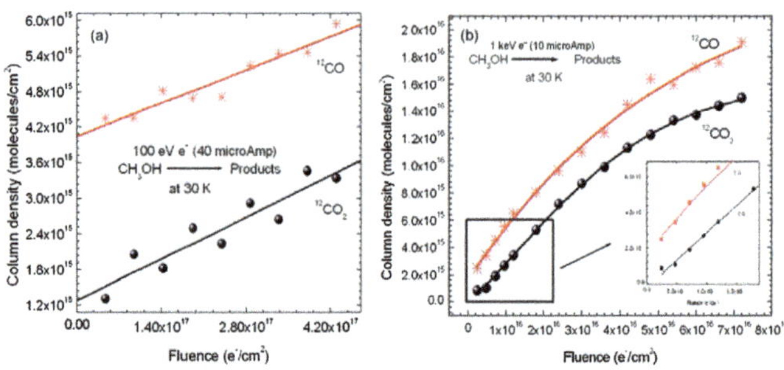

Fig. 4   Formation of CO and CO_2 by electron irradiation of a solid methanol ice film at 30 K by (a) a 100 eV and (b) a 1 keV electron beam.

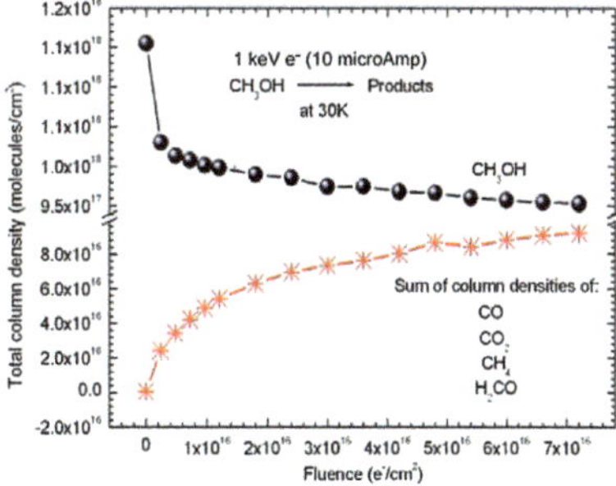

Fig. 5   Destruction rate of the CH₃OH compared with the summed formation of CO, CO₂ H₂CO and CH₄ products during irradiation of a solid methanol ice film at 30 K by 1 keV electrons.

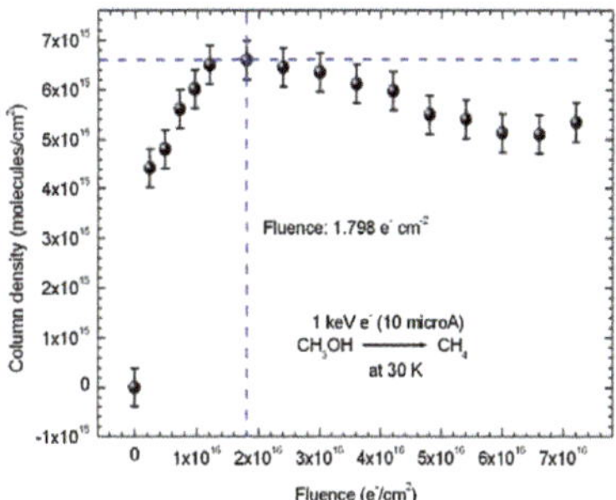

Fig. 6   The column density of CH₄ formed during electron irradiation of a methanol ice at 30 K.

measured CH_4 yield (Fig. 6), where after a fluence of $\sim$1.8 electrons cm^{-2} a reduction in the amount of methane product is observed. A very similar destruction curve for CH_4 was observed by Hudson *et al.*[18] and Baratta *et al.*[19]

3.3   Irradiation of a mixed methanol–ammonia ice

In the ISM the ice grain mantles will be a complex mixture of ices. The richer the ice composition, the more varied the species produced. As a simple example of

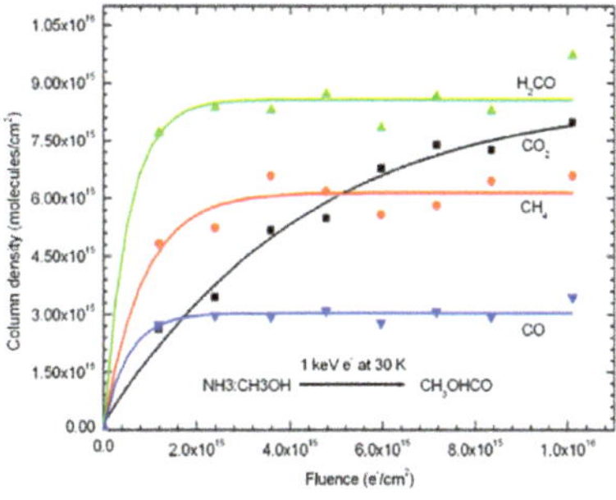

Fig. 7   Product yields measured during irradiation of a NH₃–CH₃OH ice at 30 K by 1 keV electrons.

the range and complexity of the chemistry that may occur in such an irradiated ice mixture, we examined the electron irradiation of a 1 : 1 mixture of ammonia and methanol. Ammonia is prevalent across the ISM in protostars, comets and on the surface of many planetary bodies. Indeed it is now postulated that a large part of the nitrogen available to the early solar system was in the form of highly fractionated ammonia.[20] Once again, high yields of CO and CO_2 are observed (Fig. 7), but in contrast to pure films of methanol the CO_2 column densities are higher than those of CO since the CO produced is being utilised to make other compounds, such as isocyanic acid (HNCO), formed from combining the product of NH_3 dissociation (NH_2 radical) with CO.

$$NH_3 \rightarrow NH_2 + H$$

$$NH_2 + CO \rightarrow HNCO + H$$

Prior to the irradiation of the NH_3–CH_3OH film we hypothesised that we may observe methyl amine (CH_3NH_2), but no signature of this molecule was observed. In contrast, strong yields of formamide ($HCONH_2$) were observed (Fig. 8). $HCONH_2$ can be formed from the reaction of NH_3 with formic acid (HCOOH) or methyl formate (CH_3OHCO), which is a product of the esterification of CH_3OH and HCOOH.

$$HCOOH + NH_3 \rightarrow HCONH_2 + H_2O$$

and/or

$$2CH_3OHCO + 2NH_3 \rightarrow 2HCONH_2 + 2H_2CO + H_2$$

Formic acid and methyl formate (which may also be formed by irradiation of pure methanol ice films[21]) were both observed in this experimental study during the irradiation of NH_3–CH_3OH, and once again reveal the complexity of the electron induced chemistry even in a simple binary ice. It is thought that during 'chemical evolution', $HCONH_2$ may have been used as an informational polymer, since, under certain conditions, it can form acyclonucleosides.[22] These then

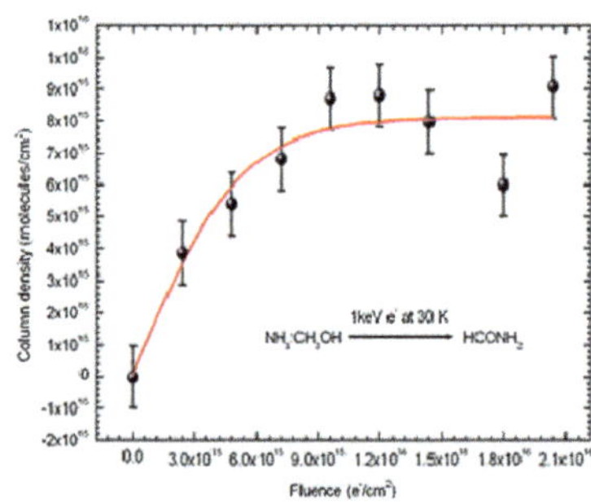

Fig. 8   The formation of formamide during irradiation of a NH_3–CH_3OH ice at 30 K by 1 keV electrons.

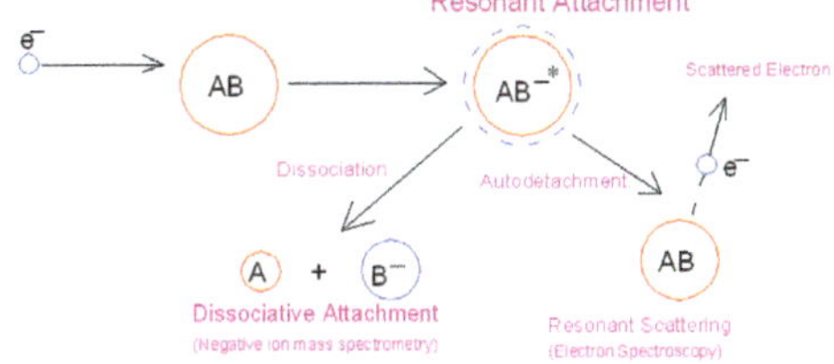

Fig. 9   Schematic of the DEA process.

combine with base purines and pyrimidines to yield acyclonucleotide, which could have played a role very similar to the part played by RNA as a repository vehicle for genetic information in the 'RNA world' hypothesis. Thus, even in a simple binary ice the ingredients for biochemistry may be readily formed by simple electron irradiation.

4.   Electron induced anion chemistry

One of the most exciting developments in recent molecular physics has been the discovery that low energy electrons (LEE) may not only dissociate the molecular target, but may do so at well-defined reaction sites, often leading to almost 100% bond selectivity and thus initiating controlled chemical processing in the local environment. The process by which LEE can induce such bond selective molecular fragmentation is known as dissociative electron attachment (DEA) (Fig. 9). An incident electron may be captured by the molecular target (XY) to form an excited state of the molecular negative ion XY*. This state, commonly called a temporary negative ion (TNI), generally decays by ejecting the excess electron within a finite time depending on its lifetime (a process called auto-detachment), but the molecular negative ion may also decay through dissociation, leading to the formation of a stable negative ion X^- and a neutral (often radical) fragment (Y), a 'reaction' summarized as $e^- + XY \rightarrow XY^{*-} \rightarrow X^- + Y$. In contrast to direct electron impact, where the incident electron must have an energy of several eV to dissociate a molecule, DEA can dissociate a molecule at energies below the dissociation energy of the ground state. In numerous cases

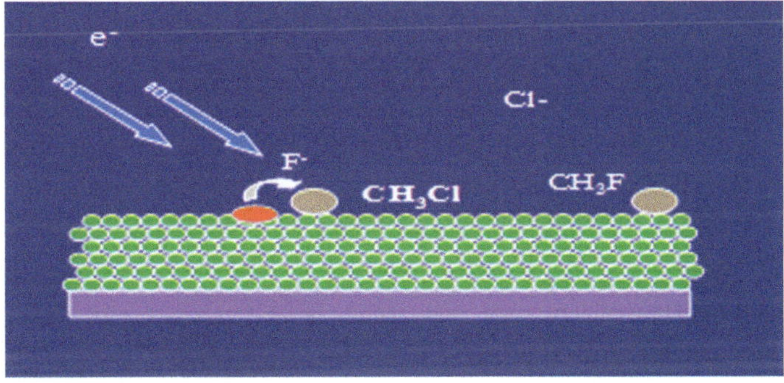

Fig. 10   Electron induced S_N2 reaction of NF_3 and CH_3Cl molecules on a surface.

DEA effectively occurs at electron impact energies of a few meV, that is at thermal energies (kT), and generally these low energy processes have very large cross sections of hundreds to thousands of Å^2, which are much larger than cross sections for direct electron (or photon) induced dissociation discussed above.

Should DEA occur in a dense medium such as in the condensed phase, within a surface layer or within a cluster, the neutral and generally more highly reactive products produced by DEA (Y) may then initiate further chemistry through reactions with neighbouring molecules (AB), *i.e.* Y + AB $\rightarrow$ AY + B. As an example, consider a mixed cluster of NF_3 and CH_3Cl (Fig. 10). The cross section for Cl^- production from CH_3Cl by direct electron impact is negligible ($<10^{-23}$ cm^2). However, F^- ions may be liberated from NF_3 and may react with CH_3Cl by the well-known nucleophilic displacement (S_N2) reaction, F^- + $CH_3Cl \rightarrow CH_3F + Cl^-$.[23] We thus form CH_3F on a surface in a mixed multilayer of co-deposited CH_3Cl and NF_3 (Fig. 10). The Cl^- ions can then be liberated/desorbed from the film and the synthesized molecular species CH_3F left on the surface. Thus DEA may lead to direct (and total) chemical transformation of the surface.

The role of DEA in ISM dust grain chemistry is yet to be quantified. Indeed, until 2006 it was not known if anions existed in the ISM. However, the possible existence of detectable abundances of anions in the ISM had been discussed for more than two decades, and even included in a chemical model of the envelope of IRC + 10216. In particular, Millar *et al.*[24] predicted an abundance of C_8H^- as large as a quarter of that of its neutral counterpart C_8H in the outer envelope of the C-star IRC +10216, a source known to be particularly rich in C-chain molecules. Recently, the first anion C_6H^- anion was detected based on laboratory work.[25] Subsequently, all linear carbon chain anions with large electron affinities: C_8H^-, C_6H^-, C_4H^-, C_5N^-, C_3N^- and CN^-, have been identified in one or more sources in the Universe, including the dark cloud core TMC-1, the prestellar core L1544, the circumstellar shell of the AGB star IRC + 10216, and the protostellar objects L1521F and L1527.

Since the binding energy of an electron to a neutral species, known as the electron affinity, is typically smaller in energy than a chemical bond, DEA is normally endothermic,[26] *e.g.*,

$$HCCCN + e^- \rightarrow CCCN^- + H$$

In order for DEA to be exothermic and so occur at the low temperatures of the ISM, there must exist weak chemical bonds in the neutral precursor. Several such processes have been discussed in the astrophysical literature. Petrie[27] realized that the formation of CN^- could occur *via* the exothermic reaction

$$MgNC + e^- \rightarrow CN^- + Mg$$

from the neutral precursor MgNC, which has been detected in IRC + 10216. Sakai *et al.*[28] calculated that the anion C_6H^- can be formed from C_6H_2:

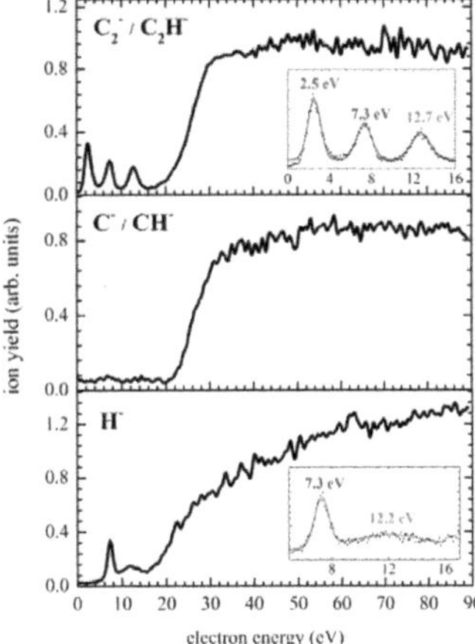

Fig. 11   Anion yields for acetylene as a function of electron energy. The inset shows the anion yield at low electron energies. The resonance peak position was determined by Gaussian fitting.

$$C_6H_2 + e^- \rightarrow C_6H^- + H,$$

a reaction that is exothermic by 16 kJ mol^{-1}. Herbst and Osamura[29] calculated that the corresponding reaction to form the anion, C_8H^-, $C_8H_2 + e^- \rightarrow C_8H^- + H$, is exothermic by 47 kJ mol^{-1}.

Electron induced anion formation processes are therefore now being included in the astrochemical models, and new data on the DEA of ISM molecular species are being assembled.[30,31] However, dipolar dissociation (DD), also known as ion pair formation, is another electron induced anion formation process and should also be included in astrochemical models:

$$e^- + AB \rightarrow AB^* + e \rightarrow A^- + B^+ + e^-,$$

To date this process has only been quantified for a very small number of hydrocarbons.[31] Fig. 11 presents anion yields for the formation of three groups of fragments: H^-, C^-/CH^-, and C_2^-/C_2H^- *via* LEE interaction with acetylene, while the C^-/CH^- anions are only formed by DD. Furthermore, DD yields may contribute >30% of the total anion yield along primary radiation (keV) tracks.

5.   Conclusions

The commissioning of the ALMA array and the next generation of space telescopes (led by JWST) heralds the dawn of a new age of astronomy, in which the role of chemistry in the interstellar medium and in star and planet formation may be quantified. A vital part of these studies will be to determine the molecular complexity in these seemingly hostile regions and explore how molecules are synthesised and survive. The current hypothesis is that many of these species are formed within the ice mantles on interstellar dust grains, with irradiation by UV light or cosmic rays stimulating chemical reactions leading to larger and more complex molecular species, including those that may be 'prebiotic' precursors. However, such irradiation releases many secondary electrons which may themselves induce chemistry. In this article we have discussed the potential role of such electron induced chemistry and illustrated, through some simple experiments, the rich molecular synthesis that electron irradiation of dust grain ice analogues may produce. It is necessary to further quantify the role of this electron chemistry, and determine whether photon and cosmic ray (ion) bombardment is better understood as chemistry induced by the secondary electrons produced from the primary (UV/ion) species during its 'absorption' in the ice mantle. Furthermore, the discovery that low energy (sub-ionisation energy) electrons can induce chemistry through the process of dissociative electron attachment needs to be reflected in astrochemistry models, particularly now that anions have been detected in space. The astrochemistry community therefore should also collaborate with the radiation chemistry community exploring DNA damage where, since the pioneering work of Sanche and co-workers[32] in 2000, it has become apparent that electron induced damage may be the major mechanism for direct damage in cellular systems.[33] The field of electron induced chemistry as an industrial tool for surface engineering is also developing rapidly, and once again the astrochemistry community can and should assimilate data from this community relevant to their research. In this year of the centenary of the Frank Hertz experiment, it is perhaps fitting to recognise the ubiquitous nature of electron collisions and how they influence the world (and universe) around us.

Acknowledgements

Much of the work presented in this article and support for the authors is derived from the European Community's Seventh Framework Programme FP7/2007-2013 (LASSIE) under grant agreement 238258 and COST Action CM0805 The Chemical Cosmos, with national support from the UK STFC and institutional support from the Open University.

References

1 A. G. G. M. Tielens, *Rev. Mod. Phys.*, 2013, **85**, 1021–1081.

2 A. G. G. M. Tielens, *IAU Symposium 135 "Interstellar Dust"*, Kluwer, Dordrecht, 1989.

3 D. P. Zaleski, N. A. Seifert, L. Amanda, *et al.*, *Astrophys. J.*, 2013, **765**, L10.

4 M. P. Bernstein, J. P. Dworkin, S. A. Sandford, G. W. Cooper and L. J. Allamandola, *Nature*, 2002, **416**, 401–3.

5 G. M. Muñoz Caro, U. J. Meierhenrich, W. A. Schutte, B. Barbier, A. Arcones Segovia, H. Rosenbauer, W. H.-P. Thiemann, A. Brack and J. M. Greenberg, *Nature*, 2002, **416**, 403–406.

6 F. J. Ciesla and S. A. Sandford, *Science*, 2012, **336**, 452.

7 Jen-Iu Lo, Sheng-Lung Chou, Yu-Chain Peng, Meng-Yeh Lin, Hsiao-Chi Lu and Bing-Ming Cheng, *J. Electron Spectrosc. Relat. Phenom.*, 2014, DOI: 10.1016/j.elspec.2013.12.014.

8 S. Pilling, D. P. P. Andrade, E. F. da Silveira, H. Rothard, A. Domaracka and P. Boduch, *Mon. Not. R. Astron. Soc.*, 2012, **423**, 2209–2221.

9 S. Pimblott and J. la Verne, *Radiat. Phys. Chem.*, 2007, **76**, 1244–1247.

10 M. H. Moore and R. L. Hudson, *IAU Colloquium 231 "Astrochemistry throughout the Universe: Recent Successes & Current Challenges"*, Cambridge, 2005.

11 K. I. Gringauz, *Nature*, 1986, **321**, 282.

12 P. Hovington, D. Drouin and R. Gauvin, *Scanning*, 1997, **19**, 1 (code download at http://www.gel.usherbrooke.ca/casino).

13 B. Sivaraman, C. Jamieson, N. J. Mason and R. I. Kaiser, *Astrophys. J.*, 2007, **669**, 1414–1421.

14 B. Sivaraman, A. M. Mebel, N. J. Mason, D. Babikov and R. I. Kaiser, *Phys. Chem. Chem. Phys.*, 2011, **13**, 421–427.

15 R. T. Garrod, S. L. W Weaver and E. Herbst, *Astrophys. J.*, 2008, **682**, 283–302.

16 C. R. Arumainayagam, Hsiao-Lu Lee, R. B. Nelson, D. R. Haines and R. P. Gunawardane, *Surf. Sci. Rep.*, 2010, **65**, 1–44.

17 M. Boyer, C. Soe, K. Chamberlain, Y. Shyur and C. Arumainayagam, *Bull. Am. Phys. Soc.*, 2011, 56.

18 R. L. Hudson and M. H. Moore, *Icarus*, 2004, **172**, 466.

19 G. A. Baratta, G. Leto and M. E. Palumbo, *Astron. Astrophys.*, 2002, **384**, 343–349.

20 S. B. Charnley and S. D. Rodgers, *Astrophys. J.*, 2002, **569**, L133–L137.

21 S. Jheeta, A. Domaracka, S. Ptasinska, B. Sivaraman and N. J. Mason, *Chem. Phys. Lett.*, 2013, **556**, 359–364.

22 R. Saladino, C. Crestini, F. Ciciriello, G. Costanzo and E. Di Mauro, *Chem. Biodiversity*, 2007, **4**(4), 694–720.

23 J. Langer, S. Matejcik and E. Illenberger, *Phys. Chem. Chem. Phys.*, 2000, **2**, 1001–1005.

24 T. J. Millar, E. Herbst and R. P. A. Bettens, *Mon. Not. R. Astron. Soc.*, 2000, **316**, 195–212.

25 M. C. McCarthy, C. A. Gottlieb, H. Gupta and P. Thaddeus, *Astrophys. J.*, 2006, **652**, L141–144.

26 K. Graupner, T. A. Field and G. C. Saunders, *Astrophys. J.*, 2008, **685**, L95.

27 S. Petrie, *Mon. Not. R. Astron. Soc.*, 1996, **281**, 137.

28 N. Sakai, T. Sakai, Y. Osamura and S. Yamamoto, *Astrophys. J.*, 2007, **667**, L65.

29 E. Herbst and Y. Osamura, *Astrophys. J.*, 2008, **679**, 1670.

30 E. Szymanska, V. S. Prabhudesai, N. J. Mason and E. Krishnakumar, *Phys. Chem. Chem. Phys.*, 2013, **15**, 998.

31 E. Szymanska, I. Cadez, E. Krishnakumar and N. J. Mason, *Phys. Chem. Chem. Phys.*, 2014, **16**, 3425.

32 B. Boudaiffa, P. Cloutier, D. Hunting, M. A. Huels and L. Sanche, *Science*, 2000, **287**, 1658–1660.

33 F. Martin, P. D. Burrow, Z. Cai, P. Cloutier, D. Hunting and L. Sanche, *Phys. Rev. Lett.*, 2004, **93**, 068101.

Faraday Discussions

PAPER

Low-energy electron-induced chemistry of condensed methanol: implications for the interstellar synthesis of prebiotic molecules

Mavis D. Boamah,[a] Kristal K. Sullivan,[a] Katie E. Shulenberger,[a] ChanMyae M. Soe,[a] Lisa M. Jacob,[a] Farrah C. Yhee,[a] Karen E. Atkinson,[ab] Michael C. Boyer,[ac] David R. Haines[a] and Christopher R. Arumainayagam*[a]

Received 23rd December 2013, Accepted 19th February 2014
DOI: 10.1039/c3fd00158j

In the interstellar medium, UV photolysis of condensed methanol (CH_3OH), contained in ice mantles surrounding dust grains, is thought to be the mechanism that drives the formation of "complex" molecules, such as methyl formate ($HCOOCH_3$), dimethyl ether (CH_3OCH_3), acetic acid (CH_3COOH), and glycolaldehyde ($HOCH_2CHO$). The source of this reaction-initiating UV light is assumed to be local because externally sourced UV radiation cannot penetrate the ice-containing dark, dense molecular clouds. Specifically, exceedingly penetrative high-energy cosmic rays generate secondary electrons within the clouds through molecular ionizations. Hydrogen molecules, present within these dense molecular clouds, are excited in collisions with these secondary electrons. It is the UV light, emitted by these electronically excited hydrogen molecules, that is generally thought to photoprocess interstellar icy grain mantles to generate "complex" molecules. In addition to producing UV light, the large numbers of low-energy (<20 eV) secondary electrons, produced by cosmic rays, can also directly initiate radiolysis reactions in the condensed phase. The goal of our studies is to understand the low-energy, electron-induced processes that occur when high-energy cosmic rays interact with interstellar ices, in which methanol, a precursor of several prebiotic species, is the most abundant organic species. Using post-irradiation temperature-programmed desorption, we have investigated the radiolysis initiated by low-energy (7 eV and 20 eV) electrons in condensed methanol at ~ 85 K under ultrahigh vacuum (5×10^{-10} Torr) conditions. We have identified eleven electron-induced methanol *radiolysis* products, which include many that have been previously identified as being formed by methanol UV *photolysis* in the interstellar medium. These

[a]Department of Chemistry, Wellesley College, Wellesley, MA 02481. E-mail: carumain@wellesley.edu; Fax: +781-283-3642; Tel: +781-283-3326

[b]Science & Engineering Department, Bunker Hill Community College, Boston, MA 02129

[c]Department of Physics, Clark University, Worcester, MA 01610

experimental results suggest that low-energy, electron-induced condensed phase reactions may contribute to the interstellar synthesis of "complex" molecules previously thought to form exclusively *via* UV photons.

1.   Introduction

Heat-, photon-, and electron-induced processing provide different routes to initiate chemical reactions. Photochemistry and thermal chemistry may differ because the former is initiated *via* an excited electronic state while the latter is initiated in the ground electronic state.[1] This difference allows the photosynthesis of products not easily accessed *via* thermal chemistry. In general, vacuum ultra-violet (VUV) photons (6–12 eV) and low-energy (< 20 eV) electrons will drive similar chemistry given the similarities between photon-induced and electron-induced excitations/ionizations, though yields and product ratios will likely differ. The possible interactions of photons with molecules, however, are restricted by selection rules governed primarily by dipole interactions and spin conservation. For example, photon-induced singlet-to-triplet transitions are nominally forbidden. Electron-induced singlet-to-triplet transitions, however, are allowed because the incident electron can be exchanged with those of the target mole-cule.[2]† Furthermore, unlike photons, electrons can be captured into resonant negative ion states that subsequently may dissociate into neutrals and negative ions.[3,4] The interactions of these negative fragments with the parent molecule or other daughter products might yield products unique to electron irradiation. In other words, low-energy electron–molecule collisions, a fundamental step that occurs during radiation chemistry, could theoretically lead to the synthesis of molecules not accessible *via* UV photochemistry. Experimental evidence partially supports this claim.[5–7]

In addition to the possibility of generating unique molecular species, low-energy, electron-induced chemistry may often predominate over UV photon-induced chemistry depending on the incident flux and the identity and phase of the target molecules. Reaction cross sections can be several orders of magnitude larger for electrons than for photons, especially at incident energies corre-sponding to resonances associated with dissociative electron attachment (DEA), making it easier for electrons to initiate chemical reactions.‡§ Moreover, electron-induced excitations, in contrast to photon-induced excitations, are not always resonant processes because an incident electron may transfer only the fraction of its energy sufficient to excite the molecule and any excess is removed by the scattered electron.

The dominance of electrons over photons in initiating condensed-phase chemical reactions has been documented for a few molecular species.[8–10] Because the chemistry ascribed to photon irradiation in thin films adsorbed on surfaces may be due to substrate photoelectrons, the role of electrons vis-à-vis photons

---

† It must be noted that selection rules in the condensed phase may differ from those in the gas phase.

‡ Reaction cross sections may, of course, change upon condensation because of intermolecular interactions and the formation of band structures.

§ DEA involves the formation of temporary negative ions that subsequently dissociate into neutral species and negatively charged ions.

may be underestimated in comparative studies. In this publication, we compare and contrast low-energy, electron-induced reactions with photon-induced reactions of condensed methanol, which is an important component of ice mantles surrounding interstellar grains. Our goal is not only to investigate fundamental differences, but also to identify how such differences impact our understanding of interstellar molecular synthesis.

In recent years, laboratory experiments and theoretical calculations have suggested that UV photon-induced surface and bulk processing of icy grain mantles containing water, carbon monoxide, methanol, and ammonia is one of the main mechanisms for the synthesis of "complex" organic molecules found in hot molecular cores and corinos, the regions within dark, dense molecular clouds warmed by one or more nearby protostars.[11–13] The source of UV light that initiates these chemical reactions is thought to be local because externally sourced UV radiation cannot penetrate these dark, dense molecular clouds. The UV photons are believed to form in these molecular clouds where cosmic rays with energies between 10 and 100 MeV ionize molecular hydrogen to generate secondary electrons, each with a mean energy around 30 eV.[14] These low-energy secondary electrons and primary cosmic rays can excite Lyman¶ and Werner band systems of molecular hydrogen.[15] UV light emission from these excited hydrogen molecules is thought to photoprocess icy grain mantles found in dark, dense molecular clouds,[16] leading to the production of radicals both light (*e.g.*, •H) and heavy (*e.g.*, $CH_3O•$). While facile light radical diffusion is possible at ~10 K,[17] the gradual warm-up from ~10 K to ~100 K in hot cores and hot corinos allows for heavy-radical diffusion. The subsequent barrierless, radical–radical reactions (*i.e.*, reactions between two open-shell species) lead to the synthesis of "complex" molecules, such as methyl formate ($HCOOCH_3$), which are potential precursors of biologically important molecules.[18] Building upon early investigations,[19,11] recent laboratory experiments that mimic photochemistry in the interstellar medium have shown that acetaldehyde (CH_3CHO), glycolaldehyde ($HOCH_2CHO$), methyl formate ($HCOOCH_3$), formic acid ($HCOOH$), acetic acid (CH_3COOH), ethylene glycol (($CH_2OH)_2$), dimethyl ether (CH_3OCH_3), and ethanol (CH_3CH_2OH) can be formed from UV photolysis of condensed methanol.[20]

In addition to producing UV photons, the non-thermal, low-energy, secondary electrons produced by cosmic rays can directly induce radiolysis reactions, a possibility which has been largely ignored in previous astrochemistry studies pertaining to the interstellar medium. Low-energy electrons may result from two processes: (1) the interaction of cosmic rays with gaseous molecular hydrogen present in the dark, dense molecular clouds,‖ and (2) the inelastic collisions that the cosmic ray experiences as it traverses through the ices. Although other high-energy radiolysis secondary products, such as excited species and ions, also cause some radiolytic changes, the inelastic collisions of the low-energy electrons with matter are hypothesized to be the primary driving force in a wide variety of radiation-induced chemical reactions.[21] The majority of these secondary electrons typically have energies below 20 eV.[22] Thus, the goal of our experiments is to investigate directly the processing of astrochemically-relevant laboratory ices by

---

¶ Not Lyman alpha, which is a spectral line of atomic hydrogen.

‖ These low-energy electrons will interact with only the top few layers of cosmic ices; our experiments simulate these interactions.

low-energy ($\leq$ 20 eV) electrons characteristic of the secondary electrons produced by the interactions of high-energy radiation, such as cosmic rays, with matter.

Our present work, involving a dedicated quasi-monoenergetic electron gun and a triple-filtered quadrupole mass spectrometer equipped with a pulse ion counting detector, builds upon previous, more limited studies conducted with $\sim$ 50 eV electrons generated from the filament of a mass spectrometer ionizer.[23,24] Only one previous study, involving post-irradiation analysis, has examined the radiolysis of condensed methanol initiated by low-energy ($\leq$20 eV) electrons relevant to high-energy radiolysis.[25,26] This study focused on the dynamics of low-energy, electron-induced synthesis of CO from condensed methanol. The results reported herein detail the formation of eleven electron-induced methanol radiolysis products and demonstrate that UV photon and low-energy ($\leq$20 eV) electron processing of methanol ices likely yield essentially the same reaction products. Consequently, our results suggest that cosmic-ray-induced, *low-energy* electrons may also play a role in interstellar molecular synthesis, previously thought to occur exclusively *via* cosmic-ray-induced UV photons.

2.   Experimental

All experiments were conducted in the Wellesley College UHV chamber at base pressures of approximately 5×10^{-10} Torr, described in detail previously.[23] A Mo(110) crystal mounted on a rotary manipulator served as the substrate for all experiments. Rotation of the crystal allowed for positioning in front of the sample doser, electron gun, and mass spectrometer. The crystal was cleaned with an oxygen dose at $\sim$$2 \times 10^{-9}$ Torr for five minutes at $\sim$1200 K, followed by heating to $\sim$2200 K for approximately 30 s.

The crystal was cooled with liquid nitrogen to $\sim$85 K prior to the introduction of gas samples into the chamber and maintained at this temperature throughout electron irradiation. At this temperature, methanol films condense as amorphous ices. We note that our inability to cool below 85 K limits our ability to detect possible low-energy, electron-induced radiolysis products that have low desorption temperatures. One such possible radiolysis product is ethane (desorption temperature $\sim$ 60 K). Other than this possibility, our inability to cool to 10 K is not a major limitation in studying the relevant chemistry pertaining to ice mantles surrounding dust grains in the interstellar medium. Lower temperatures are not critical because icy dust grains, in regions that become hot cores, are heated from 10 K to temperatures above 100 K by radiation from the ignition of a nearby new star, albeit over a very long time period.[27] Moreover, at temperatures around 10 K diffusion of heavy radicals such as $CH_3O\bullet$ is unlikely, precluding radical–radical reactions that are likely routes to "complex" molecules.

Samples were introduced into the UHV chamber through a precision leak valve. Samples other than methanol were only used for temperature-programmed desorption control experiments involving no prior electron irradiation. Samples were obtained from Sigma-Aldrich (methanol, HPLC grade 99.9%; methanol, anhydrous, 99.8%; glycolaldehyde dimer; ethylene glycol 99.8%; methyl formate, anhydrous 99%), EMD (glacial acetic acid), Fluka (formic acid, $\sim$98%) and Pharmco-AAPER (absolute ethanol, ACS/USP grade). Liquid/solid samples were transferred under a nitrogen atmosphere to Schlenk tubes, and degassed by three freeze-pump-thaw cycles. Mixtures of methanol and a suspected radiolysis product were prepared semi-quantitatively and dispensed from a single Schlenk tube. The mixture of methanol and glycolaldehyde was warmed slightly before introduction into the UHV chamber in order to ensure that the glycolaldehyde was introduced in its monomeric form.

Results of a series of temperature-programmed desorption experiments conducted in the absence of electron irradiation were used to estimate the coverage of methanol. One monolayer is defined as the maximum exposure of methanol that does not yield a multilayer peak. Film thicknesses on the Mo(110) crystal were controlled by monitoring the dosing chamber pressure drop, as measured by an MKS Baratron capacitance manometer.

Methanol films were irradiated using a Kimball Physics FRA2X1-2 flood electron gun (cathode terminal spread of 0.4 eV). The incident current on the clean crystal was set at 2 μA (flux = 2×10^{13} electrons/cm^2 s^{-1}) for all electron irradiation experiments. The crystal was grounded during electron bombardment to minimize charging of the adsorbate thin film. After irradiation, temperature-programmed desorption measurements were performed using a Hiden IDP Series 500 quadrupole mass spectrometer. To optimize the signal-to-noise ratio, no more than five masses were monitored during typical temperature-programmed desorption experiments. Temperature-programmed desorption experiments conducted in the absence of electron irradiation served as control experiments. Surface temperature was monitored using a W/5% Re *vs.* W/26% Re thermocouple spot welded to the edge of the crystal.

3.　Results and discussion

3.1.　Identification of radiolysis products

The electron-induced radiolysis products of condensed methanol were identified using the results of post-irradiation temperature-programmed desorption experiments. We have previously shown that the effects of low-energy electrons on condensed-phase molecules can be investigated by temperature-programmed desorption experiments conducted following low-energy electron irradiation of nanoscale thin films.[23] Post-irradiation temperature-programmed desorption is ideal for analyzing complex mixtures because radiolysis products can be separated based on desorption temperature. This technique is also very sensitive, allowing the detection of submonolayer quantities as low as 0.005 ML.

The experimental data were analyzed by comparing mass spectral fragments observed during thermal desorption to known mass spectra.[28] When identification of a particular radiolysis product was uncertain, temperature-programmed desorption data for methanol films containing that suspected radiolysis product were used as reference. Results of analogous experiments with methanol isotopologues ($^{13}CH_3OH$ and CD_3OD) were also used in product identification. Because there is usually a correlation between boiling points and multilayer desorption temperatures, this trend was used as tertiary evidence for product identification. Mass-to-charge ratios greater than 75 were not usually monitored because such fragments are not likely to be from nascent radiolysis products.

Because of the multitude of methanol radiolysis products whose yields were dependent on film thickness, irradiation time, and incident electron energy, several hundred post-irradiation temperature-programmed desorption experiments were conducted to help identify the electron-induced radiolysis products of

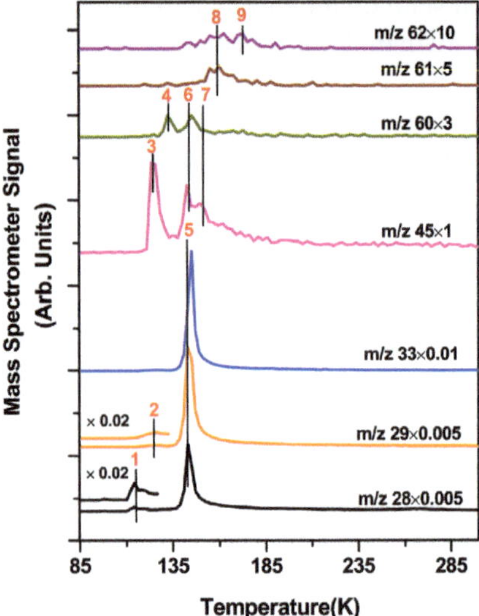

Fig. 1 Post irradiation temperature-programmed desorption data for 20 monolayers of $^{12}CH_3OH$ irradiated with 7 eV electrons for 20 min at an incident current of 2 μA (flux = 2 × 10^{13} electrons/cm^2 s^{-1} and fluence = 3 × 10^{16} electrons/cm^2) shows several desorption features: (1) CO (background), (2) formaldehyde (H_2CO), (3) dimethyl ether (CH_3OCH_3), (4) methyl formate ($HCOOCH_3$), (5) methanol (CH_3OH), (6) glycolaldehyde ($HOCH_2CHO$), (7) ethanol (CH_3CH_2OH) and acetic acid (CH_3COOH) (8) methoxymethanol (CH_3OCH_2OH), (9) ethylene glycol (($CH_2OH)_2$). Plots vertically offset for clarity.

methanol. The very small radiolysis yields and the closeness in desorption temperatures required that no more than five mass spectral fragments be monitored during temperature-programmed desorption experiments. Despite these precautions, all mass-to-charge ratios that evince peaks at a given temperature cannot be assumed to represent desorption of a single product. In addition, because of detector saturation, it was not possible to monitor mass spectral fragments (*e.g.*, m/z = 31) that were common to both methanol and some radiolysis products. Because of the reasons enumerated above, several of our identifications of low-energy electron-induced methanol radiolysis products (*e.g.*, glycolaldehyde and acetic acid) are not unambiguous.

Results of temperature-programmed desorption experiments conducted following irradiation of condensed $^{12}CH_3OH$ with 7 eV and 20 eV electrons are shown in Fig. 1 and 2, respectively.** All experiments for Fig. 1 and 2 were conducted with a film thickness of 20 ML, a flux of 2 × 10^{13} electrons/cm^2 s^{-1}, and a fluence of 3 × 10^{16} electrons/cm^2. Each figure represents a composite of several experiments conducted under identical conditions. To improve clarity, not all mass spectral fragments monitored are shown in these two figures. Results of a temperature-programmed desorption experiment conducted following irradiation of condensed $^{13}CH_3OH$ with 20 eV electrons are shown in Fig. 3.

** 7 eV is below the ionization energy of methanol. The preponderance of electrons resulting from the interaction of high-energy radiation with condensed matter have energies below 20 eV.

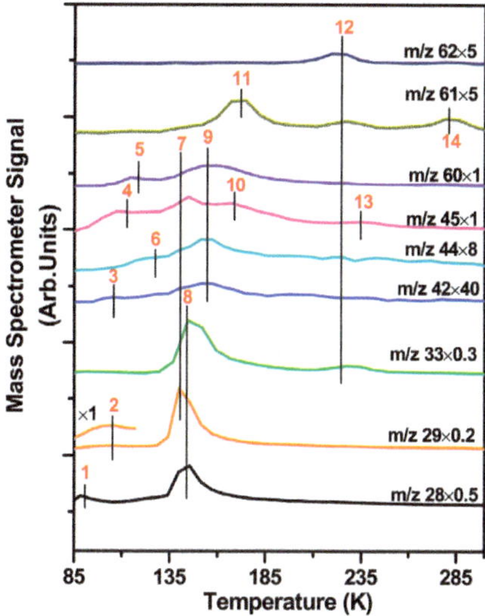

Fig. 2 Post irradiation temperature-programmed desorption data for 20 monolayers of $^{12}CH_3OH$ irradiated with 20 eV electrons for 20 min at an incident current of 2 μA (flux = 2×10^{13} electrons/cm^2 s^{-1} and fluence = 3×10^{16} electrons/cm^2) shows several desorption features: (1) CO (background), (2) formaldehyde (H$_2$CO), (3) unknown, (4) dimethyl ether (CH$_3$OCH$_3$), (5) methyl formate (HCOOCH$_3$), (6) acetaldehyde (CH$_3$CHO), (7) glycolaldehyde (HOCH$_2$CHO), (8) methanol (CH$_3$OH), (9) acetic acid (CH$_3$COOH), (10) ethanol (CH$_3$CH$_2$OH), (11) methoxymethanol (CH$_3$OCH$_2$OH), (12) ethylene glycol ((CH$_2$OH)$_2$), (13) glycolic acid (HOCH$_2$CO$_2$H), (14) 1,2,3-propanetriol (HOCH$_2$CHOH-CH$_2$OH). Plots vertically offset for clarity.

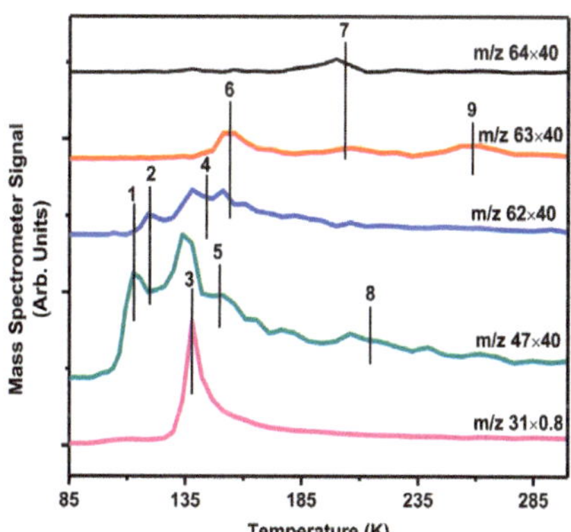

Fig. 3 Post irradiation temperature-programmed desorption data for 20 monolayers of $^{13}CH_3OH$ irradiated with 20 eV electrons for 20 min at an incident current of 2 μA (flux = 2×10^{13} electrons/cm^2 s^{-1} and fluence = 3×10^{16} electrons/cm^2) shows several desorption features: (1) dimethyl ether (CH$_3$OCH$_3$), (2) methyl formate (HCOOCH$_3$), (3) methanol (CH$_3$OH), (4) acetic acid (CH$_3$COOH), (5) ethanol (CH$_3$CH$_2$OH), (6) methoxy-methanol (CH$_3$OCH$_2$OH), (7) ethylene glycol ((CH$_2$OH)$_2$), (8) glycolic acid (HOCH$_2$CO$_2$H), (14) 1,2,3-propanetriol (HOCH$_2$CHOHCH$_2$OH). Plots vertically offset for clarity. The positioning of the vertical lines labeled 4 and 8 is somewhat arbitrary.

The identifications of the eleven electron-induced methanol radiolysis products ((A) formaldehyde (H_2CO), (B) dimethyl ether (CH_3OCH_3), (C) methyl formate ($HCOOCH_3$), (D) acetaldehyde (CH_3CHO), (E) glycolaldehyde ($HOCH_2CHO$), (F) acetic acid (CH_3COOH), (G) ethanol (CH_3CH_2OH), (H) methoxymethanol (CH_3OCH_2OH), (I) ethylene glycol (($CH_2OH)_2$), (J) glycolic acid ($HOCH_2CO_2H$), and (K) 1,2,3-propanetriol ($HOCH_2CHOHCH_2OH$)) are discussed in detail below.

A. Identification of formaldehyde (H_2CO). A very small peak identified as evidence of formaldehyde was seen for $m/z = 29$ [CHO^+] at ~115 K in post-irradiation temperature-programmed desorption experiments (Fig. 1–2). The fragment $m/z = 29$ is the dominant ion in the formaldehyde mass spectrum.[28] Results of isotopic labeling experiments with $^{13}CH_3OH$ provided additional evidence for the identification of formaldehyde (data not shown). Formaldehyde has been identified as a photolysis/radiolysis product of methanol following irradiation with UV light,[11,19,20] 55 eV electrons,[23] 1 keV electrons (tentative),[29] 5 keV electrons,[30] 1 MeV protons,[31] and 3 keV He^+ ions.[32]

B. Identification of dimethyl ether (CH_3OCH_3). Results of post-irradiation temperature-programmed desorption experiments of methanol ($^{12}CH_3OH$) displayed at ~115 K for $m/z = 45$ [$C_2H_5O^+$] (Fig. 1–2) and $m/z = 46$ [$C_2H_6O^+$] (data not shown) desorption peaks, which we assign to dimethyl ether (CH_3OCH_3). The fragments $m/z = 45$ and $m/z = 46$ are the dominant ions in the dimethyl ether mass spectrum.[28] As shown in Fig. 3, a peak for $m/z = 47$ [$^{13}C_2H_5O^+$] was observed at approximately the same temperature in post-irradiation temperature-programmed desorption experiments involving $^{13}CH_3OH$. These post-irradiation temperature-programmed desorption results were consistent with those of temperature-programmed desorption experiments involving an unirradiated 2 : 1 mixture of methanol and dimethyl ether, providing additional support for the identification of dimethyl ether as a low-energy, electron-induced radiolysis product of condensed methanol. This identification is further verified by the desorption temperatures of dimethyl ether and formaldehyde being approximately the same, consistent with their boiling points of 251 K and 254 K, respectively. Irradiation of condensed methanol with UV light[20] and 55 eV electrons[23] has been previously shown to yield dimethyl ether.

C. Identification of methyl formate ($HCOOCH_3$). The identification of methyl formate was based on results of post-irradiation temperature-programmed desorption experiments that evinced peaks at ~120 K for $m/z = 60$ [$C_2H_4O_2^+$] (Fig. 1–2). Coincident peaks were also observed in some post-irradiation experiments for $m/z = 42$ [$C_2H_2O^+$], $m/z = 29$ [HCO^+], and $m/z = 28$ [CO^+] (Fig. 2). Following $m/z = 31$[H_2CO^+],†† the fragments $m/z = 60$, 29, and 28 are the three most significant ions in the methyl formate mass spectrum.[28] As shown in Fig. 3, a peak for $m/z = 62$ [$^{13}C_2H_4O_2^+$] was observed at approximately the same temperature in post-irradiation temperature-programmed desorption experiments involving $^{13}CH_3OH$. The identification of methyl formate was further corroborated by results of temperature-programmed desorption experiments involving unirradiated 1 : 1 mixtures of methyl formate and methanol (data not shown). The desorption temperature of methyl formate is above that of dimethyl ether, consistent with their boiling points of 305 K and 251 K, respectively. However, the

---

†† Fragment $m/z = 31$, which is also the dominant ion for methanol, cannot be monitored in our experiments because of detector saturation.

observed desorption temperatures of methyl formate and acetaldehyde (boiling point 294 K) are not consistent with their boiling points. Methyl formate has been identified as a photolysis/radiolysis product of condensed methanol following irradiation with UV light,[11,20] 55 eV electrons (tentative),[24] 1 keV electrons (tentative),[29] and 5 keV electrons.[30]

   D. Identification of acetaldehyde (CH$_3$CHO). Results of temperature-programmed desorption experiments conducted following low-energy (20 eV) electron irradiation of condensed CH$_3$OH demonstrated at $\sim$ 120 K for m/z =44 [C$_2$H$_4$O$^+$] (Fig. 2), m/z =43 [C$_2$H$_3$O$^+$] (data not shown), and m/z =42 [C$_2$H$_2$O$^+$] (Fig. 2) desorption features which we attribute to acetaldehyde (CH$_3$CHO). The fragments m/z = 44, 43, and 42 are three significant ions in the published mass spectrum of acetaldehyde.[28] The desorption temperature of acetaldehyde is below that of methanol, consistent with their boiling points of 294 K and 337 K, respectively. Clear evidence was absent for the formation of acetaldehyde following 7 eV electron irradiation of methanol (Fig. 1). Acetaldehyde has been previously identified tentatively as a UV photolysis product of condensed methanol.[20]

   E. Identification of glycolaldehyde (HOCH$_2$CHO). We attribute to glycolaldehyde (hydroxy acetaldehyde) (HOCH$_2$CHO) desorption features seen for m/z = 60 [C$_2$H$_4$O$_2$$^+$], m/z = 42 [C$_2H_2O^+$], and m/z = 29 [CHO$^+$] at $\sim$140 K in post-irradiation temperature-programmed desorption experiments (Fig. 1 and 2). The fragments m/z = 60, 42, and 29 are three significant ions in the published mass spectrum of glycolaldehyde.[28] This identification of glycolaldehyde is also consistent with results (data not shown) of post-irradiation temperature-programmed desorption experiments conducted with the isotopologue ^{13}CH$_3$OH and (2) temperature-programmed desorption experiments involving no prior irradiation of methanol and glycolaldehyde mixtures. Irradiation of condensed methanol with 55 eV electrons (tentative),[24] 5 keV electrons,[30] and UV photons (tentative)[20] has been shown to yield glycolaldehyde.

   F. Identification of acetic acid (CH$_3$COOH). The identification of acetic acid is based on post-irradiation temperature-programmed desorption results that evinced desorption peaks at $\sim$ 155 K for m/z = 60 [C$_2$H$_4$O$_2$$^+$], m/z = 45 [COOH$^+$], m/z = 44 [C$_2H_4O^+$ or CO$_2$$^+$], and m/z = 42 [C$_2H_2O^+$] (Fig. 2). A desorption peak at the same temperature was also observed for m/z = 43 [C$_2$H$_3$O$^+$] (data not shown). The fragments m/z = 60, 45, and 43 are the three most significant ions in the published mass spectrum of acetic acid.[28] As shown in Fig. 3, a peak for m/z = 62 [^{13}C$_2$H$_4$O$_2$$^+$] was observed at approximately the same temperature in post-irradiation temperature-programmed desorption experiments involving ^{13}CH$_3$OH. Additional evidence to support this identification was provided by results (data not shown) of temperature-programmed desorption experiments involving unirradiated thin films containing mixtures of methanol and acetic acid. We attribute the lower desorption temperature of acetic acid (b.p. 391 K) relative to that of ethanol (b.p. 351 K) (Fig. 2) to the likely absence of acetic acid dimers in the irradiated methanol.

   G. Identification of ethanol (CH$_3$CH$_2$OH). We attribute to ethanol the peak at $\sim$ 165 K for m/z = 46 [CH$_3$CH$_2$OH$^+$] (Fig. 2) observed during temperature-programmed desorption experiments conducted following irradiation of methanol with 20 eV electrons. Desorption peaks (data not shown) at the same temperature were also observed for m/z = 43 [C$_2$H$_3$O$^+$], m/z = 44 [C$_2$H$_3$O$^+$] and m/z = 45

[$CH_3CH_2O^+$]. The fragments $m/z = 46$, 45, and 43 are three significant ions in the published mass spectrum of ethanol.[28] As shown in Fig. 3, a peak for $m/z = 47$ [$^{13}C_2H_5O^+$] was observed at approximately the same temperature in post-irradiation temperature-programmed desorption experiments involving $^{13}CH_3OH$. Ethanol has been previously identified as a low-energy electron-induced radiolysis[23] and a UV-photolysis[20] product of condensed methanol.

H. Identification of methoxymethanol (CH_3OCH_2OH). Methoxymethanol was identified as a product following irradiation of condensed methanol with low-energy electrons. This identification was based on peaks for $m/z = 61$ [$CH_3O-CH_2O^+$] and $m/z = 62$ [$CH_3OCH_2OH^+$] at ~165 K observed in the results of post-irradiation temperature-programmed desorption experiments (Fig. 1 and 2). Our results are consistent with the published mass spectrum[33] of methoxymethanol, a highly labile species for which a mass spectrum is not found in standard tables.[28] As shown in Fig. 3, a peak for $m/z = 63$ [$^{13}C_2H_5O_2{}^+$] was observed at approximately the same temperature in post-irradiation temperature-programmed desorption experiments involving $^{13}CH_3OH$. The boiling points of methoxy methanol (356 K, estimated‡‡), ethanol (351 K), and ethylene glycol (470 K) are consistent with the trends observed in the desorption temperatures for these three species. Methoxymethanol was previously identified following irradiation of condensed methanol with 55 eV electrons,[23] but not following irradiation with UV light,[11,20] 1 keV electrons,[29] or 5 keV electrons.[30]

I. Identification of ethylene glycol ($HOCH_2CH_2OH$). We attribute to ethylene glycol ($HOCH_2CH_2OH$) desorption features seen for $m/z = 33$ [CH_5O^+], $m/z = 44$ [$C_2H_3O^+$], $m/z = 61$ [$OCH_2CH_2OH^+$], and $m/z = 62$ [$HOCH_2CH_2OH^+$] at ~ 215 K in post-irradiation temperature-programmed desorption experiments (Fig. 2). These four mass-to-charge ratios are significant mass spectral fragments of ethylene glycol.[28] As shown in Fig. 3, a peak for $m/z = 63$ [$^{13}C_2H_5O_2{}^+$] was observed at approximately the same temperature in post-irradiation temperature-programmed desorption experiments involving $^{13}CH_3OH$. Ethylene glycol's boiling point (470 K) and desorption temperature are consistent with this identification. Results of previous experiments have demonstrated that γ-radiolysis,[34] low-energy electron-induced radiolysis,[23] 5 keV electron-induced radiolysis,[30] and UV photolysis[20] of condensed methanol also yield ethylene glycol. Results of temperature-programmed desorption experiments involving no prior irradiation of methanol and ethylene glycol were used in our previous identification of ethylene glycol as a radiolysis product of methanol.[23]

J. Identification of glycolic acid ($HOCH_2COOH$). Glycolic acid (hydroxyacetic acid) ($HOCH_2COOH$) was tentatively identified as a product following irradiation of condensed methanol with 20 eV electrons. This identification was based on a peak for $m/z = 45$ [$COOH^+$] at ~235 K observed in the results of post-irradiation temperature-programmed desorption experiments (Fig. 2). The absence of peaks for $m/z = 60$ and $m/z = 75$ at the same temperature is consistent with the miniscule signals for these fragments in the published mass spectrum of glycolic acid.[28] As shown in Fig. 3, a possible peak for $m/z = 47$ [$^{13}CH_2O_2{}^+$] was observed at approximately the same temperature in post-irradiation temperature-programmed desorption experiments involving $^{13}CH_3OH$. The boiling points of

---

‡‡ CSID:56311, http://www.chemspider.com/Chemical-Structure.56311.html (accessed 14 : 23, Dec 21, 2013).

glycolic acid (538 K, estimated§§) and ethylene glycol (470 K) are consistent with the trend in desorption temperatures for these two species. Glycolic acid has never been previously identified as a radiolysis/photolysis product of pure condensed methanol. Glycolic acid is likely a radiolysis product of a methanol radiolysis product, not a nascent *primary* radiolysis product of methanol. Interestingly, no evidence was seen for the formation of glycolic acid following irradiation of a 5 ML thick methanol film.

K. Identification of 1,2,3-propanetriol ($HOCH_2CHOHCH_2OH$). The tentative identification of 1,2,3-propanetriol (glycerin) (glycerol) ($HOCH_2CHOHCH_2OH$) is based on post-irradiation temperature-programmed desorption results that evinced desorption peaks at ~ 280 K for $m/z = 43\ [C_3H_7^+]$, $m/z = 61[C_2H_5O_2^+]$, and $m/z = 75\ [C_3H_7O_2^+]$ (Fig. 2). As shown in Fig. 3, a peak for $m/z = 63\ [^{13}C_2H_5O_2^+]$ was observed at approximately the same temperature in post-irradiation temperature-programmed desorption experiments involving $^{13}CH_3OH$. The boiling points of glycerol (563 K, estimated), glycolic acid (538 K), ethylene glycol (470 K) are consistent with the trends observed in the desorption temperatures for these three species. Although previously identified as a radiolysis product of methanol,[35] glycerol is likely not a nascent *primary* radiolysis product of methanol. Post-irradiation temperature-programmed desorption experiments did not show evidence for the formation of glycerol following irradiation of a 5 ML thick methanol film.

L. Non-identification of other predicted radiolysis products. Clear evidence was absent for the formation of several expected radiolysis products ((see Table 1): ethane (CH_3CH_3), dimethyl peroxide (CH_3OOCH_3), glyoxal (CHOCHO), methylene glycol ($HOCH_2OH$), methyl hydroperoxide (CH_3OOH), and formic acid (HCOOH).

3.2. Proposed reaction mechanisms for low-energy, electron-induced radiolysis of methanol

The synthesis of radiolysis products at incident electron energies as low as 7 eV indicates that electron impact ionization is not the only mechanism through which low-energy, electrons initiate reactions in the methanol thin films. Results of our recent[36] study of the low-energy electron-induced formation of ethylene glycol and methoxymethanol are consistent with a radical–radical reaction mechanism. The absence of resonant structures in plots of the yield *vs.* incident electron energy suggest that non-resonant electron impact excitation is the primary mechanism by which electrons initiate methoxy methanol and ethylene glycol formation below the ionization threshold. Moreover, the functional forms for the radiolysis yields *vs.* fluence are well fit by a quadratic function indicating that two independent dissociation events lead to the formation of both methoxymethanol and ethylene glycol.[36] Based on these results, we propose the following mechanism involving formation of methoxy ($CH_3O\bullet$) and hydroxymethyl ($\bullet CH_2OH$) radicals *via* electron impact excitation followed by radical–radical coupling:

$$CH_3OH \xrightarrow{\ e^-\ } [CH_3OH]^* \tag{1}$$

§§ CSID:737, http://www.chemspider.com/Chemical-Structure.737.html (accessed 22 : 18, Dec 21, 2013).

Table 1 Possible radiolysis products of methanol based on expected radical–radical reactions. This compendium is an extension of a previously published table.[30]

Radical 2	Radical 1					
	$\bullet CH_2OH$	$\bullet HCO$	$CH_3O\bullet$	$\bullet CH_3$	$\bullet OH$	$\bullet H$
$\bullet CH_2OH$	$HOCH_2CH_2OH$ Ethylene Glycol	CH_2OHCHO Glycolaldehyde	CH_3OCH_2OH Methoxymethanol	CH_3CH_2OH Ethanol	$HOCH_2OH$ Methylene Glycol	CH_3OH Methanol
$\bullet HCO$	CH_2OHCHO Glycolaldehyde	$CHOCHO$ Glyoxal	CH_3OCHO Methyl Formate	CH_3CHO Acetaldehyde	$HCOOH$ Formic Acid	H_2CO Formaldehyde
$CH_3O\bullet$	CH_3OCH_2OH Methoxymethanol	CH_3OCHO Methyl Formate	CH_3OOCH_3 Dimethyl Peroxide	CH_3OCH_3 Dimethyl Ether	CH_3OOH Methyl Hydroperoxide	CH_3OH Methanol
$\bullet CH_3$	CH_3CH_2OH Ethanol	CH_3CHO Acetaldehyde	CH_3OCH_3 Dimethyl Ether	CH_3CH_3 Ethane	CH_3OH Methanol	CH_4 Methane
$\bullet OH$	$HOCH_2OH$ Methylene Glycol	$HCOOH$ Formic Acid	CH_3OOH Methyl Hydroperoxide	CH_3OH Methanol	$HOOH$ Hydrogen Peroxide	H_2O Water
$\bullet H$	CH_3OH Methanol	H_2CO Formaldehyde	CH_3OH Methanol	CH_4 Methane	H_2O Water	H_2 Dihydrogen

$$[CH_3OH]^* \rightarrow \bullet CH_2OH + \bullet H \tag{2}$$

$$[CH_3OH]^* \rightarrow CH_3O\bullet + \bullet H \tag{3}$$

$$CH_3O\bullet + \bullet CH_2OH \rightarrow CH_3OCH_2OH \tag{4}$$

$$\bullet CH_2OH + \bullet CH_2OH \rightarrow HOCH_2CH_2OH \tag{5}$$

As shown in Table 1, mechanisms involving the coupling of the expected radicals ($\bullet CH_2OH$, $\bullet HCO$, $CH_3O\bullet$, $\bullet CH_3$, $\bullet OH$, and $\bullet H$) may account for the formation of all observed products except those that are likely not primary methanol radiolysis products (acetic acid (CH_3COOH), glycolic acid ($HOCH_2COOH$), and 1,2,3-propanetriol ($HOCH_2CHOHCH_2OH$)). Radical–radical reaction mechanisms are plausible given the formation of all primary radiolysis products (except perhaps acetaldehyde) at an incident electron energy of 7 eV, below the ionization energy for methanol.

3.3. Comparison to high-energy radiolysis of methanol

All four previously published high-energy (arbitrarily defined as 1 keV or higher) radiolysis studies involving methanol thin films were conducted using post-irradiation infrared-reflection absorption spectroscopy (IRAS).[29,30,32] These studies involved the irradiation of condensed methanol with either 1–5 keV electrons or 3 keV He^+ ions. As detailed in Table 2, no study identified more than six methanol radiolysis products. We attribute this low number of identified radiolysis products to the lower sensitivity of IRAS compared to TPD. Moreover, post-irradiation temperature-programmed desorption is perhaps better for analyzing complex mixtures because radiolysis products are separated by temperature. All four studies identified formaldehyde, methane, carbon dioxide, and carbon monoxide as high-energy radiolysis products of condensed methanol. Only one study identified the production of two-carbon species such as methyl formate and ethylene glycol.[30] Results of our preliminary unpublished experiments indicate that high-energy ($\sim$ 900 eV electron) radiolysis of methanol yields the same products as those reported herein, consistent with the hypothesis that high-energy radiolysis in condensed matter is mediated by low-energy electrons.

3.4. Low-energy electrons *vs.* UV photons

Of special interest is the comparison of electron-induced and photon-induced condensed-phase methanol reactions. Recent results of post-irradiation temperature-programmed desorption experiments indicate that acetaldehyde (CH_3CHO), glycolaldehyde ($HOCH_2CHO$), methyl formate ($HCOOCH_3$), formic acid ($HCOOH$), acetic acid (CH_3COOH), ethylene glycol (($CH_2OH)_2$), dimethyl ether (CH_3OCH_3), and ethanol (CH_3CH_2OH) can be formed from UV photolysis of condensed methanol.[20] All of these products except formic acid were observed in our studies following irradiation of condensed methanol with low-energy (≤ 20 eV) electrons. We attribute this difference to dissimilar experimental setups and procedures. The most conspicuous difference is the formation of methoxymethanol (CH_3OCH_2OH) *via* condensed-phase methanol reactions stimulated by low-energy electrons but not by UV photons. Resolution of this apparent

Table 2 Compendium of radiolysis/photolysis products of condensed methanol identified by post-irradiation temperature-programmed desorption (TPD), infrared spectroscopy (IR), and high resolution electron energy loss spectroscopy (HREELS). "?" symbol has been placed next to identifications that are not unambiguous

Radiolysis Products	This Work	Arumainayagam (1995)[23]	Sanche (1997)[25,26]	White (1997)[24]	Mason (2013)[29]	Kaiser (2007)[30]	Palumbo (1999)[37]	Baratta (2002)[32]	Allamandola (1988)[19]	Öberg (2009)[20]	Gerakines (1996)[11]
	20 eV electrons	55 eV electrons	0-20 eV electrons	52 eV electrons	1 keV electrons	5 keV electrons	3 keV He$^+$ ions	3 keV He$^+$ ions	UV photolysis	UV photolysis	UV photolysis
	TPD	TPD	HREELS	IR	IR	IR	IR	IR	IR	TPD & IR	IR
H_2CO	Y	Y		Y	Y	Y	Y	Y	Y	Y	Y
CH_3OCH_3	Y	Y								Y	
$HCOOCH_3$	Y			Y?		Y				Y?	Y
CH_3CHO	Y									Y?	
$HOCH_2CHO$	Y?			Y?		Y				Y?	
CH_3COOH	Y?										
CH_3CH_2OH	Y	Y								Y	Y?
CH_3OCH_2OH	Y	Y									
$(CH_2OH)_2$	Y	Y				Y				Y	Y?
$HOCH_2CO_2H$	Y?										

Table 2 (*Contd.*)

Radiolysis Products	This Work	Arumainayagam (1995)[23]	Sanche (1997)[25,26]	White (1997)[24]	Mason (2013)[29]	Kaiser (2007)[30]	Palumbo (1999)[37]	Baratta (2002)[32]	Allamandola (1988)[19]	Öberg (2009)[20]	Gerakines (1996)[11]
	20 eV electrons	55 eV electrons	0-20 eV electrons	52 eV electrons	1 keV electrons	5 keV electrons	3 keV He$^+$ ions	3 keV He$^+$ ions	UV photolysis	UV photolysis	UV photolysis
	TPD	TPD	HREELS	IR	IR	IR	IR	IR	IR	TPD & IR	IR
$HOCH_2CHOHCH_2OH$ Y?											
CO			Y		Y	Y	Y	Y	Y	Y	Y
CH_4				Y	Y	Y	Y	Y?	Y	Y	Y
CO_2					Y	Y	Y	Y	Y		Y

difference between photon- and low-energy electron-induced reactions awaits additional photochemical studies of condensed methanol. Methoxymethanol could be a potential chemical tracer of the importance of low-energy electrons for the formation of complex organics in space if UV photolysis of methanol is shown not to produce methoxymethanol. However, the UV-induced formation of methoxymethanol from condensed methanol is likely given the detection of photolysis products formed from the methoxy ($CH_3O\bullet$) and hydroxymethyl ($\bullet CH_2OH$) radicals.

Conclusions

The similarity between the products of low-energy electron and UV-photon stimulated reactions of condensed methanol is suggestive of a potentially important contributing role for cosmic ray-induced, low-energy electrons in the synthesis of "complex" organic molecules in cosmic ices. If the low-energy electron- and UV photon-induced cross sections for the formation of two products are significantly different, the relative abundance of these products in the interstellar medium may provide insight into the relative importance of low-energy electrons *vs.* UV photons in the processing of cosmic ices. Such a comparison requires knowledge of the energy-dependent electron flux whose calculation awaits Monte Carlo track simulations of cosmic ray particles traversing ices surrounding interstellar dust grains. These calculations must incorporate relativistic corrections given that galactic cosmic rays contain heavy ions of high charge and energy. While we cannot conclude with certainty that low-energy electrons contribute to the synthesis of molecules in space, it must be noted that low-energy electrons are the most abundant product of ionization radiation, such as cosmic rays, in condensed matter. However, based on what we know today, we cannot conclude with certainty whether the contributions of UV photons and low-energy electrons are equal, or if one dominates the other.

Acknowledgements

This work was supported by grants from the National Science Foundation (NSF grant number CHE-1012674 and CHE-1005032) and Wellesley College (Faculty awards and Brachman Hoffman small grants). We gratefully acknowledge several useful discussions with Professor John Yates, Dr Murthy Gudiapati, and Professor Karin Öberg. We are especially thankful to the anonymous reviewer for many useful suggestions.

References

1 B. Wardle, *Principles and Applications of Photochemistry*. Wiley: Chichester, UK, 2009.
2 K. H. Becker; C. W. McCurdy; T. M. Orlando; T. N. Rescigno, *Electron-Driven Processes: Scientific Challenges and Technological Opportunities*; August 2000.
3 E. Alizadeh and L. Sanche, Precursors of Solvated Electrons in Radiobiological Physics and Chemistry, *Chem. Rev.*, 2012, **112**(11), 5578–5602.

4 C. R. Arumainayagam, H. L. Lee, R. B. Nelson, D. R. Haines and R. P. Gunawardane, Low-energy electron-induced reactions in condensed matter, *Surf. Sci. Rep.*, 2010, **65**(1), 1–44.

5 X. L. Zhou and J. M. White, Photon-induced and electron-induced chemistry of chlorobenzene on Ag(111), *J. Chem. Phys.*, 1990, **92**(9), 5612–5621.

6 R. L. Hudson and M. H. Moore, The N-3 radical as a discriminator between ion-irradiated and UV-photolyzed astronomical ices, *Astrophys. J.*, 2002, **568**(2), 1095–1099.

7 S. C. Sparks, A. Szabo, G. J. Szulczewski, K. Junker and J. M. White, Thermal, electron, and photon induced chemistry of acetone on Ag(111), *J. Phys. Chem. B*, 1997, **101**(41), 8315–8323.

8 M. A. Henderson, R. D. Ramsier and J. T. Yates, Photon-induced *versus* electron-induced decomposition of Fe(CO)5 adsorbed on Ag(111) - iron film deposition, *J. Vac. Sci. Technol., A*, 1991, **9**(3), 1563–1568.

9 T. B. Scoggins, H. Ihm, Y. M. Sun and J. M. White, Chemistry of cyclopropane on Pt(111): Thermal, electron, and photon activation, *J. Phys. Chem. B*, 1999, **103**(32), 6791–6802.

10 E. Alizadeh, P. Cloutier, D. Hunting and L. Sanche, Soft X-ray and Low Energy Electron-Induced Damage to DNA under N-2 and O-2 Atmospheres, *J. Phys. Chem. B*, 2011, **115**(15), 4523–4531.

11 P. A. Gerakines, W. A. Schutte and P. Ehrenfreund, Ultraviolet processing of interstellar ice analogs. 1. Pure ices, *Astron. Astrophys.*, 1996, **312**(1), 289–305.

12 D. Williams and S. Viti, Modelling interstellar physics and chemistry: implications for surface and solid-state processes, *Philos. Trans. R. Soc. London, Ser. A*, 2013, **371**(1994), 20110587.

13 E. Herbst; E. F. van Dishoeck, Complex Organic Interstellar Molecules in *Annual Review of Astronomy and Astrophysics, Vol 47*, R. Blandford; J. Kormendy; E. VanDishoeck, ed. 2009; Vol. 47, pp. 427–480.

14 T. E. Cravens and A. Dalgarno, Ionization, dissociation, and heating efficiencies of cosmic-rays in a gas of molecular-hydrogen, *Astrophys. J.*, 1978, **219**(2), 750–752.

15 S. S. Prasad and S. P. Tarafdar, UV-radiation field inside dense clouds – its possible existence and chemical implications, *Astrophys. J.*, 1983, **267**(2), 603–609.

16 R. Gredel, S. Lepp, A. Dalgarno and E. Herbst, Cosmic-ray induced photodissociation and photoionization rates of interstellar-molecules, *Astrophys. J.*, 1989, **347**(1), 289–293.

17 N. Watanabe, Y. Kimura, A. Kouchi, T. Chigai, T. Hama and V. Pirronello, Direct measurements of hydrogen atom diffusion and the spin temperature of nascent H_2 molecule on amorphous solid water, *Astrophys. J.*, 2010, **714**(2), L233–L237.

18 R. T. Garrod and E. Herbst, Formation of methyl formate and other organic species in the warm-up phase of hot molecular cores, *Astron. Astrophys.*, 2006, **457**(3), 927–936.

19 L. J. Allamandola, S. A. Sandford and G. J. Valero, Photochemical and thermal evolution of interstellar/precometary ice analogs, *Icarus*, 1988, **76**(2), 225–52.

20 K. I. Öberg, R. T. Garrod, E. F. van Dishoeck and H. Linnartz, Formation rates of complex organics in UV irradiated CH3OH-rich ices I, *Astron. Astrophys.*, 2009, **504**(3), 891–U28.

21 N. J. Mason., Electron driven processes; scientific challenges and technical opportunities. *AIP Conference Proceedings* 2003, **680** (Application of Accelerators in Research and Industry), pp. 885–888.

22 S. M. Pimblott and J. A. LaVerne, Production of low-energy electrons by ionizing radiation, *Radiat. Phys. Chem.*, 2007, **76**(8–9), 1244–1247.

23 T. D. Harris, D. H. Lee, M. Q. Blumberg and C. R. Arumainayagam, Electron-Induced Reactions in Methanol Ultrathin Films Studied by Temperature-Programmed Desorption: A Useful Method to Study Radiation Chemistry, *J. Phys. Chem.*, 1995, **99**(23), 9530–5.

24 A. L. Schwaner and J. M. White, Electron-Induced Chemistry of Methanol on Ag(111), *J. Phys. Chem. B*, 1997, **101**(49), 10414–10422.

25 J. P. Jay-Gerin, M. J. Fraser, P. Swiderek, M. Michaud, C. Ferradini and L. Sanche, Evidence for CO formation in irradiated methanol and acetone: contribution of low-energy electron-energy-loss spectroscopy to γ-radiolysis, *Radiat. Phys. Chem.*, 1997, **50**(3), 263–265.

26 M. Lepage, M. Michaud and L. Sanche, Low energy electron total scattering cross section for the production of CO within condensed methanol, *J. Chem. Phys.*, 1997, **107**(9), 3478–3484.

27 D. A. Williams, W. A. Brown, S. D. Price, J. M. C. Rawlings and S. Viti, Molecules, ices and astronomy, *Astron. Geophys.*, 2007, **48**(1), 25–34.

28 S. E. Stein. "Mass Spectra". http://webbook.nist.gov/chemistry/ (accessed August 6, 2013).

29 S. Jheeta, A. Domaracka, S. Ptasinska, B. Sivaraman and N. J. Mason, The irradiation of pure CH_3OH and 1:1 mixture of NH_3:CH_3OH ices at 30 K using low energy electrons, *Chem. Phys. Lett.*, 2013, **556**, 359–364.

30 C. J. Bennett, S.-H. Chen, B.-J. Sun, A. H. H. Chang and R. I. Kaiser, Mechanistical studies on the irradiation of methanol in extraterrestrial ices, *Astrophys. J.*, 2007, **660**(2, Pt. 1), 1588–1608.

31 M. H. Moore, R. F. Ferrante and J. A. Nuth, III, Infrared spectra of proton irradiated ices containing methanol, *Planet. Space Sci.*, 1996, **44**(9), 927–935.

32 G. A. Baratta, G. Leto and M. E. Palumbo, A comparison of ion irradiation and UV photolysis of CH_4 and CH_3OH, *Astron. Astrophys.*, 2002, **384**(1), 343–349.

33 R. A. Johnson and A. E. Stanley, GC/MS and FT-IR spectra of methoxymethanol, *Appl. Spectrosc.*, 1991, **45**(2), 218–22.

34 G. Meshitsuka and M. Burton, Radiolysis of liquid methanol by CO-60 gamma-radiation, *Radiat. Res.*, 1958, **8**(4), 285–297.

35 W. J. Skraba, J. G. Burr, Jr. and D. N. Hess, Decomposition of methanol-C14 under the influence of its own radiation, *J. Chem. Phys.*, 1953, **21**, 1296.

36 M. C. Boyer, M. D. Boamah, K. K. Sullivan, C. R. Arumainayagam, M. M. Bazin, A. D. Bass and L. Sanche, To be published.

37 M. E. Palumbo, A. C. Castorina and G. Strazzulla, Ion irradiation effects on frozen methanol (CH_3OH), *Astronomy & Astrophysics*, 1999, **342**(2), 551–562.

PAPER

Stability of carbonaceous dust analogues and glycine under UV irradiation and electron bombardment

Belén Maté,* Isabel Tanarro, Miguel A. Moreno, Miguel Jiménez-Redondo, Rafael Escribano and Víctor J. Herrero

Received 13th December 2013, Accepted 4th February 2014

DOI: 10.1039/c3fd00132f

The effect of UV photon (120–200 nm) and electron (2 keV) irradiation of analogues of interstellar carbonaceous dust and of glycine were investigated by means of IR spectroscopy. Films of hydrogenated amorphous carbon (HAC), taken as dust analogues, were found to be stable under UV photon and electron bombardment. High fluences of photons and electrons, of the order of 10^{19} cm^{-2}, were needed for a film depletion of a few percent. UV photons were energetically more effective than electrons for depletion and led to a certain dehydrogenation of the HAC samples, whereas electrons led seemingly to a gradual erosion with no appreciable changes in the hydrocarbon structure. The rates of change observed may be relevant over the lifetime of a diffuse cloud, but cannot account for the rapid changes in hydrocarbon IR bands during the evolution of some proto-planetary nebulae. Glycine samples under the same photon and electron fluxes decay at a much faster rate, but tend usually to an equilibrium value different from zero, especially at low temperatures. Reversible reactions re-forming glycine, or the build-up of less transparent products, could explain this behavior. CO_2 and methylamine were identified as UV photoproducts. Electron irradiation led to a gradual disappearance of the glycine layers, also with formation of CO_2. No other reaction products were clearly identified. The thicker glycine layers (a few hundred nm) were not wholly depleted, but a film of the order of the electron penetration depth (80 nm), was totally destroyed with an electron fluence of $\sim 1 \times 10^{18}$ cm^{-2}. A 60 nm ice layer on top of glycine provided only partial shielding from the 2 keV electrons. From an energetic point of view, 2 keV electrons are less efficient than UV photons and, according to literature data, much less efficient than MeV protons for the destruction of glycine. The use of keV electrons to simulate effects of cosmic rays on analogues of interstellar grains should be taken with care, due to the low penetration depths of electrons in many samples of interest.

Instituto de Estructura de la Materia, IEM-CSIC, Serrano 123, 28006 Madrid, Spain

1   Introduction

Carbonaceous compounds, both solids and gas-phase molecules, are found in very diverse astronomical media.[1] They give the measure of the chemical complexity attainable in a given environment, and delimit the scenarios available for the chemical evolution towards life. A significant fraction of the elemental carbon is locked in large hydrocarbon structures forming small dust grains.[2] This carbonaceous dust, mostly formed in the last stages of evolution of C-rich stars, is the carrier of characteristic IR absorption bands revealing the presence of aliphatic, aromatic and olefinic functional groups in variable proportions.[3] Among the various candidate materials investigated as possible carriers of these bands, hydrogenated amorphous carbon (HAC) has led to the best agreement with the observations.[4] Carbonaceous grains are processed by H atoms, UV radiation, cosmic rays and interstellar shocks in their passage from asymptotic giant branch (AGB) stars to planetary nebulae (PN) and to the diffuse interstellar medium. In this environment, they could provide the source for clusters of polycyclic aromatic hydrocarbons (PAHs). Eventually, part of this material is incorporated into dense clouds and can end up as a component of planetary systems.[5] The mechanisms of HAC production and evolution in astronomical media are presently a subject of intensive investigation that requires not only detailed observations, but also laboratory work.[6-9]

Besides the large hydrocarbon structures just mentioned, many organic molecules, among them prebiotic species like sugars, amino acids, purines and pyrimidines have been identified in carbonaceous meteorites,[10] which are among the most primitive objects in the solar system. A rich organic inventory,[11] including the amino acid glycine,[12] has been found in comets, which are generally considered to contain some direct components from the prestellar core that gave rise to the solar nebula. A large number of organic compounds has also been recorded in observations of the interstellar medium (ISM),[13] with species of biological interest like precursors of sugars and molecules with peptidic bonds. Although the observation of glycine, the simplest amino acid, remains at best inconclusive,[14] its presence in the ISM is not unlikely since some of the molecules detected are of similar complexity. On the other hand, it is not clear whether complex and presumably fragile prestellar organic structures could survive the collapse of the protostellar core leading to the formation of a planetary system.[15] Establishing the conditions of survival of prebiotic molecules in various astronomical environments is nowadays a crucial issue for models of biogenesis.

The behavior of simple amino acids and specifically glycine in condensed phases relevant to astronomical media has drawn the attention of numerous research groups. The exchange between neutral and zwitterionic forms or the transitions between amorphous and crystalline phases as a function of temperature and chemical environment have been recently investigated in analogues of astronomical ices.[16,17] Likewise, the effects of bombarding these molecules with energetic ions,[18-21] X-Rays[22,23] or UV photons[24-29] have been the subject of many studies. A global appraisal of the different results on proton and photon irradiation is not straightforward because the experimental conditions are not always comparable and the interpretation of the data is complicated by the plurality of dissociation routes and secondary reactions.[30] Estimates based on the available data suggest that small amino acids could persist for tens or hundreds of Myr within the mantles of interstellar grains or under the surface of solar system bodies.

Energetic electrons are also known to be prevalent in astronomical media. They are present in the solar wind and in planetary magnetospheres and are also formed in the interaction of cosmic rays with matter. In interstellar grains, collision cascades induced by cosmic ray ions can lead to the production of secondary electrons with typical energies in the 0–10 keV range.[31] Although electron energies from a few eV to tens of eV are in principle the most adequate for chemical interactions, electrons with higher energies may be more efficient because they can penetrate deeper in the solids. Recent experiments by Barnett *et al.*[32] with electrons in the 100 eV–2keV interval have shown that the survival depth of organics in ices may be significantly lower than that expected from the electron penetration depth provided by current models based on Monte Carlo simulations.

The effects of electron irradiation on amino acids for astronomically relevant conditions have been less thoroughly studied than those of photons and ions. In some recent works, chemical routes for the electron stimulated formation of glycine[33,34] and even of dipeptides[35] have been explored at the surface or in the bulk of ices, but as far as we know, the survival of glycine under energetic electron bombardment has not yet been investigated. In this work we present the first results of a comparative study on the stability of HAC and glycine under bombardment by UV photons (120–200 nm) and 2 keV electrons. The investigation is based on IR spectroscopy of the samples of interest. The distinct features of photon and electron processing are discussed and compared to previous publications.

2.   Experimental details

2.1.   Plasma reactor

Carbonaceous deposits were grown by plasma enhanced chemical vapor deposition (PECVD)[36] in an inductively coupled RF discharge reactor. The reactor consists of a Pyrex tube, 30 cm length, 4 cm diameter, ended by two vacuum flanges (DIN 40KF). A 10 turns Cu coil placed externally around the central part of the tube, with a total length of 8 cm, was fed by a 13.56 MHz RF generator (Hüttinger PFG 300 RF + matchbox PFM 1500A) to produce the plasma, which was maintained at a constant power of 40 W during the deposition processes. One of the flanges of the reactor supports a piece holding an observation window and a narrow tube for gas input. This piece is easily removed to introduce the deposition substrates. The second flange is connected through a vacuum cross piece to a regulating valve and a rotary pump, to the vacuum gauges (Pirani and absolute capacitance manometers); and to a quadrupole mass spectrometer (Balzers, Prisma Plus-QMG220) used as Residual Gas Analyzer (RGA). The RGA is placed in a differentially pumped vacuum chamber connected to the reactor through a $\sim$50 μm diameter diaphragm. Typical pressures in the RGA chamber are five orders of magnitude lower than in the reactor.

The background pressure in the reactor was $\sim$3 $\times$ 10^{-3} mbar. For plasma generation, it was fed with a gas mixture of 40% CH_4 and 60% He through two fine needle valves, one for each gas, at 0.3 mbar total pressure. Gas residence time was estimated $\sim$1 s. Two silicon substrates, 2.5 cm diameter, 1 mm thickness, were placed, in horizontal orientation, roughly in the central plane of symmetry of the reactor. One of the substrates was placed $\sim$5 cm outside the coil in the direction of gas flow, and the other one in the central part of the coil. They were exposed for approximately one hour to the depositing plasma. The deposited samples had a distinct appearance (see Fig. 1). Deposits formed outside the coil (labeled HAC1) were very uniform with shiny circular interference fringes. Deposits inside the coil (labeled HAC2) were dusty, black and irregular. The IR spectra of these samples, measured for the regions marked with circles in Fig. 1, will be discussed in the next section. A thickness of $\sim$3 µm was estimated for the HAC1 samples from the interference fringes in the baseline of their IR spectra. The HAC deposits generated are stable enough to be handled and stored over a period of months without appreciable alteration.

The RGA was used to identify the species produced during the discharge and to measure the dissociation degree of the CH_4 precursor. The method was similar to that used in previous works on film deposition by DC discharges.[37,38] CH_4 dissociation, determined from the evolution of the RGA signal at $m/q^+ = 15$ a.m.u, was $\sim$70%. H_2 ($m/q^+ = 2$ a.m.u) and C_2H_2 ($m/q^+ = 26$ a.m.u) were identified as two of the main stable gas products. Ethylene and ethane are produced too, since clear increases at the $m/q^+ = 27$–30 a.m.u signals were observed, corresponding to the main peaks of the dissociation patterns of both species. C_3H_x (x = 4,6,8) formation could be also deduced from the signals at $m/q^+ = 29$, 37–41 a.m.u. Qualitative descriptions of the formation routes of C_xH_y stable products and radicals in different CH_4 rich discharges, and of the main kinetic processes involved, can be found elsewhere.[39–42]

2.2   Cryogenic chamber

The high vacuum cryogenic set up (Fig. 2) has been described elsewhere.[17,43] The stainless steel cylindrical chamber, with a residual pressure of 5×10^{-8} mbar, contains a closed-cycle helium cryostat, whose cold finger holds an IR transparent Si window in close thermal contact. The temperature of the Si substrate can be varied between 14 K and 300 K with 1 K accuracy. The system is coupled to a

Fig. 1   Samples of hydrogenated amorphous carbon deposited on Si by PECVD in the inductively coupled RF discharge reactor. Left sample: HAC1, grown outside the inductive coil. Right sample: HAC2, grown inside the inductive coil (see text). The circles indicate the regions used for spectroscopic and/or processing experiments and for the deposition of glycine.

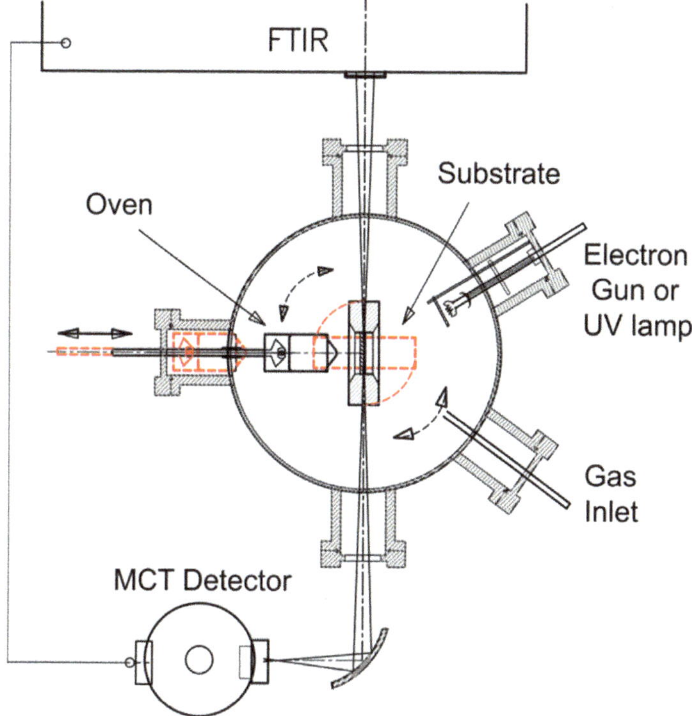

Fig. 2 Scheme of the cryogenic chamber. The UV lamp or electron gun can be interchanged to operate through the same window.

Vertex70 FTIR spectrometer in normal incidence transmission configuration. A rotatable flange allows the orientation under vacuum of the Si substrate toward different chamber ports. In one position, it faces a sublimation mini-oven that is used to evaporate glycine. This oven, of our own design, was described in more detail previously.[17,44] The Si substrate can be turned to another window to record transmission spectra, and, in a third position, it faces the UV lamp or the electron gun, depending on the experiment. The spectra shown in this work were recorded with a Mercury Cadmium Telluride detector, adding between 50 and 300 scans, with 2 cm^{-1} resolution.

The oven temperature for glycine sublimation was 140 ± 5 °C. For water deposition on top of the glycine layer, 1×10^{-5} mbar water vapor was admitted into the chamber through an independent line.

In a previous publication of our group, the amount of deposited glycine molecules was estimated from the IR spectrum *via* the absorption intensity in the 2000–800 cm^{-1} frequency range, using a theoretical value for the absorption band strength in this spectral region.[17] However, it is difficult to estimate the errors in the absolute intensities. In the present investigation we have scaled our spectra to available literature data for samples where the thickness of glycine films was experimentally measured. In particular, we adopted the values of ten Kate *et al.*[26] for vapor deposited crystalline glycine at room temperature and that of Gerakines *et al.*[20] for amorphous glycine at 20 K. The thickness of our layers varied between 80 and 900 nm, depending on the experiment.

2.3. UV lamp

The source for UV irradiation was a HAMAMATSU L10706, UV D2 lamp. The lamp is provided with a flexible pipe that allows placing its MgF_2 window 30 mm from the sample, inside the vacuum chamber. The UV beam has a diameter of 13 mm at the output window and a divergence of 7.5 degrees. The integrated emission flux in the 120 nm–200 nm wavelength range at 30 mm from the lamp window is 4×10^{14} photons $cm^{-2}\,s^{-1}$, as calculated for our particular configuration from the data provided by the manufacturer. The spectral profile of the lamp, also provided by the manufacturer, has a maximum at 155–165 nm, about 7 times more intense than the average emission of the rest of the wavelength interval. We have assumed an approximate average energy of 7.6 eV for our photons.

2.4. Electron gun

The electron source used for sample processing was built in our laboratory. It is based on the design used previously for electron-seed plasma ignition in our DC discharge reactors at low pressures.[42] It consists of a tungsten filament, 130 μm in diameter and ~20 mm long, curled to form a small spiral of about 10 mm length. The filament turns bright white and emits electrons efficiently with a current of about 2.2 A. To extract the electrons and direct them to the target, which in our case is electrically connected to ground, the filament can be negatively biased up to −2.5 kV, and a grounded grid is placed between the filament and the target, at 10 mm from the filament. The extracting grid is circular, with a 20 mm diameter, made of a very fine mesh with a separation of ~1 mm between parallel wires. The accelerated electrons going through the grid reach the target with an energy determined by the applied potential. The size of the electron beam is controlled by the size of the grid. The whole system is placed inside the vacuum chamber, and the accelerating grid is located 35 mm away from the target. In our experiments, the targets are the Si substrates with the HAC or glycine samples.

The flux of electrons reaching the samples was calibrated by replacing the Si substrate in the cryostat finger by an electrode of the same size as the sample holder (a copper disk, 1 cm diameter), placed exactly at the same place as the samples. The current from the electron gun reaching the calibration electrode was measured by means of a microamperimeter. We found that the electron flux reaching the target was proportional to the emission current and did not depend strongly on the accelerating potential. For the irradiation experiments presented in this work we have used a flux of 2.66×10^{14} electrons $cm^{-2}\,s^{-1}$, with an energy of 2.0 keV. This electron flux corresponds to an emission current of 4.6 mA (between the filament and the grid) and an electron current of 42 μA cm^{-2} at the target.

3. Results and discussion

3.1. Carbonaceus deposits

Fig. 3 shows the spectra of two of the carbonaceous deposits described in the previous section. Different deposits grown in analogous conditions had very

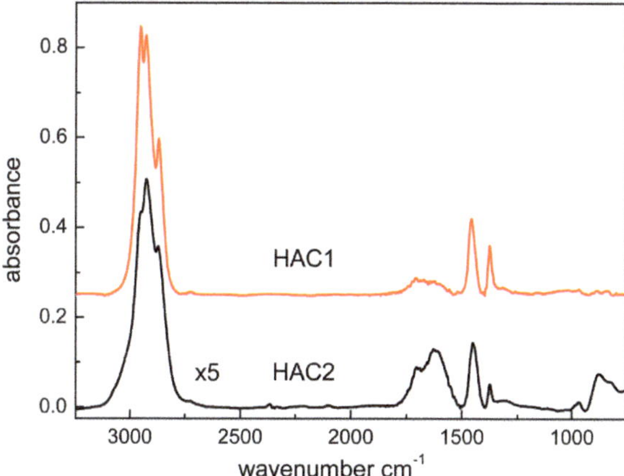

Fig. 3 IR spectra of the carbonaceous deposits shown in Fig. 1. The spectrum of the HAC2 sample has been magnified by a factor of 5. The spectrum of HAC1 is offset in the absorbance scale to facilitate the comparison.

repetitive spectra. Note that the IR absorption by the HAC2 sample is much smaller. Both spectra exhibit the characteristic features of HAC materials, but with different proportions. The following assignments are taken from ref. 7 and 9. In the region around 2900 cm^{-1} ($\sim$3.4 µm) the maximum in the HAC1 spectrum corresponds to the peak at 2959 cm^{-1}, which is assigned to the asymmetric stretching of aliphatic CH_3 groups, whereas in the HAC2 spectrum the maximum is found at 2929 cm^{-1} and is due to asymmetric stretching vibrations of the aliphatic CH_2 group. The two peaks at 1450 and 1370 cm^{-1} are attributed to deformation modes of CH_2 and CH_3 respectively. Again in this case the CH_2/CH_3 intensity ratio is larger for HAC2. The broad band in the 1600 cm^{-1} region is much stronger (in relative value) for HAC2, with a peak at $\sim$1615 cm^{-1} attributed to aromatic and olefinic C=C stretching modes. Features below 1000 cm^{-1} are also assigned to aromatic and olefinic vibrations.

The analysis of the spectra shows that both samples have aliphatic chains, but the deposit generated inside the coil of the plasma reactor, subjected to a more intensive attack by the ions and radicals in the discharge, has a smaller hydrogen content and a higher proportion of aromatic and olefinic components. This result is in agreement with intuitive expectations based on the more graphitic appearance of the HAC2 layer, as shown in Fig. 1. Dartois *et al.*[7] have produced HAC samples of variable composition by UV photolysis of different hydrocarbon precursors. Their HAC samples generated from methane or ethane produce spectra that resemble that of HAC1 in Fig. 3. On the other hand, their HAC from a xylene precursor yields a spectrum in which the 3.4 µm ($\sim$2900 cm^{-1}) band structure is close to that observed in our HAC2 spectrum. A similar 3.4 µm band, with a maximum in the central peak, is also seen in the IR spectra of HACs obtained by other groups with different methods. Mennella *et al.*[6] used hydrogenation of nano-sized carbon grains by a flux of H atoms; Kovačević *et al.*[8] employed a RF discharge of C_2H_2 and Ar; and Gadallah *et al.*[9] used ablation of graphite in a hydrogen atmosphere. The band profiles from these and other laboratory samples[4] give a good match to the shape of the 3.4 µm band observed in the diffuse ISM.[2,4,7] The ubiquity of this band shape suggests a very stable underlying structure.

In the following, we report on the distinct effects of UV light and electron bombardment on HAC1 at room temperature. The sample was irradiated for 24 h with a flux of 4×10^{14} UV photons cm^{-2} s^{-1} from the deuterium lamp ($\sim$160 nm), which corresponds to a dose of 2.63×10^{20} eV cm^{-2}, assuming an energy of 7.6 eV for the photons. The effects of this irradiation can be seen in panel a) of Fig. 4, where a difference spectrum (irradiated sample–non-irradiated sample) of the 3.4 μm band is displayed. A very small decrease in peak intensity ($\sim$3%) can be seen. The peaks in the difference spectrum are slightly displaced with respect to the HAC1 peaks because the decrease is more pronounced toward the high frequency end of the band, corresponding to the CH_3 asymmetric stretching, and less so at the central peak (CH_2 stretching). This behavior hints at an evolution towards the 3.4 μm band profile of HAC2 and thus to a loss of hydrogen. In order to visualize this effect we have represented in panel c) of the same figure the effect on the spectrum of multiplying by 20 the absorbance decrease of panel a). For the wavelength region below 2000 cm^{-1}, where the bands are much weaker, we could not derive reliable difference spectra.

If we take an estimate of 3×10^{8} eV cm^{-2} s^{-1} for the UV flux in a diffuse cloud,[9] our irradiation time corresponds to approximately 2.8×10^{4} years. Over this time, the HAC1 material should be essentially stable in the diffuse ISM, with only a slight tendency to lose hydrogen atoms and form longer chains.

Much larger irradiation doses were used by Gadallah *et al.*[9] in their study of the effect of 160 nm UV photons from a deuterium lamp on HAC samples. The IR spectra of the unprocessed deposits of Gadallah *et al.* are closer to those of HAC2 in the present work. After a dose of 1×10^{23} eV cm^{-2}, which corresponds to one

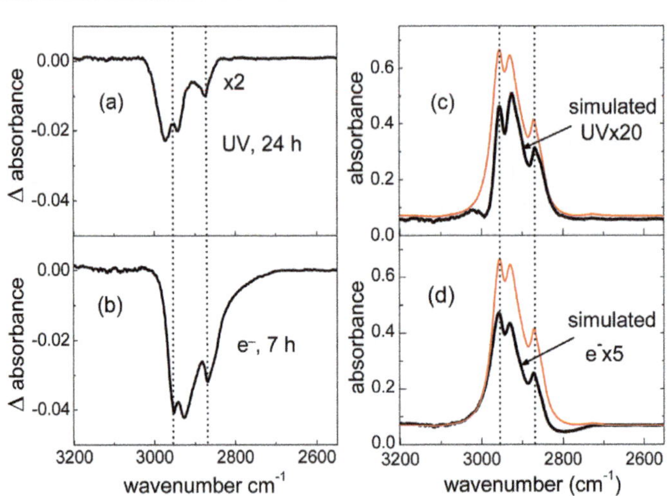

Fig. 4   Evolution of the 3.4 μm band of HAC1. Panel a) Difference spectrum (irradiated– non-irradiated) of the UV processed HAC1 sample (dose: 2.63×10^{20} eV cm^{-2}). Panel b) Id. of the HAC1 sample processed with electrons (dose: 1.34×10^{22} eV cm^{-2}). Panel c) Thin red line: HAC1 spectrum; thick black line: simulated spectrum after multiplying by 20 the absorbance decrease of panel a). Panel d) Thin red line: HAC1 spectrum; black solid line: simulated spectrum after multiplying by 5 the absorbance decrease of panel b).

third of the estimated lifetime of a diffuse cloud,[9] these authors observed a marked decrease of all aliphatic features (more than 80% at the maximum of the 3.4 µm band). The characteristic profile of the band, with a maximum in the central peak, was maintained, further corroborating the stability of this structure. Our experimental results suggest that the hydrogen-richer HAC1 material will also evolve toward this structure with sufficient UV irradiation. Gadallah *et al.* also found an increase in some of the aromatic bands and concluded that UV irradiation could explain the evolution of HAC toward PAHs during the lifetime of a diffuse cloud. Our measurements were not sensitive enough to variations in the aromatic features. However, the evolution rate observed both by Gadallah *et al.* and by us is too slow to justify the appearance and evolution of aliphatic and aromatic features in proto-planetary nebulae on time scales as short as a few hundred years.[3]

The effects of electron bombardment on HAC1 are shown in panel b) of Fig. 4. In this case, the sample was irradiated for seven hours with a $2.63 \times 10^{14}\,\mathrm{cm}^{-2}\,\mathrm{s}^{-1}$ flux of 2 keV electrons. The total processing dose was $1.34 \times 10^{22}\,\mathrm{eV\,cm}^{-2}$. After this dose, the depletion in the peaks of the 3.4 µm band was ∼6%. A comparison with the previous measurements shows that, although the depletion efficiency per photon is lower than that per electron, the energetic efficiency is much higher for photon irradiation. The difference spectrum for the electron irradiated sample does not show a preferential decrease of CH_3 bands, as compared with CH_2 bands. To make this point more clear we have represented in Fig. 4d) the simulated HAC1 spectrum after multiplying by 5 the absorbance decrease of panel b). The hints of dehydrogenation reflected in the growth of the CH_2/CH_3 ratio, commented on for the UV case, are not found here. It seems that electron bombardment gradually erodes the film rather than transforming it.

The penetration of electrons in solids is determined by a complex series of energy loss processes involving the production of UV and X-Ray bremsstrahlung, the generation of secondary electrons, ions and other reactive species, and ultimately the interaction with phonons and the dissipation of thermal energy. Penetration depths are usually estimated with the help of Monte Carlo simulations. Predictions with the CASINO code[45] for 2 keV electrons impinging on hydrogenated carbon yield penetration depths between 90 and 120 nm, depending on the assumed density and hydrogen content. For our estimated film thickness of 3 µm, all electrons will be stopped within the first layers of the HAC1 deposit and will not interact at first with the whole sample. The deeper part of the HAC1 deposit will be reached by electrons only after the depletion of the outer layers.

3.2.   Glycine

Fig. 5 shows the spectra of glycine films deposited on a cold (20 K) Si substrate and then irradiated with UV photons or electrons. Appreciable differences are found in the spectra of the differently processed samples. The spectra show that the deposited films, 600 and 900 nm thick respectively, are amorphous solids, where the glycine molecules are predominantly in a zwitterionic form[17] as shown by the small intensity of the characteristic neutral bands,[17] at 1730 and 1240 cm^{-1}. Spectra of thinner films (<400 nm) deposited at this temperature have a significant component of the neutral form. The low temperature transformation

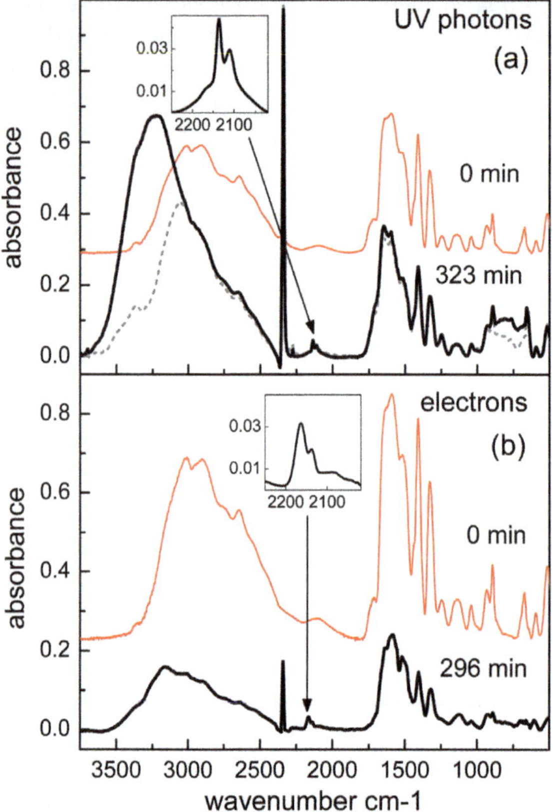

Fig. 5 IR spectra of amorphous glycine deposited at 20 K. Top: 600 nm thick deposit before (red) and after 323 min processing by UV photons. Bottom: 900 nm thick sample, before (red) and after 296 min bombardment with 2.0 keV electrons.

from neutral to zwitterionic with growing thickness is not well understood at present and will be studied in more detail in further works.

In the UV irradiated glycine film (Fig. 5a) large bands of water (3500–3000 cm^{-1}) and CO_2 (2341 cm^{-1}) grow during the experiment. These bands are attributed to deposition of background molecules. The band intensities correspond to layers of $\sim$200 nm water and $\sim$60 nm CO_2 that are consistent with the expectations for the five hour experiment. Part of the CO_2 peak is surely due to glycine decomposition. Much smaller peaks are observed at 2136 cm^{-1} (CO stretch) and at 2110 cm^{-1} (see inset in panel a) of Fig. 5). This latter peak could be due to chain oxides of the C_xO_y type.[46] Other peaks of photolysis products appear on top of the glycine bands between 1780 and 1450 cm^{-1} and will be commented on below. The small amount of neutral glycine present in the initial deposit does not disappear upon UV irradiation, as attested by the peak at 1240 cm^{-1}. The subtraction of the water contribution to the spectrum is also shown with a dotted line.

In the glycine deposit bombarded with electrons, no water is formed on top after a five hour experiment with the same chamber conditions. A peak appears at 2341 cm^{-1}, attributed to CO_2, but the absence of a water layer, and the fact that this peak is shifted by a few cm^{-1} with respect to that of pure CO_2, suggest that it corresponds to CO_2 surrounded by other molecules within the bulk of the sample.[47] It is most probably a product of glycine decomposition. Smaller peaks appear too at 2136 cm^{-1} (CO stretch) and at 2164 cm^{-1}, which could be due to OCN^-.[48] In the region between 1750 and 1000 cm^{-1} the two bands of the neutral form, at 1730 and 1240 cm^{-1}, disappear almost totally but otherwise, no significant changes are observed in the band profile. A most notable feature of this experiment was the gradual temperature increase in the sample and substrate during the intense electron irradiation. In the first four hours of the experiment the temperature increased from 20 to 76 K and then stabilized.

In an attempt to identify products of the photolysis of glycine in the 1780–1440 cm^{-1} spectral range, we intensified the changes by performing an experiment on a thinner layer (150 nm) deposited at 20 K. The results are shown in Fig. 6. In the upper trace of the top panel, corresponding to the spectrum of the unprocessed sample, the two bands of neutral glycine (1730 and 1240 cm^{-1}) characteristic of amorphous thin films are clearly visible. The lower trace of the upper panel, corresponding to the irradiated sample, shows a number of new peaks on top of glycine bands. In the lower panel, the difference spectrum (irradiated–non-irradiated) is shown.

Tentative assignments of the positive peaks in the difference spectrum of Fig. 6 are listed in Table 1. The peaks at 2136 and 2110 cm^{-1} were commented on above. The peaks at 1650, 1598 and 1482 cm^{-1} are assigned to methyilamonium ions and to methylamine. The ν_9 NH_2 wagging vibration of methylamine has been variously attributed to the 895 cm^{-1} peak[49] or to the 819 cm^{-1} peak,[33] not included in the table, but also present within the positive band at low frequency. The

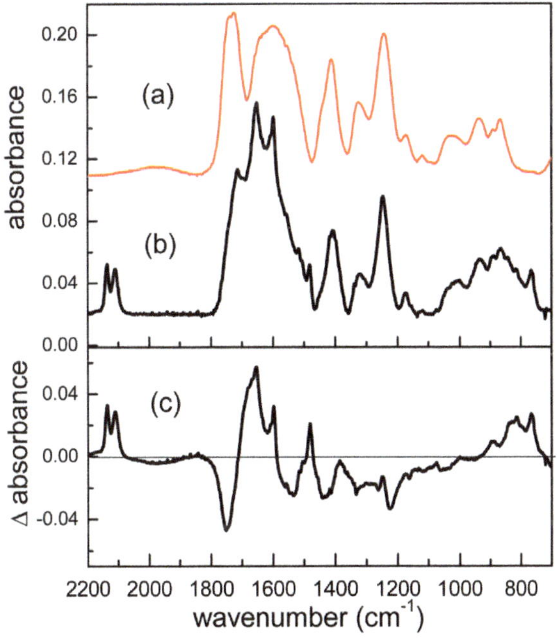

Fig. 6   IR spectrum of a thin amorphous glycine layer deposited at 20 K, a) before and b) after 280 min irradiation with UV photons. c) Difference spectrum: b) − a).

Table 1 New infrared band positions (cm^{-1}) of the processed sample at 20 K after irradiation with UV photons

Wavenumber (cm^{-1})	molecule	Assignment
2136	CO	ν_1 stretch CO
2110	CxOy oxides	ν_1 stretch CO, ref. 46
1650	$CH_3 NH_3^+$	ν_9, asymmetric bending NH_3, ref. 49
1598	CH_3NH_2	ν_4 scissor NH_2, ref. 33
1482	CH_3NH_2	ν_{12} antisymm. def. CH_3, ref. 33

production of methylamine, and the likely presence of CO_2 (see above), point to decarboxylation as one of the main mechanisms of destruction of glycine upon UV irradiation, as already noted in previous works.[24,26]

Processing experiments were also carried out with the same electron and photon fluxes, on ~500 nm thick glycine layers deposited at 300 K. At this temperature, crystalline β glycine was obtained.[19,21,26] The decay of these samples was much faster (see below), and no evidence of new products was seen in the spectra, which was not unexpected, since volatile species are not retained at this temperature. With electron processing, a 570 nm film was eliminated after 77 min, leaving a persistent residue, insoluble in water, with a broad absorption band between 1850 and 1000 cm^{-1}.

The rate of disappearance of glycine was traced by monitoring the integrated intensity of its 1400 cm^{-1} COO stretching band. The results are shown in Fig. 7.

At a given fluence, 2 keV electrons are always more efficient than 160 nm (7.6 eV) photons for the decomposition of glycine but, as in the case of the carbonaceous films discussed above, the energetic efficiency is much higher for the photons. The process of interaction of photons and electrons with glycine is also expected to be very different. Estimates of the cross section for UV absorption by glycine solid samples have been given by ten Kate et al.[26] For the 120–180 nm wavelength range, which is close to the 120–200 nm range of our deuterium lamp, these authors provided values of the dissociation cross section of ~2 × 10^{-19} cm^{-2}/molecule. Their experiments were carried out on crystalline β glycine, but we will assume that the corresponding value for our amorphous low temperature glycine is of the same order. For this value of the cross sections, all our glycine films are optically thin with respect to the UV photons.

In principle, first order kinetics could be expected for the photolysis of glycine, but this is not what is found in our experiments. After a relatively fast decay at the start, the curves in the upper panel of Fig. 7 stabilize and tend asymptotically toward values markedly different from zero. For the 20 K experiment, the total amount of glycine destroyed for a photon fluence of 8 × 10^{18} photons cm^{-2} is only 30%. Deviations from the first order kinetics law have also been observed by other authors in experiments of amino acid photolysis[29] or proton processing.[19] Gerakines et al.[19] suggested that an equilibrium value different from zero could be due to reversible processes within the sample. An alternative explanation would be the gradual shielding of the sample by a photolysis product less transparent to the UV photons.[25,50] The deposition of background water and CO_2 on the sample during the 20 K UV irradiation experiment is not expected to perturb the

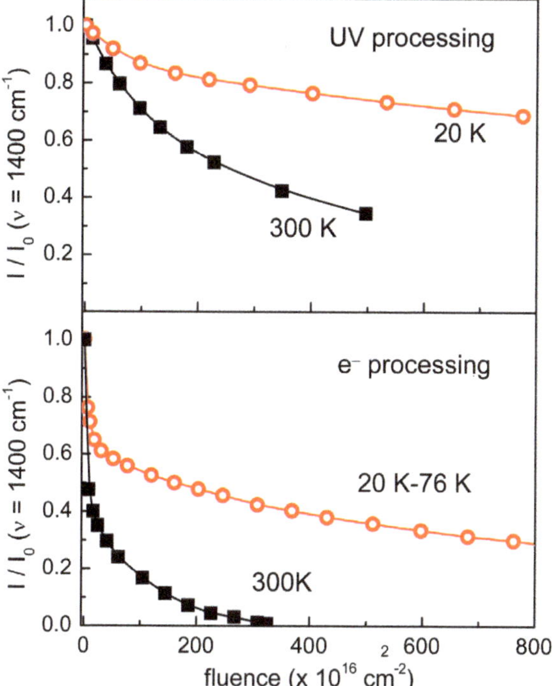

Fig. 7 Decay of the normalized intensity of the 1400 cm^{-1} spectral feature of glycine *versus* fluence of processing agent. Upper panel: UV photons (160 nm, 7.6 eV); circles: 600 nm layer, 20 K; squares: 460 nm layer, 300 K. Lower panel: electrons (2 keV); circles: 900 nm, 20–76 K; squares: 570 nm, 300 K.

photolysis of glycine, since the thin layers of these two molecules should be essentially transparent.

The hypothesis of reversible reactions could also explain the slower decay rate at low temperatures. The photolysis of glycine is expected to produce small volatile products[30] and, at low temperatures, these products would remain in the sample and be available for reactions leading back to glycine. Our experiments cannot decide on the cause of the slow decay, but in any case the observed behavior renders questionable the derivation of a meaningful half-life time from these measurements.

In the electron bombardment experiments (lower panel of Fig. 7) the situation is different. Estimates based on the CASINO code indicate that the penetration depth of 2 keV electrons in the glycine solid should be ~80 nm. The bulk of the glycine samples would thus be accessed only gradually after a layer by layer removal of the outer material. As a consequence, water and CO_2 cannot deposit since a fresh surface is continually being created by the electrons. It also seems clear that a significant amount of the energy delivered by the electrons is dissipated as heat, as shown by the temperature increase of the substrate in the course of the measurements. An experiment with a lower electron flux (5×10^{13} electrons cm^{-2} s^{-1}) was carried out to check the possible influence of the raising temperature on the observed depletion rate. In this case, the temperature stabilized at

35 K and the same decay rate was observed. The slower decay in the low temperature experiments could also be due to reversible reactions with products of glycine decomposition that do not evaporate for temperatures below 80 K. Gerakines *et al.*[19] have recently reported the decay at 15 K of a 900 nm thick glycine sample irradiated with 0.8 MeV protons. They observed a 50% decrease of the initial intensity for a fluence of 4×10^{14} protons cm^{-2}, as compared with 1.6×10^{18} electrons cm^{-2} in the present experiment (see Fig. 7). The electrons will leave most of their energy in the "opaque" solid, whereas the 0.8 MeV protons will only leave 4.2 eV ($\sim$0.5% of their energy) in the sample that has a stopping power of 4.67 eV μm^{-1}.[19] It is clear that MeV protons are much more efficient than keV electrons for the destruction of glycine, at least in samples with a thickness of hundreds of nm.

3.3   Water ice on glycine

In the final part of our work we have investigated the electron bombardment of a thin film of glycine and the shielding effect of a water ice layer on top of it. First, an amorphous glycine deposit of 80 nm thickness was prepared at 20 K on a cold substrate. In this case, the glycine was deposited on a Si window previously covered by a hydrogenated amorphous (HAC1) film. As discussed earlier, HAC is a likely component of interstellar dust. The glycine sample was then bombarded with the same electron flux used in the previous experiments (2.66×10^{14} electrons cm^{-2} s^{-1}), and the integrated intensity of the 1400 cm^{-1} band was employed to follow the evolution of glycine during electron irradiation. The results of the measurements are shown in Fig. 8 (squares). Note the heating of the samples upon electron bombardment commented on previously. The selected film thickness is of the order of the penetration length predicted by the CASINO software and the whole film is subjected to the effects of the electrons from the

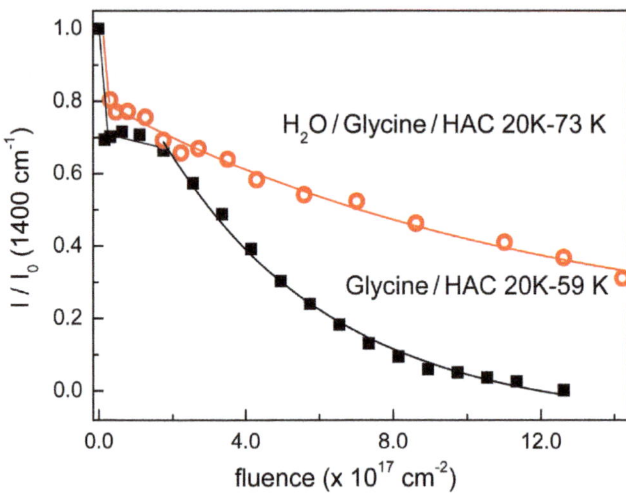

Fig. 8   Decay of the normalized intensity of the 1400 cm^{-1} spectral feature of glycine *versus* time of electron irradiation for uncovered glycine (squares) and glycine covered with a 60 nm H_2O layer (circles).

beginning of the process. After a sharp decay of about 25%, the band intensity shows a plateau associated with the transformation from the neutral to the zwitterionic form of the amino acid. When the transformation is complete, the decay continues until the total disappearance of the glycine film. A 50% decay is achieved with a fluence of $\sim 3 \times 10^{17}$ electrons cm^{-2}, much lower than the corresponding value (1.6×10^{18} electrons cm^{-2}) for the 900 nm film studied previously. The difference is consistent with the gradual penetration of electrons in the thicker film commented on above.

To study the effect of ice shielding, the experiment was repeated after growing a 60 nm water ice layer on an 80 nm glycine film deposited on HAC1 at 20 K. This ice thickness is realistic for interstellar grain mantels. The results of the electron irradiation experiment are shown in Fig. 8 (circles). Glycine is still destroyed by the electrons, but now at a slower pace. A 50% fall in intensity requires now roughly twice the electron fluence of the unshielded film. Simulations with the CASINO code showed that for vapor deposited water ice, with an assumed density of 0.67 g cm^{-3}, the electron penetration depth should be close to 200 nm. The predictions of the CASINO model further show that a significant number of electrons pierce the 60 nm water ice shield and deposit their energy within the first 30 nm of the glycine film. The relatively high permeability of ice to the penetration of electrons in the keV range was recently stressed by Barnet *et al.*[32] Further work is in progress to better quantify ice shielding effects.

In Fig. 9 we present a final comparison of the energetic efficiency of MeV protons, vacuum UV photons and keV electrons for the destruction of glycine in cold (<25 K) samples using data from this work and previous publications. This figure displays the decay of the 1400 cm^{-1} band as a function of the energy irradiated per glycine molecule. All samples chosen for this comparison are optically thin for the respective processing agents. The highest efficiency corresponds to protons and the lowest to electrons. The photon results, from different

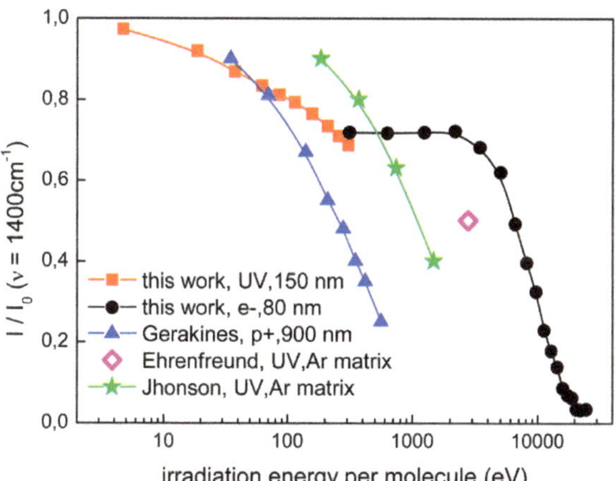

Fig. 9 Evolution of glycine 1400 cm^{-1} band as function of the energy irradiated on the sample normalized to the number eber of glycine molecules in the sample. This work: experiments (squares and circles). Literature data: triangles, from ref. 19; stars, from ref. 29; rhombus, from ref. 24. For the Ar matrices, the dilution factor has been considered.

experiments including solid layers and matrices, show certain dispersion, but lie between those of protons and electrons. Note that the plateau in the electron curve corresponds to the neutral to zwitterionic transformation commented on above (Fig. 8). The higher efficiency of protons and photons is even more marked if the actual energy deposited by the three processing agents in the samples is considered.

Kaiser *et al.* have suggested[33,35] that the effects of cosmic ray ions with energies in the MeV range could be mimicked with a bombardment of electrons in the keV range. This process should deposit a comparable amount of energy into the sample and would reproduce the secondary electron cascades expected for cosmic ray interaction. However, the comparison of the present experiments with the results of Gerakines *et al.*[19] does not support this assumption. It seems that electron cascades produced by cosmic rays within the bulk of solids are much more efficient for inducing chemical transformations than electron bombardment from the outside.

Conclusions

Hydrogenated amorphous carbon deposits are relatively stable under UV photon and electron bombardment. Decays of just a few percent in the HAC IR intensity are observed for comparatively high fluences ($\sim 10^{19}$ cm^{-2}). UV photons (160 nm, 7.6 eV) are found to induce a certain dehydrogenation of the hydrocarbon structure, probably increasing chain length. Bombardment with 2 keV electrons is energetically less efficient for depletion. It does not seem to change the structure of the hydrocarbons, but produces a layer by layer erosion of the films. Extrapolation of photon fluxes to the conditions of the diffuse interstellar medium reveals that the measured rates are significant as compared with the lifetime of a diffuse cloud, but too slow to justify the fast variation of IR hydrocarbon bands often observed in the evolution of proto-planetary nebulae.

Glycine samples subjected to comparable UV and electron fluxes are much more labile and decay at much faster rates. However, in many cases, especially in the experiments carried out at low temperatures (20–70 K), the glycine concentration tends to an equilibrium value different from zero. This equilibrium concentration could be determined by the build-up of less UV transparent products or by reversible reactions. At low temperatures, the largely volatile products of decomposition would remain in the sample and be available for reforming glycine. In accordance with previous works, decarboxylation, with production of CO_2 and methylamine, is found to be a main route of glycine UV photolysis. Carbon monoxide and possibly C_xO_y oxides are also found among the reaction products.

Electron bombardment of glycine seems to proceed in a much more indiscriminate way affecting especially the outer layers, which are gradually depleted before electrons access the bulk of the sample. The species CO_2, CO and possibly OCN$^-$ are identified in the low temperature irradiated samples. A water ice layer of 60 nm, comparable in thickness to interstellar grain mantels, is found to shield glycine only partially from electron bombardment. A comparison of our results with previous literature works indicates that, from the energetic point of view, MeV protons are more efficient than UV photons, and these, more efficient than keV electrons for the destruction of glycine. The use of keV electrons to simulate the interaction of cosmic rays with matter should be taken with care, since the effects of electrons are extremely dependent on the sample thickness in the tens of nm range.

Note added in proof

A publication on electronic processing of glycine[51] appeared in the time elapsed between the submission of this paper and the FD168 conference. Pilling and coworkers study the temperature effect on the decomposition of glycine with 2 keV electrons. Their results are in very good agreement with the ones presented in this work.

Acknowledgements

We are indebted to Y. Rodríguez-Lazcano for help with part of the measurements, to J. A. Martín Gago for advice on the construction of the electron gun, and to O. Gálvez for helpful discussions. This work was funded by the MCINN of Spain under grants FIS2010-16455 and CDS2009-00038. M. Jiménez-Redondo acknowledges also funding from the FPI program of the MICINN.

References

1 A. G. G. M. Tielens, *Rev. Mod. Phys.*, 2013, **85**, 1021.

2 J. E. Chiar, A. G. G. M. Tielens, A. J. Adamson and A. Ricca, *Astrophys. J.*, 2013, **770**, 78.

3 S. Kwok, in *The Molecular Universe*, J. Cernicharo and R. Bachiller ed, IAU Symposium 280, Cambridge University Press, 2011, p 203.

4 Y. J. Pendelton and L. J. Allamandola, *Astrophys. J. Suppl.*, 2002, **138**, 75.

5 G. M. Muñoz Caro and E. Dartois, *Chem. Soc. Rev.*, 2013, **42**, 2173.

6 V. Mennella, J. R. Brucato, L. Collangeli and M. E. Palumbo, *Astrophys. J.*, 2002, **569**, 531.

7 E. Dartois, G. M. Muñoz Caro, D. Deboffle, G. Montagnac and L. D'Hendecourt, *Astron. Astrophys.*, 2005, **432**, 895.

8 E. Kovačević, I. Stefanović, J. Berndt, Y. J. Pendelton and J. Winter, *Astrophys. J.*, 2005, **623**, 242.

9 K. A. K. Gadallah, H. Mutschke and C. Jäger, *Astron. Astrophys.*, 2012, **544**, A107.

10 O. Botha and J. L. Bada, *Surveys in Geophysics*, 2002, **23**, 411.

11 M. J. Mumma and S. B. Charnley, *Annu. Rev. Astron. Astrophys.*, 2011, **49**, 471.

12 J. E. Elsila, D. P. Glavin and J. P. Dworkin, *Meteorit. Planet. Sci.*, 2009, **44**, 1323.

13 E. Herbst and E. F. van Dishoeck, *Annu. Rev. Astron. Astrophys.*, 2009, **47**, 427.

14 L. E. Snyder, F. J. Lovas, J. M. Hollis, D. N. Friedel, P. R. Jewell, A. Remijan, V. V. Ilyushin, E. A. Alekseev and S. F. Dyubko, *Astrophys. J.*, 2005, **619**, 914.

15 P. Ehrenfreund and M. A. Sephton, *Faraday Discuss.*, 2006, **133**, 277.

16 A. Gómez-Zavaglia and R. Fausto, *Phys. Chem. Chem. Phys.*, 2003, **5**, 52.

17 B. Maté, Y. Rodríguez-Lazcano, O. Gálvez, I. Tanarro and R. Escribano, *Phys. Chem. Chem. Phys.*, 2011, **13**, 12268.

18 W. Huang, Z. Yu and Y. Zhang, *Chem. Phys.*, 1998, **237**, 223.

19 P. A. Gerakines, R. L. Hudson, M. H. Moore and J.-L. Bell, *Icarus*, 2012, **220**, 647.

20 P. A. Gerakines and R. L. Hudson, *Astrobiology*, 2013, **13**, 647.

21 S. Pilling, L. A. V. Mendes, V. Bordalo, C. F. M. Guaman, C. R. Ponciano and E. F. da Silveira, *Astrobiology*, 2013, **13**, 79.

22 S. Pilling, D. P. P. Andrade, E. M. do Nascimento, R. R. T. Marinho, H. M. Boechat-Roberty, L. H. de Coutinho, G. G. B. de Souza, R. B. de Castilho, R. l. Cavasso-Filho, A. F. Lago and A. n. de Brito, *Mon. Not. R. Astron. Soc.*, 2011, **411**, 2214.

23 A. Pernet, J. Pilmé, F. Pauzat, Y. Ellinger, F. Sirotti, M. Silly, Ph. Parent and C. Laffon, *Astron. Astrophys.*, 2013, **552**, A100.

24 P. Ehrenfreund, M. P. Bernstein, J. P. Dworkin, S. A. Sandford and L. J. Allamandola, *Astrophys. J.*, 2001, **550**, L95.

25 M. P. Bernstein, S. F. M. Ashbourn, S. A. Sandford and L. J. Allamamdola, *Astrophys. J.*, 2004, **601**, 365.

26 I. L. ten Kate, J. R. C. Garry, Z. Peeters, R. Qinn, B. Foing and P. Ehrenfreund, *Meteorit. Planet. Sci.*, 2005, **40**, 1185.

27 G. E. Orzechowska, J. D. Goguen, P. V. Johnson, A. Tsapin and I. Kanik, *Icarus*, 2007, **187**, 584.

28 A. M. Ferreira-Rodrigues, M. G. P. Homem, A. Naves de Brito, C. R. Ponciano and E. F. da Silveira, *Int. J. Mass Spectrom.*, 2011, **306**, 77.

29 P. V. Johnson, R. Hodyss, V. F. Chernow, D. M. Lipscomb and J. D. Goguen, *Icarus*, 2012, **221**, 800.

30 E. Sagstuen, A. Sanderud and E. O. Hole, *Radiat. Res.*, 2004, **162**, 112.

31 R. I. Kaiser and K. Roessler, *Astrophys. J.*, 1997, **475**, 144.

32 I. L. Barnett, A. Lignell and M. S. Gudipati, *Astrophys. J.*, 2012, **747**, 13.

33 P. D. Holtom, Ch. Bennett, Y. Osamura, N. J. Mason and R. I. Kaiser, *Astrophys. J.*, 2005, **626**, 940.

34 A. Lafosse, M. Bertin and R. Ariza, *Progr. Surf. Sci.*, **84**, p. 177.

35 R. I. Kaiser, A. M. Stockton, Y. S. Kim, E. C. Jensen and A. R. A. Mathies, *Astrophys. J.*, 2013, **765**, 111.

36 F. J. Gordillo-Vázquez, V. J. Herrero and I. Tanarro, *Chem. Vap. Deposition*, 2007, **13**, 267.

37 F. L. Tabarés, D. Tafalla, I. Tanarro, V. J. Herrero and A. M. Islyaikin, *Vacuum*, 2004, **73**, 161.

38 I. Tanarro, J. A. Ferreira, V. J. Herrero, F. L. Tabarés and C. Gómez-Aleixandre, *J. Nucl. Mater.*, 2009, **390–391**, 696.

39 M. Bauer, T. Schwarz-Selinger, W. Jacob and A. von Keudell, *J. Appl. Phys.*, 2005, **98**, 073302.

40 D. P. Liu, I. T. Martin, J. Zhou and E. R. Fisher, E R, *Pure Appl. Chem.*, 2006, **78**, 1187.

41 M. Mozetic, A. Vesel, D. Alegre, F. L. Tabares, J. Appl. Phys., 2011, **110**, pp. 053302–10.

42 I. Tanarro, V. J. Herrero, A. M. Islyaikin, I. Méndez, F. L. Tabarés and D. Tafalla, *J. Phys. Chem. A*, 2007, **111**, 9003.

43 B. Maté, A. Medialdea, M. A. Moreno, R. Escribano and V. J. Herrero, *J. Phys. Chem. B*, 2003, **107**, 11098.

44 Y. Rodríguez-Lazcano, B. Maté, O. Gálvez, V. J. Herrero, I. Tanarro and R. Escribano, *J. Quant. Spectrosc. Radiat. Transfer*, 2012, **113**, 1266.

45 D. Drouin *et al.*, 2011, *Monte Carlo Simulation of Electron Trajectory in Solids* (CASINO version 2.48) Université de Sherbrooke, Sherbrooke, Quebec, Canada., www.gel.usherbrooke.ca/casino.

46 M. J. Loeffler, G. A. Baratta, M. E. Palumbo, G. Strazzulla and R. A. Baragiola, *Astron. Astrophys.*, 2005, **435**, 587.

47 O. Gálvez, I. K. Ortega, B. Maté, B. Martín-Llorente, V. J. Herrero, R. Escribano and P. J. Gutierrez, *Astron. Astrophys.*, 2007, **472**, 691.

48 S. Jheeta, S. Ptasinka, B. Sivaraman and N. J. Mason, *Chem. Phys. Lett.*, 2012, **543**, 208.

49 J. B. Bossa, F. Borget, F. Duvernay, P. Theulé and T. Chiavassa, *J. Phys. Chem. A*, 2008, **112**, 5113.

50 O. Poch, A. Noblet, F. Stalport, J. J. Correia, N. Grand, G. Szopa and P. Coll, *Planet. Space Sci.*, 2013, **85**, 188.

51 S. Pilling, B. G. Nair, A. Escobar, H. Fraser and N. Mason, *Eur. Phys. J. D.*, 2014, **68**, 58.

DISCUSSIONS

General discussion

DOI: 10.1039/c4fd90003k

Dr Garrod opened the discussion of the paper by Anthony Jones: Your models consider atomic C and H accretion; however, atomic O and N are also likely to be present under the conditions treated in your model. What sort of competition to the dust-production process would these species introduce, and would they significantly inhibit the re-formation of carbonaceous dust?

Dr Jones answered: This critical and interesting question was already discussed, in part, in a paper last year that considered the 'doping' of hydrocarbon, a-C(:H), grains[1] and also in a soon-to-be published paper in Planetary and Space Science (PSS), from the Cosmic Dust VI meeting in Kobe in 2013, on a new framework for considering the origin of the DIBs.[2]

It appears that N atoms, in particular, can be incorporated into a-C(:H) materials at up to the 10–15% level. Additionally, O atoms are known to aid in a-C(:H) formation under CVD conditions and can incorporate into a-C(:H) as –OH and $=$O bonded configurations. So, far from inhibiting carbonaceous dust formation, heteroatoms can aid a-C(:H) formation and be incorporated as dopants, which might help to explain some unresolved astrophysical issues (*e.g.*, the origin of the DIBs, volatile Si, S and N depletion, and blue luminescence.[3]

1. A. P. Jones, *Astron. Astrophys.*, 2013, **555**, A39.
2. https://www.researchgate.net/publication/259081512_A_framework_for_resolving_the_origin_nature_and_evolution_of_the_diffuse_interstellar_band_carriers?ev=prf_pub
3. A. P. Jones, *Planet. Space Sci.*, 2014, in press.

Dr Garrod asked: Regarding the build-up of dust from individual carbon atoms: is this process strongly dependent on the dust temperature, which would likely vary according to the visual extinction? Further, how would the existence of a distribution of initial grain sizes — with smaller grains being hotter — affect the rate of grain growth and its ultimate size distribution?

Dr Jones responded: A good question to which I do not yet have a definitive answer. However, given that the carbon accretion would be onto hydrocarbon grain surfaces, the C–C bond formation process, *via* C atom insertion, should be not too dependent on T_{dust}, for typical interstellar dust temperatures (15–20 K). With regard to the dust size distribution, it now appears that, in the outer regions of molecular clouds, when dust evolution begins (see Fig. 5 in the paper, which shows that t_{acc} and t_{coag} are within a factor of two or so) it begins with small grains sticking to big grains at about the same time and under the same conditions as the onset of accretion.[1] Hence, the grain growth rate *via* accretion will depend upon the aggregate grain size distribution, which is biased towards cooler, larger grains (*e.g.*, a > 200 nm).

1. M. Koehler *et al.*, *Astron. Astrophys.*, 2012, **548**, A61.

Dr Pontoppidan commented: As you transition from the diffuse ISM to a dense cloud, the sharp aliphatic features around 3.5 microns disappear and are replaced by the mysterious "3.47 micron feature". To my knowledge, this has not been theoretically explained with any confidence. Can you expand on whether your models address this observable?

Dr Jones replied: I did wonder if the newly-derived a-C(:H) refractive index data[1–3] might be able shed some light on the origin of the 3.47 micron feature, which is assumed to be due to tertiary aliphatic CH bonds. These new a-C(:H) data do indeed show a 3.47 micron feature but it is always accompanied by other aliphatic CH_n bands and is never seen as an isolated band. In fact, for rather H-rich a-C(:H) materials with a band gap of the order of ~2 eV, the 3.47 band tends to blend into a plateau-like band in the 3.4 to 3.5 micron region with a satellite band at 3.32 to 3.35 microns, due to olefinic CH and CH_2 bands, and a weaker band at 3.25 microns due to olefinic CH_2.

The original observational evidence for the mysterious 3.47 micron band, attributed to a tertiary CH absorption on diamond,[4] was recently re-inspected by Jones.[5] It appears that the 'isolation' of this band in continuum-subtracted spectra sensitively depends upon the nature of the assumed baseline.[5] The original work (Sellgren *et al.* 1994) adopted a curved baseline that removed what looks to be a plateau-like band, in the 3.4 to 3.5 micron region, and left two apparently isolated bands, one at 3.47 microns and the other at 3.25 microns. Thus, I am not convinced by the 'detection' of an isolated feature at 3.47 microns, which appears to be too broad for a tertiary CH absorption. I therefore also have the same doubts about the 'detection' on an isolated band at 3.25 microns. With the assumption of a different baseline the detected bands at 3.25 and 3.47 microns merge into a broad plateau that looks more like that due to an a-C(:H) material with a band gap of the order of 2 eV.[5] More recent and higher resolution VLT ISAAC data in the 3 micron region towards low mass protostars appears to confirm these doubts but further work is needed to fully explain these more recent data.

1. A. P. Jones, *Astron. Astrophys.*, 2012, **540**, A1.
2. A. P. Jones, *Astron. Astrophys.*, 2012, **540**, A2.
3. A. P. Jones, *Astron. Astrophys.*, 2012, **542**, A98.
4. L. J. Allamandola, S. A. Sandford, A. G. G. M. Tielens, *Astrophys. J.*, **399**, 134.
5. A. P. Jones, Proceedings of the Dust Lifecycle meeting, Taiwan, November 2013, http://pos.sissa.it/archive/conferences/207/001/LCDU%202013_001.pdf

Professor Bergin enquired: Where is the engine of growth for carbonaceous material? Is it at the molecular cloud interface? How do you get large grains? Another question would be what is the formation mechanism?

Dr Jones answered: The engines of dust formation and growth appear to be two-fold, firstly, the circumstellar shells of evolved carbon stars (AGB, RSG, ...) and, secondly, accretion (increase in the dust mass) and coagulation (increase in the mean grain size) in the dense ISM. The latter processes probably begins in the (denser) diffuse ISM but accelerate in molecular clouds (*e.g.* see **ref. 1** and **ref. 2**). The molecular cloud interface with the diffuse ISM is probably the most important carbon dust formation site because (small) carbon grains appear to be rather fragile in shocks, HII regions and PDRs.[3–7] Large silicate grains may survive rather well in the low density ISM (*e.g.* see **ref. 7** and **ref. 8**), and so provide a substrate for hydrocarbon grains re-formation through accretion. However, the re-formation of large carbonaceous grains, if they are efficiently destroyed in SN shocks (*e.g.* see **ref.** 7), in this scenario does remain something of a problem.

1. A. P. Jones, L. Fanciullo, M. Koehler, *et al.*, *Astron. Astrophys.*, 2013, **558**, A62.
2. V. S. Parvathi *et al.*, *Astrophys. J.*, 2012, **760**, 36.
3. E. R. Micelotta, A. P. Jones and A. G. G. M. Tielens, *Astron. Astrophys.*, 2010, **510**, A36.
4. E. R. Micelotta, A. P. Jones and A. G. G. M. Tielens, *Astron. Astrophys.*, 2010, **510**, A37.
5. J. Pety, P. Gratier, V. Guzman, *et al.*, *Astron. Astrophys.*, 2012, **548**, A68.
6. J. Pety, D. Teyssier, D. Fosse, *et al.*, *Astron. Astrophys.*, 2005, **435**, 885.
7. M. Bocchio, A. P. Jones, Jonathan D. Slavin, *Astron. Astrophys.*, 2014, submitted.
8. A. P. Jones and J. A. Nuth III, *Astron. Astrophys.*, 2011, **530**, A44.

Professor van der Tak asked: How will mass estimates for interstellar clouds change if we start using your dust model, compared for example to the Ossenkopf and Henning model?

Dr Jones replied: I have not yet looked into this issue in detail but the total dust mass is comparable with most other models[1–4] because we are all bound by the same elemental abundance constraints and require similar elemental depletions. In the new model the slope of the dust emissivity varies with wavelength (steeper in the FIR than in the sub-mm), which is consistent with the latest observations from Herschel and Planck. This means that a modified blackbody approach to estimate the dust mass is not a good approximation and will not yield a 'true' mass estimate because the dust emission does not follow a single power law from FIR to mm wavelengths.

Having just compared the dust opacities/mass absorption coefficients (cm^2 g^{-1}), I find that at 250 microns the new model is about a factor of 3 higher than the Ossenkopf and Henning value (MRN model with no ice mantles) and at 1 mm it is about a factor of 7 higher. Hence, dust mass estimates with the new model will be correspondingly lower. The new dust model puts about 80% of the dust mass in canonical large grains (a ~ 50–500 nm), so mass estimates based on the cold dust emission at FIR to sub-mm wavelengths, which take into account the wavelength dependence of the dust emissivity, will measure most of the dust mass.

1. M. Compiegne *et al.*, *Astron. Astrophys.*, 2011, **525**, A103.
2. B. T. Draine, A. Li, *Astrophys. J.*, 2007, **657**, 810.
3. V. Zubko *et al.*, *Astron. Astrophys.*, 2004, **152**, 211.
4. V. Ossenkopf and Th. Henning, *Astron. Astrophys.*, 1994, **291**, 943.

Dr Semenov remarked: I can comment on uncertainty of the sub-millimeter opacities, which are commonly used to discern (dust) masses. It was studied in our old opacity paper.[1] When dust includes absorbing materials (*e.g.*, metallic iron particles, troilite, or carbonaceous matter), the sub-millimeter opacities become sensitive even to the grain topology (not talking about grain sizes, shapes, *etc.*). That is, the sub-millimeter opacities are quite uncertain (> ~ factors of several).

I also have a related question to Anthony. Do you have an idea in what form your carbonaceous material primarily exist? Which optical constants one could adopt to model its absorption and scattering properties?

1. D. Semenov, T. Henning, C. Helling, M Ilgner and E. Sedlmayr, *Astron. Astrophys.*, 2003, **410**, 611.

Dr Jones responded: I agree, the sub-mm dust opacities are rather uncertain but recent observations from Herschel and Planck indicate that the slope of the dust emissivity in the ISM depends on the wavelength (steeper in the FIR than in the sub-mm) and on the region being observed. So, I would say that not only are they somewhat uncertain but that they are also intrinsically somewhat variable. As you say the opacities are sensitive to the materials, especially to the presence of metallic Fe, troilite (FeS) and aromatic-rich carbon. With respect to Fe this also depends on how it is present in dust, as pure metallic particles or as Fe nano-inclusions with silicates, which are not equivalent. In our work, putting metallic Fe into the amorphous silicate phase in nano-inclusions gives about the same result as incorporating the Fe into the silicate as cations.[1] Our work is also showing that the sub-mm opacities are particularly sensitive to the presence of aromatic carbon as mantles and/or a separate grain population and that this may be perhaps the biggest uncertainty because we do not yet know the details of carbon cycling in the ISM. There is no reason to suppose that the nature of the (hydro)carbons and the fractional depletion of C into dust are the same every-where (*e.g.*, see **ref. 1** and **ref. 2**). In our model, depending on particle size, about 80% of the carbon is in an aromatic-rich, H-poor, hydrogenated amorphous carbon and the rest in an aliphatic-rich, H-rich phase. The optical properties for all of these materials, as a function of the H atom fraction and particle radius, have been calculated and are now available on the Strasbourg CDS website through the links given in the second[3] and third[4] of my three 2012 papers on the evolution of hydrogenated amorphous carbon materials.

1. A. P. Jones, L. Fanciullo, M. Koehler, *et al.*, *Astron. Astrophys.*, 2013, **558**, A62.
2. V. S. Parvathi *et al.*, *Astrophys. J.*, 2012, **760**, 36.
3. A. P. Jones, *Astron. Astrophys.*, 2012, **540**, A2.
4. A. P. Jones, *Astron. Astrophys.*, 2012, **542**, A98.

Dr Mennella questioned: How did you come to reconcile the formation in the ISM of an HAC mantle on grains with spectropolarimetric observations, which indicate that the carrier of the 3.4 micron band and silicates are two separate grain populations?

Dr Jones answered: This is not a problem for the new dust model because in the diffuse ISM the carbonaceous mantles on the large silicate grains are H-poor, aromatic-rich, a-C and do not exhibit any IR band in the 3 micron region (see Fig. 5 in **ref. 1**). However, in denser regions it is likely that aliphatic-rich, a-C(:H) mantles will form on all grains but once these are exposed to the interstellar radiation field in the diffuse ISM they will be rapidly UV photo-processed to a-C on a timescale of the order of 10^5 to 10^6 yr (see Fig. 1 in the Faraday Discussion paper). Hence, there appears to be no contradiction between the model and the observations.

1. A. P. Jones *et al.*, *Astron. Astrophys.*, 2013, **558**, A62.

Professor Herbst opened the discussion of the paper by Herma Cuppen: Can your inclusion of exothermicity in granular reactions and how it affects the products be used to determine the initial energy for subsequent surface reactions that occur *via* the hot atom mechanism?

Dr Cuppen answered: We use the same mechanism to thermalise adsorbing species when the temperature of the gas is different from the grain surface. We observe that this affects the sticking of atomic hydrogen, particularly in diffuse clouds where the gas is warmer than the grain.

In our current model we do not allow the exothermicity to aid surface reactions, only diffusion and desorption.

Professor Meuwly remarked: In your conclusions it is mentioned that your set-up "best describes experimental and astronomical observations as well as molecular dynamics simulations." An essential process that seems to be missing in your modelling — which is however present in both, experiments and MD simulations — is energy relaxation in any form, particularly vibrational energy relaxation. Please can you comment on this?

Dr Cuppen responded: Indeed the model does not include any vibrational relaxation, since kinetic Monte Carlo is a state-to-state method vibrational movement is not included. The advantage of this is that we can simulate the evolution of the ice system on laboratory and interstellar timescales. It is however included indirectly through our dissipation parameter B. This parameter describes on which timescale the relaxation occurs. It is obtained from molecular dynamics simulations which can simulate this process accurately but can only cover very short timescales. The MD simulations on which this was based are limited to one system and as mentioned in the conclusions, further constraints are needed. What we aimed to express with the sentence in the paper is that our model compares well with both experimental and astronomical observations, and that at the same time our dissipation parameter B mimics behaviour similar to what has been found through MD simulations.

Professor Meuwly commented: In your manuscript you mention activation energies characteristic of each of the processes that are allowed to take place (desorption, diffusion, reaction, and dissociation). In light of the discussions during the meeting it seems more appropriate to work with distributions rather than fixed values for certain of these activation energies. Please comment on how this is included in your model.

Dr Cuppen replied: In our model, we take into account the local environment. Diffusion and desorption barriers depend on the number of nearest and next-nearest neighbours and the type of neighbours. In this way, the barriers can only take certain discrete values, but they already cover a large distribution. Reaction and dissociation barriers are not affected by the local environment in this model. Part of this discussion on the distribution of binding sites originates for our earlier work,[1,2] where we explicitly propose this to be important for the formation of molecular hydrogen. The current model is based in this work.

1. Q. Chang, H. M. Cuppen and E. Herbst, *Astron. Astrophys.*, 2005, **434**, 599.
2. H. M. Cuppen and E. Herbst, *Mon. Not. R. Astron. Soc.*, 2005, **361**, 565.

Dr Garrod said: In the model, when you have produced these two excited products, do you by-pass the usual randomized Monte Carlo determination of which process will be the next to occur, to allow these species to move immediately? Or are the excited products allowed to compete naturally with other processes (to see which occurs first) according to their relative rates? If the latter is the case, then your use of a Boltzmann factor (albeit with an elevated temperature) to determine the diffusion rate of the excited species cannot be correct, because such a distribution describes the probability that a particle will achieve some specified energy as a result of *thermal* interactions, and therefore assumes that the particle is thermalized. Your model, on the other hand, assumes that the particle has already obtained extra energy, so the assignment of a temperature, and thus a Boltzmann distribution, is not a valid treatment.

Dr Cuppen responded: All processes are selected according to the usual routine. If a species is excited, its corresponding diffusion and desorption rates are increased and are therefore more likely to win the competition with reaction. Indeed just using a temperature in the Boltzmann rate is most probably not the accurate way of treating the extra energy, but we apply this as a first estimate to increase the rate. The Boltzmann factor is valid for an equilibrium situation, which this clearly is not. Here we assume that crossing barriers still proceeds through this equilibrium dependence, for lack of other information. In our simulation we see that using this approach, we obtain reasonable travelled distances as compared to Molecular Dynamics simulations for reasonable values of the dissipation timescale and excitation energy. This indicates that this treatment does not result in very unphysical changes in the diffusion rate. Follow-up Molecular Dynamics simulations should indicate how (un)valid this treatment really is.

Professor Meuwly asked: Does your simulation protocol which includes: a) directional bias and b) energy gain after formation of species lead to a thermodynamic ensemble in a strict sense? If not, how should the results in Figures 3 and 4 in the paper (as an example) be interpreted? Should the bias of a) and b) not be removed – similar in spirit to umbrella sampling simulations in Molecular Dynamics? The figures suggest that the simulations lead to an equilibrium distribution in the coverage. However, the meaning of such an equilibrium is unclear if it originates from a non-equilibrium and biased ensemble. Please clarify.

Dr Cuppen answered: It does not lead to a thermodynamic ensemble in a strict sense. One should realize that we simulate non-equilibrium processes here. This is a kinetic model. For a more in-depth discussion on the theory behind the method I refer to **ref. 1**. The constant coverage that is obtained in Figures 3 and 4 in the paper is not an equilibrium coverage, but a steady state, in which the formation of water ice is balanced by the destruction through photodissociation. If the photon flux decreases which is the case for translucent and dense clouds, one can see that this steady state is no longer obtained.

Furthermore, the directional bias (a) is an increase in the diffusion rate for the continued movement with respect to the diffusion rate in the opposite direction. This is to mimic the non-Markovian nature of the mobility of "hot species". In kinetic Monte Carlo the species would otherwise follow a random walk which would not describe the physics occurring at the surface. The directional bias is therefore introduced to obtain a more physical picture. It is thus conceptually different from the bias that is imposed during umbrella sampling MD or MC where a bias is added to the potential to lower barriers and to make it computationally feasible to simulate a certain transition.

1. H. M. Cuppen, L. J. Karssemeijer and T. Lamberts, *Chem. Rev.*, 2013, **113**, 8840.

Dr Oba said: I think you constructed models mainly based on the experimental results by Cuppen *et al.*[1] in which you assigned the bands at 3426 and 3463 cm^{-1} to be isolated OH in an O_2-matrix. You referred to a paper by Acquista *et al.*[2] for the assignment. It has long been believed that this assignment is true. However, in 1988, Cheng *et al.* reported that the isolated OH band appears at different wavenumbers (3548 cm^{-1}).[3] More recently, Langford *et al.* assigned the doublet bands, which were originally believed to be isolated OH, to be an $H_2O \cdot OH$ complex.[4] After the new assignment by Langford *et al.*, it has been experimentally and theoretically confirmed to be true.[5-7] So, do you have any other evidence that isolated OH is abundantly present in your samples?

1. H. M. Cuppen, S. Ioppolo, C. Romanzin and H. Linnartz, *Phys. Chem. Chem. Phys.*, 2010, **12**, 12077.
2. N. Acquista, L. J. Schoen and D. R. Lide Jr., *J. Chem. Phys.*, 11968, **48**, 1534.
3. B.-M. Cheng, Y.-P. Lee and J. F. Ogilvie, *Chem. Phys. Lett.*, 1988, **151**, 109.
4. V. S. Langford, A. J. McKinley and T. I. Quickenden, *J. Am. Chem. Soc.*, 2000, **122**, 12859.
5. P. D. Cooper, H. G. Kjaergaard, V. S. Langford, A. J. McKinley, T. I. Quickenden and D. P. Schofield, *J. Am. Chem. Soc.*, 2010, **125**, 6048.

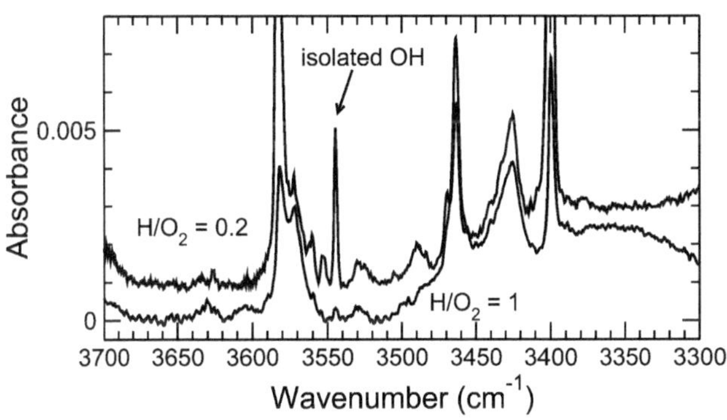

Fig. 1 Reflection Absorption Infrared Spectra of codeposition experiments of H and O_2 at 15 K. The $H/O_2 = 1$ results are taken from **ref. 1**. The other spectrum was unpublished. The $H/O_2 = 0.2$ spectrum is offset for clarity.

6. A. Engdahl, G. Karlström and B. Nelander, *J. Chem. Phys.*, 2003, **118**, 7797.
7. M. E. Jacox and W. E. Thompson, *J. Phys. Chem. A*, 2013, **117**, 9380.

Dr Cuppen replied: In our experimental paper[1] we have indeed used the $HO_2 \cdot OH$ bands at 3426 and 3463 cm^{-1} for the identification of OH in our ice. However, if we zoom in on the bottom spectrum of Fig. 2 in the paper, as shown in Fig. 1, a weak feature at 3545 cm^{-1} can be identified which should be isolated OH. For higher O_2/H ratios the feature is more prominent, as can be seen in Fig. 1 for $H/O_2 = 0.2$. This spectrum is offset for clarity.

1. H. M. Cuppen, S. Ioppolo, C. Romanzin and H. Linnartz, *Phys. Chem. Chem. Phys.*, 2010, **12**, 12077.

Professor van der Tak opened the discussion of the paper by Bérengère Parise: Given the apparent similarity between hydrogen peroxide and molecular oxygen, have you searched for H_2O_2 in Orion too?

Dr Parise answered: We did, but we did not detect it. It is however difficult to give limits on this source, because of line confusion. Moreover, the explanation for the detection of O_2 seems to be related to gas-phase chemistry after evaporation.[1] In such a case, O_2 and H_2O_2 are not necessarily related.

1. P. F. Goldsmith, R. Liseau, T. A. Bell, J. H. Black, J. H. Chen, D. Hollenbach, M. J. Kaufman, D. Li, D. C. Lis, G. Melnick, D. Neufeld, L. Pagani, R. Snell, A. O. Benz, E. Bergin, S. Bruderer, P. Caselli, E. Caux, P. Encrenaz, E. Falgarone, M. Gerin, J. R. Goicoechea, A. Hjalmarson, B. Larsson, J. Le Bourlot, F. Le Petit, M. De Luca, Z. Nagy, E. Roueff, A. Sandqvist, F. van der Tak, E. F. van Dishoeck, C. Vastel, S. Viti and U. Yildiz, *Astrophys. J.*, 2011, **737**, 96.

Professor van der Tak said: Your formaldehyde measurements indicate $T(\mathrm{rot}) = 160$ K for NGC 6334I(N), which is much higher than the dust temperature. Since this source is not a luminous protostar, could this high value be a result of optical depth?

Dr Parise answered: Yes, this is indeed certainly the case. The lower energy line of H_2CO is doubly-peaked, which is certainly a sign of large optical depth.

This implies that the upper limit we derive for HOOH based on the over-estimated rotational temperature is also overestimated. A better handle on the rotational temperature would help constraining the upper limit on the HOOH abundance to a lower value, so the value given in the paper should be seen as a very conservative one.

Professor Bergin commented: The model seems to say specific conditions give rise to H_2O_2, but 30 K gas is duplicated along many lines of sight so what is peculiar here? Herschel certainly looked for O_2 under every nook and cranny with Rho Oph the clearest case. Is this a time dependant effect?

Dr Parise replied: All observations from Herschel that I have seen so far have reached upper limits for the abundance of O_2 with respect to H_2 of at best 10^{-8} to 10^{-7}. Our gas–grain model seems to imply that this is still not sensitive enough to detect O_2 for cold ($T = 20$ K) sources at late evolutionary stage. In the case of

OphA, the Herschel O_2 detection gave an O_2 abundance of 5×10^{-8} (**ref. 1**), and our chemical model shows best agreement at a time of 6×10^5 yrs, which is not *per se* a very early time. So it would be very useful to run a detailed chemical model for the sources which have a significant amount of gas at 30 K, and towards which O_2 was searched for with Herschel, in order to give a reliable answer to your question.

1. R. Liseau, P. F. Goldsmith, B. Larsson, L. Pagani, P. Bergman, J. Le Bourlot, T. A. Bell, A. O. Benz, E. A. Bergin, P. Bjerkeli, J. H. Black, S. Bruderer, P. Caselli, E. Caux, J.-H. Chen, M. de Luca, P. Encrenaz, E. Falgarone, M. Gerin, J. R. Goicoechea, Å. Hjalmarson, D. J. Hollenbach, K. Justtanont, M. J. Kaufman, F. Le Petit, D. Li, D. C. Lis, G. J. Melnick, Z. Nagy, A. O. H. Olofsson, G. Olofsson, E. Roueff, Aa. Sandqvist, R. L. Snell, F. F. S. van der Tak, E. F. van Dishoeck, C. Vastel, S. Viti and U. A. Yıldız, *Astron. Astrophys.*, 2012, **541**, A73.

Dr Loison commented: The link between H_2O_2 detected in the gas phase and the H_2O_2 formation on grains seems obvious as H_2O_2 cannot be produced through gas phase reaction (according to the present state of the knowledge of the chemistry). You have noted that O_2 has been detected in the same region as H_2O_2, and that detection is almost the only one for O_2. You propose this cannot be a coincidence. However, if H_2O_2 is not produced through gas phase reaction, O_2 is and very efficiently. So the link should be quite complex and what information can be deduced from this double detection for the non-detection of O_2 in region where it should be very abundant (almost all dense clouds if we believe models)?

Dr Parise responded: This is a good question. I am not sure which models you are referring to.

Actually, our model (see Fig. 1 in the paper, top left panel, for the conditions $T = 10$ K and $n = 10^5$ cm^{-3}) predicts that the abundance of O_2 in the gas-phase is high at early stages, but then settles down to a value of a few 10^{-9} relative to H_2. This is well below the observational upper limits that have been published so far towards cold dense clouds (*e.g.* **ref. 1** and **ref. 2**).

Our model is also consistent with the more sensitive non-detection of O_2 towards NGC1333-IRAS4A, as discussed in the paper. So from our point of view, there is no obvious inconsistency between the model and the observations. These "high" predicted abundances are only very difficult to observe, because of the intrinsically faint transition of O_2.

1. P. F. Goldsmith *et al.*, *Astrophys. J.*, 2000, **539**, L123.
2. L. Pagani *et al.*, *Astron. Astrophys.*, 2003, **402**, L77.

Dr Semenov asked: Bérengère, could the difficulty in detecting HOOH be related to the fact that from the ground only one form (either *para* or *ortho*) of this species can be observed? Since H_2O_2 has two hydrogen atoms it should exist in two separate nuclear spin states. That is, imagine that there is a sufficient total amount of H_2O_2 in a source that could be detected with just several sigma confidence, but its particular form that we can observe is a minor fraction in the total H_2O_2.

Dr Parise replied: HOOH has two spin forms: the A-species has a spin of 1, the B-species has a spin of 3. The different ladders of the molecule are shown in Figure 1 of the detection paper by Bergman *et al.*[1] For the detection towards Oph

A, we have observed three transitions from each of the A and B-species. We clearly detected the B-transitions, while we have only a tentative detection of one of the A transitions. Because the two spin-related species are observable from the ground, I do not think the detection is impaired by a spin-related issue. You are however right, in the case of the present search, we targeted only one transition from the B-species (the brightest observed towards Oph A), so we have no information on the A-species population there.

In Oph A, the rotation diagram analysis shows that the two spin-related levels are not populated at LTE. A better understanding of the excitation of HOOH would require the computation of its collisional rates, which are presently unavailable, to the best of our knowledge.

1. P. Bergman, B. Parise, R. Liseau, B. Larsson, H. Olofsson, K. M. Menten and R. Güsten, *Astron. Astrophys.*, 2011, **531**, L8.

Professor Linnartz queried: Is there any chance that a search for HOOD could be successful and if so, would an abundance relative to H_2O_2 give enough insight in the underlying formation pathways?

Dr Parise answered: I am involved in a search for HOOD towards Oph A, which so far has been unfortunately unsuccessful. This is not very surprising, as the lines are expected to be very faint, if we assume that the D/H ratio is similar to that of water, *i.e.* of the order of 1%. A detection of HOOD (or a very low upper limit on its abundance) would indeed be a great piece of information to constrain the formation pathway of hydrogen peroxide.

Ms Caravan asked: We have recently studied the gas phase reactions of OH with ammonia and with methanol down to ~56 K and found the rate coefficients of these reactions to be orders of magnitude faster than at room temperature and than predicted by Arrhenius behavior.[1–3] One of the bimolecular products of both of these reactions is water. Given the presence of ammonia, methanol and OH in interstellar environments and your study showing H_2O_2 not to have been detected in certain environments, where according to the mechanism for production of water it would be expected to be present, I was wondering whether these gas phase reactions might be a potential alternative source of water in interstellar environments? Alternatively, if the H_2O_2 is being removed by gas phase processes as you suggest in your paper, one of these reactions could possibly be the reaction of OH with H_2O_2. The rate coefficient for this reaction is found to be quite fast at low temperatures as measured by Vakhtin *et al.*[4] If there is some contribution from the bimolecular mechanism at low temperatures, as is expected by Vakhtin *et al.*, then one of the bimolecular products of this reaction is water.

1. R. J. Shannon, M. A. Blitz, A. Goddard and D.E. Heard, *Nature Chem.*, 2013, **5**, 745.
2. J. C. Gomez Martin, R. L. Caravan, M. A. Blitz, D. E. Heard and J. M. C. Plane, *J. Phys. Chem. A*, 2014, **118**, 2693.
3. Work presented as a poster at this meeting, R.L. Caravan *et al.*, Chemical mechanisms operating at very low temperatures to enhance the rates of gas phase reactions relevant to interstellar environments, poster 32.
4. A. B. Vakhtin, D. C. McCabe, A. R. Ravishankara and S. R. Leone, *J. Phys. Chem. A*, 2003, **107**, 10642.

Dr Parise responded: Thanks for these comments. The destruction reaction in the gas phase OH + H_2O_2 is included in our model, with a rate taken from the OSU09 database. At 20 K, it corresponds to a rate of around 10^{-15} cm^3 s^{-1}.

The extrapolation of the formula of Vakhtin *et al.* outside of their nominal temperature range (96–296 K), would give a rate of around 10^{-6} cm^3 s^{-1} at 20 K, so nine orders of magnitude higher! This suggests that a direct measure of the rate at 20 K is really required to get a reliable value for this reaction.

Dr Oba returned to the discussion of the paper by Herma Cuppen: You showed us an excellent FTIR spectrum of a reaction product which supports the presence of isolated OH in an O_2 matrix (~3550 cm^{-1}). However, unfortunately, this spectrum was not shown in your previous paper.[1] In addition, you did not quantify the OH band at ~3550 cm^{-1} but the bands at 3426 and 3463 cm^{-1}, which are in fact derived from $H_2O\cdot OH$ complex. This means that, in your present and previous models,[2] you did not correctly show the variations in the band area of isolated OH in ice samples. I would suggest you to update your data and construct a new model based on the new data. I believe it would better explain the experimental results.

1. H. M. Cuppen, S. Ioppolo, C. Romanzin and H. Linnartz, *Phys. Chem. Chem. Phys.*, 2010, **12**, 12077.
2. T. Lamberts, H. M. Cuppen, S. Ioppolo and H. Linnartz, *Phys. Chem. Chem. Phys.*, 2013, **15**, 8287.

Dr Cuppen responded: For the $H/O_2 = 1$ experiments the isolated OH feature is too small to reliably integrate as a function of time, since only at the end of the co-deposition experiment can an appreciable amount of isolated OH be detected. For the $H/O_2 = 0.2$ experiment at 15 K we could make a time evolution and here the isolated OH integrated absorbance tightly follows the integrated absorbance of the bands at 3426 and 3463 cm^{-1} and the H_2O_2 bands. We therefore believe that our picture is not qualitatively altered. Furthermore, what we aimed to investigate in this paper is the role of exothermicity in general. It was initially triggered by the detection of OH, which still holds, but we took this to a broader picture.

Dr Oba remarked: You may consider that OH is formed by the reaction of HO_2 + H *via* the formation of intermediate H_2O_2 followed by the dissociation of the O–O bond in H_2O_2 by using the heat of reaction HO_2 + H. This pathway is reasonable in the gas phase because the heat of the reaction cannot be dissipated through a third body.[1-3] If the reaction HO_2 + H in an O_2-matrix proceeds similarly as in the gas phase, OH could form efficiently as well. However, assuming reactions on the actual interstellar grains, I think the heat of reaction would be dissipated through interactions with the cold grain surface to some extent, which implies that a part of the intermediate H_2O_2 remains undissociated after the formation by reaction HO_2 + H on grains. Could you give a comment on this?

1. R. Sayós, C. Oliva and M. González, *J. Chem. Phys.*, 2000, **113**, 6736.
2. Y. Ge, K. Olsen, R. I. Kaiser and J. D. Head, *AIP Conference Proceedings*, 2006, **855**, 253.
3. S. H. Mousavipour and V. Saheb, *Bull. Chem. Soc. Jpn.*, 2007, **80**, 1901.

Dr Cuppen answered: The IR spectra show very narrow features in the O_2-rich case. We therefore believe that the coupling with the matrix is limited and that indeed the reaction proceeds similarly to the gas phase. H_2O_2 will then be an intermediate that falls apart into two OH radicals. In a water-rich environment the coupling will most likely be stronger and there a larger fraction of the intermediate will remain in the form of H_2O_2. This will more be representative for the last part of the sequential hydrogenation experiments. If in our sequential simulations, we use a branching ratio of 0.91 for the H_2O_2 channel and do not allow for OH to form, the final results are very similar compared to the opposite case, since most OH radicals formed in the matrix will continue to react to form H_2O_2.

Dr Loison addressed Dr Cuppen and Miss Lamberts:You explained that the H_2O_2 formed through the H + HO_2 reaction will partly dissociate into OH + OH and the exothermicty of the reaction lead to enough kinetic energy in OH fragments to prevent recombination of OH + OH into H_2O_2. Do you think this is a very common mechanism ($C + O_2 \rightarrow CO + O$ and not to COO or CO_2) or is the adduct in general the main product, particularly on ice?

Dr Cuppen replied: This is an interesting question. I believe that one reason that H + HO_2 results in fragmentation over the formation H_2O_2 is the O_2 matrix which only weakly interacts and does not allow for fast stabilization of the adduct. In water ice, this could be a different situation. What we learn from our simulations for the first step in the reaction sequence $H + O_2 \rightarrow HO_2$, is that it cannot proceed with 100% efficiency whereas the reaction is barrierless under specific angles.[1,2] One reason for this might be that in many cases the adduct falls apart again into H and O_2. For the formation of HCOOH through the intermediate HOCO, we found experimentally that the efficiency of the reaction is dependent on the environment (CO or H_2O-rich).[3] We attributed this also to the coupling between the matrix and the HOCO complex.

1. D. Xie, C. Xu, T.-S. Ho, H. Rabitz, G. Lendvay, S. Y. Lin and H. Guo, *J. Chem. Phys.*, 2007, **126**, 074315.
2. C. Xu, D. Xie, D. H. Zhang, S. Y. Lin and H. Guo, *J. Chem. Phys.*, 2005, **122**, 244305.
3. S. Ioppolo, H. M. Cuppen, E. F. van Dishoeck and H. Linnartz, *Mon. Not. R. Astron. Soc.*, 2011, **410**, 1089.

Dr Hornekaer returned to the discussion of the paper by Anthony Jones: Superhydrogenation of PAHs to a saturated state with one excess H atom per carbon atom give chemically very stable species (*e.g.* Perhydrocoronene) with absorption cross-sections peaking at much shorter wavelength — typically in the VUV — compared to the original PAH. Do you think that these changes would enable such superhydrogenated species to survive in more areas of the ISM?

Dr Jones answered: Unfortunately the fact that the absorption of super-hydrogenated PAHs is shifted into the VUV probably does not help their survivability in the ISM. For example, Fig. 1 in our Faraday Discussion paper shows the size-dependent, hydrocarbon particle lifetime against the EUV (E > 10 eV) photo-dissociation of aliphatic/olefinic CH bonds, which is practically independent of the hydrogen content (*e.g.*, see Fig. 15 in **ref. 1**). Fig. 1 can be used to predict the survivability of super-hydrogenated PAHs (*i.e.*, polycyclic aliphatic-rich nano-particles) against CH bond photo-dissociation in the ISM, *i.e.*, the return of super-hydrogenated PAHs to an aromatic-rich PAH state *via* aromatisation. Considering

Perhydrocoronene as equivalent to an a-C(:H) sub-nm sized particle (*i.e.*, a ~ 3 nm), its aromatisation timescale would apparently be of the order of about a million years. Thus, only in regions where there is a significant attenuation of the interstellar VUV–EUV radiation field would super-hydrogenated PAHs be able to survive for longer than this. However, in such EUV-shielded regions any small grains will be accreted/coagulated onto larger grains (*e.g.*, see Fig. 5 in the paper and **ref. 2**.

1. A. P. Jones, *Astron. Astrophys.*, 2012, **540**, A2.
2. M. Koehler, B. Stepnik, A. P. Jones, V. Guillet, A. Abergel, I. Ristorcelli and J. P. Bernard, *Astron. Astrophys.*, 2012, **548**, A61.

Professor Heard commented in relation to the paper by Herma Cuppen: In Table 1 in your paper you have listed the surface and photodissociation reactions that are used in your current model. There are some other reactions involving some of the major species which might be expected to be fast, for example the recombination reaction of HO_2 ($HO_2 + HO_2$) which will generate H_2O_2, which is the same product as one of the featured reactions $H + HO_2$. Another reaction to be considered is the reaction $HO_2 + O_3 \rightarrow OH + O_2$, which will regenerate the hydroxyl radical, OH, which features in several reactions listed in Table 1.

Dr Cuppen responded: In an earlier paper[1] we have investigated the role of several other reactions in our network including $HO_2 + HO_2$ and found that introducing this reaction only leads to different results for the co-deposition simulation at 15 K which can be explained by the $HO_2 + H$ reaction as well. We therefore decided not to include this in our model, since it was hard to constrain the reaction on the basis of our experiments. We allow H atoms in our model to reside in interstitial sites; this can also be interpreted as H forming a complex with a bulk O_2 molecule. The reaction between subsurface H and HO_2 would in that scenario be equivalent to $HO_2 + HO_2$. We believe that the reaction between HO_2 and O_3 is unlikely to be important in our system since they are both transient species in a diluted medium, which effectively means that they are unlikely to exist in close vicinity.

1. T. Lamberts, H. M. Cuppen, S. Ioppolo, and H. Linnartz, *Phys. Chem. Chem. Phys.*, 2013, **15**, 8287.

Professor Schram asked in relation to the paper by Anthony Jones: Did you consider the effect of charge of clusters (typically Te in eV per nm charge)? Reasons to ask this question are: Typically findings in plasma processing plasmas (as RF) are that clusters were formed in the plasma phase up to 10–30 nm size. Then coalescence may occur resulting in large conglomerates, i.e. dust particles of 0.1–1 um size, thus exhibiting similar double humped distributions. These dust and cluster particles are both spherical in form and thus both formed in the "gas" (plasma) phase. It is certainly conceivable that these clusters contribute to deposition of amorphous layers. A second question is then what is known on the geometry of the clusters *etc.* formed in astro-chemistry?

Dr Jones replied: No we have not yet considered the effects of grain charge on the physics of the clusters, neither in terms of their intrinsic charge in the ISM nor in terms of the effects that this may have on their mutual coalescence or on their 'sticking' onto the surfaces of larger grains (*i.e.*, coagulation). Indeed, I am aware of the very interesting work on 'soot' nano-particle/cluster formation and coalescence in plasmas. In the ISM we do have to consider the effects of a dust size distribution going up to micron-sized particles. However, the exact nature of and the formation scenario for the largest carbonaceous grains around evolved stars and in the ISM is still something of an open question. The observational evidence for dust evolution in the transition from low to higher density regions seems to indicate that coagulation rather than coalescence is the more important effect, *i.e.*, the sticking of small grains (nano-particle 'clusters') onto large sub-micron sized grains, is more important than the mutual sticking of nano-particle 'clusters'. Nevertheless, we do need to consider the effects of particle charging in HII regions or photo-dissociation regions where UV photo-electron ejection effects can lead to significant grain/cluster charging and we plan to do this in the future. In our work we consider that the smallest hydrocarbon particles or clusters (containing of the order of 100 C atoms) are an intimate mix of, predominantly, aromatics with an important and key aliphatic and/or olefinic component. We have nick-named these clusters 'arophatic' structures.[1]

1. E. Micelotta *et al.*, *Astrophys. J.*, 2012, **761**, 35.

Dr Meijer queried in relation to the paper by Herma Cuppen: Your paper does not state which spin state for oxygen molecule and oxygen atom was used. Can you please clarify that?

Dr Cuppen answered: The model does not explicitly include spin states. However only reactions that can occur with ground state (molecular) oxygen are included in our reaction network, since we simulate conditions where in principle all species should be in the ground state. Experimentally, the oxygen molecules are deposited at room temperature.

Dr Garrod returned to the discussion of the paper by Bérengère Parise: You suggest that, for the single source in which you have detected hydrogen peroxide, a large fraction of the material may be at the 'goldilocks' temperature for grain-surface H_2O_2 formation, while in other sources it would not. You further suggest that this may be due to external heating of Oph A. Is there any reason to expect, *a priori*, that an externally-heated source should have such a uniform temperature?

Dr Parise replied: I have not done the detailed radiative transfer modelling, but I guess that an externally heated source may have a flatter temperature gradient than an internally heated source, because of the geometry of the illumination. Our prediction for the temperature of Oph A could be tested against the observational determination of the dust temperature from recent Herschel observations.

Mr Li commented: First, thanks for the nice talk. I was wondering, are the temperatures and densities fixed in your model? Are the simulations as a function of time? According to my experience, one can indeed extract some useful information about the physical chemistry processes from a single point model.

However, a complete model considering all variables may be more useful and suggestive. Sometimes these two types of models may give quite different (at least not exactly the same) predictions. Could you comment on this with regard to your simulations?

Dr Parise responded: You are right, and for this exact reason we ran two types of models. First, we ran models with static conditions (that is to say, constant temperature and density, but we follow the evolution of the chemistry with time). Then we also ran models with varying temperature and density. And it is indeed clear that the variation with time of the temperature is a key aspect for the chemistry, as one has to start in cold enough conditions for ices to form, and then warm up for oxygen atoms to become mobile enough to react.

Miss Lamberts asked: You mentioned that the production of HOOH is extremely sensitive to the temperature, and that outside of the 20–30 K temperature range the expected HOOH abundance is very low. Did you check the influence of the exact values of the reaction rates that are relevant for the formation of HOOH on the grains, or, in other words, did you consider performing a sensitivity analysis such as those performed by Wakelam *et al.*[1] and Lamberts *et al.*?[2]

1. V. Wakelam *et al.*, *Astron. Astrophys.*, 2010, **517**, A21.
2. Lamberts *et al.*, *Phys. Chem. Chem. Phys.*, 2013, **15**, 8287.

Dr Parise replied: We have not done a sensitivity analysis yet, but that would definitely be an interesting study. For the moment, I think we are very limited by the fact that the gas-phase destruction channels for HOOH are incomplete, and therefore I do not think a sensitivity study based on the present chemical network would be very relevant.

I would assume that the exact value of the temperature at which HOOH is abundant is highly dependent on the oxygen mobility on the grains (to form O_2 in the first place).

Professor Kaiser opened the discussion of the paper by Jonathan Rawlings: First, you state that detailed modeling studies on the formation of propylene in TMC-1 propose that propylene (CH_3CHCH_2) cannot be formed by conventional interstellar gas phase chemistry. However, a closer look at references 15 and 17 in your paper suggests that this conclusion is based on incomplete models. The models speculate without scientific evidence that ill-defined ion-molecule reactions involving $C_3H_7^+$ might play a role in the formation of propylene *via* a dissociative recombination process. However, it is well known that the gas phase reaction of the methylidyne radical (CH) with ethane (C_2H_6) is very fast with rate constants in the order of 10^{-10} cm^3 s^{-1} even at 23 K;[1] recent computational studies indicate that propylene is the exclusive reaction product.[2] Therefore, objective gas phase models have to incorporate the rapid neutral–neutral reaction of methylidyne radicals with ethane to objectively evaluate the role of this reaction in the gas phase formation of propylene without any unjustified bias on speculative ion–molecule pathways.

Second, the proposed reaction sequence of methyl radicals (CH_3) with methylidyne (CH) forming triplet methylcarbene (CH_3CH), which then reacts with triplet carbene (CH_2) to singlet propylene is purely speculative. There is no experimental evidence. It is very dangerous to disseminate reaction pathways based on guesswork. It is well documented in the astrophysical literature — mainly from the ion–molecule and gas–grain communities — that once speculative reaction models have been 'proposed,' after a few years, the astronomy community assumes that these speculative pathways have been 'well established,' although there is no scientific evidence! It can take years for physical chemists to verify and/or validate a reaction sequence, which can be written down in a matter of seconds.

1. A. Canosa, I. R. Sims, D. Travers, I. W. M. Smith, B. R. Rowe, *Astron. Astrophys.*, 1997, **323**, 644.
2. A. M. Mebel, private communication, 2014.

Professor Rawlings answered: In response to your first question, we agree that our reaction scheme may not be completely correct in the details, but it is at least viable.

So far, efficient gas-phase mechanisms for propylene formation, as applicable to the conditions in the interstellar medium, have not been identified in the literature. In addition, some of the gas-phase channels for propylene formation that exist in the astrochemical reaction data files are spurious and incorrect (*e.g.* as a photodissociation product of 1,3-butadiene).

In any case, we emphasise that a mechanism is required which is capable of generating a range of complex organic molecules, and not just propylene.

In response to your second question, we agree that the proposed reaction pathways are highly speculative, but they are at least plausible, on grounds of valency matching *etc.* It must be remembered that we are talking about three-body (and perhaps even higher order) reactions in an extremely high density environment, so the issue of spin conservation in bimolecular reactions is not strictly relevant.

It is right and proper that existing chemical data is used, wherever it is available, and we have done this to the best of our ability. However, to suggest that the availability of well-defined and quantified chemical pathways denies the possible existence of other, speculative, mechanisms is very strange. If other authors then incorporate these alternatives without question, then they are at fault – but that is certainly not the approach that we advocate.

Professor Kamp asked: I did not understand the explosive mechanism you describe in your paper. Can you please expand on the details of how accreting H atoms on a dust grain (maybe with an ice mantle) in space can lead to an explosion of the grain and/or mantle?

Professor Rawlings responded: There are various sources of latent chemical energy in both the dust grains themselves and in/on the ices. Thus, radicals can be produced in both of these components by the action of external radiation fields, the passage of cosmic rays and electrons *etc.* This source of chemical potential energy is well-known (*e.g.* **ref. 1**). In addition, accreting radicals — most notably hydrogen atoms — will release significant energy on recombination. In the mechanism that we are proposing, once such reactions are triggered, the exothermicity (and enhanced radical mobility) drives a runaway process in which a significant fraction of all of the energy that is stored in radicals is released.

1. C. D. Gay, P. C. Stancil, S. Lepp and A. Dalgarno, *Astrophys. J.*, 2011, **737**, L44.

Dr Garrod commented: In the paper, you suggest that, for the mechanism of run-away heating and explosive desorption to be effective, the atomic hydrogen content of the combined dust-grain and ice-mantle material must exceed around 5%. Assuming that most free (*i.e.* not chemically-bound) atomic H is likely to be fixed in the ice rather than within the dust grain itself, this further suggests that the required atomic-hydrogen content of the ice mantle alone would need to be upwards of 10%.

This seems implausible, given that such a proportion would require that, on average, every atomic hydrogen in the ice mantle would be in contact with another atomic H, while not reacting with it. How could such a structure even physically exist, or survive long enough for an appreciable ice mantle to form with this extreme atomic-hydrogen content?

Professor Rawlings answered: Firstly, I think that we need to make it clear that the proposed energy source is not just the recombination of hydrogen atoms (and other radicals) to form H_2, but also the chemical energy stored in the hydrogenated amorphous carbon (HAC) substrate. It is probably energetic events in the HAC substrate that get the explosion started in the mantle.

Typical interstellar grains/ices have a very high surface density of binding sites. There is also direct laboratory evidence (scanning tunnelling microscopy) which shows that extremely high coverage and clustering of H-atoms on graphite can occur.[1] Finally, please note that the radicals do not just exist in the form of surface accreted atoms, but are generated within the ices by UV radiation, cosmic rays *etc.* – these radicals are locked into a solid matrix and so will not be chemically active until mobilized.

1. L. Hornekaer *et al.*, *Phys. Rev. Lett.*, 2006, **97**, 186102.

Professor Herbst asked: The radicals on the grain react exothermically either to cause the explosion in the first place, or possibly as the explosion occurs. How then can the radicals be available for synthetic chemistry in the gas after the explosion?

Professor Rawlings replied: There is no reason why (a) reactions on grains should go to completion and (b) only non-radical species are produced in these reactions. The explosions occur on such rapid timescales that incomplete conversion and the production of non-equilibrium product species are likely.

Mr Ligterink said: The laboratory work cited in the paper addresses only hydrogen atoms on carbonaceous surfaces which result in a release of energy/explosion at a certain critical density. This is of course not like an interstellar dust grain, where you also expect the grain to be covered in H_2O, CO, CO_2, CH_4, NH_3

and a large number of other species. Do you think this different environment will still give a grain explosion? Could it be hindered by for example the porosity of the ice, different sticking coefficients compared to carbon or trapping of H atoms?

Professor Rawlings responded: We do, of course, employ an idealised model of the nature and morphology of the dust grains, but that is a criticism that can be made of any model of gas-grain interactions and surface chemistry.

The principal source of the explosion energy comes from the recombination of hydrogen atoms – and that is independent of the nature of the substrate.

However, if we were to make the assumption that the heat that initiates an explosion in the mantle comes from energetic events in the grain core then, again the nature of the mantle composition is not likely to be very important.

Mr Fedoseev remarked: Short laboratory time scales promote the chain explosion mechanisms described in your work, since on the short time scale adsorbed atoms and radicals have less time to find each other and recombine prior to the critical density being achieved. In addition effective heat dissipation is challenged on short time scales and this in turn can promote the chain mechanisms described. However, in space adsorbed atoms and radicals have days, months, or years to find each other, recombine and dissipate the energy produced in the reactions. Therefore it would be difficult to achieve critical density. Would the process of chain recombination and catastrophic explosion described in the experiments be still realistic in space on such long time scales?

Professor Rawlings answered: The laboratory evidence of explosions seems incontrovertible, although — as you state — the long timescales that are available in the interstellar medium will mean that some of the radicals will be lost. What one has to remember is that the processing of ices (by UV, cosmic rays, electrons *etc.*) need not be accompanied by prompt migration or even recombination on long timescales if the radicals and defects so-generated are not free to migrate. Tunnelling of large chemical species is inhibited, even on very long timescales.

Dr Ellinger commented: The relative abundances of isomeric COMs are important data that should help in the constraining of chemical models. All three isomers of $C_2H_4O_2$ chemical formula have been observationally identified. Their relative stabilities are such that CH_3COOH (AA) is the most stable with $HCOOCH_3$ (MF) and $HOCH_2CHO$ (GA) being 0.74 and 1.2 eV higher on the energy scale, respectively. In the explosion mechanism, there is so much energy released that three-body interactions are considered and energy barriers can be overpassed. At least for some time these conditions resemble those of a Local Thermodynamic Equilibrium; then a minimum energy principle applies and the abundances should parallel the relative stabilities, $n(AA) > n(MF) > n(GA)$. Observations provide a different order, $n(MF) > n(GA) > n(AA)$. However, since grain ice mantles are involved in the process, the adsorption energy of each isomer may be worth considering. Both experiments and theory[1,2] show that AA is the most strongly bound to the ice with an energy that is greater than the desorption of the ice itself, whereas MF is the most weakly bound and will desorb well before the other isomers. Could this influence your models and help rationalizing the observations?

1. M. Lattelais *et al.*, *Astron. Astrophys.*, 2011, **532**, A12.
2. D. Burke *et al.*, poster 60.

Professor Rawlings replied: This is an interesting question, although there are a number of unknowns in the model.

If the explosions result in a series of chemical reactions that occur on chemical timescales that are typically very much shorter than the physical/dilution time-scale then it may be that the isomers may be produced in 'normal' ratios, reflective of a thermochemical equilibrium. But, if that doesn't hold and/or non-equilibrium chemical conditions pertain then it is possible that unusual isomeric ratios may result.

In addition, the chemistry in the lengthy quiescent phases between explosions will be dominated by freeze-out and desorption processes which will further skew the ratios in the way that you describe.

What is known is that there are significant source-to-source variations in the abundance ratios and it may well be that, with a little more careful thinking, these could be used to diagnose the relative importance of the explosion mechanism.

Professor Bergin stated: I am afraid I disagree with this model of grain surface ice desorption. Detailed models can tell you how many radicals build up on a grain surface. If you are right, every quiescent dark cloud should have evaporated complex molecules in their centre, and they don't. Moreover we see demonstrative depletion holes for CO in pre-stellar cores, with the depletion rings larger for less volatile CS. Thus there is clear evidence for some volatility control in the evapo-ration process, which will not be the case for explosive desorption. Can you comment on this?

Professor Rawlings responded: I disagree with the first point; firstly, the models that you refer to tend to be highly simplistic with regards to their assumptions about the composition and morphology of both the grain cores and the ices. We know that typical grains actually have a very high surface density of binding sites that can accommodate radicals. The models are also incomplete; *e.g.* as we have heard at this meeting, the contribution to radical production by low energy electrons has probably been severely under-estimated. There are essentially no empirical constraints on the abundance of radicals in true inter-stellar ices. The statement concerning the ubiquity of complex organic molecules (COMs) is too simplistic; in propitious environments our model predicts the presence of COMs at, or near, current detection limits. Moreover, the efficiency of the mechanism is dependent on the local conditions; the hydrogen atom abun-dance, gas and dust temperatures, the UV and cosmic ray ionization rates *etc.* It is therefore to be expected that there will be substantial source-to-source variations.

Significantly, we know that COMs are detected in some dark, quiescent sources (*e.g.* TMC-1, L1689b and B1-b) but not others. This is an issue that is at least as problematic for 'traditional' surface chemistries (most of which fail to include formation-enthalpy-driven desorption) as it is for our more hypothetical model. In the darker/denser regions of cores — where the hydrogen atom abundance is insignificant — the explosion mechanism will not operate and the regular freeze-out and desorption mechanisms will yield the types of depletion morphologies to which you refer.

Professor Linnartz commented: The Greenberg experiments you are referring to were performed under regular high vacuum conditions, typically with thousands of monolayers (MLs) of ice. More recent experiments use ultra-high vacuum, and the investigated ices are as thin as a few tens of MLs. On one hand, this allows controlling experimental settings in a much better way, but also, such a setting is closer to the thickness of interstellar ices. So what is the consequence for a picture in which complex species are formed during an explosion? Is this picture based on laboratory experiments using thick ices, or is it still an option for thin ices, and if so, how should one look at such an explosion; as a major catastrophic event, in which the full ice is evaporated, or a more local event, involving a few MLs only?

Professor Rawlings answered: The efficiency of the process will depend on whether the explosion is initiated in the ice mantle or *via* heat transfer from energetic events in the substrate (that is to say, the grain core – composed of hydrogenated amorphous carbon and silicates). In the former case, the effect may indeed be localized, but in the latter situation, it is possible that more catastrophic heating may occur.

Dr Semenov opened the discussion of the paper by Catherine Walsh: According to current global 3D MHD disk models, the transport and accretion of matter due to redistribution of angular momentum is essentially a 3D phenomenon and does not have a preferred flow direction. Therefore, for us, disk chemistry modelers, it will be hard to perform an MHD, realistically modeling the disk chemistry, unless the time-dependent chemistry is fully coupled with a feasible 3D disk dynamical model.

Dr Walsh replied: This is good point which should certainly be highlighted. The exploratory models presented here were intended as a step beyond static disk models, in which the chemistry is computed for a fixed set of physical conditions (representative of a particular location within the disk) as a function of time only. The calculations show that, for the simplest possible case (a net accretion flow), dynamical timescales can be important, in particular for determining the composition (and origin) of the material in the planet-forming regions of proto-planetary disks.

Coupling the chemistry with a more realistic model of mass transport in protoplanetary disks would be challenging but nonetheless very worthwhile. Generating a global map of the chemical composition of the disk would be computationally very demanding, but one could certainly calculate the chemical evolution for those particular parcels of gas which are delivered to the planet-forming region. This approach would be analogous to that recently conducted for coupled physiochemical models of collapsing envelopes, the physical structure for which is determined using either a semi-analytical approach[1–3] or full 3D magnetohydrodynamics.[4]

A similar calculation has been done for a hydrodynamic simulation of a gravitationally unstable protoplanetary disk around a young solar-mass star.[5] The authors found that individual parcels of material followed somewhat chaotic paths and thus were subjected to a range of different physical conditions over time which subsequently affected the resulting chemical composition of the disk.

1. R. Visser, E. F. van Dishoeck, S. D. Doty and C. P. Dullemond, *Astron. Astrophys.*, 2009, **495**, 881.
2. R. Visser, S. D. Doty and E. F. van Dishoeck, *Astron. Astrophys.*, 2011, **534**, A132.
3. M. N. Drozdovskaya, C. Walsh, R. Visser, D. Harsono and E. F. van Dishoeck, *Mon. Not. R. Astron. Soc.*, 2014, submitted.
4. U. Hincelin, V. Wakelam, B. Commercon, F. Hersant and S. Guilloteau, *Astrophys. J.*, 2013, **775**, 44.
5. J. D. Ilee, A. C. Boley, P. Caselli, R. H. Durisen, T. W. Hartquist and J. M. C. Rawlings, *Mon. Not. R. Astron. Soc.*, 2011, **417**, 2950.

Dr Woods commented: Jes Jørgensen and his team recently detected glycolaldehyde and other organics in the gas phase around the low-mass Class 0/I protostar IRAS16293-2422 out to distances of 60 AU or so.[1] In your model, which follows the chemistry to a later evolutionary stage, large organics such as glycolaldehyde only appear in the gas phase within a radius ~10 AU from the central star. Can you comment on the difference in the distribution of large organics here? Is it solely a matter of the evolution of the object?

1. J. Jørgensen *et al.*, *Astrophys. J.*, 2012, **757**, L4.

Dr Walsh answered: I concur with Dr Semenov's comment that IRAS 16293 is a relatively young source which likely has a significantly higher mass accretion rate ($>10^{-7}$ $M_\odot$ yr^{-1}) than that seen for disks around T Tauri stars (~10^{-8} $M_\odot$ yr^{-1}).[1] The resulting heating due to viscous dissipation can push the 'snow line' outwards, for example, the water (and complex organics) snow line ($T \approx 100$ K) can be pushed out to a few 10's of AU.[2,3] High spatial and spectral resolution observation should help determine whether the glycolaldehyde emission indeed arises from an embedded disk in IRAS 16293.

1. N. Calvet, L. Hartmann and S. E. Strom, in *Protostars and Planets IV*, 2000, 377.
2. S. S. Davis, *Astrophys. J.*, 2005, **620**, 994.
3. G. M. Kennedy and S. J. Kenyon, *Astrophys. J.*, 2008, **673**, 502.

Dr Semenov said: I have a comment about a recent detection of glycolaldehyde in a solar-mass protostar by Jes Jorgensen. The reason that this COM sits in the gas phase within inner 20 AU may be related to the fact that this star is still being built and thus accretion rates are relative large, leading to efficient viscous accretion heating of these inner regions of the infalling envelope. The enhanced inner temperatures keep glycolaldehyde in the gas phase.

Professor Kaiser commented: It is a clear misconception that it is well established that the formation of COMs happens on the surface of ices. The fact that astrochemical models predict that some COMs can be formed on the surface of grains at elevated temperatures is simply an artifact because only a few percent (!) of the material of the grain is considered to be important. Models completely ignore that COMs are formed at temperatures as low as 10 K within the 'bulk' of interstellar ices *via* non-equilibrium chemistry. You cannot expect that models which only account for a few percent of the chemistry provide objective results. Laboratory experiments have demonstrated that the interaction of ionizing radiation with low temperature ices does form COMs as complex as hydrocarbons,[1–4] (polycyclic) aromatic hydrocarbons (PAHs),[5,6] aldehydes and ketones,[7–10] carboxylic acids,[10–13] sugars,[14,15] amides,[16] amino acids,[17] dipeptides,[18] and amines.[19] In summary, there are many feasible pathways which lead to COM formation. Only complete astrochemical models, which account for all feasible pathways lead to realistic results.

1. Interaction of ionization radiation with 'bulk' ices and formation of COMs at 10 K *via* suprathermal (non-equilibrium, non-Arrhenius) chemistry.
2. Formation of COMs *via* radical–radical recombination within the 'bulk' of the ices upon warming up (thermal chemistry).
3. Formation of COMs *via* radical–radical recombination on the surfaces of the ices upon warming up (thermal chemistry).

1 R. I. Kaiser and K. Roessler, *Astrophys. J.*, 1998, **503**, 959.
2. C. J. Bennett, C. S. Jamieson, Y. Osamura and R. I. Kaiser, *Astrophys. J.*, 2006, **653**, 792.
3. Y. S. Kim, C. J. Bennett, L. Sheng, O'Brian and R. I. Kaiser, *Astrophys. J.*, 2010, **711**, 744.
4. B. M. Jones and R. I. Kaiser, *J. Phys. Chem. Lett.*, 2013, 4, 1965.
5. R. I. Kaiser and K. Roessler, *Astrophys. J.*, 1997, **475**, 144.
6. L. Zhou, W. Zheng, R. I. Kaiser, A. Landera, A. M. Mebel, M. C. Liang and Y. L. Yung, *Astrophys. J.*, 2010, **718**, 1243.
7. C. J. Bennett, C. S. Jamieson, Y. Osamura and R. I. Kaiser, *Astrophys. J.*, 2005, **624**, 1097.
8. C. J. Bennett, Y. Osamura, M. D. Lebar and R. I. Kaiser, *Astrophys. J.*, 2005, **634**, 698.
9. L. Zhou, R. I. Kaiser, L. G. Gao, A. H. H. Chang, M. C. Liang and Y. Y. Yung, *Astrophys. J.*, 2008, **686**, 1493.
10. R. I. Kaiser, S. Maity and B. M. Jones, *Phys. Chem. Chem. Phys.*, 2014, **16**, 3399.
11. Y. S. Kim and R. I. Kaiser, *Astrophys. J.*, 2010, **725**, 1002.
12. C. J. Bennett, Y. S. Kim, R. I. Kaiser, T. Hama and M. Kawasaki, *Astrophys. J.*, 2011, **727**, 27.
13. C. J. Bennett and R. I. Kaiser, *Astrophys. J.* 2007, **660**, 1289.
14. C. J. Bennett, S. H. Chen, B. J. Sun, A. H. H. Chang and R. I. Kaiser, *Astrophys. J.*, 2007, **660**, 1588.
15. C. J. Bennett and R. I. Kaiser, *Astrophys. J.*, 2007, **661**, 899.
16. B. M. Jones, C. J. Bennett and R. I. Kaiser, *Astrophys. J.*, 2011, **734**, 1.
17. P. D. Holtom, C. J. Bennett, Y. Osamura, N. J. Mason and R. I. Kaiser, *Astrophys. J.*, 2005, **626**, 940.
18. R. I. Kaiser, Y. S. Kim, A. Stockton, E. Jensen and R. A. Mathies, *Astrophys. J.*, 2013, **765**, 111.
19. Y. S. Kim and R. I. Kaiser, *Astrophys. J.*, 2011, **729**, 1.

Dr Walsh responded: I agree with Professor Kaiser that the wording in the submitted version of the paper is too strong regarding the formation of COMs in interstellar and circumstellar environments: we will reword this appropriately in the proofs and are happy to have the opportunity to do so.

I would like to clarify the methods used to calculate the ice abundances. The models presented here assume two phases only (gas and ice) and we do not distinguish between 'bulk' and 'surface' ice. We assume the bulk diffusion rates are the same as the surface diffusion rates and relate the surface/bulk diffusion barrier to each species' desorption barrier (or binding energy). We also allow dissociation and ionisation *via* energetic photons and particles to occur throughout the bulk ice. Hence, radicals and atoms are produced throughout the ice mantle and can diffuse and react with other radicals, atoms, or species present within the ice. We also include many different pathways to COMs; however, because only a handful of ice systems have been investigated in the laboratory, we acknowledge the models do suffer from a lack of quantitive data for many reactions (*e.g.*, binding energies, reaction barriers, diffusion barriers, branching ratios) and we use extrapolated or estimated parameters. As I commented during one of the discussions, perhaps the answer is to simplify the astrochemical models and use these same models to simulate laboratory experiments to quantify the critical parameters listed above. This would help significantly improve the larger chemical networks. I concur that thermal grain surface chemistry alone cannot explain the formation of complex organic molecules *via* radical–radical association reactions at low temperatures (~10 K). The modelling community should now work towards including non-thermal ice chemistry triggered by irradiation by energetic particles and photons, especially considering the recent detections of gas-phase complex organic molecules in dark clouds and presellar cores.[1]

1. A. Bacmann, V. Taquet, A. Faure, C. Kahane and C. Ceccarelli, *Astron. Astrophys.*, 2012, **541**, L12.

Dr Garrod commented: Models that treat both the ice-surface and bulk ice-mantle chemistry do indeed exist now; see **ref. 1**.

1. R. T. Garrod, *Astrophys. J.*, 2013, **765**, 60.

Dr Jones asked: It seems likely that in these regions the grain surfaces will be coated with an aliphatic-rich amorphous hydrocarbon mantle. In your modelling what do you think would be the likely effects of a such a chemically rich surface underlying the ice mantle?

Dr Walsh responded: A source of carbon-rich material between the 'bare' refractory grain surface and the water-rich ice mantle could certainly instigate an interesting chemistry at the surface–water ice interface. Given that energetic particles and photons can penetrate the ice mantle and that ice mantles are believed to be layered (water ice forms first and is later covered by a layer of, *e.g.*, CO/CH_3OH mixed ice), this may help to build chemical complexity deep within the ice mantle.

In the model presented here, we treat the ice mantle as a single phase, *i.e.*, we do not treat the bulk and surface ice separately and nor do we take into consideration the layering of ices. It would nonetheless be interesting to add an aliphatic component to the ice mantle in our simple two-phase model to see the effect on the generation of COMs. It would also be interesting to explore this in models with a more robust treatment of the ice layering, for example, the GRAINOBLE model[1] or macroscopic Monte Carlo models,[2] both of which have the ice mantle chemistry coupled to the gas-phase chemistry.

1. V. Taquet, C. Ceccarelli and C. Kahane, *Astron. Astrophys.*, 2012, **548**, A42.
2. A. Vasyunin and E. Herbst, *Astrophys. J.*, 2013, **762**, 86.

Dr Garrod asked in relation to the paper by Jonathan Rawlings: Are there any clear observational tests that would demonstrate that the proposed mechanism is occurring in any astronomical environment?

Professor Rawlings replied: This is a good question and a definitive answer would indeed help to validate the model. One possible test could be based on methanol production. In 'traditional' models CH_3OH is generated from CO in the solid state, whilst in our model CH_4 plays a more important role. If CH_3OH can be detected in regions where the ices are believed to be CO-poor (perhaps due to enhanced cosmic ray desorption) then that would be a good test of the hypothesis. There is another — quite different — possible test; it has been postulated[1,2] that the heat liberated from the crystallization of an amorphous carbon outer layer of a grain may be able to heat the underlying silicate core to a temperature of ~1000 K. If this happens, then this would result in the annealing/crystallization of inter-stellar silicate grains. The widespread presence of crystalline silicates in cold, quiescent regions would also give support to our theory.

1. C. Kaito *et al.*, *Astrophys. J.*, 2007, **666**, L57.
2. K. K. Tanaka *et al.*, *Astrophys. J.*, 2010, **717**, 586.

Dr Heays returned to the discussion of the paper by Catherine Walsh: Would it be useful to study the sensitivity of the observable predictions of your model to the various reaction rates and desorption energies within?

Dr Walsh answered: The sensitivity of the chemistry to the gas-phase rate coefficients has been studied for dark cloud conditions,[1] and also for proto-planetary disks.[2] This exercise enabled both sets of authors to identify those species sensitive to the adopted reaction rate coefficients and also those reactions for which a better constraint on the rate coefficient is required. These calculations are made possible because the gas-phase chemical databases provide an esti-mation of the level of error for each rate coefficient.

It would certainly be useful to do a similar exercise for thermal grain-surface chemistry, although this is governed somewhat by more uncertain global parameters (such as grain-size distribution, diffusion barriers, *etc.*) rather than individual reaction rate coefficients. In previous work, we looked at the effects of varying the diffusion barrier height and the reactive desorption rate.[3] Both of these parameters have a global effect: a larger diffusion barrier shifts the region in which COMs are formed inwards towards the star (where the temperature is higher) and a faster rate for reactive desorption increases the abundances of gas-phase COMs in the disk midplane. A natural extension of this work would be to consider whether this variation leads to an observable change in molecular line emission allowing a comparison with observations.

1. V. Wakelam, E. Herbst, J. Le Bourlot, F. Hersant., F. Selsis and S. Guilloteau, *Astron. Astrophys.*, 2010, **517**, A21.
2. A. I. Vasyunin, D. Semenov, Th. Henning, V. Wakelam, E. Herbst and A. M. Sobolev, *Astrophys. J.*, 2008, **672**, 629.
3. C. Walsh, T. J. Millar, H. Nomura, S. Widicus Weaver, Y. Aikawa, J. C. Laas and A. Vasyunin, *Astron. Astrophys.*, 2014, **563**, A33.

Dr Semenov commented: Concerning uncertainties in the calculated molecular abundances in disk chemical models. Back in 2008 my student, Anton Vasyunin, and I, studied in detail the error bars in the modeled abundances caused by the uncertainties of gas-phase reactions only (see **ref. 1**). In summary, we found that abundances of simple molecules such as H_2O, HCO^+, HCN, H_2CO, *etc.* are uncertain by factors of 3–4. The exception is H_2 and CO for which the chemistry is well known; their uncertainties are within ~30%. The error bars for more complex molecules (>4–5 atoms) are uncertain roughly by a factor of 10 or more. If one then takes into account uncertainties in surface processes (*e.g.*, in binding energies), the error bars become even larger.

1. A. I. Vasyunin *et al.*, *Astrophys. J.*, 2008, **672**, 629.

Faraday Discussions

PAPER

The cycling of carbon into and out of dust

Anthony P. Jones,* Nathalie Ysard, Melanie Köhler, Lapo Fanciullo,
Marco Bocchio, Elisabetta Micelotta, Laurent Verstraete
and Vincent Guillet

Received 9th December 2013, Accepted 11th February 2014
DOI: 10.1039/c3fd00128h

Observational evidence seems to indicate that the depletion of interstellar carbon into dust shows rather wide variations and that carbon undergoes rather rapid recycling in the interstellar medium (ISM). Small hydrocarbon grains are processed in photo-dissociation regions by UV photons, by ion and electron collisions in interstellar shock waves and by cosmic rays. A significant fraction of hydrocarbon dust must therefore be re-formed by accretion in the dense, molecular ISM. A new dust model (Jones *et al.*, *Astron. Astrophys.*, 2013, **558**, A62) shows that variations in the dust observables in the diffuse interstellar medium ($n_H \leq 10^3$ cm^{-3}), can be explained by systematic and environmentally-driven changes in the small hydrocarbon grain population. Here we explore the consequences of gas-phase carbon accretion onto the surfaces of grains in the transition regions between the diffuse ISM and molecular clouds (*e.g.*, Jones, *Astron. Astrophys.*, 2013, **555**, A39). We find that significant carbonaceous dust re-processing and/or mantle accretion can occur in the outer regions of molecular clouds and that this dust will have significantly different optical properties from the dust in the adjacent diffuse ISM. We conclude that the (re-)processing and cycling of carbon into and out of dust is perhaps the key to advancing our understanding of dust evolution in the ISM.

1 Introduction

Many lines of observational evidence and much modelling work have indicated that interstellar carbon grains and, in particular, carbonaceous nanoparticles are rather 'volatile' dust species that seemingly undergo rapid processing in the interstellar medium (ISM).[1-8] Recent work strongly suggests that these grains most likely consist of hydrogenated amorphous carbons, a-C(:H), which encompass H-poor, aromatic-rich a-C through to H-rich, aliphatic-rich a-C:H materials.[9-15] Such hydrocarbon nanoparticles are processed in the intense radiation fields of photo-dissociation regions (PDRs) by UV photon-induced destruction[6,9] and by ion and electron collisions. This processing can occur in interstellar shock waves, in a hot gas ($T \geq 10^6$ K) and by cosmic ray interactions.[1-3,7,8,16] Given that a

Institut d'Astrophysique Spatiale, CNRS/Université Paris Sud, Orsay, 91405, France. E-mail: Anthony.Jones@ias.u-psud.fr; Fax: +33 1 6985 8675; Tel: +33 1 6985 8647

significant fraction of carbon is in dust in the diffuse ISM, rapid hydrocarbon dust destruction implies that it must be re-formed by accretion in the dense, molecular ISM.[4]

Here we adopt a new dust model,[9] which can explain many of the variations in the dust observables (UV-NIR extinction and IR-mm emission) in the diffuse interstellar medium ($n_H \leq 10^3$ cm^{-3}) in terms of systematic and environmentally-driven changes in the small hydrocarbon grain population. Further, the photo-processing of newly-exposed a-C:H nanoparticles in the shells around evolved stars could provide a viable mechanism for fullerene formation.[17,18] We study the accretion of gas-phase carbon and briefly consider the formation of aggregates by the coagulation of small a-C(:H) grains onto the surfaces of large grains in the transition regions between the diffuse ISM and molecular clouds.[19-22] We conclude that the processing and cycling of carbon into and out of dust appears to be the key to advancing our understanding of dust evolution in the ISM.

2 The evolution of carbonaceous dust in the ISM

In the laboratory, a-C(:H) solids are known to darken upon UV photon irradiation, thermal annealing and ion irradiation.[23-25] In a-C(:H) materials, this process, known as photo-darkening,† leads to a decrease in the band gap or optical gap energy, E_g. The band gap for a-C(:H) materials ($E_g = -0.1$ to 2.7 eV)‡ is directly related to X_H, the H atom fraction ($E_g = 4.3X_H$), and also to the ratio, R, of the sp^3 and sp^2 C atomic fractions, X_{sp^3} and X_{sp^2}, respectively, $i.e.$,[11,12]

$$R = \frac{X_{sp^3}}{X_{sp^2}} \approx \frac{(8X_H - 3)}{(8 - 13X_H)} \sim \frac{(0.6E_g - 1.0)}{(2.7 - E_g)}. \tag{1}$$

The above equation is an approximation to the exact relationship between E_g, X_H and R, which depends upon the sp^2 aromatic domain sizes, the –CH$_3$ methyl group concentration and the particle size.[11-13] The band gap is therefore a proxy for the optical properties of an a-C(:H) with given R and X_H.[11,12]

It is therefore the evolution of the band gap that can be used to trace and characterise the inherent variability in the a-C(:H) optical properties, which is of prime importance in unravelling the evolution of hydrocarbon grains in the ISM.[9,26-28] Recent modelling work[11-13] shows that the optical properties of a-C(:H) materials can be completely determined by two parameters, the band gap, E_g, and the particle size, a, which turn out to be coupled for $a < 30$ nm, in that the minimum possible band gap is given by:[14]

$$E_g(a)_{\text{min}} = \left(\frac{1}{a \, [\text{nm}]} - 0.2 \right) \text{eV}, \tag{2}$$

$i.e.$, the effective band gap, $E_{g,\text{eff}}$, for a given size particle is

$$E_{g,\text{eff}} = \max[E_{g,\text{bulk}}, E_g(a)_{\text{min}}], \tag{3}$$

---

† Photo-darkening refers to an increase in the *dark*, graphite-like or aromatic content. Hence, the term photo-darkening is used to describe the aromatisation of a-C(:H) materials.

‡ Here we adopt the Tauc gap for the optical band gap, E_g, $i.e.$, the energy axis intercept in a plot of $(E\alpha)^{0.5}$ vs. E, where $\alpha = 4\pi k/\lambda$ is the absorption coefficient.[11]

where $E_{g,bulk}$ is the expected bulk a-C(:H) material band gap. For particles with radii larger than $\sim$30 nm, this minimum band gap limit does not apply.

In this work we explore the likely consequences of interstellar a-C(:H) dust evolution in the transition between the diffuse ISM and the outer regions of molecular clouds ($n_H = 10^2$–10^4 cm^{-3}).

2.1 Hydrocarbon dust photo-processing

The (photo-)dissociation of CH bonds within a-C(:H) grains leads to the loss of H atoms from the aromatic, aliphatic and olefinic hydrocarbon structures in the chemical network, a progressive aromatisation and a closing of the optical band gap. This process is at the heart of the evolution of the physical and optical properties of a-C(:H) materials. Quantifying this fundamental process is therefore key to advancing our understanding of the evolution of these materials in the ISM. Following earlier work,[11,15] the photo-darkening rate can be expressed as:

$$\Lambda_{UV,pd}(a) = F_{EUV}\, \sigma_{CH}\, Q_{abs}(a,E)\varepsilon \ [s^{-1}], \tag{4}$$

where $F_{EUV} \simeq 3 \times 10^7$ photons cm^{-2} s^{-1} is the CH bond-dissociating EUV photon flux in the local ISM[29] (for the interstellar radiation field, ISRF, in the solar vicinity, i.e., $G_0 = 1$), $\sigma_{CH} \simeq 10^{-19}$ cm^2 is the CH bond photo-dissociation cross-section,[30–33] $Q_{abs}(a,E)$ is the particle absorption efficiency factor in the optical cross-section determination, i.e., $\sigma(a,E) = \pi a^2 Q_{abs}(a,E)$, and ε is the photo-dissociation or photo-darkening efficiency.

The a-C(:H) particle photo-processing time-scale, as a function of grain radius and the local ISRF, can then be expressed as:

$$\tau_{UV,pd}(a, G_0) = \frac{1}{\Lambda_{UV,pd}(a)\ G_0}. \tag{5}$$

Fig. 1 shows the a-C(:H) particle photo-processing time scale as a function of ε and grain radius. As can be seen in the figure, the earlier estimates for $\tau_{UV,pd}(a)$ with the assumption that $\varepsilon = 0.1$ (shown as diamonds)[15] predict decreasing EUV processing time scales with increasing particles size up to $a \simeq 30$ nm, and are then approximately constant time scales at around $(5 \times 10^4/G_0)$ years for $a > 30$ nm. The increased photo-processing time scales for $a < 30$ nm are due to the decrease in the grain photon absorption efficiency with decreasing particle size, i.e., $Q_{abs} \propto a$ for particles smaller than the wavelength.

We now introduce a size-dependent photo-dissociation or photo-darkening efficiency, $\varepsilon(a)$, into our earlier derivation,[11,15] and adopt the following:

$$\varepsilon(a) = \left(\frac{2}{a\ [nm]}\right) \ \text{for} \ a > 2 \ nm, \quad \text{otherwise} \quad \varepsilon(a) = 1 \ \text{for} \ a \leq 2 \ nm. \tag{6}$$

For a-C(:H) grains, such a size-dependent CH bond photo-dissociation efficiency appears to be a physically more realistic scenario than a fixed and size-independent ε. With increasing particle size, other photon-driven processes will begin to compete with CH photo-dissociation and will likely become more important in large grains ($a > 10$ nm), i.e., EUV photon absorption leading to thermal excitation and/or fluorescence. For small particles, $a \leq 2$ nm, the CH

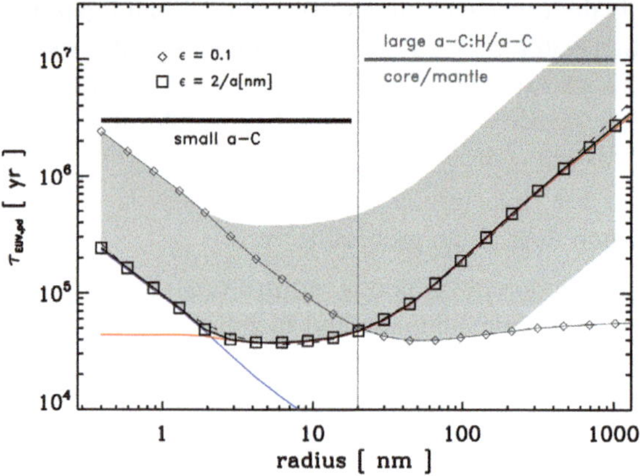

Fig. 1 The a-C(:H) photo-processing time scale for $G_0 = 1$ as a function of particle radius, a, and the CH bond photo-darkening efficiency, ε ($\varepsilon = 0.1$, shown as diamonds; $\varepsilon = 2/a$ for $a > 2$ nm, eqn (6), shown as squares). The vertical grey line indicates the critical radius between small homogeneous and larger core–mantle particles. The grey band indicates an order of magnitude uncertainty on the derived time scales; for grains with radii $\leq$ 200 nm, the derived photo-processing time scales are lower limits. The blue and red lines indicate the small and large grain ($\varepsilon \times Q_{abs})^{-1}$ behaviours, *i.e.*, $\propto 1/Q_{abs}$ and $\propto a/(2Q_{abs})$, respectively. The mostly-hidden dashed grey line is the analytical fit using eqn (7) to the derived a-C(:H) photo-processing time-scales (squares).

bond photo-dissociation is determined by the absorption efficiency factor Q_{abs} (see Fig. 1), and adopting $\varepsilon(a) = 1$ therefore implies that every absorbed photon leads to CH bond breaking, as might be expected for particles in the 'molecular domain'. If we now consider the effects of including the size-dependent photo-darkening efficiency, $\varepsilon(a)$, given by eqn (6) and shown by the squares in Fig. 1, we find a minimum in the processing time scale of the order of $(4 \times 10^4/G_0)$ years for particles with $a \approx 3$–10 nm. For smaller and larger grains in the modelled dust size distribution the processing time scale is $>(10^5/G_0)$ years. For particles smaller than $\sim$3 nm, the same trend as in the constant ε processing time scales is apparent, but it is shifted down by an order of magnitude because for $a < 2$ nm we now assume $\varepsilon(a) = 1$, rather than $\varepsilon = 0.1$ as in the previous work. For particles with radii larger than 10 nm, the processing time scales increase with radius because of the assumed size-dependence of the photo-darkening efficiency, $\varepsilon(a) \propto a^{-1}$.

The a-C(:H) grain photo-processing time scales in the ISM, for a size-dependent photo-dissociation efficiency $\varepsilon(a)$, can be analytically expressed as a function of the grain radius, a, in nm;

$$\tau_{UV,pd}(a) = \frac{10^4}{G_0}\left\{2.7 + \frac{6.5}{(a\,[\text{nm}])^{1.4}} + 0.04\,(a\,[\text{nm}])^{1.3}\right\}[\text{years}]. \tag{7}$$

This fit to the predicted a-C(:H) photo-processing time scales is shown by the somewhat hidden dashed line fit to the square data points in Fig. 1.

2.2 Hydrocarbon dust erosion

In addition to EUV photo-processing in PDRs, which leads to the erosion of small carbonaceous particles,[6] a-C(:H) dust will also be highly susceptible to processing, erosion and destruction in supernova (SN)-generated shock waves in the warm inter-cloud medium.[1,34,35] Recent work[2,3,7,8,16] shows that small a-C(:H) and polyaromatic particles (*e.g.*, polycyclic aromatic hydrocarbons, PAHs) are rather easily destroyed in SN shock waves, in a hot gas and by cosmic rays. This occurs primarily through the effects of electronic excitation and dissociation following electron and ion collisions, a process that is more important than direct "knock-on" sputtering in sub-nanometer- and nanometer-sized particles. The important conclusion of these works is that most of the small carbon population in energetic regions will be destroyed. The lifetime of carbonaceous grains in the ISM that is derived from these studies is rather short, *i.e.*, ≤ 100 million years. This catastrophic situation implies that carbonaceous dust must be efficiently re-formed in dense regions of the ISM, where re-accretion re-forms carbonaceous matter principally in the form of mantles on the surviving silicate and any surviving carbonaceous grains.[4] The logical conclusion of this is that large pre-solar a-C(:H) grains should be rather rare. However, pre-solar carbon-rich grains of nanodiamond, SiC and graphite with anomalous isotopic compositions typical of evolved stars and supernovae are found in meteorites,[36] but they are probably not the dominant carrier of the solid carbon phase.

2.3   Hydrocarbon dust (re-)accretion

Here we consider in detail the (re-)accretion of gas phase C and H atoms to form a-C(:H) carbonaceous mantles on dust in the transition between the diffuse and dense molecular ISM. In particular, we find that the nature of the mantles formed on dust in the transition to denser regions is critically determined by the extinction at UV wavelengths and the hydrogenation of aromatic-rich carbona-ceous grain materials.[37–40]

Here we explore the qualitative effects of mantle (re-)accretion in the outer regions of molecular clouds. We assume a semi-inifinite planar cloud and adopt the simple density profile into the cloud from the surface, *e.g.*, similar to that derived for dense filaments[21] and photo-dissociation regions,[41]

$$n_{\mathrm{H}}(l) = n_{\mathrm{H}}(0)[1 + 10^5 l^2], \tag{8}$$

where $n_{\mathrm{H}}(0)$ is the cloud surface density in cm^{-3}, which we assume to be typical of the diffuse ISM, *i.e.*, $n_{\mathrm{H}}(0) = 40\ \mathrm{cm}^{-3}$, and l is the distance into the cloud from its surface in pc. The cloud density profile is shown in Fig. 2 along with the column density, N_{H}, the optical depths, A_{V} and A_{EUV} ($E_{\mathrm{EUV}} = 10$ eV), and the attenuation of V and EUV band photons as a function of distance into the cloud.

Recently, a completely new dust modelling approach was proposed, which does not include polycyclic aromatic hydrocarbons (PAHs), graphite grains or "astronomical silicate".[9] Instead, it uses the size-dependent optical and thermal properties for hydrogenated amorphous carbons, a-C(:H),[11–15] and the optical properties of an amorphous silicate with metallic iron inclusions.[9] The new aspects of this model are:

- the continuum of a-C(:H) optical properties,
- the FUV photo-processing of aliphatic-rich a-C:H into aromatic-rich a-C,
- the a-C(:H) core–mantle structure of all carbon grains with radii > 20 nm,
- a-C mantled amorphous forsterite-type silicates, and
- the incorporation of iron as metal nano-inclusions into the silicate (as a result of reduction by mantle carbon diffusion into the silicate[42]).

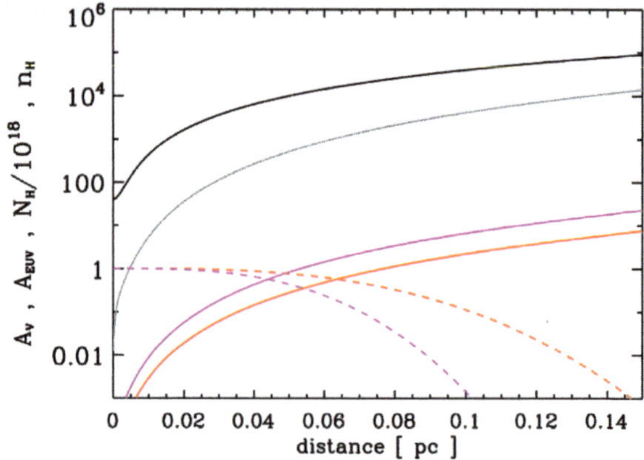

Fig. 2 The cloud parameters: density n_H (H cm^{-3}, black), column density $N_H/10^{18}$ (H cm^{-2}, grey), A_V (red), the attenuation at the V band wavelength $I/I_0 = e^{-A_V}$ (dashed red), $A_{EUV} = 3A_V$ ($E_{EUV} = 10$ eV $\equiv 124$ nm, violet, see Fig. 4) and the attenuation at the EUV wavelength $I/I_0 = e^{-A_{EUV}}$ (dashed violet), as a function of distance into the cloud from its surface.

This new model satisfactorily explains the dust extinction, scattering and emission in the diffuse ISM and predicts the evolution of the carbonaceous dust properties in response to local conditions.[9]

In this work we apply the Jones *et al.* dust model,[9] using the extinction and mass size distributions for the model derived using the DustEM tool,[43] as shown in Fig. 3 and 4. In this model the gas phase C atom accretion time scale to form a-C(:H) mantles is given by

$$t_{acc} = [\Sigma_{total} n_H X_C \nu_C S_C]^{-1},\tag{9}$$

where Σ_{total} is the total dust cross-section per H atom, X_C is the relative abundance of gas phase carbon atoms and $\nu_C = [8k_B T_{kin}/(\pi m_C)]^{0.5}$ is their thermal velocity

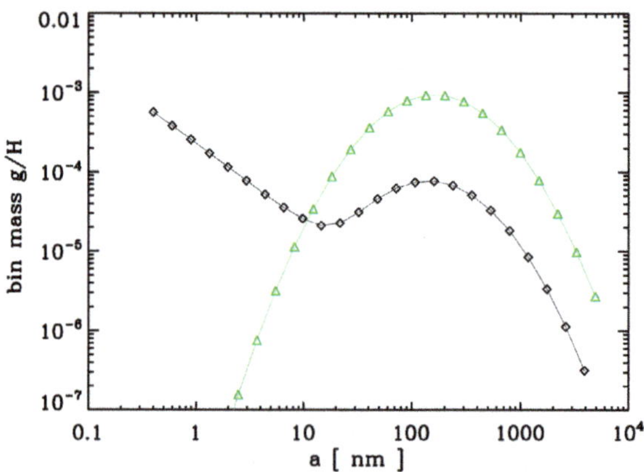

Fig. 3 The model dust mass–size distribution in the diffuse ISM: a-C(:H) grains (black) and a-Sil$_{Fe}$–a-C core–mantle grains (green).

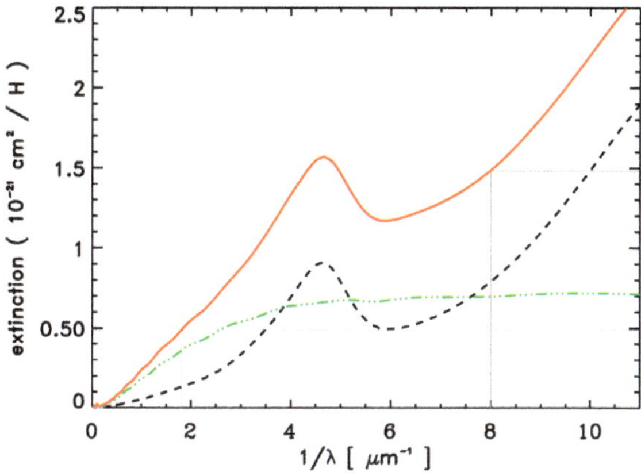

Fig. 4　The modelled dust extinction in the diffuse ISM: a-C(:H) grains (black) and a-Sil$_{Fe}$– a-C core–mantle grains (green). The total extinction is given by the red line. The grey lines indicate the extinction in the V band and for 10 eV EUV photons.

(k_B is the Boltzmann constant, $T_{kin} = 80$ K is the gas kinetic temperature and m_C is the C atom mass), and S_C is the C atom sticking coefficient (assumed to be unity). We assume a cosmic abundance of $\sim$400 ppm for carbon;[5] the dust model requires 233 ppm[9] therefore leaving $\sim$167 ppm available for accretion onto dust in the form of a-C(:H) mantles.

It has been shown that H atom incorporation into carbonaceous materials at low temperature ($T_{kin} = 80$ K) can lead to their hydrogenation,[37–40] and we therefore consider this possibility within the framework of our qualitative model. The H atom sticking time scale onto all grains is

$$t_H = [\Sigma_{total} n_H \nu_H S_H]^{-1}, \tag{10}$$

where $\nu_H = [8k_B T_{kin}/(\pi m_H)]^{0.5}$ is the H atom thermal velocity, and S_H is the H atom sticking coefficient. Here we assume the canonical value $S_H = 0.3$. We define the low temperature a-C(:H) grain (re-)hydrogenation rate as

$$t_{Hin} = [\xi \Sigma_{total} n_H \nu_C S_H]^{-1}, \tag{11}$$

where ξ is the efficiency for H atom incorporation into the a-C(:H) structure. Given that most incident H atoms will combine with other incident H atoms to form H_2, which is then ejected, ξ should be significantly less than unity.

For the size distribution in the Jones *et al.* standard diffuse ISM dust model,[9] we derive $\Sigma_{total} = 7.3 \times 10^{-21}$ cm^2 per H atom and find that the accretion time scale and mantle thickness are very dependent upon the lower grain size limit. For example, the accretion of $\sim$170 ppm of carbon onto all grain surfaces in the standard diffuse ISM dust model yields $t_{acc} \approx 10^7$ years for $n_H = 10^4$ cm^{-3} and a mantle thickness of $\sim$3 nm on all grains. However, in the outer regions of molecular clouds, the small grains ($a \leq 5$ nm) are accreted/coagulated onto large grains.[19,21,22] As an illustration of the important effects of small particles, if we remove all grains with radii < 5 nm from the dust size distribution, we find a carbonaceous mantle thickness of 160 nm, a reduction in the total grain cross-section by more than an order of magnitude, to $\Sigma_{total} = 1.4 \times 10^{-22}$ cm^2 per H atom, and a corresponding increase in the C atom accretion time scale to $t_{acc} \geq 10^9$ years for $A_V < 3$. Thus, if carbon mantle accretion onto nanoparticles is inhibited by stochastic heating events or if the nanoparticles themselves are coagulated/accreted onto or into large particles, the accreted mantle thickness could be $\gg 3$ nm, but the accretion time scales would then be $>10^7$ years.

Interestingly, analysis of the $\sim$100 nm thick "organic" coatings on the mineral grains in interplanetary dust particles (IDPs) indicates that this primitive matter is the result of the condensation of C-bearing "ices" onto grain surfaces and the formation of refractory matter by subsequent UV or other ionising irradiation.[44] These "organic" mantles, which carry H and N isotopic anomalies consistent with molecular cloud or outer Solar System material, resemble the a-C(:H) mantles on the grains in the Jones *et al.* diffuse ISM dust model[9] and could be the result of the direct accretion of the remaining gas phase carbon as a-C:H, as discussed here, rather than requiring a carbon-rich ice precursor phase. Such "organic" coatings on carbonaceous grains would lead to large aliphatic-rich a-C:H grains that resemble the "organic globules" detected in primitive meteorites and comets, which show D/H and ^{15}N/^{14}N enrichments, and that are likely the product of low temperature ($\sim$10 K) chemical reactions in cold molecular clouds or the outer regions of the proto-solar nebula.[45]

In Fig. 5 we show the time scale for C atom accretion into a-C(:H) on all grains, t_{acc}, and the a-C(:H) grain EUV photo-processing time scale, t_{pd}, for 1, 10 and 100 nm radius particles, as a function of distance into the cloud and A_V. We deduce from Fig. 5 that the accretion and EUV photo-processing time scales are similar

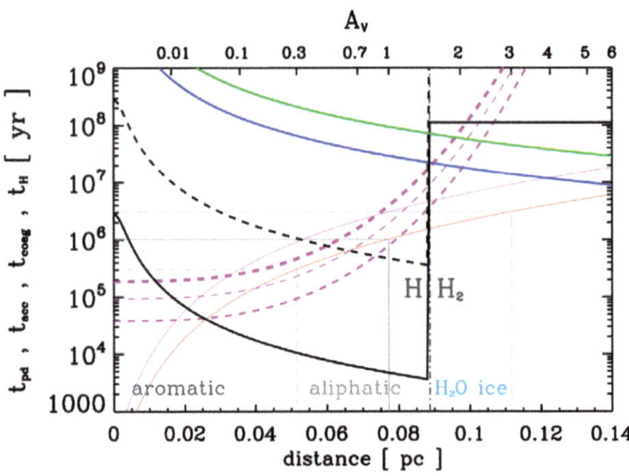

Fig. 5   The dust processing time scales as a function of A_V and distance into the cloud: the C atom accretion time scale t_{acc} (blue), the small a-C grain–large silicate grain coagulation time scale t_{coag} (green), the H atom sticking time scale, t_H (black), the H atom incorporation time scale, t_{Hin} (black dashed), and the EUV photo-processing time scale for 1, 10 and 100 nm radius a-C(:H) grains t_{pd} (thin, medium and thick dashed violet lines, respectively). Also shown are A_V (red) and $A_{EUV} = 3A_V$ (violet), multiplied by 10^6. The grey lines indicate the depth into the cloud at which $A_V \sim 0.3$, 1 and 3 (short-dashes, solid and long-dashes, respectively). The vertical dash-dotted line indicates the atomic-to-molecular hydrogen transition (here assumed to occur at $A_V = 1.5$). The approximate aromatic, aliphatic and H₂O ice accreted mantle regimes are also indicated.

for cloud depths of the order of $\sim$0.09–0.1 pc, equivalent to an A_V of $\sim$2. Nearer to the cloud surface, $A_V < 2$ where $t_{acc} \gg t_{pd}$, it is evident that grain mantling can only occur on timescales of $\geq 10^7$ years. In the absence of any incident H atom hydrogenation of the grains, the outer grain surfaces/mantles will be EUV-photolysed into H-poor, aromatic-rich materials, *i.e.*, low band gap a-C with $E_g < 0.2$ eV. However, deeper into the cloud, the CH bond photo-dissociating EUV photons are attenuated and the cloud interiors, $A_V > 2$, are UV-poor regions (UVPRs) where a-C:H grains and mantles will not be EUV photo-processed, because the photolysis time scale is significantly longer than the C (and H) atom accretion time scale. Thus, in UVPRs the accreted mantles will be H-rich, aliphatic-rich materials, *i.e.*, wide band gap a-C:H with $E_g \geq 2$ eV.

We now consider the possibility that H atom hydrogenation is efficient at low temperatures.[37–40] Fig. 5 indicates that in the low density outer regions of clouds, with $A_V < 0.01$, the H atom collision rate is insufficient to counteract the EUV photo-dissociation of the CH bonds within a-C(:H) grain, even if every H atom that sticks to the grain is incorporated into the a-C(:H) structure, *i.e.*, $\xi = 1$. Adopting a significantly lower efficiency, $\xi = 0.01$, leads to the same result, but the depth for the onset of hydrogenation shifts deeper into the cloud ($A_V \gtrsim 0.7$). These results imply that in the low extinction diffuse ISM and outer cloud regions ($A_V < 0.01$ for $\xi = 1$ or $A_V < 0.7$ for $\xi = 0.01$, with $n_H \approx 40$ cm^{-3} and $T_{kin} \approx 80$ K), small a-C(:H) particles and any a-C(:H) on the surfaces/mantles of large grains must be made of predominantly H-poor, aromatic-rich, a-C materials. Deep within the clouds, hydrogen is predominantly in molecular form for $A_V \gtrsim 1.5$,[46] except for a background H atom abundance of $\sim$1 cm^{-3} due to the effect of cosmic rays, and so the hydrogenation process will be reduced by orders of magnitude in a molecular gas. However, as Fig. 5 shows, at intermediate optical depths, 0.01–$0.7 \lesssim A_V \lesssim 1.5$ (the lower limit depends on the assumed value for ξ), $t_{Hin} < t_{pd} < t_{acc}$, *i.e.*, the H atom incorporation into a-C(:H) is faster than both CH bond photo-dissociation and mantle accretion. Thus, the transformation of aromatic-rich grain materials into H-rich, aliphatic-rich materials could occur on time scales of the order of 10^6 years. Somewhat deeper into the cloud, $\sim$1.5 $\lesssim A_V \lesssim 2.5$, and immediately after the transition from atomic to molecular hydrogen, H atom incorporation is again dominated by photo-dissociation (albeit at a very low level), and any remaining gas-phase carbon in the clouds would accrete along with the ice mantles in regions with $A_V \gtrsim 1.5$.

2.4   Grain coagulation effects

Following earlier work[22] the coagulation time scale between dust species 1 and 2 can be expressed as

$$t_{coag} = \{\pi(a_1 + a_2)^2 X_1 n_H \Delta v\}^{-1}, \tag{12}$$

where a_i is the grain radius, X_1 is the relative abundance of dust species 1 and Δv is the relative velocity between the particles. We are interested in the time scale for large grains to sweep up small aromatic-rich grains, in an accretion-type coagulation process[19–22] that leads to the formation of aromatic particle mantles. In order to estimate the coagulation time scale we set $a_1 = 5$ nm and $a_2 = 160$ nm, similar to previous estimates,[19] and based on the dust masses in the new model.[9] The small–large grain coagulation timescale is shown in Fig. 5 and indicates that small grain coagulation always lags behind C atom accretion as a viable a-C(:H) mantling process in cloud interiors.

3   Astrophysical implications

Fig. 6 gives a schematic view of the evolution of the accreted/transformed a-C(:H) grain–mantle composition as a function of the optical depth, A_V, and the H atom incorporation efficiency into a-C(:H), ξ. Fig. 5 and 6 show that in the diffuse ISM and in the outer regions of molecular clouds, the time scale for H atom incorporation into a-C(:H) grains is significantly longer than the CH photo-dissociation time scale. In low extinction regions ($A_V < 0.01$–0.7) H atom incorporation into a-C(:H) is therefore seemingly not fast enough to transform a-C into a-C:H. Thus, given that the EUV CH photo-dissociation depth is of the order of 20 nm,[11] this work indicates that small a-C(:H) grains and the outer surfaces of large a-C(:H) grains will be rapidly transformed ($t_{pd} \geq 4 \times 10^4$ years) to aromatic-rich a-C (see Fig. 5), as in the recently-proposed dust model.[9] The cores of large a-C(:H) particles ($a \gg 20$ nm) being shielded from the effects of CH bond EUV photo-dissociation can therefore consist of H-rich, aliphatic-rich a-C:H material.[11] Clearly, the cores of these large a-C(:H) grains can only be H-rich in the diffuse ISM if they were formed as such, because they cannot be re-hydrogenated there. However, in slightly higher extinction regions ($A_V > 0.01$–0.7), any grain core a-C material could be transformed into a-C:H by the effects of H atom incorporation,[37–40] if this process is efficient and if they are not already H-rich. Interestingly, as shown in Fig. 5, somewhat deeper into a cloud than the atomic-to-molecular hydrogen interface, at $A_V \approx 2$, the time scale for H atom incorporation into a-C increases above that for CH bond photo-dissociation, and so any accreting a-C(:H) mantles would be aromatic-rich rather than aliphatic-rich. This would seemingly occur with the onset of ice mantle formation for $A_V \geq 1.5$.

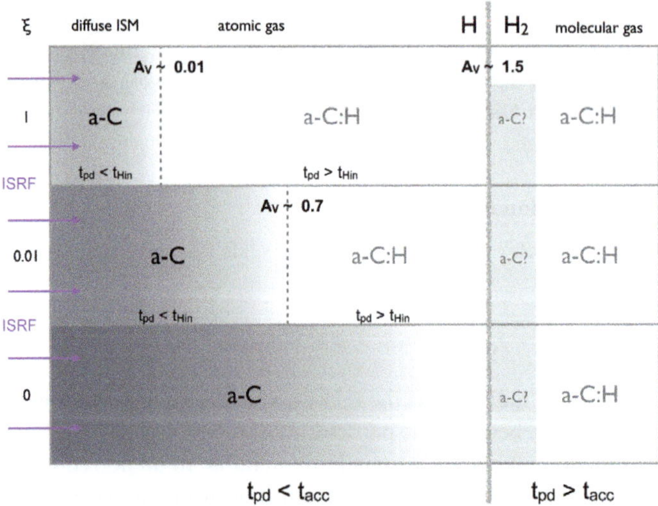

Fig. 6   A schematic view of the accreted/transformed carbonaceous grain–mantle composition as a function of the optical depth, A_V, and the H atom incorporation efficiency into a-C(:H), ξ.

It has been proposed that in the ISM the carbonaceous grains consist of an aliphatic-rich mantle overlying an aromatic-rich core.[47] This is the inverse of the Jones *et al.* dust model,[9] and is at odds with our results for dust processing in the diffuse ISM. Nevertheless, within molecular clouds it is indeed likely that the carbonaceous grain surfaces will be aliphatic-rich due to the effects of a-C:H mantle accretion and/or the incident H atom aliphatisation of aromatics.

The formation of H-rich, aliphatic-rich mantles and grain surfaces, through a-C:H mantle accretion or H atom incorporation into a-C grains, leads to a slight decrease in the visible extinction, a diminution of the UV bump but little change in the FUV extinction.[9] This is consistent with the high carbon depletion ($\sim$400 ppm of C in dust) along the line of sight towards HD 207198,[5,9] but is inconsistent with the extinction curve along lines of sight through the denser diffuse ISM, with $R_V \sim 5$, where the UV extinction is flatter than the average, *i.e.*, for lines of sight with $R_V \sim 3$. The flatter UV extinction along high R_V lines of sight implies that the coagulation of small grains onto large grains occurs before or is contemporaneous with carbonaceous mantle accretion. However, this preliminary study seems to indicate that accretion occurs before small grains coagulate onto large grains, *i.e.*, t_{coag} is always > t_{acc}. The extinction curve variations and far-IR to sub-mm dust emission generally seem to imply that small grains stick onto big grains before significant accretion occurs.[19] An exception to this seems to be the HD 207198 line of sight[5,9] where the formation of a-C:H mantles, or the transformation of a-C to a-C:H by H atom incorporation,[37] is strongly implied. Detailed studies of the evolution of dust evolution in the low density ISM, through the effects of coagulation and accretion, will be required in order to unravel the exact sequence of events as a function of the local conditions.

4   Summary and conclusions

Carbonaceous dust in the ISM is apparently a much more fragile dust component than previously thought and must therefore be efficiently re-formed in the ISM. In this work we have investigated the composition and evolution of a-C(:H) dust in the transition from the low-density diffuse ISM to the outer reaches of molecular clouds. The primary processes that drive this evolution are the CH bond EUV photo-dissociative aromatisation, a-C:H $\rightarrow$ a-C, and the reverse process of H atom incorporation and aliphatisation, a-C $\rightarrow$ a-C:H, leading to (re-)hydrogenation. We find that in the low-density diffuse ISM ($A_V < 0.01$–0.7) photo-processing dominates over aliphatisation, and that small carbonaceous grains/thin mantles and the outer surfaces of large carbonaceous grains should consist predominantly of low band gap (E_g of $\sim$0 eV) aromatic-rich a-C materials. However, in the diffuse ISM, where H atom (re-)hydrogenation is slower than EUV photo-aromatisation, the carbonaceous grain cores may consist of H-rich aliphatic-rich material if they were formed as such. In slightly higher extinction regions ($A_V > 0.01$–0.7) we find that H atom incorporation into a-C leading to aliphatisation and an opening of the band gap could dominate. Interestingly, somewhat deeper into a cloud than the H–H_2 interface ($A_V \approx 2$), any accreting gas phase carbon will tend to form a-C rather than a-C:H mantles because of the low H atom abundance. This unusual effect seemingly occurs at about the same optical depth as the accretion of ice mantles (at $A_V \gtrsim 1.5$, assuming a semi-infinite slab cloud model), which merits a more detailed analysis than has been possible here.

In order to fully quantify the effects studied here a much more detailed study of the equilibrium composition of a-C(:H) materials in the ISM is required. This will necessitate a much deeper understanding of the spatio-temporal evolution of the dust composition (*e.g.*, band gap, E_g) and the dust size distribution resulting from the effects of CH bond photo-dissociation and a-C(:H) grain photo-fragmentation in the low-density ISM, and the effects of accretion and coagulation in denser regions. In particular, a much deeper understanding of the interplay between accretion and coagulation is required. Allied to this must be a study of the likely effects of turbulence and radiation pressure on the accretion onto dust and grain coagulation in the outer regions of molecular clouds. Also, given that C^+ remains the most abundant form of gas phase carbon to optical depths A_V of ~ 3,[46] and that the grains in molecular clouds are negatively charged, the collision cross-sections for accretion could be somewhat enhanced within clouds. However, as is often assumed, ion accretion onto negatively-charged grains leads to the recombination and desorption of the incident ion. Thus, any accretion gain could be offset by C^+ recombination upon collision. These processes could play an important role in dust evolution in these transition regions because they would tend to enhance the gas–grain and grain–grain collision velocities and therefore reduce the relevant time scales. However, it should be noted that small grain coagulation onto large grains leads to a reduced total grain surface and therefore thicker mantles, but also to significantly longer accretion time scales.

The carbonaceous mantle accretion phenomena investigated here ought to have important observable consequences. Thus, the outer regions of molecular clouds ($\sim 0.5 \leq A_V \leq 3$) merit deeper investigation because they will provide strong constraints on the nature and evolution of the matter that accretes onto dust and on the coagulation process.

Acknowledgements

This research was, in part, made possible through the financial support of the Agence National de la Recherche (ANR) through the program CIMMES (ANR-11-BS56-029-02).

References

1 L. Serra Díaz-Cano and A. P. Jones, *Astron. Astrophys.*, 2008, **492**, 127–133.

2 E. R. Micelotta, A. P. Jones and A. G. G. M. Tielens, *Astron. Astrophys.*, 2010, **510**, A36.

3 E. R. Micelotta, A. P. Jones and A. G. G. M. Tielens, *Astron. Astrophys.*, 2010, **510**, A37.

4 A. P. Jones and J. A. Nuth, *Astron. Astrophys.*, 2011, **530**, A44.

5 V. S. Parvathi, U. J. Sofa, J. Murthy and B. R. S. Babu, *Astrophys. J.*, 2012, **760**, 36.

6 P. Pilleri, J. Montillaud, O. Berné and C. Joblin, *Astron. Astrophys.*, 2012, **542**, A69.

7 M. Bocchio, E. R. Micelotta, A.-L. Gautier and A. P. Jones, *Astron. Astrophys.*, 2012, **545**, A124.

8 E. R. Micelotta, A. P. Jones and A. G. G. M. Tielens, *Astron. Astrophys.*, 2011, **526**, A52.

9 A. P. Jones, L. Fanciullo, M. Köhler, L. Verstraete, V. Guillet, M. Bocchio and N. Ysard, *Astron. Astrophys.*, 2013, **558**, A62.

10 A. P. Jones, *Astron. Astrophys.*, 2013, **555**, A39.

11 A. P. Jones, *Astron. Astrophys.*, 2012, **540**, A2.

12 A. P. Jones, *Astron. Astrophys.*, 2012, **540**, A1.

13 A. P. Jones, *Astron. Astrophys.*, 2012, **542**, A98 (paper III).

14 A. P. Jones, *Astron. Astrophys.*, 2012, **545**, C3.

15 A. P. Jones, *Astron. Astrophys.*, 2012, **545**, C2.

16 M. Bocchio, A. P. Jones, L. Verstraete, E. M. Xilouris, E. R. Micelotta and S. Bianchi, *Astron. Astrophys.*, 2013, **556**, A6.

17 J. Bernard-Salas, J. Cami, E. Peeters, A. P. Jones, E. R. Micelotta and M. A. T. Groenewegen, *Astrophys. J.*, 2012, **757**, 41.

18 E. R. Micelotta, A. P. Jones, J. Cami, E. Peeters, J. Bernard-Salas and G. Fanchini, *Astrophys. J.*, 2012, **761**, 35.

19 M. Köhler, B. Stepnik, A. P. Jones, V. Guillet, A. Abergel, I. Ristorcelli and J.-P. Bernard, *Astron. Astrophys.*, 2012, **548**, A61.

20 M. Köhler, V. Guillet and A. Jones, *Astron. Astrophys.*, 2011, **528**, A96.

21 N. Ysard, A. Abergel, I. Ristorcelli, M. Juvela, L. Pagani, V. Könyves, L. Spencer, G. White and A. Zavagno, *Astron. Astrophys.*, 2013, **559**, A133.

22 B. Stepnik, A. Abergel, J.-P. Bernard, F. Boulanger, L. Cambrésy, M. Giard, A. P. Jones, G. Lagache, J.-M. Lamarre, C. Meny, F. Pajot, F. Le Peintre, I. Ristorcelli, G. Serra and J.-P. Torre, *Astron. Astrophys.*, 2003, **398**, 551–563.

23 S. Iida, T. Ohtaki and T. Seki, in *Optical Effects in Amorphous Semiconductors*, AIP Conf. Proceedings No. 120, ed. P. C. Taylor and S. G. Bishop, AIP, New York, 1985, p. 258.

24 F. W. Smith, *J. Appl. Phys.*, 1984, **55**, 764–771.

25 M. Godard, G. Féraud, M. Chabot, Y. Carpentier, T. Pino, R. Brunetto, J. Duprat, C. Engrand, P. Bréchignac, L. D'Hendecourt and E. Dartois, *Astron. Astrophys.*, 2011, **529**, A146.

26 W. W. Duley, *Mon. Not. R. Astron. Soc.*, 1996, **283**, 343–346.

27 A. P. Jones, *Cosmic Dust – Near and Far*, 2009, pp. 473–481.

28 A. P. Jones, W. W. Duley and D. A. Williams, *Q. J. R. Astron. Soc.*, 1990, **31**, 567–582.

29 R. C. Henry, *Astrophys. J.*, 2002, **570**, 697–707.

30 A. R. Welch and D. L. Judge, *J. Chem. Phys.*, 1972, **57**, 286–290.

31 Y. A. Gruzdkov, K. Watanabe, K. Sawabe and Y. Matsumoto, *Chem. Phys. Lett.*, 1994, **227**, 243–247.

32 V. Mennella, G. M. Muñoz Caro, R. Ruiterkamp, W. A. Schutte, J. M. Greenberg, J. R. Brucato and L. Colangeli, *Astron. Astrophys.*, 2001, **367**, 355–361.

33 G. M. Muñoz Caro, R. Ruiterkamp, W. A. Schutte, J. M. Greenberg and V. Mennella, *Astron. Astrophys.*, 2001, **367**, 347–354.

34 A. P. Jones, A. G. G. M. Tielens, D. J. Hollenbach and C. F. McKee, *Astrophys. J.*, 1994, **433**, 797–810.

35 A. P. Jones, A. G. G. M. Tielens and D. J. Hollenbach, *Astrophys. J.*, 1996, **469**, 740–764.

36 E. Anders and E. Zinner, *Meteoritics*, 1993, **28**, 490–514.

37 V. Mennella, *Astrophys. J.*, 2010, **718**, 867–875.

38 V. Mennella, J. R. Brucato, L. Colangeli and P. Palumbo, *Astrophys. J.*, 2002, **569**, 531–540.

39 V. Mennella, *Astrophys. J.*, 2006, **647**, L49–L52.

40 V. Mennella, *Astrophys. J.*, 2008, **682**, L101–L104.

41 H. Arab, A. Abergel, E. Habart, J. Bernard-Salas, H. Ayasso, K. Dassas, P. G. Martin and G. J. White, *Astron. Astrophys.*, 2012, **541**, A19.

42 C. Davoisne, Z. Djouadi, H. Leroux, L. D'Hendecourt, A. Jones and D. Deboffle, *Astron. Astrophys.*, 2006, **448**, L1–L4.

43 M. Compiègne, L. Verstraete, A. Jones, J.-P. Bernard, F. Boulanger, N. Flagey, J. Le Bourlot, D. Paradis and N. Ysard, *Astron. Astrophys.*, 2011, **525**, A103.

44 G. Flynn, *38th COSPAR Scientific Assembly*, 2010, p. 3242.

45 S. Messenger, K. Nakamura-Messenger and L. Keller, *37th COSPAR Scientific Assembly*, 2008, p. 2018.

46 A. G. G. M. Tielens and D. Hollenbach, *Astrophys. J.*, 1985, **291**, 722–754.

47 J. E. Chiar, A. G. G. M. Tielens, A. J. Adamson and A. Ricca, *Astrophys. J.*, 2013, **770**, 78.

PAPER

The formation of ice mantles on interstellar grains revisited – the effect of exothermicity

T. Lamberts,[ab] X. de Vries[a] and H. M. Cuppen*[a]

Received 13th December 2013, Accepted 4th February 2014

DOI: 10.1039/c3fd00136a

Modelling of grain surface chemistry generally deals with the simulation of rare events. Usually deterministic methods or statistical approaches such as the kinetic Monte Carlo technique are applied for these simulations. All assume that the surface processes are memoryless, the Markov chain assumption, and usually also that their rates are time independent. In this paper we investigate surface reactions for which these assumptions are not valid, and discuss what the effect is on the formation of water on interstellar grains. We will particularly focus on the formation of two OH radicals by the reaction H + HO$_2$. Two reaction products are formed in this exothermic reaction and the resulting momentum gained causes them to move away from each other. What makes this reaction special is that the two products can undergo a follow-up reaction to form H$_2$O$_2$. Experimentally, OH has been observed, which means that the follow-up reaction does not proceed with 100% efficiency, even though the two OH radicals are formed in each other's vicinity in the same reaction. This can be explained by a combined effect of the directionality of the OH radical movement together with energy dissipation. Both effects are constrained by comparison with experiments, and the resulting parametrised mechanism is applied to simulations of the formation of water ice under interstellar conditions.

1   Introduction

Water is one of the molecules most vital for life on Earth. How water was delivered to Earth is not completely clear. Different pathways have been suggested, including scenarios where Earth's water is of extraterrestrial origin.[1] Water frozen on interstellar dust particles could have been trapped during the accretion of dust to form our planet, or water may have been transported by comets at a later stage. In both cases, interstellar ices are at the origin of the mechanism.[2,3] In the cold

Theoretical Chemistry, Institute for Molecules and Materials, Radboud University Nijmegen, Heyendaalseweg 135, 6525 AJ Nijmegen, The Netherlands. E-mail: hcuppen@science.ru.nl

bRaymond and Beverly Sackler Laboratory for Astrophysics, Leiden Observatory, University of Leiden, P. O. Box 9513, NL 2300 RA Leiden, The Netherlands

regions of molecular clouds, water is formed through grain surface reactions. On the grain surface, water can be formed through simple exothermic addition reactions, where the grain serves as a third body to take up the excess energy.[4]

In 2007, we performed the first microscopic astrochemical simulations on the surface formation of water in different environments: diffuse, translucent, and dense clouds.[5] This model was based on the physical and chemical information that was available at the time, which was rather poor, but it triggered many experimental studies by several different laboratories.[6–9] Because of these new experiments the surface reaction network is now much better constrained and the understanding of the formation and structure of solid water and its precursors has improved in general. Recently, a new microscopic kinetic Monte Carlo model was developed[10] that is capable of simulating different types of water formation experiments[11,12] and that puts constraints on the reaction rates of many of the reactions within the surface network. Table 1 gives an overview of this network as it is currently understood.[9–13] In the present discussions, the new kinetic Monte Carlo model (KMC) will be the basis for simulations under interstellar conditions. We further aim to show the importance of exothermicity of reactions, and how this leads to non-Markovian behaviour and chemical desorption.

Table 1 List of surface and photodissociation reactions used in the current model. For their corresponding rates we refer to Cuppen and Herbst,[5] and Lamberts et al.[10]

Temperature-independent reactions

H	$+$	H	$\rightarrow$	H_2	
H	$+$	O_2	$\rightarrow$	HO_2	
H	$+$	HO_2	$\rightarrow$	products	
				OH	$+ OH$
				H_2O_2	
				H_2	$+ O_2$
				H_2O	$+ O$
H	$+$	O	$\rightarrow$	OH	
O	$+$	O	$\rightarrow$	O_2	
H	$+$	O_3	$\rightarrow$	O_2	$+ OH$
H	$+$	OH	$\rightarrow$	H_2O	

Temperature-dependent reactions

H	$+$	H_2O_2	$\rightarrow$	H_2O	$+ OH$
H_2	$+$	O	$\rightarrow$	OH	$+ H$
H_2	$+$	HO_2	$\rightarrow$	H_2O_2	$+ H$
H_2	$+$	OH	$\rightarrow$	H_2O	$+ H$
OH	$+$	OH	$\rightarrow$	products	
				H_2O_2	
				H_2O	$+ O$
O	$+$	O_2	$\rightarrow$	O_3	

Photodissociation reactions

OH	$\rightarrow$	O	$+ H$
H_2O	$\rightarrow$	OH	$+ H$
O_2	$\rightarrow$	O	$+ O$
O_3	$\rightarrow$	O_2	$+ O$

Grain surface chemistry is generally simulated by either rate equations or some stochastic method that solves the master equation. The master equation describes the change in probability to be in a certain state at a certain time. In these simulations a state is often represented by, *e.g.*, the species on the grain, their position, the temperature of the grain, *etc.* One of the assumptions at the root of the derivation of the master equation is that the events bringing the system from one state to the next are memoryless: the Markov chain assumption. Typical grain surface processes such as diffusion and desorption of thermalised species are effectively memoryless: they occur at much longer time scales than the (lattice) vibrations and all history about which states were previously visited is lost due to the vibrations between two transitions. For a more in-depth discussion, see ref. 14. Newly formed reaction products can behave differently, however. Most grain surface reactions are highly exoergic. Part of this exothermicity will be immediately transferred to the icy mantle, but a substantial fraction likely remains in the reaction products, making them "hot". For reactions with two or more reaction products the energy gain is distributed as momentum over the products. This will have two effects: the hot species are more likely to move and desorb than thermalised species, and because they have some momentum, their movement will not be memoryless and their trajectory will therefore not follow a random walk. The energy responsible for this behaviour eventually dissipates through collisions with the grain and/or ice mantle. The extent of this effect will therefore strongly depend on the time scale for energy dissipation, which in turn depends on the local environment. Collisions with molecules making up the ice mantle are much more effective than collisions with the harder grain material, which does not absorb the energy as easily.

A clear example of a reaction where non-Markovian behaviour becomes important is the formation of two OH radicals by the reaction $H + HO_2$. This is one of the few significant surface reactions in the water network that leads to two reaction products. The formed radicals will move away from each other due to their opposite relative momenta. If this reaction occurs on top of an ice or a grain surface, the species will continue to diffuse on the surface. If the reaction occurs in the bulk phase, however, they can loose their directionality through collisions with neighbouring molecules. What makes this reaction special is that the two products can undergo a follow-up reaction to form H_2O_2. In a co-deposition experiment of O_2 and H with an overabundance of O_2, the most likely pathway to OH formation is through this $H + HO_2$ reaction. Experimentally, OH has been observed under these conditions, which means that the follow-up reaction of OH + OH does not proceed with 100% efficiency, even though the two OH radicals are formed close together in the same reaction.

In ref. 10 we have already suggested that one of the reasons why the kinetic Monte Carlo simulations cannot reproduce an appreciable amount of OH without artificially changing the physical parameters, is their inherent Markovian behaviour. Because of the assumption that the processes are memoryless, the OH radicals can travel in all directions, including towards the location where the OH radicals were initially formed. It is relatively straightforward to give reaction products some extra energy in kinetic Monte Carlo simulations, but to alter their trajectory and not make it a random walk is much harder. Without this change, however, newly formed OH radicals can easily meet again and react.

To constrain the time scale for energy dissipation is not straightforward. Some information is available for the photoproducts of water ice photodissociation by molecular dynamics simulations,[15–17] which are determined by Newton's equations of motion and do not assume Markovian behaviour. However, these results are limited to a water-rich environment, and they may not be applicable to the formation of the first monolayers of the water ice mantle. Dulieu *et al.*[18] showed in a combined experimental and simulation study that the chemical desorption of water-related species can be much higher in the monolayer regime. Garrod *et al.*[19] have already suggested the importance of chemical desorption for, *e.g.*, the gas phase detection of methanol in cold dark clouds. This work was based on the Rice–Ramsperger–Kessel (RRK) theory, which relates the excess energy and the binding energy of species to a desorption probability. They modified this theory by adding an unconstrained a parameter, which they chose to be 0.1. Later, Cazaux *et al.*[20] studied the influence of chemical desorption on the gas phase composition by applying a KMC model, similar to one used by Cuppen and Herbst,[5] but specifically focusing on the impact on the gas phase and on fractionation. In the present paper, we vary the time scale for thermalisation and we will show how this affects the overall evolution of the grain mantle. We will not only focus on chemical desorption, but also on its influence on the grain surface chemistry and ice structure. The parameters that correspond best to our experiments will be applied in Section 4 to model the evolution of a grain in diffuse cloud, translucent cloud and dense cloud conditions.

2   Methodology

The formation of water is simulated by microscopic kinetic Monte Carlo simulations. For a detailed description of the method we refer to ref. 14. The program and its parametrisation are described in detail in ref. 5 and 10. Here, we briefly describe the main characteristics of these models. This is followed by a description of the adaptations that were implemented to overcome the use of empirically fitted parameters, replacing them by more physically relevant values. Finally we highlight the differences between the experimental and astrochemical simulation runs.

The grain is represented by a lattice model in which each lattice site can be occupied by one of the species shown in Table 2. Each species in the lattice has 6

Table 2   Species in the model and their corresponding binding energy contribution, E, in Kelvin, depending on the specific environment

Species	Grain	H, H_2, O	Rest
H	105	10	70
H_2	80	10	50
O_2	240	240	240
OH	210	20	210
O_2H	630	60	630
H_2O_2	1370	140	1370
H_2O	1260	130	1260
O	260	30	260
O_3	630	60	630

neighbours and 12 nearest neighbours, corresponding to a primitive cubic lattice. The total binding energy, $E_{\text{tot,bind}}$, for each species to a site is calculated by additive contributions of its neighbours. Nearest neighbours add a contribution of E, and next-nearest neighbours a contribution of $E/8$. Values for E are quoted in Table 2. The neighbour below the particle adds a double contribution ($2E$). Small species, $i.e.$, H, H_2, OH, and O, are allowed to occupy interstitial sites in the ice. This is included to account for the experimentally observed penetration of H atoms into solid O_2. Diffusion of these species to subsurface positions only occurs when an O_2 or a HO_2 molecule is atop the final position, since penetration in water-like structures has not been observed experimentally. Larger species can be present in the intermediate layer, though, as a result of the positioning of reaction products.

After the deposition of a species on the grain, it can desorb, diffuse, react or dissociate. Each event is assumed to be thermally activated, and the event rate is calculated using

$$k = \nu \exp\left(-\frac{E_{\text{a},i}}{T}\right) \tag{1}$$

where $E_{\text{a},i}$ is the activation energy (or barrier) for process i in Kelvin, and ν is the attempt frequency, which is approximated by the standard value for physisorbed species, $\dfrac{kT}{h} = 10^{12}\ \text{s}^{-1}$. Desorption can only occur if a species is positioned in the top layer and depends on the total binding energy of the site. Diffusion can occur to each of the 18 neighbouring sites, provided that the site is empty. If the site is full, the small species have a probability of moving to a corresponding interstitial site. Hopping events are calculated taking into account the binding energy contribution to the site, as well as a term to ensure microscopic reversibility. Reactions can occur with co-reactants occupying one of the 6 nearest-neighbour and interstitial sites. Photodissociation only occurs in the interstellar simulations under the influence of the interstellar radiation field with

$$k = \alpha_{\text{photo}} \exp(-\gamma_{\text{photo}} A_{\text{V}}), \tag{2}$$

or through cosmic-ray induced photons

$$k = \alpha_{\text{cr photo}} \zeta, \tag{3}$$

where ζ is the cosmic ray ionization rate which is taken to be $1.3 \times 10^{-17}\ \text{s}^{-1}$. Table 1 presents an overview of the chemical and photodissociation reactions included in our model. Surface reaction rates can be found in Lamberts $et\ al.$,[10] and photodissociation rates in van Dishoeck $et\ al.$[21] Deviations from these literature values are discussed in Section 2.1.

2.1 Adaptations of the Monte Carlo routine

The kinetic Monte Carlo simulation study of Lamberts $et\ al.$[10] could only reproduce the experimental observation of significant OH abundances if the mobility of OH radicals and the barrier for the OH + OH reaction were artificially increased. The values required were not consistent with other independent estimates of the OH mobility and the OH + OH reaction barrier,[22] but were necessary to prevent immediate recombination of the two OH radicals to form H_2O_2. Moreover, to ensure the correct ratio between formed OH and H_2O_2, a fixed branching ratio for the reaction $H + HO_2$ leading to 2 OH and to H_2O_2 was chosen. The latter might, however, be the result of a two step process: $H + HO_2 \rightarrow 2\ OH \rightarrow H_2O_2$.

In the present paper, both the diffusion and activation barrier have been adjusted to values of 210 and 0 K, respectively, and the product channel $H + HO_2 \rightarrow H_2O_2$ is removed. A physical explanation for the experimental detection of OH is the initial opposite relative momentum that two reaction products obtain upon reaction. To simulate this, we made two additional adaptations to the program described by Lamberts *et al.*[10] The two reaction products in the first reaction will be allowed to react or move with a directional bias, such that a higher probability is given to moving in the opposite direction to the other OH radical. The total hopping rate remains the same. The directionality is retained as long as the species is excited, *i.e.*, when the temperature of the species is higher than the temperature of the surface. Typically this allows for one to three directional hops to be performed.

Furthermore, species that are formed during an exothermic reaction resulting in two or more reaction products receive a certain energy gain that can be utilized to enhance their mobility: increased hopping and desorption rates. The excess energy is expressed in terms of temperature, T_{ex}, and is distributed over the produced species in a ratio inversely proportional to their masses, following conservation of momentum.[17] The energy gained by species A in a reaction with products A and B is therefore $T_{gain,A} = T_{ex}\dfrac{m_B}{m_A + m_B}$. We use the following notations: T_{ex} is the excess energy of the reaction, T_{gain} is the energy given to a specific reaction product directly after reaction, and $T_{species}$ is the temperature of a species at any given time. For the experimental simulations, the energy is distributed (almost) equally between the two products for the dominant reactions, since these products are of (nearly) equal mass. For the interstellar conditions where photodissociation reactions and the reaction $H_2 + OH \rightarrow H_2O + H$ determine the chemistry to a large extent, T_{gain} becomes very different for the two products.

In the original simulations by Lamberts *et al.*[10] an arbitrary, small excess energy of only 100 K for each reaction product, regardless of their number, was included in the model. The event time of the next event for the reaction product was then calculated using

$$\Delta t = -\frac{\ln X}{k_{tot}\left(T_{species}\right)} \qquad (4)$$

where X is a pseudo-random number between 0 and 1 and k_{tot} is the sum of all rates (diffusion, desorption, and reaction) for this molecule. These rates were considered constant during Δt for a temperature $T_{species}$, that was set equal to $T_{gain} = 1/2\ T_{ex}$ after reaction. This energy was dissipated through hopping events, where each diffusion step reduces $T_{species}$ by an arbitrary factor of 1.6. After 10^{-8} s, the local temperature of the excited species was set back to the temperature of the surface. This is based on the assumption that a molecule on the surface will be thermalised after 10 ns. The species were found to either react immediately to form a new molecule, or remain in their initial configuration for times much longer than 10 ns. The outcome of the simulation results is therefore not too sensitive to the exact choice of this time scale.

The assumption that the rates remain stationary during Δt is a rather crude approximation. To be able to study the effect of thermalisation, we have changed the algorithm which determines Δt to accommodate the effect of changing rates, *i.e.*, including energy dissipation. The dissipation of the energy is expected to be exponential.[23] To reduce the computational complexity of an exponential energy loss, a simpler expression (eqn (5)) was suggested by Cuppen and Hornekær.[24] The sensitivity of the model to the exact functional form was checked and found to be small. The initial temperature gain is assumed to decay following

$$T'(t) = \max\left(T_{\text{surf}}, \frac{T_{\text{gain}}}{(1 + B(t_{\text{r}} - t_{\text{a}}))^2} \right) \tag{5}$$

where t_{r} is the time at which the reaction occurred. A similar model was applied in ref. 24. The parameter B can be chosen freely. Here two types of B parameters are used: B, to describe the decay in the presence of an ice mantle, and B_{grain}, the decay on a bare grain surface. The varying temperature leads to non-stationary rates, and the kinetic Monte Carlo time step cannot be determined in the usual way for this situation (see also ref. 14). We applied the method of Jansen[25] to obtain the time step, using eqn (5) instead of a linearly increasing temperature.

Here we assume that the total T_{ex} that is distributed over the two reaction products is never larger than the maximum exothermicity of the reactions included in the network. The main effect that we believe the excess energy has on the evolution of the ice species is to move the reactants over a large enough distance that they will not be able to react together. A large part of the energy is thought to be dissipated through (internal) rotational and vibrational modes and interaction with the substrate, but the time scales involved in this dissipation are expected to be considerably smaller than those for the translational excitation. Rotational excitation might change the reaction rates because of a change in incoming angle and, moreover, rotational and/or vibrational excitation will most likely affect the energetics involved in the reaction. However, most ice reactions are diffusion limited and we expect these effects not to dominate.

2.2 Experimental *vs.* interstellar simulations

There are a few differences between the simulations of the experimental conditions and the simulations of interstellar ices. One of the differences is the composition of the gas phase (see Section 4). The H-atom flux for experimental conditions is chosen to be 5×10^{12} atoms cm^{-2} s^{-1}, in agreement with the experimental fluxes used in ref. 11 and 12 multiplied by a sticking fraction of 0.2. The other species arriving at the surface are H_2 and O_2 for an experimental simulation, or H_2 and O under interstellar conditions. Under interstellar conditions the flux is determined by

$$f_{\text{A}} = \frac{v_{\text{A}} n(\text{A}) n_{\text{H}}}{4} \tag{6}$$

where $n(\text{A})$ is the relative gas phase abundance of species A, n_{H} is the total hydrogen density, and v_{A} is the mean velocity of species A in the gas. For a cloud with a temperature of 20 K and a hydrogen density $n_{\text{H}} = 10^4$ cm^{-3}, with half of its hydrogen in the form of H atoms, the H atom flux is 1.6×10^8 atoms cm^{-2} s^{-1}. The initial oxygen atom abundance is $3 \times 10^{-4} n_{\text{H}}$, which slowly diminishes over the course of the simulation to account for the freeze-out of the O atoms in terms of the newly formed surface species.

Besides the large difference in order of magnitude between the fluxes, there is also a change in the time scale over which the simulation is run. Experimental simulations last up to the length of the experiment (an hour), whereas astrochemical simulations are associated with time scales typical for dense clouds, *i.e.*, 10^5–10^6 years.[26] Here we can cover between 4×10^4 and 2.5×10^5 years. Finally, the photodissociation events occur only in the interstellar simulations, where H_2O, H_2O_2, O_3, HO_2, and OH are allowed to photodissociate with a rate similar to their gas phase rate.

3 Simulations of experiments

Here, we present and discuss kinetic Monte Carlo simulations of the following four experiments: sequential hydrogenation of O_2 at 15 K and 25 K, and codeposition of O_2 and H at 15 K and 25 K. Following the approach in ref. 10 we have chosen to reproduce these experimental results since they are representative for different experiments of the O_2 + H reaction route. The experimental surface abundances are reproduced as closely as possible, but using parameters based on chemical and physical arguments rather than empirical ones, which was done in the previous study. As mentioned above, one of the crucial parameters was found to be the OH mobility rate combined with the OH + OH reaction barrier. These, as well as different choices for T_{ex}, B and B_{grain} (eqn (5)) are changed in the current study to obtain a set of physico-chemically acceptable, best fit parameters that can be implemented in simulations run over astrochemically relevant time scales. The parameters chosen for all simulations are summarized in Table 3. The experimental simulations were performed on a smooth surface, analogous to the polished gold substrate used in the experimental set-up. Since the experimental studies are all bulk studies, the exact nature of the initial substrate is only of marginal influence.

Fig. 1 shows a comparison of the time evolution of the surface abundances that are experimentally obtained[12] with simulated results for runs I, II and V. Changing

Table 3 Summary of the parameters chosen for all experimental simulations

Run	$H + H_2O_2{}^a$ (%)	$E_{a,OH\ diff}$ (K)	$E_{a,OH + OH\ react}$ (K)	B (s^{-1})	B_{grain} (s^{-1})	T_{ex} (K)	Directionality
I[10]	56 : 32 : 2 : 7	105	600	–	–	100^a	off
II	91 : 0 : 2 : 7	210	0	–	–	100^a	off
III	91 : 0 : 2 : 7	210	0	10^{12}	10^{12}	200	on
IV	91 : 0 : 2 : 7	210	0	10^{12}	10^{12}	1000	on
V	91 : 0 : 2 : 7	210	0	10^{12}	10^{12}	1400	on
VI	91 : 0 : 2 : 7	210	0	10^{12}	10^{12}	2000	on
VII	91 : 0 : 2 : 7	210	0	10^{11}	10^{11}	1400	on
VIII	91 : 0 : 2 : 7	210	0	10^{13}	10^{13}	1400	on
IX	91 : 0 : 2 : 7	210	0	10^{13}	10^{13}	2000	on

[a] Branching ratios for the four product channels listed in Table 1. [b] A $T_{gain} = \frac{1}{2} T_{ex}$ of 100 K is given to each product, regardless of the number of products and their masses.

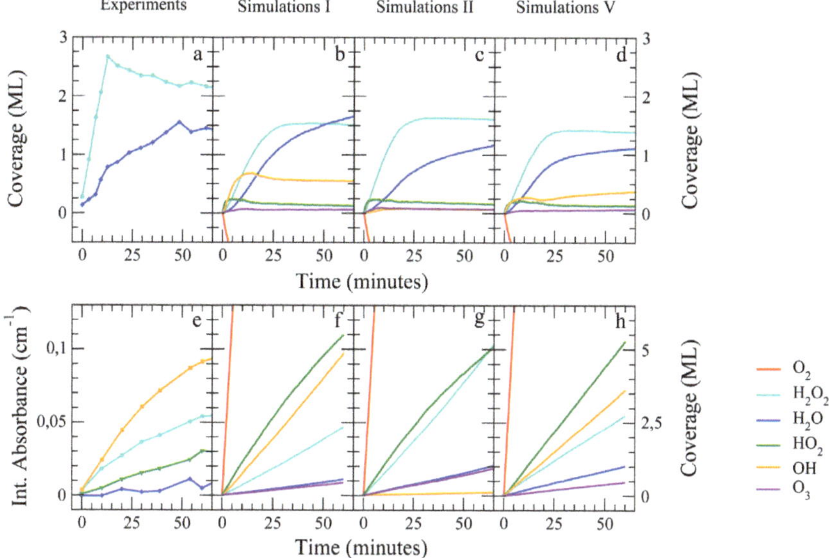

Fig. 1 Evolution of the surface abundances of O_2, OH, HO_2, H_2O, H_2O_2 and O_3 as a function of time; (a) and (e) experimentally,[12] (b) and (f) simulations I from Table 3, using the parameters from ref. 10, (c) and (g) simulations II, using only physical parameters, (d) and (g) simulations V, using best fit parameters from this work (d).

the set of parameters from simulation I to II encompasses the use of more realistic values. For instance, the barrier for the OH + OH reaction is removed following the recommended gas phase value from Atkinson *et al.*[22] The new value, however, results in a poorer agreement with the experimental results for the codeposition simulations. In general, this is due to the fact that all OH radicals produced by the reaction H + HO_2 are used in a follow-up reaction to subsequently produce H_2O_2, since there is no preventing mechanism. This is clear from the decrease of the OH surface abundance by 5 ML and the simultaneous increase of the H_2O_2 abundance by 2.5 ML. For the sequential hydrogenation at 15 K, a similar effect is observed, and all H_2O is subsequently formed through the reactions H + H_2O_2 → H_2O + OH and H + OH → H_2O. H_2O is slightly under-produced since there are not enough OH radicals available for reaction, but no large discrepancies with respect to the experiment are observed.

To improve the reproduction of the experimental codeposition surface abundances, directional hopping is included as well as the use of non-stationary excess energy. In Fig. 2, co-deposition simulations III and V–VIII are presented for 15 K and 25 K.

Firstly, the dependence of the surface abundances on T_{ex} is manifested mainly through the difference in the OH *versus* H_2O_2 production. A higher mobility of the reaction products from the reaction H + HO_2 allows them to move away from each other and prevent subsequent reaction to H_2O_2. This can also be seen in Table 4, where the ratio of follow-up reactions over the initial H + HO_2 → 2OH reaction is given. This ratio decreases with T_{ex}. The overall $\dfrac{[2OH]}{[H_2O_2]}$ abundance ratio is also

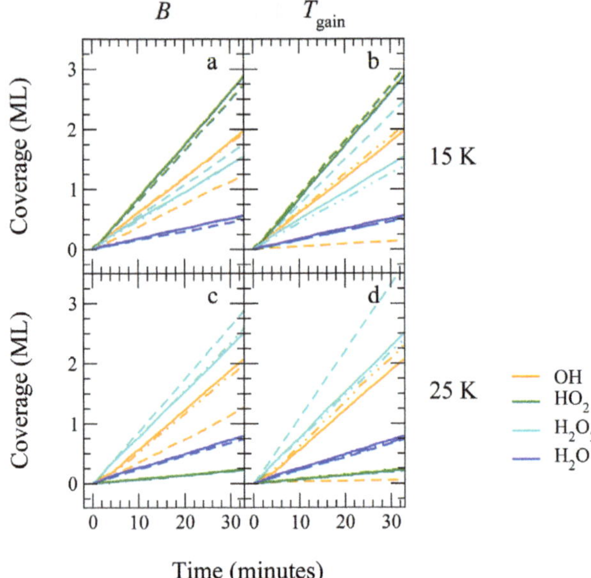

Time (minutes)

Fig. 2 Evolution of the surface abundances of OH, HO$_2$, H$_2$O, H$_2$O$_2$ and O$_3$ as a function of time for 15 K and 25 K with different values for B and T_{ex}. Solid lines indicate $B = 10^{12}$ s^{-1} and $T_{ex} = 1400$ K. Dashed lines indicate the lower value for B or T_{ex}, and dashed-dotted lines indicate the higher value. Simulations V, VII and VIII are depicted in panels (a) and (c), and runs III, V and VI in panels (b) and (d).

Table 4 Summary of the effect of the B and T_{ex} parameters on the formation of OH and its subsequent behaviour in the experimental simulations

Run	T_{ex} (K)	B (s^{-1})	T_{surf} (K)	$\dfrac{2OH \rightarrow products}{HO_2 + H \rightarrow 2OH}$	$\dfrac{[2OH]}{[H_2O_2]}$	Diffusion ('hot' events/species)
III	200	10^{12}	15	0.90	0.03	1.0
IV	1000	10^{12}	15	0.62	0.49	1.1
V	1400	10^{12}	15	0.55	0.66	1.4
VI	2000	10^{12}	15	0.50	0.75	2.2
VII	1400	10^{11}	15	0.62	0.35	4.4
VIII	1400	10^{13}	15	0.55	0.66	1.1
IX	2000	10^{13}	15	0.50	0.79	1.1
III	200	10^{12}	25	0.90	0.01	1.0
IV	1000	10^{12}	25	0.69	0.32	1.2
V	1400	10^{12}	25	0.63	0.42	1.4
VI	2000	10^{12}	25	0.60	0.49	2.2
VII	1400	10^{11}	25	0.72	0.22	4.8
VIII	1400	10^{13}	25	0.64	0.40	1.1
IX	2000	10^{13}	25	0.62	0.42	1.1

affected by other reactions and desorption, and can therefore not be directly derived from the first ratio. It is, however, more easily comparable to experimentally measurable quantities. Cuppen *et al.*[12] determined this ratio to be 1.6 ± 0.2 for 15 K and less than 0.5 for 25 K. The assumption was made that the bandstrength of the OH stretching mode of H_2O_2 is twice the bandstrength of that of the OH radical. If this assumption is indeed valid, a T_{ex} of 200 K is obviously not enough to allow for sufficient build-up of OH in the ice. A value of 1400 K or 2000 K appears to best describe this experimental observation.

Panels (a) and (c) in Fig. 2 give the results of runs V, VII, and VIII and show the influence of B. Similar trends are observed when simulations VI and IX are compared. Again, the results are also summarized in Table 4. The $\dfrac{2OH \rightarrow products}{HO_2 + H \rightarrow 2OH}$ ratio remains, as expected, more or less unaffected with varying B since this is almost purely determined immediately after the formation of the two OH radicals. The influence of the B parameter becomes most apparent if one considers the number of diffusion events during thermalisation of a single species, which is presented in the last column of Table 4. These diffusion events can bring the OH radicals to nearest-neighbour sites over a distance of approximately 3 Å, using a site density of 10^{15} sites cm^{-2}, or to next-nearest-neighbour sites over a distance of 4.5 Å. The number of diffusion events increases with decreasing B for the same T_{ex}, since B controls how fast thermalisation occurs. The initial momentum of the particle is, however, determined by T_{ex}, and the number of diffusion events is therefore a function of both. The average diffusion of a single species with an initial T_{ex} value of 1400 K ranges between 1.1 and 4.4 hops. Since we use a directionality for the mobility, OH will have more or less travelled in a straight line, in the opposite direction of the other OH radical formed. The OH radicals have therefore covered a spatial range of 3.3 to 20 Å. Molecular dynamics simulations by Andersson et $al.$[15] and Arasa et $al.$[17] show that the distance travelled by an excited OH radical inside bulk, non-porous ice is approximately 2 Å. The distance travelled on top of a surface can range up to 80 Å. In our simulations, species are always restricted by either bulk molecules or neighbours residing on step edges and islands. Therefore we expect the distance travelled by a hot reaction product to be larger than 2 Å, but considerably smaller than 80 Å. We expect a value of 10^{12} s^{-1} for the B parameter therefore to be the most realistic.

Experimentally, a large drop in the HO_2 and OH surface abundances accompanied by a large increase in H_2O_2 is observed upon increasing the temperature from 15 to 25 K. The decreasing OH abundance is only reproduced by simulation III; the decrease in HO_2 and increase in H_2O_2 are seen in all simulations. Experimentally, at 15 K OH is more abundant, whereas at 25 K H_2O_2 is more abundant. Taking this into account leads to best fit parameters of $B = 10^{12}$ in combination with $T_{ex} = 1400$ or 2000 K. In conclusion, our simulations of the codeposition experiments show that thermalisation is indeed an important effect which can help explain the observed behaviour.

Simulations of sequential hydrogenation experiments show, however, that the produced $H_2O : H_2O_2$ ratio for the low T_{ex} value reflects the experimental results better. Moreover, the slight decrease in H_2O_2 abundance around 10–20 minutes of experiment is best reproduced by an excess energy of 1400 K. For this reason, we will continue to use the parameters of simulation V throughout the remainder of the paper, but we will comment in the text as to how the other parameter choices affect the obtained experimental simulation results.

Finally, as mentioned in the introduction, Dulieu et $al.$[18] studied the importance of chemical desorption of reaction products through sequential O_2 hydrogenation experiments, where the amount of deposited O_2 remained in the (sub) monolayer regime. Their underlying substrate was an amorphous silicate or a graphite surface. They found substantial desorption of the formed H_2O molecules. This is caused, at least in part, by the lack of binding with the surrounding molecules. Their (sub)monolayer system is fundamentally different from our bulk studies. Using our optimised model we indeed find a similar trend, albeit only concerning the desorption of the formed OH radicals. Desorption of OH radicals for a surface covered with $\sim$0.5 ML O_2 is a factor of 2.5–5 larger than on a surface covered with $\sim$3.5 ML O_2 for T_{ex} values of 1400 and 2000 K. Desorption of H_2O is not prominent, as expected, since a water molecule atop a water surface with one H_2O neighbour has a total binding energy of 5020 K in our simulations, comparable to the value reported by Dulieu *et al.*[18]

4   Simulations of interstellar conditions

We have chosen to run simulations under interstellar conditions, using a T_{ex} of 1400 K, $B_{ice} = 10^{12}$ s^{-1} and $B_{grain} = 10^{11}$ or 10^{12} s^{-1}. The physical conditions and relative fluxes of the different species are given in Table 5 along with the values used in the laboratory. The physical conditions are chosen equal to those in Cuppen and Herbst[5] for A2, D2, E2 and F2, representing a diffuse, translucent and two dense clouds, respectively. Also, the same initial grain surface was used as in this previous study. This initial grain surface possesses a high degree of surface roughness and is therefore thought to be representative for the irregularly shaped interstellar grains. For dense cloud condition II, H-atom diffusion is rapid and H-atom desorption is slow. Simulations can therefore not be run for astrophysically relevant timescales. We can however comment on the relative contribution of the H_2O production channels, since we find that these do not change beyond the production of one monolayer. Fig. 3 shows the surface abundances for all species for two simulation runs of a typical diffuse cloud, and Fig. 4 for the translucent and dense cloud I conditions. Fig. 5 gives cross sections of the resulting three simulated grains. Tables 6 and 7 summarize the contributions of the different reaction routes to OH and H_2O formation, respectively, for all four conditions.

4.1 Diffuse clouds

The simulations of the diffuse cloud conditions within 2×10^4 years reach an almost steady state situation of a more or less empty grain with a H_2O surface

Table 5 Physical conditions and initial fluxes of the reactants in the simulations under laboratory and interstellar conditions

Parameter	Lab	Diffuse	Translucent	Dense I	Dense II
A_V (mag)	–	0.5	3	5	10
n_H (cm^{-3})	–	1×10^2	1×10^3	5×10^3	2×10^4
T_{gas} (K)	300	80	40	20	10
T_{grain} (K)	15	18	14	12	10
f_A (cm^{-2} s^{-1})					
H	2.5×10^{13}	3.2×10^6	2.3×10^5	3.2×10^4	2.3×10^4
H_2	2.5×10^{13}	–	8.0×10^6	2.9×10^7	8.0×10^7
O	–	2.4×10^2	1.7×10^3	6.1×10^3	1.7×10^4
O_2	2.5×10^{13}	–	–	–	–

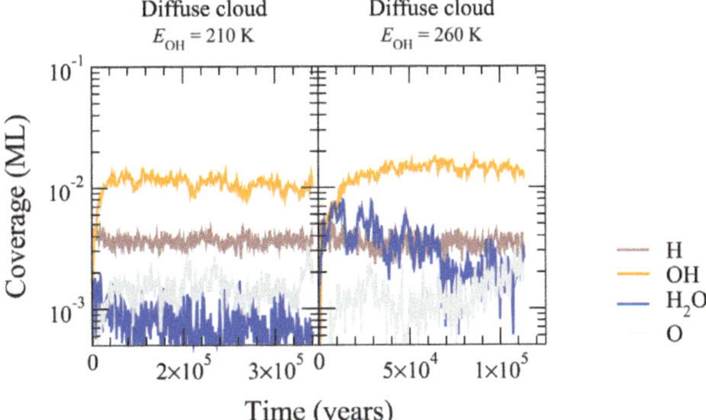

Fig. 3   Simulated surface abundances of ice species in monolayers as a function of time under two diffuse cloud conditions. Panel (a) shows the abundances for a standard run with $E_{OH} = 210$ K, and panel (b) for a run with $E_{OH} = 260$ K.

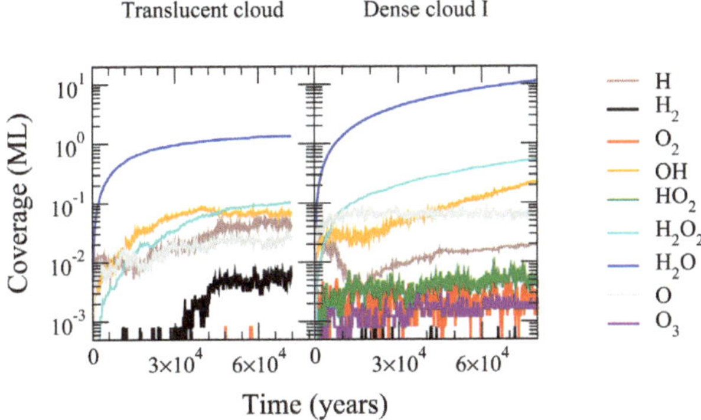

Fig. 4   Simulated surface abundances of ice species in monolayers as a function of time under (a) translucent and (b) dense cloud I conditions.

Fig. 5   Cross sections of simulated grains for the three cloud conditions: diffuse, translucent, and dense I, respectively. The grain sites are shown in black, orange is H_2, dark blue is H_2O, light blue is H_2O_2, yellow is OH, and grey is O.

abundance of 1×10^{-3} ML and an OH surface abundance of 0.01 ML. Deposited oxygen atoms are efficiently converted into H_2O through reactions with H atoms. Due to the strong radiation field for these conditions, all formed water molecules are photodissociated into OH which desorbs in 48% of the photodissociation events during the steady state. The remaining OH either reacts again with H atoms to form H_2O or is further dissociated into O and H. This cycle results in the low surface coverage of H_2O and OH; the total ice coverage is lower than the

Table 6 Contributions of the different surface reaction routes to OH formation in percentage

	T_{ex} (K)	H + HO$_2$	H + O	H + O$_3$	H + H$_2$O$_2$
Diffuse	1400	0.0	100.0	0.0	0.0
	2000	0.0	100.0	0.0	0.0
Translucent	1400	0.5	89.7	0.0	9.8
	2000	0.7	92.2	0.0	7.1
Dense I	1400	13.2	82.7	0.9	3.2
Dense II	1400	31.6	55.1	6.6	6.6

Table 7 Contributions of the different reaction routes to H$_2$O formation in percentage

	T_{ex} (K)	H + OH	H$_2$ + OH	H + H$_2$O$_2$	OH + OH	H + HO$_2$
Diffuse	1400	99.5	0.0	0.0	0.5	0.0
	2000	99.8	0.0	0.1	0.1	0.0
Translucent	1400	85.4	2.4	10.4	1.8	0.0
	2000	87.6	2.2	8.7	1.5	0.0
Dense I	1400	76.7	17.6	4.0	1.0	0.7
Dense II	1400	22.5	61.6	11.3	2.4	2.2

equivalent simulations in Cuppen and Herbst,[5] (0.05 ML) and in the previous simulations H$_2$O was the dominant species. The reason for this discrepancy is the high OH photodesorption rate. We will come back to this issue later. The species that remain on the surface fill the vacancies of the grain surface or reside near the step edges, as can be seen in the left panel of Fig. 5, which shows cross sections of the simulated grain surfaces. The photodesorption rate of OH radicals on top of a grain fluctuates as a function of time, and appears to depend on the time-dependent, local configuration of the grain. If the water molecules are mostly present in a local configuration where the photodesorption rate is lower, their oxygen atoms will repeatedly undergo the formation–dissipation cycle, resulting in a lower average desorption rate. Once these atoms have desorbed, the time-dependent photodesorption rate increases again until such a favourable configuration is again obtained. However, considering all the dust grains in a typical diffuse cloud, these fluctuations will average out.

The photodesorption efficiency (amount of hot desorbed OH radicals divided by the number of H$_2$O photodissociation events) fluctuates between 28 and 80% with a median value of 48%. This is significantly higher than the values obtained by Andersson et al.:[15] 9 and 2% for OH desorption in the first monolayer of crystalline and amorphous ice, respectively. In their case, the efficiency also depends on the local configuration. One would expect the desorption rate on top of a grain to be higher than from bulk ice, because excess energy is probably easier to transfer to a molecular environment than to a more brittle grain surface. For this reason we ran additional simulations with a B_{grain} parameter of 10^{11} s^{-1}. This B_{grain} parameter is used for energy dissipation of species that do not have any molecules in their immediate vicinity. For these simulations the photodesorption efficiency of OH upon H_2O photodissociation behaves in a very similar manner (51%). The steady state coverages of OH and H_2O remain the same in this case, as can be seen in Fig. 3. Our values may be much higher, but concern an almost empty grain surface. The binding energy of an OH radical on top of a surface is most likely lower than that of an OH embedded in a water layer. We ran some additional simulations in which we increased the binding energy of OH with the surface from $E = 210$ K to $E = 260$ K. In this case the photodesorption efficiency of OH ranges between 5 and 17% with an average efficiency of 9%, which is comparable to the results of Andersson et $al.$[15] The steady state surface abundance is now 6×10^{-3} ML for H_2O and 0.02 ML for OH. Using this increased binding energy, the experimental simulations remain very similar and still reproduce the experimental observations. Finally, another parameter that influences the photodesorption efficiency is T_{ex}. Increasing its value to 2000 K also increases the photodesorption. In the following, all discussed simulations are parametrised according to $T_{ex} = 1400$ K, $B_{ice} = 10^{12}$ s^{-1} and $B_{grain} = 10^{11}$ s^{-1}, and using the standard OH binding energy.

During the simulations, not only is H_2O formed but also H_2. We see here that because of our rough surface, the residence time of the hydrogen atoms is not the rate limiting step, but rather the sticking probability of the hydrogen atoms to the surface. Since under these conditions the gas phase has a relatively high temperature of 80 K, the sticking probability of H atoms to the surface is relatively low ($\sim$33%), but more than 47% of all H atoms remaining on the surface react to form H_2, leading to a H_2 formation efficiency of 15%. This is in good agreement with the results of Chang et $al.$[27], who found a recombination efficiency of $\sim$10% for temperatures as high as 21.8 K, which is enough to account for significant H_2 production in diffuse interstellar clouds. Again we can observe a time-dependent H_2 formation rate in the beginning of the simulation when the ice develops. This is because of similar arguments as for the time-dependent OH photodesorption rate. E values for H binding to the bare grain lower than 105 K result in lower H_2 formation rates, mainly because of the reduced sticking. Since we know observationally that H_2 has to be formed under these conditions, we have chosen to use this value for E. Lower values of E result in a similar production of H_2O and all other species, except H_2.

4.2   Translucent clouds

In general we observe a similar behaviour to the results of Cuppen and Herbst,[5] $i.e.$, the competition between photodissociation and surface reaction is key. The new photodissociation rates, following van Dishoeck et $al.$,[21] are higher than those previously used, but attenuate faster with visual extinction, which results in lower rates at $A_V = 3$ mag.

In our present simulations the growing ice layer reaches a thickness of slightly more than a monolayer after 7×10^4 years, where a steady state seems to have been reached. Cuppen and Herbst[5] obtained surface coverages of a few monolayers under equivalent circumstances. They found that the bottom layers are heavily processed and contain more oxygen-rich species like OH, O_2, and O_3, since these layers lose some of their hydrogen due to photoprocessing. Here, we stay around the monolayer regime and we therefore do not see the same effect. One of the reasons for this is the higher photodesorption rate. The ice structure found in the previous work was very porous due to the heavy processing. The middle panel in Fig. 5 depicts a cross section of the ice surface, where only some small pores can be found. The lack of strong porosity upon processing is in agreement with experimental work on ion irradation of water ices by Palumbo,[28] and is probably caused by the exothermicity of the photoproducts that allows for restructuring.

The contributions of the various reaction routes (Tables 6 and 7) show that most of the OH is formed through the reaction O + H, although the dissociation of H_2O also plays a role (see below). Even though the amount of H_2 is high, the H_2 + OH route is less efficient in the formation of water ice since the reaction is not barrierless. We observe H_2O_2 as a result of the reaction route OH + OH. The follow-up reaction H + H_2O_2 is consequently also more important for both OH and H_2O formation than in the previous 2007 work.[5] The H_2O production rate decreases with increasing time. We also observe that the contributions of the various reaction routes change with time: before the steady state is reached, OH is formed only through H + O, and H_2O only through H + OH. At later times, however, the contribution of H + H_2O_2 becomes stronger. Both the evaporation and dissociation rate of the various species also increase up to the point where the total surface coverage is $\sim$0.6 ML, and then the rates stabilize.

To distinguish between desorption originating from photodissociation events and from reaction heat is not trivial, hence we will only comment on it qualitatively. The largest part of the desorbing OH radicals originates from photodissociation of water. Changing the exothermicity to 2000 K leads to a larger amount of chemical desorption events, whereas increasing the binding energy of OH to the surface decreases this amount. Even though the relative contributions of the various reaction routes do not change upon changing these parameters, the final amount of (water) ice produced does. An increased desorption, regardless of the origin, decreases the ice thickness since fewer radicals are available for reaction.

4.3   Dense clouds

The right panel of Fig. 4 shows the surface abundance of a water covered grain for dense cloud condition I. For both dense cloud conditions I and II, water ice is efficiently formed, more efficiently than in the Cuppen and Herbst[5] simulations. The main reasons for this are that H_2 does not stick to the surface as easily, which prevented many surface reactions in the earlier work, and that the ice is more compact because of the use of excess energy. The latter can be seen in the right panel of Fig. 5, where a very compact ice is formed; much more compact than that in Cuppen and Herbst,[5] where towers of H_2O were formed, typical of ballistic deposition.[29] In the present simulations, the formed species have some momentum to move around and find a more favourable binding site, which leads to a smoothing of the formed ice. Oba *et al.*[30] indeed found that ice that is formed on a surface shows more characteristics of a compact ice than ice that is deposited at similar low surface temperatures.

Under dense cloud conditions, the O_2 route becomes increasingly important, since the O : H flux ratio increases on going from diffuse, to translucent, to dense I and finally to dense II conditions. This can be seen in Table 6. The contribution of the H + HO_2 reaction to the formation of the hydroxyl radical increases. The intermediate HO_2 is formed from H + O_2. At the same time, we see that the related reaction O_3 + H increases as well. A clear difference with the Cuppen and Herbst[5] simulations is that now the O_2 channel does not exclusively proceed with H_2O_2 as an intermediate for H_2O formation. Table 7 shows that its contribution increases with density (O : H flux ratio), but not to the same extent as in the previous simulations. In the present work, most of the OH formed through the O_2 formation channel is transformed into water through the follow-up reactions OH + H and OH + H_2. Since the first has no barrier, this reaction is preferred under most circumstances. However, under dense cloud II conditions water ice is predominantly formed through the latter reaction, since the H_2 flux becomes much higher than the H flux and the surface residence time of H_2 increases at the same time because of the low surface temperature. Indeed, the main differences between the two dense cloud conditions are the density and the grain temperature. The percentage of intermediate species is higher for dense cloud II than for dense cloud I. This is probably because H atoms cannot convert these intermediate species into stable species as easily under these conditions. A lower temperature results in less H diffusion, and the H-atom flux relative to the H_2 flux and O-atom flux is lower than for dense cloud I.

Photodesorption is not important here due to the high visual extinctions involved. Part of the formed water ice is returned to the gas phase, however, by chemical desorption. The 1400 K excess energy can be used for desorption when two reaction products are formed, and therefore we see that chemical desorption is more efficient for the dense cloud II simulations, since the contribution of reactions with two products (H_2 + OH and H + H_2O_2) is larger here. Roughly, the desorption of OH is 10 and 29% respectively, and for H_2O it amounts to 5 and 13%.

In our current model, it is not possible for the intermediate species HO_2 and H_2O_2 to desorb, although these species have been observed in the gas phase.[31–33] Desorption mechanisms other than chemical desorption could be responsible for these observations. These mechanisms include photodesorption through cosmic ray photons or cosmic ray desorption, where a cosmic ray hits a grain and results in the desorption of most of the mantle material.[34,35] This rare stochastic event is rather hard to simulate considering a single grain, but integrated over a large cloud with many dust grains the effect can be significant. Another possible desorption mechanism that is not included here is the so-called kick-out mechanism, where a cold species desorbs as a result of being hit by a hot species. This was observed in molecular dynamics simulations to be important.[15–17] Experimentally, this mechanism was confirmed by time-of-flight measurements of the desorption fragments of water ice photodissociation.[36] Similar conclusions were drawn for the photodesorption of N_2 in the presence of CO, where the excited CO kicks out N_2 which has no absorbance in the applied photon range.[37]

4.4 Comparison to observations and other models

Much of the work done since Cuppen and Herbst[5] has focussed on the importance of the molecular oxygen route for water formation. Here we found that this route has the largest implications in dense clouds, as expected. A particularly interesting region in which to study the influence of O_2 in dense clouds is ρ Oph A, where O_2, HO_2 and H_2O_2 have been detected in recent years.[31–33,38,39] The derived abundances with respect to hydrogen are 5×10^{-8}, 1×10^{-10} and 1×10^{-10}, respectively. The hybrid moment equation approach was applied by Du and

Parise[33] to model the production of interstellar H_2O_2 on the surface of dust grains. They ran a model for a range of different physical conditions and found reasonable agreement with observations for a time of 6×10^5 years, with abundance ratios of the three species of $60 : 3 : 1$ in the gas phase for $O_2 : HO_2 : H_2O_2$. They used the chemical desorption mechanism introduced by Garrod et al.[19] and found that the exact setting of the a parameter has a large influence on the gas phase composition. Furthermore, the reaction $H + H_2O_2$ has a lower barrier than that adopted here. This factor could increase their H_2O_2 abundance.

In our model we do not calculate any gas phase abundances, but as outlined by Öberg et al.[40] the gas phase abundance seems to reflect the composition of the co-existing ice mantles (see also Fig. 1 in Du and Parise[33]). The dense cloud simulation runs do indeed produce O_2, HO_2 and H_2O_2 next to H_2O, where we do not see that the O_2 abundance is much higher than for HO_2 and H_2O_2, which is likely caused by the much lower densities considered. The physical parameters determined for ρ Oph A ($T = 21$ K and $n_H = 6 \times 10^5$ cm^{-3}) are not typical for dense clouds and lead to an atomic oxygen abundance of $3 \times 10^{-4} n_H$, corresponding to a flux of 7.5×10^5 atoms cm^{-2} s^{-1}. Comparing this to our values of 6.1×10^3 and 1.7×10^4 atoms cm^{-2} s^{-1}, it is clear that much more O_2 and HO_2 can be produced. The results of our optimised model run with parameter settings as close as possible to those of ρ Oph A show the following behaviour for the ice abundances: $O_3 \gg O_2 > HO_2 > H_2O_2$. If we assume that the species H_2O_2 and HO_2 are predominantly formed on the grain surface and that the gas phase is a good reflection of the ice composition, our $H_2O_2 : HO_2$ abundance ratio of 0.6 is in good agreement with the observed ratio presented by Parise et al.[32] of 1.

As compared to the previous simulations of Cuppen and Herbst,[5] our final amount of produced H_2O is similar and varies as a function of A_V. Since the 2007 results were in good agreement with the observations by Whittet et al.[41] of water ice in Taurus dark clouds, in terms of the water ice threshold value of $A_V = 3.2$ K and linear growth of the column density of water ice above this value. The present results show a similar agreement. The main differences between the studies are the formation routes for water ice. We expect this to have consequences for the deuterium fractionation of the ice species, although this was not explored in the present paper. Since some of the channels such as $H_2 + OH$ posses a barrier which can be overcome through quantum chemical tunneling, the formation of H_2O will be favoured over HDO. For other channels this might have the reverse trend. Cazaux et al.[20] included deuteration reactions in their water ice network to study this effect, but unfortunately there is not enough experimental and quantum chemical evidence available to back up this network and to make it predictive.

5   Discussions and conclusions

In the present paper, exothermicity of reactions is used to increase the momentum of the final products by increasing the hopping and desorption rates. Garrod et al.[42] showed that chemical desorption can play an important role in explaining the observed abundance of different gas phase chemical species. Later, Cazaux et al.[20] and Dulieu et al.[18] made similar conclusions. Here we again see the same effect. But exothermicity does not only lead to desorption but can also lead to a more compact ice, which has an effect on many diffusion properties and can allow reactants to meet. In our simulations, we indeed see this compaction.

Here the exothermicity is only considered for reactions with two or more reaction products, and the excess energy is only employed for diffusion and desorption and not to overcome chemical reactions. Also allowing single reaction products to desorb with a higher probability leads to too much desorption under interstellar conditions: even in dense clouds hardly any ice is formed. Applying the excess energy to overcome reactions as well leads to too much $H + H_2O_2$ under experimental conditions. If the energy is not partitioned between the two products according to their relative masses, the desorption rate of OH upon H_2O photodissociation becomes near unity. Our set-up in the current paper therefore best describes experimental and astronomical observations as well as molecular dynamics simulations. Moreover, we believe this to be a rather accurate description of the underlying physics and chemistry, where the exothermicity is transferred into kinetic energy. Because of conservation of momentum this can only be done for reactions with multiple products, and the energy is distributed considering their relative masses. How excess energy can be applied to overcome chemical barriers is not clear and this mechanism is currently missing in our models. With this discussion we hope to trigger new dedicated studies and discussions on this topic and on the role of exothermicity in general. The role of a kick-out mechanism also remains unexplored in this work.

Acknowledgements

H.M.C. is grateful for support from the VIDI research program 700.10.427, which is financed by The Netherlands Organisation for Scientific Research (NWO) and from the European Research Council (ERC-2010-StG, grant agreement no. 259510-KISMOL). T.L. is supported by the Dutch Astrochemistry Network financed by The Netherlands Organisation for Scientific Research (NWO).

References

1 M. J. Drake, *Meteorit. Planet. Sci.*, 2005, **40**, 519.

2 K. Muralidharan, P. Deymier, M. Stimpfl, N. H. de Leeuw and M. J. Drake, *Icarus*, 2008, **198**, 400–407.

3 D. Bockelée-Morvan, D. C. Lis, J. E. Wink, *et al.*, *Astron. Astrophys.*, 2000, **353**, 1101–1114.

4 E. F. Dishoeck, E. Herbst and D. A. Neufeld, *Chem. Rev.*, 2013, **113**, 9043 and references therein.

5 H. M. Cuppen and E. Herbst, *Astrophys. J.*, 2007, **668**, 294–309.

6 N. Miyauchi, H. Hidaka, T. Chigai, A. Nagaoka, N. Watanabe and A. Kouchi, *Chem. Phys. Lett.*, 2008, **456**, 27–30.

7 E. Matar, E. Congiu, F. Dulieu, A. Momeni and J. L. Lemaire, *Astron. Astrophys.*, 2008, **492**, L17–L20.

8 S. Ioppolo, H. M. Cuppen, C. Romanzin, E. F. van Dishoeck and H. Linnartz, *Astrophys. J.*, 2008, **686**, 1474–1479.

9 Y. Oba, N. Watanabe, T. Hama, K. Kuwahata, H. Hidaka and A. Kouchi, *Astrophys. J.*, 2012, **749**, 67.

10 T. Lamberts, H. M. Cuppen, S. Ioppolo and H. Linnartz, *Phys. Chem. Chem. Phys.*, 2013, **15**, 8287.

11 S. Ioppolo, H. M. Cuppen, C. Romanzin, E. F. van Dishoeck and H. Linnartz, *Phys. Chem. Chem. Phys.*, 2010, **12**, 12065.

12 H. M. Cuppen, S. Ioppolo, C. Romanzin and H. Linnartz, *Phys. Chem. Chem. Phys.*, 2010, **12**, 12077.

13 F. Dulieu, *IAU Symposium*, 2011, pp. 405–415.

14 H. M. Cuppen, L. J. Karssemeijer and T. Lamberts, *Chem. Rev.*, 2013, **113**, 8840.

15 S. Andersson, A. Al-Halabi, G.-J. Kroes and E. F. van Dishoeck, *J. Chem. Phys.*, 2006, **124**, 064715.

16 C. Arasa, S. Andersson, H. M. Cuppen, E. F. van Dishoeck and G. Kroes, *J. Chem. Phys.*, 2010, **132**, 184510.

17 C. Arasa, S. Andersson, H. M. Cuppen, E. F. van Dishoeck and G. J. Kroes, *J. Chem. Phys.*, 2011, **134**, 164503.

18 F. Dulieu, E. Congiu, J. Noble, S. Baouche, H. Chaabouni, A. Moudens, M. Minissale and S. Cazaux, *Sci. Reports*, 2013, **3**, 1338.

19 R. Garrod, I. H. Park, P. Caselli and E. Herbst, *Faraday Discuss.*, 2006, **133**, 51.

20 S. Cazaux, V. Cobut, M. Marseille, M. Spaans and P. Caselli, *Astron. Astrophys.*, 2010, **522**, A74.

21 E. F. van Dishoeck, B. Jonkheid and M. C. van Hemert, *Faraday Discuss.*, 2006, **133**, 231.

22 R. Atkinson, D. L. Baulch, R. A. Cox, J. N. Crowley, R. F. Hampson, R. G. Hynes, M. E. Jenkin, M. J. Rossi and J. Troe, *Atmos. Chem. Phys.*, 2004, **4**, 1461–1738.

23 D. V. Shalashilin and B. Jackson, *J. Chem. Phys.*, 1998, **109**, 2856–2864.

24 H. M. Cuppen and L. Hornekær, *J. Chem. Phys.*, 2008, **128**, 174707.

25 A. P. J. Jansen, *Comput. Phys. Commun.*, 1995, **86**, 1.

26 E. A. Bergin and M. Tafalla, *Annu. Rev. Astron. Astrophys.*, 2007, **45**, 339–396.

27 Q. Chang, H. M. Cuppen and E. Herbst, *Astron. Astrophys.*, 2005, **434**, 599–611.

28 M. E. Palumbo, *Astron. Astrophys.*, 2006, **453**, 903–909.

29 G. A. Kimmel, Z. Dohnálek, K. P. Stevenson, R. S. Smith and B. D. Kay, *J. Chem. Phys.*, 2001, **114**, 5295–5303.

30 Y. Oba, N. Miyauchi, H. Hidaka, T. Chigai, N. Watanabe and A. Kouchi, *Astrophys. J.*, 2009, **701**, 464–470.

31 P. Bergman, B. Parise, R. Liseau, B. Larsson, H. Olofsson, K. M. Menten and R. Güsten, *Astron. Astrophys.*, 2011, **531**, L8.

32 B. Parise, P. Bergman and F. Du, *Astron. Astrophys.*, 2012, **541**, L11.

33 F. Du and B. Parise, *Astron. Astrophys.*, 2011, **530**, A131.

34 T. I. Hasegawa and E. Herbst, *Mon. Not. R. Astron. Soc.*, 1993, **261**, 83–102.

35 E. Herbst and H. M. Cuppen, *Proc. Natl. Acad. Sci. U. S. A.*, 2006, **103**, 12257–12262.

36 A. Yabushita, T. Hama, M. Yokoyama, M. Kawasaki, S. Andersson, R. N. Dixon, M. N. R. Ashfold and N. Watanabe, *Astrophys. J. Lett.*, 2009, **699**, L80–L83.

37 K. I. Öberg, E. F. van Dishoeck and H. Linnartz, *Astron. Astrophys.*, 2009, **496**, 281–293.

38 B. Larsson, R. Liseau, L. Pagani, P. Bergman, P. Bernath, N. Biver, J. H. Black, R. S. Booth, V. Buat, J. Crovisier, C. L. Curry, M. Dahlgren, P. J. Encrenaz, E. Falgarone, P. A. Feldman, M. Fich, H. G. Florén, M. Fredrixon, U. Frisk, G. F. Gahm, M. Gerin, M. Hagström, J. Harju, T. Hasegawa, Å. Hjalmarson, L. E. B. Johansson, K. Justtanont, A. Klotz, E. Kyrölä, S. Kwok, A. Lecacheux, T. Liljeström, E. J. Llewellyn, S. Lundin, G. Mégie, G. F. Mitchell, D. Murtagh, L. H. Nordh, L.-Å. Nyman, M. Olberg, A. O. H. Olofsson,

G. Olofsson, H. Olofsson, G. Persson, R. Plume, H. Rickman, I. Ristorcelli, G. Rydbeck, A. A. Sandqvist, F. V. Schéele, G. Serra, S. Torchinsky, N. F. Tothill, K. Volk, T. Wiklind, C. D. Wilson, A. Winnberg and G. Witt, *Astron. Astrophys.*, 2007, **466**, 999–1003.

39 R. Liseau, P. F. Goldsmith, B. Larsson, L. Pagani, P. Bergman, J. Le Bourlot, T. A. Bell, A. O. Benz, E. A. Bergin, P. Bjerkeli, J. H. Black, S. Bruderer, P. Caselli, E. Caux, J.-H. Chen, M. de Luca, P. Encrenaz, E. Falgarone, M. Gerin, J. R. Goicoechea, Å. Hjalmarson, D. J. Hollenbach, K. Justtanont, M. J. Kaufman, F. Le Petit, D. Li, D. C. Lis, G. J. Melnick, Z. Nagy, A. O. H. Olofsson, G. Olofsson, E. Roueff, A. Sandqvist, R. L. Snell, F. F. S. van der Tak, E. F. van Dishoeck, C. Vastel, S. Viti and U. A. Yıldız, *Astron. Astrophys.*, 2012, **541**, A73.

40 K. I. Öberg, S. Bottinelli and E. F. van Dishoeck, *Astron. Astrophys.*, 2009, **494**, L13–L16.

41 D. C. B. Whittet, P. A. Gerakines, J. H. Hough and S. S. Shenoy, *Astrophys. J.*, 2001, **547**, 872–884.

42 R. T. Garrod, V. Wakelam and E. Herbst, *Astron. Astrophys.*, 2007, **467**, 1103–1115.

Faraday Discussions

PAPER

Characterizing the chemical pathways for water formation – a deep search for hydrogen peroxide†

Bérengère Parise,[*ab] Per Bergman[c] and Karl Menten[b]

Received 29th November 2013, Accepted 11th February 2014

DOI: 10.1039/c3fd00115f

In 2011, hydrogen peroxide (HOOH) was observed for the first time outside the solar system (Bergman *et al.*, *Astron. Astrophys.*, 2011, **531**, L8). This detection appeared *a posteriori* to be quite natural, as HOOH is an intermediate product in the formation of water on the surface of dust grains. Following up on this detection, we present a search for HOOH in a diverse sample of sources in different environments, including low-mass protostars and regions with very high column densities, such as Infrared Dark Clouds (IRDCs). We do not detect the molecule in any other source than Oph A, and derive 3σ upper limits for the abundance of HOOH relative to H_2 lower than that in Oph A for most sources. This result sheds a different light on our understanding of the detection of HOOH in Oph A, and shifts the question of why this source seems to be special. Therefore we rediscuss the detection of HOOH in Oph A, as well as the implications of the low abundance of HOOH, and its similarity with the case of O_2. Our chemical models show that the production of HOOH is extremely sensitive to temperature, and is favored only in the range 20–30 K. The relatively high abundance of HOOH observed in Oph A suggests that the bulk of the material lies at a temperature in the range 20–30 K.

1   Introduction

Water has long been known to exist in star-forming regions, both in the gas phase[1] and in the form of ices.[2] The Herschel Space Observatory has recently shown that water is present virtually everywhere it was looked for, and showed detections in new environments such as cold prestellar cores[3,4] and protoplanetary disks[5] (see, for example, the results of the WISH Key Project[6]).

[a]*School of Physics and Astronomy, University of Cardiff, The Parade, Cardiff, CF24 3AA, UK. E-mail: Berengere.Parise@astro.cf.ac.uk; Fax: +44 (0)29 208 74056; Tel: +44 (0)29 208 74649*

[b]*Max-Planck-Institut für Radioastronomie, Auf dem Hügel 69, 53121 Bonn, Germany*

[c]*Onsala Space Observatory, Chalmers University of Technology, 439 92 Onsala, Sweden*

† Based on observations with the Atacama Pathfinder EXperiment (APEX) telescope. APEX is a collaboration between the Max Planck Institute for Radio Astronomy, the European Southern Observatory, and the Onsala Space Observatory.

The relatively low abundance of water in the gas phase of cold regions, as well as its high abundance in the ices coating dust grains, has led to the hypothesis that water forms very efficiently on the surface of dust grains. Recent laboratory experiments have investigated the details of the different formation pathways.[7-10] In these experiments, water was shown to form through three different routes, starting with the hydrogenation of O, O_2 and O_3, respectively. The relative importance of the three routes in the different environments is not yet fully understood, but can have a huge impact on the resulting composition of the formed ices.

HOOH is formed on grains as a precursor in the formation of water by the O_2 route:

$$O_2 + H \rightarrow HO_2 \tag{1}$$

$$HO_2 + H \rightarrow HOOH \tag{2}$$

$$HOOH + H \rightarrow H_2O + OH \tag{3}$$

Reactions (1) and (2) on dust grains were hypothesized by Allen et $al.$[11] based on the related gas-phase reactions, and the addition of reaction (3) to the models was proposed by Tielens et $al.$[12] Laboratory experiments recently showed that these reactions indeed proceed.[8,10]

In view of the ubiquity of water, and the central role of HOOH in its formation, it came somewhat as a surprise that HOOH was only recently detected for the first time in the interstellar medium,[13] namely in ρ Oph A. Subsequent detailed chemical modelling, solving the gas-phase and grain surface chemistry with the Hybrid Moment Equation (HME) method,[14] showed that HOOH is indeed a major precursor of H_2O in that source.[15] Further confidence was brought to the modelling by the (predicted) detection of HO_2,[16] again raising the question of why HOOH had not been detected earlier. No dedicated observational search had been reported previously though, although the frequencies were long known,[17,18] so the lack of detection may simply be attributed to the lack of search efforts.

Following the detection of HOOH towards Oph A, we intend here to clarify the observational picture, and present a search for HOOH in a sample of ten sources of different natures and in different environments.

2   Observations

2.1   Technical details

Using the APEX telescope,[19] we targeted the HOOH $3_{0,3}-2_{1,1}$ transition towards a sample of sources. The energy of the upper level for this transition is 31 K. This line was the brightest of a set of transitions observed towards ρ Oph A.[13] The sources are listed in Table 1 and described in detail in Section 2.2. The observations were made on August 6th and 7th 2011, under acceptable (PWV = 1–3 mm) weather conditions. The APEX1 receiver[20] was tuned at the frequency of the HOOH line (219.166860 GHz) and was connected to the XFFTS spectrometer.[21] The main beam efficiency at this frequency is 0.75, and the angular resolution is $28''$ (FWHM).

Table 1 Source list

Source	RA (J2000)	DEC (J2000)	Distance (pc), and reference
ρ Oph-A SM1	16:26:27.2	−24:24:04	120[23]
ρ Oph-B2-MM8	16:27:28.0	−24:27:06.9	120[23]
G15.01-0.67	18:20:21.22	−16:12:42.2	2100[24]
G018.82-00.28MM1	18:25:56.1	−12:42:48	4800[25]
G018.82-00.28MM4	18:26:15.5	−12:41:32	4800[25]
G028.53-00.25MM1A	18:44:18.08	−03:59:34.33	5700[25]
NGC6334I(N)	17:20:54.63	−35:45:08.9	1600[26]
G1.6-0.025	17:49:43.6	−27:33:52	8000[27,28]
NGC1333-IRAS4A	03:29:10.3	+31:13:32	235[29]
L1527	04:39:53.9	+26:03:10	140[30]
RCrA-IRS7B	19:01:56.4	−36:57:27	130[31]

2.2 Source sample

In the following we give a brief description of our source sample (Table 1). The sources have been selected based on different criteria. An important condition was high column densities, to increase the sensitivity of the detection. Some sources were selected because they share some characteristics with Oph A: either proximity (Oph B) to test the role of the local conditions, or average temperature conditions around 20–25 K (IRDC sources, envelopes of low-mass protostars). Finally, to span as many different chemical environments as possible, we also selected one prototypical source representing each of the following chemical classes: hot cores (NGC6334), hot corinos (NGC1333-IRAS4A), warm carbon chain chemistry (L1527), and "hot chemistry without hot cores"[22] (G1.6).

Relevant information (dust and gas temperatures, as well as H_2 column densities) for our sample is listed in Table 2, and references for these values are given in the following text. We have converted H_2 column densities found in the literature to correspond to the value averaged in a 28″ beam, similar to our HOOH observations. The conversion is discussed source by source. For comparison, we also include Oph A, where HOOH was detected.[13]

ρ Oph-B2-MM8 is located in the same dark cloud system as ρ Oph A. It is the brightest clump in the Oph B2 region at 1.3 mm, as observed by Motte *et al.*[32] These authors estimated a mass of 1.5 $M_\odot$, assuming a dust temperature of 12 K.[33] The peak column density in an 11″ beam for the Oph B2 region is 4.1×10^{23} cm^{-2} (Motte *et al.*[32]), and that should correspond to the value centered on MM8. They quote a source size of 4000 × 4000 AU, which corresponds to 25″ at the distance they adopted at that time (160 pc). We can therefore extrapolate the H_2 column density in a 28″ beam to be 2.2×10^{23} cm^{-2}.

G15.01-0.67 is located in the M17 nebula (also known as the Omega Nebula). It is a SCAMPS source (high-mass pre/protocluster clump detected in the SCUBA Massive Pre/Protocluster Core Survey, Thompson *et al.*[34]) studied by Pillai *et al.*[35] It was detected on the edge of one of the SCUBA fields from Thompson *et al.*[34] The $C^{18}O$ excitation temperature is 32 K, and the NH_3 rotational temperature is 26 K.[35] The total $N(H_2)$ column density is 16.6×10^{23} cm^{-2}, as derived from the 850 μm dust continuum flux smoothed to a resolution of 20″.[35] In the extreme case of a

Table 2 Source properties

Source	v_{lsr} (km s^{-1})	Linewidth (km s^{-1})	T_{dust} (K)	T_{gas} (K)	T_{H_2CO}[c] (K)	$N(H_2)$[e] (cm^{-2})
ρ Oph-A SM1				24	33 ± 3	1.5 × 10^{23}[f]
ρ Oph-B2-MM8	3.9[a]	1.1 ± 0.1[a]	12		≤16	2.2 × 10^{23}
G15.01-0.67	18.4[b]	4.9 ± 0.1[a]		26–32	64 ± 11	1.1 × 10^{24}
G018.82-00.28MM1	40.4[a]	5.7 ± 0.1[a]	26		61 ± 10	1.6 × 10^{23}
G018.82-00.28MM4	64.9[a]	4.4 ± 0.3[a]	17		≤29	9.3 × 10^{22}
G028.53-00.25MM1	86.3[a]	6.34 ± 0.2[a]	17		57 ± 9	3.6 × 10^{23}
NGC6334I(N)	−3.8[b]	5.7 ± 0.1[b]	30–35		160 ± 67	8.5 × 10^{23}
G1.6-0.025	51.7[b]	5.3 ± 0.3[b]		60	195 ± 107	4 × 10^{22}
NGC1333-IRAS4A	7.0[a]	≤3.3[d]		24	—[d]	1.3 × 10^{23}
L1527	5.9[a]	1.1 ± 0.1[a]		16	23 ± 2	4.1 × 10^{22}
RCrA-IRS7B	5.7[a]	2.5 ± 0.2[a]		22–40	47 ± 6	5.9 × 10^{22}

[a] Measured on the H_2CO 218.222 GHz low-energy line. [b] Measured on the H_2CO 218.475 GHz line, because the low-energy line is double peaked (NGC6334I(N)) or shows other signs of high opacity (G1.6-0.025). [c] H_2CO rotational temperature deduced from the three observed H_2CO lines (this study). Upper limits are given when the high-energy lines are not detected. The values tabulated when the three lines are detected should also be considered as upper limits, as the low-energy line is likely to be optically thick. [d] The lines have a non-Gaussian shape, the emission being dominated by the outflow. We therefore derive only an upper limit on the linewidth for the envelope, and refrain from giving a rotational temperature from H_2CO. [e] Beam-averaged H_2 column densities in the 28″ APEX beam. [f] This value refers to the column density of the whole cloud, as traced by the C^{18}(3–2) line, corrected for opacity.[38] The central core, traced from H_2CO and CH_3OH lines, accounts for a column density of 3×10^{22} cm^{-2}. This latter value was used to derive the detected HOOH abundance.[13]

point source, the column density should be reduced by a factor of 2 to obtain it within a 28″ beam. If the source has a 20″ size, the decrease would be by a factor of 1.5, and this is what we adopt here.

G018.82-00.28 MM1 and MM4 are the two most massive cores within the IRDC MSXDC G018.82-00.28. They were detected and characterised by Rathborne et al.[36] using the MAMBO 1.2 mm continuum receiver. Their angular sizes are 23″ and 31″, respectively.[36] Their dust temperatures were derived from broadband SEDs (26 and 17 K, respectively) by Rathborne et al.,[37] and their masses were then determined from the 1.2 mm flux (495 M$_\odot$ and 228 M$_\odot$, respectively). The H_2 column densities in the MAMBO 11″ beam are 3.2 × 10^{23} cm^{-2} and 1.5 × 10^{23} cm^{-2}, respectively. Taking into account their angular sizes, this leads to column densities of 1.6 × 10^{23} cm^{-2} and 0.93 × 10^{23} cm^{-2}, respectively, when averaged in a 28″ beam. We find that the LSR velocity from the MM1 core as measured from the H_2CO lines (40.4 km s^{-1}) is surprisingly different from that of the rest of the complex, as measured by Rathborne et al.[36] (65.8 km s^{-1}). This could be a sign that the MM1 core, which already appears spatially separated from the rest of the complex on MSX images,[36] is actually at a different distance than the rest of the complex.

G028.53-00.25 MM1 is the most massive core within the IRDC G028.53-00.25, another IRDC studied by Rathborne et al.[37] They derive a temperature of 17 K from their broadband SED study, a mass of 1088 M$_\odot$ from the 1.2 mm data, and a H_2 column density of 5.6 × 10^{23} cm^{-2} averaged in the MAMBO 11″ beam. Its angular size is 33″.[36] This translates into a H_2 column density of 3.6 × 10^{23} cm^{-2} in a 28″ beam.

NGC6334I(N) is a star formation site located north of the more developed NGC6334I ultracompact HII region and molecular core. Sandell[39] mapped the region at five wavelengths in the range 350 μm to 1.3 mm. Although the temperature cannot be constrained independently from the dust opacity at these wavelengths, they found plausible fits in the temperature range $T_{dust} = 30\text{--}35$ K. This is consistent with the temperature estimates of the extended gas by Kuiper *et al.*[40] (from low-energy NH_3 lines) and McCutcheon *et al.*[41] (from ^{12}CO lines). Note that Beuther *et al.*[42] found evidence for higher gas temperatures in the compact ($\sim$2–3$''$) core by means of NH_3 (5,5) and (6,6) inversion lines, but we are likely not sensitive to this hotter gas. We derive $N(H_2) = 4.2 \times 10^{24}$ cm^{-2} averaged on the source from the average H_2 density and the source size ($11'' \times 8''$) tabulated by Sandell.[39] This translates into 8.5×10^{23} cm^{-2} averaged in a 28$''$ beam.

G1.6-0.025 is one of the molecular clouds within the Central Molecular Zone (CMZ) surrounding the galactic center. It has been studied in detail by Menten *et al.*[27] We targeted here position 3 from Menten *et al.*,[27] for which they derived kinetic temperatures from a detailed analysis of CH_3OH excitation. The extended cloud, whose emission peaks at $v_{lsr} = 51$ km s^{-1}, was shown to have a temperature of 60 K, and a H_2 column density of 4×10^{22} cm^{-2} (from ^{13}CO measurements, in a 2$'$ beam).[27] We assume here that this extended gas has the same column density at smaller scales.

NGC1333-IRAS4A is a Class 0 low-mass protostar located in the NGC1333 complex in the Perseus molecular cloud. The distance of this complex is 235 pc.[29] At this distance, 28$''$ represents 6580 AU. We compute the H_2 column density averaged in this beam from the density power law derived by Kristensen *et al.*[43] to be 1.3×10^{23} cm^{-2}. Maret *et al.*[44] and Maret *et al.*[45] studied the emission of formaldehyde (H_2CO) and methanol (CH_3OH) towards this source, and derived rotational temperatures of 24 K for both species.

L1527 is a young low-mass protostar located in the Taurus-Auriga molecular cloud, at a distance of 140 pc.[30] Its evolutionary stage is still debated.[46] As for NGC1333-IRAS4A, we compute the H_2 column density from the density power law of Kristensen *et al.*[43] Maret *et al.*[44] derived a rotational temperature of 16 K from the study of H_2CO lines.

RCrA-IRS7B is a Class 0 protostar located in the R Coronae Australis complex, at a distance of 130 pc.[31] Several molecular lines were observed using APEX towards this source by Schöier *et al.*[47] The kinetic temperatures derived from H_2CO and CH_3OH lines are 40 and 22 K, respectively.[47] Their study derives an H_2 column density of 3×10^{23} cm^{-2} in the APEX2a 18$''$ beam, based on the $C^{34}S$ line. Lindberg and Jørgensen[48] derived the density profile of the source based on SCUBA and Herschel continuum data. Computation of the column density into a 18$''$ beam using their profile leads to 8×10^{22} cm^{-2}, which, compared to the value derived from the $C^{34}S$ line, gives an idea of the uncertainty of the derivation of column densities from lines and dust (here a factor of 4). Adopting the density profile,[48] we derive a H_2 column density of 5.9×10^{22} cm^{-2} in a 28$''$ beam.

2.3 Observational results

The HOOH $3_{0,3}\text{--}2_{1,1}$ line was not detected towards any of the sources in our sample. The noise rms values reached towards each source are listed in Table 3. The rms levels reached are in the range of 16–25 mK at 0.52 km s^{-1} resolution.

Table 3 Observational results

Source	HOOH rms[a] (mK)	HOOH[b] $\int T_a dv$ (mK km s^{-1})	H$_2$CO 3_{03}–2_{02} (E_{up} = 21.0 K) $\int T_a dv$ (K km s^{-1})	H$_2$CO 3_{21}–2_{20} (E_{up} = 68.2 K) $\int T_a dv$ (K km s^{-1})	H$_2$CO 3_{22}–2_{21} (E_{up} = 68.2 K) $\int T_a dv$ (K km s^{-1})
ρ Oph-A[c]		125	4.55 ± 0.04	0.60 ± 0.04	0.62 ± 0.04
ρ Oph-B2-MM8	22.9	≤45	1.83 ± 0.02	≤0.02[b]	≤0.02[b]
G15.01-0.67	19.2	≤92	20.57 ± 0.03	5.64 ± 0.03	5.28 ± 0.03
G018.82-00.28MM1	21.1	≤109	6.48 ± 0.04	1.81 ± 0.04	1.54 ± 0.03
G018.82-00.28MM4	21.0	≤96	0.87 ± 0.03	≤0.03[b]	≤0.03[b]
G028.53-00.25MM1	17.7	≤96	2.17 ± 0.03	0.61 ± 0.03	0.42 ± 0.04
NGC6334I(N)	18.6	≤97	44.18 ± 0.06	18.94 ± 0.04	17.72 ± 0.07
G1.6-0.025	16.1	≤80	1.41 ± 0.05	0.69 ± 0.03	0.55 ± 0.03
NGC1333-IRAS4A	24.6	≤75	—[d]	—[d]	—[d]
L1527	18.4	≤32	1.63 ± 0.01	0.11 ± 0.01	0.13 ± 0.01
RCrA-IRS7B	23.1	≤61	9.21 ± 0.02	1.94 ± 0.02	1.84 ± 0.02

[a] At resolution 0.52 km s^{-1}. [b] The tabulated upper limits on the integrated intensity are 3σ. [c] From Bergman et al.[13] for HOOH and Bergman et al.[38] for H$_2$CO. [d] See footnote d of Table 2.

Three H$_2$CO lines are present within the large bandwidth of the XFFTS. Their fluxes are also listed in Table 3, as they will be useful for the interpretation of the HOOH upper limits (Section 3).

3 Analysis

3.1 Upper limits on the HOOH column density

In order to derive upper limits on the HOOH column densities from the rms noise of the observations, we need knowledge of both the typical linewidth in each source, as well as of the excitation temperature of the considered HOOH transition.

The typical linewidth for each source can be inferred from the observations, as other lines are present in the XFFTS range. For this purpose, we use the lower excitation line of H$_2$CO, which is detected towards all sources, and derive its width using a Gaussian fit to the line. For sources where the higher energy lines of H$_2$CO are also detected, and the lower energy line is obviously broadened through opacity effects, we measure the width of the higher excitation lines. This linewidth is listed in the third column of Table 2. The resulting 3σ upper limits on the integrated flux are listed in column 3 of Table 3.

Estimating the excitation temperature of the line is more difficult. On the one hand, the dust temperature has been measured for some of the sources (Table 2). The dust temperature is expected to be equal to the gas temperature in dense regions, where the two phases thermalize through collisions. But the average densities in our beam may not always be high enough to reach this state.

On the other hand, we can compute the rotational temperature of H$_2$CO from the three para-lines which are in the observed band. In the case where the three lines are detected, this rotational temperature should be seen as an upper limit to the kinetic temperature, as the lower energy line is likely to be optically thick. We checked this approach on the SM1 core of Oph A (also called the D-peak position), where HOOH was first detected.[13] We computed the T_{rot} based on only those three H_2CO lines, as observed by Bergman *et al.*[38] We found $T_{rot} = 33 \pm 3$ K, which is indeed higher than the kinetic temperature that was derived for this source from more detailed modelling, using more transitions and taking line opacities into account: 22.5 K (modified rotation diagram technique) and 24 K (ALI technique using H_2CO and CH_3OH).[38]

When the upper energy lines are not detected, we can likewise derive an upper limit on the rotational temperature from their non-detection.

The excitation temperature of the HOOH line should lie somewhere in the range bracketed by the dust temperature and our derived H_2CO excitation temperature. In the cases where credible gas temperatures have been derived in previous studies (see column 5 of Table 2), we take their values for the HOOH excitation temperature. This avoids using the likely overestimated H_2CO rotational temperature when the H_2CO lines are very optically thick. This is, for example, the case with G15, where we get a 3σ upper limit on HOOH abundance of 4×10^{-12} when assuming $T_{ex} = 32$ K. Taking the likely overestimated temperature of 64 K would lead to a value of 7×10^{-12}. In cases where we do not have a good gas temperature estimate from the literature, we adopt the temperature we derived from H_2CO. This should result in an extremely conservative value (*i.e.* an overestimated upper limit), as described above.

The 3σ upper limits on the HOOH column densities are listed in Table 4. For the sake of clarity, the T_{ex} adopted for deriving the upper limit is also given. The 3σ limits are of the same order or lower than the detected column density towards ρ Oph A, except for NGC6334(N) where the higher temperature pulled the limit on the column density towards a much higher value. This is consistent with the fact that the 3σ upper limits of the integrated flux listed in Table 3 are 1–4 times lower than the detected flux towards ρ Oph A.

3.2 HOOH abundances

We derived the upper limits for the HOOH abundance in each source by using the H_2 column density averaged over the APEX beam (see Section 2.2 for a detailed

Table 4 3σ upper limits on the HOOH column densities and abundances relative to H_2

Source	$T_{ex}{}^c$ (K)	$N(HOOH)$ (cm^{-2})	[HOOH] : [H$_2$]
ρ Oph-A	22	$(3–8) \times 10^{12}$	$(1–3) \times 10^{-10\ a}$ $(2–6) \times 10^{-11\ b}$
ρ Oph-B2-MM8	16	$\leq 1.5 \times 10^{12}$	$\leq 7 \times 10^{-12}$
G15.01-0.67	32	$\leq 3.8 \times 10^{12}$	$\leq 4 \times 10^{-12}$
G018.82-00.28MM1	61	$\leq 8.1 \times 10^{12}$	$\leq 5 \times 10^{-11}$
G018.82-00.28MM4	29	$\leq 3.7 \times 10^{12}$	$\leq 4 \times 10^{-11}$
G028.53-00.25MM1A	57	$\leq 6.7 \times 10^{12}$	$\leq 2 \times 10^{-11}$
NGC6334I(N)	160	$\leq 2.4 \times 10^{13}$	$\leq 3 \times 10^{-11}$
G1.6-0.025	60	$\leq 5.9 \times 10^{12}$	$\leq 1.5 \times 10^{-10}$
NGC1333-IRAS4A	24	$\leq 2.6 \times 10^{12}$	$\leq 2 \times 10^{-11}$
L1527	16	$\leq 1.0 \times 10^{12}$	$\leq 2.5 \times 10^{-11}$
RCrA-IRS7B	40	$\leq 3.0 \times 10^{12}$	$\leq 5 \times 10^{-11}$

[a] Using the compact core H_2 column density.[38] [b] Using the full H_2 column density. [c] The tabulated T_{ex} is the value used to derive the column densities.

discussion of each source). The resulting 3σ upper limits are tabulated in Table 4. We revisited the case of Oph A by estimating the abundance of HOOH under the assumption that the emission comes from the full gas along the line-of-sight (case (b) in Table 4, where the full H_2 column density is traced from $C^{18}O(3-2)$ observations), while Bergman *et al.*[13] assumed it originated in the dense core traced in H_2CO and CH_3OH (case (a) in Table 4).

All 3σ upper limits (except for the case of G1.6, where the relatively high upper limit stems from a combination of high excitation temperature and low H_2 column density) are well under the detected abundance of HOOH in OphA, when the detected molecule is assumed to be located in the compact core.[13] All 3σ upper limits are of the same order as or slightly lower than the HOOH abundance in Oph A, if the full column density traced by $C^{18}O(3-2)$ is taken into account.

The derived upper limits are therefore significant, and constrain the abundance of HOOH to a strictly lower value than in ρ Oph A in all sources.

4   Discussion

The main result of our observational search is that HOOH is very rare in the interstellar medium. This sheds some new light on why the molecule was only detected very recently.[13] Under the specific physical conditions of ρ Oph A however, HOOH is abundant, so the puzzle remains as to what makes this source so different from all other sources in our sample.

In the following, we discuss the direct implications of our observations, as well as the implications in terms of chemical modelling.

4.1   HOOH and O_2, similarly elusive

We note that the detection of HOOH towards a sole source (ρ Oph A) is very similar to the case of O_2. O_2 has been searched for towards many sources[49,50] but detected so far in only two sources,[51-53] the strongest case being ρ Oph A, for which the O_2 abundance relative to H_2 is 5×10^{-8}.[52] It is certainly not a simple coincidence that O_2 and HOOH have been detected towards the same source, and that they are both otherwise elusive, as their chemistry is tightly linked *via* the grain surface reactions (1) and (2).

4.2   The role of the environment

Our source sample allows us to investigate the role of the environment and of possible local chemical anomalies on the abundance of HOOH. ρ Oph B belongs to the same molecular cloud as ρ Oph A, and is situated therefore also at the same short distance from the sun (120 pc). The H_2 column density that we derive for ρ Oph B is somewhat higher than that of ρ Oph A. The 3σ upper limit on the HOOH abundance in ρ Oph B is very low (3 to 50 times lower, depending on the assumption for the location of HOOH in ρ Oph A) compared to the HOOH abundance observed in ρ Oph A.

The lack of detection towards ρ Oph B seems to discard the possibility that the detection in ρ Oph A is due to anomalous initial elemental abundances. Instead, the comparison of these two sources seems to indicate that the temperature is important: ρ Oph B is rather cold (≤16 K), while ρ Oph A has a temperature of 24 K. We address this point in more detail in the following sections.

4.3 The role of the present average temperature and density conditions

Du *et al.*[15] modelled in detail the abundance of HOOH in Oph A, as well as that of many other observed molecules also believed to form on dust grains, using a fully coupled gas–grain model, taking into account the latest experimental results[10] for the reaction rates on the grains. Their model was able to reproduce the abundance of HOOH in the source, and even successfully predicted the abundance of HO_2, which was detected subsequently.[16] Gas-phase HOOH was found to originate mainly from the desorption of HOOH formed on the grains through reaction (2). The model of Du *et al.*[15] assumed a constant temperature of 21 K, and a constant density of 6×10^5 cm^{-3}. The best match for the abundance of all considered molecules observed (O_2, HOOH, HO_2, H_2CO and CH_3OH) was obtained for an age of 6×10^5 years. At earlier times, HOOH was found to be overabundant. They showed that the temperature plays an important role in the HOOH abundance, whereas the role of density is not as significant. A temperature change from 20 to 22 K was shown to increase the HOOH abundance by an order of magnitude, while the abundance did not vary much between 22 K and 30 K (see their Fig. 4). At the given evolutionary time of 6×10^5 years, the abundance of HOOH decreased with increasing density in the range 10^5–10^6 cm^{-3} (see their Fig. 6). Their modelling implies that the HOOH abundance should be at least as high as that in Oph A for sources younger than 6×10^5 years, with temperatures in the range 21–30 K and densities lower than or equal to 6×10^5 cm^{-3} (with the assumption of constant temperature and density).

In our sample, several sources have an average temperature in the range 21–30 K. Among them, NGC1333-IRAS4A is certainly the best source to compare to Oph A (although IRAS4A is almost twice as distant, *cf.* Table 1). Its present temperature is similar to that of Oph A. An LVG study of the H_2CO emission[44] led to an estimated density of $(3–4) \times 10^5$ cm^{-3}, while the density at a radius equal to FPBW/2 is 8×10^5 cm^{-3} according to the density profile of Kristensen *et al.*[43] The average density of IRAS4A is therefore also very similar to that of Oph A. Unless this former source is older than $(6–10) \times 10^5$ years, the model of Du *et al.*[15] would therefore predict an HOOH abundance at least as high in this source as that in Oph A, with the assumption of constant temperature and density. The lack of detection of HOOH towards this source is therefore puzzling.

The low-mass Class 0 protostar IRAS4A was much colder in the past, having, as has already been pointed out by Yıldız *et al.*,[54] likely evolved through a long ($\geq 8 \times 10^5$ years) cold precollapse phase ($T \approx 10$ K). Yıldız *et al.*[54] derived a low 3σ upper limit for the abundance of O_2 towards IRAS4A, based on Herschel observations. Their interpretation for the low abundance of O_2 was that most of the O_2 formed in the gas in the early stages was hydrogenated into water on the grains during the long cold precollapse phase. As the hydrogenation of O_2 leads to the formation of HOOH, our non-detection of HOOH might add some new constraints to this interpretation. Furthermore, their study shows that the temporal evolution of the physical conditions plays an important role in the non-detection of O_2.

In order to further investigate what could be the key difference between Oph A and the other sources, we performed new chemical model calculations, including models with variable temperature and density.

4.4 Chemical modelling

a. Description of the chemical model. The model is based on the same assumptions as that of Du *et al.*,[15] and takes into account the correction from Du *et al.*[55] The model has been described in detail in Du *et al.*,[15] and here we summarize its main characteristics. It solves the coupled gas-phase and grain surface chemistry using the Hybrid Moment Equation (HME) method,[14] which was developed to address correctly the stochasticity of the grain surface chemistry. The HME code was benchmarked against Monte Carlo simulations, and showed very good agreement in the results.[14] No layering of the ices is considered in this version of the code. The surface reactions considered in the model are listed in Appendix B of Du *et al.*[15] (but see also Du *et al.*[55]). They are based on a combination of selected reactions from Allen and Robinson,[11] Tielens and Hagen,[12] and Hasegawa *et al.*[56] Some of these earlier reaction rates have been updated according to recent experimental results.[10,57-59] We assume quantum tunneling for the reactions having a barrier. The reaction probabilities depend on the product $a\sqrt{E_a}$, where a is the barrier width, and E_a is the barrier height. The absolute values of these parameters are therefore not needed, and we can without introducing any limitation for our purposes assume $a = 1$ Å for all reactions, and then derive the corresponding barrier height E_a from the experimental results. In particular, the barrier heights of reactions (2) and (3) are estimated based on Cuppen *et al.*,[10] while that of reaction (1) is assumed to be 600 K (a value which is intermediate between those of Tielens and Hagen[12] and Ioppolo *et al.*[59]). Photodissociation reactions induced by cosmic rays and chemical desorption reactions are also included.

b. Stationary physical conditions. The time evolution of physical parameters in observed sources is in general very difficult to constrain. For example, it is unknown how steep the increase of temperature is during the formation of a protostar. It is therefore useful as a first step to look at chemical models with stationary physical conditions, to understand the first order effects of the temperature and density on the chemistry.

Fig. 1 shows the predictions for three different temperatures (10 K, 15 K and 21 K), all other parameters being identical. We fixed the density to 10^5 cm^{-3} (the value chosen by Yıldız *et al.*[54] for the precollapse phase for IRAS4A). The predicted abundance of HOOH in the late stages increases with temperature in the presented range, as does the abundance of O_2 (as already observed by Du *et al.*[15] at a higher density). This might already give the reason for the abundance of HOOH in Oph B (at temperature 12 K) being lower than in Oph A. Our model predicts that HOOH is more than an order of magnitude less abundant at 12 K than at 21 K at late stages.

It is interesting to note that O_2 forms on grain surfaces at 10 K in our model, even with our assumption that the diffusion barrier is a relatively high fraction (0.77) of the desorption barrier compared to other models (Yıldız *et al.*[54] used 0.5). This surface formation outweighs the contribution of the freeze-out of gas-phase O_2 to the abundance of O_2 ice until about 10^5 years. This is in contradiction to the conclusion one reaches based on the very low mobility of O atoms compared to H atoms at these temperatures. This might be due to a relatively high O : H accreting ratio at the beginning of the evolution, leading to a high concentration of O on the surface.

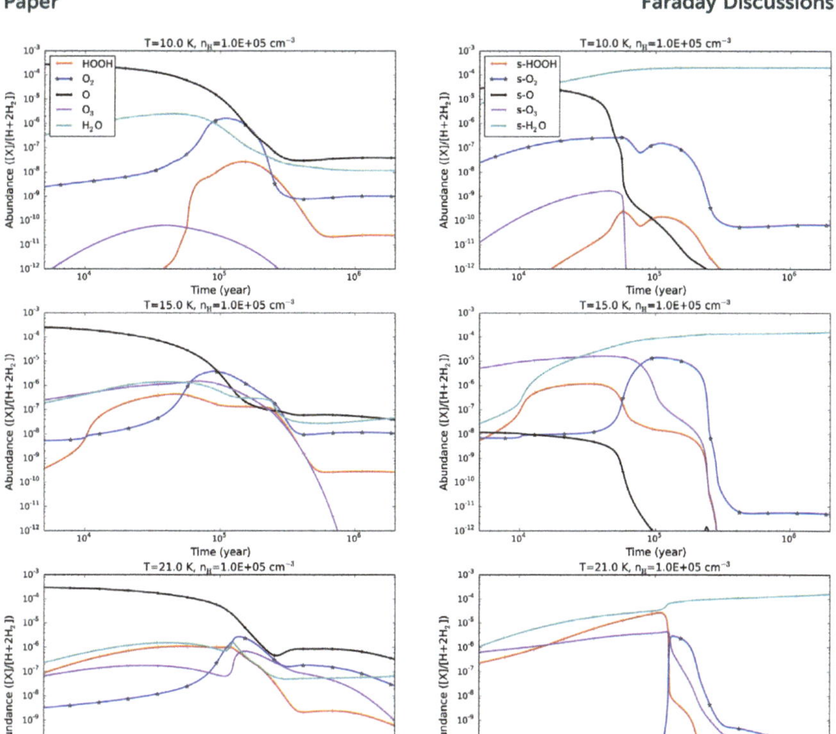

Fig. 1　Evolution with time of the gas-phase (left panel) and solid (ice, right panel) abundances of O, O_2, O_3, HOOH and H_2O for the fixed density of 10^5 cm^{-3}, and temperatures of 10 K (upper panels), 15 K (central panels) and 21 K (lower panels).

At early stages, the HOOH abundance is higher than at $t > 6 \times 10^5$ years. As Du *et al.*[15] already mentioned, the gaseous destruction mechanisms of HOOH are likely not complete in the present chemical networks, and therefore the HOOH abundance may be overestimated overall. It is therefore possible that the rarity of HOOH is due to an observational bias, and that Oph A corresponds to the rare example of a young object. Such an interpretation could only be confirmed after the gas-phase destruction mechanisms of HOOH have been reviewed, and the thorough modelling study from Du *et al.*[15] is updated, to take into account the agreement of the abundance of many species simultaneously. From now on, we will focus on other possible interpretations for the detection of HOOH in Oph A.

　c.　IRAS4A: is HOOH consistent with the O_2 upper limit? We can now return to the problem of HOOH in IRAS4A. Our model with $T = 10$ K and $n_H = 10^5$ cm^{-3} corresponds to the conditions assumed by Yıldız *et al.*[54] for the precollapse phase of IRAS4A.

　Our model predicts an $[O_2] : [H+2H_2]$ abundance ratio of $\sim 10^{-9}$ (*i.e.* $[O_2] : [H_2] \approx 2 \times 10^{-9}$ at times longer than 8×10^5 years (times beyond which the model settles to constant values), which is consistent with the upper limit on the O_2 abundance towards IRAS4A.[54]

The predicted abundance of HOOH is $[HOOH] : [H_2] \approx 5 \times 10^{-11}$. This is hardly consistent with our non-detection ($[HOOH] : [H_2] \leq 2 \times 10^{-11}$, 3σ). A slightly higher density has the effect of lowering the expected HOOH abundance (for $n_H = 2 \times 10^5$ cm^{-3}, the predicted abundance decreases to 3×10^{-11}). Here again, the model might be reconciled with these observations if missing gaseous destruction mechanisms for HOOH were added. A more detailed test would involve doing a shell modelling as Yıldız et al.[54] did, to account for the increased density and temperature after the embedded protostar formed, but this is beyond the scope of this paper.

d.   The warming-up phase. We want to further investigate the role of the temperature in the formation of HOOH. However, one cannot realistically model the conditions of star-forming regions using stationary conditions. Indeed, star-forming regions are evolving from cold cloud conditions to warm conditions. At the start, the cold conditions ensure that grain-surface chemistry plays a key role, whereas it would not in models which already start with warm conditions. We here ran chemical models with a warming-up phase, with the aim of further elucidating the impact of the temperature on the abundance of HOOH.

All models have a constant density of 10^5 cm^{-3}. The early-time temperature is taken to be 12 K, and the warm-up phase starts at 2×10^5 years. The temperature is then increased by 1 K every 2×10^4 years. The models differ from each other by the temperature at which the warm-up phase is stopped. The temperature is then kept constant until the end of the evolution (10^7 years). The resulting HOOH abundances are shown in Fig. 2.

The abundance of HOOH in the model that remains at 12 K has a similar behaviour to that of the 10 K model (from the previous section), but with a final steady-state abundance about twice as high as that at 10 K. When increasing the temperature from 12 K, the HOOH abundance firstly increases and then remains consistently higher than that of the 12 K model, as long as the temperature reached remains under 30 K. In the case of a final temperature of 21 K, the HOOH abundance is more than one order of magnitude larger than that at 12 K at all times. For final temperatures of 28 to 30 K, the enhancement of the HOOH abundance can reach up to two orders of magnitude, and then slowly decays to less than one order of magnitude enhancement within a few 10^6 years.

However, for sources in which the warm-up reaches temperatures above 31 K, the situation changes. A steep decrease in the HOOH abundance is observed very quickly after the initial enhancement, appearing at around 28–30 K. For $T > 33$ K, the abundance decreases quickly to values several orders of magnitude lower than the abundance for 12 K.

We also ran similar models in which the warm-up phase was started later (at 7×10^5 years), and found that the main facts listed above still apply, and that the H_2O_2 abundance reached at later times is unchanged. The duration of the cold phase therefore does not seem to be of much significance.

The conclusion of this study is that enhancement of the HOOH abundance does occur in a very limited range of temperatures, around 20–30 K. Any further warm-up above 30 K will result in the rapid destruction of HOOH.

4.5   The special case of ρ Oph A

The detection of HOOH in Oph A, in stark contrast to the non-detection in the other sources, could therefore reflect the fact that the bulk of the material in

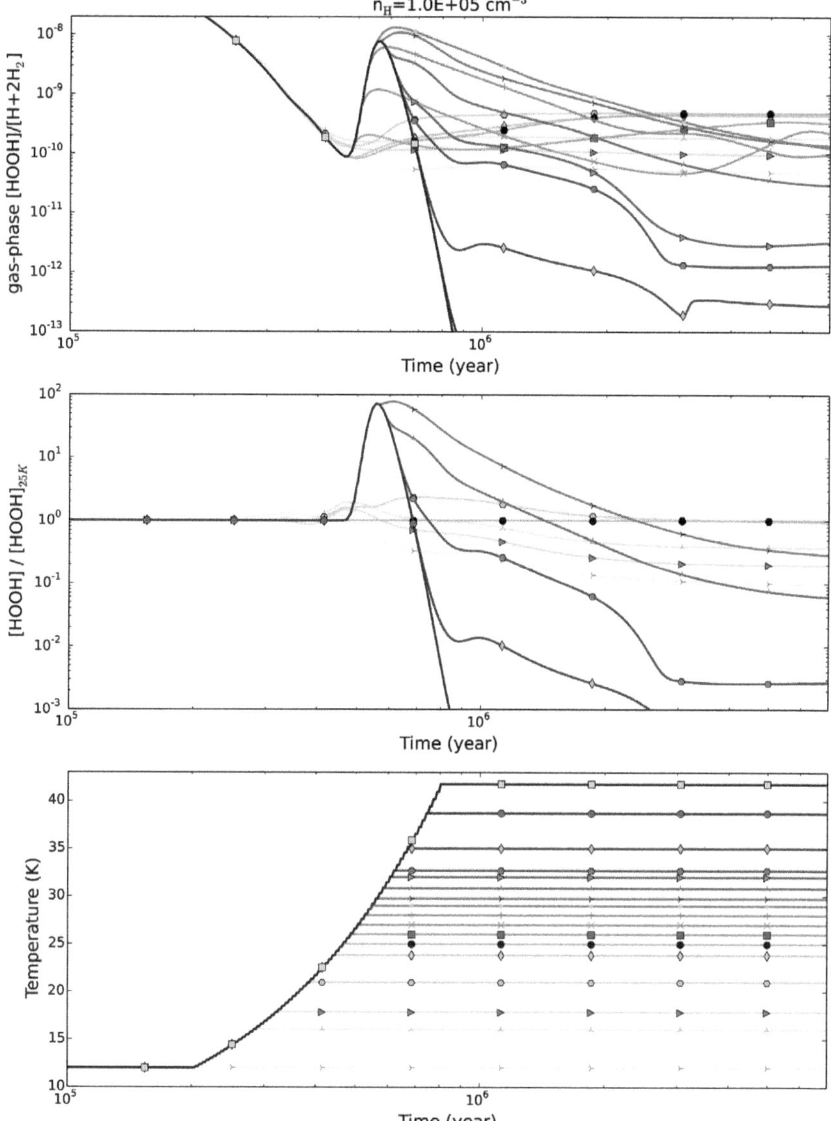

Fig. 2 Evolution with time of the abundance of gas-phase HOOH (upper panel) in the warming-up models. The temperature evolution is shown in the lower panel, while the middle panel shows the abundance of HOOH normalized to that in the 25 K model.

Oph A is within the favorable temperature range (20–30 K), whereas this is not the case for the other sources. In Oph A, the estimates of the gas temperature at different positions of the clump lead to temperatures in the range 24–30 K (see *e.g.* Table 8 from Bergman *et al.*[38]). The fraction of the mass in the 20–30 K range to the total mass may be close to unity.

For the IRAS4A protostar, the average temperature is 24 K, as for Oph A. But here the protostar is internally heating its envelope, causing a steep temperature gradient. As a result, only a small portion of the gas is actually at the average temperature. Using the density profile from Kristensen *et al.*,[43] and approximating the temperature profile with a power law, we find that only ~15% of the total envelope mass in IRAS4A is in the temperature range 20–30 K. Therefore, the HOOH abundance enhancement should be strongly suppressed in this object. Considering only the H_2 in the range 20–30 K, the upper limit on the HOOH abundance in this gas becomes 1.3×10^{-10} cm^{-3}, a value that does not conflict with the detection in Oph A. This interpretation could be further put to the test by integrating much more deeply in IRAS4A.

Similarly, the other sources in our sample may not have the bulk of their mass in the 20–30 K range. The IRDC sources might be warmer than ~35 K (as suggested by the high rotational temperatures for H_2CO), in which case the predicted abundance for HOOH would fall to a few 10^{-12} or even lower. Additionally, these sources are much further away, so the material that might be at the favorable temperature is heavily beam-diluted. On the contrary, Oph B is too cold to have a significant enhancement of HOOH. Our model with the stationary conditions $T = 12$ K and $n = 10^6$ cm^{-3} predicts a HOOH abundance of ~2×10^{-12} at late times, a value well below our observed upper limit. Finally, the case of the two other low-mass protostars L1527 and RCrA-IRS7B is certainly similar to that of IRAS4A, where the internal heating by the protostar may cause a steep temperature gradient.

How could ρ Oph A achieve this particular condition? A closer look at the environment of Oph A shows that it is externally heated by the S1 source,[60] which is in fact a close binary system (B4 + K).[61] The slightly curved morphology to the east of the main ridge of Oph A (see *e.g.*, Fig. 1 from Larsson *et al.*[51]) seems to coincide with the edge of the ISOCAM bright emission surrounding the S1 source.[62] This might be an indication that Oph A was formed from compression under the radiative pressure from the B4 star. Compression and external heating may be the cause of the unusually warm conditions within Oph A.

Finally, we remark that because of the close chemical link between O_2 and HOOH, we expect the abundance of both molecules to show some degree of correlation in astronomical sources. At longer times ($t > 8 \times 10^5$ years), we get from our (static) chemical models $[O_2] : [HOOH] \approx 40$ for a range of temperatures between 10 to 21 K, at the fixed density of 10^5 cm^{-3}. The search for O_2 has proven to be very difficult because of the atmospheric opacity at the frequencies of its intrinsically weak magnetic dipole rotational transitions, requiring deep observations from satellites (SWAS, Odin, Herschel). We suggest here that the search for sources with high O_2 content could be easily approached by searching first for HOOH, a molecule that is much easier to target with ground-based telescopes.

5 Conclusions

Following our detection of HOOH towards Oph A, we have searched for HOOH in a sample of ten sources, of different nature and in different environments. HOOH was not detected towards any of the sources in our sample, and significant upper limits for the HOOH abundance could be obtained. These negative results shed new light on key parameters in the O_2/HOOH chemistry. We ran new gas–grain chemical models, taking into account a warm-up phase after a cold early cloud phase. The models show that the production of HOOH is extremely sensitive to the temperature, and that outside of the 20–30 K temperature range the expected HOOH abundance is very low. We conclude that the key difference between Oph A and the other sources is that the bulk of the material in Oph A is likely to be at this favorable temperature, whereas most of the mass in the other sources may lie outside this range. The peculiar conditions of Oph A may be caused by external heating. This interpretation could explain the scarcity of detections of O_2 and HOOH in the ISM, and could be tested by observing other externally heated sources.

Acknowledgements

We thank F. Du for interesting discussions during the early phases of this project, as well as R. Liseau for discussions regarding the stellar environment of ρ Oph A.

References

1 E. A. Bergin, G. J. Melnick, J. R. Stauffer, M. L. N. Ashby, G. Chin, N. R. Erickson, P. F. Goldsmith, M. Harwit, J. E. Howe, S. C. Kleiner, D. G. Koch, D. A. Neufeld, B. M. Patten, R. Plume, R. Schieder, R. L. Snell, V. Tolls, Z. Wang, G. Winnewisser and Y. F. Zhang, *Astrophys. J.*, 2000, **539**, L129–L132.

2 D. C. B. Whittet, W. A. Schutte, A. G. G. M. Tielens, A. C. A. Boogert, T. de Graauw, P. Ehrenfreund, P. A. Gerakines, F. P. Helmich, T. Prusti and E. F. van Dishoeck, *Astron. Astrophys.*, 1996, **315**, L357–L360.

3 P. Caselli, E. Keto, L. Pagani, Y. Aikawa, U. A. Yıldız, F. F. S. van der Tak, M. Tafalla, E. A. Bergin, B. Nisini, C. Codella, E. F. van Dishoeck, R. Bachiller, A. Baudry, M. Benedettini, A. O. Benz, P. Bjerkeli, G. A. Blake, S. Bontemps, J. Braine, S. Bruderer, J. Cernicharo, F. Daniel, A. M. di Giorgio, C. Dominik, S. D. Doty, P. Encrenaz, M. Fich, A. Fuente, T. Gaier, T. Giannini, J. R. Goicoechea, T. de Graauw, F. Helmich, G. J. Herczeg, F. Herpin, M. R. Hogerheijde, B. Jackson, T. Jacq, H. Javadi, D. Johnstone, J. K. Jørgensen, D. Kester, L. E. Kristensen, W. Laauwen, B. Larsson, D. Lis, R. Liseau, W. Luinge, M. Marseille, C. McCoey, A. Megej, G. Melnick, D. Neufeld, M. Olberg, B. Parise, J. C. Pearson, R. Plume, C. Risacher, J. Santiago-García, P. Saraceno, R. Shipman, P. Siegel, T. A. van Kempen, R. Visser, S. F. Wampfler and F. Wyrowski, *Astron. Astrophys.*, 2010, **521**, L29.

4 P. Caselli, E. Keto, E. A. Bergin, M. Tafalla, Y. Aikawa, T. Douglas, L. Pagani, U. A. Yıldız, F. F. S. van der Tak, C. M. Walmsley, C. Codella, B. Nisini, L. E. Kristensen and E. F. van Dishoeck, *Astrophys. J.*, 2012, **759**, L37.

5 M. R. Hogerheijde, E. A. Bergin, C. Brinch, L. I. Cleeves, J. K. J. Fogel, G. A. Blake, C. Dominik, D. C. Lis, G. Melnick, D. Neufeld, O. Panić, J. C. Pearson, L. Kristensen, U. A. Yıldız and E. F. van Dishoeck, *Science*, 2011, **334**, 338.

6 E. F. van Dishoeck, L. E. Kristensen, A. O. Benz, E. A. Bergin, P. Caselli, J. Cernicharo, F. Herpin, M. R. Hogerheijde, D. Johnstone, R. Liseau, B. Nisini, R. Shipman, M. Tafalla, F. van der Tak, F. Wyrowski, Y. Aikawa, R. Bachiller, A. Baudry, M. Benedettini, P. Bjerkeli, G. A. Blake, S. Bontemps, J. Braine, C. Brinch, S. Bruderer, L. Chavarría, C. Codella, F. Daniel, T. de Graauw, E. Deul, A. M. di Giorgio, C. Dominik, S. D. Doty, M. L. Dubernet,

P. Encrenaz, H. Feuchtgruber, M. Fich, W. Frieswijk, A. Fuente, T. Giannini, J. R. Goicoechea, F. P. Helmich, G. J. Herczeg, T. Jacq, J. K. Jurgensen, A. Karska, M. J. Kaufman, E. Keto, B. Larsson, B. Lefloch, D. Lis, M. Marseille, C. McCoey, G. Melnick, D. Neufeld, M. Olberg, L. Pagani, O. Panić, B. Parise, J. C. Pearson, R. Plume, C. Risacher, D. Salter, J. Santiago-García, P. Saraceno, P. Stäuber, T. A. van Kempen, R. Visser, S. Viti, M. Walmsley, S. F. Wampfler and U. A. Yıldız, *Publ. Astron. Soc. Pac.*, 2011, **123**, 138–170.

7 K. Hiraoka, T. Miyagoshi, T. Takayama, K. Yamamoto and Y. Kihara, *Astrophys. J.*, 1998, **498**, 710.

8 N. Miyauchi, H. Hidaka, T. Chigai, A. Nagaoka, N. Watanabe and A. Kouchi, *Chem. Phys. Lett.*, 2008, **456**, 27–30.

9 H. Mokrane, H. Chaabouni, M. Accolla, E. Congiu, F. Dulieu, M. Chehrouri and J. L. Lemaire, *Astrophys. J.*, 2009, **705**, L195–L198.

10 H. M. Cuppen, S. Ioppolo, C. Romanzin and H. Linnartz, *Phys. Chem. Chem. Phys.*, 2010, **12**, 12077.

11 M. Allen and G. W. Robinson, *Astrophys. J.*, 1977, **212**, 396–415.

12 A. G. G. M. Tielens and W. Hagen, *Astron. Astrophys.*, 1982, **114**, 245–260.

13 P. Bergman, B. Parise, R. Liseau, B. Larsson, H. Olofsson, K. M. Menten and R. Güsten, *Astron. Astrophys.*, 2011, **531**, L8.

14 F. Du and B. Parise, *Astron. Astrophys.*, 2011, **530**, A131.

15 F. Du, B. Parise and P. Bergman, *Astron. Astrophys.*, 2012, **538**, A91.

16 B. Parise, P. Bergman and F. Du, *Astron. Astrophys.*, 2012, **541**, L11.

17 P. Helminger, W. C. Bowman and F. C. de Lucia, *J. Mol. Spectrosc.*, 1981, **85**, 120–130.

18 D. Petkie, *J. Mol. Spectrosc.*, 1995, **171**, 145–159.

19 R. Güsten, L. Å. Nyman, P. Schilke, K. Menten, C. Cesarsky and R. Booth, *Astron. Astrophys.*, 2006, **454**, L13–L16.

20 V. Vassilev, D. Meledin, I. Lapkin, V. Belitsky, O. Nyström, D. Henke, A. Pavolotsky, R. Monje, C. Risacher, M. Olberg, M. Strandberg, E. Sundin, M. Fredrixon, S.-E. Ferm, V. Desmaris, D. Dochev, M. Pantaleev, P. Bergman and H. Olofsson, *Astron. Astrophys.*, 2008, **490**, 1157–1163.

21 B. Klein, S. Hochgürtel, I. Krämer, A. Bell, K. Meyer and R. Güsten, *Astron. Astrophys.*, 2012, **542**, L3.

22 M. A. Requena-Torres, J. Martín-Pintado, A. Rodríguez-Franco, S. Martín, N. J. Rodríguez-Fernández and P. de Vicente, *Astron. Astrophys.*, 2006, **455**, 971–985.

23 L. Loinard, R. M. Torres, A. J. Mioduszewski and L. F. Rodríguez, *Astrophys. J.*, 2008, **675**, L29–L32.

24 R. Chini and V. Hoffmeister, in *Star Formation in M17*, ed. B. Reipurth, Astronomical Society of the Pacific, 2008, p. 625.

25 R. Simon, J. M. Rathborne, R. Y. Shah, J. M. Jackson and E. T. Chambers, *Astrophys. J.*, 2006, **653**, 1325–1335.

26 P. Persi and M. Tapia, in *Star Formation in NGC 6334*, ed. B. Reipurth, Astronomical Society of the Pacific, 2008, p. 456.

27 K. M. Menten, R. W. Wilson, S. Leurini and P. Schilke, *Astrophys. J.*, 2009, **692**, 47–60.

28 M. J. Reid, *Annu. Rev. Astron. Astrophys.*, 1993, **31**, 345–372.

29 T. Hirota, T. Bushimata, Y. K. Choi, M. Honma, H. Imai, K. Iwadate, T. Jike, O. Kameya, R. Kamohara, Y. Kan-Ya, N. Kawaguchi, M. Kijima, H. Kobayashi, S. Kuji, T. Kurayama, S. Manabe, T. Miyaji, T. Nagayama, A. Nakagawa, C. S. Oh, T. Omodaka, T. Oyama, S. Sakai, T. Sasao, K. Sato, K. M. Shibata, Y. Tamura and K. Yamashita, *Publ. Astron. Soc. Jpn.*, 2008, **60**, 37.

30 S. J. Kenyon, D. Dobrzycka and L. Hartmann, *The Astronomical Journal*, 1994, **108**, 1872–1880.

31 R. Neuhäuser and J. Forbrich, in *The Corona Australis Star Forming Region*, ed. B. Reipurth, Astronomical Society of the Pacific, 2008, p. 735.

32 F. Motte, P. Andre and R. Neri, *Astron. Astrophys.*, 1998, **336**, 150–172.

33 I. Ristorcelli, J. M. Lamarre, G. Serra, F. Pajot, M. Giard, J. P. Bernard and J. P. Torre, *Diffuse Infrared Radiation and the IRTS*, 1997, p. 144.

34 M. A. Thompson, J. Hatchell, A. J. Walsh, G. H. MacDonald and T. J. Millar, *Astron. Astrophys.*, 2006, **453**, 1003–1026.

35 T. Pillai, F. Wyrowski, J. Hatchell, A. G. Gibb and M. A. Thompson, *Astron. Astrophys.*, 2007, **467**, 207–216.

36 J. M. Rathborne, J. M. Jackson and R. Simon, *Astrophys. J.*, 2006, **641**, 389–405.

37 J. M. Rathborne, J. M. Jackson, E. T. Chambers, I. Stojimirovic, R. Simon, R. Shipman and W. Frieswijk, *Astrophys. J.*, 2010, **715**, 310–322.

38 P. Bergman, B. Parise, R. Liseau and B. Larsson, *Astron. Astrophys.*, 2011, **527**, A39.

39 G. Sandell, *Astron. Astrophys.*, 2000, **358**, 242–256.

40 T. B. H. Kuiper, W. L. Peters, III, J. R. Forster, F. F. Gardner and J. B. Whiteoak, *Astrophys. J.*, 1995, **446**, 692.

41 W. H. McCutcheon, G. Sandell, H. E. Matthews, T. B. H. Kuiper, E. C. Sutton, W. C. Danchi and T. Sato, *Mon. Not. R. Astron. Soc.*, 2000, **316**, 152–164.

42 H. Beuther, A. J. Walsh, S. Thorwirth, Q. Zhang, T. R. Hunter, S. T. Megeath and K. M. Menten, *Astron. Astrophys.*, 2007, **466**, 989–998.

43 L. E. Kristensen, E. F. van Dishoeck, E. A. Bergin, R. Visser, U. A. Yıldız, I. San Jose-Garcia, J. K. Jørgensen, G. J. Herczeg, D. Johnstone, S. F. Wampfler, A. O. Benz, S. Bruderer, S. Cabrit, P. Caselli, S. D. Doty, D. Harsono, F. Herpin, M. R. Hogerheijde, A. Karska, T. A. van Kempen, R. Liseau, B. Nisini, M. Tafalla, F. van der Tak and F. Wyrowski, *Astron. Astrophys.*, 2012, **542**, A8.

44 S. Maret, C. Ceccarelli, E. Caux, A. G. G. M. Tielens, J. K. Jørgensen, E. van Dishoeck, A. Bacmann, A. Castets, B. Lefloch, L. Loinard, B. Parise and F. L. Schöier, *Astron. Astrophys.*, 2004, **416**, 577–594.

45 S. Maret, C. Ceccarelli, A. G. G. M. Tielens, E. Caux, B. Lefloch, A. Faure, A. Castets and D. R. Flower, *Astron. Astrophys.*, 2005, **442**, 527–538.

46 S. J. Kenyon, M. Gómez and B. A. Whitney, in *Low Mass Star Formation in the Taurus-Auriga Clouds*, ed. B. Reipurth, Astronomical Society of the Pacific, 2008, p. 405.

47 F. L. Schöier, J. K. Jørgensen, K. M. Pontoppidan and A. A. Lundgren, *Astron. Astrophys.*, 2006, **454**, L67–L70.

48 J. E. Lindberg and J. K. Jørgensen, *Astron. Astrophys.*, 2012, **548**, A24.

49 P. F. Goldsmith, G. J. Melnick, E. A. Bergin, J. E. Howe, R. L. Snell, D. A. Neufeld, M. Harwit, M. L. N. Ashby, B. M. Patten, S. C. Kleiner, R. Plume, J. R. Stauffer, V. Tolls, Z. Wang, Y. F. Zhang, N. R. Erickson,

D. G. Koch, R. Schieder, G. Winnewisser and G. Chin, *Astrophys. J.*, 2000, **539**, L123–L127.

50 L. Pagani, A. O. H. Olofsson, P. Bergman, P. Bernath, J. H. Black, R. S. Booth, V. Buat, J. Crovisier, C. L. Curry, P. J. Encrenaz, E. Falgarone, P. A. Feldman, M. Fich, H. G. Floren, U. Frisk, M. Grin, E. M. Gregersen, J. Harju, T. Hasegawa, Å. Hjalmarson, L. E. B. Johansson, S. Kwok, B. Larsson, A. Lecacheux, T. Liljeström, M. Lindqvist, R. Liseau, K. Mattila, G. F. Mitchell, L. H. Nordh, M. Olberg, G. Olofsson, I. Ristorcelli, A. Sandqvist, F. von Scheele, G. Serra, N. F. Tothill, K. Volk, T. Wiklind and C. D. Wilson, *Astron. Astrophys.*, 2003, **402**, L77–L81.

51 B. Larsson, R. Liseau, L. Pagani, P. Bergman, P. Bernath, N. Biver, J. H. Black, R. S. Booth, V. Buat, J. Crovisier, C. L. Curry, M. Dahlgren, P. J. Encrenaz, E. Falgarone, P. A. Feldman, M. Fich, H. G. Florén, M. Fredrixon, U. Frisk, G. F. Gahm, M. Gerin, M. Hagström, J. Harju, T. Hasegawa, Å. Hjalmarson, L. E. B. Johansson, K. Justtanont, A. Klotz, E. Kyrölä, S. Kwok, A. Lecacheux, T. Liljeström, E. J. Llewellyn, S. Lundin, G. Mégie, G. F. Mitchell, D. Murtagh, L. H. Nordh, L.-Å. Nyman, M. Olberg, A. O. H. Olofsson, G. Olofsson, H. Olofsson, G. Persson, R. Plume, H. Rickman, I. Ristorcelli, G. Rydbeck, A. A. Sandqvist, F. V. Schéele, G. Serra, S. Torchinsky, N. F. Tothill, K. Volk, T. Wiklind, C. D. Wilson, A. Winnberg and G. Witt, *Astron. Astrophys.*, 2007, **466**, 999–1003.

52 R. Liseau, P. F. Goldsmith, B. Larsson, L. Pagani, P. Bergman, J. Le Bourlot, T. A. Bell, A. O. Benz, E. A. Bergin, P. Bjerkeli, J. H. Black, S. Bruderer, P. Caselli, E. Caux, J.-H. Chen, M. de Luca, P. Encrenaz, E. Falgarone, M. Gerin, J. R. Goicoechea, Å. Hjalmarson, D. J. Hollenbach, K. Justtanont, M. J. Kaufman, F. Le Petit, D. Li, D. C. Lis, G. J. Melnick, Z. Nagy, A. O. H. Olofsson, G. Olofsson, E. Roueff, A. Sandqvist, R. L. Snell, F. F. S. van der Tak, E. F. van Dishoeck, C. Vastel, S. Viti and U. A. Yıldız, *Astron. Astrophys.*, 2012, **541**, A73.

53 P. F. Goldsmith, R. Liseau, T. A. Bell, J. H. Black, J.-H. Chen, D. Hollenbach, M. J. Kaufman, D. Li, D. C. Lis, G. Melnick, D. Neufeld, L. Pagani, R. Snell, A. O. Benz, E. Bergin, S. Bruderer, P. Caselli, E. Caux, P. Encrenaz, E. Falgarone, M. Gerin, J. R. Goicoechea, Å. Hjalmarson, B. Larsson, J. Le Bourlot, F. Le Petit, M. De Luca, Z. Nagy, E. Roueff, A. Sandqvist, F. van der Tak, E. F. van Dishoeck, C. Vastel, S. Viti and U. Yıldız, *Astrophys. J.*, 2011, **737**, 96.

54 U. A. Yıldız, K. Acharyya, P. F. Goldsmith, E. F. van Dishoeck, G. Melnick, R. Snell, R. Liseau, J.-H. Chen, L. Pagani, E. Bergin, P. Caselli, E. Herbst, L. E. Kristensen, R. Visser, D. C. Lis and M. Gerin, *Astron. Astrophys.*, 2013, **558**, A58.

55 F. Du, B. Parise and P. Bergman, *Astron. Astrophys.*, 2012, **544**, C4.

56 T. I. Hasegawa, E. Herbst and C. M. Leung, *Astrophys. J. Suppl.*, 1992, **82**, 167–195.

57 S. Ioppolo, H. M. Cuppen, C. Romanzin, E. F. van Dishoeck and H. Linnartz, *Astrophys. J.*, 2008, **686**, 1474–1479.

58 G. W. Fuchs, H. M. Cuppen, S. Ioppolo, C. Romanzin, S. E. Bisschop, S. Andersson, E. F. van Dishoeck and H. Linnartz, *Astron. Astrophys.*, 2009, **505**, 629–639.

59 S. Ioppolo, H. M. Cuppen, C. Romanzin, E. F. van Dishoeck and H. Linnartz, *Phys. Chem. Chem. Phys.*, 2010, **12**, 12065.

60 R. Liseau, G. J. White, B. Larsson, S. Sidher, G. Olofsson, A. Kaas, L. Nordh, E. Caux, D. Lorenzetti, S. Molinari, B. Nisini and F. Sibille, *Astron. Astrophys.*, 1999, **344**, 342–354.

61 M. Gagné, S. L. Skinner and K. J. Daniel, *Astrophys. J.*, 2004, **613**, 393–415.

62 A. Abergel, J. P. Bernard, F. Boulanger, C. Cesarsky, F. X. Desert, E. Falgarone, G. Lagache, M. Perault, J.-L. Puget, W. T. Reach, L. Nordh, G. Olofsson, M. Huldtgren, A. A. Kaas, P. Andre, S. Bontemps, M. Burgdorf, E. Copet, J. Davies, T. Montmerle, P. Persi and F. Sibille, *Astron. Astrophys.*, 1996, **315**, L329–L332.

Faraday Discussions

PAPER

The formation of glycine and other complex organic molecules in exploding ice mantles

J. M. C. Rawlings,*[a] D. A. Williams,[a] S. Viti,[a] C. Cecchi-Pestellini[b] and W. W. Duley[c]

Received 21st December 2013, Accepted 14th February 2014

DOI: 10.1039/c3fd00155e

Complex Organic Molecules (COMs), such as propylene (CH_3CHCH_2) and the isomers of $C_2H_4O_2$ are detected in cold molecular clouds (such as TMC-1) with high fractional abundances (Marcelino *et al.*, *Astrophys. J.*, 2007, **665**, L127). The formation mechanism for these species is the subject of intense speculation, as is the possibility of the formation of simple amino acids such as glycine (NH_2CH_2COOH). At typical dark cloud densities, normal interstellar gas-phase chemistries are inefficient, whilst surface chemistry is at best ill defined and does not easily reproduce the abundance ratios observed in the gas phase. Whatever mechanism(s) is/are operating, it/they must be both efficient at converting a significant fraction of the available carbon budget into COMs, and capable of efficiently returning the COMs to the gas phase. In our previous studies we proposed a complementary, alternative mechanism, in which medium- and large-sized molecules are formed by three-body gas kinetic reactions in the warm high density gas phase. This environment exists, for a very short period of time, after the total sublimation of grain ice mantles in transient co-desorption events. In order to drive the process, rapid and efficient mantle sublimation is required and we have proposed that ice mantle 'explosions' can be driven by the catastrophic recombination of trapped hydrogen atoms, and other radicals, in the ice. Repeated cycles of freeze-out and explosion can thus lead to a cumulative molecular enrichment of the interstellar medium. Using existing studies we based our chemical network on simple radical addition, subject to enthalpy and valency restrictions. In this work we have extended the chemistry to include the formation pathways of glycine and other large molecular species that are detected in molecular clouds. We find that the mechanism is capable of explaining the observed molecular abundances and complexity in these sources. We find that the proposed mechanism is easily capable of explaining the large abundances of all three isomers of $C_2H_4O_2$ that are observationally inferred for star-forming regions. However, the model currently does not provide an obvious

[a]*University College London, Department of Physics and Astronomy, Gower Street, London WC1E 6BT, United Kingdom. E-mail: jcr@star.ucl.ac.uk*

[b]*INAF – Osservatorio Astronomico di Palermo, P.za Parlamento 1, I-90134 Palermo, Italy*

[c]*University of Waterloo, Department of Physics and Astronomy, Waterloo, Ontario, Canada N2L 3G11*

explanation for the predominance of methyl formate, suggesting that some refinement to our (very simplistic) chemistry is necessary. The model also predicts the production of glycine at a (lower) abundance level, that is consistent with its marginal detection in astrophysical sources.

1   Introduction

Organic molecules containing more than a few carbon atoms, known collectively as Complex Organic Molecules (or COMs), are found in a wide variety of inter-stellar astrophysical environments, from dynamically active star-forming regions, to quiescent molecular clouds. Unlike many of the smaller molecules that have been identified in molecular clouds, which are often exotic unsaturated radicals, detected COMs include organic species that are well known in the laboratory (*e.g.* alcohols, acids, esters, ketones, saturated and unsaturated hydrocarbons, simple sugars, *etc.*). However, the mechanisms by which they are formed remains largely unknown. Simple bimolecular reactions between neutral species and/or between ions and neutrals (that are driven by ultraviolet radiation and ionization by cosmic rays) are known to be efficient for the formation of the smaller species, but cannot explain the observed abundances of COMs.

It is believed that many of the COMs are formed on the surfaces of dust grains, in ice mantles, *via* solid-state reactions that are perhaps catalysed by the presence of ultraviolet photons and/or thermal processing in star-forming regions. However, even these mechanisms fail to reproduce some of the observed abun-dances. Notable examples of instances where the models struggle are in pre-dicting: (a) the presence of significant quantities of propylene (CH_3CHCH_2); (b) the abundances and ratios of the three isomers of $C_2H_4O_2$, glycolaldehyde (CH_2OHCHO), methyl formate ($HCOOCH_3$) and acetic acid (CH_3COOH); and (c) the possible presence of the simplest amino acid, glycine (NH_2CH_2COOH).

The three isomers of $C_2H_4O_2$ have all been detected in star-forming regions, but – to date – no detections of these species have been made in molecular clouds. However, where detected, methyl formate is found to be much more abundant than the other isomers. For example, the relative abundances of acetic acid-: glycolaldehyde : methyl formate in one well-studied source (Sgr B2(N)) are ~1 : 4 : 26. This is consistent with the observation that interstellar molecules with a C–O–C backbone structure are preferred over those with a C–C–O structure.[16] In the context of 'large molecule chemistry', methyl formate is considered to be less important that the other two (C–C–O structure) isomers; glycolaldehyde is the simplest sugar, whilst acetic acid is probably involved in the main synthesis channels for the simplest amino acid, glycine (NH_2CH_2COOH), so it may be an important precursor to biomolecule formation.[26]

In this paper we review the results that we have obtained from an alternative scenario, based on proposed gas-phase reactions that may occur in the high-density gas following catastrophic ice mantle sublimation. We then present and discuss an extension to the model that is capable of providing plausible expla-nations for cases (b) and (c) above – noting, for example, that our previous reaction scheme only included the glycolaldehyde isomer of $C_2H_4O_2$.

In Section 2 we describe the fundamental hypothesis that forms the basis for our studies; that the ice mantles on dust grains can be catastrophically sublimated and that a rapid and efficient three-body chemistry between radicals can take place in the gas-phase. Section 3 gives an extended summary of our previous work and provides the context for the new work presented in this paper. In section 4 we describe how the chemistry has been extended, being driven by the inclusion of additional ice mantle species, so as to describe the formation of glycine, the isomers of $C_2H_4O_2$ and other COMs. Section 5 describes the physical and chemical model. Our computational model and results are presented in section 6, and our conclusions discussed in section 7.

2   Explosive chemistry in sublimating ice mantles

Laboratory experiments involving the slow warming of chemically-mixed ices[8] demonstrate that desorption occurs in several distinct and narrow temperature bands, of which the most important to our studies is the so-called co-desorption band. This band is where the major component of the ice, H_2O, desorbs and carries with it all other species.

This understanding of desorption has been shown to be consistent with current observations of so-called 'hot cores'. These are the 'remnants' of the natal molecular envelopes out of which high mass stars form. They are heated by the young protostar and the ice mantles sublimate. By transferring molecular material back into the gas phase, they effectively reveal the composition of the ice mantles. The reasonable assumption is made that the formation of a hot core is a time-dependent process.[28] In a typical hot core, the material is relatively warm ($\sim$200 K) and is exceptionally rich in molecules, some of which are hydrogen-poor, such as vinyl cyanide CH_2CHCN, ethyl cyanide CH_3CH_2CN, methyl formate $HCOOCH_3$, dimethyl ether $(CH_3)_2O$, acetone $(CH_3)_2CO$, acetic acid CH_3COOH, and ethanol CH_3CH_2OH (see, $e.g.$ ref. 26). None of these species is readily formed in standard interstellar gas-phase chemistries. The conventional wisdom is therefore that the source of these relatively complex species is solid-state chemistry occurring in the ice mantles on dust grains. Highly detailed computational models by Herbst, Garrod, and their collaborators ($e.g.$ ref. 11) have investigated the chemistry that may arise from the creation of (photolytically or cosmic-ray generated) mobile radicals within the ices. Their models predict a rich and relatively complex chemistry. These species are subsequently returned to the gas phase when the ice mantles are (thermally) sublimated.

At the same time, laboratory evidence shows that the catastrophic recombination of hydrogen atoms and other accumulated radicals in a solid may abruptly raise the temperature of the solid to $\sim$10^3 K. For abrupt temperature excursions of this magnitude, any ices that are present will be completely converted to gas and sublimated explosively.

In this paper, we develop and extend our proposal that COMs may be formed in the high-density gas phase that pertains immediately after the ice mantle on a dust grain is catastrophically desorbed ('explodes'). In this model, hydrogen atoms are accreted from the gas phase and accumulate in the ices until a critical hydrogen atom density is reached, whereupon a runaway explosion occurs.[10] This process releases all of the chemical energy stored in the grain. This can include both the hydrogen recombination energy and the energy stored in other radicals, as well as energy released due to phase changes in the grain substrate. The theory of radical recombination-driven mantle explosions is very similar to that described in ref. 14, except that the mechanism for mantle explosion is the spontaneous internal recombination of trapped hydrogen atoms, rather than an external heating source. Unlike the laboratory situation, there is a large excess of hydrogen in the interstellar medium. Hydrogen–ice bonding energies are of the order of $\sim$1000 K, and ices typically possess a high density of binding sites. It therefore seems highly plausible that any explosion will be hydrogen-driven, although other radicals can and will contribute. Note that as this mechanism does not require the presence of an external heating source, and it may therefore be as applicable to quiescent dark clouds as it is to dynamically active regions.

It was shown in ref. 10 that the number of H atoms required to cause this explosion is equivalent to about 5% of the total number of atoms in the grain plus mantle. An instantaneous conversion from solid to gas would create a gas with a number density similar to that of the solid, *i.e.* about 10^{23} cm^{-3}. This is unlikely, but it is nevertheless possible that the density is initially extremely high, if only for a very short period of time. In our models we postulate that three-body reactions occurring in this extremely dense and fairly warm gas (with a temperature that is sufficiently high to overcome any activation barriers) can create molecules of considerable complexity. The reactants are the radicals that are produced in the ices by the action of cosmic rays and UV photons.

The sublimated gas is subsequently assumed to undergo a free expansion which occurs on a time scale on the order of nanoseconds. However, the density in this sublimate is so high that many collisions occur before the gas has relaxed to more normal interstellar conditions. In this picture, the episodic explosions therefore enrich the interstellar gas with the products of the three-body reactions, and the material undergoing this enrichment is accumulated during the interval between explosions. Ideas of a similar kind (but operating on much longer time scales) were first explored in ref. 9, which suggested that amino acids, peptides, and a variety of organometallic compounds could be created from evaporating ices confined within cavities inside aggregate grains.

3   Summary of the results from our previous models

In ref. 7 [Paper I], we presented the basic model, and considered the gas-phase reactions that occur between the primary (saturated) constituents of the ice mantles, once released into the gas phase. In that study there was no attempt to identify the nature of the products of the reactions – the purpose was to establish whether or not the proposed mechanism is viable, *i.e.* that it results in significant abundances of COMs.

The chemistry that we adopted was entirely hypothetical, since there is no information available about three-body gas-phase chemistry at extremely high number densities. We considered various types of possible three-body reactions. For these, one normally considers the third body as being chemically inert, but whose purpose is to collisionally stabilise the excited product of the reaction. However, we also postulated reactions in which all three species are chemically active. Whilst some of the product species can be formed in a single-stage reaction, many require two stages, involving intermediate species after the first stage.

The physical model is very simple and idealised: we consider a situation in which a sphere of ice is instantaneously sublimated into the gas phase. This gas then is assumed to freely expand into a vacuum at some fraction, ε, of the sound speed v_s. The parameter ε makes allowance for real (non-spherical) grain morphologies, and/or the effects of trapping in cavities. Unhindered spherical expansion corresponds to $\varepsilon = 1$.

If the sphere of gas has an initial radius r_0 and density n_0, then, by mass conservation, at any time t after mantle sublimation, the density n is given by

$$\frac{n}{n_0} = \frac{1}{\left(1 + 10^9 \varepsilon t\right)^3} \tag{1}$$

where we have assumed that r_0 is comparable to the typical thickness of an ice mantle ($r_0 = 10^{-5}$ cm), and the local sound speed $v_s = 10^4$ cm s^{-1}. At the instant of sublimation, the gas density (n_0) may be comparable to the density in the solid-state.

On inspection of the results from that model, we found that the most significant of the free parameters are the initial density (n_0) and the value(s) of the rate coefficients (k_{3B}). There is, however, a wide range of the (n_0, k_{3B}) parameter space which results in the formation of significant abundances of COMs. If activation barriers can be overcome, then the chemistry will be fast and efficient.

Observational studies (*e.g.* ref. 1 and 2) emphasise that detected molecules (such as amino acetonitrile, NH_2CH_2CN, and ethyl formate, C_2H_5OCHO) tend to be hydrogen-poor, which is consistent with the mechanism that we postulate; H atoms will be ejected in the bond-breaking and bond-making processes of three-body chemistry.

In ref. 23 [Paper II], we took a more holistic and stochastic view of the process, by developing a much more complex model of a population of grains that successively cycle through phases of ice mantle accretion, explosion and mixing of the chemically rich gas into the ISM. This can lead to a cumulative molecular enrichment of the interstellar medium and allowed us to obtain more realistic, time-averaged abundances. We also postulated a more specific reaction scheme (involving both one-stage and two-stage formation processes) for identifiable reaction products, many of which have been clearly detected in the interstellar medium.

Importantly, we addressed a limitation of Paper I, which only considered reactions between (saturated) primary ice components. In this and subsequent work, we recognised that reactions between radicals are likely to be much more important and significant to the chemical evolution of the exploding mantles. Thus, an inspection of databases for known three-body reactions (*e.g.* ref. 29) shows that the rate coefficients for reactions between radicals may be quite large ($>10^{-26}$ cm^6 s^{-1}), as compared to reactions involving saturated species (for which $k \sim 10^{-33}$–10^{-31} cm^6 s^{-1}). To try and produce a self-consistent chemistry, we assumed that any saturated molecules produced in either stage are unreactive and that only the radicals are capable of undergoing further stages of association.

In the quiescent phase, we followed the dark cloud gas-phase and freeze-out chemistry of some 81 gas-phase and 25 solid-state species, assuming that full hydration of atoms and simple hydrides to CH_4, NH_3, H_2O and H_2S occurs on grain surfaces. During this phase we assumed that reactive radicals are created in the ice mantles due to the action of impinging cosmic rays. We note that there may be contributions both from direct cosmic ray impact and photolysis by the cosmic ray induced radiation field.[22] This is a cumulative effect and we assume that it is limited to the stripping of a single hydrogen atom from saturated species. Thus, H_2O may give rise to OH, CH_4, CH_3, *etc.*, and the population of these radicals is associated with the bulk of the ice. The relative abundance of the radical to the parent saturated species will therefore be proportional to the cosmic ray ionization rate and the period of exposure. The time dependence of the chemistry was followed, together with the deposition of ices and the accumulation of weakly bound H atoms, until the atomic hydrogen abundance in the ices reaches some threshold value (f_H). At this point we assume that the hydrogen explosively recombines (with 100% efficiency) to H_2 and all components of the ice mantles are instantaneously heated and fully sublimated.

In the explosion phase we considered the chemistry in the high density, rapidly expanding gas in the immediate vicinity of a dust grain following ice mantle sublimation. As explained above, the radical–radical–H_2O reactions are expected to be faster than the radical–neutral–H_2O reactions, and so we adopted larger values for k_{3B} than we did in Paper I.

We then followed the chemical evolution through a number of cycles (typically 5), resulting in a total duration of $\sim$1.5 Myr, which is comparable to the average lifetime for a dense cloud. In those cases where the choice of parameters leads to shorter inter-explosion times, we cycled the chemistry though a larger number of explosions. The frequency of the episodic temperature excursions is controlled by the cloud chemistry, so that there will be some optimal value for n_H; if it is too high and the molecular content of the gas is low, then this implies efficient desorption–dissociation is taking place, which will not be conducive to the formation of ices. If it is too low, then the inter-explosion time scale becomes too long for the processes described here to be important.

As described above, this model comprises two phases for each cycle. Phase I models the (standard) dark cloud chemistry, with freeze-out and (limited) surface chemistry, whilst Phase II models the chemistry in the high density, rapidly expanding gas, in the immediate vicinity of a dust grain following ice mantle sublimation. The two phases are physically and chemically distinct from each other, but the output from each phase feeds into the other as material cycles between the two.

In our standard model (Model 1), the temperature, density and extinction in the dark cloud phase (Phase I) are $n_I = 10^4$ cm^{-3}, $T_I = 10$ K, and $A_v = 3$ magnitudes, respectively. The cosmic-ray ionization rate of H_2 is $\zeta = 1.3 \times 10^{-17}$ s^{-1} and the initial atomic hydrogen density $n_{H,0} = 1.0$ cm^{-3}. The initial density and

Table 1 Parameter values used in the models described in Paper II. See text for details

Model	Parameter values
1	Standard
2	$n_I = 10^5$ cm^{-3}, $A_v = 10$
3	$n_I = 10^6$ cm^{-3}, $A_v = 10$
4	$n_I = 10^7$ cm^{-3}, $A_v = 10$
5	$\zeta = 1.3 \times 10^{-16}$ s^{-1}, $n_{H,0} = 10$ cm^{-3}, $n_{cyc} = 50$
6	$\zeta = 1.3 \times 10^{-16}$ s^{-1}, $n_I = 10^5$ cm^{-3}, $n_{H,0} = 10$ cm^{-3}, $A_v = 10$, $n_{cyc} = 50$

temperature in the explosion phase (Phase II) are $n_{\mathrm{II}} = 10^{20}$ cm^{-3} and $T_{\mathrm{II}} = 1000$ K, respectively. The 'universal' three-body reaction rate-coefficient is set at $k_{3B} = 10^{-28}$ cm^6 s^{-1} and the chemistry is cycled through 5 explosions ($n_{\mathrm{cyc}} = 5$). The variations of these parameter values used in the other models are given in Table 1. Results from these models are illustrated in Table 2, which gives the time-averaged abundances of selected species and illustrates the sensitivity of the results to variations in the free parameters.

Although there are a large number of poorly-constrained free parameters in the model, the results are found not to be strongly sensitive to the values used. Key results from the model included the prediction of (i) the presence of detectable gas-phase abundances of COMs in cold molecular clouds, (ii) significant abundances of undetectable molecules, such as C_2H_6, which may play a role in the formation of other larger species, and (iii) a mechanism for the formation of larger molecules of biochemical importance in molecular clouds.

We found that this mechanism may be an important source of precursors to biomolecule formation as well as smaller organic species, such as methanol and formaldehyde. However, some caution is necessary here; our models would seem to predict excessively high gas-phase abundances of methanol which could imply that we have overestimated either the CH_4 abundance in the ices, or the rate for the CH_3 + OH association. Nevertheless, the most important prediction of our model was the gas-phase presence of these larger molecular species in quiescent molecular clouds and not just in dynamically active regions such as hot cores. This is not predicted by the standard solid-state chemistries, which require a physical mechanism to return the chemically enriched ice mantles to the gas phase. The detection of COMs in quiescent clouds would therefore give strong observational support for the proposed mechanism.

Table 2 Results from Paper II: time-averaged fractional abundances of selected species in the final cycle for models 1–6. The nomenclature $a(b)$ implies a value of $a \times 10^b$

Species	Model 1	Model 2	Model 3	Model 4	Model 5	Model 6
NH_3	6.3(−8)	1.9(−6)	3.3(−7)	3.5(−8)	1.1(−7)	5.0(−8)
H_2O	4.1(−6)	1.6(−5)	1.6(−6)	1.6(−7)	7.3(−6)	1.2(−5)
H_2CO	1.9(−9)	2.5(−8)	4.5(−10)	2.4(−11)	1.0(−9)	1.8(−9)
H_2S	6.4(−10)	4.3(−9)	5.2(−10)	5.2(−11)	7.9(−10)	4.3(−9)
CH_3OH	1.6(−7)	5.5(−8)	5.5(−9)	5.5(−10)	3.2(−7)	8.7(−7)
NH_2OH	2.9(−8)	1.3(−8)	1.7(−9)	1.7(−10)	9.2(−8)	3.9(−8)
HCOOH	1.3(−10)	7.7(−11)	1.3(−12)	6.9(−14)	4.7(−11)	4.7(−9)
C_2H_6	4.9(−8)	2.1(−8)	2.1(−9)	2.1(−10)	9.9(−8)	2.2(−7)
CH_3NH_2	7.5(−9)	4.7(−9)	6.3(−10)	6.5(−11)	1.3(−8)	1.2(−8)
CH_3CH_3O	1.5(−11)	2.2(−11)	2.3(−12)	2.2(−13)	4.4(−11)	2.0(−9)
C_2H_5OH	1.5(−11)	2.2(−11)	2.3(−12)	2.2(−13)	4.4(−11)	2.0(−9)
CH_3CHO	1.4(−11)	2.9(−11)	4.8(−13)	2.6(−14)	6.6(−12)	1.1(−9)
$HCONH_2$	2.8(−12)	6.3(−12)	1.4(−13)	8.1(−15)	9.1(−13)	1.9(−11)
$(CH_2OH)_2$	1.8(−13)	3.6(−14)	3.5(−15)	3.5(−16)	8.5(−13)	2.1(−10)
CH_2OHCHO	6.5(−14)	4.3(−14)	7.1(−16)	3.7(−17)	1.6(−14)	6.8(−12)
CH_2OHNH_2	5.7(−12)	4.9(−12)	6.7(−13)	6.9(−14)	3.3(−11)	7.2(−11)
CH_3OCH_3O	1.8(−13)	3.6(−14)	3.5(−15)	3.5(−16)	8.5(−13)	2.1(−10)
CH_3OOH	2.6(−10)	5.1(−11)	5.1(−12)	5.0(−13)	1.6(−9)	2.1(−8)
CH_3OCH_2OH	1.5(−13)	2.7(−14)	2.7(−15)	2.6(−16)	8.3(−13)	1.8(−10)

In ref. 24 [Paper III], we retained and expanded the chemistry of Paper II, but reverted to the simpler one-grain physical model of paper I to study the specific issue of propylene formation.

Propylene, CH_3CHCH_2, has been detected in dense gas towards the cyanopolyyne peak of the low mass star-forming region TMC-1 with a substantial fractional abundance of $\sim 2 \times 10^{-9}$ relative to hydrogen.[18] Detailed modelling studies[15,17] suggest that propylene cannot be readily formed by conventional interstellar gas phase chemistry.

We again proposed formation by simple two-stage radical addition:

$$CH_3 + CH + H_2O \rightarrow CH_3CH + H_2O,$$

followed by

$$CH_3CH + CH_2 + H_2O \rightarrow CH_3CHCH_2 + H_2O.$$

Although hypothetical, there is some laboratory evidence to justify that this type of reaction scheme is viable (*e.g.* ref. 20).

In Table 3 we give the final (asymptotic) abundances for propylene as a function of two free parameters: the fraction of the ice that is converted to radicals (F_{rad}) and the value of the product of the three-body reaction rate coefficient with the square of the initial, post-explosion density ($k_{3B}n_0^2$). In these calculations, the gas is allowed to expand freely following the ice mantle sublimation.

With this assumption, the free parameters are found to be the ice composition, the branching ratios for radical production, F_{rad}, and $k_{3B}n_0^2$.

The efficiency of the conversion of the radicals trapped in the ice into complex organic molecules is effectively determined by the ratio of the dynamical (geometrical dilution) time scale to the chemical time scale and, as can be seen from Table 3, a saturation limit is seen to apply to those situations where $t_{chem} \ll t_{dyn}$; in this case a robust limiting value for the propylene abundance (relative to H_2O) of $Y_{sat} \sim 1.9 \times 10^{-6}$ is obtained. Thus, provided that 0.1% or more of the ice is converted to radicals and the product of the reaction rate coefficients with the square of the initial, post-sublimation gas density is $>10^{14}$ s^{-1}, the abundances of the COMs may approach saturation levels. We found that the value of this

Table 3 Results from Paper III: log of the final abundances (relative to H_2O) for propylene (CH_3CHCH_2)

$\log_{10}(k_{3B}n_0^2)$	$\log_{10} F_{rad}$				
	-4.00	-3.50	-3.00	-2.50	-2.00
10.0	-16.86	-15.36	-13.86	-12.37	-10.87
11.0	-14.86	-13.37	-11.87	-10.39	-8.94
12.0	-12.87	-11.39	-9.94	-8.57	-7.34
13.0	-10.94	-9.57	-8.34	-7.29	-6.39
14.0	-9.34	-8.29	-7.39	-6.60	-5.90
15.0	-8.39	-7.60	-6.90	-6.28	-5.73
16.0	-7.90	-7.28	-6.73	-6.22	-5.72
17.0	-7.73	-7.22	-6.72	-6.22	-5.72
18.0	-7.72	-7.22	-6.72	-6.22	-5.72

saturation limit is primarily dependent on the chemical initial conditions (*i.e.* the ice mantle composition and the degree of processing to radicals), and is essentially independent of the physical parameters.

Using these results we estimated an optimal injection rate for propylene into a dense cloud core like those seen in TMC-1. Comparing this time scale to that for the destruction of propylene by gas-phase chemistry in the molecular cloud, we used a simple argument to estimate the time-averaged abundance of propylene that could result from this mechanism: $X_{prop} \sim 2 \times 10^{-10}$–$2 \times 10^{-9}$. This is comparable to the observationally inferred value in TMC-1.

We therefore found that rather special conditions may be required to produce the observationally inferred values of X_{prop}. The dependence on the abundance of atomic carbon and oxygen in the gas phase, as well as the properties (such as the temperature) of the dust grains may help to explain why propylene is detected in TMC-1, but not in other sources, such as Orion-KL.

4　Extending the chemistry

In this section we discuss how we can extend our model to describe possible formation routes for glycine and the isomers of $C_2H_4O_2$, *via* the high density radical association scheme.

One possible way of forming glycine in such a high density regime would be *via* a three-stage process involving four radicals. The first step would form COOH by simple association of CO and OH:

$$CO + OH + H_2O \rightarrow COOH + H_2O.$$

Then:

$$NH_2 + CH_2 + H_2O \rightarrow NH_2CH_2 + H_2O$$

followed by

$$NH_2CH_2 + COOH + H_2O \rightarrow NH_2CH_2COOH + H_2O.$$

Or, alternatively,

$$CH_2 + COOH + H_2O \rightarrow CH_2COOH + H_2O$$

followed by

$$CH_2COOH + NH_2 + H_2O \rightarrow NH_2CH_2COOH + H_2O.$$

However, it is unlikely that these mechanisms can produce glycine with the requisite efficiency: In our previous studies we limited the chemical network to reactions between radicals of matching valency, on the grounds that unsaturated product species would be rapidly converted into more complex molecules. In those studies, we did not wish to investigate the formation of species more complex than propylene, so that was a satisfactory approximation. It is exactly those reactions involving radicals formed in the second stage to form more complex species, such as glycine, that are central to this formation mechanism. Thus, consideration of all possible reactions (and also including CO) would lead to a network of over 1400 reactions and a substantial enhancement in the number of chemical species. The combined effects of requiring three stages for the formation of glycine, plus the fact that the distribution of C, N and O would be over a much larger number of products implies that the process would only yield a small fractional abundance of glycine.

However, recent observations of hot core sources implies that a much simpler, one-stage process, may be viable: specifically, formic acid (HCOOH), glyco-laldehyde (CH_2OHCHO), methyl formate ($HCOOCH_3$), and acetic acid (CH_3COOH) are seen to be present in at least some hot core environments.[30,3,25] Whilst not providing conclusive proof of the fact, this would strongly suggest that they (and their associated radicals) are also present in the ice mantles, thus enriching the variety of molecules present in interstellar ices. If so, then the first two stages of the glycine mechanism described could be bypassed, due to the likely presence of the COOH and CH_2COOH radicals in the sublimated gas. Glycine could then be formed by just one single-stage reaction:

$$CH_2COOH + NH_2 + H_2O \rightarrow NH_2CH_2COOH + H_2O.$$

Similar one-step reactions could also lead to the formation of glycolaldehyde (CH_2OHCHO), methyl formate ($HCOOCH_3$) and the re-formation of acetic acid (CH_3COOH):

$$CH_2OH + CHO + H_2O \rightarrow CH_2OHCHO + H_2O$$

$$CH_3 + HCOO + H_2O \rightarrow HCOOCH_3 + H_2O$$

$$CH_3 + COOH + H_2O \rightarrow CH_3COOH + H_2O.$$

In our previous studies we assumed that the dust ice mantles were composed of H_2O, CH_4, NH_3, H_2CO and CH_3OH. The primary radicals in the first-stage chemistry were the photolysis products of these species, resulting from the abstraction of a hydrogen atom: OH, CH_3, CH_2, CH, NH_2, NH, CHO, CH_3O and CH_2OH.

Here we now assume that since formic acid (HCOOH) and acetic acid (CH_3COOH) are also present in the ice then the following additional photolytically generated radicals would also be present: COOH, CH_2COOH, CH_3COO and HCOO. To implement this we have augmented the chemistry of our previous models (first stage only) to include the reactions between these additional radicals and our original set (given above).

With these additions, a set of 46 possible additional reactions involving 47 possible additional species could result. The additional species are given in Table 4 and the additional reactions are listed in Table 5. With these augmentations to the propylene chemistry, the chemistry consists of a total of 185 reactions between 165 species.

The inclusion of these extra species introduces some new variables into the model: the fractional abundances of HCOOH and CH_3COOH (relative to water) in

Table 4 Extra chemical species

$COOH$	$CO(OH)_2$	CH_2COOH	$CH_2OHCOOH$	CH_3COO
CH_3COOOH	$HCOO$	$HCOOOH$	CH_3COOH	C_2H_5COOH
CH_3COOCH_3	$HCOOCH_3$	$(CH_2)_2COOH$	CH_3COOCH_2	$HCOOCH_2$
$CHCOOH$	$CH_2CHCOOH$	CH_3COOCH	$HCOOCH$	NH_2COOH
CH_2NH_2COOH	CH_3COONH_2	$HCOONH_2$	$NHCOOH$	$CH_2NHCOOH$
CH_3COONH	$HCOONH$	$CHOCOOH$	$CH_2CHOCOOH$	$CH_3COOCHO$
$HCOOCH_3$	CH_3OCOOH	CH_2CH_3OCOOH	CH_3COOCH_3O	$CH_2CH_2OHCOOH$
CH_3COOCH_2OH	$HCOOCH_2OH$	$(COOH)_2$	$CH_2(COOH)_2$	$CH_3COOCOOH$
$HCOOCOOH$	$(CH_2COOH)_2$	CH_3COOCH_2COOH	$CH_2(HCOO)COOH$	$(CH_3COO)_2$
$CH_3COOHCOO$	$(HCOO)_2$			

the canonical ice mantle and the branching ratios for the formation of the radicals by photolysis;

$$HCOOH \rightarrow COOH + H$$

$$HCOOH \rightarrow HCOO + H$$

and

$$CH_3COOH \rightarrow CH_2COOH + H$$

$$CH_3COOH \rightarrow CH_3COO + H.$$

In our models we assume that the $HCOOH$ and CH_3COOH solid-state abundances, relative to H_2O, lie in the range of 0.001–1%, consistent with observational constraints. For the branching ratios, we assume that there is no preferred channel in either case – *i.e.*, the ratios are 0.5 for each branch. These values, as well as the values of the other parameters in our model, are given in Table 6.

5   The physical and chemical model

5.1   Physical model of the explosion

In this study, we are looking for a proof-of-concept result, rather than accurate quantitative predictions for the abundances of the COMs that we are investigating. We therefore employ the simple physical model of a single grain explosion that was used in Papers I and III, rather than the complex multi-phase cyclic model described in Paper II – which would not be justified, bearing in mind the speculative nature of our extension to the chemistry.

As stated above, we assume that a purely gas-phase chemistry occurs in the high density gas that is produced as a result of the sudden and complete sublimation of the ice mantle. The cause of that process is not the subject of this paper, but it could be an externally driven sudden heating event, or result from the catastrophic recombination of radicals trapped in the ices and/or phase changes within the substrate. In either case we simply note that any sudden and efficient desorption process is potentially capable of driving the three-body gas-phase chemistry that we describe above. As in Papers I and III, we again consider an idealised situation in which a sphere of ice is instantaneously sublimated into the gas phase.

Table 5 Reactions added to the propene chemistry. These are all three-body reactions, with H_2O as the third reactant. The products of the first 12 reactions are all radicals with a valency of 1 or 2

No.	Reactant 1	Reactant 2	Product	Name
1	NH	COOH	NHCOOH	
2	NH	CH_2COOH	$CH_2NHCOOH$	
3	NH	CH_3COO	CH_3COONH	
4	NH	HCOO	HCOONH	
5	CH_2	COOH	CH_2COOH	
6	CH_2	CH_2COOH	$(CH_2)_2COOH$	
7	CH_2	CH_3COO	CH_3COOCH_2	
8	CH_2	HCOO	$HCOOCH_2$	
9	CH	COOH	CHCOOH	
10	CH	CH_2COOH	$CH_2CHCOOH$	
11	CH	CH_3COO	CH_3COOCH	
12	CH	HCOO	HCOOCH	
13	NH_2	COOH	NH_2COOH	Carbamic acid
14	NH_2	CH_2COOH	CH_2NH_2COOH	Glycine
15	NH_2	CH_3COO	CH_3COONH_2	
16	NH_2	HCOO	$HCOONH_2$	
17	OH	COOH	$CO(OH)_2$	
18	OH	CH_2COOH	$CH_2OHCOOH$	
19	OH	CH_3COO	CH_3COOOH	
20	OH	HCOO	HCOOOH	
21	CH_3	COOH	CH_3COOH	Acetic acid
22	CH_3	CH_2COOH	C_2H_5COOH	Propionic acid
23	CH_3	CH_3COO	CH_3COOCH_3	Methyl acetate
24	CH_3	HCOO	$HCOOCH_3$	Methyl formate
25	CHO	COOH	CHOCOOH	
26	CHO	CH_2COOH	$CH_2CHOCOOH$	
27	CHO	CH_3COO	$CH_3COOCHO$	
28	CHO	HCOO	$HCOOCH_3O$	
29	CH_3O	COOH	CH_3OCOOH	
30	CH_3O	CH_2COOH	CH_2CH_3OCOOH	
31	CH_3O	CH_3COO	CH_3COOCH_3O	
32	CH_3O	HCOO	$HCOOCH_3O$	
33	CH_2OH	COOH	$CH_2OHCOOH$	
34	CH_2OH	CH_2COOH	$CH_2CH_2OHCOOH$	
35	CH_2OH	CH_3COO	CH_3COOCH_2OH	
36	CH_2OH	HCOO	$HCOOCH_2OH$	
37	COOH	COOH	$(COOH)_2$	Oxalic acid
38	COOH	CH_2COOH	$CH_2(COOH)_2$	
39	COOH	CH_3O	$CH_3COOCOOH$	
40	COOH	HCOO	HCOOCOOH	
41	CH_2COOH	CH_2COOH	$(CH_2COOH)_2$	
42	CH_2COOH	CH_3COO	CH_3COOCH_2COOH	
43	CH_2COOH	HCOO	$CH_2(HCOO)COOH$	
44	CH_3COO	CH_3COO	$(CH_3COO)_2$	
45	CH_3COO	HCOO	$CH_3COOHCOO$	
46	HCOO	HCOO	$(HCOO)_2$	

5.2   Chemistry produced from exploding ices

On empirical grounds – based on observations along lines of sight towards low mass stars – we assume that ice consists mainly of H_2O, CO, CO_2, CH_4, NH_3,

Table 6 Physical and chemical parameters defining the explosion. Where a variable, the values used in our standard model are given in square brackets. See the text for further description

Parameter	Value
$CH_4 : H_2O(ice)$	0.04
$NH_3 : H_2O(ice)$	0.01
$H_2CO : H_2O(ice)$	0.03
$CH_3OH : H_2O(ice)$	0.03
$HCOOH : H_2O(ice)$	0.0001–0.01 [0.001]
$CH_3COOH : H_2O(ice)$	0.0001–0.01 [0.001]
Fraction converted into radicals (F_{rad})	0.01–1% [1%]
Initial density (n_0)	10^{19}–10^{23} cm^{-3} [10^{22}]
Three-body rate coefficients (k_{3B})	10^{-32}–10^{-26} cm^6 s^{-1} [10^{-28}]
Trapping parameter (ε)	0.1–1.0 [1.0]

Molecule $\rightarrow$ radicals	Branching ratio
$H_2O \rightarrow OH$	0.5
$CH_4 \rightarrow CH_3, CH_2, CH$	0.33, 0.33, 0.33
$NH_3 \rightarrow NH_2, NH$	0.5, 0.5
$H_2CO \rightarrow CHO$	0.5
$CH_3OH \rightarrow CH_3O, OH, CH_3, CH_2OH$	0.1, 0.2, 0.2, 0.5
$HCOOH \rightarrow COOH, HCOO$	0.5, 0.5
$CH_3COOH \rightarrow CH_2COOH, CH_3COO$	0.5, 0.5

H_2CO, CH_3OH (*e.g.*, ref. 21), together with the recently-identified additions of $HCOOH$ and CH_3COOH. Other species (such as OCN^-), which are not relevant to our reaction scheme, may also be present, but at much lower abundance levels. In addition, we also do not include CO_2, which, although it has a typical relative abundance of 21%, is tightly bound and probably unreactive under the assumed conditions.

The relative abundances of the ice mantle constituents show significant variations along different lines of sight, (*e.g.* ref. 4 and 13) which reflect the different physical and chemical conditions when the ices were formed. These are key free parameters in the model and we speculate that they are a major factor in the observed variations in the abundances of COMs. In our standard model, the canonical values for the ice abundances that we have used are based on a variety of observational studies (*e.g.* ref. 5) and are given in Table 6.

The chemistry that we propose is driven by the presence of radicals. Following previous studies we assume that molecules in the ice mantles are subject to photodissociation driven by the secondary radiation field that is generated by cosmic ray ionization and the excitation of ambient H_2.[22]

As in previous studies, we assume that the process of radical production is limited to the stripping of just one hydrogen atom per molecule. Thus, H_2O may give rise to OH, CH_4 to CH_3, *etc.*, and the population of these radicals is associated with the bulk of the ice. We omit from the photodissociation products the atoms such as O, C, and N on the grounds that the overabundance of hydrogen will tend to enhance the hydrogenation. Hot hydrogen atoms in the explosion will tend to establish a population of hydrides.

In Paper II we argued that an upper limit for the fraction of mantle species that is photodissociated is of the order of 1%. So, as in Paper III, we have considered a (realistic) range of $F_{rad} = 0.01$–1%. Of course, the products and branching ratios for these photolysis reactions are highly uncertain. We have followed the practice of our previous studies and used values (given in Table 6) that are consistent with laboratory and theoretical determinations. As described in our basic model, it is assumed that these radicals are trapped within the ices until 'mobilised' when an explosion occurs.

As the ice composition is dominated by H_2O, and to a lesser extent by CO, we again assume that the third body in the three-body reactions is usually H_2O, although we have also included some reactions where CO is the (chemically passive) third body.

The rate coefficients for these various reactions are entirely hypothetical as no detailed information is available for any of the reactions in our list. As with previous studies and for the sake of simplicity, we assume that all reactions have the same basic rate coefficients (k_{3B}). By referring to existing databases of three-body reactions (*e.g.*, RATE12[19]) it is evident that for saturated species, most of the values of k_{3B} for the reactions lie in the range 10^{-27}–10^{-32} cm^6 s^{-1}, but they are significantly larger for reactions involving radicals.

At the densities that we are considering, the chemistry is completely dominated by three-body reactions. We therefore do not include any two-body, photochemical or cosmic ray induced reactions in our chemical network. Also, we note that there are no clear observations of sulfur-bearing species in ice mantles and so we do not include any complex sulfur-bearing species.

The first stage of our updated radical addition chemistry considered here involves association between the thirteen radical species identified in Table 6. Note that $\sim$99% of the molecules from the ices are not involved, except as third bodies stabilising the products. The first stage generates a set of associations, giving products that are either molecules (formed when the initial radicals have equal valences) or radicals (formed where the initial radicals have unequal valences). The product molecules are often familiar, including some of the extra species in Tables 4 and 5, and many of them are detected interstellar species. It is assumed that these product molecules take no further part in the radical chemistry. However, product radicals from the first stage may undergo further stages of association. Just as in the first stage, associations in the second stage may give rise either to saturated molecules or to new radicals. The new radicals may, if the expansion time scale permits, give rise to a further stage of chemistry.

6   Computational models and results

As with our models of propylene formation, we have developed two different types of models to investigate the efficiency of the formation of the isomers of $C_2H_4O_2$ and glycine *via* the proposed mechanism:

1. Model A, which uses a single set of parameters to determine the time-dependence of the chemistry.

2. Model B, which performs a grid of calculations and determines the final (asymptotic) abundance of a selected species (*e.g.* glycine) as a function of variations in several free parameters.

The numerical calculation of the time-dependencies of the chemical species utilises the LSODE integration package. As the chemical rate coefficients are taken to be the same, the chemical time scales in the model are similar and the differential equations that describe the time-dependencies of the abundances are not numerically stiff.

We also follow the practice of Papers I–III and adopt a single value for the rate coefficient (k_{3B}) for all reactions. This incorporates any implicit dependence on the temperature. Our previous work (Paper II) has shown that the temperature-dependence of the rate coefficients is relatively unimportant, although at high temperatures the chemistry would tend to yield a thermochemical equilibrium composition. An inspection of databases for known three-body reactions (*e.g.* ref. 29) reveals that for reactions between radicals, the rate coefficients can be quite large ($>10^{-26}$ cm^6 s^{-1}), whereas for reactions involving saturated species they may be significantly smaller ($\sim 10^{-33}$–10^{-31} cm^6 s^{-1}). Although we do not include the latter, we have considered values of k_{3B} that cover the range 10^{-32}–10^{-26} cm^6 s^{-1}.

Table 6 specifies the other parameters and the range of values that we have investigated in our studies. The first part gives the composition of the ice, with abundances relative to the dominant component, H_2O. The second part gives the

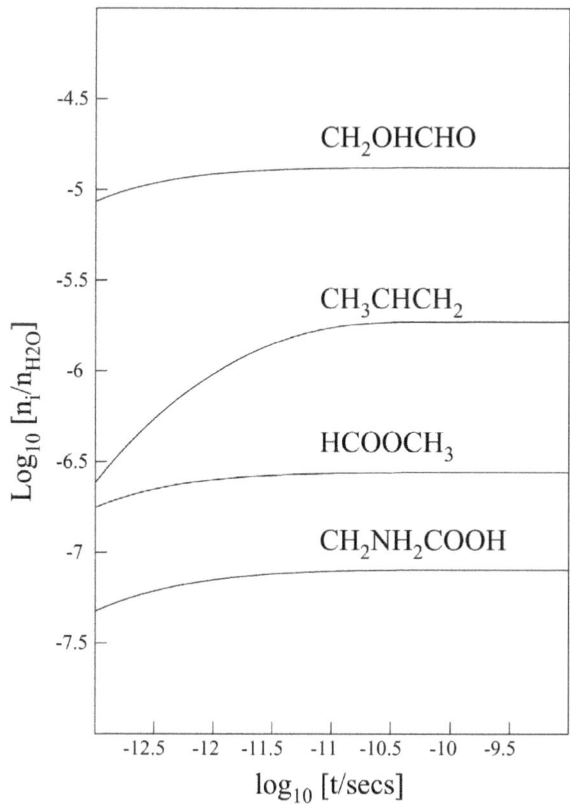

Fig. 1   Example of the results obtained from Model A, showing the time-dependencies of methyl formate (HCOOCH$_3$), glycolaldehyde (CH$_2$OHCHO), propylene (CH$_3$CHCH$_2$) and glycine (NH$_2$CH$_2$COOH).

physical characteristics of the explosion phase, and the third part identifies the radicals produced in the ices, together with the branching ratios.

An example of the results that are obtained from Model A are presented in Fig. 1 and Table 7.

In Paper I we very roughly estimated the expected time scale for the significant conversion of reactant species by three-body reactions into complex organics on the assumptions that (i) there are no significant activation energy barriers to the reactions in question, and (ii) the reactions are generally constructive. This was found to be of the order of $\sim 10^{-13}$–10^{-12} s. This is very rapid indeed and is typically much shorter than the dynamical time scale. In this case, the details of the expansion and cooling of the gas are of minimal relevance; essentially the chemistry takes place very shortly after the sublimation of the ice mantles when the gas density is highest; the relative abundances of the product molecules are then "frozen in" to the flow. These results are confirmed in Fig. 1 which shows results that are obtained from our 'standard' model in which the initial density (n_0) is 10^{22} cm^{-3}, the 'universal' rate coefficient (k_{3B}) is 10^{-28} cm^6 s^{-1}, and the HCOOH and CH$_3$COOH abundances in the ice are both 0.1%.

Table 7 gives the final (asymptotic) abundances relative to H$_2$O (for selected species) obtained from the same model. As in Paper II, we see that the modelled abundances are all quite large, showing that effective conversion of carbon to COMs is taking place. Note, however, that these abundances are relative to H$_2$O and so need to be scaled down by a factor of the H$_2$O abundance relative to hydrogen ($\sim 10^{-4}$) to obtain something like a mean gas-phase abundance. It should also be remembered that these results were obtained for a single ice mantle explosion event and so cannot be compared directly to values determined from observations. To do that requires a knowledge of the frequency of the explosions and how the chemistry evolves in the ISM in the inter-explosion periods. In most cases, the gas-phase chemistry of these species is very poorly understood. An approximate estimate of the time-averaged abundances could be estimated from assumptions about the explosion frequency and the destruction channels/rates, but that is beyond the scope of this paper and is probably not justified by the crudeness of our assumptions about the chemistry.

As in previous studies, we can see from Table 6 that there are several free parameters in the model, most importantly the ice composition, the branching ratios for radical production, the fraction of the ice that is converted to radicals (F_{rad}), the density (n_0), the rate-coefficients (k_{3B}) and the trapping parameter (ε).

Table 7 Example of the final abundances (relative to H$_2$O) for selected species obtained from Model A. The nomenclature $a(b)$ implies a value of $a \times 10^b$

Species	Abundance	Species	Abundance
CH$_3$NH$_2$	2.5(−6)	CH$_3$OCH$_3$	1.5(−6)
C$_2$H$_5$OH	6.2(−6)	CH$_2$OHCHO	1.3(−5)
CH$_3$CHCH$_2$	1.9(−6)	CH$_3$COOH	2.8(−7)
C$_2$H$_5$COOH	2.9(−7)	CH$_3$COOCH$_3$	2.8(−7)
HCOOCH$_3$	2.8(−7)	CH$_2$CHCOOH	2.6(−7)
NH$_2$COOH	7.4(−8)	CH$_2$NH$_2$COOH	7.9(−8)
HCOONH	7.9(−8)	(COOH)$_2$	1.0(−8)

To show how the modelled abundances depend on these various parameters, we show in Table 8 examples of the results obtained from Model B. For this particular calculation, the 'universal' rate coefficient is 10^{-27} cm^6 s^{-1}, and the HCOOH and CH$_3$COOH abundances in the ice are both 0.1%. The initial gas density (n_0) and the fractional conversion to radicals in the ice (F_{rad}) are left as free parameters.

Despite the limitations mentioned above, we can see from this table that (a) significant abundances of glycine are predicted, that are comparable to observationally inferred limits, and (b) the 'saturation' effect that was seen in the case of the propylene chemistry that was the subject of Paper III is clearly apparent here as well. Thus, in Table 8, the abundance levels (of glycine) in the bottom half of the table show obvious evidence of saturation. Since this regime of the (F_{rad}, n_0) parameter space is quite feasible, it would seem reasonable to believe that, in the model that we postulate, abundances of propylene and other COMs may be expected to be in this saturation limit. A similar conclusion can be made for the isomers of C$_2$H$_4$O$_2$.

As with propylene, the abundances of these COMs are therefore very insensitive to the physical parameters (such as n_0 and k_{3B}). Also, we again find that the saturation levels are defined by the ice composition, and not the physical characteristics of the mantle explosion.

We find obvious near-linear dependencies on the efficiency of radical formation F_{rad} and, for methyl formate, acetic acid and glycine, a similar dependence on the ice abundances of formic acid and acetic acid. This is for obvious reasons – the formation channels for these species all involve the radicals produced from the acids.

Glycolaldehyde, on the other hand, has unrelated formation channels and so is insensitive to the abundances of the acids in the ice.

7   Discussion and conclusions

In our model of high density gas-phase radical association we find that – in optimal conditions and after exploring the full volume of parameter space that we have defined – glycine can attain a 'saturation' abundance of $\sim 10^{-6}$ relative to H$_2$O. This corresponds to $\sim 10^{-10}$ relative to hydrogen in the fully-mixed gas phase and is comparable to the observationally inferred limits.

Table 8   Example of the results obtained from Model B: log of the final abundances (relative to H$_2$O) for glycine (NH$_2$CH$_2$COOH) as a function of the initial gas density (n_0) and the fractional conversion to radicals in the ice (F_{rad})

| | $\log_{10} F_{rad}$ | | | | |
$\log_{10}(n_0^2)$	−4.00	−3.50	−3.00	−2.50	−2.00
19.0	−12.30	−11.30	−10.31	−9.32	−8.35
19.5	−11.31	−10.32	−9.35	−8.43	−7.61
20.0	−10.35	−9.43	−8.61	−7.89	−7.25
20.5	−9.61	−8.89	−8.25	−7.67	−7.13
21.0	−9.25	−8.67	−8.13	−7.61	−7.10
21.5	−9.13	−8.61	−8.10	−7.60	−7.10
22.0	−9.10	−8.60	−8.10	−7.60	−7.10

In the case of glycolaldehyde, the saturation limit abundance (relative to H_2O) is large ($\sim 10^{-4}$–10^{-5}) and is insensitive to the free parameters, other than the fractional conversion to radicals in the ice (F_{rad}). The modelled abundances of the other isomers of $C_2H_4O_2$ are also typically large ($\sim 10^{-6}$–10^{-8}) and are consistent with their clear detection in astrophysical sources, even if for the examples that we have shown, the abundance ratios for the isomers do not match observations. Methyl formate and acetic acid are both sensitive to the proportion of the acid radicals in the ices, so that variation in these values can explain the observed variations in the $HCOOCH_3 : CH_2OHCHO$ and $CH_3COOH : CH_2OHCHO$ ratios.

However, in our reaction scheme, both methyl formate and acetic acid are formed from radicals created by the dissociation of formic acid in the ice (HCOO and COOH respectively). As the branching ratio for these radicals is taken to be 50 : 50 (and in any case is unlikely to be variable) our model fails to reproduce variations in the $HCOOCH_3 : CH_3COOH$ ratio. The most likely explanation for this is that our reaction scheme is incomplete, although other factors may also be important.

Although speculative, our model is based on plausible assumptions. We again note that a 'normal' gas-phase chemistry is unable to explain the observed abundances of COMs, and a viable alternative mechanism for their formation is required. This has to be both highly efficient at converting carbon to COMs, and – if it is assumed that they are formed from material contained in dust ice mantles – the COMs must be efficiently transmitted to the gas-phase. Our proposed 'explosion chemistry' is completed on extremely short time scales. The products of this chemistry may not be directly observed, but are injected into the dense gas of a molecular cloud core. Observational effects are therefore not necessarily defined by the explosion chemistry, but are determined by the injection and subsequent chemistry in the dark cloud.

As we have noted in our previous studies, there are clearly a large number of simplifications and approximations in our model; the reaction network that we have adopted, and the specific extensions to model the chemistry of glycine and the isomers of $C_2H_4O_2$, are very speculative, if plausible. The adoption of a single value for the three-body reaction rate coefficients is a major simplification. In reality there could be large differences between the rate coefficients which would result in strong variations in the formation efficiencies and abundance ratios. In addition, we do not know the branching ratios in the ice photolysis reactions, nor do we have very strong constraints on the relative abundances of the ice constituents (which may vary from place to place). But at least this seems to suggest that there is a way forward to providing a viable mechanism for the formation of glycine and for understanding the relative abundances of the isomers of CH_2OHCHO in cold clouds. The model provides a possible explanation for certain observationally-inferred abundance anomalies in the interstellar medium. In particular, we have proposed a mechanism that can provide a chemical link between the three isomers of $C_2H_4O_2$, and that can explain some of the variations in the ratios of their abundances towards different sources.

Finally, we note that several alternative mechanisms have been proposed for the formation of COMs. In the models presented in ref. 27, radicals are created in ice mantles by UV photolysis. It is then assumed that the grains are heated by grain–grain collisions and the radicals react with each other in the mantles. Chemical explosions release the COMs, including glycine and glycolaldehyde, into the gas phase. Specific mechanisms have also been postulated for the formation of methyl formate, including the solid-state reaction of methanol and CO in cosmic ray irradiated ices, the solid-state reaction of formic acid and methanol,[16] and the gas-phase reaction of methanol and formaldehyde in hot cores.[6] In addition, ref. 12 has investigated glycine formation in the solid-state, following a similar radical reaction mechanism to that which we propose is operating in the gas phase. There is no reason why the mechanism that we are proposing and some, or all, of these alternatives should not also be possible, and may indeed complement each other.

References

1 A. Belloche, K. M. Menten, C. Comito, H. S. P. Müller, P. Schilke, J. Ott, S. Thorwirth and C. Hieret, *Astron. Astrophys.*, 2008, **482**, 179.

2 A. Belloche, R. T. Garrod, H. S. P. Müller, K. M. Menten, C. Comito and P. Schilke, *Astron. Astrophys.*, 2009, **499**, 215.

3 M. T. Beltrán, C. Codella, S. Viti, R. Neri and R. Cesaroni, *Astrophys. J.*, 2009, **690**, L93.

4 A. C. A. Boogert, *et al.*, *Astrophys. J. Suppl.*, 2004, **154**, 359.

5 F. A. van Broekhuizen, K. M. Pontoppidan, H. J. Fraser and E. F. van Dishoeck, *Astron. Astrophys.*, 2005, **441**, 249.

6 P. Caselli, T. I. Hasegawa and E. Herbst, *Astrophys. J.*, 1993, **408**, 548.

7 C. Cecchi-Pestellini, J. M. C. Rawlings, S. Viti and D. A. Williams, *Astrophys. J.*, 2010, **725**, 1581 [Paper I].

8 M. P. Collings, M. A. Anderson, R. Chen, J. W. Dever, S. Viti, D. A. Williams and M. R. S. McCoustra, *Mon. Not. R. Astron. Soc.*, 2004, **354**, 1133.

9 W. W. Duley, *Mon. Not. R. Astron. Soc.*, 2000, **319**, 791.

10 W. W. Duley and D. A. Williams, *Astrophys. J.*, 2011, **737**, L44.

11 R. T. Garrod, S. L. Widicus Weaver and E. Herbst, *Astrophys. J.*, 2008, **682**, 283.

12 R. T. Garrod, *Astrophys. J.*, 2013, **765**, 60.

13 E. L. Gibb, D. C. B. Whittet, A. C. A. Boogert and A. G. G. M. Tielens, *Astrophys. J. Suppl.*, 2004, **151**, 35.

14 J. M. Greenberg, *Astrophys. Space Sci.*, 1976, **39**, 9.

15 E. Herbst, E. Roueff and D. Talbi, *Mol. Phys.*, 2010, **108**, 2171.

16 J. M. Hollis, F. J. Lovas and P. R. Jewell, *Astrophys. J.*, 2000, **540**, L107.

17 Z. Lin, *et al.*, *Astrophys. J.*, 2013, **765**, 80.

18 N. Marcelino, J. Cernicharo, M. Agúndez, E. Roueff, M. Gerin, J. Martín-Pintado, R. Mauersberger and C. Thum, *Astrophys. J.*, 2007, **665**, L127.

19 D. McElroy, C. Walsh, A. J. Markwick, M. A. Cordiner, K. Smith and T. J. Millar, *Astron. Astrophys.*, 2013, **550**, A36.

20 K. Norinaga and O. Deutschmann, *Ind. Eng. Chem. Res.*, 2007, **46**, 3547.

21 K. I. Öberg, A. C. A. Boogert, K. M. Pontoppidan, S. van den Broek, E. F. van Dishoeck, S. Bottinelli, G. Blake and N. J. Evans II, *Astrophys. J.*, 2011, **740**, 109.

22 S. S. Prasad and S. P. Tarafdar, *Astrophys. J.*, 1983, **267**, 603.

23 J. M. C. Rawlings, D. A. Williams, S. Viti, C. Cecchi-Pestellini and W. W. Duley, *Mon. Not. R. Astron. Soc.*, 2013, **430**, 264 [Paper II].

24 J. M. C. Rawlings, D. A. Williams, S. Viti and C. Cecchi-Pestellini, *Mon. Not. R. Astron. Soc.: Lett.*, 2013, **436**, L59 [Paper III].

25 A. Remijan, Y.-S. Shiao, D. N. Friedel, D. S. Meier and L. E. Snyder, *Astrophys. J.*, 2004, **617**, 384.

26 E. Snyder, *Proc. Natl. Acad. Sci. U. S. A.*, 2006, **103**, 12243.

27 W. H. Sorrell, *Astrophys. J.*, 2001, **555**, L129.

28 S. Viti, M. P. Collings, J. W. Dever, M. R. S. McCoustra and D. A. Williams, *Mon. Not. R. Astron. Soc.*, 2004, **354**, 1141.

29 J. Woodall, M. Agúndez, A. J. Markwick-Kemper and T. J. Millar, *Astron. Astrophys.*, 2007, **466**, 1197.

30 P. M. Woods, H. Calcutt, S. Viti, C. Codella, M. T. Beltrán and G. Kelly, *Mon. Not. R. Astron. Soc.*, 2014, submitted.

PAPER

Complex organic molecules along the accretion flow in isolated and externally irradiated protoplanetary disks

Catherine Walsh,[*a] Eric Herbst,[b] Hideko Nomura,[c] T. J. Millar[d] and Susanna Widicus Weaver[e]

Received 13th December 2013, Accepted 10th February 2014

DOI: 10.1039/c3fd00135k

The birth environment of the Sun will have influenced the physical and chemical structure of the pre-solar nebula, including the attainable chemical complexity reached in the disk, important for prebiotic chemistry. The formation and distribution of complex organic molecules (COMs) in a disk around a T Tauri star is investigated for two scenarios: (i) an isolated disk, and (ii) a disk irradiated externally by a nearby massive star. The chemistry is calculated along the accretion flow from the outer disk inwards using a comprehensive network which includes gas-phase reactions, gas-grain interactions, and thermal grain-surface chemistry. Two simulations are performed, one beginning with complex ices and one with simple ices only. For the isolated disk, COMs are transported without major chemical alteration into the inner disk where they thermally desorb into the gas reaching an abundance representative of the initial assumed ice abundance. For simple ices, COMs can efficiently form on grain surfaces under the conditions in the outer disk. Gas-phase COMs are released into the molecular layer *via* photodesorption. For the irradiated disk, complex ices are also transported inwards; however, they undergo thermal processing caused by the warmer conditions in the irradiated disk which tends to reduce their abundance along the accretion flow. For simple ices, grain-surface chemistry cannot efficiently synthesise COMs in the outer disk because the necessary grain-surface radicals, which tend to be particularly volatile, are not sufficiently abundant on the grain surfaces. Gas-phase COMs are formed in the inner region of the irradiated disk *via* gas-phase chemistry induced by the desorption of strongly bound molecules such as methanol; hence, the abundances are not representative of the initial molecular abundances injected into the outer disk. These

[a]*Leiden Observatory, Leiden University, P. O. Box 9513, 2300 RA Leiden, The Netherlands. E-mail: cwalsh@leidenuniv.nl; Fax: +31 71 527 5819; Tel: +31 71 527 6287*

[b]*Departments of Chemistry, Astronomy, and Physics, University of Virginia, Charlottesville, VA 22904, USA*

[c]*Department of Earth and Planetary Sciences, Tokyo Institute of Technology, 2-12-1 Ookayama, Meguro-ku, Tokyo, 152-8551, Japan*

[d]*Astrophysics Research Centre, School of Mathematics and Physics, Queen's University Belfast, University Road, Belfast, BT7 1NN, UK*

[e]*Department of Chemistry, Emory University, Atlanta, GA 30322, US*

results suggest that the composition of comets formed in isolated disks may differ from those formed in externally irradiated disks with the latter composed of more simple ices.

1   Introduction

The current accepted paradigm of isolated low-mass star formation involves the gravitational collapse of a dense condensation within a molecular cloud forming a young star surrounded by a torus of dust and gas, the *protoplanetary disk*.[1] The dust and gas contained within the disk provide the ingredients for planets and other planetary system objects such as comets.[2] However, star formation rarely occurs in isolation: stars generally form in multiple systems and within or nearby stellar clusters.[1,3] One of the closest and best-studied disks, TW Hydrae, is associated with a small stellar cluster.[4] There is also recent evidence that the Sun may have been born within a stellar cluster which has long since dispersed.[5] The nearest active high-mass star-forming region, the Orion molecular cloud (at a distance of ≈ 400 pc), also contains many young low-mass stars which host protoplanetary disks, so-called *proplyds*.[6] The proplyds in Orion are not only irradiated by the central star, they are also heavily irradiated and indeed photo-evaporated by the external radiation field originating from nearby massive OB stars ($T_{\mathrm{eff}} \gtrsim 10\ 000$ K).[7]

Given a large proportion of low-mass stars form within extreme environments, this raises questions regarding the influence of the external conditions on the physical and chemical structure of the disk. The birth environment may also have an impact on the planet-formation process as it is believed that planetesimals form *via* the coagulation of dust grains in the dense midplane.[8] Ice-covered dust grains can help this process: experiments have demonstrated that the presence of the ice mantle can significantly increase the sticking probability upon collision at low relative velocities.[9–11] Ice mantles also help increase the molecular complexity of the gas by providing a substrate for grain-surface chemistry to build so-called *complex organic molecules* (henceforth referred to as COMs) which do not often have efficient gas-phase routes to formation under typical interstellar or circumstellar conditions.[12] The chemistry of COMs in protoplanetary disks is of significance because these molecules may provide the 'building blocks' of larger, more complex, *prebiotic* molecules, for example, amino acids.[13–15] Is it possible for such species to form within protoplanetary disks and survive assimilation into planets and comets? How does the external environment of the protoplanetary disk influence the resulting abundances and distribution of COMs? Externally irradiated disks are expected to be significantly warmer than isolated disks because the heating is dominated by the absorption of stellar and interstellar far-ultra-violet (FUV; 91.2 nm $< h\nu < 210$ nm) photons by dust grains.[16–18] Does this warmer environment facilitate or impede the production and transport of COMs in the disk? What variation in chemical complexity is expected in the planet- and comet-forming zones ($\lesssim 50$ AU) of externally irradiated protoplanetary disks compared with isolated protoplanetary disks?

The formation of COMs in a static model of an isolated T Tauri disk was investigated in previous work.[19] COMs were found to form efficiently *via* grain-surface association reactions on dust grains reaching peak fractional abundances $\sim 10^{-6}$–10^{-4} that of the H nuclei number density. The physical and chemical structure of a protoplanetary disk externally irradiated by a nearby massive star was also investigated.[20] Here, it was found that the disk atmosphere remains predominantly molecular in nature and the surface density is sufficient to effectively shield the midplane from the intense external FUV radiation. Consequently, the midplane temperature in the outer disk is sufficiently low (<100 K) for non-volatile species (*e.g.*, H_2O and CH_3OH) to exist as ice on dust grains, potentially facilitating dust-grain coagulation and the formation of icy planetesimals.

In this work, the abundance and distribution of COMs within a typical T Tauri disk are investigated for two local environments: (i) external irradiation by the interstellar radiation field only, and (ii) external irradiation by a nearby massive O-type star. The aim is to determine the influence of the external environment on the disk structure and the resulting implications on the formation, survival, and transport of COMs. In previous work, the chemistry evolved at each grid point as a function of time, independent from neighbouring grid points.[19,20] Here, the chemistry is calculated along streamlines which follow the accretion flow (and transport of material) from the outer disk into the planet- and comet-forming zone ($\lesssim 50$ AU). The influence of the accretion flow on the chemistry and transport of molecules within protoplanetary disks has been investigated previously.[21-24] However, this is the first investigation of the grain-surface formation and transport of COMs along the accretion flow in both an isolated disk and an externally irradiated disk.

The remainder of the paper is structured as described. In Section 2 the current observational and theoretical understanding of the chemical structure of protoplanetary disks is briefly discussed. In Section 3, the protoplanetary disk model is described, including the methods used to determine the disk physical structure (Section 3.1), the chemical network used (Section 3.2), and the method used to calculate the chemical evolution along the accretion flow (Section 3.3). In Section 4 the results of the simulations are presented and in Section 5 the main conclusions are stated.

2   Chemistry in protoplanetary disks

The complex interplay of radiation and physics in protoplanetary disks leads to disks that possess a chemically layered structure where the dominant chemical species reflect the dominant chemistry.[25] Photodissociation and ionisation by UV photons and X-rays create a surface layer abundant in radicals, atoms, and ions. Shielding of the UV and X-ray radiation by dust and gas allows ion–molecule reactions to build molecular complexity in the gas phase creating a warm, molecule-rich layer deeper in the atmosphere. Towards the disk midplane, the increasing density and decreasing temperature facilitate the formation (or preservation) of ice mantles on dust grains. There is also radial stratification due to the steady increase in density and temperature moving inwards along the midplane towards the central star. As the sublimation temperature of each ice species is surpassed, the molecule desorbs from the grain mantle and is returned to the gas phase. The radius at which this occurs, known as the *snowline*, depends on the molecular binding energy of each distinct ice species. For example, the CO snowline is expected to reside further out in the disk than the H_2O snowline because CO ice is more volatile than H_2O ice ($T_{CO} \approx 20$ K *versus* $T_{H_2O} \approx 100$ K) which is postulated to have implications on the C/O ratios in forming planetesimals.[26]

The physical conditions in the cold, outer disk (>10 AU) appear favourable to the production of molecules at least as complex as those observed in dark, molecular clouds.[27] This region is best observed at far-IR to (sub)mm wavelengths where many molecules emit *via* pure rotational transitions. However, to date, small, simple molecules only have been detected and have been limited to those species which are relatively abundant and which possess simple rotational spectra, leading to observable line emission. The molecules detected in multiple disks, thus far, include CO, HCO^+, CN, HCN, CS, N_2H^+, SO, C_2H, and associated ^{13}C-, ^{18}O-, and deuterium-containing isotopologues.[4,28–35] The detection of more complex species is hindered by the apparent size of protoplanetary disks: in astronomical terms, disks are tiny objects, on the order of a few arcseconds only in diameter. Nevertheless, current interferometric observatories have allowed the detection of several relatively complex molecules, H_2CO,[36] HC_3N,[37] and c-C_3H_2.[38] Fortunately, the Atacama Large Millimeter/Submillimeter Array (ALMA), which is currently in its 'Early Science' phase, will have the necessary specifications to reach the sensitivities required to detect complex molecules in nearby protoplanetary disks, provided these species exist in the disk atmosphere with a sufficiently high abundance.

It is now generally accepted that most COMs form within or on ice mantles on the surfaces of dust grains.[12] The most famous example is the simplest alcohol, methanol (CH_3OH). Originally proposed some decades ago,[39] it is now known that the main route to gas-phase methanol is *via* the desorption of methanol ice which itself forms *via* the sequential hydrogenation of CO ice,

$$CO \xrightarrow[(1)]{H} HCO \xrightarrow[(2)]{H} H_2CO \xrightarrow[(3)]{H} \begin{array}{c} CH_3O \\ CH_2OH \end{array} \xrightarrow[(4)]{H} CH_3OH. \qquad (1)$$

Steps (2) and (4) in the above reaction scheme are assumed to possess no reaction barrier since they involve atom addition to a radical. However, steps (1) and (3) are postulated to possess large reaction barriers (≈ 2500 K), determined using quantum calculations.[40] Laboratory studies performed at 10 K have shown that steps (1) and (3) have *effective* barriers ≈ 400 K, suggesting the reaction sequence is assisted by the quantum tunnelling of H atoms through the reaction barriers.[41,42] This mechanism efficiently forms methanol ice under dark cloud conditions reaching abundances on the order of 1% to 10% that of water ice, in line with astrophysical observations.[43]

Modern grain-surface chemical networks now include a plethora of atom–radical and radical–radical reaction pathways to COMs and also precursor molecules necessary for forming COMs in the gas.[44–46] The continuing development of these networks has been driven by the detection of rotational line emission from multiple COMs in so-called *hot cores* and *hot corinos*.[12] These objects are thought to be the remnant molecular material leftover from the process of high-mass and low-mass star formation, respectively. One proposed mechanism for forming COMs in these sources is that simple ices formed on grain surfaces at 10 K (*e.g.*, CO, H_2O, NH_3, CH_4, and CH_3OH) undergo warming to ≈ 20–30 K (caused by the ignition of the embedded star) where the grain-surface molecules achieve sufficient mobility for radical–radical association to form more complex species, *e.g.*, methyl formate ($HCOOCH_3$).[44,45] The necessary radicals are produced on the grain *via* dissociation by UV photons generated internally *via* the interaction of cosmic rays with H_2 molecules.[47] The importance of radiation processing for generating complex molecules in ices has been known for some time, and can be efficient at low temperatures (10 K).[48–50] Further warming to ≥ 100 K allows those COMs formed on the grain surface to return to the gas phase thus 'seeding' the gas with complex molecules. Line emission from COMs in hot cores and hot corinos is often characterised by a gas temperature ≥ 100 K, and abundance estimates range between $\sim 10^{-10}$–10^{-6} that of the H_2 number density.[12]

The formation of complex organic molecules in isolated protoplanetary disks may occur *via* the same mechanism given that a significant fraction of the disk material is ≥ 20 K. A fundamental difference between hot cores and protoplanetary disks is the presence of UV photons and X-rays which permeate the disk surface. In the cold, outer disk, the primary mechanism for releasing non-volatile molecules, such as, H_2O and CH_3OH, into the gas phase is desorption triggered by the absorption of UV photons, a process termed *photodesorption*.[51] In the disk mid-plane, this process is triggered by the absorption of cosmic-ray-induced UV photons. Higher in the disk atmosphere, photodesorption by stellar UV photons and X-rays and interstellar UV photons dominates.[52,53] Photodesorption has been well-studied in the laboratory and is understood to be an indirect process. For pure CO ice, desorption from the ice surface is triggered by the electronic excitation of CO molecules in the bulk ice.[54] For pure H_2O ice, surface desorption is induced by photodissociation of H_2O molecules deeper in the ice.[55,56] In a protoplanetary disk, this generates a *photodesorbed* gas-phase layer of those species otherwise expected to reside on the grain at the temperatures in the outer disk, including H_2O and CH_3OH. Indeed, photodesorption is the favoured explanation for the detection of cold H_2O in the disk of TW Hya.[57] For a protoplanetary disk externally irradiated by a nearby massive star, the disk material is significantly warmer (>50 K); hence, the mechanism for complex molecule formation and release into the gas phase may differ from that described above for isolated systems.

Whether COMs can form within disks also has implications on the chemical significance of comets, considered pristine material left over from the process of planet formation.[58] If COMs do not form efficiently in disks, then it is possible that comets originally consisted of simple ices which were subsequently altered (either *via* UV photons or thermal processes) at a more advanced phase of the evolution of the Solar System. Whether the Sun's natal protoplanetary disk was externally irradiated by nearby massive stars may also play a role in determining the attainable chemical complexity. A further consideration, not addressed in our previous work, is the composition of the material *entering* the protoplanetary disk. Is it possible for complex molecules to have formed at an earlier stage in the evolution of the star? If so, how are their abundances altered as they are transported through the disk? The work presented in this discussion aims to address these fundamental questions.

3 Protoplanetary disk model

The physical and chemical structure of a protoplanetary disk around a typical T Tauri star is simulated for two scenarios: (i) the *isolated* case, in which the disk is irradiated by FUV and X-ray photons originating from the star and interstellar medium (ISM) only, and (ii) the *irradiated* case, in which the disk undergoes additional irradiation by FUV photons from a nearby (<0.1 pc) massive O-type star.

3.1 Physical model

The physical structure of each protoplanetary disk model is calculated using the methods outlined in previous work.[59,60] Here, the important parameters only are highlighted.

The central star for both cases is a typical classical T Tauri star with mass, $M_* = 0.5\ M_\odot$, radius, $R_* = 0.5\ R_\odot$, and an effective temperature, $T_* = 4000$ K.[61] The disk is axisymmetric and in Keplerian rotation about the parent star. The physical structure of the disk is assumed to be steady (*i.e.*, unchanging in time) with a constant mass accretion rate, $\dot{M} = 10^{-8}\ M_\odot\ \mathrm{yr}^{-1}$. The kinematic viscosity in the disk, ν, is parameterised using the dimensionless α parameter which scales the maximum expected size of turbulent eddies by the sound speed (c_s) and scale height (H) of the disk, $\nu \approx \alpha c_\mathrm{s} H$.[62] For a typical T Tauri disk, $\alpha \sim 0.01$.

The dust temperature, gas temperature, and density structure of the disk are determined in a self-consistent manner by iteratively solving the necessary equations.[63] Initially, the physical structure of an optically thick viscous accretion disk in hydrostatic equilibrium is adopted.[64] The specific intensity everywhere in the disk, $I_\nu(r,\ \phi)$, is calculated by solving the axisymmetric two-dimensional radiative transfer equation for the propagation of radiation originating from the central star and external sources such as the ISM and/or a nearby massive star. The dust temperature at each position in the disk is then recalculated assuming local radiative equilibrium between the absorption and reemission of radiation by dust grains. The new gas density and temperature are determined by solving the equations of hydrostatic equilibrium in the vertical direction and detailed balance between heating and cooling of the gas, respectively. The new disk structure is adopted and the above steps performed until the convergence criterion is reached. In this way, the FUV and X-ray mean intensities are also determined at each position in the disk.

Heating sources included are radiation from the parent star and the ISM (and the nearby massive O-type star for the irradiated disk case) and the radiative flux generated by viscous dissipation in the disk midplane.[62] Gas heating mechanisms included are grain photoelectric heating induced by FUV photons and X-ray heating due to the X-ray ionisation of H atoms. The gas is allowed to cool *via* gas–grain collisions and line transitions.

UV excess emission is commonly observed towards T Tauri stars.[65] The radiation field from the central star is modelled as a combination of black body radiation at the star's effective temperature, optically thin hydrogenic bremsstrahlung radiation at a higher temperature ($T_\mathrm{br} = 25\ 000$ K), and Lyman-α line emission. The total FUV luminosity from the central star is $\sim 10^{31}\ \mathrm{erg\ s}^{-1}$. For the isolated disk case, external irradiation by the interstellar radiation field is also included ($G_0 \approx 1.6 \times 10^{-3}\ \mathrm{erg\ cm}^{-2}\ \mathrm{s}^{-1}$). The irradiated disk experiences additional irradiation by a nearby O-type star. The O-type star is assumed to radiate as a black body with effective temperature, $T_\mathrm{eff} = 45\ 000$ K. The FUV flux at the disk surface is set to $G_\mathrm{ext} = (4 \times 10^5) \times G_0$. This corresponds to a distance between the disk and the massive star of $\lesssim 0.1$ pc. Young stars are also bright in X-rays.[66] The X-ray emission from the central star for both cases is modelled as a fit to the observed *XMM-Newton* spectrum from the nearby classical T Tauri star, TW Hya. The total X-ray luminosity is $\sim 10^{30}$ erg s^{-1}.

The main FUV opacity is absorption and scattering by dust grains. The dust grains are assumed to be well mixed with the gas with a gas-to-dust mass ratio ~ 100 and the dust-grain size distribution adopted is that which reproduces the extinction curve of dense clouds.[67] In the computation of the chemical reaction rates, for simplicity, a single dust-grain size and number density are used (see Section 3.2). The extinction of X-rays occurs *via* two processes: absorption *via* photoionisation of all elements and Compton scattering by H atoms.[68]

3.2   Chemical model

The physical conditions encountered in protoplanetary disks cover a wide range in density, temperature, and radiation field strength. To compute the chemistry in all regions of the disk in an appropriate manner, a network which includes all possible chemical processes must be used. The network used is the same as in previous work[19] which originates from the expanded version of the *Ohio State University* (OSU) network.[44–46] The methods used to compute the majority of the reaction rate coefficients are the same as those outlined in previous work.[53] The full chemical network consists of ≈ 9350 reactions involving ≈ 800 species. A brief description is given here.

3.2.1   Gas-phase reactions. The gas-phase network includes two-body reactions, photodissociation and photoionisation, direct cosmic-ray ionisation, and cosmic-ray-induced photodissociation and photoionisation. The core network is expanded to include direct X-ray ionisation and X-ray-induced dissociation and ionisation of elements and molecules. This is simulated by replicating the set of cosmic-ray reactions and scaling the dissociation and ionisation rates by the total X-ray ionisation rate calculated at each grid point in the disk, $\zeta_{xr}(x,z)$. The photoreaction rates in the core network are normalised to the average strength of the unshielded interstellar radiation field, G_0.

Under astrophysical conditions, many molecules dissociate primarily *via* line absorption rather than *via* continuum absorption.[69,70] For those species which achieve sufficiently large column densities, self shielding can dominate over shielding by dust grains.[71] Self shielding occurs when foreground molecular material absorbs the photons necessary for photodissociation to occur deeper into the atmosphere. The photodissociation cross sections can also overlap with those for H_2. In this work, the self- and mutual-shielding of H_2, CO, and N_2 are taken into account.[72–74]

3.2.2   Gas–grain interactions. The gas-phase network is supplemented with gas–grain interactions to allow the condensation (freezeout) and sublimation (desorption) of molecules onto, and from, dust grains, respectively. The freezeout rate onto dust grains is dependent upon the geometrical cross section of dust grains and the velocity of the gas. It is assumed that sticking occurs upon each collision with a dust grain, *i.e.*, a sticking coefficient equal to 1 is used for all species. H atoms are known to stick less efficiently to grain surfaces than other species.[75] Here, the temperature dependence of the sticking coefficient for H atoms is taken into account.[76,77]

The thermal desorption rate is calculated assuming the bound molecule oscillates in a harmonic potential well.[78] The thermal desorption rate depends on the binding energy of the molecule to the grain surface, the temperature of the dust, and the characteristic frequency of vibration of the molecule. Volatile molecules, *e.g.*, CO, will desorb from the grain mantle at lower temperatures than less volatile species, *e.g.*, H_2O. Thermal desorption triggered by cosmic-ray heating of dust grains is also included.[79] This process is significant for volatile molecules only.

For species which are strongly bound to the grain mantle, *e.g.*, H_2O, photodesorption is the primary mechanism for releasing molecules into the gas phase in the cold, outer disk. The photodesorption rate of each species depends on the flux of UV photons, the geometric cross section of a dust-grain binding site, the surface coverage of the molecule, and the molecular yield per UV photon. This process is well studied in the laboratory and photodesorption yields are now determined for CO, N_2, CO_2, H_2O, and CH_3OH.[50,55,80,81] For all other species, a yield of 10^{-3} molecules photon^{-1} is adopted, in line with those values already constrained in the laboratory. Recent experiments and simulations have also determined that photodesorption occurs from the top two monolayers of the ice mantle only, and this is accounted for in the calculation of the photodesorption rates.[54] Photodesorption triggered by all sources of UV photons are included: external stellar and interstellar UV photons, and UV photons generated internally by the interaction of cosmic rays with H_2 molecules.

3.2.3 Grain-surface reactions. A comprehensive grain-surface network is adopted to simulate the formation of COMs.[44–46] The grain-surface association reactions are assumed to occur *via* the Langmuir–Hinshelwood mechanism only, *i.e.*, two adsorbed species on the grain surface diffuse and react. The reaction rate coefficients are calculated using the rate equation method.[78] This method is appropriate due to the relatively high densities, $\gtrsim 10^7$ cm^{-3}, in the regions of the disk where grain-surface chemistry is important. Because of the discrete nature of dust grains, processes occurring on grain surfaces are stochastic: however, comparisons of stochastic models of grain-surface chemistry with models employing the rate equation method show that stochastic effects are most important at lower densities ($\lesssim 10^5$ cm^{-2}).[82–84]

Classical thermal diffusion is assumed for all species, except H and H_2 which are also allowed to quantum tunnel through the diffusion barrier. This is the 'optimistic' case: recent experiments investigating the surface mobility of H atoms on water ice have proved inconclusive regarding the exact contribution of quantum diffusion to the total diffusion rate.[85] The ratio of the diffusion barrier to the binding energy of each species is assumed to be 0.3. Again, this is an 'optimistic' value which allows the efficient diffusion and reaction of relatively volatile species at $\sim$20 K. For each exothermic grain-surface association reaction which leads to a single product, there is a probability that the product will be returned to the gas phase. This is the process of 'reactive' or 'chemical' desorption and provides an additional desorption mechanism for non-volatile molecules. A probability of 1% is adopted in this work.[86]

3.2.4 Dust-grain model. A fixed dust grain size of 10^{-5} cm is adopted and the number density of dust grains is assumed to equal $\sim 10^{-12}$ that of H nuclei. These values correspond to so-called classical grains which possess $\approx 10^6$ surface binding sites per grain.

3.2.5 H_2 formation rate. H_2 molecules also form on dust-grain surfaces and the conversion from H to H_2 and *vice versa* must also be taken into account. Here, the most optimistic case is adopted: it is assumed that the rate of formation of gas-phase H_2 equates to half the rate of arrival of H atoms onto the grain surfaces. This allows efficient formation of H_2 on warm dust grains (≥ 25 K) in line with astrophysical observations towards diffuse clouds and photon-dominated regions (PDRs).[87,88] If the explicit grain-surface route to H_2 is included, the rate of formation of gas-phase H_2 drops significantly at warm temperatures (≥ 25 K) due to the volatile nature of H atoms. In reality, the formation rate likely lies between these two extrema. Several solutions have been suggested such as allowing both the physisorption and chemisorption of H atoms on dust grains, and including H_2 formation *via* both the Eley–Rideal and Langmuir–Hinshelwood mechanisms.[89–91] Certainly, these solutions should be explored in future protoplanetary disk models.

Note that this work does include the temperature-dependent behaviour of the sticking coefficient of H atoms which is found to decrease with increasing temperature.[76,77]

3.3   Accretion flow

In previous work, the chemistry was computed at each point in the disk as a function of time, using initial abundances generated using a simple molecular cloud model.[20,20,53,92] The chemistry was set up to evolve at each grid point independent of that in neighbouring grid points, *i.e.*, the rate of change in abundance of species i at a point(x, z) was given by,

$$\frac{\mathrm{d}n_i(x,z)}{\mathrm{d}t} = P_i - D_i \ \mathrm{cm}^{-3}\mathrm{s}^{-1}, \tag{2}$$

where P_i and D_i are the production and destruction rates of species, i, respectively. Here, chemical abundances are computed along the accretion flow by integrating the chemical evolution of a fluid parcel along streamlines from the outer edge of the disk inwards towards the central star. The rate of change in abundance of species i, is derived assuming,

$$\frac{\partial(n_i v_x)}{\partial x} = \frac{\mathrm{d}n_i}{\mathrm{d}t} = P_i - D_i \ \mathrm{cm}^{-3}\mathrm{s}^{-1}, \tag{3}$$

where v_x is the radial velocity along a streamline. Using the continuity equation,

$$\frac{\partial(n_T v_x)}{\partial x} = 0 \ \mathrm{cm}^{-3} \ \mathrm{s}^{-1}, \tag{4}$$

where n_T is the total number density, the rate of change in abundance of species i with radius is given by

$$\frac{\mathrm{d}n_i(x,z)}{\mathrm{d}x} = \frac{1}{v_x(x)} \frac{\mathrm{d}n_i(x,z)}{\mathrm{d}t} + \frac{n_i(x,z)}{n_T(x,z)} \frac{\mathrm{d}n_T(x,z)}{\mathrm{d}x} \ \mathrm{cm}^{-2}. \tag{5}$$

The radial velocity of the flow is given by

$$v_x(x) = -\frac{\dot{M}}{2\pi\Sigma(x)x} \ \mathrm{cm} \ \mathrm{s}^{-1}, \tag{6}$$

where $\dot{M}$ is the assumed mass accretion rate and $\Sigma(x)$ is the surface density of the disk at a radius, x. Streamlines, l, are defined as fractions of the disk scale height, *i.e.*, $l = fH$, with $0 \leq f \leq 4$. The disk scale height is determined at the disk midplane for all streamlines,

$$H_0(x) = \frac{c_{s0}(x)}{\Omega(x)} \propto T_0^{1/2} x^{3/2},$$ (7)

where $c_{s0}(x) = \sqrt{kT_0(x)/m_0(x)}$, $T_0(x)$, and $m_0(x)$ are the sound speed, gas temperature, and mean molecular mass in the disk midplane, and $\Omega(x) = \sqrt{GM_*/x^3}$ is the Keplerian angular velocity at a radius, x. G, M_*, and k represent the gravitational constant, stellar mass, and Boltzmann's constant, respectively. The disk scale height, H, and the geometrical height, z, are related *via* the assumption of hydrostatic equilibrium which defines the density structure, $\rho(x, z)$,

$$\rho(x, z) = \rho_0(x)\exp\left(\frac{-z^2}{2H(x)^2}\right),$$ (8)

where $\rho_0(x)$ is the density in the midplane, $z = 0$, at a radius, x. Physical conditions, such as density, temperature, and radiation field strength, are extrapolated along the flow by assuming each parameter follows a power law behaviour $(\propto \alpha x^{-\beta})$ between grid points.

3.4 Initial conditions

The initial abundances injected into the outer edge of each streamline are extracted at a time of 10^5 years from a simple time-dependent molecular cloud model with a fixed density of 10^5 cm^{-3} and a high visual extinction of $\gg 10$ mag. The chemical model used is the same as that described in Section 3.2. A cosmic-ray ionisation rate of 1.3×10^{-17} s^{-1} is assumed. The time-dependent cloud model calculation is begun assuming all species are in atomic form except hydrogen which is in molecular form. The assumed elemental abundances for H : He : C : N : O are $1.0 : 9.75(-2) : 1.4(-4) : 7.5(-5) : 3.2(-4)$. For the heavier elements, H : Na : Mg : Si : S : Cl : Fe, the ratios are $1.0 : 2.0(-9) : 7.0(-9) : 8.0(-9) : 8.0(-8) : 4.0(-9) : 3.0(-9)$.

It is assumed that the material entering the outer region of the protoplanetary disk has remained shielded from the central star. Two sets of initial abundances are generated: one for the 'cold' case assuming a constant gas and dust temperature of 10 K, and one for the 'warm' case assuming a constant temperature of 30 K. In this way, either 'pristine' dark cloud initial abundances are used, or it is assumed a degree of thermal processing has occurred before injection into the disk. The latter value of 30 K was chosen based on preceding work investigating the formation and distribution of COMs in a static model of an isolated protoplanetary disk.[19] In that work it was found that thermal grain-surface chemistry efficiently operates and builds COMs in the ice when the dust temperature exceeds ≈ 20–30 K, depending on the assumed barrier to surface diffusion (0.3–0.5 times the desorption energy). For the lower value of the diffusion barrier, a peak in the abundances of grain-surface COMs was seen at a temperature of ≈ 30 K; hence, this temperature is adopted for the 'warm' case. In addition, the maximum dust temperature reached at the outer edge of the isolated protoplanetary disk model is ≈ 30 K. This higher temperature also affects the relative abundances and subsequent chemistry of volatile species, for example, CO. In the cold model, only those COMs which can form efficiently *via* atom-addition reactions on grain surfaces reach appreciable fractional abundances because at 10 K atoms alone have sufficient mobility to diffuse and react. However, at 30 K, weakly bound molecules, such as molecular radicals, also have sufficient mobility, allowing greater complexity to build within the ice mantle *via* radical–radical association reactions.[44] Hence, for the 'cold' case, the calculations begin with simple ices only, whereas for the 'warm' case, the calculations begin with simple and complex ices. For the warm case, cosmic-ray-induced photo-desorption releases a proportion of COMs from the grain mantle so that the calculations also begin with appreciable fractional abundances of COMs in the gas phase.

The initial molecular fractional abundances (with respect to total H nuclei density) for COMs of interest in this work are listed in Table 1. The species highlighted here are those which have been observed in the gas phase in cold molecular clouds and/or hot cores/corinos and/or require methanol ice as a parent molecule. Also listed are the assumed binding energies for each species.[44] Dimethyl ether, CH_3OCH_3, methyl formate, $HCOOCH_3$, and glycolaldehyde, $HOCH_2CHO$, are all thought to form *via* the association of radicals produced either *via* the hydrogenation of CO ice, or the photodissociation of methanol ice,[45,46,50]

$$CH_3OH + h\nu \rightarrow CH_3 + OH, \tag{9}$$

$$\rightarrow CH_3O + H \tag{10}$$

$$\rightarrow CH_2OH + H. \tag{11}$$

The relative branching ratios for these photodissociation pathways are thought to influence the resulting abundances of CH_3CHO, CH_3OCH_3, $HCOOCH_3$, and $HOCH_2CHO$, *via* the surface-association reactions,

$$CH_3 + HCO \rightarrow CH_3CHO, \tag{12}$$

$$CH_3O + CH_3 \rightarrow CH_3OCH_3, \tag{13}$$

$$CH_3O + HCO \rightarrow HCOOCH_3, \tag{14}$$

Table 1   Initial molecular fractional abundances and binding energies, $E_B{}^a$

| Molecule | Cold | | Warm | | E_B |
	Gas	Ice	Gas	Ice	(K)
CH_3OH	9.8(−10)	1.8(−06)	1.3(−09)	3.0(−06)	5530
$HCOOH$	7.0(−11)	2.8(−11)	1.3(−09)	1.7(−06)	5570
CH_3CHO	8.7(−12)	9.8(−11)	7.0(−11)	1.2(−08)	2780
CH_3OCH_3	3.1(−14)	1.5(−11)	9.0(−11)	8.1(−08)	3680
$HCOOCH_3$	1.2(−15)	2.3(−15)	5.6(−11)	1.2(−07)	4100
$HOCH_2CHO$	4.4(−21)	1.8(−18)	1.1(−11)	6.6(−08)	6680

^{a} Note: $a(b) = a \times 10^b$. Binding energies from Garrod *et al.*, (2008).

$$CH_2OH + HCO \rightarrow HOCH_2CHO, \tag{15}$$

respectively. In this work, a ratio of $3:1:1$ for pathways (9), (10), and (11) is assumed. This corresponds to the 'standard' case in the study of methyl formate formation in hot cores by Laas *et al.*, (2011).[46]

The faster diffusion rate of grain-surface radicals is apparent in the increased abundances of $HCOOH$, CH_3CHO, CH_3OCH_3, $HCOOCH_3$, and $HOCH_2CHO$ on the grain surface for the warm case. This leads to a higher abundance in the gas: grain-surface species formed on the grain are released to the gas *via* non-thermal desorption (see Section 3.2). Hence, the calculations for the warm case begin with significantly higher abundances of COMs more complex than methanol than for the cold case.

4   Results

4.1   Physical structure

The locations and paths of the streamlines over which the chemistry is calculated are shown in Fig. 1 for the isolated case (top panel) and the irradiated case (bottom panel). The chemistry evolves along each streamline from outside to inside in the direction of increasing time (shown in the top axes of each panel). The disk scale height at large radii is larger for the irradiated disk due to the higher gas temperature in the disk midplane. The streamlines considered are those which are deep enough in the disk such that ices remain frozen out in the outer disk.

The number density (red lines) and gas temperature (blue lines) along each streamline as a function of disk radius is displayed in Fig. 2. The integrated FUV flux (green lines) and X-ray flux (gold lines) as a function of radius are also shown in Fig. 3. The data for the isolated disk are shown in the left-hand panel and those for the irradiated disk are shown in the right-hand panel.

The density structure in both disk models is similar along streamlines at the same scale height because the assumed surface density is the same in both models. The X-ray fluxes in both models are similar for the same reason *i.e.*, the X-rays 'see' a similar attenuating column of material. The temperature in the outer region ($\gtrsim 10$ AU) of the irradiated disk (right-hand panel) is significantly higher than for the isolated disk. This is due to the increased FUV flux from the nearby O-type star which is shown in Fig. 3. However, the temperature in the inner region ($\lesssim 1$ AU) of the isolated disk is higher than that of the irradiated disk for the higher streamlines ($\gtrsim 2\,H$). This is because at small radii, the disk scale height for the isolated disk is slightly larger than for the irradiated disk; hence, the higher streamlines for the isolated disk flow through regions of higher temperature (and thus lower density). This is also demonstrated by the increased FUV flux for these streamlines (see Fig. 3).

4.2   Chemical evolution

The fractional abundance (with respect to H nuclei density) of each species listed in Table 1 is shown in Fig. 4–13 for both the isolated and irradiated disk models. We do not show the results for CH_3OCH_3 because the behaviour is the same as that for $HCOOCH_3$. The blue and red lines in each plot represent the results for which the 'cold' and 'warm' sets of initial molecular abundances are assumed,

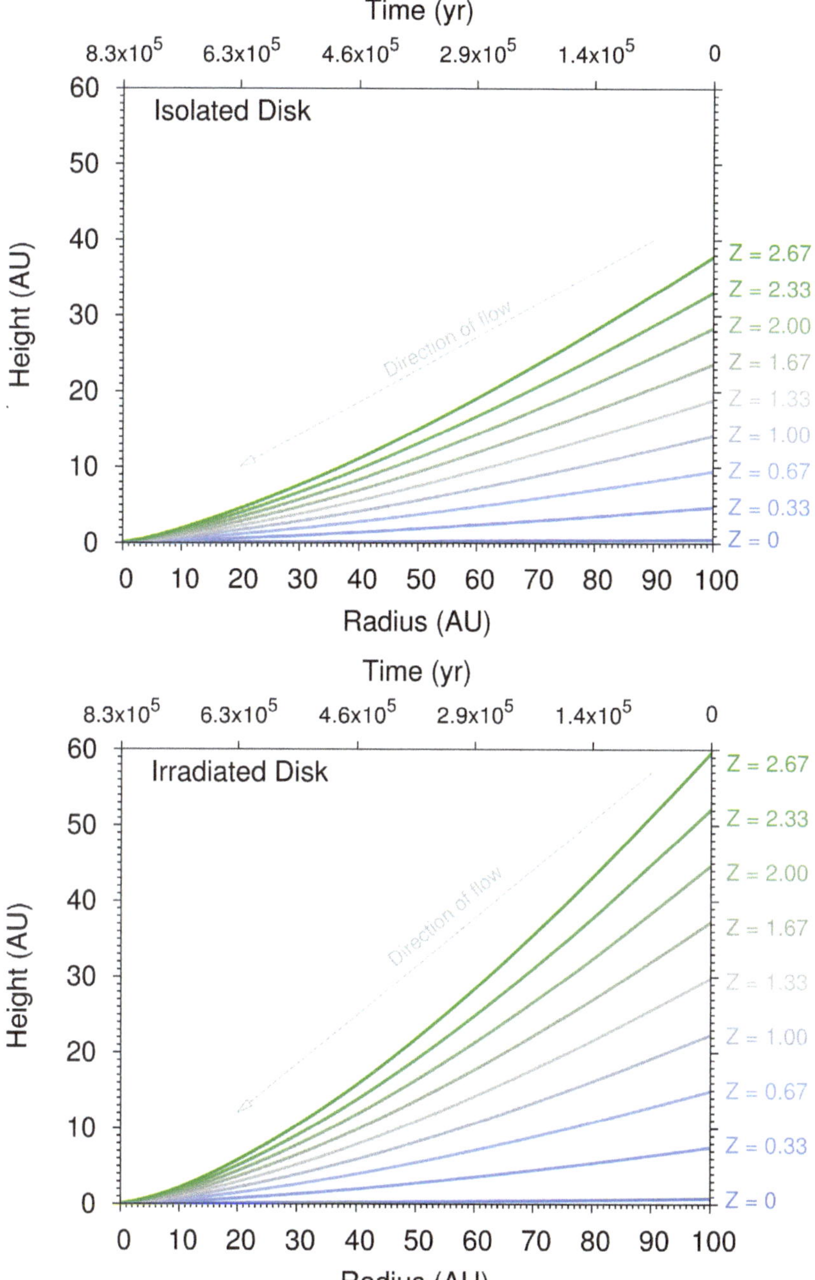

Fig. 1 Location and paths of streamlines along which the chemistry is calculated for the isolated disk (top panel) and irradiated disk (bottom panel). Time (shown on the top axes) increases from right to left.

respectively. There are several general trends. The ice mantle survives to larger scale heights in the isolated disk than in the irradiated disk ($Z \lesssim 2\,H$ *versus* $Z \lesssim 0.67\,H$). The stronger UV field in the outer regions of the irradiated disk efficiently

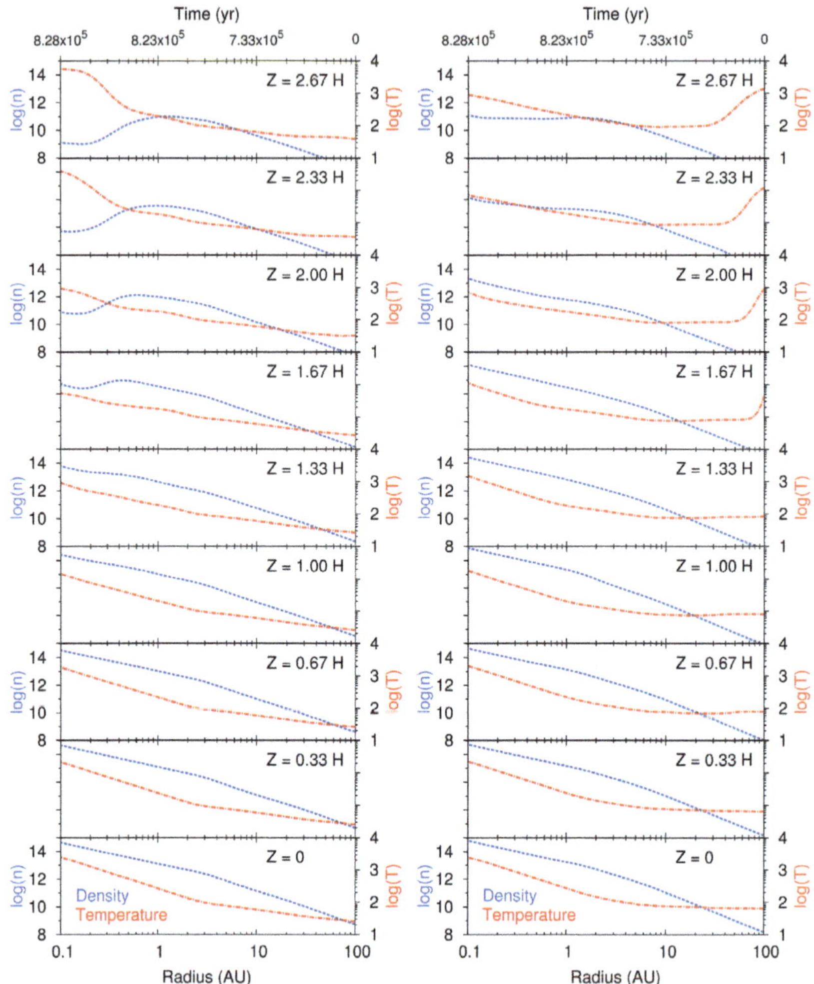

Fig. 2　Number density (blue lines) and gas temperature (red lines) along each streamline as a function of radius for the isolated disk (left-hand panel) and irradiated disk (right-hand panel). Time (shown on the top axes) increases from right to left.

strips the ice from the grain surface *via* photodesorption. Gas-phase COMs are subsequently destroyed by photodissociation. The results suggest gas-phase COMs do not survive in the molecular layers of externally irradiated disks. The gas-phase COMs in both models reach their highest fractional abundance in the inner disk midplane. The origin of the gas-phase COMs is either thermal desorption from the ice mantle or gas-phase formation. The dominant process depends on the composition of the ice mantle entering the inner disk. The initial abundances (*i.e.*, cold or warm) are not as critical for the isolated disk as for the irradiated disk. For those COMs which begin with a negligible abundance on the grain for the cold case, these species are formed efficiently on the grain surface under the conditions in the outer isolated disk (see, *e.g.*, Fig. 8, 10, and 12). On the other hand, for the irradiated disk, the initial composition is critical.

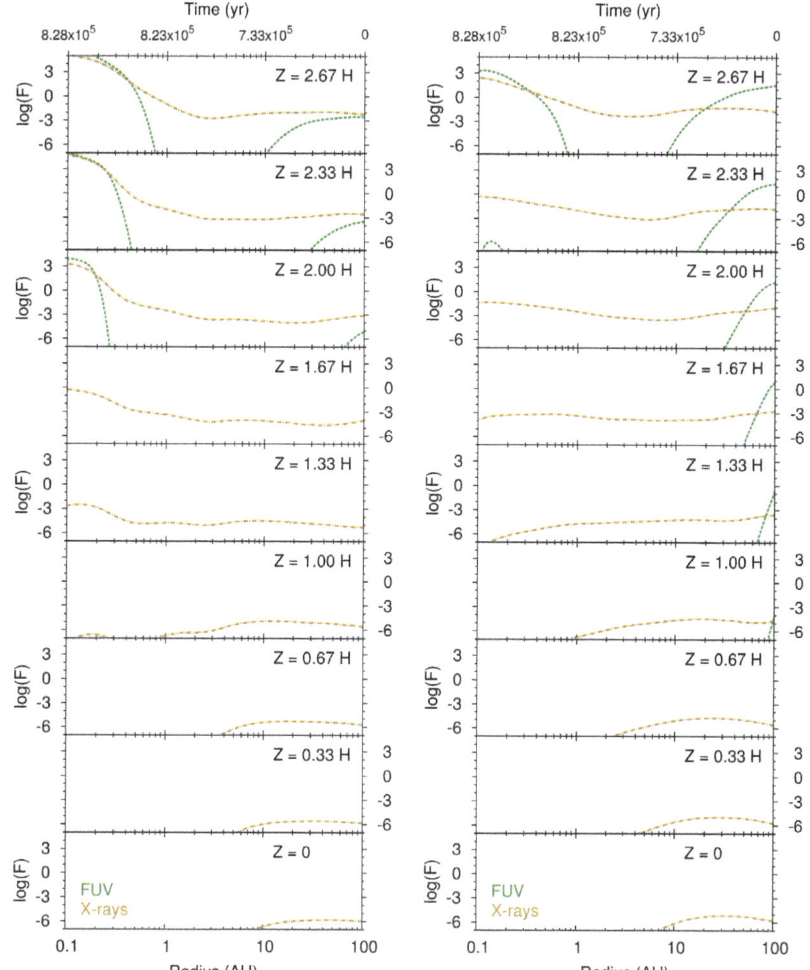

Fig. 3   FUV flux (green lines) and X-ray flux (gold lines) along each streamline as a function of radius for the isolated disk (right-hand panel) and irradiated disk (left-hand panel). The units are erg cm^{-2} s^{-1}. Time (shown on the top axes) increases from right to left.

The specific behaviour and chemistry of the species predicted by reactions (12) to (15) are now discussed.

4.2.1   Methanol. In Fig. 4 and 5 the fractional abundance of gas-phase (left-hand panel) and grain-surface (right-hand panel) methanol, CH_3OH, is shown as a function of disk radius for the isolated and irradiated disk models, respectively.

The methanol abundance is preserved along the accretion flow in the mid-plane in both models. At ≈ 2 AU, the temperature is sufficiently high for methanol ice to thermally desorb from the grain surface into the gas phase. This transition region is known as the *snowline* and resides at the same radius for both models. This is because the heating in the inner disk is dominated by the central star and viscous dissipation rather than external irradiation. The abundance reached in the gas phase is the same as that for the ice injected into both models. This

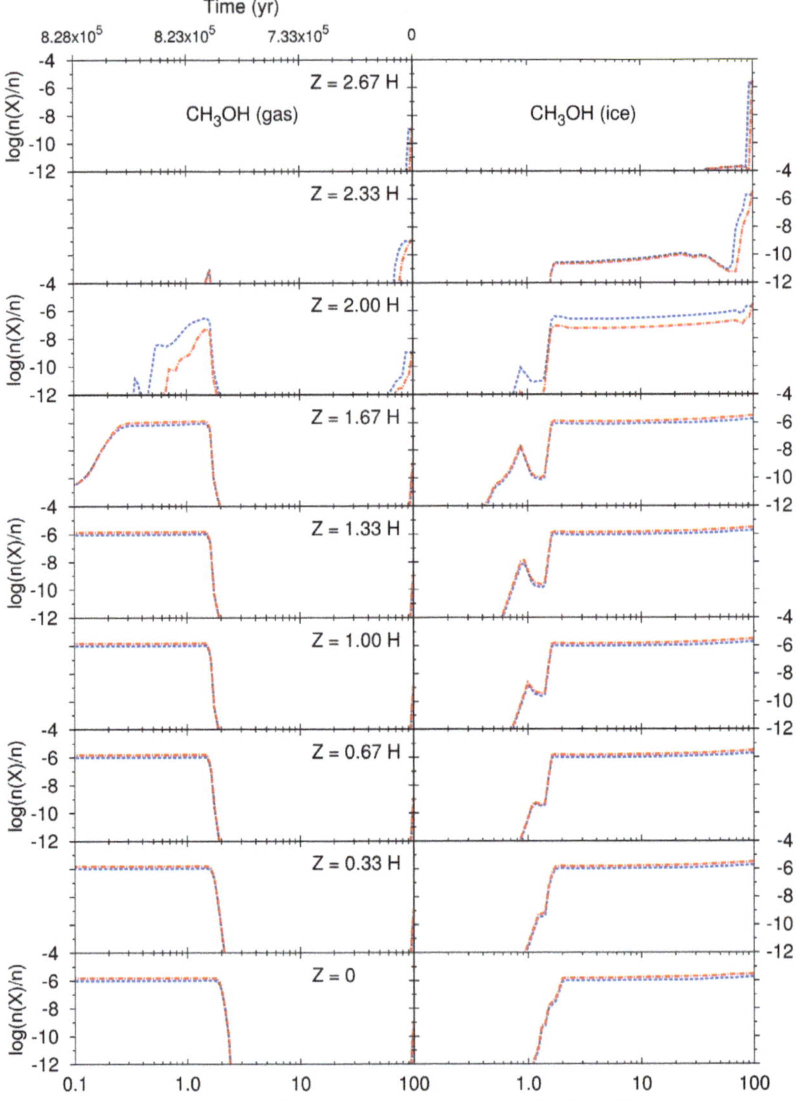

Fig. 4 Fractional abundance (with respect to number density) of CH$_3$OH gas and ice along each streamline as a function of disk radius for the isolated disk model. The blue lines show results from the model using the 'cold' (10 K) set of initial abundances and the red lines from the model using the 'warm' (30 K) set of initial abundances (see, Table 1).

suggests that the methanol entering the planet- and comet-forming region in both irradiated and isolated protoplanetary disks has an interstellar origin.

At higher scale heights in the isolated disk, methanol is photodesorbed into the gas phase reaching a fractional abundance $\sim 10^{-9}$ with respect to number density at radii ≥ 50 AU. This photodesorbed layer is not present in the irradiated case. Whether gas-phase molecules survive in the photodesorbed layer requires a delicate balance between photodesorption and photodissociation in the

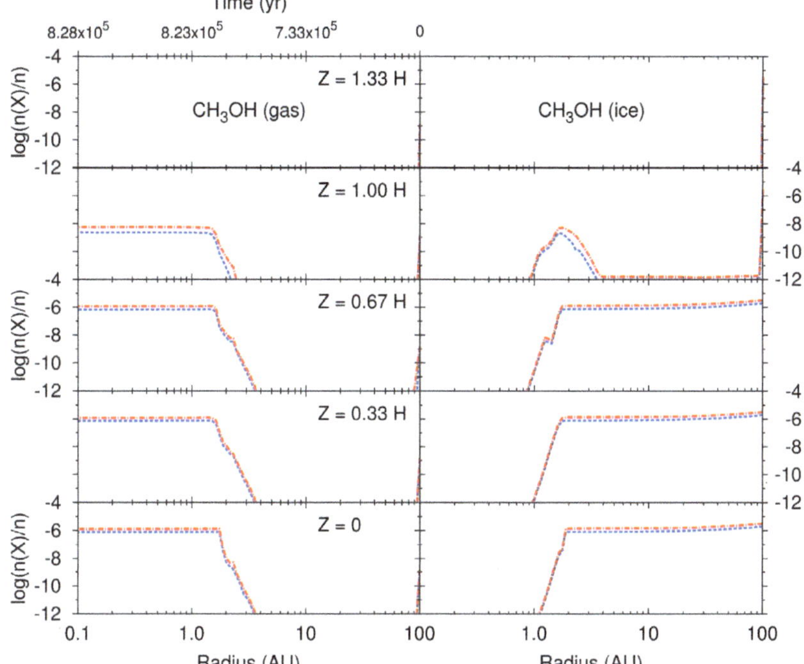

Fig. 5 Fractional abundance (with respect to number density) of CH_3OH gas and ice along each streamline as a function of disk radius for the irradiated disk model. The blue lines show results from the model using the 'cold' (10 K) set of initial abundances and the red lines from the model using the 'warm' (30 K) set of initial abundances (see, Table 1).

molecular layer of the disk. The results suggest that gas-phase methanol may be observable in isolated protoplanetary disks as was found in the static model.[19]

There are only minor differences between the cold and warm sets of abundances for both disk models because methanol ice reaches similar fractional abundances in the cold and warm cloud models ($\approx 10^{-6}$).

4.2.2 Formic acid. In Fig. 6 and 7 the fractional abundance of gas-phase (left-hand panel) and grain-surface (right-hand panel) formic acid, HCOOH, is shown as a function of disk radius for the isolated and irradiated disk models, respectively. The story for formic acid is different from that for methanol. Formic acid is postulated to possess several routes to formation on grain surfaces at low temperatures, including the radical–radical association reaction,

$$HCO + OH \rightarrow HCOOH. \tag{16}$$

HCOOH has also been postulated to form *via* hydrogenation of the 'HOCO' complex, [93] *i.e.*,

$$CO + OH \rightarrow HOCO, \tag{17}$$

$$HOCO + H \rightarrow HCOOH. \tag{18}$$

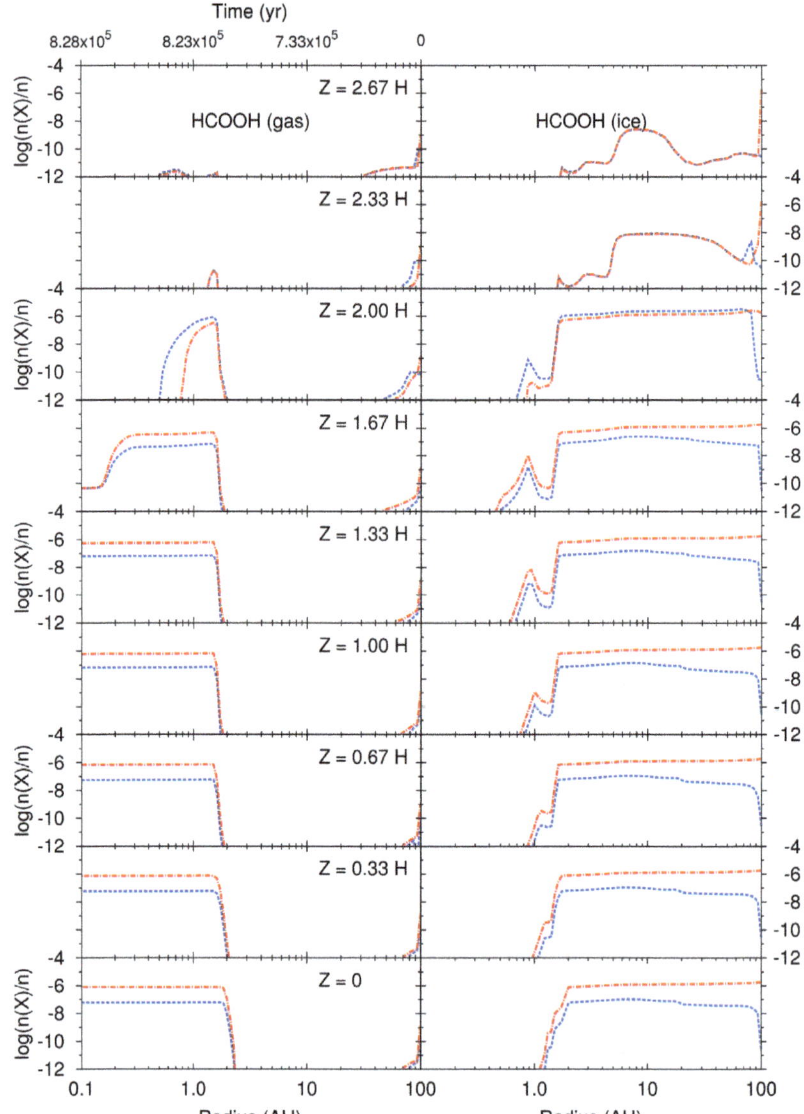

Fig. 6 Fractional abundance (with respect to number density) of HCOOH gas and ice along each streamline as a function of disk radius for the isolated disk model. The blue lines show results from the model using the 'cold' (10 K) set of initial abundances and the red lines from the model using the 'warm' (30 K) set of initial abundances (see, Table 1).

Öberg *et al.* studied the formation of HCOOH *via* the first reaction during the irradiation of pure methanol ice. Upper limits only were determined due to the difficulty in uniquely identifying HCOOH bands in the RAIRS (reflection–absorption infrared spectroscopy) spectrum.[50] Ioppolo *et al.* later studied the formation of HCOOH *via* the second pathway and determined it was efficient at low temperatures $\lesssim 20$ K.[93] Both pathways are included in the network used here; however, the reaction, $CO + OH \rightarrow HOCO$, has a large reaction barrier, ≥ 1500 K,

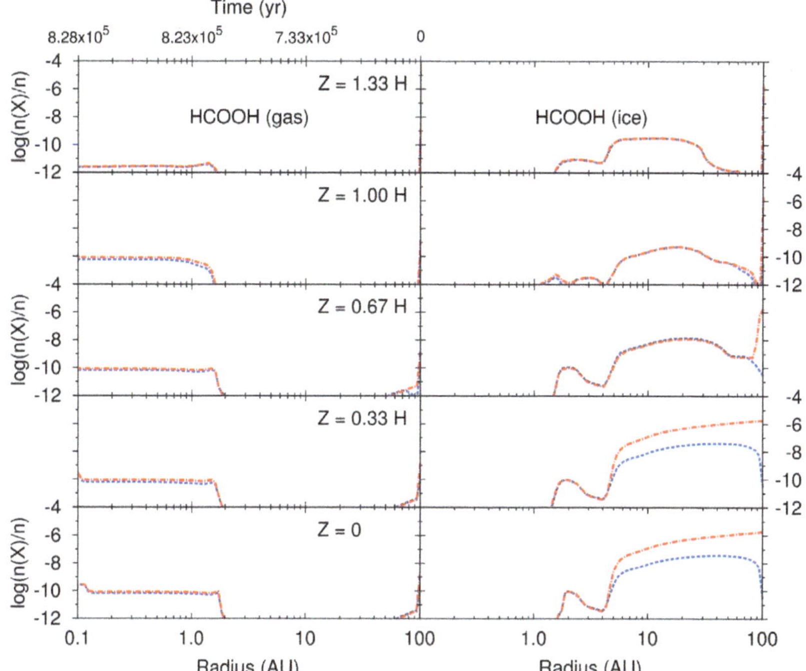

Fig. 7 Fractional abundance (with respect to number density) of HCOOH gas and ice along each streamline as a function of disk radius for the irradiated disk model. The blue lines show results from the model using the 'cold' (10 K) set of initial abundances and the red lines from the model using the 'warm' (30 K) set of initial abundances (see, Table 1).

originating from studies of the gas-phase reaction potential energy surface.[94] Because thermal grain-surface chemistry is used, this reaction is slow at low temperatures and HCOOH is not efficiently formed at $\sim$10 K. As a consequence, the distribution of HCOOH along the accretion flow is sensitive to the assumed initial abundance.

In the isolated disk, the abundance of HCOOH ice and gas for the warm case is around two orders of magnitude higher than for the cold case. A comparison with the initial abundances shows that grain-surface HCOOH is formed in the outer disk for the cold case in both models albeit reaching a lower abundance than that achieved for the warm case. In the isolated disk, as found for methanol, the formic acid fractional abundance, $\sim$$10^{-7}$ – 10^{-6}, is preserved along the accretion flow. The snowline for HCOOH also resides at a similar radius to that for methanol ($\approx$2 AU) reflecting their similar binding energies (see Table 1). At higher scale heights in the isolated disk, formic acid is photodesorbed reaching a gas-phase fractional abundance $\sim$ 10^{-9}. Formic acid ice transported into the inner regions of isolated protoplanetary disks may have an interstellar origin provided the initial abundance entering the disk is sufficiently high. HCOOH ice has been observed in the envelopes of several low-mass protostars with abundances $\sim$1–5% of the water ice abundance.[95]

In the irradiated disk, the gas-phase formic acid abundance in the inner region of the disk does not reflect that injected into the outer disk. The higher

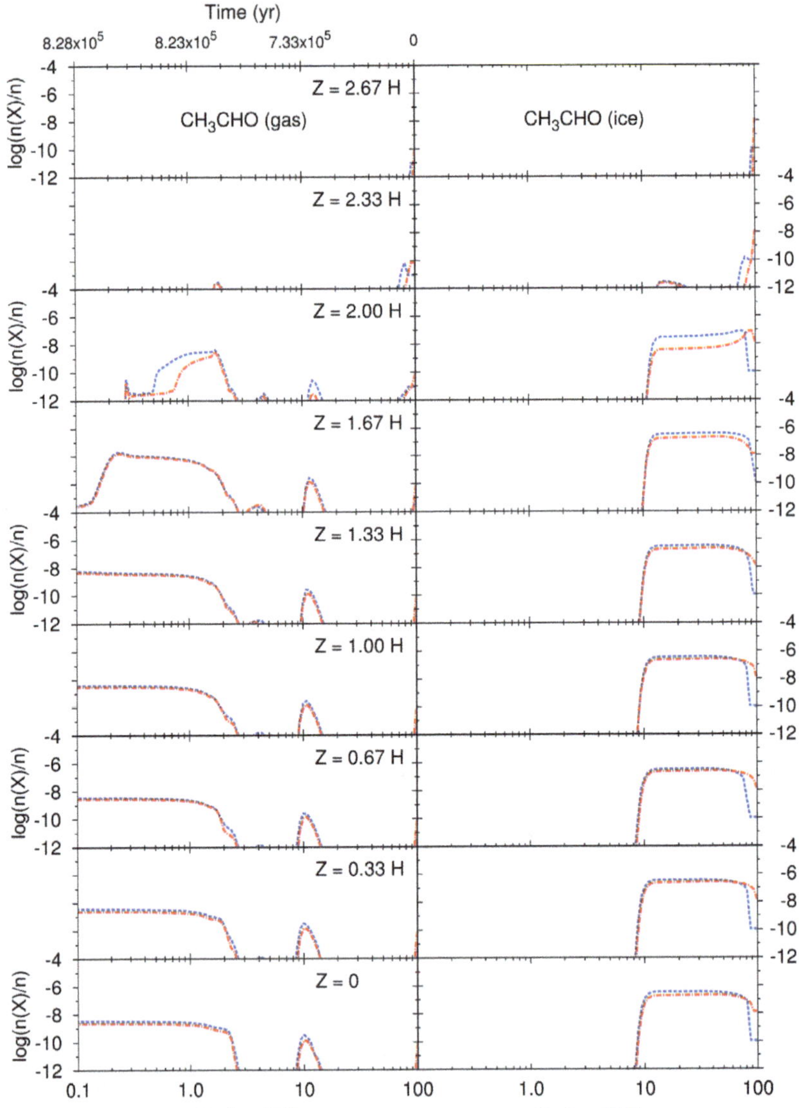

Fig. 8 Fractional abundance (with respect to number density) of CH_3CHO gas and ice along each streamline as a function of disk radius for the isolated disk model. The blue lines show results from the model using the 'cold' (10 K) set of initial abundances and the red lines from the model using the 'warm' (30 K) set of initial abundances (see, Table 1).

temperature (>60 K) in the midplane helps thermally process the ice such that the cold and warm models converge at a radius of ≈ 5 AU, *i.e.*, outside the snowline at ≈ 2 AU. This thermal processing leads to a drop in the fractional abundance of formic acid ice from $\sim 10^{-8}$–10^{-6} to 10^{-10} at ≈ 5 AU. Hence, the thermally desorbed formic acid inside of the snowline reaches a peak fractional abundance $\sim 10^{-10}$. Molecules which rely on radical–radical association reactions are sensitive to the higher temperature because radicals generally possess a lower binding

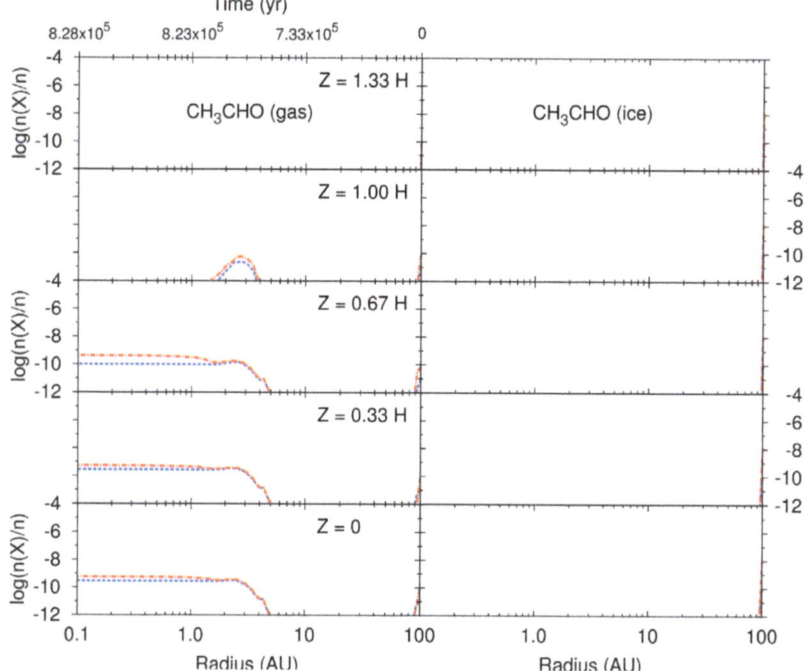

Fig. 9 Fractional abundance (with respect to number density) of CH₃CHO gas and ice along each streamline as a function of disk radius for the irradiated disk model. The blue lines show results from the model using the 'cold' (10 K) set of initial abundances and the red lines from the model using the 'warm' (30 K) set of initial abundances (see, Table 1).

energy to the grain surface than molecules, *e.g.*, the binding energy assumed here for OH and HCO are 2850 K and 1600 K, respectively. Whether a molecule can form efficiently *via* this process relies on a sufficient supply of precursor radicals on the grain surface. At higher temperatures, the thermal desorption of radicals begins to compete with the thermal grain-surface association rates. For the case of formic acid in the irradiated disk midplane, this occurs at ≈ 80 K (at ≈ 5 AU).

4.2.3   Acetaldehyde. In Fig. 8 and 9 the fractional abundance of gas-phase (left-hand panel) and grain-surface (right-hand panel) acetaldehyde, CH_3CHO, is shown as a function of disk radius for the isolated and irradiated disk models, respectively.

In contrast to formic acid, acetaldehyde is not sensitive to the assumed initial abundances. In the isolated disk for the cold case, CH_3CHO forms efficiently in the outer disk midplane reaching a similar ice fractional abundance as for the warm case, $\sim 10^{-7}$, for radii ≥ 10 AU. Acetaldehyde ice can from on the grain *via* the association of the CH_3 and HCO radicals. This reaction is barrierless and so can proceed at the low temperatures in the outer regions of the isolated disk. The snowline for acetaldehyde resides at ≈ 10 AU reflecting its lower binding energy (2780 K). There is a narrow thermally desorbed region of gas-phase acetaldehyde at ≈ 10 AU. Within this radius, the temperature is sufficiently high for efficient thermal desorption of both the CH_3 and HCO radicals so that the radical–radical association formation pathway is no longer effective. Acetaldehyde can also form

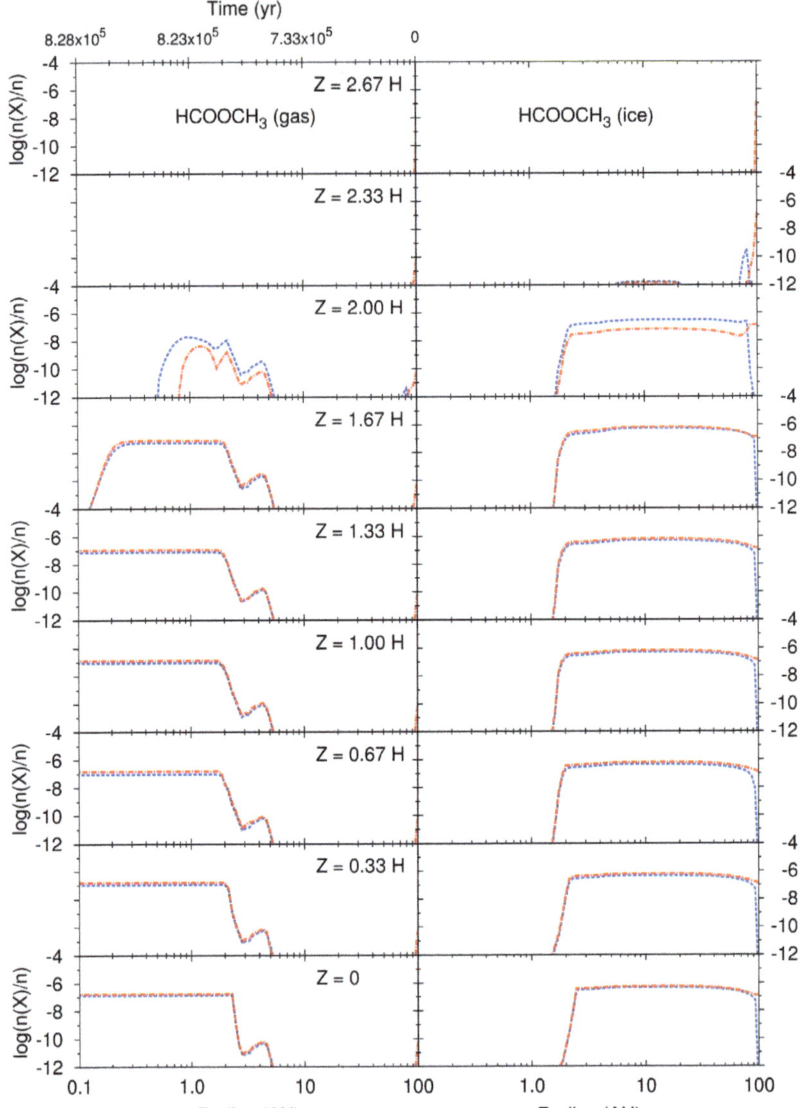

Fig. 10 Fractional abundance (with respect to number density) of HCOOCH₃ gas and ice along each streamline as a function of disk radius for the isolated disk model. The blue lines show results from the model using the 'cold' (10 K) set of initial abundances and the red lines from the model using the 'warm' (30 K) set of initial abundances (see, Table 1).

via hydrogenation of CH_3CO which itself forms *via* the association of CH_3 and CO; however, this reaction has a large reaction barrier (≈ 3500 K) and CO is also a volatile species with a binding energy of 1150 K. The origin of gas-phase acetaldehyde in the inner disk midplane ($\lesssim 2$ AU) is thus gas-phase chemistry induced by the thermal desorption of strongly bound COMs, for example, C_2H_5OH. Gas-phase acetaldehyde forms primarily *via* dissociative electron recombination of the protonated form, $CH_3CH_2O^+$, which can form *via* reaction of cations with C_2H_5OH.

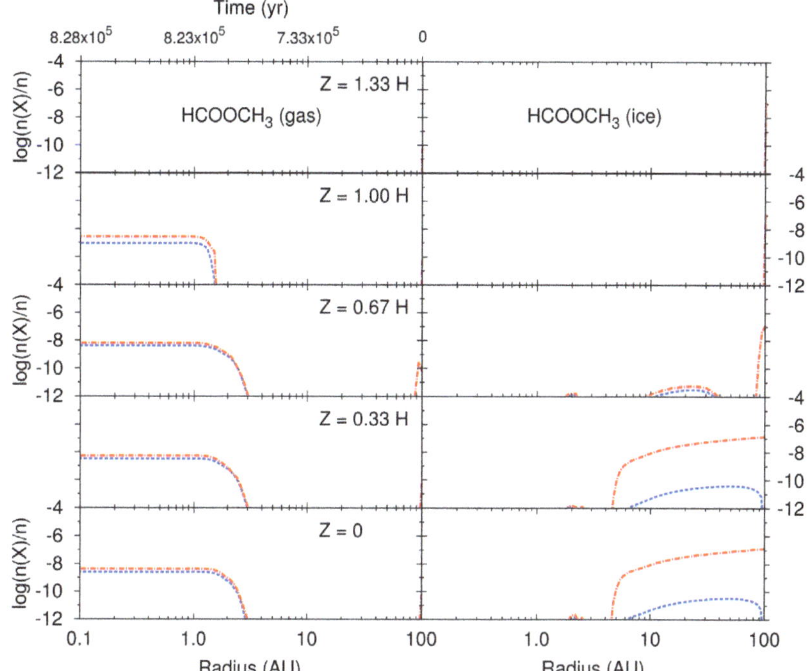

Fig. 11 Fractional abundance (with respect to number density) of HCOOCH$_3$ gas and ice along each streamline as a function of disk radius for the irradiated disk model. The blue lines show results from the model using the 'cold' (10 K) set of initial abundances and the red lines from the model using the 'warm' (30 K) set of initial abundances (see, Table 1).

Acetaldehyde ice in the outer region of the disk may have an interstellar origin, again, if the injected abundance is sufficiently high. There are no reported detections of acetaldehyde ice in molecular clouds or protostellar envelopes. However, gas-phase acetaldehyde has been detected towards several dark clouds and prestellar cores with a fractional abundance $\sim10^{-11}$–10^{-10} with respect to the H$_2$ gas density.[96–98] The relatively constant abundance across sources suggests the source of CH$_3$CHO is grain-surface formation followed by non-thermal desorption into the gas phase.

In the irradiated disk, the temperature is sufficiently high to thermally desorb CH$_3$CHO (and its precursor radicals) such that the molecule survives neither on the grain, nor in the gas. CH$_3$CHO molecules destroyed in the gas phase cannot be replenished *via* grain-surface chemistry due to the volatile nature of CH$_3$ and HCO. This result suggests CH$_3$CHO ice which is formed in comets within externally irradiated disks may have a secondary origin and is not pristine. The origin of gas-phase acetaldehyde in the inner disk is gas-phase formation following the thermal desorption of strongly bound COMs. Thus the fractional abundance of gas-phase acetaldehyde in the inner regions of the isolated disk and the irradiated disk are similar, $\sim10^{-9}$.

4.2.4 Methyl formate. In Fig. 10 and 11 the fractional abundance of gas-phase (left-hand panel) and grain-surface (right-hand panel) methyl formate, HCOOCH$_3$, is shown as a function of disk radius for the isolated and irradiated disk models, respectively.

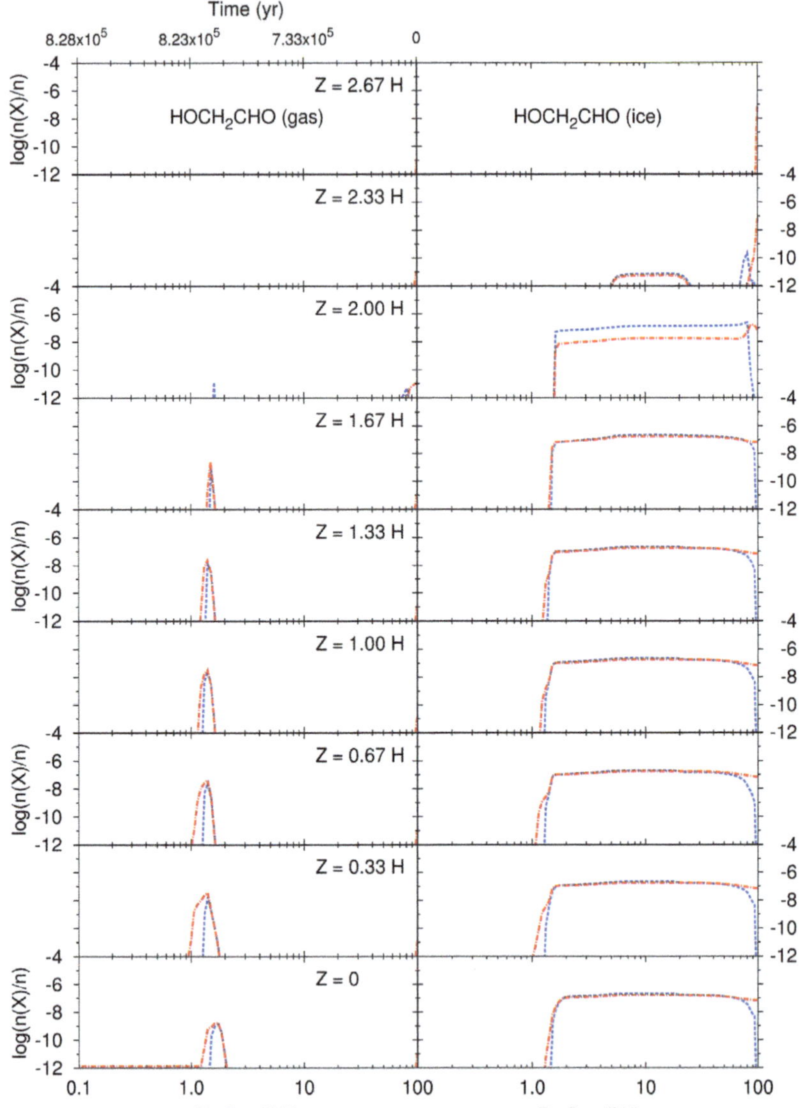

Fig. 12 Fractional abundance (with respect to number density) of HOCH$_2$CHO gas and ice along each streamline as a function of disk radius for the isolated disk model. The blue lines show results from the model using the 'cold' (10 K) set of initial abundances and the red lines from the model using the 'warm' (30 K) set of initial abundances (see, Table 1).

For the isolated disk, the abundance of methyl formate is not sensitive to the assumed initial abundances: methyl formate is efficiently formed in the outer disk *via* the association of the CH$_3$O and HCO radicals, reaching a peak fractional abundance, $\sim 10^{-7}$. This reaction is barrierless and can proceed at the temperatures in the outer disk, ≈ 20 K. This agrees with the result found for the static model.[19] The snowline for methyl formate lies slightly beyond the methanol snowline, between 2 and 3 AU. This reflects the slightly lower binding energy

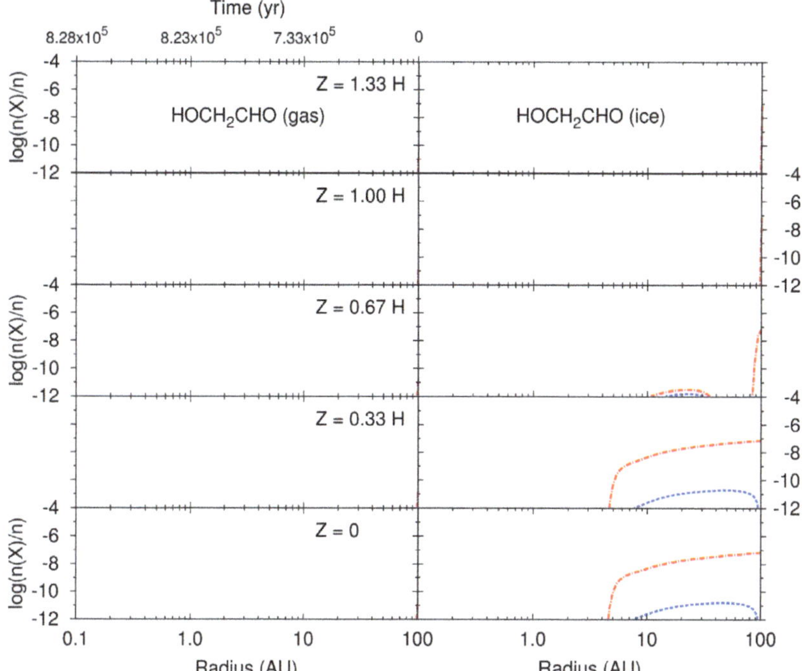

Fig. 13 Fractional abundance (with respect to number density) of HOCH$_2$CHO gas and ice along each streamline as a function of disk radius for the irradiated disk model. The blue lines show results from the model using the 'cold' (10 K) set of initial abundances and the red lines from the model using the 'warm' (30 K) set of initial abundances (see, Table 1).

assumed for methyl formate (4100 K). The abundance of methyl formate is preserved in the inner disk region. The origin of methyl formate in the planet- and comet-forming regions of isolated disks may have an interstellar origin depending on the initial abundance entering the disk. Gas-phase methyl formate has been observed in dark clouds and prestellar cores with a fractional abundance similar to that for acetaldehyde $\sim 10^{-11}$–10^{-10}.[96–98]

In the irradiated disk model, the abundance of methyl formate ice is sensitive to the assumed initial abundances. For the cold case, methyl formate ice reaches a peak fractional abundance $\sim 10^{-10}$ contrasted with $\sim 10^{-7}$ for the warm case. Again, the higher temperature in the irradiated disk leads to efficient desorption of the necessary precursor radicals, CH$_3$O and HCO, which have assumed binding energies, 2500 K and 1600 K, respectively. Methyl formate thermally desorbs at a radius ≈ 5 AU whereby it is destroyed by gas-phase chemistry. Within ≈ 2 AU, gas-phase methyl formate is replenished *via* gas-phase chemistry induced by the thermal desorption of strongly bound COMs. Methyl formate can form in the gas *via* the dissociative recombination of the protonated form of the molecule, HCOOCH$_4{}^+$. This species, in turn, can form *via* the following ion-neutral reactions,[46]

$$CH_3OH_2^+ + HCOOH \rightarrow HCOOCH_4^+ + H_2O, \tag{19}$$

$$HCOOH_2^+ + CH_3OH \rightarrow HCOOCH_4^+ + H_2O. \tag{20}$$

Thus, the formation of gas-phase methyl formate is triggered by the thermal desorption of CH_3OH and $HCOOH$ ice within ≈ 2 AU. The peak fractional abundance attained is $\sim 10^{-9}$ compared with $\sim 10^{-7}$ for the isolated case. This result correlates well with previous work which concluded gas-phase formation of methyl formate contributes, at most, to $\sim 1\%$ of the total gas-phase abundance in hot cores.[46]

The scenario described here for methyl formate, $HCOOCH_3$, is similar to that for dimethyl ether, CH_3OCH_3.

4.2.5   Glycolaldehyde. In Fig. 12 and 13 the fractional abundance of gas-phase (left-hand panel) and grain-surface (right-hand panel) glycolaldehyde, $HOCH_2$-CHO, is shown as a function of disk radius for the isolated and irradiated disk models, respectively. Glycolaldehyde is an interesting complex molecule because it is the simplest molecule which possesses both an aldehyde and a hydroxyl group. It has also recently been detected in the gas-phase towards the low-mass protostar, IRAS-16293 with a fractional abundance, $\sim 10^{-8}$ with respect to H_2.[99]

Similar to methyl formate, glycolaldehyde is not sensitive to assumed initial abundances in the outer regions of the isolated disk and can form efficiently *via* the barrierless reaction between CH_2OH and HCO. Glycolaldehyde thermally desorbs within a radius of ≈ 2 AU and exists in the gas phase in a narrow region between ≈ 1 and 2 AU. Within this region, the conditions are such that grain-surface reformation cannot compete with gas-phase destruction on the timescales of the accretion flow; however, it is possible that the gas-phase chemical network is incomplete for larger COMs such as glycolaldehyde.

The glycolaldehyde ice abundances in the outer region of the irradiated disk are very sensitive to the assumed initial abundances, again, requiring an appreciable abundance of precursor radicals on the grain. Once the glycolaldehyde desorbs, it is efficiently destroyed in the gas phase.

4.3   Comparison with previous disk models

The influence of the accretion flow (or advection) on the gas-grain balance in protoplanetary disks has been investigated previously.[21-24] Aikawa *et al.* computed the chemistry along streamlines from the outer disk (≈ 400–600 AU) to the inner disk (≈ 10–70 AU) using a method similar to that outlined here. Ilgner *et al.* and Heinzeller *et al.* later investigated the chemical evolution of parcels of gas moving inwards along the accretion flow along multiple trajectories to build two-dimensional 'snapshots' of the chemical structure of the planet-forming region of the disk ($\lesssim 10$ AU). In these works grain-surface chemistry was not included which restricted the discussion to small, simple species. Nomura *et al.* considered the chemistry and transport of COMs in a protoplanetary disk for the first time. However, as in previous works, a grain-surface network was not explicitly included and the calculations followed the subsequent gas-phase chemistry upon thermal desorption of the parent COMs assumed to be already present in the ice mantle. The discussion was also restricted to the inner planet-forming zone ($\lesssim 12$ AU). Hence, the work presented here is the first investigation of grain-surface chemistry and the subsequent formation and destruction of COMs along the accretion flow in a protoplanetary disk.

Recently, the chemical evolution along parcels of gas in a hydrodynamical simulation of a gravitationally unstable embedded disk were reported for the first time.[100] The parcels were found to follow erratic paths encountering a range of physical conditions and spikes in temperature and density due to shocks. The resulting molecular abundances trace density enhancements in spiral arms and a subsequent calculation of the molecular line emission suggested these structures may be observable with modern interferometric facilities such as ALMA.[101] Again, grain-surface chemistry was not included in these calculations, and the influence on the abundance and distribution of COMs in hydrodynamic simulations of protoplanetary disks is yet to be investigated.

There are several additional physical processes thought to be important in disks which can influence the physical structure and resulting chemistry.[102] The effects of turbulent mixing on the composition of protoplanetary disks has been studied by numerous groups.[22,24,103–106] Early work suggested that vertical mixing does not change the chemically stratified nature of the disk; however, mixing can significantly increase the depth of the molecular layer and thus the vertical column densities.[104] Of these works, only one has included the formation of complex organic molecules *via* thermal grain-surface chemistry.[106] It was found that the abundances of COMs are very sensitive to the treatment of turbulent mixing, a similar conclusion to that presented here for the case of advection. A chemical model including both processes would be ideal; however, such models remain computationally challenging.

The dust evolution in the disk can have a profound effect on the disk physical structure and resulting chemistry. Grain growth (coagulation) skews the dust-grain size distribution towards larger grains whereas settling (sedimentation) depletes the surface layers of large grains and increases the relative abundance of large grains in the disk midplane.[107,108] Several groups have investigated the influence of dust-grain evolution on disk chemistry.[109–113] The inclusion of grain growth tended to push the molecular layer deeper into the disk with increasing grain size; however, the vertical column densities of molecules remained unaffected.[110] Grain settling was found to lead to a smaller freeze out zone and a larger molecular layer due to the increased penetration of UV radiation.[111] Recent models have coupled both dust evolution processes in a disk model with full gas-grain chemistry.[112,113] In these models the chemically stratified nature of the disk is retained, albeit with a molecular layer which resides deeper into the disk. Both works also found that the gas-phase column densities are enhanced (at the expense of the grain-surface column densities) relative to the results using pristine interstellar dust. Grain coagulation may affect the formation rate of COMs due to the reduced surface area available for freeze out of the necessary precursor species. On the other hand, grain settling increases the dust-to-gas mass ratio in the disk midplane relative to that higher in the disk atmosphere, somewhat counteracting the reduction in total surface area. In addition, the deeper penetration of UV radiation may increase the efficiency of both thermal and non-thermal desorption of COMs formed on the grain into the gas phase. The combined effect of accretion flow and dust evolution on the abundance and distribution of COMs in protoplanetary disks should be explored in the future.

5 Conclusions

The results presented here suggest that COMs which have efficiently formed on grain surfaces under cold, *i.e.*, prestellar, conditions survive the transport along the accretion flow in the midplane of isolated disks; hence, the composition of icy planetesimals in isolated disks may be representative of the pristine interstellar ice entering the outer shielded regions of the disk, provided these species reach appreciable abundances under prestellar conditions. This conclusion is similar to that reached for simple ice species, *e.g.*, H_2O, in investigations on the influence of collapse, infall, and subsequent disk formation in low-mass protostars.[114] Currently, a similar investigation is underway for more complex ice species.[115] The successful identification of gas-phase COMs such as CH_3OH, CH_3CHO, $HCOOCH_3$, and CH_3OCH_3 in dark clouds and prestellar cores also suggest that grain-surface chemistry can operate effectively in dense, cold environments.[96–98] However, it should be noted that, of the species considered in this work, only methanol ice and formic acid ice have been detected in prestellar environments.[43,95,116–119]

If simple ices only are injected into the outer disk, the physical conditions facilitate the efficient production of COMs such as those considered here: CH_3CHO, $HCOOCH_3$, CH_3OCH_3, and $HOCH_2CHO$. Hence, there is an increase in molecular complexity from cloud to disk, producing abundances of complex molecules similar to those observed in cometary comae.[120] This conclusion is similar to that reached in the static disk model.[19] The chemical timescales are sufficiently short in the outer disk that the results from the static model and accretion model (presented here) are similar but not identical. The fractional abundances of grain-surface COMs along the disk midplane in the static model vary with radius: the long lifetime of the disk allows the chemistry to settle to the local physical conditions. Hence, the assumed initial abundance of COMs are not preserved in the static model as is seen in the accretion flow model presented here. This work has shown that the grain-surface chemistry is sensitive to both the initial abundances adopted in the model and the accretion flow.

Once the snowline of each species is breached, the COMs can thermally desorb with the gas-phase abundance reflecting that injected from the ice mantle as also found in previous work investigating the transport of methanol ice in a disk.[23] This is in stark contrast to that found in the static model in which the extreme densities and temperatures encountered in the inner midplane are able to destroy gas-phase COMs over the lifetime of the disk, $\sim 10^6$ years. In this model, the gas spends only ≈ 5000 years within 1 AU of the star. In order to properly treat the transport and chemistry of complex molecules within the planet-forming region of isolated protoplanetary disks, the influence of the accretion flow must be taken into account. In the disk molecular layer, COMs photodesorb into the gas phase, reaching peak fractional abundances similar to those found in the static model.[19]

Conversely, the radiation field in the irradiated disk is too strong for photo-desorbed molecules to survive in the gas phase. It is also found that the assumed initial abundances are more important for this model due to the much higher temperatures in the outer disk. If simple ices are injected, thermal grain-surface chemistry is unable to form complex molecules because the necessary precursor radicals, which tend to be particularly volatile, can desorb efficiently. If more complex ices are injected into the outer disk, the grain-surface COMs undergo thermal processing such that the gas phase abundances in the inner region do not reflect those injected at the outer edge of the disk. The origin of gas-phase COMs in the inner region of the irradiated disk is gas-phase formation induced by the thermal desorption of strongly bound molecules, such as CH_3OH, which do not rely on radical–radical formation. This could mean that icy planetesimals which form in the midplane of irradiated disks are likely to be more thermally processed and also composed of more simple ices than those in isolated disks. This is interesting given that the Sun has been postulated to have formed in a stellar cluster which has long since dispersed.[5] If the comets in the Solar System were originally composed of more simple ices, their current composition could reflect post-formation processing of the cometary surface.[121]

The work presented here does not address one of the outstanding issues in grain-surface chemistry, that is, the efficient formation at 10 K of those complex molecules which are thought to form *via* radical–radical association reactions. Under the current paradigm, molecular radicals do not have sufficient mobility to diffuse within or on the ice mantle. There are several processes which have been postulated to help build chemical complexity at 10 K including the radiation-processing of ices *via* UV photons and cosmic rays,[50,122] and a high efficiency of reactive desorption.[123,124] These issues will be explored, in the context of proto-planetary disk chemistry, in future work.

Acknowledgements

C.W. acknowledges support from the European Union A-ERC grant 291141 CHEMPLAN and financial support (program number 639.041.335) from the Netherlands Organisation for Scientific Research (NWO). E.H. wishes to acknowledge the support of the National Science Foundation for his astro-chemistry program, and his program in chemical kinetics through the Center for the Chemistry of the Universe. He also acknowledges support from the NASA Exobiology and Evolutionary Biology program through a subcontract from Rensselaer Polytechnic Institute. H.N. acknowledges the Grant-in-Aid for Scientific Research 23103005 and 25400229. She also acknowledges support from the Astrobiology Project of the CNSI, NINS (Grant Number AB251002, AB251012). Astrophysics at QUB is supported by a grant from the STFC. S.L.W.W. acknowledges support from startup funds provided by Emory University.

References

1 F. H. Shu, F. C. Adams and S. Lizano, *Annu. Rev. Astron. Astrophys.*, 1987, **25**, 23.

2 J. P. Williams and L. A. Cieza, *Annu. Rev. Astron. Astrophys.*, 2011, **49**, 67.

3 C. J. Lada and E. A. Lada, *Annu. Rev. Astron. Astrophys.*, 2003, **41**, 57.

4 J. H. Kastner, B. Zuckerman, D. A. Weintraub and T. Forveille, *Science*, 1997, **277**, 67.

5 F. C. Adams, *Annu. Rev. Astron. Astrophys.*, 2010, **48**, 47.

6 C. R. O'dell, Z. Wen and X. Hu, *Astrophys. J.*, 1993, **410**, 696.

7 J. Bally, R. S. Sutherland, D. Devine and D. Johnstone, *Astron. J.*, 1998, **116**, 293.

8 J. Blum and G. Wurm, *Annu. Rev. Astron. Astrophys.*, 2008, **46**, 21.

9 F. G. Bridges, K. D. Supulver, D. N. C. Lin, R. Knight and M. Zafra, *Icarus*, 1996, **123**, 422.

10 K. D. Supulver, F. G. Bridges, S. Tiscareno, J. Lievore and D. N. C. Lin, *Icarus*, 1997, **129**, 539.

11 B. Gundlach, S. Kilias, E. Beitz and J. Blum, *Icarus*, 2011, **214**, 717.

12 E. Herbst and E. F. van Dishoeck, *Annu. Rev. Astron. Astrophys.*, 2009, **47**, 427.

13 M. P. Bernstein, J. P. Dworkin, S. A. Sandford, G. W. Cooper and L. J. Allamandola, *Nature*, 2002, **416**, 401.

14 G. M. Muñoz Caro, U. J. Meierhenrich, W. A. Schutte, B. Barbier, A. Arcones Segovia, H. Rosenbauer, W. H.-P. Thiemann, A. Brack and J. M. Greenberg, *Nature*, 2002, **416**, 403.

15 D. E. Woon, *Astrophys. J.*, 2002, **571**, L177.

16 S. J. Kenyon and L. Hartmann, *Astrophys. J.*, 1987, **323**, 714.

17 E. I. Chiang and P. Goldreich, *Astrophys. J.*, 1997, **490**, 368.

18 P. D'Alessio, J. Canto, N. Calvet and S. Lizano, *Astrophys. J.*, 1998, **500**, 411.

19 C. Walsh, T. J. Millar, H. Nomura, E. Herbst, S. Widicus Weaver, Y. Aikawa, J. C. Laas and A. I. Vasyunin, *Astron. Astrophys.*, 2014, **563**, A33.

20 C. Walsh, T. J. Millar and H. Nomura, *Astrophys. J.*, 2013, **766**, L23.

21 Y. Aikawa, T. Umebayashi, T. Nakano and S. M. Miyama, *Astrophys. J.*, 1999, **519**, 705.

22 M. Ilgner, T. Henning, A. J. Markwick and T. J. Millar, *Astron. Astrophys.*, 2004, **415**, 643.

23 H. Nomura, Y. Aikawa, Y. Nakagawa and T. J. Millar, *Astron. Astrophys.*, 2009, **495**, 183.

24 D. Heinzeller, H. Nomura, C. Walsh and T. J. Millar, *Astrophys. J.*, 2011, **731**, 115.

25 E. A. Bergin, Y. Aikawa, G. A. Blake and E. F. van Dishoeck, *Protostars and Planets V*, 2007, 751.

26 K. I. Öberg, R. Murray-Clay and E. A. Bergin, *Astrophys. J.*, 2011, **743**, L16.

27 P. Caselli and C. Ceccarelli, *Astron. Astrophys. Rev.*, 2012, **20**, 56.

28 A. Dutrey, S. Guilloteau and M. Guelin, *Astron. Astrophys.*, 1997, **317**, L55.

29 G.-J. van Zadelhoff, E. F. van Dishoeck, W.-F. Thi and G. A. Blake, *Astron. Astrophys.*, 2001, **377**, 566.

30 C. Qi, J. E. Kessler, D. W. Koerner, A. I. Sargent and G. A. Blake, *Astrophys. J.*, 2003, **597**, 986.

31 E. F. van Dishoeck, W.-F. Thi and G.-J. van Zadelhoff, *Astron. Astrophys.*, 2003, **400**, L1.

32 W.-F. Thi, G.-J. van Zadelhoff and E. F. van Dishoeck, *Astron. Astrophys.*, 2004, **425**, 955.

33 S. Guilloteau, V. Piétu, A. Dutrey and M. Guélin, *Astron. Astrophys.*, 2006, **448**, L5.

34 C. Qi, D. J. Wilner, Y. Aikawa, G. A. Blake and M. R. Hogerheijde, *Astrophys. J.*, 2008, **681**, 1396.

35 K. I. Öberg, C. Qi, J. K. J. Fogel, E. A. Bergin, S. M. Andrews, C. Espaillat, T. A. van Kempen, D. J. Wilner and I. Pascucci, *Astrophys. J.*, 2010, **720**, 480.

36 Y. Aikawa, M. Momose, W.-F. Thi, G.-J. van Zadelhoff, C. Qi, G. A. Blake and E. F. van Dishoeck, *Pub. Astron. Soc. Japan*, 2003, **55**, 11.

37 E. Chapillon, A. Dutrey, S. Guilloteau, V. Piétu, V. Wakelam, F. Hersant, F. Gueth, T. Henning, R. Launhardt, K. Schreyer and D. Semenov, *Astrophys. J.*, 2012, **756**, 58.

38 C. Qi, K. I. Öberg, D. J. Wilner and K. A. Rosenfeld, *Astrophys. J.*, 2013, **765**, L14.

39 M. Allen and G. W. Robinson, *Astrophys. J.*, 1977, **212**, 396.

40 D. E. Woon, *Astrophys. J.*, 2002, **569**, 541.

41 N. Watanabe and A. Kouchi, *Astrophys. J.*, 2002, **571**, L173.

42 G. W. Fuchs, H. M. Cuppen, S. Ioppolo, C. Romanzin, S. E. Bisschop, S. Andersson, E. F. van Dishoeck and H. Linnartz, *Astron. Astrophys.*, 2009, **505**, 629.

43 E. L. Gibb, D. C. B. Whittet, A. C. A. Boogert and A. G. G. M. Tielens, *Astrophys. J. Suppl.*, 2004, **151**, 35.

44 R. T. Garrod and E. Herbst, *Astron. Astrophys.*, 2006, **457**, 927.

45 R. T. Garrod, S. L. W. Weaver and E. Herbst, *Astrophys. J.*, 2008, **682**, 283.

46 J. C. Laas, R. T. Garrod, E. Herbst and S. L. Widicus Weaver, *Astrophys. J.*, 2011, **728**, 71.

47 S. S. Prasad and S. P. Tarafdar, *Astrophys. J.*, 1983, **267**, 603.

48 L. J. Allamandola, S. A. Sandford and G. J. Valero, *Icarus*, 1988, **76**, 225.

49 P. A. Gerakines, W. A. Schutte and P. Ehrenfreund, *Astron. Astrophys.*, 1996, **312**, 289.

50 K. I. Öberg, R. T. Garrod, E. F. van Dishoeck and H. Linnartz, *Astron. Astrophys.*, 2009, **504**, 891.

51 M. S. Westley, R. A. Baragiola, R. E. Johnson and G. A. Baratta, *Planet. Space Sci.*, 1995, **43**, 1311.

52 K. Willacy and W. D. Langer, *Astrophys. J.*, 2000, **544**, 903.

53 C. Walsh, T. J. Millar and H. Nomura, *Astrophys. J.*, 2010, **722**, 1607.

54 M. Bertin, E. C. Fayolle, C. Romanzin, K. I. Öberg, X. Michaut, A. Moudens, L. Philippe, P. Jeseck, H. Linnartz and J.-H. Fillion, *Phys. Chem. Chem. Phys.*, 2012, **14**, 9929.

55 K. I. Öberg, H. Linnartz, R. Visser and E. F. van Dishoeck, *Astrophys. J.*, 2009, **693**, 1209.

56 C. Arasa, S. Andersson, H. M. Cuppen, E. F. van Dishoeck and G.-J. Kroes, *J. Chem. Phys.*, 2010, **132**, 184510.

57 M. R. Hogerheijde, E. A. Bergin, C. Brinch, L. I. Cleeves, J. K. J. Fogel, G. A. Blake, C. Dominik, D. C. Lis, G. Melnick, D. Neufeld, O. Panić, J. C. Pearson, L. Kristensen, U. A. Yıldız and E. F. van Dishoeck, *Science*, 2011, **334**, 338.

58 M. J. Mumma, P. R. Weissman and S. A. Stern, *Protostars and Planets III*, 1993, p. 1177.

59 H. Nomura and T. J. Millar, *Astron. Astrophys.*, 2005, **438**, 923.

60 H. Nomura, Y. Aikawa, M. Tsujimoto, Y. Nakagawa and T. J. Millar, *Astrophys. J.*, 2007, **661**, 334.

61 S. J. Kenyon and L. Hartmann, *Astrophys. J. Suppl.*, 1995, **101**, 117.

62 J. E. Pringle, *Annu. Rev. Astron. Astrophys.*, 1981, **19**, 137.

63 H. Nomura, *Astrophys. J.*, 2002, **567**, 587.

64 D. N. C. Lin and J. Papaloizou, *Mon. Not. R. Astron. Soc.*, 1980, **191**, 37.

65 G. H. Herbig and R. W. Goodrich, *Astrophys. J.*, 1986, **309**, 294.

66 E. D. Feigelson and T. Montmerle, *Annu. Rev. Astron. Astrophys.*, 1999, **37**, 363.

67 J. C. Weingartner and B. T. Draine, *Astrophys. J.*, 2001, **548**, 296.

68 P. R. Maloney, D. J. Hollenbach and A. G. G. M. Tielens, *Astrophys. J.*, 1996, **466**, 561.

69 E. F. van Dishoeck, *Astrochemistry*, 1987, p. 51.

70 E. F. van Dishoeck, B. Jonkheid and M. C. van Hemert, *Faraday Discuss.*, 2006, **133**, 231.

71 A. E. Glassgold, P. J. Huggins and W. D. Langer, *Astrophys. J.*, 1985, **290**, 615.

72 H.-H. Lee, E. Herbst, G. Pineau des Forets, E. Roueff and J. Le Bourlot, *Astron. Astrophys.*, 1996, **311**, 690.

73 R. Visser, E. F. van Dishoeck and J. H. Black, *Astron. Astrophys.*, 2009, **503**, 323.

74 X. Li, A. N. Heays, R. Visser, W. Ubachs, B. R. Lewis, S. T. Gibson and E. F. van Dishoeck, *Astron. Astrophys.*, 2013, **555**, A14.

75 E. Matar, H. Bergeron, F. Dulieu, H. Chaabouni, M. Accolla and J. L. Lemaire, *J. Chem. Phys.*, 2010, **133**, 104507.

76 X. Sha, B. Jackson, D. Lemoine and B. Lepetit, *J. Chem. Phys.*, 2005, **122**, 014709.

77 H. M. Cuppen, L. E. Kristensen and E. Gavardi, *Mon. Not. R. Astron. Soc.*, 2010, **406**, L11.

78 T. I. Hasegawa, E. Herbst and C. M. Leung, *Astrophys. J. Suppl.*, 1992, **82**, 167.

79 T. I. Hasegawa and E. Herbst, *Mon. Not. R. Astron. Soc.*, 1993, **261**, 83.

80 K. I. Öberg, G. W. Fuchs, Z. Awad, H. J. Fraser, S. Schlemmer, E. F. van Dishoeck and H. Linnartz, *Astrophys. J.*, 2007, **662**, L23.

81 K. I. Öberg, E. F. van Dishoeck and H. Linnartz, *Astron. Astrophys.*, 2009, **496**, 281.

82 T. Stantcheva, P. Caselli and E. Herbst, *Astron. Astrophys.*, 2001, **375**, 673.

83 R. T. Garrod, A. I. Vasyunin, D. A. Semenov, D. S. Wiebe and T. Henning, *Astrophys. J.*, 2009, **700**, L43.

84 A. I. Vasyunin, D. A. Semenov, D. S. Wiebe and T. Henning, *Astrophys. J.*, 2009, **691**, 1459.

85 N. Watanabe, Y. Kimura, A. Kouchi, T. Chigai, T. Hama and V. Pirronello, *Astrophys. J.*, 2010, **714**, L233.

86 R. T. Garrod, V. Wakelam and E. Herbst, *Astron. Astrophys.*, 2007, **467**, 1103.

87 C. Gry, F. Boulanger, C. Nehmé, G. Pineau des Forêts, E. Habart and E. Falgarone, *Astron. Astrophys.*, 2002, **391**, 675.

88 E. Habart, F. Boulanger, L. Verstraete, G. Pineau des Forêts, E. Falgarone and A. Abergel, *Astron. Astrophys.*, 2003, **397**, 623.

89 S. Cazaux and A. G. G. M. Tielens, *Astrophys. J.*, 2004, **604**, 222.

90 W. Iqbal, K. Acharyya and E. Herbst, *Astrophys. J.*, 2012, **751**, 58.

91 J. Le Bourlot, F. Le Petit, C. Pinto, E. Roueff and F. Roy, *Astron. Astrophys.*, 2012, **541**, A76.

92 C. Walsh, H. Nomura, T. J. Millar and Y. Aikawa, *Astrophys. J.*, 2012, **747**, 114.

93 S. Ioppolo, H. M. Cuppen, E. F. van Dishoeck and H. Linnartz, *Mon. Not. R. Astron. Soc.*, 2011, **410**, 1089.

94 W.-C. Chen and R. A. Marcus, *J. Chem. Phys.*, 2005, **123**, 094307.

95 A. C. A. Boogert, K. M. Pontoppidan, C. Knez, F. Lahuis, J. Kessler-Silacci, E. F. van Dishoeck, G. A. Blake, J.-C. Augereau, S. E. Bisschop, S. Bottinelli, T. Y. Brooke, J. Brown, A. Crapsi, N. J. Evans, II, H. J. Fraser, V. Geers, T. L. Huard, J. K. Jørgensen, K. I. Öberg, L. E. Allen, P. M. Harvey, D. W. Koerner, L. G. Mundy, D. L. Padgett, A. I. Sargent and K. R. Stapelfeldt, *Astrophys. J.*, 2008, **678**, 985.

96 K. I. Öberg, S. Bottinelli, J. K. Jørgensen and E. F. van Dishoeck, *Astrophys. J.*, 2010, **716**, 825.

97 A. Bacmann, V. Taquet, A. Faure, C. Kahane and C. Ceccarelli, *Astron. Astrophys.*, 2012, **541**, L12.

98 J. Cernicharo, N. Marcelino, E. Roueff, M. Gerin, A. Jiménez-Escobar and G. M. Muñoz Caro, *Astrophys. J.*, 2012, **759**, L43.

99 J. K. Jørgensen, C. Favre, S. E. Bisschop, T. L. Bourke, E. F. van Dishoeck and M. Schmalzl, *Astrophys. J.*, 2012, **757**, L4.

100 J. D. Ilee, A. C. Boley, P. Caselli, R. H. Durisen, T. W. Hartquist and J. M. C. Rawlings, *Mon. Not. R. Astron. Soc.*, 2011, **417**, 2950.

101 T. A. Douglas, P. Caselli, J. D. Ilee, A. C. Boley, T. W. Hartquist, R. H. Durisen and J. M. C. Rawlings, *Mon. Not. R. Astron. Soc.*, 2013, **433**, 2064.

102 T. Henning and D. Semenov, *Chem. Rev.*, 2013, **113**, 9016.

103 D. Semenov, D. Wiebe and T. Henning, *Astrophys. J.*, 2006, **647**, L57.

104 K. Willacy, W. Langer, M. Allen and G. Bryden, *Astrophys. J.*, 2006, **644**, 1202.

105 Y. Aikawa, *Astrophys. J.*, 2007, **656**, L93.

106 D. Semenov and D. Wiebe, *Astrophys. J. Suppl.*, 2011, **196**, 25.

107 C. Dominik and A. G. G. M. Tielens, *Astrophys. J.*, 1997, **480**, 647.

108 C. P. Dullemond and C. Dominik, *Astron. Astrophys.*, 2004, **421**, 1075.

109 B. Jonkheid, F. G. A. Faas, G.-J. van Zadelhoff and E. F. van Dishoeck, *Astron. Astrophys.*, 2004, **428**, 511.

110 Y. Aikawa and H. Nomura, *Astrophys. J.*, 2006, **642**, 1152.

111 J. K. J. Fogel, T. J. Bethell, E. A. Bergin, N. Calvet and D. Semenov, *Astrophys. J.*, 2011, **726**, 29.

112 A. I. Vasyunin, D. S. Wiebe, T. Birnstiel, S. Zhukovska, T. Henning and C. P. Dullemond, *Astrophys. J.*, 2011, **727**, 76.

113 V. Akimkin, S. Zhukovska, D. Wiebe, D. Semenov, Y. Pavlyuchenkov, A. Vasyunin, T. Birnstiel and T. Henning, *Astrophys. J.*, 2013, **766**, 8.

114 R. Visser, E. F. van Dishoeck, S. D. Doty and C. P. Dullemond, *Astron. Astrophys.*, 2009, **495**, 881.

115 M. N. Drozdovskaya, C. Walsh, R. Visser, D. Harsono and E. F. van Dishoeck, *Mon. Not. R. Astron. Soc.*, 2014, submitted.

116 W. A. Schutte, A. C. A. Boogert, A. G. G. M. Tielens, D. C. B. Whittet, P. A. Gerakines, J. E. Chiar, P. Ehrenfreund, J. M. Greenberg, E. F. van Dishoeck and T. de Graauw, *Astron. Astrophys.*, 1999, **343**, 966.

117 E. L. Gibb, D. C. B. Whittet, W. A. Schutte, A. C. A. Boogert, J. E. Chiar, P. Ehrenfreund, P. A. Gerakines, J. V. Keane, A. G. G. M. Tielens, E. F. van Dishoeck and O. Kerkhof, *Astrophys. J.*, 2000, **536**, 347.

118 K. M. Pontoppidan, E. Dartois, E. F. van Dishoeck, W.-F. Thi and L. d'Hendecourt, *Astron. Astrophys.*, 2003, **404**, L17.

119 A. C. A. Boogert, T. L. Huard, A. M. Cook, J. E. Chiar, C. Knez, L. Decin, G. A. Blake, A. G. G. M. Tielens and E. F. van Dishoeck, *Astrophys. J.*, 2011, **729**, 92.

120 D. Bockelée-Morvan, J. Crovisier, M. J. Mumma and H. A. Weaver, in *The composition of cometary volatiles*, ed. M. C. Festou, H. U. Keller and H. A. Weaver, University of Arizona Press, Tucson, 2004, p. 391.

121 M. J. Mumma and S. B. Charnley, *Annu. Rev. Astron. Astrophys.*, 2011, **49**, 471.

122 C. J. Bennett and R. I. Kaiser, *Astrophys. J.*, 2007, **661**, 899.

123 A. I. Vasyunin and E. Herbst, *Astrophys. J.*, 2013, **769**, 34.

124 F. Dulieu, E. Congiu, J. Noble, S. Baouche, H. Chaabouni, A. Moudens, M. Minissale and S. Cazaux, *Sci. Rep.*, 2013, **3**, 1338.

Faraday Discussions

DOI: 10.1039/c4fd90004a

DISCUSSIONS

General discussion

Professor Kamp opened the discussion of the paper by Cornelia Jäger: When you re-condense your dust back into the Ne matrix, you note that your product is Mg poor. Where is the Mg going? How do we get back to the kind of compositions that we usually observe in space?

Dr Jäger replied: In the interstellar medium (ISM), the dust destruction rate is faster than the stardust injection rate. However, measurements of the interstellar depletion indicate that 99.9% of all refractory materials reside in solid grains, and therefore there must be an efficient dust reformation process. This condensation process has to occur at low temperature and low density. Until now, experimental proof of the cold condensation process was lacking. In a series of experiments, we wanted to prove the condensation of complex silicates at temperatures of about 10 K. The main goal of our study, the final proof of the cold condensation, has been successfully demonstrated in the presented experiments. The dust components produced in circumstellar environments and destroyed in shock waves represent the reservoir for atoms and molecules that accrete on surviving grains and may form silicates or carbonaceous molecules. In order to produce relevant molecular and atomic species, we have used amorphous silicate materials with olivine and pyroxene stoichiometry produced in our laboratory, which have successfully been used as a cosmic dust analog material.[1] An efficient method to evaporate amorphous silicate materials and to produce the molecular precursors for such a condensation process is laser ablation. In our laboratory, this method has been successfully used to produce amorphous silicates by gas-phase condensation processes. Previous laser ablation experiments and subsequent condensation of silicates in a quenching gas atmosphere at high temperatures have clearly shown that the composition of the condensate represents the composition of the target.[2,3] In these condensation experiments, energies of between 100 and 250 mJ pulse^{-1}, corresponding to power densities between 4×10^8 and 9×10^9 W cm^{-2}, were used. For the cold condensation experiments, a Nd:YAG laser with a pulse energy of 20 to 25 mJ pulse^{-1} has been employed. The MgO depletion process in the evaporation is very probably an experimental result owing to the low power density of the laser used in the ablation process. The power density of the laser is not sufficient to vaporize the material completely. The evaporation process is accompanied by heating and partial melting of the target material. This leads to a phase separation within the silicate material and the formation of an MgO phase in the silicate. MgO is a very heat-resistant material that is difficult to evaporate. Consequently, the laser ablation results in a fractionation of the target silicate.

The final condensate is depleted in MgO. In our next experiments on cold condensation, we will adjust our experimental conditions in order to achieve the same stoichiometry of the condensed and target material. The application of a laser with higher energy can avoid such a fractionation process of the silicate target material.

1 J. Dorschner *et al.*, *Astron. Astrophys.*, 1995, **300**, 503.
2 C. Jäger *et al.*, in *Cosmic Dust – Near and Far*, ASP Conference Series, 2009, vol. 414, p. 319.
3 T. Sabri *et al.*, *Astrophys. J.*, 2014, **780**, 180.

Mr Rab asked: How efficient is the process of cold condensation in an astronomical environment, especially considering the low densities? How fast can you grow particles from, for example, nanometer to micrometer sized particles in the interstellar environment (diffuse ISM and dark clouds)?

Dr Jäger responded: What we can say at the moment is that this cold condensation process is very efficient in the laboratory. The formation of the solid silicates occurs within minutes without any additional energy. In the ISM, of course, the formation rate depends on the accretion rate of the silicate-forming molecules and atoms. In the diffuse ISM, silicate grains are not expected to reach 1 µm in size. Following Mathis *et al.*,[1] an upper limit of 0.25 µm is generally applied. Nevertheless, silicate grains with sizes of 1–1.5 µm are present in dense regions,[2] and cause the coreshine phenomenon.[3,4] Hirashita modelled the growth of dust grains in dense molecular clouds with solar metallicity, where destructive processes can be neglected.[5] He found that the growth of silicate grains by accretion of gas-phase species at their surface is efficient only for sizes up to ~0.001 µm with a time scale shorter than ~10 Myr. In contrast, grains larger than 0.01 µm are formed by coagulation of nanometer sized grains over periods longer than ~10 Myr. In his model, however, Hirashita kept the upper limit of 0.25 µm set for silicate grains in the diffuse ISM by Mathis *et al.* In order to model the formation of grains as large as 1 µm, this limit must be removed. By this means, Hirashita and Voshchinnikov found that silicate grains can grow up to 1 µm over ~100–200 Myr in dense clouds.[6] It was noted that such a duration is of the order of the lifetime of a cloud.

1 J. S. Mathis, W. Rumpl and K. H. Nordsieck, *Astrophys. J.*, 1977, **217**, 425.
2 M. Andersen, J. Steinacker, W.-F. Thi, L. Pagani, A. Bacmann and R. Paladini, *Astron. Astrophys.*, 2013, **559**, A60.
3 J. Steinacker, L. Pagani, A. Bacmann and S. Guieu, *Astron. Astrophys.*, 2010, **511**, A9.
4 L. Pagani, J. Steinacker, A. Bacmann, A. Stutz and T. Henning, *Science*, 2010, **329**, 1622.
5 H. Hirashita, *Mon. Not. R. Astron. Soc.*, 2012, **422**, 1263.
6 H. Hirashita and N. V. Voshchinnikov, *Mon. Not. R. Astron. Soc.*, 2014, **437**, 1636.

Professor Plane addressed Dr Jäger: I have two comments on this very interesting study. Firstly, in your laser ablation technique you observe the same phenomenon of differential ablation that occurs when meteoroids ablate in the Earth's upper atmosphere – Mg ablates less efficiently than Fe from what are essentially olivine particles.[1] Secondly, it is striking that you make pyroxene during the deposition in the Ne matrix rather than olivine. We have done experiments where we make Fe–Mg–Si–O nanoparticles photochemically in the gas phase, by irradiating a mixture of $Fe(OH)_5$, $Mg(OC_2H_5)_2$ and $Si(OC_2H_5)_4$ in the presence of O_3.[2] EDX analysis shows that only olivines – not pyroxenes – form, and by altering the ratios of the Fe and Mg precursors, we can manufacture particles with any stoichiometry between forsterite and fayalite. In these experiments O_3 is present, which I assume is not the case in your experiment. It would therefore be interesting if you could introduce O_3 into the chamber and/or the Ne matrix, to see if olivine rather than pyroxene is produced under your experimental conditions.

1 T. Vondrak, J. M. C. Plane, S. L. Broadley and D. Janches, *Atmos. Chem. Phys.*, 2008, **8**, 7015.
2. R. W. Saunders and J. M. C. Plane, *Icarus*, 2011, **212**, 373.

Dr Jäger replied: Thank you very much for the interesting comments and hints. The change of stoichiometry we have observed in the laser ablation and condensation process is probably related to the applied experimental conditions. In our laboratory, we have already produced silicates as cosmic dust analogs by gas-phase condensation techniques. The laser ablation setup that we use to simulate the gas-phase condensation of amorphous silicates at high temperatures (as it may occur in circumstellar shells) consists of a target that is made of either a silicate, or metal powders mixed in a stoichiometric ratio to form pyroxene-type (Me : Si = 1) or olivine-type (Me : Si = 2) silicates. The target material evaporated by the Nd:YAG laser is finally condensed in a quenching gas atmosphere that is composed of either pure He, or He and oxygen. The silicates produced in this setup have nearly the same composition as the target used for the ablation. In the case where we realize a high power density for the laser evaporation, we are able to avoid differential ablation.[1,2] However, in astrophysical environments, the complete evaporation of silicates may not be expected in the same way as that observed in the Earth's upper atmosphere by meteorite ablation. I agree with you that systematic studies using varying laser energies and target compositions should be performed in order to learn more about the preferential condensation of silicates. In our next experiments on cold silicate condensation, we will focus on this problem. However, in our cryogenic matrix, we have also observed the presence of ozone molecules (see the feature at 1039 cm^{-1} in Fig. 3 and 5 in the paper) that have probably been formed in the ablation process. Based on preliminary calculations of theoretical IR spectra, the amount of ozone isolated in the matrices is at least as large as the amount of Si-bearing species. However, in our experiments, we did not form olivine. This could be related to the non-equilibrium conditions in our condensation process.

1 C. Jäger *et al.*, in *Cosmic Dust – Near and Far*, ASP Conference Series, 2009, vol. 414, p. 319.
2 T. Sabri *et al.*, *Astrophys. J.*, 2014, **780**, 180.

Dr Jones commented: In some very nice work, Mathieu Roskosz has shown that out-of-equilibrium diffusion processes can quite naturally explain the formation of a pyroxene-type silicate from an olivine-type silicate stoichiometry at temperatures below the annealing temperature.[1,2] While this may not be the origin of the effect that you observe, because of the very different nature of your samples and their formation, it may be of general interest. Your current experiments are "clean", and hence you form silicates. However, do you think that if there were some "pollutants" present, as in the ISM (*e.g.*, C, N or S atoms), that you would still form silicates, and what type of silicates might they be? For example, the presence of C atoms can lead to the reduction of Fe to metallic iron during the annealing of amorphous Mg–Fe silicates.[3]

1 M. Roskosz *et al.*, *Astron. Astrophys.*, 2011, **529**, A111.
2 M. Roskosz *et al.*, *Astrophys. J.*, 2009, **707**, L174.
3 C. Davoisne *et al.*, *Astron. Astrophys.*, 2006, **448**, L1.

Dr Jäger replied: Thank you very much for pointing to the work of Roskosz *et al.* Indeed, these are two very interesting papers clearly showing how complicated the silicate chemistry is. However, in our cold condensation process, we are not under equilibrium conditions, and therefore I would not expect reactions such as the formation of pyroxene from olivine-type silicates. As discussed in previous comments, the reason for obtaining a different stoichiometry to that in the laser ablated target is the low laser power density, which leads to phase separation and fractionation of the silicate material. However, in cold astrophysical environments, the type of silicate formed depends on the ratio of silicate-forming molecules and atoms that accrete on the surface of a grain, and this may depend on many factors, such as the destruction process of circumstellar grains, the site of destruction and recondensation, turbulence that may mix the precursors, and so on. For instance, selective sputtering of silicates by ion bombardment due to shock waves may also lead to changes in the ratio of the silicate-forming species, since light elements (O, Mg) are preferentially sputtered compared to the heavier components Si or Fe.[1] The simultaneous accretion of many species of silicon-, carbon-, nitrogen- and sulfur-bearing molecules on the surface of a grain may of course influence the condensation process. In any case, they would separate the silicate-forming species from each other. However, we do not believe that this process would prevent the growth of solid layers. First of all, the accreted species may diffuse on the surface and finally find a partner for reaction. In the condensation process studied in the neon matrix, the molecules are also isolated from each other since they are embedded in the cryogenic matrix at the beginning. In the second step, the condensation of silicates is stimulated by the diffusion of silicate-forming species due to annealing of the cryogenic matrix up to 13 K. The formation of the broad 10 μm band of amorphous silicates has already been observed in this phase. In addition, there are other scenarios that may eventually influence the condensation. The accretion of carbon atoms and clusters and the final formation of carbonaceous solids, such as hydrogenated amorphous carbon, may be slower since carbon in the solid form as well as in clusters may react with O_2 and H_2O ice to form CO and CO_2 molecules. In addition, carbon can also react with hydrogen to form volatile materials that can more easily desorb than condensed solid layers. Nitrogen or small nitrogen-bearing species may also react with oxygen and hydrogen. In contrast, a solid layer of silicates is resistant to oxidation and hydrogenation. One may change the surface of the solid, but erosion as in the case of carbon is impossible. However, even silicates may change their composition and structure. As you already mentioned, reduction processes in the presence of carbon or hydrogen are, of course, possible scenarios that may strongly influence the final composition of silicates condensed at low temperature in the interstellar medium.

1 K. Demyk *et al.*, *Astron. Astrophys.*, 2001, **368**, L38.

Professor van der Tak opened the discussion of Dr Ioppolo's paper: How much column density of ice is needed for an observable feature? Can we see or have we already seen such features?

Dr Ioppolo replied: In the laboratory, we see our best features arising from a laboratory column density of the order of $\sim 10^{18}$ cm^{-2}. Indeed, for all but the weakest features, we are likely quite sensitive to at least an order of magnitude less than that.

In the case of water, this aligns very well with previously observed ice column densities, which typically vary from 10^{17} to 10^{18} cm^{-2}.[1,2]

Malfait *et al.* reproduced the THz spectral profile of a Herbig Ae star HD142527 by comparing ISO data with some laboratory spectra of silicates, montmorillonite, and crystalline water ice. This is a further indication that the interstellar features of solid molecules are detectable in the THz.

1 A. G. Tielens *et al.*, *Astrophys. J.*, 1991, **381**, 181.
2 M. Tanaka *et al.*, *Astrophys. J.*, 1994, **430**, 779.
3 K. Malfait *et al.*, *Astron. Astrophys.*, 1999, **345**, 181.

Dr Pontoppidan addressed Dr Ioppolo: The 63 μm water ice feature is very broad, and detections with Herschel are difficult since it blends with the overall SED shape. The 43 μm band is potentially much sharper, but was unfortunately not covered by Herschel. The next abundant ice species that may have a sharp feature in the Herschel/THz range is CO_2, for which old laboratory work suggested the presence of a 150 μm feature. Did you look for this in your experiments, or are you planning to?

Dr Ioppolo replied: In 1999, Malfait *et al.* reproduced the far-infrared spectral profile of a Herbig Ae star HD142527 by comparing ISO data with some laboratory spectra of silicates, montmorillonite, and crystalline water ice.[1] Although the 63 μm band of water ice is quite broad, detection of water ice should still be possible with Herschel PACS data. We are currently providing new THz laboratory spectra of water ice at different temperatures to be compared to several different star-forming regions observed by the Herschel Space Observatory.

As you mentioned, CO_2 ice is the second most important candidate to be detected in the THz spectral region. Moore and Hudson (1994) showed that CO_2 ice has a sharp strong peak at ~ 85 μm.[2] We recently deposited layers of crystalline CO_2 ice at 75 K. Our THz spectra confirmed the presence of the 85 μm band, and showed that CO_2 has a second slightly less intense band at higher wavelength. To the best of our knowledge, this last feature has never been reported before in the literature, because far-infrared laboratory spectra acquired with traditional FTIR spectrometers usually range between 20 and 100 μm. In our spectra, both these features appear to be very narrow and intense. Therefore, they are good candidates for the detection of interstellar solid CO_2 in the far-infrared with Herschel data. Our CO_2 THz spectra will be incorporated in a future publication.

1 K. Malfait *et al.*, *Astron. Astrophys.*, 1999, **345**, 181.
2 M. H. Moore and R. L. Hudson, *Astron. Astrophys., Suppl. Ser.*, 1994, **103**, 45.

Professor Zacharias asked: How does the THz set-up compare with an IR free-electron laser like FELIX in Nijmegen with respect to sensitivity, tunability and temporal resolution?

Dr Ioppolo answered: Our spectrometer has good spectral coverage from 0.3 to 7.5 THz in one pulse, which is almost the entire THz region. None of the individual FEL experiments in Nijmegen covers the entire region accessible to our instrument. The sensitivity of any experiment is based on the signal-to-noise ratio of the spectrometer, and this ratio is influenced by both the source brightness and the detection sensitivity. Given the numbers provided by Wijnen *et al.*,[1] there are detection schemes for FEL experiments that allow for the detection of single pulses of THz light from the FEL at the nJ level. As such, one could construct a spectrometer with excellent sensitivity using a FEL. However, a bench-top spectrometer such as ours offers adequate sensitivity for the spectroscopy of astrochemical ice analogs.

1 F. J. P. Wijnen, G. Berden and R. T. Jongma, *Opt. Express*, 2010, **18**, 26517.

Professor van Dishoeck addressed Professor van der Tak: This is a comment on your question about the detection of far-infrared ice features and the column densities needed to see them.

As pointed out by Dr Ioppolo, the 45 and 63 µm water ice features have been detected by ISO in a few sources, mostly in emission in protoplanetary disks.[1,2] They have also been seen in absorption toward two embedded massive protostars.[3] With improved data reduction methods, the 63 µm water ice feature is now also robustly detected in a handful of disks around Herbig Ae stars with the Herschel PACS instrument.[4,5] For the disk around HD 142527, the 63 µm feature seen by Malfait *et al.* is confirmed, but the hydrated silicate feature (ascribed to a clay, montmorillionite) is not, illustrating the difficulty in detecting these weak broad features on top of a broad continuum.

The detected emission features have been interpreted with models that indicate that a substantial fraction of the available oxygen (on the order of 25–50%) must be locked up in water ice in the emitting layer of the disk. To detect minor ice features, the signal-to-noise ratio on the continuum needs to be significantly higher than the current values achieved with ISO and PACS of, at most, 100; this implies that the instrumental characteristics, in particular the stability and relative spectral response function, need to be well understood and characterized.

1 K. Malfait *et al.*, *Astron. Astrophys.*, 1999, **345**, 181.
2 E. I. Chiang *et al.*, *Astrophys. J.*, 2001, **547**, 1077.
3 E. Dartois *et al.*, *Astron. Astrophys.*, 1998, **338**, L21.
4 M. K. McClure *et al.*, *Astrophys. J.*, 2012, **759**, L10.
5 J. Bouwman *et al.*, in preparation.

Professor Linnartz continued the discussion of Dr Ioppolo's paper: One common problem in IR ice spectroscopy is the overlap of generally unstructured features, making it hard to unambiguously identify the constituents of the ice. On the one hand, similar vibrational modes of different species have similar spectra (*e.g.*, CH stretches for different polycyclic aromatic hydrocarbons (PAHs) in water ice), and on the other hand different modes of different molecules can also coincide. In the THz domain this is definitely also an issue, especially as the accessible energy domain is even smaller. Therefore, to what extent does the THz domain offer a molecule-selective tool to identify ice in space unambiguously?

Dr Ioppolo replied: THz vibrational modes are fingerprints of intra- and mostly intermolecular forces. These modes are therefore very much linked to the composition, temperature, and structure of the ice, which means that species that share the same functional groups do not necessarily present absorption features at the same wavelengths. Moreover, one of the most important advantages of using THz spectroscopy *versus* spectroscopy in the mid-IR to detect interstellar solid species is that ices can potentially be detected in both absorption and emission along any line of sight in the THz range.

Our laboratory spectra showed that water ice presents some distinct broad features between 4 and 7 THz. These features do not overlap with the sharper bands of CO_2 (see my previous answer to Dr Pontoppidan's question), or even with some bands from more complex species, such as methanol. This is especially true in the case that the ice has experienced some thermal processing. Our new laboratory THz spectra of high-resolution (0.1 cm^{-1}) crystalline methanol ice at 10 K revealed a forest of sharp features in the spectral region between 2 and 4 THz. In Fig. 4 of our paper, we only observed three absorption bands due to the lower spectral resolution used in this work. Therefore, high-resolution THz spectra can be used to detect interstellar species in the solid phase, even in regions where mid-IR data of ices are not available because of a lack of a background light source.

Professor Kamp asked: Can you comment on whether different layered ice structures could be distinguished from the emission signatures we can observe from ices in space?

Dr Ioppolo communicated in reply: The THz laboratory spectra of pure ices presented in our paper indicate that different species have different absorption features that are structure- and temperature-dependent. In a previous work, we showed that as long as species are not mixed together, the unique signature of each component is still visible in layered ices.[1] On the other hand, when different species are mixed together, the THz spectrum will depend on the interaction of the different species within the ice (as in the case of CO_2 in water ice in ref. 1).

If we consider that interstellar ices can be divided in polar and non-polar layers, and we assume that the main components of these ices are water, carbon monoxide, carbon dioxide, and sometimes even methanol, Fig. 8a in our paper shows that water should still be detectable even if layered or mixed with a small percentage of methanol. This could also be the case for other species, such as CO–CO_2 mixed or layered ices (see my previous answer to Dr Pontoppidan's question).

1 M. A. Allodi, S. Ioppolo, M. J. Kelley, B. A. McGuire and G. A. Blake, *Phys. Chem. Chem. Phys.*, 2014, **16**, 3442.

Professor Arumainayagam opened the discussion of the paper by Ralf-Ingo Kaiser: Are you able to monitor the species desorbing during electron irradiation?

In other words, are you able to do electron-stimulated desorption (ESD) experiments?

Professor Kaiser responded: Yes, in principle we can do that. But in the present experiments, we do not see any evidence that the 5 keV electrons desorb any species.

Professor Dulieu commented: The set-up you have built is impressive and at the cutting edge of what can be done. I am sure that you will provide a lot of new, rich information to the community thanks to the multi-detection possibilities. I was wondering what your strategy has been for the choice of the detection system you use. For example, why did you discard some possibilities, like very high m/z QMS detection with tunable electron energy for ionisation, or even gas chromatography? Putting energy into mixed ices induces a very rich chemistry. How do you deconvolve the branching ratios of the different products?

Professor Kaiser replied: With respect to the strategy, we are attempting to elucidate reaction mechanisms of the formation of new molecules in the condensed phase (solid state) at temperatures as low at 5 K, and also during the warm up of the irradiated samples, in which newly formed molecules can sublime into the gas phase. There is no single analytical tool capable of tracing all newly formed molecules on-line and *in situ* under contamination-free ultrahigh vacuum (UHV) conditions. Therefore, it is necessary to exploit highly complementary detection schemes in the condensed phase (infrared, Raman, UV/Vis), and for those molecules subliming into the gas phase, like single photon fragment-free vacuum ultraviolet (VUV) photoionization coupled with reflectron time-of-flight mass spectrometry (ReTOF-MS). Note that the on-line and *in situ* operation of the spectrometers is a crucial prerequisite to elucidate the formation of new molecules and functional groups over the experimental period. By following the temporal evolution of the absorption bands, we can extract time-dependent concentration profiles of newly formed species. These profiles can be fitted kinetically to derive the reaction mechanisms for the formation of even bio-relevant molecules like glycolaldehyde.[1-9]

For the condensed phase, infrared, Raman, and UV/Vis are capable of probing (small) individual molecules,[10-36] and most importantly functional groups. However, as the molecules get more complex, it is not feasible that infrared, Raman, and UV/Vis can discriminate between complex organic molecules with identical functional groups, such as acetaldehyde (CH_3CHO), propanal (C_2H_5CHO), butanal (C_3H_7CHO) and higher molecular weight aldehydes.[37-39] To discriminate those, the sample is warmed up to allow the newly synthesized molecules to sublime. These are identified *via* fragment-free single photon ionization coupled with reflectron time-of-flight mass spectrometry (ReTOF-MS).[40-41] It is important to note that these molecules can be photoionized according to their known ionization energies and then mass-analyzed. Since different isomers have different ionization energies, the correlation of the ionization energy with the mass-to-charge ratio of the product helps to uniquely identify individual molecules. The photoionization cross sections of organic molecules are mostly known, so it is also feasible to derive branching ratios from the ion signals. Since in some cases the ionization potentials of structural isomers are within about 0.1

eV of each other, ionization with tunable low energy electrons is problematic; energy spreads down to 0.1 eV are rarely achieved with low energy electrons, so structural isomers with similar ionization energies can both be ionized. This prevents isomer specific detection.

With respect to GC-MS, it is important to stress that, upon warm up, organic residues might be formed. The analysis of these residues is not trivial. Upon venting of the machine, traces of oxygen can never be fully prevented from potentially oxidizing labile molecules or functional groups in the residues. Further, the residues must be dissolved under mild conditions without hydrolyzing the molecules formed.[42] This is not always feasible and requires sophisticated techniques.

Infrared spectroscopy: Utilizing a Nicolet Fourier Transform Infrared (FTIR) 6700 spectrometer with a liquid nitrogen cooled MCTB-type detector, the ices are probed in the 10 000–500 cm^{-1} region. This includes the astronomically relevant range from 1.0–2.5 μm (10 000–4000 cm^{-1}), so that the laboratory spectra can be archived and compared with prospective data from space missions. Infrared spectra are also utilized to identify (small) individual molecules based on excitation of vibrations associated with changes in the permanent dipole moment of the molecule and also functional groups. Additionally, a non-volatile residue is typically formed whenever carbon is present in the ice. The residue may consist of a complex mixture of chemical components, leading to significant overlap of individual peaks in their respective spectra. In this particular case, a deconvolution technique is applied to the infrared spectra to discern individual species. The success of this method has been demonstrated previously and led to the detection of functional groups of dipeptides.

Raman spectroscopy: We exploit a novel custom built Raman system for 532 nm laser excitation, a super notch filter, and an intensified CCD detector. To generate the 532 nm light, we utilize a frequency doubled Nd:YAG laser. Raman spectroscopy identifies vibrations of a molecule causing a change in the molecular polarizability, *e.g.* homonuclear, infrared inactive diatomic species such as molecular hydrogen, oxygen and nitrogen. Raman analysis has a number of advantages compared to infrared spectroscopy, such as the sharpness of the spectral features. Considering that our ice samples also contain water, which is a very strong infrared absorber, the Raman spectrum of water is weak; this allows the identification of functional groups of astrobiologically important molecules such as alkylphosphonic acids in water-bearing ices. We demonstrated that this instrument obtains a very good Raman signal.[43–45]

Ultraviolet–visible spectroscopy: UV/Vis spectra are recorded by a Thermo Nicolet Evolution 600 Spectrometer operating in the range of 190–1100 nm. UV/Vis spectroscopy helps in the identification of electronic transitions of functional groups. We demonstrated that this instrument can collect excellent UV/Vis data.[46]

Reflectron time-of-flight mass spectrometry (ReTOF) & photoionization: Products released into the gas phase are sampled after ionization *via* a reflectron time-of-flight mass spectrometer, which monitors the complete product spectrum simultaneously based on the mass-to-charge ratios of the ionized neutral molecules. Although electron impact is a widely used approach to ionize gas phase molecules, electron impact ionization cannot be always applied to discriminate large molecules of astrobiological importance: electron impact often leads to significant fragmentation of the parent, which can severely reduce or even eliminate entirely the signal of the parent molecule. Instead, fragment-free single photon soft photoionization with vacuum ultraviolet photons (VUV) is implemented to observe the molecules as they sublimate into the gas phase. We are exploiting two of the key advantages of soft photoionization. First, fragment-free photo ionization combined with the detection of the parent ion at a well-defined mass-to-charge ratio helps to determine explicitly the molecular mass of the molecule. Secondly, by tuning the photoionization wavelength from 5 to 10 eV we can produce a photoionization efficiency (PIE) curve, which reports the intensity of the parent ion *versus* the photon energy at a well-defined mass-to-charge ratio of the prospective alkylphosphonic acid. These curves can be utilized to determine the ionization energy of the molecule of interest and can be compared with known PIE curves. Since distinct isomers have different PIEs, we can even determine the branching ratios of both isomers by fitting the experimentally determined PIE with a linear combination of the PIEs of distinct isomers. Therefore, the use of tunable fragment-free single photon soft photoionization is crucial in determining to what extent structural isomers are formed.

1 C. J. Bennett and R. I. Kaiser, *Astrophys. J.*, 2007, **661**, 899.
2 L. Zhou, R. I. Kaiser, L. G. Gao, A. H. H. Chang, M. C. Liang and Y. Y. Yung, *Astrophys. J.*, 2008, **686**, 1493.
3 C. J. Bennett, Y. S. Kim, R. I. Kaiser, T. Hama and M. Kawasaki, *Astrophys. J.*, 2011, **727**, 27.
4 B. M. Jones, C. J. Bennett and R. I. Kaiser, *Astrophys. J.*, 2011, **734**, 1.
5 C. J. Bennett and R. I. Kaiser, *Astrophys. J.*, 2007, **660**, 1289.
6 C. J. Bennett, C. S. Jamieson, Y. Osamura and R. I. Kaiser, *Astrophys. J.*, 2005, **624**, 1097.
7 P. D. Holtom, C. J. Bennett, Y. Osamura, N. J. Mason and R. I. Kaiser, *Astrophys. J.*, 2005, **626**, 940.
8 C. J. Bennett, Y. Osamura, M. D. Lebar and R. I. Kaiser, *Astrophys. J.*, 2005, **634**, 698.
9 C. J. Bennett, S. H. Chen, B. J. Sun, A. H. H. Chang and R. I. Kaiser, *Astrophys. J.*, 2007, **660**, 1588.
10 C. J. Bennett, C. P. Ennis and R. I. Kaiser, *Astrophys. J.*, 2014, **782**, 63.
11 B. Sivaraman, A. M. Mebel, N. J. Mason, R. I. Kaiser and D. Babikov, *Phys. Chem. Chem. Phys.*, 2011, **13**, 421.
12 C. P. Ennis, C. J. Bennett and R. I. Kaiser, *Phys. Chem. Chem. Phys.*, 2011, **13**, 9469.
13 C. P. Ennis, C. J. Bennett, B. M. Jones and R. I. Kaiser, *Astrophys. J.*, 2011, **733**, 1.
14 C. P. Ennis and R. I. Kaiser, *Astrophys. J.*, 2012, **745**, 103.
15 C. J. Bennett, C. S. Jamieson and R. I. Kaiser, *Astrophys. J., Suppl. Ser.*, 2009, **182**, 1.
16 C. J. Bennett, C. S. Jamieson and R. I. Kaiser, *Phys. Chem. Chem. Phys.*, 2010, **12**, 4032.
17 W. Zheng and R. I. Kaiser, *J. Phys. Chem. A*, 2010, **114**, 5251.
18 C. J. Bennett, C. S. Jamieson and R. I. Kaiser, *Phys. Chem. Chem. Phys.*, 2009, **11**, 4210.
19 W. Zheng, D. C. Jewitt, Y. Osamura and R. I. Kaiser, *Astrophys. J.*, 2008, **674**, 1242.
20 C. S. Jamieson, A. M. Mebel and R. I. Kaiser, *Chem. Phys. Lett.*, 2008, **450**, 312.
21 W. Zheng and R. I. Kaiser, *Chem. Phys. Lett.*, 2007, **450**, 55.
22 B. Sivaraman, C. S. Jamieson, N. J. Mason and R. I. Kaiser, *Astrophys. J.*, 2007, **669**, 1414.
23 C. S. Jamieson, A. M. Mebel and R. I. Kaiser, *Chem. Phys. Lett.*, 2007, **443**, 49.
24 C. J. Bennett and R. I. Kaiser, *Astrophys. J.*, 2005, **635**, 1362.
25 R. I. Kaiser, G. Eich, A. Gabrysch and K. Roessler, *Astron. Astrophys.*, 1999, **346**, 340.
26 C. J. Bennett, C. Jamieson, A. M. Mebel and R. I. Kaiser, *Phys. Chem. Chem. Phys.*, 2004, **6**, 735.
27 C. S. Jamieson, C. J. Bennett, A. M. Mebel and R. I. Kaiser, *Astrophys. J.*, 2005, **624**, 436.
28 C. S. Jamieson, A. M. Mebel and R. I. Kaiser, *Phys. Chem. Chem. Phys.*, 2005, 7, 4089.
29 W. Zheng, D. Jewitt and R. I. Kaiser, *Astrophys. J.*, 2006, **639**, 534.
30 C. S. Jamieson, A. M. Mebel and R. I. Kaiser, *Astrophys. J., Suppl. Ser.*, 2006, **163**, 184.
31 W. Zheng, D. Jewitt and R. I. Kaiser, *Astrophys. J.*, 2006, **648**, 753.
32 C. S. Jamieson, A. M. Mebel and R. I. Kaiser, *Chem. Phys. Lett.*, 2007, **440**, 105.
33 C. S. Jamieson, A. M. Mebel and R. I. Kaiser, *ChemPhysChem*, 2006, 7, 2508.
34 W. Zheng, D. C. Jewitt and R. I. Kaiser, *Chem. Phys. Lett.*, 2007, **435**, 289.
35 C. S. Jamieson and R. I. Kaiser, *Chem. Phys. Lett.*, 2007, **440**, 98.
36 W. Zheng, D. C. Jewitt and R. I. Kaiser, *Phys. Chem. Chem. Phys.*, 2007, **20**, 2556.
37 Y. S. Kim and R. I. Kaiser, *Astrophys. J.*, 2011, **729**, 1.

38 Y. S. Kim and R. I. Kaiser, *Astrophys. J.*, 2010, **725**, 1002.
39 C. J. Bennett, B. Jones, E. Knox, J. Perry, Y. S. Kim and R. I. Kaiser, *Astrophys. J.*, 2010, **723**, 641.
40 R. I. Kaiser, S. Maity and B. M. Jones, *Phys. Chem. Chem. Phys.*, 2014, **16**, 3399.
41 B. M. Jones and R. I. Kaiser, *J. Phys. Chem. Lett.*, 2013, **4**, 1965.
42 R. I. Kaiser, Y. S. Kim, A. Stockton, E. Jensen and R. A. Mathies, *Astrophys. J.*, 2013, **765**, 111.
43 C. J. Bennett, S. J. Brotton, B. M. Jones, A. K. Misra, S. Sharma and R. I. Kaiser, *Anal. Chem.*, 2013, **85**, 5659.
44 S. Brotton and R. I. Kaiser, *Rev. Sci. Instr.*, 2013, **84**, 055114.
45 S. Brotton and R. I. Kaiser, *J. Phys. Chem. Lett.*, 2013, **4**, 669.
46 B. M. Jones, R. I. Kaiser and G. Strazzulla, *Astrophys. J.*, 2014, **781**, 85.

Dr McGregor addressed Sergio Ioppolo: My research focuses on heterogeneous catalysis, and hence deals with both gas–solid and liquid–solid interactions. In the latter case, working with experts in the spectroscopy field, we have also studied structuring in methanol–water and acetone–water mixtures using THz time-domain spectroscopy. For the former we identify a number of critical concentrations at which the structuring changes (namely 10, 30 and 70 mol% methanol), ranging from a system dominated by water networks, through systems dominated by mixed water–methanol networks, to a system dominated by methanol–methanol interactions. This has practical implications in changing properties such as electron solvation (or in my case reactions occurring in the liquid when used as a solvent). Much of this structuring derives from dynamical hydrogen-bond forming/breaking processes, which will not be present in solid-phase ices. We similarly observed strong interactions in acetone systems (could this be carbonyl–carbonyl interactions?). Despite the ices discussed in this paper being in the solid-phase, do they posses an analogous type of structuring to liquid-phase systems and does this have a concentration dependence? If so, can we apply any of the knowledge gained from liquid-phase systems to astro-chemically-relevant ices?

Dr Ioppolo replied: Certainly, amorphous ices present more similarities with liquids than crystalline ices. Fig. 8 in our paper shows three different mixtures of water and methanol, and another three mixtures of water and acetaldehyde. All these ices were deposited in crystalline form in order to investigate the interactions of different crystals and to enhance the probability of observing sharp distinct features from the different components of the mixtures. Our choice to study 2 : 1, 1 : 1, and 0.5 : 1 mixtures was also dictated, as in your studies, by the curiosity to investigate different regimes dominated by water networks, by mixed water–methanol (or water–acetaldehyde) networks, and by methanol–methanol (acetaldehyde–acetaldehyde) interactions. From a first look at these different regimes, it is clear that they all differ from each other. Moreover, trying to fit these spectra with pure crystal components was quite difficult, because even when the profile of the water-rich mixtures looked qualitatively similar to the spectrum of pure crystalline water ice, all the features were shifted. This is a clear indication of the interaction between different species within an ice. In this respect, the knowledge gained from liquid-phase systems in the past few years can be easily used to identify interesting cases that can potentially be investigated in the solid phase in the THz, regardless of their interstellar relevance.

Professor Rawlings remarked: The discussion of the effects of ice layering is fascinating and possibly very important. However, real interstellar dust grains do not look like smooth spheroids. How severely will the real morpholgies and natures of the surfaces affect these conclusions?

Dr Ioppolo responded: The thickness of our laboratory ices ranges between a fraction of a micron and several microns. In the current configuration, our THz spectrometer is not sensitive to thicknesses lower than a few thousand mono-layers (ML), where 1 ML = 10^{15} molecules cm^{-2}. Several studies showed that the contribution from the interaction between the substrate and the ice is negligible for thicknesses equal to or higher than a few tens of ML.[1-3] Therefore, the THz modes seen in our spectra are mostly bulk modes. We are aware of the fact that the nature and shape of substrates may influence peak positions and profiles of THz modes of ices a few ML thick. However, THz spectra of interstellar ices, that are generally a few tens of ML thick, should resemble the THz spectra obtainable with our spectrometer. In the near future, we plan to investigate THz surface *vs.* bulk modes more systematically, as well as using reflection–absorption THz spectroscopy to enhance our sensitivity to thinner layers.

1 J. He and G. Vidali, *Astrophys. J.*, 2014, **788**, 50.
2 F. Dulieu *et al.*, *Nat. Sci. Rep.*, 2013, **3**, 1338.
3 S. Ioppolo *et al.*, *Phys. Chem. Chem. Phys.*, 2010, **12**, 12065.

Dr Pontoppidan remarked: From an astronomer's perspective, the mid- to far-infrared wavelength region is threatened. It requires expensive space or airborne telescopes, but few, if any remain. SOFIA has an uncertain future and Herschel is no longer operational. Ground-based observatories often put lower priorities on mid-infrared facilities. The one bright light on the horizon is JWST, but that is also a mission with a limited lifetime. We astronomers need important science cases to support the development of new infrared telescopes. There seems to be a tendency by laboratory astrochemists to wait for astronomers to provide exciting science cases in need of laboratory work, but it has to go the other way as well. We need a more proactive role from laboratory astrochemists to help us provide "slam dunk" science cases. Only if we work together as equal partners can we compete with other astrophysical science cases for the limited funds to develop astronomical facilities.

Professor Linnartz said: It may be interesting for many of you to know that we have started a new coordinated initiative to visualize the ongoing efforts within European Laboratory Astrophysics. To this end a new site has been initiated (www.labastro.eu), and for interested scientists it is possible to join a new mail distribution list. The focus of the site is to show (1) the research that is being performed within Europe, (2) which groups are involved, and (3) the European infrastructure (also in terms of databases) that is available. An application has been prepared to provide access to European laboratories performing astro-physical research, in order to stimulate on-topic projects in which clear questions can be addressed with highly specialized equipment. For more details, contact: labastro@strw.leidenuniv.nl.

Professor Mason said in reply: Similarly, the European Planetary Science community (http://www.europlanet-eu.org) has developed a collaborative research network under the auspices of FP6 and FP7, and it is coordinating several bids to Horizon 2020 including a bid for an Advanced Research Infrastructure. Planetary science research includes many studies that overlap with this meeting (*e.g.*, the irradiation of planetary ices leading to molecular synthesis on icy bodies and the study of chemistry of planetary atmospheres). The underlying atomic, molecular data (rate constants, cross sections, spectroscopy) are very similar to the data used in astrochemistry, and use the same databases. VAMDC is an EU led project to provide access to such databases through a single portal (www.vamdc.eu and www.vamdc.org). For further information and engagement in either project, contact nigel.mason@open.ac.uk.

Regarding databases, there are several international activities, such as the International Virtual Observatory Project IVOA (www.ivoa.net) and VAMDC (www.vamdc.eu), which are collating both data and databases to provide a resource for the astronomy/astrochemistry community. An important part of these activities is the development of a "roadmap" to provide protocols for data assembly, validation and assessment, which will be important for the astrochemistry community (*e.g.* in modelling). The astrochemistry community should participate more strongly in such activities.

Professor McCoustra continued the discussion of the paper by Sergio Ioppolo: Following up on Professor Mason's comments, I note the very rich and complex spectrum presented by Dr Ioppolo for solid species. However, we should be careful in not just blindly measuring spectra, as we may be accused of stamp collecting, and think more on the question of whether we can generate new physical understanding of these solid state materials from these spectra. No doubt this means working very closely with the theoretical and computational chemistry community. However, density functional theory (DFT) of an isolated molecule or dimer is only a starting point. Even within the limit of an isolated molecule there are issues in blindly using DFT codes, especially in this spectral region, where vibrations are highly anharmonic and difficult to accurately predict computationally. The collective motions of the solid and the coupling of the collective motions with the molecular motions must be considered. Perhaps the molecular dynamics community have something to add to the interpretation of these spectra?

Dr Ioppolo responded: In our paper, we used some MP2 calculations to find a simple explanation for the differences between the THz spectra of acids and the other investigated species. We understand the importance of using more refined simulation methods that can better resemble more realistic cases. In this respect, molecular dynamics can be a very useful tool to understand mechanisms such as the diffusion of species within an ice. In order to achieve this goal, however, we need to first assign and study the nature of the THz modes of solid species.

Recently, we used CRYSTAL09 code to generate predictions of crystalline modes in the THz spectral range. CRYSTAL09 is an *ab initio* program for the study of crystalline solids, and therefore does not simulate gas-phase single molecules or dimers. It expands the single molecule functions as a linear combination of Bloch functions and is able to predict energies and properties that are unique to the crystalline form of a studied molecule. In particular, the spectral modes predicted by CRYSTAL09 are all optical phonon modes. Some of our preliminary results show that the synthetic spectrum of crystalline methanol resembles the laboratory spectrum of the same ice quite well. Moreover, the anharmonicity in the vibrational modes of a solid molecule can be modelled from the spectral shift of THz features visible in laboratory spectra of crystalline ices acquired at different temperatures. We are currently working on comparing these anharmonic contributions observed experimentally to the CRYSTAL09 predicted modes in order to retrieve more information on the nature of the THz modes of molecules in the solid phase.

Dr Oba asked: Have you ever analysed deuterated molecules like D_2O? Is the THz spectroscopy sensitive to such deuterated molecules?

Dr Ioppolo replied: Yes, our laboratory results show that THz spectroscopy is sensitive to deuterated species. Recently, we deposited crystalline D_2O at 150 K and cooled it down to 10 K. The D_2O THz spectrum looks qualitatively similar to the spectrum of H_2O, except it is shifted by 0.3 THz toward lower frequencies. The shift is large enough to distinguish the two molecules from each other.

Professor Linnartz addressed Cornelia Jäger: In your introduction you mentioned experiments performed both in a matrix and in a helium droplet environment. Whereas Ne matrix isolation spectroscopy is considered to provide a highly inert environment, more and more evidence is being found to show that dynamical effects in He droplets may not be completely negligible. Moreover, there is a difference in temperature between He droplets and Ne matrices. As you performed experiments in both environments, I was wondering whether you found systematic differences in the cold dust condensates formed in your experiments?

Dr Jäger responded: Indeed, we performed experiments on the cold condensation of $(SiO)_x$ in helium clusters and in neon matrices. Liquid helium droplets provide an ultralow temperature of $T = 0.37$ K, and they are proven to be superfluid. All species embedded inside the He droplets are able to move freely, but they are confined in the droplet, thus resembling the situation of accreting molecules or atoms on the surface of dust grains. In addition, helium droplets provide an ideal opportunity to study chemical reactions and the formation of intermediates in a low temperature condensation process. In both environments (helium clusters and neon matrices), the formation of SiO clusters could be observed by characterizing the intermediates either *via* mass spectrometry, or by UV/Vis spectroscopy. The oligomerization of SiO molecules was found to be barrierless. The reaction rate does not follow the Arrhenius equation and consequently, it does not show strong temperature dependence. Therefore, this reaction is expected to be fast in the low-temperature environment of the interstellar medium on the surface of dust grains. Condensation of $(SiO)_x$ and the formation of nanoscale amorphous SiO grains could be observed at a temperature of 0.37 K by the incorporation of numerous SiO molecules into a helium cluster and by annealing and evaporation of SiO-doped Ne matrices at about 10 K. The formation of the silicate in the neon matrix at this low temperature has been monitored by *in situ* MIR spectroscopy. The disappearance of the molecular bands and the occurrence of the 10 μm band typical of amorphous silicates are direct evidence for the condensation process. Both condensates are very similar in their internal structure. The condensed grains show an amorphous structure and both are characterized by the typical 10 μm band. We did not observe any differences, such as the formation of special defects.

Professor Dulieu opened the discussion of the paper by Gianfranco Vidali: Unfortunately the data you used for D_2 on compact water does not meet all the expected conditions; in particular, the data used in Amiaud *et al.* (2007)[1] are at low coverages. However, we have more suitable sets of unpublished data that we would be happy to provide to you for testing your model. The disadvantage of the TPD is that this technique mixes up many processes. Do you envisage using isothermal desorptions (as we did[2]) to separate the different processes of diffusion, desorption and even reactivity?

1 L. Amiaud *et al.*, *J. Chem. Phys.*, 2007, **127**, 144709.
2 J. A. Noble *et al.*, *Astron. Astrophys.*, 2012, **543**, A5.

Professor Vidali communicated in reply: Indeed, the data of Amiaud *et al.*[1] of D_2 adsorption/desorption on/from nonporous water ice are for a coverage that is much less than a monolayer. In such a case, only a fraction of the sites are occupied. Therefore, the distribution of sites obtained from those data is skewed towards the deep sites, that is, the sites that are preferentially occupied by D_2 molecules. The saturation coverage of the D_2 on water ice is 0.55 exposed monolayers (EML), which should be equivalent to about 0.15 ML instead of 1 ML. This means that even if the sample temperature is as low as 10 K, only the deepest 15% of sites are occupied at 10 K. The distribution obtained from the direct inversion represents deep adsorption sites only, since the shallow sites are not occupied. One cannot obtain the full distribution unless the exposure

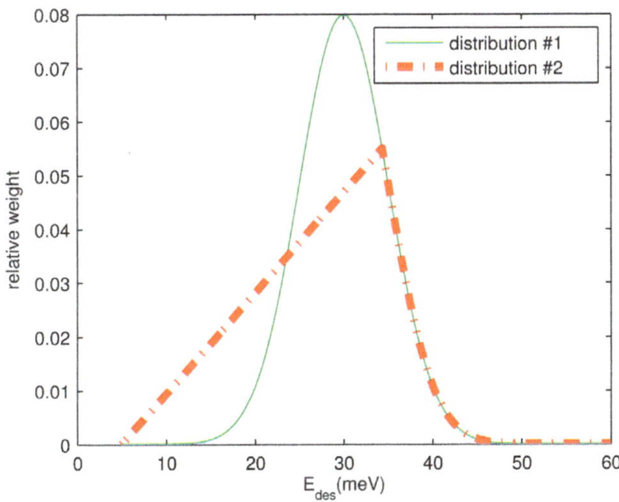

Fig. 1   Two model desorption energy distributions sharing the same tail.

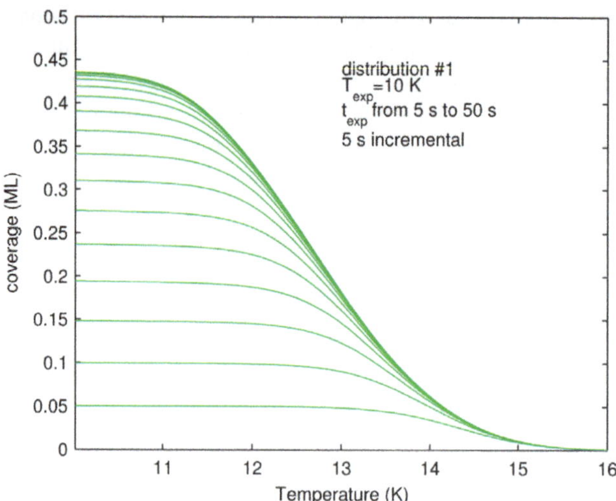

Fig. 2 Simulated coverage using the desorption energy distribution #1 in Fig. 1.

temperature is low enough that desorption from the shallower sites is negligible during exposure.

The question arises of whether useful information about diffusion can be extracted in instances where the data don't give the complete distribution of adsorption energy sites. This is the case when the surface temperature is not low enough that a full layer of adsorbate can be built. To make some progress on this point, we rephrase the question as how an unknown distribution of shallow sites affects the TPD traces. Since the shallow sites are always empty, the exact distribution might not matter and we can use a guessed shallow-site distribution. We now create two model distributions, having the same distribution of deep sites

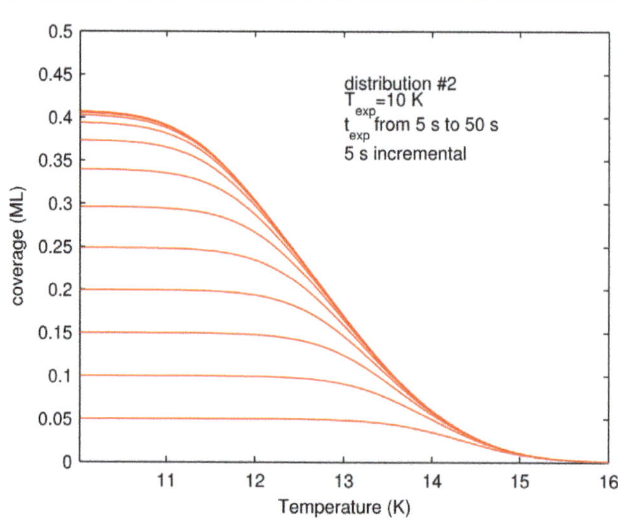

Fig. 3 Simulated coverage using the desorption energy distribution #2 in Fig. 1.

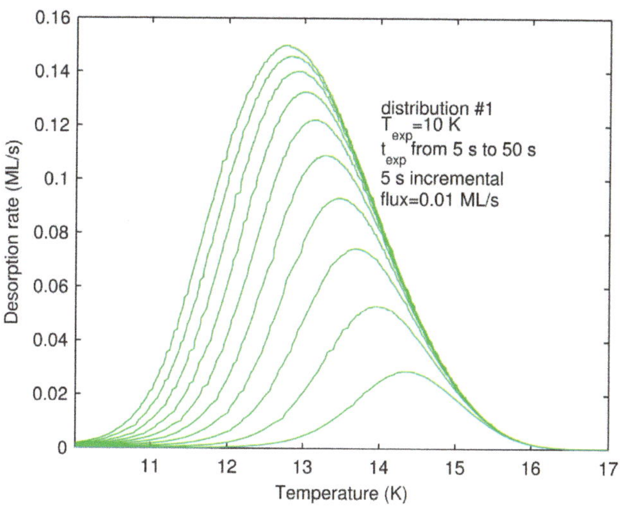

Fig. 4   Simulated TPD spectra using the desorption energy distribution #1 in Fig. 1.

but with different shallow-site distributions, as shown in Fig. 1. The exposure temperature in the model is chosen as 10 K, and the maximum exposure time is such that the coverage is about four tenths of a monolayer (see Fig. 2 and 3). The effective exposure rate (flux times sticking coefficient) is 0.01 ML s^{-1}; exposure duration is 5 s to 50 s, in 5 s increments. The ΔE value[2] is set to be 10 meV. A heating rate of 10 K min^{-1} is applied 100 seconds after exposure; this gives enough time for molecules in the shallow sites to desorb or move into the deep sites.

The simulated TPD spectra are shown in Fig. 4 and Fig. 5. It can be seen that the distribution of the shallow sites affects only the beginning of the traces, but

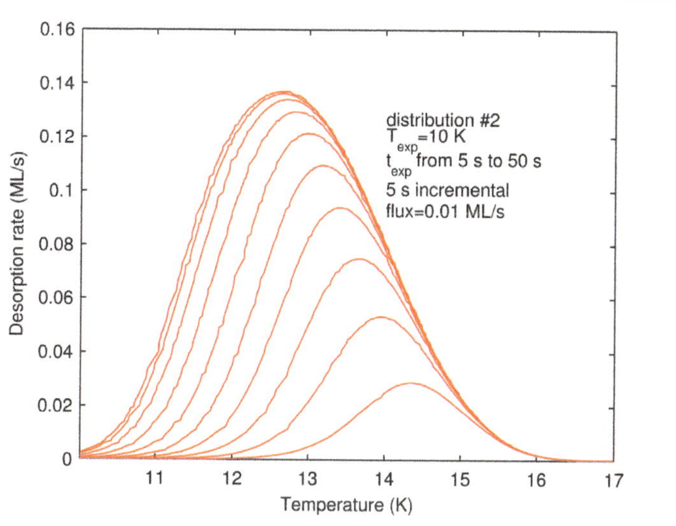

Fig. 5   Simulated TPD spectra using the desorption energy distribution #2 in Fig. 1.

overall there isn't a significant change in the TPD spectra. Had we chosen a lower saturation coverage, such as a quarter of a layer, there would have been hardly any differences in the TPD traces. Thus, one should be able to use a guessed desorption energy distribution for the unknown shallow sites without significantly affecting the TPD traces that are most dependent on the deeper sites.

As for your question about the techniques of TPD and isothermal desorption, I agree that TPD is a blunt but indispensible instrument. Because of the challenges of working at low temperature and with surfaces that are not characterized, or cannot easily be characterized as one would do in routine surface science experiments, TPD is often the only technique that has been widely used in studies of astrochemical processes on surfaces. Isothermal desorption is an attractive tool that was used in the 1980s to study phase transitions in two dimensions and quasi-two dimensions on the surfaces of metals, graphite and semiconductors. It has its own challenges; for example, it requires careful monitoring of conditions to ensure that measurements are done under equilibrium conditions. We have done a study of Hg adsorption on Cu(001) and have shown how equilibrium isobars can be used to obtain useful information about adsorption energetics, such as the isosteric heat of adsorption.[3]

1 L. Amiaud *et al.*, *J. Chem. Phys.*, 2007, **127**, 144709.
2 J. He and G Vidali, *Faraday Discuss.*, 2014, **168**, DOI: 10.1039/c3fd00113j.
3 P. A. Dowben *et al.*, *Surf. Sci.*, 1990, **230**, 113.

Mr Doronin addressed Professor Vidali: Thank you for your article. Your method will certainly be useful for treating our experimental data: we observe similar behaviour in our TPD studies on both amorphous and crystal water ice.

Another point that may be added to Section 4.4, suggestions for experimental investigations, is to systematically perform the TPD experiments at a set of different heating ramp ratios, varying by at least a factor of 10. This might allow the differentiation of diffusion and desorption processes. Also, it may help to obtain a more accurate value of the desorption pre-factor. In the best case, the model should fit all the experiments performed at different ramp ratios.

Professor Vidali replied: What you describe is indeed what is done in many studies in surface science that use the TPD technique. In those cases, the starting temperature is much higher and the temperature of the samples can be increased rather quickly. Therefore, heating rates from $1\ \mathrm{K\ s^{-1}}$ to $100\ \mathrm{K\ s^{-1}}$ and sometimes beyond are used. Unfortunately, operating at low temperature requires a good thermal contact with a liquid helium bath or equivalent. This means that heat generated at the sample can disperse to other parts of the sample holder, causing unwanted desorption and lower heating rates. Eventually however, a technical solution will be found because the easiest way to determine the pre-exponential in the Polanyi–Wigner expression is to change the heating rate, as you suggest.

Professor Linnartz enquired: You mentioned ice porosity, and during this Discussion we have not spoken much about ice structure. Generally, it is assumed that ices in space are amorphous, largely consisting of water, and mainly compact (c-ASW). This is in line with physical-chemical intuition (upon heating or UV processing ice molecules can rearrange to fill pores), and is consistent with the lack of detection of dangling OH bonds in space. However, in a number of recent studies,[1,2] it was found that an ice that does not exhibit dangling bonds still can be compacted, which is indicative of at least some level of remaining porosity.

From a modelling perspective, would it be possible to draw any conclusions on the level of porosity in ice in space? That is, are the differences in the chemical processes in (non-)porous ices in the models so great that this may be indicative of the (non-)porosity of ice in space ?

1 K. Isokoski, J.-B. Bossa, T. Triemstra and H. Linnartz, *Phys. Chem. Chem. Phys.*, 2014, **16**, 3456.
2 J.-B. Bossa, K. Isokoski, D. M. Paardekooper, M. Bonnin, E. P. van der Linden, T. Triemstra, S. Cazaux, A. G. G. M. Tielens and H. Linnartz, *Astron. Astrophys.*, 2014, **561**, A136.

Professor Vidali replied: As presented, our diffusion–desorption rate equation model is best used for gas–surface kinetics. The introduction of pores or cracks into the model brings about two complications. First, we know little about pores in the actual ISM. One could argue, paraphrasing what you said, that because of how ices are built up, with the release of condensation energy, and how they are subjected to energetic processes, they might have a less porous structure than many like to think. Second, there is no direct desorption from pores, while the direct desorption from surface deep sites is possible. However, without a better model available, one can still use the model (at least for interpreting laboratory data; a separate problem is how to translate this information to the ISM) and treat the pores as deep adsorption sites. Porous ices differ from non-porous ices by three characteristics: 1) porous ices have broader desorption energy distributions; 2) the desorption energy distribution from porous ices is centered at a higher energy value than that from non-porous ices; and 3) if one observes the transition from submonolayer coverage to multilayer coverage, the transition happens at a higher coverage for porous ices than for non-porous ices. More studies (or breakthroughs!) are needed to characterize the level of porosity in ISM ices. There aren't many models of atom–surface processes in pores, at least in the astro-chemistry literature. I recall a model by Awad *et al.*,[1] where the ice was modelled as a collection of cracks; more recently, a model has been reported by Kalvāns and Shmeld.

1 Z. Awad, T. Chigai, Y. Kimura, O. M. Shalabiea and T. Yamamoto, *Astrophys. J.*, 2005, **626**, 262.
2 J. Kalvāns and I Shmeld, *Astron. Astrophys.*, 2010, **521**, A37.

Dr Congiu addressed Professor Vidali: This is a comment meant to link the model you presented to the exciting discussion we had yesterday about the diffusion of O atoms, from which we learned that it would be advisable that other groups contribute to the understanding of the physical processes behind the mobility of heavy atoms. If I look at the rate equations of your model, it seems that the diffusion term accounts for diffusion of a given species between empty sites only. All the occupied sites are actually skipped, and thus any possible reaction event is discarded *a priori*. I believe that it would be very interesting if you could improve your current model by adding the possibility that two species react if they come to the same adsorption site. This would allow you to infer, for example, diffusion energy barriers of oxygen atoms, as we did in our paper.[1]

1 E. Congiu, M. Minissale, S. Baouche, H. Chaabouni, A. Moudens, S. Cazaux, G. Manicò, V. Pirronello and F. Dulieu, *Faraday Discuss.*, 2014, **168**, DOI: 10.1039/c4fd00002a, and references therein.

Professor Vidali responded: Ours is just a first step in addressing the question of how we can extract useful information on the diffusion of adsorbed species using data from TPD experiments. Here, we have applied our model to the adsorption of molecules, where the assumption that there can be no adsorption on or diffusion to an already occupied site is a reasonable one. Certainly, as we move to consider reactions between radicals on surfaces we need to incorporate your suggestions into our model.

Professor McCoustra opened the discussion of the paper by Jean-Hugues Fillion: We have previously reported on the adsorbate-mediated desorption of water from the benzene-water system.[1] In that system, we too observe a wavelength dependence of both the benzene itself and the substrate water on which it is sitting, reflecting the first electronic transition in benzene at around 250 nm. The mechanism we proposed for that process involved photon absorption by the benzene (water is transparent at and around 250 nm), non-radiative relaxation from the excited electronic state to the ground state surface, and desorption of the benzene and water molecules and clusters associated with unimolecular decay of the weakly bound benzene–water surface adsorption complex. Could Professor Fillion enlighten us to the mechanism pertinent in this instance?

1 (a) J. D. Thrower, D. J. Burke, M. P. Collings, A. Dawes, P. J. Holtom, F. Jamme, P. Kendall, W. A. Brown, I. P. Clark, H. J. Fraser, M. R. S. McCoustra, N. J. Mason and A. W. Parker, *J. Vac. Sci. Technol., A*, 2008, **26**, 919; (b) J. D. Thrower, D. J. Burke, M. P. Collings, A. Dawes, P. J. Holtom, F. Jamme, P. Kendall, W. A. Brown, I. P. Clark, H. J. Fraser, M. R. S. McCoustra, N. J. Mason and A. W. Parker, *Astrophys. J.*, 2008, **673**, 1233.

Professor Fillion responded: In your experiment, it appears that an indirect benzene-mediated desorption mechanism, in which the mid-UV energy absorbed by the benzene molecules is transferred to neighbouring water molecules, inducing them to desorb, is indeed operative. Such a process is similar to what we observed in the VUV with CO, N_2 and CO_2 molecules. It would be of great interest to confirm the mechanism by recording a photodesorption spectrum together with the determination of the yield. In that case, the indirect desorption from the hydrogen-bonded system could be validated.

Professor Herbst asked: Are there any examples other than CO in which a discrete photodesorption spectrum is measured? Does the photodesorption of CO occur *via* the triplet or singlet Π state?

Professor Fillion replied: The photodesorption of CO occurs *via* the singlet Π state. A similar discrete spectrum has been obtained for N_2, where a singlet Π state vibronic progression has been observed above 12 eV in the photodesorption spectrum. No other examples have been found to date, due to the fact that most solid molecular ices do not present discrete absorption spectra. It would also be of interest to find a counter-example, where the discrete absorption spectrum is not

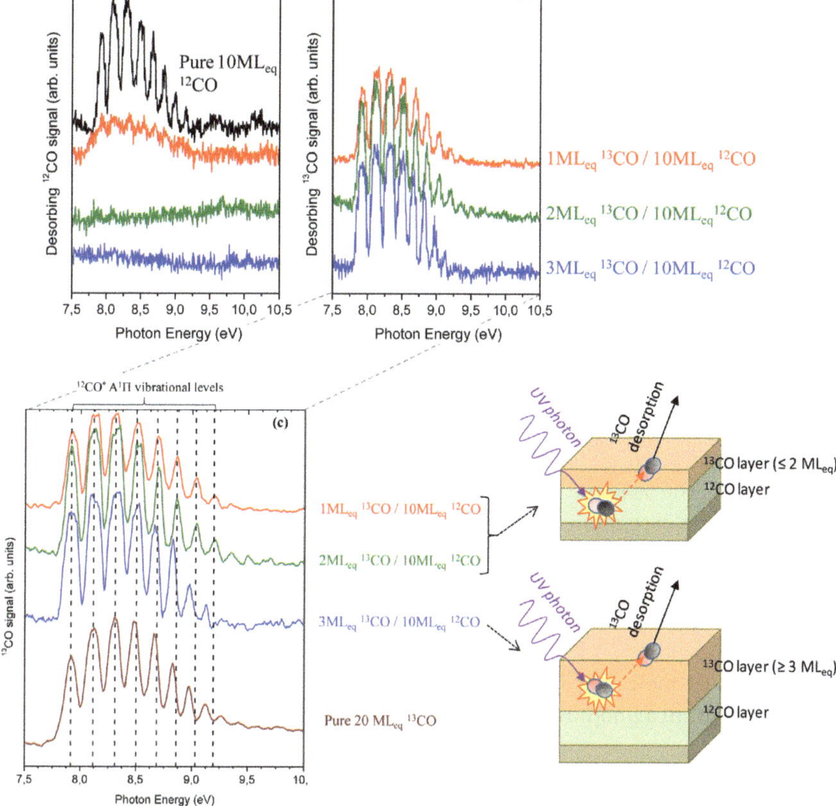

Fig. 6 ^{12}CO (a) and ^{13}CO (b) photon-stimulated desorption spectra at 14 K of 10 ML ^{12}CO ice, covered with an increasingly thick layer of ^{13}CO. (c) Enlarged image of the ^{13}CO PSD spectra of the ^{13}CO/^{12}CO layered ice compared to the spectrum obtained from pure ^{13}CO ice at 14 K. The dashed lines correspond to the energies of the first vibronic levels of pure ^{12}CO ice, from Lu *et al.*[1] Figure reproduced from Bertin *et al.*[2]

reproduced by the photodesorption spectrum. In that case, one could perhaps identify more precisely the origin of the energy transfer.

Professor Watanabe commented: I am interested in the details of the desorption induced by electronic transition (DIET) process you observed in solid CO. Is it similar to cavity ejection or excimer dissociation for rare gas solids? I would like to know how CO molecules in underlayers work for DIET. The dependence of the CO thickness on the photodesorption rate may provide more information on the DIET process. That is, the photodesorption rate may depend on the CO thickness. This is also important for astrochemical implications. Have you measured the CO thickness dependence on the photodesorption rate? Do you have any comment on the details of CO DIET?

Professor Fillion responded: We have done a detailed analysis of the thickness dependence of the CO DIET. Photon-stimulated desorption (PSD) spectra recorded for several thicknesses ranging from 5 to 250 ML are identical, and do not depend on the solid substrate on which the ices are grown. In order to gain more insight into the origin of the CO DIET from thin ices, we have performed detailed spectral analysis of isotopically labelled CO ices: thin layers of ^{13}CO ice, deposited with different thicknesses on top of a 10 ML thick ^{12}CO ice. As observed in Fig. 6, 1 ML of ^{13}CO deposited on ^{12}CO reduces the ^{12}CO desorption yield, but does not completely quench it, and a strong ^{13}CO signal is observed. The addition of a second layer of ^{13}CO largely blocks all ^{12}CO desorption, while the ^{13}CO desorption rate becomes comparable to that of ^{12}CO from 10 ML of pure ^{12}CO ice. Only molecules from the two topmost layers are thus photodesorbed.

Complementary information on the excitation process can be extracted from the details of the PSD spectra; the peak positions in the PSD spectra for different ^{13}CO ice coverages are presented in Fig. 6(c). For ^{13}CO coatings thinner than 2 ML, the PSD spectrum matches the vibronic structure of the $A^1\Pi$ $(v' = 0) \rightarrow X^1\Sigma^+$ $(v = 0)$ electronic transition for condensed ^{12}CO. The majority of the initial photon absorption events therefore must take place in the underlying ^{12}CO ice and not in the isotopic top layer, from which the photodesorbing molecules originate; the excess energy deposited in the ^{12}CO is transferred to the surface ^{13}CO molecules. This is an indication that CO ice photodesorption is an indirect process, where surface molecules are desorbed following the electronic excitation of sub-surface molecules. As 3 ML of ^{13}CO are needed to observe ^{13}CO-related absorption features, it is likely that the initial excitation step occurs within the 2 to 3 topmost molecular layers. UV absorption in the top ice layer does not efficiently result in a photodesorption event; an extended discussion can be found in Bertin et al.[2]

1 H.-C. Lu, H.-K. Chen, B.-M. Cheng, Y.-P. Kuo and J. F. Ogilvie, *J. Phys. B: At. Mol. Opt. Phys.*, 2005, **38**, 3693.
2 M. Bertin, E. C. Fayolle, C. Romanzin, K. I. Öberg, X. Michaut, A. Moudens, L. Philippe, P. Jeseck, H. Linnartz and J.-H. Fillion, *Phys. Chem. Chem. Phys.*, 2012, **14**, 9929.

Professor Zacharias addressed Professor Fillion and Professor Herbst: Previously we observed the resonant desorption of molecules from several 100 ML thick molecular ices of methane and deuterated methane,[1] N_2O,[2] and CD_3F,[3] adsorbed at about 25 K on NaCl(100). In all cases a desorption of the monomer species was observed when IR-active vibrations were excited by tunable radiation of the free-electron laser FELIX, then in Nieuwegein/Utrecht. For a layered system of methane and deuterated methane it was found that upon excitation of the methane C–H stretching vibration, the desorption of CD_4 is observed when the CD_4 layers were on top of the CH_4 layers. On the other hand, when CH_4 was on top of CD_4, the excitation of methane yielded only the desorption of CH_4 molecules and no CD_4 desorption.

1 B. Redlich, H. Zacharias, G. Meijer and G. von Helden, *J. Chem. Phys.*, 2006, **124**, 044704.
2 B. Redlich, L. van der Meer, H. Zacharias, G. Meijer and G. von Helden, *Nucl. Instrum. Meth. A*, 2003, **507**, 556.
3 B. Redlich, H. Zacharias, G. Meijer and G. von Helden, *Phys. Chem. Chem. Phys.*, 2002, **4**, 3448.

Professor Herbst opened the discussion of the paper by Helmut Zacharias: Do you have a sense of what the efficiency or cross section is for the resonant infrared desorption of methane from NaCl(100)? Long ago, F. Dzegilenko and I published a desorption mechanism[1] in which the radiative excitation of water leads, *via* a dipole–dipole interaction, to transfer of energy into the librational mode of nearby CO molecules, which then desorb. The efficiency calculated is on the order of 10^{-4}.

1 F. Dzegilenko and E. Herbst, *Astrophys. J.*, 1995, **443**, L81.

Professor Zacharias replied: For the desorption of CD_3F from NaCl(100) at $T_s = 40$ K *via* the IR resonant excitation of the ν_3 mode at 958 cm^{-1} ($\lambda \sim 10.44$ μm), we measured the velocity distribution of desorbing CD_3F molecules as a function of applied laser pulses and at different fluences without re-dosing the surface.[1] In these experiments we prepared molecular films with thicknesses of several hundred monolayers. When plotting the observed intensities semi-logarithmically *vs.* the applied photon numbers we can estimate the desorption cross section. Taking the published data from ref. 1, we arrive at an IR resonant desorption cross section of about $\sigma_{des} \sim 1 \times 10^{-21}$ cm^2. This value is more than three orders of magnitude smaller than typical non-resonant UV-induced desorption cross sections of small molecules from oxide surfaces.[2] The microscopic nature of the desorption mechanism – dipole–dipole interactions or phonon-induced, as studied in ref. 3, or *via* collision cascades – is presently not known, because dilution experiments with rare gases or other molecules have not yet been performed.

It might additionally be noted that for monolayer coverages the internal stretching mode of diatomic molecules does not couple efficiently to the desorption coordinate. When the CO or NO stretching modes at $\lambda \sim 4.7$ μm and 5.3 μm, respectively, of molecules adsorbed on a NiO(100) substrate were directly excited by IR free-electron laser pulses, no desorption signal could be detected.[4]

For atomic hydrogen desorption from graphite using XUV radiation at $h\nu = 38.8$ eV, a desorption cross section of $\sigma \sim 1.3 \times 10^{-19}$ cm^2 and a yield per pulse of about 3×10^{-5} ML were measured.[5] This is about a factor of five larger than for desorption with $\lambda = 400$ nm radiation. There, initial cross sections of $\sigma = (2.5 \pm 1.0) \times 10^{-20}$ cm^2 and $(4.9 \pm 1.0) \times 10^{-20}$ cm^2 have been derived for H and D atoms, respectively.[6] These cross sections describe the decrease of the surface coverage monitored at the H and D atomic signals. They therefore also include other possible desorption channels, like associative desorption. The atomic desorption yield per pulse at $\lambda = 400$ nm was found to be about 2×10^{-5} ML and 5×10^{-5} ML for H and D atoms, respectively.

1 B. Redlich, H. Zacharias, G. Meijer and G. von Helden, *Phys. Chem. Chem. Phys.*, 2002, **4**, 3448.
2 (a) M. Menges, B. Baumeister, K. Al-Shamery, H.-J. Freund, C. Fischer and P. Andresen, *J. Chem. Phys.*, 1994, **101**, 3318; (b) G. Eichhorn, M. Richter, K. Al-Shamery and H. Zacharias, *J. Chem. Phys.*, 1999, **111**, 386.
3 F. Dzegilenko and E. Herbst, *Astrophys. J.*, 1995, **443**, L81.
4 B. Redlich and H. Zacharias, unpublished results, 2002.
5 B. Siemer, T. Olsen, T. Hoger, M. Rutkowski, C. Thewes, S. Düsterer, J. Schiøtz and H. Zacharias, *Chem. Phys. Lett.*, 2010, **500**, 291.
6 R. Frigge, T. Hoger, B. Siemer, H. Witte, M. Silies, H. Zacharias, T. Olsen and J. Schiøtz, *Phys. Rev. Lett.*, 2010, **104**, 256102.

Dr Hornekær commented: It is surprising that you do not see any fast component for atomic hydrogen desorption from graphite. Have you looked for molecular hydrogen desorption? Regarding the explanation for the two-peak velocity distribution of H atoms desorbing from graphite, you propose that this may be due to H atoms sitting in a uniform "*para*-distribution" on the surface. Such a distribution may well work out in theoretical calculations, however, in reality we never observe such uniform distributions of H on the surface in scanning tunnelling microscopy studies. Also, even if it did exist, how does it explain the fact that we get a higher velocity from this distribution?

Professor Zacharias replied: We also expected fast atoms, and therefore in the XUV desorption experiments we carefully monitored the delay time range corresponding to fast atoms, with a null result. Only afterwards did we extended the delay range up to 70 μs for a flight path of only 7 mm, which revealed the slow H atoms reported in Fig. 8 in our paper. In these experiments we did not look for hydrogen nor deuterium molecules.

We agree that a uniform *para*-configuration over the whole surface would be quite exotic. However, it seems to be conceivable that patches of such a configuration may exist. Unlike desorption with femtosecond pulses in the near UV ($\lambda = 400$ nm), where fast atoms (up to 15 km s^{-1}) from *ortho*-pairs[1] were observed,[2] such fast atoms were not found for the desorption with XUV pulses.

The faster velocity of H atoms from the uniform *para*-configuration can be rationalized even in a simple DIET picture by inspection of Fig. 9 in our paper. In the theoretical calculation, this configuration not only shows a larger binding energy per H atom, but the potential is also shifted closer towards the graphite surface. Assuming the same potential for the electronically excited state, V_1, a Franck–Condon excitation of the H–C bond by the hot substrate electrons leads to an energetically higher starting point for the motion along the reaction coordinate than the excitation out of a *para*-pair configuration. The form of the potential curve for the electronically excited state displayed in Fig. 9 is, however, only a good guess. From the simulation we can only retrieve the derivative of this potential.[2]

1 L. Hornekær, E. Rauls, W. Xu, Z. Sljivancanin, R. Otero, I. Stensgaard, E. Laegsgaard, B. Hammer and F. Besenbacher, *Phys. Rev. Lett.*, 2006, **97**, 186102.
2 R. Frigge, T. Hoger, B. Siemer, H. Witte, M. Silies, H. Zacharias, T. Olsen and J. Schiøtz, *Phys. Rev. Lett.*, 2010, **104**, 256102.

Professor Linnartz asked: You mentioned in your presentation the formation of D_3O^+, and I was wondering whether you also found evidence for an ionic $[D_5O_2]^+$ (or $[H_5O_2]^+$) species – basically a proton sandwiched between two water molecules. The reason I ask is that charge stabilized van der Waals-like complexes have a high binding energy, can be studied spectroscopically very well,[1] and may be of astronomical interest, as long as an acceptable formation pathway can be provided. In the gas phase this will not work out, but the solid phase may provide an alternative route.

1 H. Linnartz, D. Verdes and J. P. Maier, *Science*, 2002, **297**, 1166.

Professor Zacharias replied: In these studies the time-of-flight mass spectrometer was designed to resolve the kinetic energy distribution of directly desorbing ions of predominantly low masses. Therefore, the mass resolution rapidly decreased at higher masses. Due to this low resolution we hesitate to clearly assign signals observed at 42 amu, which could be due to $D_3O^+(D_2O)_1$, and at 62 amu, which could be due to $D_3O^+(D_2O)_2$, to these water cluster ions, although such an assignment is likely to be correct. We would expect the formation of water cluster ions, and such protonated (or deuterated) clusters have high binding energies of 1370 meV and 850 meV, respectively.[1] The signal observed at 42 amu shows a time dependence very similar to that shown for $D_2O_2^+$ and D_3O^+ in Fig. 5 and 6 in our paper.

1 Y. K. Lau, S. Ikuta and P. Kebarle, *J. Am. Chem. Soc.*, 1982, **104**, 1462.

Professor Dulieu addressed Professor Zacharias and Professor Fillion: I am especially interested in the return to the gas phase of adsorbed molecules, after an energetic event, especially after a chemical reaction.

In your experiments, once absorbed, how is the energy shared among the possible channels? What could be the exact role of the substrate? Do you have any idea of the efficiency of the gas phase release or phonon transfer and the effect of the type of surface?

Professor Fillion responded: In the case of the desorption induced by electronic transition (DIET) in the VUV as observed for CO_2, it appears that the vibrational relaxation in the lattice actually competes with the desorption efficiency. Experiments on CO_2 ice at 75 K performed by C. Yan and T. Yates[1] indicate that the coupling of CO_2 vibrational states in the matrix is inversely correlated with the CO_2 desorption rates: a low phonon-mediated energy relaxation enhances the photodesorption. We have observed a similar effect by comparing the desorption behaviour from layered samples at 10 K,[2] and this could be important in the astrophysical context discussed at this conference. High CO photodesorption rates are obtained for 1 ML of CO condensed onto graphite, N_2 or CO_2, whereas an amorphous water ice substrate will quench CO photodesorption. This strong difference arises from the very different nature of the intermolecular $CO–H_2O$ interactions compared to other lattices; it has been shown that dangling O–H bonds of surface H_2O show a unique ability to rapidly and efficiently evacuate excess vibrational energy from the surface to the bulk. This opens a new relaxation pathway for excess energy originating from the surrounding CO excitation, which competes with the desorption of other CO molecules. In conclusion, it is important to keep the energy provided by electronic excitation very localised in the (sub)surface region for efficient desorption.

1 C. Yan and T. Yates, *J. Chem. Phys.*, 2013, **138**, 154303.
2 M. Bertin, E. C. Fayolle, C. Romanzin, K. I. Öberg, X. Michaut, A. Moudens, L. Philippe, P. Jeseck, H. Linnartz and J.-H. Fillion, *Phys. Chem. Chem. Phys.*, 2012, **14**, 9929.

Professor Zacharias responded: The energy redistribution depends on the thickness of the molecular ice layer and on the energy of the incoming or reactive photon or particle. For monomolecular layers and energies above the bandgap of the substrate, most electrons will be primarily excited. Within a picosecond they form a hot electron cloud which subsequently cools down *via* electron–phonon collisions. In principle, this could be calculated if the electron–phonon coupling constant, λ, is known. Unfortunately, for the mixture of materials which makes up a grain particle, this coupling constant is unknown. Even for graphite and graphene the value is currently under debate.[1-3] Direct phonon excitation by an atomic or molecular fragment can also take place. This can be estimated using the hard cubes model.[4,5] With hydrogen as one of the collision partners, the direct phonon excitation in a grain surface will amount to only a few percent of the energy lost by the hydrogen.

For thick molecular ice layers which absorb the incoming photons completely, the process is more complicated. Again, in the UV and VUV the primary process excites electrons, but this excitation will first lead to a localized vibrational motion *via* Franck–Condon pumping. These vibrations then couple to neighboring molecules and, in the solid state language, excite phonons. Presently, we are tackling this problem for the water ice layer using XUV radiation in collaboration with a theory group. We are optimistic that will have the first results this year.

1 T. Kampfrath, L. Perfetti, F. Schapper, C. Frischkorn and M. Wolf, *Phys. Rev. Lett.*, 2005, **95**, 187403.
2 H. Yan, D. Song, K. F. Mak, I Chatzakis, J. Maultzsch and T. F. Heinz, *Phys. Rev. B*, 2009, **80**, 121403.
3 J. C. Johannsen *et al.*, *Phys. Rev. Lett.*, 2013, **111**, 027403.
4 R. M. Logan and J. C. Keck, *J. Chem. Phys.*, 1968, **49**, 860.
5 E. K. Grimmelmann, J. C. Tully and M. J. Cardillo, *J. Chem. Phys.*, 1980, 72, 1039.

Dr Hornekær addressed Professor Zacharias: Do you observe any changes in the graphite substrate after the radiation exposures used in your H atom desorption experiments?

Professor Zacharias replied: Unfortunately, we did not monitor the surface of the graphite by STM or AFM. In damage experiments on amorphous carbon films on silicon, we focused the XUV radiation up to fluences of 19 mJ cm^{-2}. Even after 20 000 shots on the same spot no damage was found, with either a profilometer or with an AFM.[1] It is, however, evident that crystalline graphite is definitely a different surface than an amorphous carbon film.

In the desorption experiments of H (D) atoms from graphite with femtosecond pulses of 400 nm radiation,[2] we varied the desorption fluences over a wide range, from 5 to more than 50 mJ cm^{-2}. In this range a uniformly increasing non-linear yield dependence , $Y \sim I^n$, was observed, with values of the exponent n of $n \sim (2.0 \pm 0.5)$ and $n \sim (2.6 \pm 0.5)$ for deuterium and hydrogen, respectively, both for p-polarized light. For fluences between 10 and 28 mJ cm^{-2}, we measured the velocity distribution of H and D atoms. In all cases the observed velocity distribution did not change with fluence. It should also be noted that we exposed a specific surface area to only five laser pulses before moving the surface to a new spot.

1 B. Siemer, T. Hoger, M. Rutkowski, M. Menneken, S. Düsterer and H. Zacharias, *Proc. SPIE*, 2013, 8777, 87770F.
2 R. Frigge, T. Hoger, B. Siemer, H. Witte, M. Silies, H. Zacharias, T. Olsen and J. Schiøtz, *Phys. Rev. Lett.*, 2010, **104**, 256102.

PAPER

Cold condensation of dust in the ISM

Gaël Rouillé,[a] Cornelia Jäger,[*a] Serge A. Krasnokutski,[a] Melinda Krebsz[b] and Thomas Henning[c]

Received 12th February 2014, Accepted 13th February 2014

DOI: 10.1039/c4fd00010b

The condensation of complex silicates with pyroxene and olivine composition under conditions prevailing in molecular clouds has been experimentally studied. For this purpose, molecular species comprising refractory elements were forced to accrete on cold substrates representing the cold surfaces of surviving dust grains in the interstellar medium. The efficient formation of amorphous and homogeneous magnesium iron silicates at temperatures of about 12 K has been monitored by IR spectroscopy. The gaseous precursors of such condensation processes in the interstellar medium are formed by erosion of dust grains in supernova shock waves. In the laboratory, we have evaporated glassy silicate dust analogs and embedded the released species in neon ice matrices that have been studied spectroscopically to identify the molecular precursors of the condensing solid silicates. A sound coincidence between the 10 μm band of the interstellar silicates and the 10 μm band of the low-temperature siliceous condensates can be noted.

1 Introduction

AGB stars and supernovae (SNs) are main cosmic dust factories. The formed stardust is finally distributed into the interstellar medium (ISM), and eventually becomes a part of cold and dense molecular clouds. Interstellar dust is exposed to destructive processes caused by supernova-induced shock waves. It was estimated that only a few percent of the total mass of stardust survive these destructive processes in the ISM.[1,2] However, observations of refractory elements in the ISM clearly show a depletion of these elements from the gas phase. Consequently, an efficient condensation process of dust grains in the ISM is required to balance the discrepancy between the stellar formation of dust grains and their interstellar destruction.[2] Recently, Jones & Nuth[3] discussed a compatible injection and

[a]Laboratory Astrophysics Group of the Max Planck Institute for Astronomy at the Friedrich Schiller University Jena, Institute of Solid State Physics, Helmholtzweg 3, 07743 Jena, Germany. E-mail: cornelia.jaeger@uni-jena.de; Fax: +49-3641-9-47308; Tel: +49-3641-9-47354

[b]Institute for Geological and Geochemical Research, Research Centre for Astronomy and Earth Sciences, Hungarian Academy of Sciences, 45 Budaörsi street, 1112 Budapest, Hungary

[c]Max Planck Institute for Astronomy, Königstuhl 17, 69117 Heidelberg, Germany

destruction time scale for silicate dust particles, an assumption based on inherent uncertainties in the determination of destruction efficiencies.

In the ISM, refractory and other atoms and molecules generated by the erosion of grains in supernova shocks slowly accrete onto surfaces of surviving grains at low temperatures. This process is suggested to occur in the very dense cores of molecular clouds (MCs) where the temperature of dust grains is between 10 and 20 K. However, such processes may already occur in the outer, less dense regions of molecular clouds. The accreted species can finally react among each other to form solid layers on the grains. In the very dense cores of MCs, other abundant gaseous species such as CO, H_2O, and small carbon-rich molecules simultaneously accrete with refractory species and may prevent the formation of refractory silicate material. However, fast desorption processes of carbon-based condensates may also lead to a preferred formation of silicate condensates and to a separation of siliceous and carbonaceous dust in the ISM.

Very recently, the condensation of SiO molecules at low temperature using neon matrix and helium droplet isolation techniques has been studied.[4] SiO represents the major component of the interstellar silicates. Reactions between SiO molecules were found to be barrierless. The energy of SiO polymerization reactions has been determined experimentally using a calorimetric method, and theoretically with calculations based on the coupled-cluster and density functional theories. The experiments have clearly revealed the efficient formation of SiO_x condensates at temperatures of about 10 K.[4]

In the present work, we extend our recently performed experimental studies on the cold condensation process of SiO to more complex systems containing magnesium and iron. This step is necessary to obtain more insight into the low-temperature condensation process of realistic interstellar silicates. We apply laser vaporization to silicate samples in an attempt to produce the precursor species involved in the condensation of interstellar silicates. In a first stage, the vaporized species are deposited and isolated in a Ne matrix where they are cooled down to temperatures relevant to the ISM. This stage gives us the opportunity to identify the laser-vaporized species by absorption spectroscopy in the UV and mid-IR wavelength domains. The second stage consists of annealing and warming the matrix up to 13 K until the complete evaporation of the Ne atoms, in order to cause the accretion of the laser-vaporized species at low temperature.

2 Experimental

Chunks of two amorphous silicates, synthesized in-house by melting and quenching,[5] were used as targets. The formula of one silicate was Mg_2SiO_4, corresponding to the stoichiometry of forsterite, the Mg endmember of the olivine group. The formula of the other silicate was $Mg_{0.4}Fe_{0.6}SiO_3$, i.e., the composition of a pyroxene.

Laser vaporization was carried out using a pulsed laser source (Continuum Minilite II) emitting photons with a wavelength of 532 nm. The laser was operated with a repetition rate of 10 pulses per second. Each pulse lasted 5 ns and carried an energy of 20 to 25 mJ. The laser beam was focused at the surface of the target and, during the experiments, it was shifted every minute to vaporize a fresh part of the target and also to avoid drilling through it. Holes 0.4 to 0.5 mm in diameter were created during the experiments. Assuming this diameter coincides with the diameter of the laser beam at the target position, a fluence of 2 to 4 GW cm^{-2} is inferred.

Matrix isolation spectroscopy was performed with Ne (Linde, purity 99.995%) as the matrix material. Each matrix was grown on a KBr substrate (Korth Kristalle GmbH). This material transmits photons in the mid-IR wavelength domain and also at useful UV wavelengths, with a lower limit of 205 nm. A compressed-He closed-cycle cryocooler (Advanced Research Systems Inc. DE-204SL) was employed to cool down the substrate and to maintain it at low temperature. In order to form a Ne matrix, the substrate was kept at ~6 K and the Ne mass flow rate was set to 5 standard cubic centimeters per minute. The target for laser vaporization and the substrate were separated by a distance of ~55 mm.

The cryocooler could be rotated while being operated, allowing the substrate to face three directions and the corresponding pairs of opposite ports that equipped the vacuum chamber. These ports allowed us, in turn, to deposit the Ne atoms and laser-vaporized species on the substrate, to measure IR spectra and UV spectra in transmission. A Fourier transform IR spectrometer (Bruker VERTEX 80v) and a UV spectrometer (JASCO V-670 EX) were optically coupled to the vacuum chamber. The IR beam of the FTIR spectrometer, guided along an evacuated optical path by means of gold-coated mirrors, passed through the vacuum chamber to reach an external detector. Optical fibers fitted with collimating optics were used to carry the photons from the UV spectrometer to the vacuum chamber and back.

The IR spectra were measured by averaging 64 scans carried out at a speed of 10 kHz with a resolution of 1 cm^{-1}. The UV spectra were measured with equal step and resolution of 0.2 nm at a rate of ~11 nm min^{-1}.

3   Results

3.1   Analysis of the evaporated species

The isolation of the vaporized species in the Ne matrices gives us the opportunity to identify them using absorption spectroscopy, and also to verify the presence of contaminants. For instance, our spectra show bands due to the Ne matrix-isolated contaminants CO (2141 cm^{-1}),[6] CO$_2$ (2348 cm^{-1}),[7] H$_2$O (line systems at 1631 and 3783 cm^{-1}),[8] and (H$_2$O)$_2$ (lines at 3590 and 3734 cm^{-1}).[8]

3.1.1   Mg$_2$SiO$_4$ target. Magnesium atoms are easily detected in the UV region. In Fig. 1, the line of atomic Mg at 275 nm is very strong, indicating the efficient vaporization of this element. Allowing for a matrix-induced wavelength shift, this line corresponds to that measured at 285.30 nm in vacuum.[9] Despite the large amount of Mg atoms, dimers are not observed. They would give rise to a band at ~257 nm corresponding to the $B^1\Pi_u \leftarrow X^1\Sigma_g^+$ transition.[10,11] A spectrum of Mg atoms deposited in a Ne matrix to serve as a reference supports the assignment.

Still at UV wavelengths, the peaks in a very weak, structured feature originating at 234 nm correspond to bands of the $A^1\Pi \leftarrow X^1\Sigma^+$ transition of the SiO molecule.[4,12] Thus only a very little amount of this species was deposited in the Ne matrix. Accordingly, features recently attributed to SiO oligomers, which are produced by barrierless reactions between the SiO molecules, are not found.[4]

In the 205 to 350 nm range, Si atoms in their ground state give medium to strong absorption lines at 220.87, 243.95, and 251.51 nm in vacuum.[9] These lines

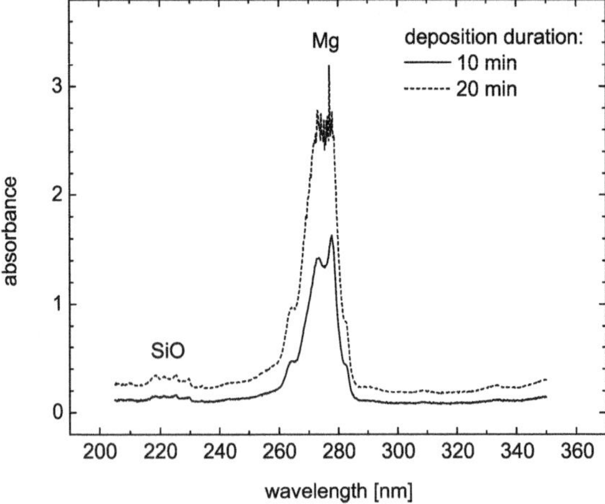

Fig. 1   Ultraviolet absorption features of species isolated in a Ne matrix at 6.3 K following laser vaporization of amorphous Mg_2SiO_4. The features are attributed to SiO and Mg. The absorption line of Mg is saturated after accumulating material for 20 min.

are not seen in our spectrum, even considering matrix-induced wavelength shifts. In the same region, O I atoms have only one very weak transition.

In the mid-IR domain, shown in Fig. 2, weak absorption bands arise at 913, 1039, 1107, 1164, 1228, 1754, 2152, and 2192 cm^{-1}. We assign the absorption bands seen at 1039 and 1107 cm^{-1} to the ν_3 and ν_1 modes of O_3, respectively.[13] Another band due to O_3 is not seen at 2109 cm^{-1} $(\nu_1 + \nu_3)$,[13] most likely because being a combination band it is too weak. The identification of O_3 could have been confirmed by the observation of the broad Hartley band in the UV spectrum, near 253 nm.[14] This band, however, is not observed. The comparison between the

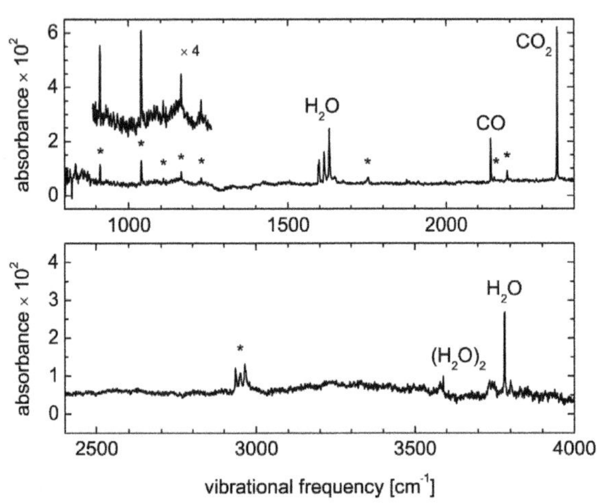

Fig. 2   Infrared absorption features of species isolated in a Ne matrix at 6.3 K following laser vaporization of amorphous Mg_2SiO_4. Asterisks mark the features of interest. The material had been accumulated for 20 min.

absorption cross-sections of the UV and IR transitions would clarify our assignment. The peak at 1164 cm^{-1} is attributed to O_4^+.[15,16] Those at 913 and 2192 cm^{-1} were recently reported and discussed by Jacox and Thompson.[17] It was found that these absorptions may be caused by a single species, which would be a molecular complex involving H_2 and possibly H_3O^+ or $H_2O_5^+$.

The line found at 1228 cm^{-1} is attributed to SiO.[18,19] As expected from the analysis of the UV spectrum, the IR bands that characterize the oligomers Si_2O_2 and Si_3O_3 are not seen.[18-20] Absorptions that would possibly reveal the presence of magnesium oxides lie at wavelengths longer than 10 µm, outside the range we have scanned.[21]

Finally, a tight group of bands that arises at 2937, 2951, and 2966 cm^{-1} resembles the bands of the ν_2 mode of H_2O near 1620 cm^{-1},[8] with a positive shift of 1337 cm^{-1}. The bands at 1754 and 2152 cm^{-1} remain unassigned.

3.1.2 $Mg_{0.4}Fe_{0.6}SiO_3$ target. In the UV region, shown in Fig. 3, the lines of the Fe and Mg atoms are strong, demonstrating again the efficient vaporization of magnesium, as observed with the Mg_2SiO_4 target, and also that of iron. Only a weak contribution of SiO is possibly detected. The lines assigned to Fe I are favorably compared with the vacuum wavelengths.[9] A spectrum of Fe atoms deposited in a Ne matrix was measured to serve as a reference, which supports the assignment. One can note that there is no evidence of the broad Hartley band of O_3.

In the IR spectrum displayed in Fig. 4, we find bands already observed when using the Mg_2SiO_4 target. Thus O_3 (1039 cm^{-1}), O_4^+ (1164 cm^{-1}), and SiO (1228 cm^{-1}) are present in the matrix. The band assigned above to the ν_1 mode of O_3 is absent. On the other hand, new bands of O_4^+ isomers are detected at 1321 and 2808 cm^{-1}.[15,16] The absorptions caused by a complex possibly involving H_2 and possibly H_3O^+ or $H_2O_5^+$ are also detected.[17]

There is no evidence again of the Si_2O_2 and Si_3O_3 oligomers. The unassigned peaks found at 1754 and 2152 cm^{-1} in the experiment with the Mg_2SiO_4 target are absent, like the group of bands previously observed around 2950 cm^{-1}.

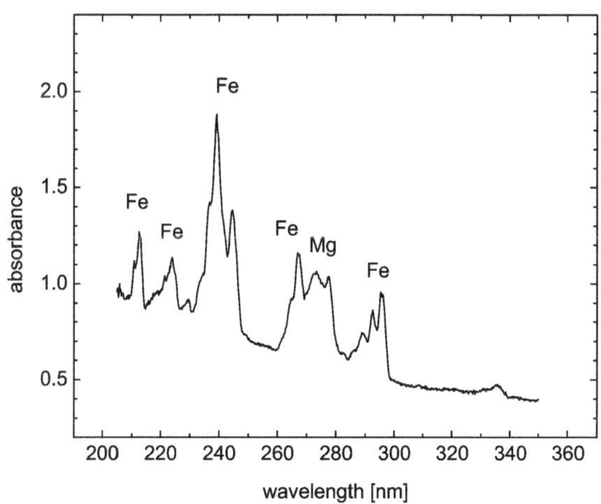

Fig. 3 Ultraviolet absorption features of species isolated in a Ne matrix at 6.3 K following laser vaporization of amorphous $Mg_{0.4}Fe_{0.6}SiO_3$. The features are attributed to Fe and Mg atoms. The material had been accumulated for 20 min.

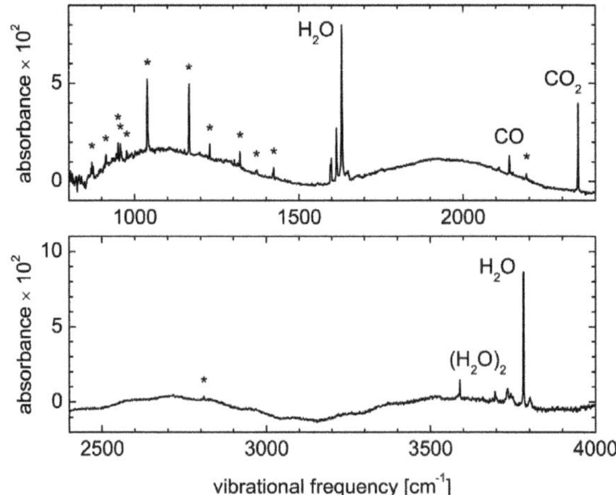

Fig. 4 Infrared absorption features of species isolated in a Ne matrix at 6.3 K following laser vaporization of amorphous $Mg_{0.4}Fe_{0.6}SiO_3$. Asterisks mark the features. The material had been accumulated for 40 min.

On the other hand, there are additional bands at the positions 870, 950, 958, 976, 1369, and 1424 cm^{-1}. After comparison with measurements on species isolated in Ar matrices, we tentatively assign the bands at 1369 and 1424 cm^{-1} to SiO_3 (1363.5 cm^{-1} in Ar matrix) and SiO_2 (1416 cm^{-1} in Ar matrix), respectively.[22] Other tentative assignments would be to FeO (870 cm^{-1} here, 873.08 cm^{-1} in Ar matrix),[23] and OFeO (950 cm^{-1} here, 945.8 cm^{-1} in Ar matrix).[24]

The absorption at 976 cm^{-1} has not been assigned.

3.2 Analysis of the low-temperature condensates

The condensates were studied by IR spectroscopy, high-resolution transmission electron microscopy (HRTEM), and energy dispersive X-ray (EDX) spectroscopy. Generally, IR spectroscopy is the best method to follow up the silicate condensation *in situ* that means independence of the temperature. The formation of the broad silicate bands at about 10 and 20 μm corresponding to the Si–O stretching and bending modes of amorphous silicates is a clear indicator for the first appearance of a solid layer. However, due to the experimental requirements (study of the gaseous precursors and the condensate of one sample *in situ*), the final layer thickness of the condensed silicate was rather small. Therefore, the typical silicate bands were weak and the measurement of the 20 μm band was not possible without having a very low signal to noise ratio.

In each experiment, after annealing and complete evaporation of the Ne atoms, the species constituting the background gas have deposited on the still cold substrate giving various features. Fig. 5 shows spectra obtained with the Mg_2SiO_4 target after evaporation of the Ne atoms and cooling to 6.5 K. Beside the absorption bands due to CO, CO_2, and H_2O, a feature is seen near 1000 cm^{-1}, which we attribute to a solid condensate. The profile of this band is likely affected by the presence of the broad water ice feature that peaks near 780 cm^{-1}.[25] After warming to room temperature, the absorption band attributed to the condensate

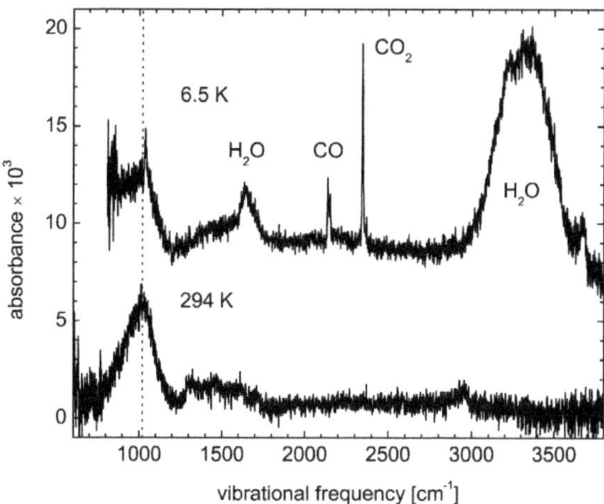

Fig. 5 Infrared spectra of the condensate obtained with the Mg_2SiO_4 target. (Top) Spectrum measured at 6.5 K after warming to 13 K. (Bottom) Spectrum measured at room temperature.

shows an asymmetric profile, which is steeper on the higher frequency side. The maximum of the band is located at $\sim$1020 cm^{-1} (9.8 μm).

Working with the $Mg_{0.4}Fe_{0.6}SiO_3$ target, we did not observe immediately the expected IR absorption near 1000 cm^{-1}, neither at low nor at room temperature. Nonetheless, after the experiment, a solid deposit on the substrate has been detected that has been measured using a clean KBr substrate as reference. A distinct band was observed near 1000 cm^{-1} clearly pointing to the formation of a silicate material. This finding is certainly caused by small problems with the reference measurements of pure KBr during the cooling and warming up phase. Previous temperature-dependent IR measurements of low-temperature SiO_x condensates revealed the formation of the solid phase already at about 10 K.[19] Similarly, the magnesium silicate condensate shown in Fig. 5 has also been formed at low temperature (13 K). Therefore, we act on the assumption that the solid condensate was already formed at low temperature.

Fig. 6 shows IR spectra of final condensates obtained by evaporation of a Mg_2SiO_4 and a $Mg_{0.4}Fe_{0.6}SiO_3$ target, respectively. Both spectra were taken at room temperature. A small shift of the 10 μm band is observed. The bands have their maximum at $\sim$990 cm^{-1} (10.1 μm) and $\sim$1020 cm^{-1} (9.8 μm), respectively. A lower polymerization degree in the condensate produced from a pyroxene-like target ($Mg_{0.4}Fe_{0.6}SiO_3$) should give a band that is shifted to smaller wavelengths compared to those produced from the olivine target (Mg_2SiO_4). The comparison of the bands show that both condensates have to have similar stoichiometry. This has been confirmed by EDX analysis (see Table1).

In addition, the final condensates were studied by high-resolution transmission electron microscopy (HRTEM). HRTEM micrographs of the magnesium iron silicate condensed at low temperatures show fluffy aggregates that are composed of nanometer-sized grains. The sizes of individual primary grains vary between 3 and 15 nm. The internal grain structure is found to be completely amorphous and the final condensate shows a clear homogeneity in structure and

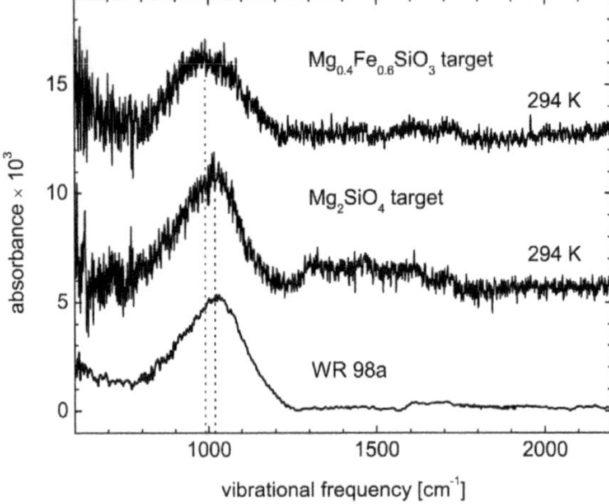

Fig. 6 Infrared spectra of the final condensates obtained by laser-vaporization of amorphous $Mg_{0.4}Fe_{0.6}SiO_3$ and Mg_2SiO_4 targets measured at room temperature. The third spectrum shows the normalized silicate absorption feature toward WR 98a representing the local ISM.[26] The spectra are vertically shifted for clarity.

Table 1 Composition of the condensed silicates by EDX analysis

Condensate	At% Mg	At% Fe	At% Si	At% O
MgFe-silicate	4.9	15.8	17.3	62.0
Mg-silicate	20.8	—	20.9	58.3

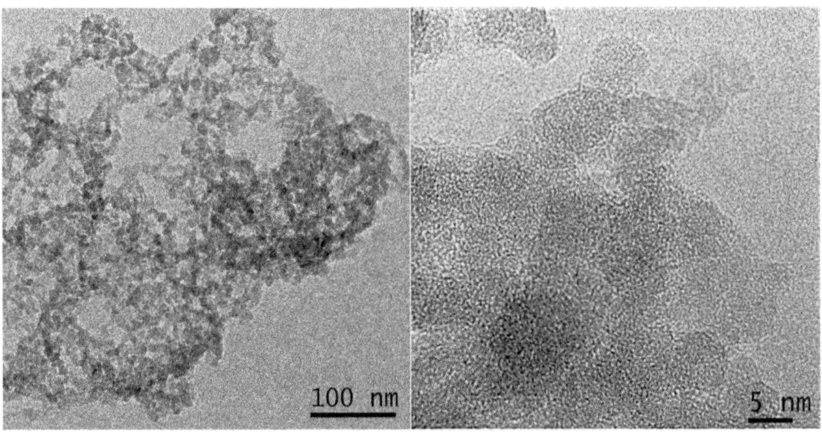

Fig. 7 HRTEM images of the final condensate prepared from the evaporation and recondensation of a magnesium iron silicate at low temperatures. The left image shows an overview of a big cluster. The right image presents a direct view inside the amorphous grains. Please note the irregular structure of the interior of the grains.

composition (see Fig. 7). No hints at phase separations of the MgFe-silicate into individual oxides such as FeO, MgO, and/or SiO_2 can be observed. The morphology and structure of the grains is very similar to the low-temperature condensate produced from SiO molecules.[4] In addition, very similar results have been obtained for the pure magnesium silicate prepared by laser ablation from a target with the composition Mg_2SiO_4. The formed fluffy aggregates are composed of small primary particles with a completely amorphous structure.

The analytical characterization was complemented by energy-dispersive X-ray spectroscopy (EDX). The method allows the determination of the final composition of the silicate material. The analysis revealed an iron-rich magnesium silicate for the condensate produced from the $Mg_{0.4}Fe_{0.6}SiO_3$ target, the composition of which is given in Table 1. Compared to the composition of the target, the material became poor in magnesium. This is probably due to a selective evaporation process. In an amorphous $Mg_{0.4}Fe_{0.6}SiO_3$ silicate, the MgO represents the most heat-resistant component of the silicate. The composition of the final condensate produced from the Mg_2SiO_4 target is also shown in Table 1. Compared to the target, the condensate is depleted in Mg and shows a pyroxene stoichiometry.

4 Astrophysical discussion

In the ISM, gaseous species may accrete on cold surfaces of pre-existing small particles. The repeated erosion of silicate stardust in SN-induced shock waves is the source of these gaseous species in the ISM. To simulate such processes in the laboratory, we have evaporated silicate dust analogs that have been produced by melting and quenching. These materials have successfully been used to model spectral energy distributions of many astrophysical sources.[27–29] The applied material represents a realistic source for the release of astrophysically relevant gaseous species.

The isolation of the evaporated species in condensed rare gases cools them down before they can interact among each other. This step is necessary as laser-vaporized molecules are hot and their internal energy may affect the chemistry of accretion. The analytical characterization of the species released from silicates and finally embedded in the Ne matrix has shown that beside SiO mainly Mg, Fe, and small molecules such as O_3, O_4^+, SiO_2, SiO_3, FeO, and FeO_2 are precursors of the silicates. It is their accretion that resulted in the formation of silicate grains. While SiO and FeO have been discovered in the ISM,[30–32] the other molecules have not been reported. A model for the chemistry of silicon in dense interstellar clouds includes SiO_2 in its network.[33] The production of SiO by interstellar shocks in molecular outflows predicts the presence of SiO_2.[34]

The condensation of pure silicon monoxide and of more complex silicates with olivine and pyroxene composition by accretion of molecules and atoms on cold surfaces, and subsequent reactions between them at temperatures between 10 and 20 K, has been proven experimentally. The grains were formed by reactions of the embedded molecules during the annealing (10–12 K) and evaporation of the Ne matrix (13 K). The applied conditions are comparable to those prevailing in molecular clouds. The final condensates are fluffy aggregates consisting of small nanometer-sized primary grains. All low-temperature silicates possess amorphous structures and form fluffy aggregates.

The dust condensation at low temperature and low density in the ISM is a process that is discussed to be necessary to keep the balance between dust destruction and formation. So far, there is no exact description of where the cold condensation process may take place. Turbulence in the ISM can quickly distribute the refractory elements such as Mg, Si, and other species produced in SN shocks into the surrounding medium. In SN ejecta, synthesized materials are mixed by turbulence with the surrounding ISM on a time scale of 100 Myr.[35] One can assume a similar time scale for the mixing of elements liberated from destroyed grains. The process of cold condensation may already be active in diffuse clouds with low temperature and in low-density regions of molecular clouds. In molecular cloud environments with maximum density, where many non-refractory and refractory species and elements may accrete simultaneously on the grains, very complex ices should be formed. There are two competitive processes influencing the growth of the solid layer which are the desorption and the sticking processes. The desorption process is strongly determined by the bonding energy between the species and the surface. According to Draine,[2] for binding energies of 0.1 eV, the lifetime is about 5×10^5 yr. Accreted species that may form strong bonds, which are typical for refractory solids, may grow very fast and remain on the surface for a long time. Diffusion and desorption processes on grain surfaces may finally trigger the formation of more stable siliceous and carbonaceous solids in addition to less stable complex ices. Furthermore, the interstellar UV field and UV photons from young stars inside the clouds, and cosmic rays, can penetrate the interior of such MCs and trigger reactions between accreted molecules and clusters.

Selection processes that may lead to the formation of spatially separated interstellar dust components have to be addressed in upcoming experimental studies.

The condensation of species produced by laser vaporization of silicates has already been studied, albeit under different conditions. The vaporized species were condensed in the gas phase in a quenching gas atmosphere at high temperatures.[36–41] The particles formed in this process were deposited on various substrates before being investigated. The structure and general morphology of the high-temperature and low-temperature siliceous condensates turned out to be remarkably similar. Both are characterized by amorphous structures. In addition, both types of condensates show very similar spectral properties well comparable to the 10 μm profile of the observed interstellar silicates. So far, distinct structural and spectral differences between high- and low-temperature condensates cannot be observed.

5   Conclusions

The condensation of complex silicates with pyroxene composition at temperatures between 10 and 20 K by accretion of molecules and atoms on cold surfaces, and subsequent reactions between them, has been proven experimentally. The experiments clearly demonstrate an efficient silicate formation at low temperatures. The final condensates are fluffy aggregates consisting of small nanometer-sized primary grains. All low-temperature silicates condense in amorphous form and they were found to be homogeneous in structure and composition. To study the gaseous precursors of such condensation processes that can be formed by erosion of dust grains in SN schock waves, we have evaporated glassy silicate materials with a pulsed Nd:YAG laser. The liberated gaseous species were embedded in solid neon matrices that have been studied by spectroscopy in the UV/VIS and IR range. The IR spectral properties of low-temperature siliceous condensates do not much differ from silicates produced in high-temperature condensation processes. A sound coincidence between the 10 μm band of the interstellar silicates measured by Chiar & Tielens[26] and the 10 μm band of the low-temperature siliceous condensate can be noted.

Acknowledgements

The authors acknowledge the support of the Deutsche Forschungsgemeinschaft through project No. He 1935/26-1 within the framework of the Priority Program 1573 "Physics of the Insterstellar Medium". M. K. is grateful for the award of an Eötvös Scholarship of the Hungarian State.

References

1 S. Zhukovska, H.-P. Gail and M. Trieloff, *Astron. Astrophys.*, 2008, **479**, 453–480.
2 B. T. Draine, *Cosmic Dust - Near and Far*, 2009, p. 453.
3 A. P. Jones and J. A. Nuth, *Astron. Astrophys.*, 2011, **530**, A44.
4 S. A. Krasnokutski, G. Rouillé, C. Jäger, F. Huisken, S. Zhukovska and T. Henning, *Astrophys. J.*, 2014, **782**, 15.
5 J. Dorschner, B. Begemann, T. Henning, C. Jaeger and H. Mutschke, *Astron. Astrophys.*, 1995, **300**, 503.
6 H. Dubost, *Chem. Phys.*, 1976, **12**, 139–151.
7 L. Wan, L. Wu, A.-W. Liu and S.-M. Hu, *J. Mol. Spectrosc.*, 2009, **257**, 217–219.
8 D. Forney, M. E. Jacox and W. E. Thompson, *J. Mol. Spectrosc.*, 1993, **157**, 479–493.
9 A. Kramida, Yu. Ralchenko, J. Reader and NIST ASD Team, *NIST Atomic Spectra Database (ver. 5.1)*, [Online]. Available: http://physics.nist.gov/asd[2014, February 7]. National Institute of Standards and Technology, Gaithersburg, MD., 2013.
10 J. G. McCaffrey and G. A. Ozin, *J. Chem. Phys.*, 1988, **88**, 2962–2971.
11 B. Healy, P. Kerins and J. G. McCaffrey, *Low Temp. Phys.*, 2012, **38**, 679–687.
12 J. Hormes, M. Sauer and R. Scullman, *J. Mol. Spectrosc.*, 1983, **98**, 1–19.
13 P. Brosset, R. Dahoo, B. Gauthier-Roy, L. Abouaf-Marguin and A. Lakhlifi, *Chem. Phys.*, 1993, **172**, 315–324.
14 A. Jaye, W. Laasch and P. Gütler, *Investigations of the Hartley band of ozone isolated in rare gas matrices*, in DESY Photon Science HASYLAB Annual Report, Part 1, Contribution 21-241, 1998.
15 W. E. Thompson and M. E. Jacox, *J. Chem. Phys.*, 1989, **91**, 3826–3837.
16 M. E. Jacox and W. E. Thompson, *J. Chem. Phys.*, 1994, **100**, 750–751.
17 M. E. Jacox and W. E. Thompson, *J. Phys. Chem. A*, 2013, **117**, 9380–9390.
18 R. K. Khanna, D. D. Stranz and B. Donn, *J. Chem. Phys.*, 1981, **74**, 2108–2115.
19 G. Rouillé, S. A. Krasnokutski, M. Krebsz, C. Jäger, F. Huisken and T. Henning, Cosmic dust formation at cryogenic temperatures, in *The Life Cycle of Dust in the Universe: Observations, Theory, and Laboratory Experiments*, ed. A. Andersen, M. Baes, H. Gomez, C. Kemper and D. Watson, PoS(LCDU 2013)047, 2014.
20 J. W. Hastie, R. H. Hauge and J. L. Margrave, *Inorg. Chim. Acta*, 1969, **3**, 601–606.
21 L. Andrews, E. S. Prochaska and B. S. Ault, *J. Chem. Phys.*, 1978, **69**, 556–563.

22 B. Tremblay, P. Roy, L. Manceron, M. E. Alikhani and D. Roy, *J. Chem. Phys.*, 1996, **104**, 2773–2781.

23 D. W. Green, G. T. Reedy and J. G. Kay, *J. Mol. Spectrosc.*, 1979, **78**, 257–266.

24 G. V. Chertihin, W. Saffel, J. T. Yustein, L. Andrews, M. Neurock, A. Ricca and C. W. Bauschlicher, *J. Phys. Chem.*, 1996, **100**, 5261–5273.

25 K. I. Öberg, H. J. Fraser, A. C. A. Boogert, S. E. Bisschop, G. W. Fuchs, E. F. van Dishoeck and H. Linnartz, *Astron. Astrophys.*, 2007, **462**, 1187–1198.

26 J. E. Chiar and A. G. G. M. Tielens, *Astrophys. J.*, 2006, **637**, 774–785.

27 C. Gielen, J. Bouwman, H. van Winckel, T. Lloyd Evans, P. M. Woods, F. Kemper, M. Marengo, M. Meixner, G. C. Sloan and A. G. G. M. Tielens, *Astron. Astrophys.*, 2011, **533**, A99.

28 M. Min, L. B. F. M. Waters, A. de Koter, J. W. Hovenier, L. P. Keller and F. Markwick-Kemper, *Astron. Astrophys.*, 2007, **462**, 667–676.

29 F. J. Molster, L. B. F. M. Waters and A. G. G. M. Tielens, *Astron. Astrophys.*, 2002, **382**, 222–240.

30 R. W. Wilson, A. A. Penzias, K. B. Jefferts, M. Kutner and P. Thaddeus, *Astrophys. J.*, 1971, **167**, L97–L100.

31 C. M. Walmsley, R. Bachiller, G. Pineau des Forêts and P. Schilke, *Astrophys. J.*, 2002, **566**, L109–L112.

32 R. S. Furuya, C. M. Walmsley, K. Nakanishi, P. Schilke and R. Bachiller, *Astron. Astrophys.*, 2003, **409**, L21–L24.

33 E. Herbst, T. J. Millar, S. Wlodek and D. K. Bohme, *Astron. Astrophys.*, 1989, **222**, 205–210.

34 P. Schilke, C. M. Walmsley, G. Pineau des Forêts and D. R. Flower, *Astron. Astrophys.*, 1997, **321**, 293–304.

35 M. S. Oey, *A Massive Star Odyssey: From Main Sequence to Supernova*, 2003, p. 620.

36 J. R. Stephens and R. W. Russell, *Astrophys. J.*, 1979, **228**, 780–786.

37 J. R. Stephens, A. Blanco, E. Bussoletti, L. Colangeli, S. Fonti, V. Mennella and V. Orofino, *Planet. Space Sci.*, 1995, **43**, 1241–1246.

38 J. R. Brucato, L. Colangeli, V. Mennella, P. Palumbo and E. Bussoletti, *Astron. Astrophys.*, 1999, **348**, 1012–1019.

39 J. R. Brucato, V. Mennella, L. Colangeli, A. Rotundi and P. Palumbo, *Planet. Space Sci.*, 2002, **50**, 829–837.

40 C. Jäger, H. Mutschke, T. Henning and F. Huisken, *Cosmic Dust - Near and Far*, 2009, p. 319.

41 T. Sabri, L. Gavilan, C. Jäger, J. L. Lemaire, G. Vidali, H. Mutschke and T. Henning, *Astrophys. J.*, 2014, **780**, 180.

PAPER

THz and mid–IR spectroscopy of interstellar ice analogs: methyl and carboxylic acid groups

S. Ioppolo,*[ab] B. A. McGuire,†[c] M. A. Allodi†[c] and G. A. Blake[ac]

Received 21st December 2013, Accepted 7th February 2014

DOI: 10.1039/c3fd00154g

A fundamental problem in astrochemistry concerns the synthesis and survival of complex organic molecules (COMs) throughout the process of star and planet formation. While it is generally accepted that most complex molecules and prebiotic species form in the solid phase on icy grain particles, a complete understanding of the formation pathways is still largely lacking. To take full advantage of the enormous number of available THz observations (*e.g.*, *Herschel Space Observatory*, SOFIA, and ALMA), laboratory analogs must be studied systematically. Here, we present the THz (0.3–7.5 THz; 10–250 cm^{-1}) and mid–IR (400–4000 cm^{-1}) spectra of astrophysically-relevant species that share the same functional groups, including formic acid (HCOOH) and acetic acid (CH$_3$COOH), and acetaldehyde (CH$_3$CHO) and acetone ((CH$_3$)$_2$CO), compared to more abundant interstellar molecules such as water (H$_2$O), methanol (CH$_3$OH), and carbon monoxide (CO). A suite of pure and mixed binary ices are discussed. The effects on the spectra due to the composition and the structure of the ice at different temperatures are shown. Our results demonstrate that THz spectra are sensitive to reversible and irreversible transformations within the ice caused by thermal processing, suggesting that THz spectra can be used to study the composition, structure, and thermal history of interstellar ices. Moreover, the THz spectrum of an individual species depends on the functional group(s) within that molecule. Thus, future THz studies of different functional groups will help in characterizing the chemistry and physics of the interstellar medium (ISM).

1   Introduction

Much astrochemical research in recent years has focused on understanding the chemical and physical pathways that potentially lead to the formation of biotic

[a]*Division of Geological and Planetary Sciences, California Institute of Technology, 1200 E. California Blvd., Pasadena, CA 91125, USA. E-mail: ioppolo@caltech.edu; Fax: +1 (626) 568-0935; Tel: +1 (626) 395-6962*
[b]*Institute for Molecules and Materials, Radboud University Nijmegen, PO Box 9010, Nijmegen, NL 6500 GL, The Netherlands*
[c]*Division of Chemistry and Chemical Engineering, California Institute of Technology, 1200 E. California Blvd., Pasadena, CA, 91125, USA*
† These authors contributed equally to this work.

and prebiotic molecules in the interstellar medium (ISM). This focus has only intensified with the detection of glycine, the simplest amino acid, in cometary samples recently returned by the STARDUST mission.[1] The quest to understand the possible cosmic origins of life as we know it is of great importance to the wider scientific community and in the eyes of the public at large. Therefore, the laboratory, modeling, and observational communities must work together to achieve this ambitious goal. Yet, despite having detected amino acids in meteorites and comets, we lack a complete understanding of the formation pathways for even these relatively simple prebiotic precursors.

For instance, recent laboratory experiments, corroborated by astrochemical models, strongly support the hypothesis that hydroxylamine (NH_2OH), a possible precursor of glycine, is efficiently formed in the solid state under dense dark cloud conditions.[2-6] Deep within dark interstellar clouds, where temperatures drop as low as 10 K and densities are on the order of 10^5 cm^{-3}, gas-phase species accrete onto grains, forming icy mantles. Interstellar grains act as a third body capable of carrying off the excess energy released from newly formed bonds, thus promoting surface reactions. During the early stages of star and planet formation, solid-state chemistry is largely dominated by barrierless reactions involving free atoms – H, D, C, N, and O – landing and diffusing on the ice surface of the grains. Later, energetic processing (*e.g.*, heating, UV-irradiation) becomes important. In this scenario, NH_2OH has been shown to be formed through the barrierless hydrogenation of NO and NO_2. However, single-dish observations by Pulliam *et al.*[7] failed to detect NH_2OH in various warm astronomical environments. The non-detection of NH_2OH can be explained by efficient photodissociation pathways or thermally-activated surface reactions within the ice.

Another relevant example is methyl formate ($HCOOCH_3$), an abundant molecule detected toward a plethora of interstellar sources (see, *e.g.*, Nummelin *et al.*[8] and refs. therein). Until the last decade, $HCOOCH_3$, as well as most other complex organic molecules (COMs), was presumed to form primarily through gas-phase ion-molecule reactions. However, both experimental and theoretical work has shown that the reactions that were thought to give rise to $HCOOCH_3$ are inefficient in the gas-phase.[9] Instead, recent gas-grain chemical models have shown that $HCOOCH_3$, and indeed many COMs, are likely to be formed from radical–radical and radical–neutral reactions on and within the icy mantles of dust grains (see, *e.g.*, Garrod *et al.*[10] and Garrod[11]). Often, these radicals are small organic functional groups (*e.g.* CH_3, OH, NH_2, and CHO) formed from the UV photodissociation of simpler organic species such as methanol (CH_3OH) (see, *e.g.*, Öberg *et al.*,[12] Laas *et al.*,[13] Laas *et al.*[14] and refs. therein). This suggests that many observed COMs may share common formation pathways in these ices. Thus, examining sets of interstellar species which share common functional groups offers essential insight into how these species form in interstellar environments.

Mid-infrared (mid-IR) spectroscopy (400–4000 cm^{-1}) has been fundamental in improving our understanding of ice and dust in the ISM over the past few decades. As a result, the most abundant simple molecules have been detected in the solid phase *via* mid-IR astronomical observations (see Öberg *et al.*[15] for a review). However, transitions at IR frequencies correspond to intramolecular vibrational modes of the molecules in the ice. Therefore, complex species that share functional groups may present similar IR spectra with absorption features that overlap with each other and with those of the most abundant species in the solid phase (*e.g.*, H_2O, CO, CO_2, H_2CO, CH_3OH, CH_4, HCOOH, and NH_3). Mid-IR observations are also limited by the fact that we can observe solid-phase species only in absorption, rather than in emission, greatly limiting the sample set of astronomical sources.

In contrast, TeraHertz (THz), or far-IR, spectroscopy, covering the region from 0.1 to 10 THz (3–330 cm^{-1}), directly probes lower frequency vibrations that correspond to low energy intra- and, especially, intermolecular modes. THz spectroscopy is therefore particularly sensitive to the long range interactions between molecules and allows for direct measurements of large-scale structural changes.[16] Moreover, unlike the case in the mid-IR spectral region, at THz wavelengths, photon energies are sufficiently small, so temperatures less than ~150 K can provide a detectable emission signal, and absorption can occur against the background continuum. Thus, THz spectroscopy provides the unique opportunity to detect ices in disks and clouds both in emission and in absorption against the widespread dust continuum, through a number of distinct features and, theoretically, along any line of sight.

To take full advantage of the enormous number of available THz observations (*e.g.*, *Herschel Space Observatory*, SOFIA, and ALMA), laboratory analogs must be studied systematically. Some THz lab data on ices are presently available.[17,18] Here, we present the THz and mid-IR spectra of perhaps the simplest series of astrophysically-relevant molecules which share common, and progressively more complex functional groups (see Fig. 1), as well as the abundant ice species H_2O, CH_3OH, and CO.

2   Experimental methods

The experimental apparatus has been described in detail elsewhere.[16] Briefly, a high-resistivity intrinsic Si substrate, located inside a high-vacuum chamber (5×10^{-6} Torr at 300 K), is cooled to ~8 K using a He-cooled cryostat. A heating coil can control the temperature of the substrate from 8 K to 300 K. Gas samples are

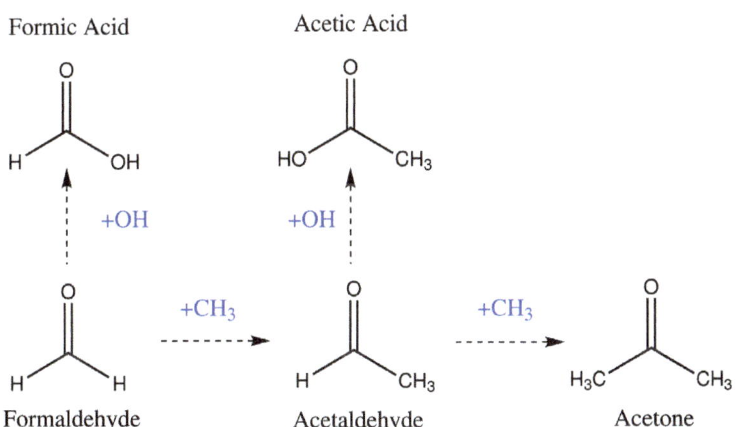

Fig. 1 Cartoon demonstrating the increasing complexity achievable with the addition of a single functional group (in this case OH or CH_3 radicals) to a simpler, neutral species. Arrows do not represent reaction pathways or mechanisms.

prepared in a metal dosing line and deposited onto the Si substrate to form an ice *via* a capillary stainless steel pipe normal to the Si surface. The end of the pipe is capped with a metal mesh with a 38 μm hole size to ensure a uniform, one-sided ice deposition. Transmission spectra can be collected in both the mid-IR from 400 to 4000 cm^{-1}, and the THz from 0.3 to 8 THz (10 to 265 cm^{-1}). Mid-IR spectra were taken using a Nicolet 6700 FTIR spectrometer externally coupled to a mercury-cadmium-telluride (MCT) detector. THz spectra were collected using a home-built time-domain (TD) THz spectrometer which is described in more detail below. The beams from the two spectrometers enter the vacuum chamber along separate paths and are incident on the substrate at 45° relative to the surface. This configuration is chosen to allow for the simultaneous acquisition of THz and mid-IR spectra. Unless otherwise specified, 256 scans are co-added to produce an average when acquiring mid-IR spectra using the FTIR (~10 min of acquisition time), while 64 scans are co-added for the THz spectra (~2 h of acquisition time). Each spectrum was ratioed with the background scan of a blank substrate held at 250 K.

2.1 THz instrumentation

There have been some changes to the THz spectrometer since the publication of Allodi *et al.*[16] (see Fig. 2). The output of the ultrafast regenerative amplifier (Legend Elite, Coherent Inc.) seeds an optical parametric amplifier (OPA) (Light Conversion Inc. TOPAS-C) capable of generating <40 fs pulses in the near-IR. Taking the idler output of the OPA at 1745 nm, and focusing that beam through a beta-barium borate (BBO) crystal creates the two-color plasma needed for THz pulse generation. This plasma generates THz pulses in a similar manner to the two-color plasma created with an 800 nm pulse and its second harmonic; however, the THz generation mechanism in plasma is more efficient.[19] Previously, pulses with roughly 3.5 mJ of energy at 800 nm created a plasma to generate THz pulses. Now, pulses with roughly 350 μJ of energy at 1745 nm generate a plasma that produces THz pulses with a power comparable to the previous plasma at 800 nm. The wavelength of 1745 nm was chosen because it provided the largest peak THz electric fields for the current optical alignment of the spectrometer.

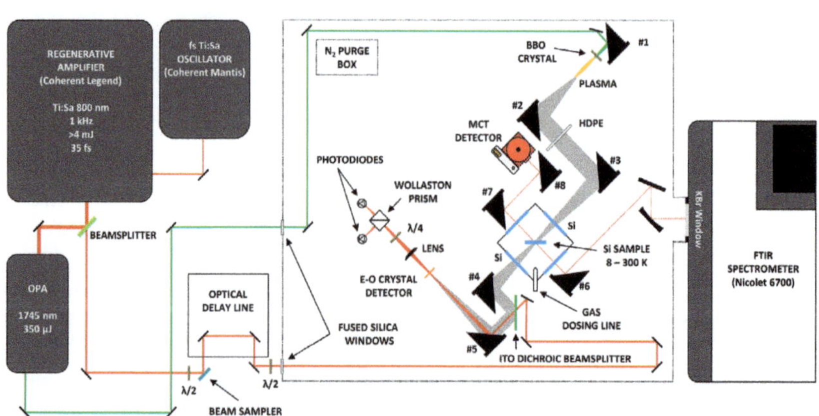

Fig. 2 Schematic of the Caltech astronomical ice analog experimental setup, including the home-build time domain THz spectrometer (not to scale).

The THz field is detected *via* Electro-optic (EO) sampling in a GaP crystal (Del Mar Photonics Inc.) using a small percentage of the output of the ultrafast amplifier at 800 nm that is not used to seed the OPA. This crystal is different from the one used by Allodi *et al.*[16] in that it has a thin EO active layer optically contacted to a thicker inactive layer. Specifically, a GaP (110) layer 200 μm thick, and capable of EO sampling, is optically contacted to a GaP (100) layer 4 mm thick that is EO inactive. This provides the advantage of pushing etalon features, or recurrences of the pulse in the time domain, outside the window being scanned. This ensures that any changes between the sample scan and the reference scan in the time domain will not lead to artificial signals in the frequency domain.

2.2   Data processing

The THz waveforms were collected over a 12 ps window in the time domain, starting 3 ps before the THz peak and continuing until 9 ps after the THz peak. A lock-in amplifier (Stanford Research Systems SR830) set for a 100 ms time constant was used to acquire the THz signals, which are modulated by an optical chopper set to 500 Hz. The waveforms contain 119.92 points per ps to ensure adequate sampling of the frequency content from 0.3 to 8 THz. Absorption spectra of the samples studied in the frequency domain are found by taking a fast Fourier transform (FFT) of the time domain data and dividing by a reference scan of the bare Si substrate.

The resolution of the spectra collected is approximately equal to the inverse of the length of scan in the time domain. For these experiments, this yields a ~90 GHz resolution in the frequency domain. With the optically contacted GaP crystal described in the section above, it is possible to increase the resolution of these spectra on our instrument to ~10 GHz by scanning a larger window in the time domain. However, this time window was chosen so as to quickly collect high signal-to-noise spectra of many different ices. The following data analysis procedure allows us to extract the maximum information from the collected data.

Apodization of the time-domain data prevents noise from spectral leakage. When Fourier transforming a time-domain signal collected in a window of finite duration, a boxcar function becomes convolved with the true time-domain signal, resulting in each line in the frequency domain being convolved with a sinc function.[20] Spectral leakage arises because a sinc function will add many sidelobes to the main peak, and so distinguishing between separate peaks becomes more difficult. Choosing a different apodization function from the boxcar will suppress the sidelobes and prevent spectral leakage that would arise from the sidelobes of one feature overlapping with the peak of another nearby feature. However, there is a balance to be struck between suppressing the sidelobes, but not distorting the lineshape drastically. In addition, asymmetric apodization functions must be used given the asymmetric nature of the THz signal in the collection window.[21] All the data in this paper are processed using an asymmetric Hann window peaked at the maximum of the THz signal. The Hann window[22] provides a good balance between sidelobe suppression and lineshape concerns.

After apodization, zeros were added to the apodized data to both interpolate between points in the frequency domain and ensure that the length of the data set equals a power of two. Zero padding in the time domain before taking an FFT is mathematically identical to interpolation in the frequency domain.[23]

Interpolation can help to identify the center of a feature with high accuracy. This can be especially useful for high signal-to-noise data since increasing the signal-to-noise ratio increases the accuracy of the measurement of peak centers. In addition to interpolation, zero padding to create a data set of length 2^N, where N is a positive integer, maximizes the speed of the FFT calculation.

2.3 Computational methodology

Ab Initio quantum chemical calculations were performed on dimers of formic acid and acetic acid molecules to provide a molecular interpretation of some of the unassigned modes present in the THz spectral region as shown and discussed in the next sections. All calculations were performed using the MP2/6-311+G(d,p) level of theory to adequately model the hydrogen-bonding interactions between molecules and were carried out using Gaussian 09.[24] Since harmonic frequencies calculated with MP2 are known to be systematically biased toward higher frequencies than are observed,[25] the calculated frequencies in wavenumbers are scaled by a value of 0.9646 as used by Pople *et al.*[26]

3 Results

A listing of the mid-IR and THz spectra of simple and more complex pure ices, as well as some selected mixtures, are presented in Table 1. The water used for the ice samples is deionized and purified (Milli-Q water purifier, EMD Millipore) before being degassed by several freeze-pump-thaw cycles. In general, all the liquid samples used in this work (CH_3OH from Omni Solv 99.9%, HCOOH from Sigma Aldrich 98–100%, CH_3CHO from Fluka/Sigma Aldrich 99.5%, CH_3COOH from J. T. Baker 99.7%, and $(CH_3)_2CO$ from Macron Chemicals 99.5%) are degassed by freeze-pump-thaw cycles. Gaseous CO is used as received from Air Liquide (99.97%).

Crystalline ices are deposited close to their desorption temperature under our experimental conditions, while all the amorphous ices are formed at 10 K. After deposition, crystalline ices are cooled down in temperature, while amorphous ices are generally heated or annealed to higher temperatures as shown in Table 1. Our selected crystalline and amorphous films are in all cases less than 10 μm thick and are generally condensed at a rate of $\sim$30 μm hr^{-1}.

3.1 Mid-IR spectra of selected pure ices

Fig. 3 shows the mid-IR crystalline spectra of the species investigated in this work. All the spectra are acquired during deposition to obtain non-saturated IR absorption bands. Here, 64 scans are generally co-added to produce a non-saturated spectrum.

H_2O. Fig. 3 shows a typical mid-IR spectrum of cubic-crystalline water ice deposited at 140 K.[27–29] The water IR spectrum exhibits four broad bands between 400 and 4000 cm^{-1}: (*i*) the libration of water molecules at 12.5 μm (800 cm^{-1}); (*ii*) the bending mode at $\sim$6 μm (1650 cm^{-1}); (*iii*) a combination band at 4.5 μm (2200 cm^{-1}); (*iv*) and the OH-stretching mode at $\sim$3 μm (3250 cm^{-1}). Peak position and band profile of these bands are strongly dependent on the type of ice (amorphous *vs.* crystalline) and on thermal history of the ice as will be discussed in more detail later in the text.

Table 1 List of experiments performed in this work

Experiment	Ratio	T_{dep}/K	Thermal History/K
Pure ice			
H_2O		175	140, 10
H_2O		150	150, 75, 10
H_2O		140	140, 75, 10, 10[a]
H_2O		140	10
H_2O		125	10
H_2O		10	10, 125, 150
CO		30	25, 10
CO		30	10
CO		10	10, 25[b]
CH_3OH		140	100, 75, 10
CH_3OH		140	100, 75, 10
CH_3OH		10	10, 75, 100[c]
$HCOOH$		150	100, 75, 10
$HCOOH$		10	10, 75, 75[d], 10
CH_3CHO		125	100, 75, 10
CH_3CHO		125	10, 10, 10
CH_3CHO		10	10, 75, 100, 100[e]
CH_3CHO		10	10, 40, 60, 80, 10[e]
CH_3COOH		150	100, 75, 10
CH_3COOH		180	100, 75, 10
CH_3COOH		180	100, 75, 10
CH_3COOH		10	10, 75, 100, 100[f]
$(CH_3)_2CO$		150	100, 75, 10
$(CH_3)_2CO$		10	10, 75, 100
$(CH_3)_2CO$		10	10, 75, 100, 100[d]
Mixed ice			
$H_2O{:}CH_3OH$	2 : 1	140	10
$H_2O{:}CH_3OH$	1 : 1	140	10
$H_2O{:}CH_3OH$	0.5 : 1	140	10
$H_2O{:}CH_3CHO$	2 : 1	125	10
$H_2O{:}CH_3CHO$	1 : 1	125	10
$H_2O{:}CH_3CHO$	0.5 : 1	125	10
$H_2O{:}(CH_3)_2CO$	2 : 1	150	10
$H_2O{:}(CH_3)_2CO$	1 : 1	150	10
$H_2O{:}(CH_3)_2CO$	0.5 : 1	150	10

[a] Annealed to 175 K for 10 min. [b] Annealed to 32 K for 5 min. [c] Annealed to 140 K for 10 min and 145 K for <1 min. [d] Annealed to 150 K for 10 min. [e] Annealed to 125 K for 5 min. [f] Annealed to 200 K for 5 min; T_{dep} is the substrate temperature during deposition; Thermal History notes the temperatures that the ice have been sequentially exposed to. The heating ramp rate is in all cases 20 K min^{-1}.

CO. The spectrum of CO ice deposited at 30 K is largely flat with one strong absorption feature at 4.6 μm (2140 cm^{-1}). ^{13}CO is also present in the ice as a small fraction of the total ^{12}CO according to its ^{13}C≡O stretch mode band visible in Fig. 3 at 4.8 μm (2090 cm^{-1}).

CH$_3$OH. Methanol, the final ice mantle product created by the hydrogenation of CO ice, has a somewhat more complex mid-IR spectrum than its precursor: (*i*) a torsion mode at 4.8 μm (710 cm^{-1}); (*ii*) the strong CO-stretch at 9.7 μm (1030 cm^{-1}); (*iii*) the CH$_3$-rock at 8.8 μm (1130 cm^{-1}); (*iv*) the OH-bend at ~7 μm

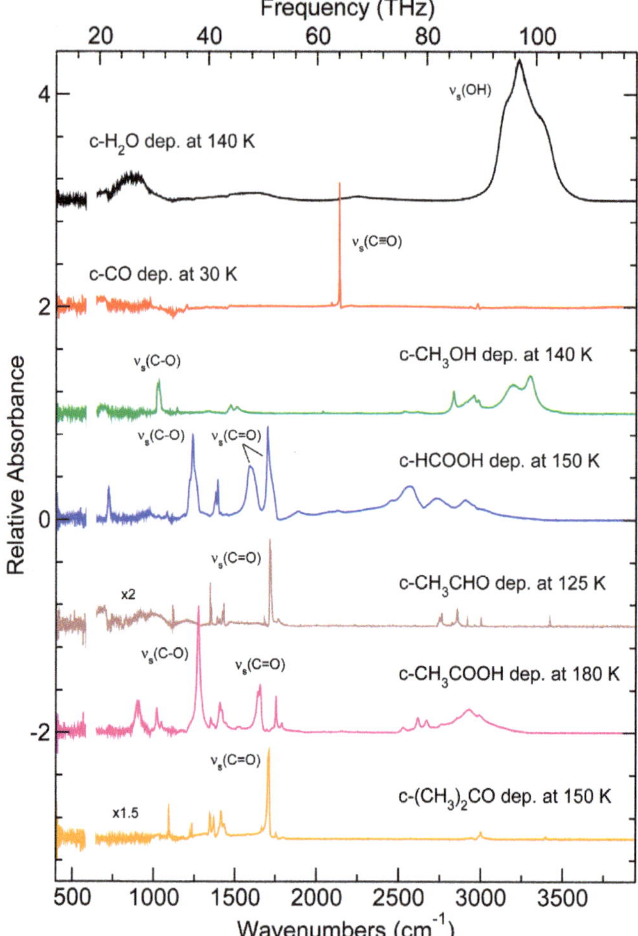

Fig. 3 Mid-IR FTIR spectra of non-saturated crystalline pure ices. Spectra are acquired during deposition of thicker ices. The strongest vibrational modes are assigned in the figure.

(1415 cm^{-1}); (*v*) the CH$_3$-deformation at 6.8 μm (1470 cm^{-1}); (*vi*) a series of CH-stretch modes between 3.5 and 3.3 μm (2800 and 3000 cm^{-1}); (*vii*) OH-stretch mode at ∼3 μm (3300 cm^{-1}).[30,31]

HCOOH. Pure HCOOH deposited at 150 K presents several absorption bands in the mid-IR: (*i*) the OCO-bend mode at 14.2 μm (700 cm^{-1}); (*ii*) a OH-bend mode at 10.7 μm (930 cm^{-1}); (*iii*) the CH-bend mode at 9.3 μm (1070 cm^{-1}); (*iv*) the strong C–O stretching mode at 8.3 μm (1210 cm^{-1}); (*v*) the OH-bend and the CH-bend modes at 7.2 μm (1390 cm^{-1}); (*vi*) two strong C=O stretching modes at 6 and 5.8 μm (1650 and 1715 cm^{-1}); (*vii*) and several OH-stretch modes and a CH-stretch mode in the spectral region between 3.9 and 3.2 μm (2500 and 3100 cm^{-1}).[32]

CH$_3$CHO. Crystalline acetaldehyde deposited at 125 K has several sharp peaks in the mid-IR: (*i*) the CH$_3$-rock at 9 μm (1115 cm^{-1}); (*ii*) the symmetric CH$_3$ deformation at 7.4 μm (1350 cm^{-1}); (*iii*) the in-plane CH wag at 7.1 μm (1400 cm^{-1}); (*iv*) the asymmetric CH$_3$ deformation at 7 μm (1430 cm^{-1}); (*v*) the CO-stretch at 5.8 μm (1710 cm^{-1}); (*vi*) the carbonyl CH-stretch at 3.7 μm (2710 cm^{-1}); (*vii*) and a series of methyl CH-stretch between 3.4 and 2.7 μm (2900 and 3600 cm^{-1}).[33]

CH$_3$COOH. Acetic acid has a desorption temperature higher than water. Therefore, we deposited acetic acid at 180 K to obtain a crystalline film. Acetic acid presents a forest of features in the mid-IR. Thus, here we name only the strongest absorption bands shown in Fig. 3: (*i*) the γ(OH) at 10.8 μm (920 cm^{-1}); (*ii*) the ρ(CH$_3$) at 9.5 μm (1050 cm^{-1}); (*iii*) the very strong C–O stretching mode at 7.7 μm (1300 cm^{-1}); (*iv*) the OH at 7.1 μm (1410 cm^{-1}); (*v*) the C=O stretching mode at 6 μm (1650 cm^{-1}); (*vi*) a combination mode at 3.7 μm (2650 cm^{-1}); (*vii*) and the OH-stretch at 3.4 μm (2900 cm^{-1}).[34]

(CH$_3$)$_2$CO. Acetone deposited at 150 K has several absorption bands, among the more prominent are: (*i*) the CH$_3$-rock at 9.1 μm (1090 cm^{-1}); (*ii*) the CC-stretch at 8.2 μm (1210 cm^{-1}); (*iii*) a series of CH$_3$ deformation modes between 7.4 and 6.6 μm (1350 and 1500 cm^{-1}); (*iv*) the CO-stretch at 5.8 μm (1710 cm^{-1}); (*v*) and the CH-stretch at 3.3 μm (3000 cm^{-1}).[30]

Fig. 3 clearly shows that molecules that share the same functional group present similar mid-IR absorption features (*e.g.*, similar intramolecular vibrations). For instance, acetaldehyde and acetone have the methyl group in common and show almost identical IR features; formic acid and acetic acid share the carboxylic acid group and have similar IR spectra as well. This can be potentially a cause of spectral confusion when mixtures of species containing the same functional groups are investigated in the mid-IR.

The THz spectral range is dominated by intermolecular forces (lattice modes). These modes involve the relative motion of molecules as a whole and can be either translational or librational modes, and are therefore sensitive to the structure of the hydrogen bond network in general and the nature of the unit cell in particular for crystalline solids. Moreover, THz features often give information on the lattice of the ice or transitions between different phases (*e.g.*, amorphous *vs.* crystalline).[18] Therefore, THz spectroscopy can potentially give more information on the long range structure of the ice than mid-IR spectroscopy does, especially when pure ices of complex species or complex mixtures are investigated. In the next section, we present the THz spectra of the ices shown in Fig. 3. Because the intermolecular modes potentially involve many monomers, and are distinct for various topological combinations of the individual molecules, theoretical predictions of the resulting THz spectra are much more involved than are those for mid-infrared spectra. To date, little such work has been performed on species beyond water and CO/CO$_2$, and so the following discussion is necessarily qualitative.

3.2 THz spectra of selected pure ices

Fig. 4 shows some selected crystalline (left panel) and amorphous (right panel) pure ices in the spectral region between 0.3 and 7.5 THz. Although crystalline ices are deposited at temperatures close to the desorption temperature of each species, all the spectra shown in Fig. 4 are acquired at 10 K. Moore and Hudson[18] studied the THz spectra of crystalline and amorphous pure water, carbon monoxide, and methanol ices at 13 K in the spectral range between 3 and 15 THz

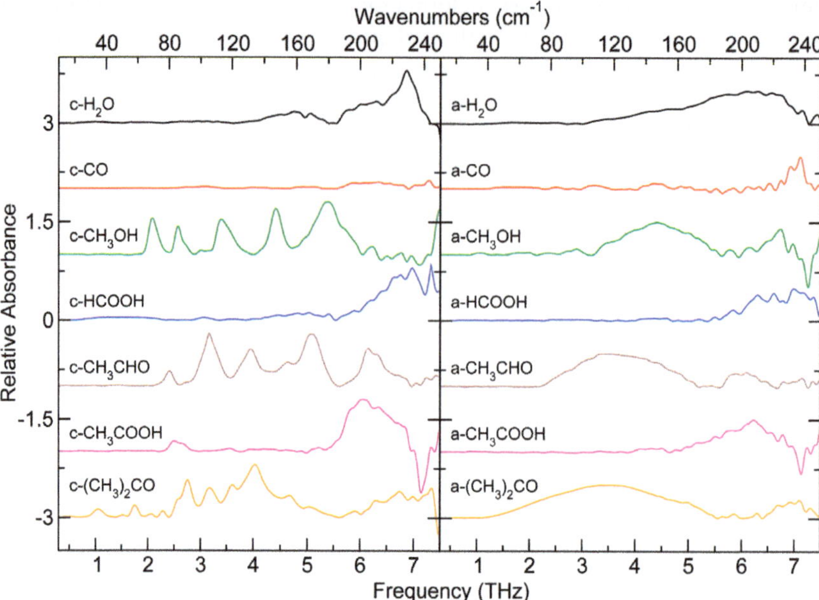

Fig. 4 THz spectra of selected crystalline (left panel) and amorphous (right panel) pure ices. Although crystalline ices are deposited at higher temperatures, all the spectra are acquired at 10 K.

by means of a standard FTIR technique. In their study, crystalline samples were formed by annealing ices that were pre-deposited at 13 K. Our results are in good agreement with those from Moore and Hudson,[18] *i.e.*, peak position and band profiles are consistent with those found in literature. In general, amorphous features are found to be few, very broad, and strong, while crystalline bands are numerous, narrow, and often very strong. One of the novelties of our work is that the spectra shown in Fig. 4 include a spectral region that was not previously covered by FTIR techniques (*i.e.*, between 0.3 and 3 THz). In this region, several new features are identified (see below). Moreover, our work includes a series of species never investigated in the THz range: HCOOH, CH_3CHO, CH_3COOH, and $(CH_3)_2CO$.

H_2O. The THz spectrum of water ice is the most studied in the literature.[18,35–37] Amorphous water ice presents a strong and very broad band peaked at 6.2 THz, and a second broad band at 4.6 THz that is a shoulder of the previous one. Other features are all very weak. Crystalline water ice deposited at 140 K and cooled to 10 K presents distinct peaks at 6.9 THz (H-bond stretch between bilayers), 5.7 THz (out-of-phase vibration within a bilayer), 4.9 THz (proton disordered motion), 4.3 THz, 1.9 THz (corresponding to an O–O–O bend), and 1 THz (as assigned in one of our previous works).[16,38,39]

CO. The THz spectrum of CO is nearly flat.[18] The spectrum of crystalline CO deposited at 30 K and cooled to 10 K indicates that there may be a very weak and broad feature between 6 and 7 THz. The amorphous spectrum of CO at 10 K shows very weak peaks at 3 and 4.5 THz. The presence of weak features cannot be due to contaminations since the mid-IR spectra of CO ice do not show the presence of major impurities. Our THz–TD spectra are less sensitive to thin ices than the mid-

IR FTIR measurements are. The stronger peak at 7.3 THz in the amorphous spectrum could be due to baseline subtraction artifacts. At those frequencies, the signal-to-noise ratio is quite low and a flat spectrum is therefore harder to reduce than a spectrum with clear THz features.

CH_3OH. Amorphous methanol shows a broad feature at 4.4 THz and perhaps the beginning of another broad band at 6.8 THz as also suggested in Ref. 18. Apart from the three peaks (5.4, 4.4, and 3.4 THz) shown in Ref. 18, crystalline methanol deposited at 140 K and cooled to 10 K presents two other strong and sharp peaks at 2.6 and 2.1 THz. These peaks have not been reported before.

$HCOOH$. Although the mid-IR spectrum of formic acid is quite complex with a forest of bands in absorption, the THz spectrum looks quite flat. Amorphous and crystalline formic acid have only a broad and strong band at $\sim$7 THz. The crystalline spectrum deposited at 150 K and cooled to 10 K shows some very weak features at lower frequencies, while the strong band is blue-shifted compared to the amorphous case.

CH_3CHO. Similar to methanol, amorphous acetaldehyde displays a broad band at 3.5 THz and a smaller one at 6 THz. Crystalline acetaldehyde deposited at 125 K and cooled at 10 K present several strong peaks at 6.2, 5.1, 4, 3.2, and 2.4 THz.

CH_3COOH. As for formic acid, acetic acid spectra are quite flat except for a strong band at $\sim$6 THz. Crystalline acetic acid has also another band at 2.5 THz. As opposed to the other species studied here, acetic acid desorbs above the desorption temperature of water ice and therefore acetic acid is deposited crystalline at 180 K in this work.

$(CH_3)_2CO$. Crystalline acetone has perhaps the most complex THz spectrum studied here. Small and sharp bands are visible even at low frequencies where the other molecules do not have any absorption bands. There are two strong peaks at 4 and 2.8 THz. The other numerous bands are medium or weak. Amorphous acetone has a strong broad band at 3.5 and another band at 7 THz.

The THz spectra of pure methanol, acetaldehyde, and acetone are quite similar: the amorphous spectra all have a broad and strong band and a second broad band at higher frequencies; those bands split in several distinct features in the crystalline spectra of those molecules cooled at 10 K. All the bands are strong and sharp. However, they do not perfectly overlap with each other. Formic acid and acetic acid present several spectral similarities as well: the spectra are generally quite flat, both amorphous and crystalline, and have a strong and broad feature around 7 THz. This may result from the strong dimer interactions expected for these species where a cyclic hydrogen bonded geometry is expected in both the gas and solid state. Finally, as noted above, the THz spectra of different molecules do not exactly overlap with each other. Thus, THz features can potentially provide a unique fingerprint of certain ice structures and compositions.

3.3 The structure effect: the case of water ice

Fig. 5 shows the full spectral range of our current setup (from 0.3 to 120 THz or 10 to 4000 cm^{-1} with a few gaps). We plot water ice data here as example since it is the best studied and understood at both mid-IR and THz wavelengths. As is true for the other species in this work, the intermolecular forces and hydrogen-bonded network that characterizes solid state water enables a series of reversible and irreversible transformations to be accessed at low temperature. Allodi *et al.*[16] showed that crystalline water ices subjected to different temperature cycles present the same spectral profile in the THz at the same temperatures. This is an example of reversible transformation. However, when amorphous water ice is heated to higher temperatures the ice goes through a series of irreversible transformations such as the transition between amorphous solid porous water and amorphous solid compact water (above ~100 K), or amorphous solid compact water and crystalline water. However, crystallization of an annealed ice has been proven to be not completed even above 140 K.[29]

Fig. 5 investigates the degree of crystallization of the water ice deposited at 175, 140, 125, and 10 K, respectively. Here, THz data clearly show substantial differences under different conditions. All the spectra are cooled to 10 K to better compare them with each other. The spectrum deposited at 175 K shows the full crystallization of the ice with the strongest H-bond stretch between bilayers at 6.9 THz. At 140 K, water is clearly crystalline but a small fraction of it is still amorphous. At 125 K, water shows the first signs of reorganization with a distinct separation between the bands at 4.9 and 6.9 THz, and a sharp tip at 6.8 THz. For comparison a spectrum of water ice deposited at 10 K is also shown. Mid-IR spectra show differences from each other, however, the THz range gives more direct signs of the transformation within the ice.

3.4 The temperature effect: crystalline ices

Allodi *et al.*[16] showed that the THz band profiles are temperature dependent in the case of crystalline water ice. Here we extend this to all the species studied in this

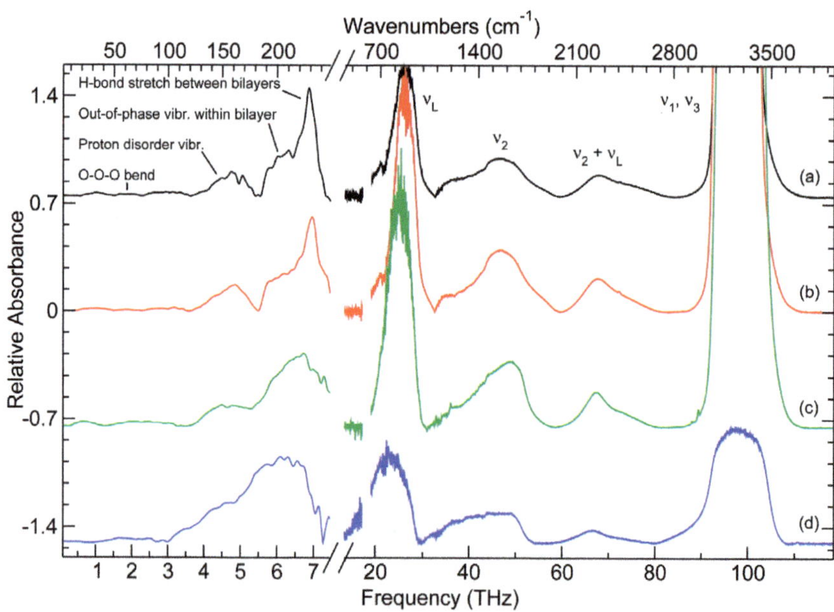

Fig. 5 Spectra of water ice deposited at different temperatures: (a) 175 K; (b) 140 K; (c) 125 K; and (d) 10 K. The figure shows the complete spectral window covered by the setup as used in this work.

work. Fig. 6 shows the THz spectra of crystalline ices at 100 K (black curves), 75 K (red curves), and 10 K (green curves), with the exception of one spectrum of water ice that is at 140 K (black curve) instead of 100 K. The deposition temperatures for all the species are the same as described in Fig. 3. Apart from the acids, which behave differently, all the other species have strong changes that are dependent on the temperature variations of the ice. At 100 K, several sharp peaks are visible for water, methanol, acetaldehyde, and acetone. These peaks become sharper at 75 K and clearly stronger at 10 K. At low temperatures, new peaks are detected. Moreover, these peaks reveal a temperature dependent blueshift at lower temperatures. The peak positions at all the temperatures investigated are summarized in Table 2.

As discussed before, these transformations are reversible. The peak shifting and broadening in the THz spectra results from the anharmonicity of the vibrational potential, and thus provide a spectral fingerprint of the intermolecular forces. Since the anharmonicity of the Morse potential decreases the spacing

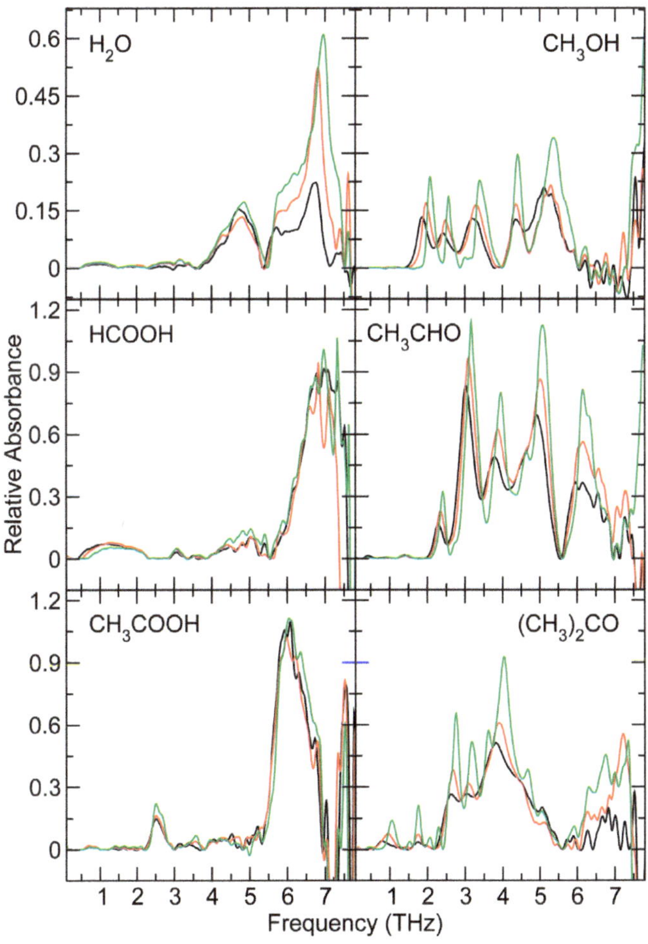

Fig. 6 THz spectra of crystalline ices at different temperatures: (black curves) 100 K; (red curves) 75 K; and (green curves) 10 K. The black curve of water ice is acquired at 140 K.

between adjacent vibrational levels, hot bands exhibit redshifts, *i.e.*, appear at lower frequencies than the corresponding fundamental transitions. The vibrationally averaged structure of the crystal is then temperature dependent since the wavefunctions of the hot bands are different from each other and the ground state. At low temperature, only the lower lying states on the vibrational potential are populated and the absorption bands appear blueshifted and sharper with respect to the absorption bands due to hot bands.

Table 2 List of absorption bands of the molecules studied at different temperatures in the THz spectral region between 0.3 and 7.5 THz

c-H_2O			a-H_2O	c-CH_3OH			a-CH_3OH
140 K	75 K	10 K	10 K	100 K	75 K	10 K	10 K
1.0 w	1.0 w	1.0 w					
1.9 vw	1.9 vw	1.9 vw	1.7 w	1.85 s	1.95 s	2.1 s	1.0 w
2.7 vw	2.7 vw	2.7 vw		2.4 s	2.5 s	2.6 s	1.9 w
3.1 w	3.1 w	3.1 w	3.5 m			3.0 w	2.3 w
4.3 m	4.3 m	4.3 m		3.2 s	3.3 s	3.4 s	3.5 m
4.7 s	4.8 s	4.9 s	4.6 s, sh, br	4.37 s	4.37 s	4.43 vs	4.4 s, br
5.7 m, sh	5.7 s, sh	5.7 s, sh	6.2 vs, br	5.1 s	5.3 s	5.4 vs	6.8 s
6.7 s	6.8 vs	6.9 vs					

c-HCOOH			a-HCOOH	c-CH_3CHO			a-CH_3CHO
100 K	75 K	10 K	10 K	100 K	75 K	10 K	10 K
1.2 w	1.4 w	1.6 w		0.4 vw	0.4 vw	0.4 vw	
3.0 w	3.0 w	3.0 w		0.9 vw	0.9 vw	0.9 vw	
3.6 vw	3.6 vw	3.6 vw		1.4 vw	1.4 vw	1.4 vw	
4.8 w	4.8 w	4.8 m	4.3 vw	2.28 m	2.36 m	2.43 m	
7.0 vs	7.0 vs	7.0 vs	6.8 vs, br	3.0 s	3.1 vs	3.19 vs	
				3.8 m s^{-1}	3.88 s	3.96 s	3.5 s, br
						4.6 w, sh	
				4.9 s	5.0 s	5.1 vs	
				6.0 m s^{-1}	6.1 s	6.2 s	6.0 m, br
				7.3 m	7.3 m	7.3 m	

c-CH_3COOH			a-CH_3COOH	c-$(CH_3)_2CO$			a-$(CH_3)_2CO$
100 K	75 K	10 K	10 K	100 K	75 K	10 K	10 K
0.8 vw	0.8 vw	0.8 vw		0.48 vw	0.48 vw	0.48 vw	
1.8 vw	1.8 vw	1.8 vw		0.8 vw	0.96 vw	1.05 w	
2.5 m	2.5 m	2.5 m		1.3 vw	1.45 vw	1.53 vw	
3.4 w	3.5 w	3.6 w	3.4 m	1.75 vw	1.75 vw	1.75 w	
4.3 w	4.3 w	4.3 w	4.3 w	2.0 vw	2.02 vw	2.06 vw	
5.2 w	5.2 w	5.2 w				2.3 w	
5.9 vs	6.0 vs	6.1 vs	6.2 vs, br			2.57 vw, sh	
				2.63 m	2.68 m	2.76 s	
				3.04 m	3.1 m	3.18 s	
						3.61 m, sh	3.5 vs, br
				3.8 s	3.9 s	4.04 vs	
				4.5 vw, sh	4.5 vw, sh	4.7 m, sh	
				5.05 w, sh	5.05 w, sh	5.05 w, sh	5.0 w, sh
				5.6 w	5.6 w	5.6 w	
				5.8 w	5.8 w	5.9 w	5.8 w
				6.3 w	6.3 w	6.3 m	6.3 w
				6.8 w	6.8 w	6.8 m s^{-1}	7.0 s

[a] vw = very weak, w = weak, m = medium, s = strong, vs = very strong, sh = shoulder, br = broad.

Moreover, it is possible for features not present at high temperatures to grow in at lower temperatures. If the shape of the vibrational potential is such that there are multiple local minima, corresponding to different crystal structures, at higher temperatures, different vibrational states corresponding to these different structures are populated. Based on the different processing of the ice, it is possible to freeze-out the different structures and thus have one feature split into several at lower temperature. Thus, THz spectra of crystalline ices have the potential to give direct information on the temperature of the studied system.

THz spectroscopy is not only sensitive to intermolecular forces, it also probes low energy intramolecular modes such as torsional motions of individual molecules. Some torsion–vibration spectra of methanol, acetaldehyde, acetic acid, and acetone are already available in the literature as reviewed by Shimanouchi.[40] These data are in good agreement with our THz laboratory spectra: methanol has a torsional mode at 6 THz that corresponds to the very strong mode at 5.4 THz of our spectrum of crystalline methanol at 10 K as shown in Table 2; the 4.6 THz mode of crystalline acetaldehyde found in our spectrum at 10 K corresponds to the torsional mode assigned at 4.5 THz; acetic acid has a torsional mode at 2.8 THz that matches well with a band at 2.5 THz as seen in our crystalline spectrum at 10 K; the acetone torsional mode at 3.2 THz overlaps with a strong band at 3.18 THz as found in our spectrum of crystalline acetone at 10 K.

3.5   The temperature effect: amorphous ices

As previously discussed, amorphous ice goes through irreversible transformations if annealed to higher temperatures. Fig. 7 shows the THz spectra of amorphous ices as deposited at 10 K (black curves) and heated to higher temperatures. In the left-panel of Fig. 7, amorphous water ice is heated to 125 K (red curve) and 150 K (green curve). Comparing these spectra to those acquired at the same temperatures and shown in Fig. 5 gives information on the degree of crystallinity of the ice. The red and green curves in Fig. 5 represent ices directly deposited at 140 and 125 K, respectively. These ices clearly present the highest degree of crystallization at the selected temperatures, whereas amorphous ices heated to the same temperatures show an amorphous component underneath the crystalline line-shapes. This indicates that the ice is partially reorganized but still largely amorphous. The degree of crystallinity increases with temperature as expected. This is also the case for methanol and acetone. Amorphous methanol is heated to 75 K (red curve), and annealed to 140 K for ~10 min and cooled to 100 K (green curve). Spectra at 10 and 75 K do not present major differences indicating that the ice is still amorphous. After annealing, the ice shows a clear crystalline structure. The features are however broader than the ones shown in Fig. 6, indicating that the crystallization process is not yet completed. Amorphous acetone heated to 75 K (red curve), 100 K (green curve), and annealed to 150 K and cooled to 100 K (blue curve), shows an even clearer amorphous component even when the ice is annealed for ~10 min.

Amorphous acetaldehyde approaches a fully crystalline structure at lower temperatures than the other species studied here. In the right-center panel of Fig. 7, amorphous acetaldehyde is heated to 40 K (red curve), 60 K (green curve), 80 K (blue curve), and annealed to 125 K for ~5 min and cooled back to 10 K (brown curve). Acetaldehyde goes through crystallization at temperatures above

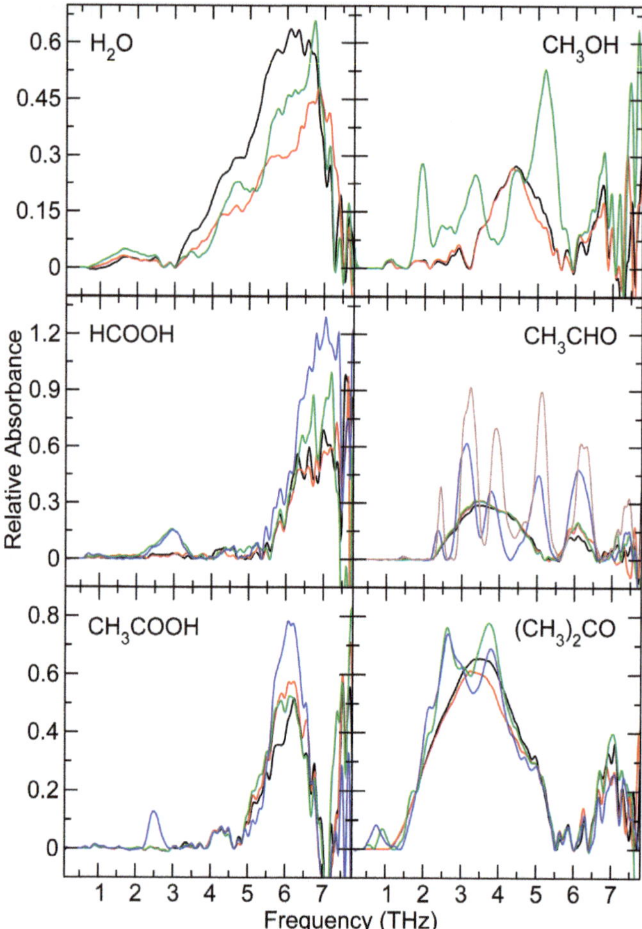

Fig. 7   THz spectra of amorphous ices at different temperatures: see text for the assignment of the different temperatures.

70 K under our experimental conditions, and at 80 K shows already an almost perfect crystalline structure comparable to the one shown in Fig. 6. The annealing process does not greatly change the degree of crystallinity. The difference here is the temperature of acquisition of the spectra (*i.e.*, 80 K *vs.* 10 K). As shown in Fig. 6, crystalline acetaldehyde ice shows blueshifted, sharper, and more intense features at low temperatures. This is also the case for the spectra in Fig. 7.

Amorphous formic acid is heated to 75 K (red curve), annealed to 150 K for ~10 min and cooled back to 75 K (green curve), and cooled to 10 K (blue curve). The acid presents small changes in the band profile of the strongest feature at ~6.8 THz when the ice temperature is increased. However, after the ice is annealed to 150 K for ~10 minutes a new feature appears at 3 THz. Cooling the now crystalline ice to 10 K allows for an increase of the intensity of the main feature at 6.8 THz. This is consistent with the behavior of the other crystalline ices as previously discussed, but is not consistent with the almost complete lack of changes seen in crystalline formic acid as shown in Fig. 6. It is interesting to note that the new crystalline band at 3 THz does not change with changes in temperature, while only the main band increases in intensity at lower temperatures. However the band profile of the main band at 10 K is now similar to the one of the crystalline ice as shown in Fig. 6 indicating that the ice is fully crystalline. Further experiments are needed to show whether the crystalline band at 6.8 THz is temperature dependent as shown in Fig. 7 or is temperature independent (see Fig. 6). A possible explanation is that the ice was not fully crystalline when the second spectrum at 75 K was acquired. Since the acquisition of a THz spectrum lasts for $\sim$2 h with our current setup, the ice may have had enough time to reorganize at this temperature and the spectrum at 10 K shows that more than a temperature dependence.

A similar effect is seen in the acetic acid spectra. Amorphous acetic acid is heated to 75 K (red curve), 100 K (green curve), and annealed to 200 K for $\sim$5 min before being cooled back to 100 K (blue curve). Also for acetic acid, a new band at 2.5 THz is present in the annealed spectrum and the main band at 6.2 THz looks sharper and more intense. The fact that this band looks sharper already at 100 K is an indication that the ice is fully crystalline, adding support to our idea that the changes of the main band profile at 6.8 THz in formic acid are due to further crystallization more than to the energy state distribution of the fundamental and hot band features whose intensities are a strong function of temperature.

In an attempt to provide molecular insight into why the structures of the amorphous and crystalline acid ices present similar THz spectra, *ab initio* quantum chemistry calculations were performed on a dimer of each species. The cyclic dimer structure is such that the acidic proton of one molecule is donating to the carbonyl group of the second molecule and vice-versa. While these are gas-phase dimers, they can provide some insight into solid-phase structures as well. The hydrogen-bonded interaction that forms a dimer is indeed quite strong and dimers can therefore dominate the structure of pure ices. Our calculations provide evidence in support of this. In the case of the formic acid dimer, the calculations reveal a strong mode corresponding to a rocking of the two molecules, which stretches their hydrogen bonds, at 235 cm^{-1} (7.0 THz). This is in good agreement with the formic acid mode observed in the experimental spectrum at $\sim$7 THz. A similar mode is found in the calculated vibrations for the acetic acid dimer at 160 cm^{-1} (4.8 THz). According to our experimental data, however, the strongest mode of acetic acid peaks at $\sim$6 THz. Differences between MP2 simulations and THz laboratory results can be explained by the more constrained nature of the molecules in the ice that can cause the frequency of this rocking vibration to be higher in energy in the solid phase than in the gas phase. Overall, while the predicted mode of acetic acid does not agree perfectly with the experimentally-measured transition, the calculations produce the same red shift in frequency, relative to the formic acid mode, as observed in the experiments. This offers evidence to support our assignments of these transitions, since a model of harmonic vibrations predicts a decrease in frequency when mass is increased, as is the case when a heavier functional group is added ($-CH_3$ *vs.* H). Therefore, similarities between THz spectra of amorphous and crystalline acidic ices can be explained by the direct deposition of acidic dimers onto the substrate. As a result of the strong hydrogen-bonding interaction between dimers, the features around 6 and 7 THz remain largely unchanged when amorphous ices are annealed.

3.6 The composition effect: binary mixtures

Although there are some far-IR spectra of mixed binary and more complex ices reported in the literature, there are no data available at low frequencies (between 0.3 and 3 THz). For instance, Moore and Hudson[18] mixed water with other astrophysically-relevant species with a ratio of 10 : 1 and 2 : 1. All their amorphous ices were deposited at 13 K, while the crystalline mixtures were obtained by annealing amorphous ices to 155 K and cooling it back to 13 K. This procedure may lead to some desorption, diffusion, and reorganization. However, we showed that annealed ices are hardly fully crystalline unless the ice is annealed to temperatures close to their desorption temperature for several minutes. This can also cause desorption of part of the ice. Allodi *et al.*[16] showed that CO_2 mixed in water ice segregates at higher temperatures between the bilayers of water ice, disrupting the crystalline structure of water ice. Spectra of these mixtures are hard to interpret if not supported by proper THz ice-database that includes spectra of pure ices at different temperatures and different mixtures deposited at high temperatures. Therefore, our goal here is the investigation of the spectral changes of crystalline ice when binary mixtures are deposited at different ratios and at high temperatures.

The top panels of Fig. 8 show the spectra of water and methanol mixed ices with a ratio of 2 : 1, 1 : 1, and 0.5 : 1 deposited at 140 K. By looking at both the THz and the mid-IR regions, it is possible to conclude that although the ices are deposited at the same temperature and under the same experimental conditions, the composition of the ice has a strong effect on the spectral profile. It should be noted that the mid-IR spectra shown in the figure are acquired during deposition to have non-saturated infrared bands (64 co-added scans). This allows us to monitor the profile of the fundamental vibrational modes of water and methanol in the mid-IR and retrieve information from both the THz and the mid-IR regions.

The 2 : 1 mixture (Fig. 8a) shows the clear signs of crystalline water ice with some impurities due to the presence of methanol. As opposed to the case of CO_2 that has been shown to affect strongly the water modes in the THz, methanol does not seem to have a strong impact on the H-bond stretch mode between bilayers or the proton disorder mode of water ice.[16] When the amount of methanol is comparable to the amount of water (1 : 1 mixture, Fig. 8b), however, the aforementioned water modes are affected by the methanol and a single broad band appears at ~6 THz. Three water modes and two methanol modes lay on the region covered by this broad band. The last mixture (0.5 : 1, Fig. 8c) shows four peaks that partially overlap with each other and most likely are due to the single peaks seen in the spectra of pure crystalline water and methanol ices. However, they are slightly shifted. This indicates that water affects the structure of methanol more than the other way around and that the interaction of the two molecules with each other can be investigated in the THz.

The bottom panels of Fig. 8 shows the spectra of water and acetaldehyde mixed ices with a ratio of 2 : 1, 1 : 1, and 0.5 : 1 deposited at 125 K. In this case, water is not yet crystalline, but our THz spectra show that the degree of reorganization of the ice is substantially higher than for amorphous water ice at 10 K. By comparing the three mixtures we can conclude that the water structure is affected by the presence of acetaldehyde in a way that is not strongly dependent on the amount of acetaldehyde in the ice: two broad and strong peaks due to water ice are visible in

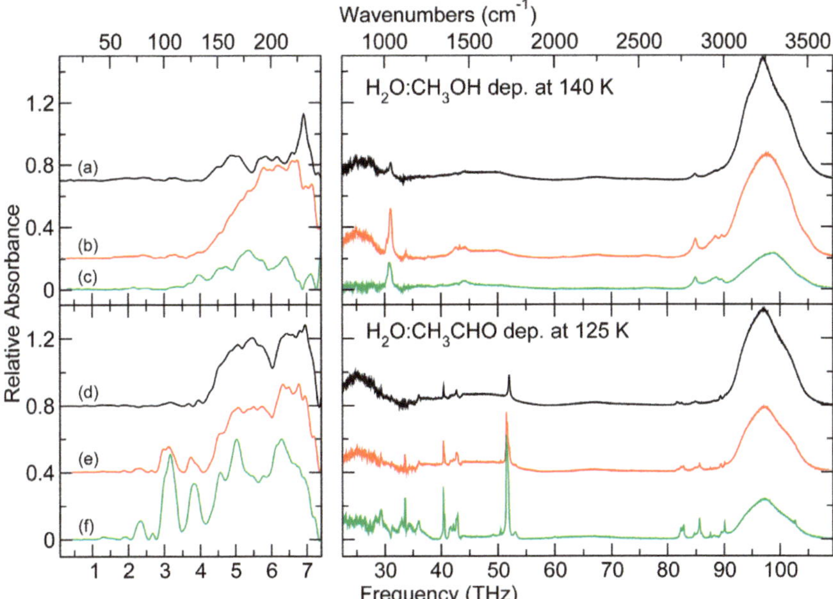

Fig. 8 THz spectra of crystalline mixed ices at 10 K. The left panels show THz spectra of $H_2O : CH_3OH$ 2 : 1 (a), 1 : 1 (b), and 0.5 : 1 (c) mixtures (top) and $H_2O : CH_3CHO$ 2 : 1 (d), 1 : 1 (e), and 0.5 : 1 (f) mixtures (bottom). The right panels show the same ices in the mid-IR. In this case, mid-IR spectra are acquired during deposition to obtain non-saturated features.

all the investigated mixtures and their profile does not change in the different mixtures. Another important conclusion is that peaks due to crystalline acetaldehyde grow in proportionally with the amount of acetaldehyde in the mixture and are visible even when acetaldehyde is half of the total amount of water in the ice. These two conclusions indicate that acetaldehyde and water are not extensively mixed at 125 K and that the crystalline structure of acetaldehyde is preserved. It is possible that acetaldehyde is segregating and crystallizing in the pores of water ice that are still present at 125 K.

4   Astrophysical implications

All the species studied in this work are astrochemically-relevant and have been identified in the ISM: water was first detected in 1969 in Sgr B2, Orion, and W49;[41] carbon monoxide was detected in 1970 in Orion;[42] methanol was first detected in 1970 in Sgr B2;[43] formic acid was detected in 1971 in Sgr B2;[44] acetaldehyde was detected in 1973 in Sgr B2;[45,46] acetic acid was first detected in 1997 in Sgr B2;[47] and acetone was detected in 1987 in Sgr B2.[48]

The astrochemical origin of the species studied here is only well understood in some cases. For instance, CO forms efficiently in the gas phase; water has been proven to be formed efficiently at cold temperatures in the solid phase through surface reactions (for a review see van Dishoeck *et al.*[49]); the surface hydrogenation of CO ice leads to the formation of formaldehyde and methanol at low temperatures;[31,50] formic acid is formed through the surface hydrogenation of the HOCO complex.[51] However, little is known about the formation of other species such as acetaldehyde, acetic acid, and acetone. Although non-energetic surface reaction pathways leading to the formation of the more complex species studied in this work have not yet been investigated in the laboratory, several models indicate that those species should be formed in the solid phase and could participate in the formation of even more complex species.[11] For instance, Bisschop *et al.*[52] studied the surface hydrogenation of acetaldehyde and found that it converts to ethanol (C_2H_5OH) or to CH_4, H_2CO, and CH_3OH.

Interstellar ices are mainly composed of water, carbon monoxide, carbon dioxide, methane, ammonia, methanol, and formaldehyde. The spectroscopic identification of other less abundant species, such as formic acid and acetaldehyde studied here, relies on weak bands in the mid-IR that do not overlap with the stronger bands of the main ice components. The OH and CH bending modes of HCOOH at 7.25 μm and the CH_3 deformation of CH_3CHO at 7.41 μm are good examples. Solid-state abundances of formic acid are between 1 and 5% in both low and high mass star forming regions with respect to H_2O.[53-56] The detection of acetaldehyde is less certain, but abundances comparable with those of formic acid have been reported.[55,56]

This work shows that THz features from molecules such as formic acid, acetaldehyde, acetic acid, and acetone are distinct and retain important information on the composition and structure of the ice. Therefore, a spectral investigation of these species in the THz range in space has the potential to contribute to the identification of the composition of interstellar ices. This can be used to better understand: (*i*) the spatial distribution of such species in the ISM since THz spectra can be acquired ideally along any line of sight; (*ii*) the chemistry that governs the formation of such species, since THz features are a unique fingerprint of the chemical composition of the ice; (*iii*) and the physical conditions that the ice experiences during the evolution of star-forming regions, since THz modes are sensitive to the structure of the ice.

5   Conclusions

We present the first THz results (0.3–7.5 THz; 10–250 cm^{-1}) from astrophysically-relevant species that share the same functional groups, such as formic acid and acetic acid, and acetaldehyde and acetone, compared to more abundant interstellar molecules like water, CO, and methanol. Some pure and mixed binary ices are also discussed here. The effect of the composition and the structure of the ice at different temperature is shown. Below we enumerate our conclusions:

1. Our spectra of pure water, CO, and methanol ices at different temperatures agree with previous literature studies. Since our spectral coverage extends to a range previously uninvestigated (0.3–3 THz), we were able to identify several new features in the methanol spectrum as we did for water ice in a previous study.[16]

2. THz spectra of crystalline formic acid and acetic acid deposited at high temperatures present a strong broad feature around 7 THz and are quite flat elsewhere. THz spectra of crystalline methanol, acetaldehyde, and acetone have a forest of sharp, often strong features. In general, the THz spectrum of an individual molecule depends on the functional groups present in the molecule.

3. THz spectra of amorphous species that share the same functional group are similar. Methanol, acetaldehyde, and acetone have a broad feature around 4 THz, while formic acid and acetic acid have a broad feature around 7 THz.

4. *Ab initio* quantum chemical calculations on dimers of formic acid and acetic acid suggest the assignment of the strong band observed at $\sim$7 THz to a rocking mode of two molecules. The similarities between THz spectra of amorphous and crystalline acidic ices can then be explained by the direct deposition of dimers from the gas phase and the strong hydrogen-bonding interaction between dimers.

5. Crystalline ices undergo reversible transformations when the ice is heated or cooled to different temperatures. Amorphous ices undergo irreversible transformations when the ice is heated. THz spectra are sensitive to both reversible and irreversible changes within the structure of the ice.

6. THz spectra retain information of the composition of the ice, as can be seen in our studies of binary mixtures. This is shown when binary mixtures are studied. This information can be used to understand dynamics within the ice such as diffusion, reaction, and desorption.

7. Our work demonstrates the importance of THz spectroscopy in the study of astrochemical ices and contributes to the body of experimental laboratory data needed to interpret astronomical observations from *Herschel Space Observatory*, SOFIA, and ALMA.

Acknowledgements

This work was supported by the NSF CRIF:ID and CSDM programs and the NASA Exobiology and Laboratory Astrophysics programs. The authors thank Coherent, Inc. for the loan of the Optical Parametric Amplifier used in these experiments. S.I. acknowledges support from a Niels Stensen Fellowship and a Marie Curie Fellowship (FP7-PEOPLE-2011-IOF-300957). B.A.M. gratefully acknowlwdges funding by an NSF Graduate Research Fellowship. M.A.A. was supported by the Department of Defense (DoD) Air Force Office of Scientific Research, National Defense Science and Engineering Graduate (NDSEG) Fellowship, 32 CFR 168a.

References

1 J. E. Elsila, D. P. Glavin and J. P. Dworkin, *Meteorit. Planet. Sci.*, 2009, **44**, 1323–1330.

2 E. Congiu, G. Fedoseev, S. Ioppolo, F. Dulieu, H. Chaabouni, S. Baouche, J. L. Lemaire, C. Laffon, P. Parent, T. Lamberts, H. M. Cuppen and H. Linnartz, *Astrophys. J.*, 2012, **750**, L12.

3 E. Congiu, H. Chaabouni, C. Laffon, P. Parent, S. Baouche and F. Dulieu, *J. Chem. Phys.*, 2012, **137**, 054713.

4 G. Fedoseev, S. Ioppolo, T. Lamberts, J. F. Zhen, H. M. Cuppen and H. Linnartz, *J. Chem. Phys.*, 2012, **137**, 054714.

5 M. Minissale, G. Fedoseev, E. Congiu, S. Ioppolo, F. Dulieu and H. Linnartz, *Phys. Chem. Chem. Phys.*, 2014, **16**, 8257.

6 S. Ioppolo, G. Fedoseev, M. Minissale, E. Congiu, F. Dulieu and H. Linnartz, *Phys. Chem. Chem. Phys.*, 2014, **16**, 8270.

7 R. L. Pulliam, B. A. McGuire and A. J. Remijan, *Astrophys. J.*, 2012, **751**, 1.

8 A. Nummelin, P. Bergman, Å. Hjalmarson, P. Friberg, W. M. Irvine, T. J. Millar, M. Ohishi and S. Saito, *Astrophys. J. Suppl.*, 2000, **128**, 213–243.

9 A. Horn, H. Møllendal, O. Sekiguchi, E. Uggerud, H. Roberts, E. Herbst, A. A. Viggiano and T. D. Fridgen, *Astrophys. J.*, 2004, **611**, 605–614.

10 R. T. Garrod, S. L. W. Weaver and E. Herbst, *Astrophys. J.*, 2008, **682**, 283–302.

11 R. T. Garrod, *Astrophys. J.*, 2013, **765**, 60.

12 K. I. Öberg, S. Bottinelli and E. F. van Dishoeck, *Astron. Astrophys.*, 2009, **494**, L13–L16.

13 J. C. Laas, R. T. Garrod, E. Herbst and S. L. Widicus Weaver, *Astrophys. J.*, 2011, **728**, 71.

14 J. C. Laas, B. M. Hays and S. L. Widicus Weaver, *J. Phys. Chem. A*, 2013, **117**, 9548–9554.

15 K. I. Öberg, A. C. A. Boogert, K. M. Pontoppidan, S. van den Broek, E. F. van Dishoeck, S. Bottinelli, G. A. Blake and N. J. Evans, II, *Astrophys. J.*, 2011, **740**, 109.

16 M. A. Allodi, S. Ioppolo, M. J. Kelley, B. A. McGuire and G. A. Blake, *Phys. Chem. Chem. Phys.*, 2014, **16**, 3442–3455.

17 J. E. Bertie, *Appl. Spectrosc.*, 1968, **22**, 634–640.

18 M. H. Moore and R. L. Hudson, *Astron. Astrophys. Suppl. Ser.*, 1994, **103**, 45–56.

19 M. Clerici, M. Peccianti, B. E. Schmidt, L. Caspani, M. Shalaby, M. Giguère, A. Lotti, A. Couairon, F. Légaré, T. Ozaki, D. Faccio and R. Morandotti, *Phys. Rev. Lett.*, 2013, **110**, 253901.

20 P. R. Griffiths and J. A. de Haseth, *Fourier Transform Infrared Spectroscopy*, 2nd ed., Wiley, 2007.

21 R. K. H. Galvao, S. Hadjiloucas, A. Zafiropoulos, G. C. Walker, J. W. Bowen and R. Dudley, *Opt. Lett.*, 2007, **32**, 3008–3010.

22 R. B. Blackman and J. W. Tukey, *The Measurement of Power Spectra, From the Point of View of Communications Engineering*, Dover, 1959.

23 J. O. Smith, *Mathematics of the Discrete Fourier Transform (DFT), with Audio Applications*, 2nd ed., W3K Publishing, 2007.

24 M. J. Frisch, G. W. Trucks, H. B. Schlegel, G. E. Scuseria, M. A. Robb, J. R. Cheeseman, G. Scalmani, V. Barone, B. Mennucci, G. A. Petersson, H. Nakatsuji, M. Caricato, X. Li, H. P. Hratchian, A. F. Izmaylov, J. Bloino, G. Zheng, J. L. Sonnenberg, M. Hada, M. Ehara, K. Toyota, R. Fukuda, J. Hasegawa, M. Ishida, T. Nakajima, Y. Honda, O. Kitao, H. Nakai, T. Vreven, J. A. Montgomery, Jr., J. E. Peralta, F. Ogliaro, M. Bearpark, J. J. Heyd, E. Brothers, K. N. Kudin, V. N. Staroverov, R. Kobayashi, J. Normand, K. Raghavachari, A. Rendell, J. C. Burant, S. S. Iyengar, J. Tomasi, M. Cossi, N. Rega, J. M. Millam, M. Klene, J. E. Knox, J. B. Cross, V. Bakken, C. Adamo, J. Jaramillo, R. Gomperts, R. E. Stratmann, O. Yazyev, A. J. Austin, R. Cammi, C. Pomelli, J. W. Ochterski, R. L. Martin, K. Morokuma, V. G. Zakrzewski, G. A. Voth, P. Salvador, J. J. Dannenberg, S. Dapprich, A. D. Daniels, FarkasJ. B. Foresman, J. V. Ortiz, J. Cioslowski and D. J. Fox, *Gaussian 09 Revision D.01*, Gaussian Inc., Wallingford CT, 2009.

25 C. J. Cramer, *Essentials of Computational Chemistry*, 2nd ed., Wiley, 2004.

26 J. A. Pople, A. P. Scott, M. W. Wong and L. Radom, *Isr. J. Chem.*, 1993, **33**, 345–350.

27 A. H. Hardin and K. B. Harvey, *Spectrochim. Acta, Part A*, 1973, **29**, 1139–1151.

28 G. A. Baratta, G. Leto, F. Spinella, G. Strazzulla and G. Foti, *Astron. Astrophys.*, 1991, **252**, 421–424.

29 P. Jenniskens, S. F. Banham, D. F. Blake and M. R. S. McCoustra, *J. Chem. Phys.*, 1997, **107**, 1232–1241.

30 W. Hagen, A. G. G. M. Tielens and J. M. Greenberg, *Astron. Astrophys. Suppl. Ser.*, 1983, **51**, 389–416.

31 G. W. Fuchs, H. M. Cuppen, S. Ioppolo, S. E. Bisschop, S. Andersson, E. F. van Dishoeck and H. Linnartz, *Astron. Astrophys.*, 2009, **505**, 629.

32 S. E. Bisschop, G. W. Fuchs, A. C. A. Boogert, E. F. van Dishoeck and H. Linnartz, *Astron. Astrophys.*, 2007, **470**, 749–759.

33 H. Hollenstein and H. H. Günthard, *Spectrochim. Acta, Part A*, 1971, **27**, 2027–2060.

34 M. Haurie and A. Novak, *Spectrochim. Acta*, 1965, **21**, 1217–1223.

35 J. E. Bertie and S. M. Jacobs, *J. Chem. Phys.*, 1977, **67**, 2445–2448.

36 R. L. Hudson and M. H. Moore, *Astrophys. J.*, 1993, **404**, L29–L32.

37 D. M. Hudgins, S. A. Sandford, L. J. Allamandola and A. G. G. M. Tielens, *Astrophys. J. Suppl.*, 1993, **86**, 713–870.

38 S. Krishnamurthy, R. Bansil and J. Wiafe-Akenten, *J. Chem. Phys.*, 1983, **79**, 5863–5870.

39 G. Profeta and S. Scandolo, *Phys. Rev. B: Condens. Matter Mater. Phys.*, 2011, **84**, 024103.

40 T. Shimanouchi, *Nat. Stand. Ref. Data Scr., Nat. Bur. Stand. (US)*, 1972, **39**, 1–164.

41 A. C. Cheung, D. M. Rank, C. H. Townes, D. D. Thornton and W. J. Welch, *Nature*, 1969, **221**, 626–628.

42 R. W. Wilson, K. B. Jefferts and A. A. Penzias, *Astrophys. J.*, 1970, **161**, L43.

43 J. A. Ball, C. A. Gottlieb, A. E. Lilley and H. E. Radford, *Astrophys. J.*, 1970, **162**, L203.

44 B. Zuckerman, J. A. Ball and C. A. Gottlieb, *Astrophys. J.*, 1971, **163**, L41.

45 N. Fourikis, M. W. Sinclair, B. J. Robinson, P. D. Godfrey and R. D. Brown, *Australian Journal of Physics*, 1974, **27**, 425.

46 W. Gilmore, M. Morris, P. Palmer, D. R. Johnson, F. J. Lovas, B. E. Turner and B. Zuckerman, *Astrophys. J.*, 1976, **204**, 43–46.

47 D. M. Mehringer, L. E. Snyder, Y. Miao and F. J. Lovas, *Astrophys. J.*, 1997, **480**, L71.

48 F. Combes, M. Gerin, A. Wootten, G. Wlodarczak, F. Clausset and P. J. Encrenaz, *Astron. Astrophys.*, 1987, **180**, L13–L16.

49 E. F. van Dishoeck, E. Herbst and D. A. Neufeld, *Chem. Rev.*, 2013, **113**, 9043–9085.

50 N. Watanabe and A. Kouchi, *Astrophys. J.*, 2002, **571**, L173–L176.

51 S. Ioppolo, H. M. Cuppen, E. F. van Dishoeck and H. Linnartz, *Mon. Not. R. Astron. Soc.*, 2011, **410**, 1089–1095.

52 S. E. Bisschop, G. W. Fuchs, E. F. van Dishoeck and H. Linnartz, *Astron. Astrophys.*, 2007, **474**, 1061–1071.

53 W. A. Schutte, J. M. Greenberg, E. F. van Dishoeck, A. G. G. M. Tielens, A. C. A. Boogert and D. C. B. Whittet, *Astrophysics and Space Science*, 1998, **255**, 61–66.

54 W. A. Schutte, A. C. A. Boogert, A. G. G. M. Tielens, D. C. B. Whittet, P. A. Gerakines, J. E. Chiar, P. Ehrenfreund, J. M. Greenberg, E. F. van Dishoeck and T. de Graauw, *Astron. Astrophys.*, 1999, **343**, 966–976.
55 E. L. Gibb, D. C. B. Whittet, A. C. A. Boogert and A. G. G. M. Tielens, *Astrophys. J. Suppl.*, 2004, **151**, 35–73.
56 J. V. Keane, A. G. G. M. Tielens, A. C. A. Boogert, W. A. Schutte and D. C. B. Whittet, *Astron. Astrophys.*, 2001, **376**, 254–270.

Faraday Discussions

PAPER

Infrared and reflectron time-of-flight mass spectroscopic study on the synthesis of glycolaldehyde in methanol (CH₃OH) and methanol–carbon monoxide (CH₃OH–CO) ices exposed to ionization radiation†

Surajit Maity, Ralf I. Kaiser* and Brant M. Jones*

Received 6th December 2013, Accepted 6th February 2014

DOI: 10.1039/c3fd00121k

We present conclusive evidence on the formation of glycolaldehyde (HOCH₂CHO) synthesized within astrophysically relevant ices of methanol (CH₃OH) and methanol–carbon monoxide (CH₃OH–CO) upon exposure to ionizing radiation at 5.5 K. The radiation induced chemical processes of the ices were monitored on line and *in situ via* infrared spectroscopy which was complimented by temperature programmed desorption studies post irradiation, utilizing highly sensitive reflectron time-of-flight mass spectrometry coupled with single photon fragment free photoionization (ReTOF-PI) at 10.49 eV. Specifically, glycolaldehyde was observed *via* the ν_{14} band and further enhanced with the associated frequency shifts of the carbonyl stretching mode observed in irradiated isotopologue ice mixtures. Furthermore, experiments conducted with mixed isotopic ices of methanol–carbon monoxide (^{13}CH₃OH–CO, CH₃^{18}OH–CO, CD₃OD–^{13}CO and CH₃OH–C^{18}O) provide solid evidence of at least three competing reaction pathways involved in the formation of glycolaldehyde *via* non-equilibrium chemistry, which were identified as follows: (i) radical–radical recombination of HCO and CH₂OH formed *via* decomposition of methanol – the "two methanol pathway"; (ii) *via* the reaction of one methanol unit (CH₂OH from the decomposition of CH₃OH) with one carbon monoxide unit (HCO from the hydrogenation of CO) – the "one methanol, one carbon monoxide pathway"; and (iii) formation *via* hydrogenation of carbon monoxide resulting in radicals of HCO and CH₂OH – the "two carbon monoxide pathway". In addition, temperature programmed desorption studies revealed an increase in the amount of glycolaldehyde formed, suggesting further thermal chemistry of trapped radicals within the ice matrix. Sublimation of glycolaldehyde during the warm up was also monitored *via* ReTOF-PI and validated *via* the mutual agreement of the associated isotopic frequency shifts within the infrared band positions and the identical sublimation profiles obtained from the ReTOF spectra and infrared spectroscopy of the

Department of Chemistry, W. M. Keck Research Laboratory in Astrochemistry, University of Hawaii at Manoa, Honolulu, Hawaii, HI, 96822, USA. E-mail: ralfk@hawaii.edu; brantmj@hawaii.edu

† Electronic supplementary information (ESI) available. See DOI: 10.1039/c3fd00121k

corresponding isotopes. In addition, an isomer of glycolaldehyde (ethene-1,2-diol) was tentatively assigned. Confirmation of the identified pathways based on infrared spectroscopy was also obtained from the observed ion signals corresponding to isotopomers of glycolaldehyde. These coupled techniques provide clear, concise evidence of the formation of a complex and astrobiologically important organic, glycolaldehyde, relevant to the icy mantles observed in the interstellar medium.

1. Introduction

During the last decade, the glycolaldehyde molecule ($HCOCH_2OH$) has received considerable attention in the astronomy, astrochemistry, and astrobiology communities. Glycolaldehyde was first detected in the richest molecular source in the Galaxy, Sgr B2(N),[1] located in the galactic center molecular cloud Sagittarius B2, also known as the Large Molecular Heimat.[2] Chemical differentiation was later observed in the structural isomers of glycolaldehyde,[3] *i.e.* methyl formate ($HCOOCH_3$) and acetic acid (CH_3COOH), whose formation routes are unknown.[2] Consequently, an understanding of the chemistry behind the formation of glycolaldehyde and its isomers may help to explain the observed differentiation, since structural isomers can be exploited as a chemical clock to unravel the underlying chemistry in extraterrestrial environments such as star forming regions. A subsequent extensive line search of glycolaldehyde allowed the determination of the fractional abundance of $f(H_2) \approx 5.9 \times 10^{-11}$ in Sgr B2 (N).[4] From here, a tentative identification of glycolaldehyde outside of the galactic center was made in 2005 towards the hot core G31.40+0.31,[5] and later confirmed by the same authors in 2009,[6] with the glycolaldehyde emission originating from the hottest ($\geq$300 K) section with a rather large (estimated) fractional abundance of $f(H_2) \approx 5 \times 10^{-8}$. This yielded the only confirmed detection outside our galaxy. In addition to glycolaldehyde, other complex organic molecules (COM) have been detected in the galactic center, such as methyl formate ($HCOOCH_3$), acetaldehyde (CH_3COH), ketene (H_2CCO), and ethylene oxide (c-C_2H_4O).[7] The most recent detection of glycolaldehyde has been with the Atacama Large Millimeter Array (ALMA) toward the protostellar binary system IRAS 16293-2422.[8] Here, glycolaldehyde was found in the warm (200–300 K) gas close to the individual components of the binary, with an estimated fractional abundance of $f(H_2) \approx 5 \times 10^{-9}$, which the authors note correlates with an origin related to the energetic processing *via* UV photolysis of an icy grain mantle consisting of methanol and carbon dioxide followed by sublimation during the warm up phase.

As glycolaldehyde is the simplest monosaccharide sugar detected in numerous astrophysical environments ranging from hot cores to the galactic center molecular cloud, the detection of glycolaldehyde immediately prompted interest within the astrobiological community, as the presence of glycolaldehyde directly pertains to the 'origin of life' question. Here, glycolaldehyde is one of the key precursors in the formation of ribonucleic acid (RNA).[9-13] Since the 'RNA World' origin of life hypothesis[14-17] is currently one of the dominating theories within the community,[18,19] glycolaldehyde has been heavily implicated in the literature on the premise that it correlates well with the abiotic formation of ribose, which is a central back bone of RNA. Yet, one of the main difficulties in considering an 'RNA World' relationship to the origin of life is the complication of forming RNA prebiotically.[20,21] There has been experimental evidence however, showing that abiotic formation of nucleotides from basic starting materials is possible, and thus not an entirely impractical chemical route. In particular, glycolaldehyde can react *via* an aldol condensation to ultimately produce ribose.[22] Additionally the polymerization of formaldehyde (the formose reaction[23]) has been suggested as plausible prebiotic route to glycolaldehyde and ribose (Shapiro, 2000).[24] However, as pointed out by Shapiro (2000),[24] no prebiotic route exists that preferentially leads to the synthesis of ribose specifically over a conglomerate of other sugars, despite many efforts. Some success has been demonstrated in generating ribose 2,4-diphosphate in a possibly prebiotic reaction of monophosphate glyco-laldehyde and formaldehyde.[25] Also, it should be noted that ribose has been shown to be produced in elevated yields following a formose type mechanism in mixtures containing aldehyde, glycolaldehyde, and minerals such as pyrite, borate and silicates.[26] In addition, ribonucleotides were abiotically synthesized in a multistep reaction utilizing only cyanamide (CN_2H_2), cyanoacetylene (C_3HN), glyceraldehyde ($C_3H_6O_3$) and glycolaldehyde ($HOCH_2CHO$).[27] Consequently, the abiotic synthesis of nucleotides is feasible. Furthermore, as pointed out recently, an upper limit of 10^8 kg year^{-1} of glycolaldehyde may have been deposited on an early Earth *via* comets and interplanetary dust, in addition to the 10^8 kg year^{-1} that may have been formed *in situ* on an early Earth.[28] Additionally, the detection of glycolaldehyde around the newly formed star IRAS 16293B implies that this molecule may be involved in planet forming processes.[8]

Due to the importance of glycolaldehyde, over the last few decades, studies exploring the radiation induced synthesis of glycolaldehyde pertaining to ices of methanol have been numerous. Of the first papers, most were focused on spectroscopy of radicals; the photolysis ($E_{hv} = 145$ nm, 8.4 eV) of methanol isolated in an argon matrix produced the hydroxymethyl (CH_2OH) radical.[29] The production of this CH_2OH radical in addition to the methyl (CH_3) and the methoxy radical (CH_3O) *via* X-ray radiolysis of crystalline methanol was later confirmed using electron pair resonance spectroscopy at 4.2 K.[30] Products of γ-irradiated methanol ice were found to be molecular hydrogen (H_2), ethylene glycol ($HOCH_2CH_2OH$), methane (CH_4), carbon monoxide (CO), dimethyl ether (CH_3OCH_3), and ethanol (CH_3CH_2OH).[31] From here, pure methanol and binary mixtures of methanol with water and carbon monoxide were exposed to distinct sources of ionizing radiation (summarized in Table S1 of the ESI†). All of the previous experimental results predominantly agree that simple molecules (CO, CO_2, CH_4, H_2CO), with some degree of overlapping with the detection of ethanol, ethylene glycol, water, and the formyl radical (HCO), are formed *in situ via* energetic processing. Crucial discrepancies do exist however, such as the assignment of acetone, which was first suggested as a result of 3 keV He$^+$ bombardment of pure methanol,[32] and was subsequently tentatively assigned in later studies.[33–36] However, the assignment of acetone was refuted in a study with 0.8 MeV H$^+$ bombardment of a methanol-water mixture;[37] from this study, the assignment of the observed vibrational bands that were ascribed to acetone were subsequently attributed to the $HCOO^-$ ion. Far IR spectroscopy was used to examine the phase change of crystalline to amorphous methanol induced by 0.7 MeV H$^+$ irradiation. Although no products were specifically identified, the authors noted that the irradiated methanol would not re-crystallize with thermal annealing, which was attributed to additional products synthesized within the methanol ice matrix.[38] Higher order alcohols were attributed to an unidentified feature at 1088 cm^{-1} after broad band UV photolysis of pure methanol ice *via* a hydrogen microwave discharge lamp at 10 K in addition to the standard group of daughter molecules (CO, CO_2, CH_4, H_2CO). This assignment of a higher mass alcohol was further verified by the observation that the IR band started to disappear (indicating sublimation) at 120 K and was completely gone at 230 K, suggesting a molecule with a similar volatility to methanol.[39] After the first detection of glycolaldehyde in the galactic center,[1] a resurgence in methanol irradiation experiments occurred in the hopes of explaining the origin of this sugar, along with its isomers, methyl formate and acetic acid. One of the first irradiation experiments to identify glycolaldehyde as a product of methanol ice exposed to ionizing radiation utilized 5 keV electrons, thereby simulating the track of secondary electrons generated within the trajectory of galactic cosmic rays[40] and a binary mixture of methanol and carbon monoxide.[41] Here, glycolaldehyde was identified in irradiated methanol ices at 10 K and also following warm up, whereas in the binary mixture (CH_3OH–CO), glycolaldehyde in conjunction with methyl formate was immediately identified *in situ* at 10 K with a total applied dose of 0.7 eV per 16 amu. In addition, glycolaldehyde and methyl formate were distinguished by utilizing temperature programmed desorption coupled with gas phase detection *via* a quadrupole mass spectrometer based on the unique fragmentation pattern (*e.g.* the methyl formate cation – $HCOO^+$) at distinct sublimation temperatures. Following the studies of Bennett *et al.*,[40,41] similar work was conducted utilizing broad band UV photons produced *via* a hydrogen microwave discharge lamp, in which glycolaldehyde was only inferred as a potential product after photolysis of CH_3OH and CH_3OH–CO ices. Recently, glycolaldehyde was only briefly mentioned as an endogenous product upon exposure to 200 keV protons at relatively high doses of 34 eV[42] and 18 eV[43] per 16 amu. In the most recent experiment, Chen *et al.* reported the formation of glycolaldehyde in irradiated methanol ices utilizing soft X-rays with peak energies of 300 eV and 550 eV over a broad band spectrum (250–1200 eV)[44] following warm up, in agreement with the previous observation of Bennett *et al.*;[41] methyl formate, acetic acid, formic acid and ethylene glycol were also reported. No mention of glycolaldehyde was made in the experiments involving various heavy cosmic ray analogs,[45] or in the irradiation of methanol at very low doses with soft X-rays.[46] To summarize, numerous experiments have been conducted over recent decades regarding the chemical modification of methanol ices upon exposure to ionizing radiation relevant to astrophysical conditions. Within these studies, a general consensus on the formation of small molecules, which can be detected easily *via* infrared spectroscopy (FTIR) (CO, CO_2, H_2CO, CH_4) and for the most part ethanol, ethylene glycol, and methyl formate, has been ascertained. However, the main difference thus far is the ambiguity in the assignment of glycolaldehyde in the majority of the processed ices.

In this study, we present compelling evidence from infrared spectral data correlated for the first time with temperature programmed desorption studies exploiting single photon ionization (10.49 eV) reflectron time-of-flight mass spectrometry (ReTOF) gas phase detection of glycolaldehyde synthesized in irradiated ices of CH_3OH and CH_3OH–CO along with selected isotopologues, exposed to doses of up to 6.5 eV per molecule relevant to the life time of an interstellar icy grain within a cold molecular cloud prior to the warm up (star formation) phase; at this point the grain begins to warm, causing radicals to diffuse and the endogenous radiolytically synthesized molecules to sublimate into the gas phase, allowing for direct observation *via* radio astronomical observations.

2. Experimental

The experiments were carried out in a novel, contamination-free ultra-high vacuum (UHV) chamber at the W. M. Keck Research Laboratory in Astrochemistry (Fig. 1). The main chamber was evacuated down to a base pressure typically of a few 10^{-11} Torr using oil-free magnetically suspended turbomolecular pumps backed with dry scroll pumps. A cold finger assembled from oxygen free high conductivity copper (OFHC) was coupled to a UHV compatible closed-cycle helium refrigerator (Sumitomo Heavy Industries, RDK-415E). A polished silver mirror was then mounted to the cold finger insulated with 0.1 mm thick indium foil to ensure thermal conductivity, and subsequently cooled to a final temperature of 5.5 ± 0.1 K; the entire ensemble was freely rotatable within the horizontal center plane, and translatable in the vertical (z-axis) *via* a UHV compatible bellow (McAllister, BLT106) and differential pumped rotational feed through (Thermoionics Vacuum Products, RNN-600/FA/MCO). From here, the corresponding gases were then deposited through a glass capillary with a background (uncorrected for ion gauge sensitivity) pressure reading in the main chamber of 5×10^{-8} Torr for approximately 3 minutes, yielding an ice sample with final thicknesses of 510 ± 10 nm for pristine methanol ices, and 495 ± 10 nm for mixed ices of

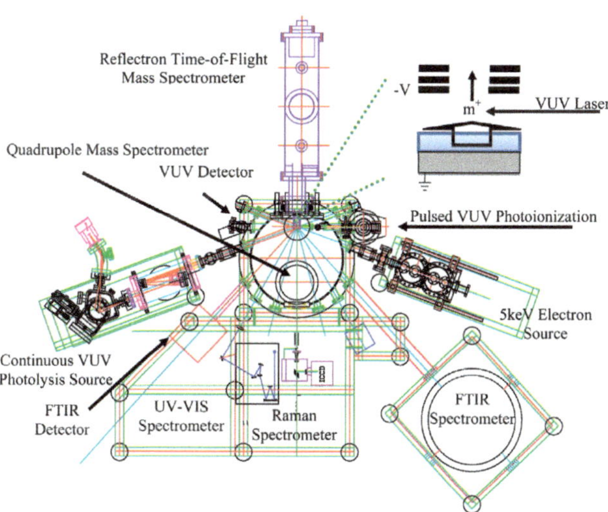

Fig. 1 Schematic top view of the main chamber including the analytical instruments, radiation sources, and the cryogenic target (point of converging lines). The alignment of the cryogenic target, radiation sources, infrared, Raman, and UV−vis spectrometer allows simultaneous online and *in situ* measurements of the modification of the targets upon the exposure to irradiation. After the irradiation, the cold head can be rotated 180° to face the ReTOF mass spectrometer; the target can then be warmed up allowing the newly formed products to sublimate, whereupon they are photoionized and mass analyzed. The inset (top right) shows the geometry of the ReTOF ion source lenses with respect to the target and ionization laser.

methanol–carbon monoxide. The thickness of the sample was determined *in situ* using laser interferometry. Here, the cooled silver target was rotated to face a HeNe laser (CVI Melles-Griot; 25-LHP-230, 632 nm), which struck the target at an incident angle of 4° relative to the sample normal, and was reflected towards a photodiode (CVI Melles-Griot Si Photodiode; 13DAS005 with a 632.8 nm narrow band pass filter, CVI Xll-632.8-12.6M). The induced current in the photodiode was monitored as a function of time with a picoammeter (Keithley Model 6485), while the gas was introduced into the chamber at a constant rate through a precision leak valve, whereupon it finally condensed onto the low temperature silver target. During the deposition, the HeNe laser was reflected off the surface of the silver target and the freshly deposited ice sample, causing an interference pattern. The relation between the period of the interference curve between two maxima or minima relates to a change in thickness, Δd, as described by Heavens,[47] where λ_0 is the laser wavelength (632.8 nm), θ_i is the incident angle (4°), and n_f is the refractive index of the ice. Here, *a priori* knowledge of the index of refraction (n_f) is required for accurate determination of the deposited ice thickness. Taking into consideration the known index of refraction for the substrate,[48–50] the index of refraction can also be derived by measuring the intensity ratio between the maximum and minimum of the measured interference curve.[49,50] From this technique, we derived an index of refraction $n_f = 1.34 \pm 0.02$ for pure methanol, in agreement with published data of 1.33,[51] and $n_f = 1.35 \pm 0.02$ for the binary CH_3OH–CO mixture. The column densities were calculated utilizing a modified Lambert–Beer relationship[52–54] with the absorption coefficients of 1.3×10^{-17} cm molecule^{-1} and 1.1×10^{-17} cm molecule^{-1} for the 1028 cm^{-1} (ν_8; CH_3OH) and 2090 cm^{-1} (ν_1; ^{13}CO) bands, respectively.[41,55,56] Accordingly, the mixed methanol–carbon monoxide ice was found to be in the ratio of $(4.0 \pm 0.2) : (5.0 \pm 0.2)$. In addition, isotopically-labeled CD_3OD, $^{13}CH_3OH$, and $CH_3^{18}OH$ ices for pure methanol irradiation experiments and CD_3OD–CO, $^{13}CH_3OH$–CO, $CH_3^{18}OH$–CO, CD_3OD–^{13}CO, $CH_3^{18}OH$–$C^{18}O$, and CH_3OH–$C^{18}O$ mixed ices for methanol–carbon monoxide irradiation experiments were also used to confirm infrared assignments *via* isotope shifts, and in the reflectron time-of-flight data *via* shifts their in mass-to-charge ratios.

Each of the amorphous ices was irradiated with 5 keV electrons isothermally at 5.5 ± 0.1 K for one hour at 30 nA over a square area of 1.0 ± 0.1 cm^2 and at an angle of 70° relative to the surface normal. The emission current was measured prior to irradiation on line and *in situ* utilizing a Faraday cup (Kimball Physics, FC-71) mounted inside a differentially pumped chamber on an ultra-high vacuum compatible translation state. The total dose deposited into the ice sample was determined from Monte Carlo simulations (CASINO)[57] taking into consideration the scattering coefficient and the energy deposited from the back scattered electrons. The total energy deposited into the amorphous ice was 6.5 ± 0.8 eV per CH_3OH molecule, and 5.2 ± 0.8 eV per molecule on average for the irradiation experiments of the binary CH_3OH–CO (4 : 5) ice mixture. It should be noted here that in determining the applied dose of the isotopic analogs, that both the index of refraction and density were assumed to be that of their respective normal counterparts, as most of these data are not empirically available. Furthermore, the density of the CH_3OH–CO ice mixture was calculated to be 1.026 g cm^{-3} based on the column density weighted fraction of their respective pure densities (1.020 g cm^{-3} CH_3OH, 1.029 g cm^{-3} CO[41]) in the limit of volume additivity.[58]

For the on line and *in situ* identification of new molecular band carriers of the ices during irradiation, a Fourier transform infrared spectrometer (Nicolet 6700) monitored the samples throughout the duration of the experiment, with an IR spectrum collected every two minutes in the range of 6000–400 cm^{-1} at a resolution of 4 cm^{-1}. Each FTIR spectrum was recorded in the absorption–reflection–absorption mode (reflection angle = 45°) for two minutes, resulting in a set of 30 infrared spectra during the radiation exposure for each system. We recognize that integrated band areas can be altered by optical interference effects inherit in absorption–reflection–absorption FTIR spectroscopy, as demonstrated by Teolis *et al.*;[59] however, this issue is circumvented by integration of weak bands, whose absorbance remains linear with respect to the amount of ice deposited.[60] After the irradiation, the sample was kept at 5.5 K for one hour, then temperature programmed desorption (TPD) studies were conducted by heating the irradiated ices at a rate of 0.5 K min^{-1} to 300 K. Throughout the thermal sublimation process, the ice samples were monitored *via* infrared spectroscopy and single photon ionization reflectron time-of-flight mass spectroscopy[61] separately, *i.e.* each experiment was conducted twice. For the gas phase detection, the products were ionized upon sublimation *via* single photon ionization exploiting pulsed (30 Hz) coherent vacuum ultraviolet (VUV) light at 118.2 nm (10.49 eV), generated *via* non-linear four wave mixing. The ions were then extracted into a reflectron time-of-flight mass spectrometer, whereupon the ions were mass resolved according to their arrival times. Please see the ESI† for more detailed description regarding CASINO calculations, coherent VUV light generation, and reflectron time-of-flight mass spectroscopy.

3. Results

3.1 Infrared spectroscopy

3.1.1 Qualitative analysis. The infrared spectra of pure methanol ice and mixed ices of methanol (CH_3OH)–carbon monoxide (CO) recorded before and after irradiation along with the assignments are shown in Fig. 2A and 2B. During the irradiation, multiple new absorption features emerged (summarized in the ESI, Tables S2 and S3†) alongside the absorptions of the pristine ices. In the irradiated methanol (Fig. 2A, ESI Table S2†) and methanol–carbon monoxide ices (Fig. 2B, ESI Table S3†), the hydroxymethyl radical (CH_2OH) was detected *via* the ν_4 absorption band at 1192 cm^{-1} and 1193 cm^{-1}, respectively, in good agreement with previously reported values of 1197 cm^{-1} in UV photolyzed methanol ices by Gerakines *et al.*,[39] 1197 cm^{-1} and 1192 cm^{-1} in methanol[40] and methanol–carbon monoxide ices[41] exposed to electron irradiation, and 1193 cm^{-1} in methanol ices irradiated with soft X-rays by Chen *et al.*[44] Formaldehyde (H_2CO) was observed *via* the ν_2, ν_3, and ν_4 absorption bands at 1246 cm^{-1}, 1499 cm^{-1} and 1726 cm^{-1} in methanol ice, and at 1249 cm^{-1}, 1497 cm^{-1} and 1726 cm^{-1} in methanol–carbon monoxide ice; again, these absorptions agree with previous studies.[37,39–41] Methyl formate ($HCOOCH_3$) was identified *via* its ν_{14} fundamental at 1714 cm^{-1}, which is in good agreement with the literature value of 1718 cm^{-1} in electron irradiation experiments with methanol[40] and methanol–carbon monoxide ices[41] and UV photolysis of methanol ices by Gerakines *et al.*,[39] and 1720 cm^{-1} by Modica *et al.* with pure methyl formate spectra at 16 K.[42,43] The formation of a formyl radical (HCO) was identified *via* the ν_3 fundamental at 1842 cm^{-1} in both the irradiated

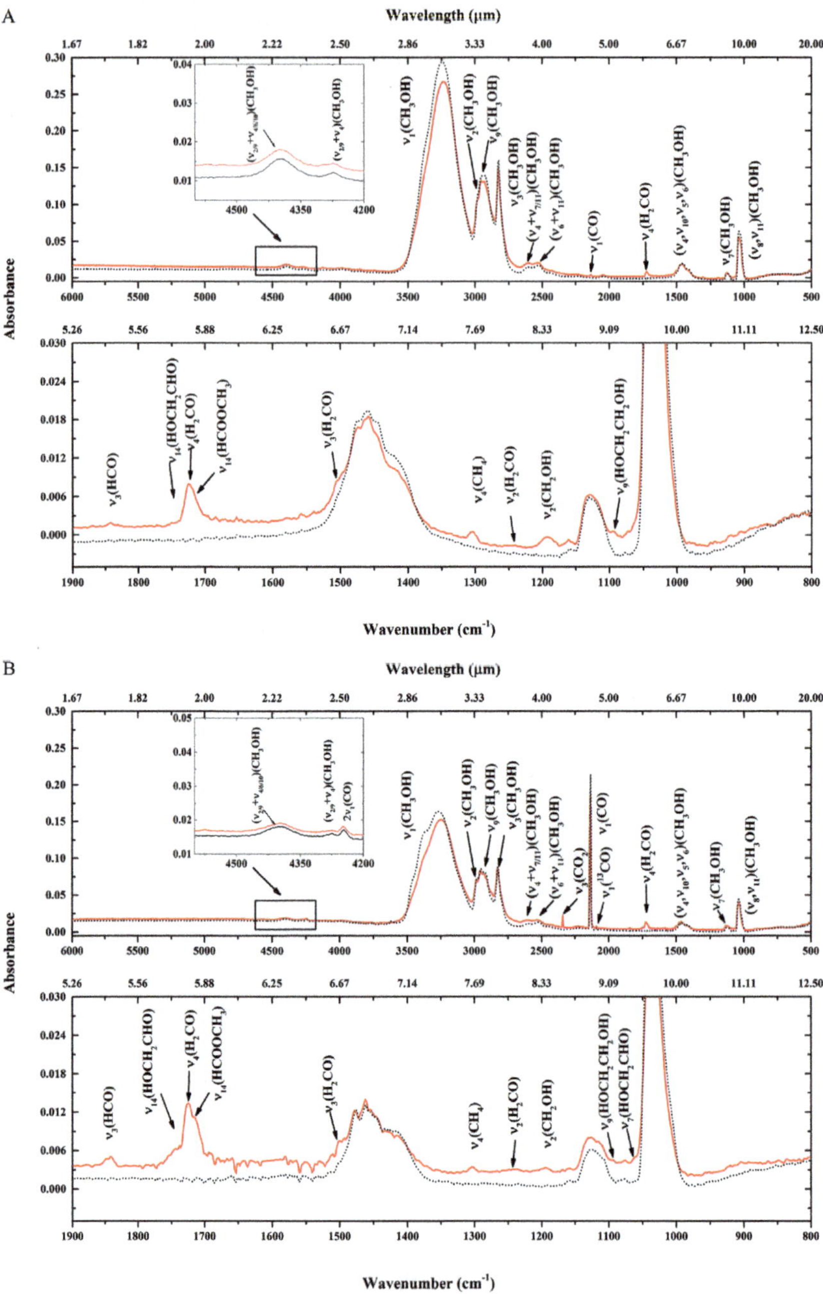

Fig. 2   (A) (Top) Infrared spectra of the methanol ices at 5.5 K before (dotted trace) and after (solid trace) irradiation. (Bottom) The newly formed products as labeled are shown in the 1900–800 cm^{-1} region. (B) (Top) Infrared spectra of the methanol–carbon monoxide ices at 5.5 K before (dotted trace) and after (solid trace) irradiation. (Bottom) The newly formed products as labeled are shown in the 1900–800 cm^{-1} region.

ices, as previously assigned at 1848 cm^{-1} by Hudson *et al.*,[37] and at 1842 cm^{-1} by Bennett *et al.*[40,41] and Chen *et al.*[39,44] Methane (CH_4) was detected *via* the ν_4 fundamental at 1304 cm^{-1} and 1303 cm^{-1} in the methanol and methanol–carbon monoxide ices, respectively. Formation of carbon monoxide (CO) was confirmed *via* the ν_1 fundamental at 2135 cm^{-1} in methanol ice. In addition, carbon dioxide was detected *via* the ν_3 mode at 2339 cm^{-1} in irradiated methanol ice and 2342 cm^{-1} in the irradiated methanol–carbon monoxide ice. Ethylene glycol was identified in both irradiated ices *via* the ν_9 fundamental at 1094 cm^{-1}, based on the assignment of this molecule in previous studies of irradiated methanol ices[37,40] at 1090 cm^{-1} and 1088 cm^{-1}. It should be mentioned that all other relatively strong infrared absorption bands of ethylene glycol coincidentally overlap with the methanol absorptions[40,41] and consequently are masked. Glycolaldehyde ($HOCH_2$-CHO) was identified by the ν_{14} (CO of HCO stretching) fundamental at 1743 cm^{-1} in both methanol and methanol–carbon monoxide ices. The assignment of the observed bands to glycolaldehyde are based on previous studies of irradiated methanol[40] ices at 1747 cm^{-1}, irradiated methanol–carbon monoxide ices[41] at 1757 cm^{-1}, and matrix isolation studies of glycolaldehyde at 13 K at 1747 cm^{-1}.[62,63] A second band at 1062 cm^{-1} in the methanol–carbon monoxide ice can be attributed to the ν_7 fundamental of glycolaldehyde, in agreement with previous studies.[41–43,62,63] The assignment of these absorptions was also confirmed *via* their isotopic shifts in irradiated ices consisting of CD_3OD, $^{13}CH_3OH$ and $CH_3{}^{18}OH$, as compiled in Table S2 in the ESI,† and irradiated binary mixed ices consisting of CD_3OD–CO, $^{13}CH_3OH$–CO, $CH_3{}^{18}OH$–CO, CD_3OD–^{13}CO, $CH_3{}^{18}OH$–$C^{18}O$ and CH_3OH–$C^{18}O$ as compiled in Table S3 in the ESI.†

3.1.2 **Deconvolution of the carbonyl absorption features.** The new absorption features connected to the carbonyl functional group in the 1800–1600 cm^{-1} region are very broad (Fig. 3a and 3b), implying the presence of multiple carriers. Further evidence is obtained from the additional infrared absorption features of the carbonyl band regions in the irradiated mixed isotopic ices, as shown in Figure 3b. These findings suggest the presence of multiple underlying molecular carriers, and possibly different isotopomers (in the mixed isotopic ices) as well. Therefore, the absorption features in 1800–1600 cm^{-1} were deconvoluted with the peak positions, and their associated assignments are shown in Fig. 3 and listed in Tables 1–3. The rationale behind the assignments is discussed below.

In the case of irradiated methanol (CH_3OH) ices, the deconvolution (Fig. 3a) identified four distinct bands centered at 1743 cm^{-1}, 1726 cm^{-1}, 1714 cm^{-1} and 1697 cm^{-1}. The band at 1743 cm^{-1} is assigned to the ν_{14} band of glycolaldehyde ($HOCH_2CHO$), as discussed previously based on earlier experimental work.[40,41,62,63] The absorptions at 1726 cm^{-1} and 1714 cm^{-1} can be linked to the ν_4 fundamental of formaldehyde (H_2CO) and the ν_{14} of methyl formate ($HCOOCH_3$), which are in good agreement with the literature data at 1726 cm^{-1} and 1718 cm^{-1}, respectively.[39–43] Finally, the band observed at 1697 cm^{-1} can be attributed to the Fermi resonance splitting of the ν_{14} fundamental and the $2\nu_6$ overtone band of glycolaldehyde.[41,44,63] To support these assignments, deconvolution of the bands in 1800–1600 cm^{-1} region for the isotopically labeled ices of CD_3OD, $^{13}CH_3OH$, and $CH_3{}^{18}OH$ were also completed. The ν_{14} band of glycolaldehyde ($HOCH_2CHO$) at 1743 cm^{-1} is expected to show an isotopic shift of 26 cm^{-1} in CD_3OD ices,[64] 37 cm^{-1} in $^{13}CH_3OH$ ices,[64] and 30 cm^{-1} in $CH_3{}^{18}OH$ ices.[65] In the present investigation, the corresponding isotopomers of glycolaldehyde, $DOCD_2CDO$ in CD_3OD

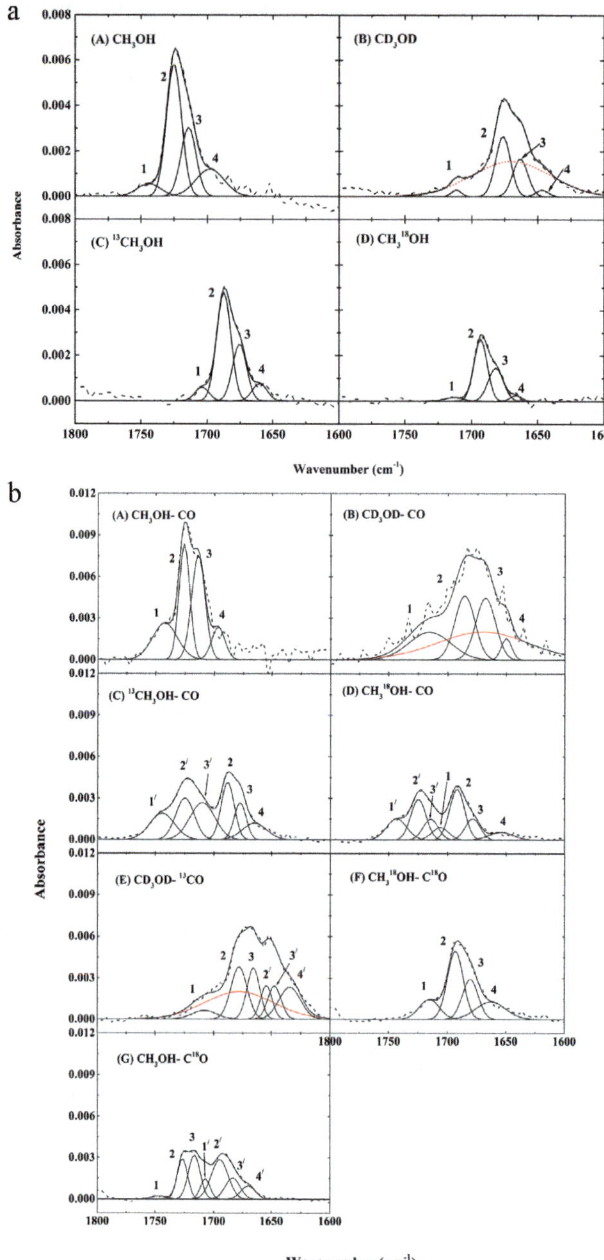

Fig. 3   (a) Deconvoluted infrared absorption features in the region of the carbonyl functional group in (A) CH_3OH, (B) CD_3OD, (C) $^{13}CH_3OH$, and (D) $CH_3{}^{18}OH$ ices. The bands marked as (1) and (4) are assigned as the ν_{14} and $2\nu_6$ bands of glycolaldehyde ($HOCH_2CHO$), respectively. The bands marked as (2) and (3) are assigned to formaldehyde (H_2CO) and methyl formate ($HCOOCH_3$). The dotted line in (B) corresponds to $2\nu_8$ of CD_3OD as in the pristine ice. (b) Deconvoluted infrared absorption features in the region of the carbonyl functional group in (A) $CH_3OH–CO$, (B) $CD_3OD–CO$, (C) $^{13}CH_3OH–CO$, (D) $CH_3{}^{18}OH–CO$, (E) $CD_3OD–^{13}CO$, (F) $CH_3{}^{18}OH–C^{18}O$, and (G) $CH_3OH–C^{18}O$ ices. The bands marked as (1), (2), (3) and (4) are assigned to ν_{14} of glycolaldehyde, ν_4 of formaldehyde and ν_{14} of methyl formate, and $2\nu_6$ bands of glycolaldehyde and their isotopomers (Table 3), respectively.[41] In ices, $HO^{13}CH_2^{13}CHO$ in $^{13}CH_3OH$ ices, and $H^{18}OCH_2CH^{18}O$ in $CH_3^{18}OH$ ices, are assigned based on peak positions of 1711 cm^{-1}, 1703 cm^{-1} and 1713 cm^{-1}, respectively, with corresponding isotopic shifts of 32 cm^{-1}, 40 cm^{-1} and 30 cm^{-1} (Table 1), in agreement with the expected frequency shifts.[64,65] The formaldehyde identification was based on the ν_4 band, observed at 1676, 1687 and 1693 cm^{-1} in the irradiated CD_3OD, $^{13}CH_3OH$ and $CH_3^{18}OH$ ices, respectively. The corresponding isotopic shifts of 50, 39 and 33 cm^{-1} agree well with the respective literature values[66,67] of 45, 38 and 33 cm^{-1}. Methyl formate was confirmed *via* the ν_{14} band at 1676 cm^{-1} and 1682 cm^{-1} in $^{13}CH_3OH$ and $CH_3^{18}OH$ ices, with isotopic shifts of 38 cm^{-1} and 32 cm^{-1} respectively, in close agreement with typical shifts of carbonyl stretching frequencies upon ^{13}C and ^{18}O substitution of about 40 cm^{-1} and 30 cm^{-1}, respectively.[65] In addition, the band assigned to the Fermi resonance (of the ν_{14} fundamental and an overtone from $2\nu_6$) also shows relevant isotopic shifts (Table 1).[65]

The deconvoluted absorption bands of the carbonyl functional group in the 1800–1600 cm^{-1} region reveal four distinct bands (Figure 3a) in the case of isotopically pure (only one isotope of each atom) binary ices consisting of CH_3OH–CO, CD_3OD–CO and $CH_3^{18}OH$–$C^{18}O$, similar to that of the irradiated pure methanol ices. Here, the glycolaldehyde isotopomers $HOCH_2CHO$, $DOCD_2CDO$ and $H^{18}OCH_2CH^{18}O$ were identified at 1743 cm^{-1}, 1714 cm^{-1} and 1715 cm^{-1} (Table 2), respectively, in excellent agreement with the assigned bands of the corresponding isotopomers observed in irradiated methanol ices (Table 1). However, in the mixed isotopic ices, *i.e.*, $^{13}CH_3OH$–CO, $CH_3^{18}OH$–CO, CD_3OD–^{13}CO and CH_3OH–$C^{18}O$, where two different atomic isotopes ($^{12}C/^{13}C$ and $^{16}O/^{18}O$) are present together, the profile of the carbonyl band was rather unique. Here, evidence of two different isotopomers of glycolaldehyde is observed in the irradiated ices of the mixed isotopes. Fig. 4 presents the isotopomers of glycolaldehyde in these mixed isotopic ices, showing that at least two different carbonyl functional groups could be formed and subsequently detected in the relevant infrared spectral region. For example, $H^{13}C^{16}O$ and $H^{12}C^{16}O$ following irradiation of the $^{13}CH_3OH$–CO system, $H^{12}C^{18}O$ and $H^{12}C^{16}O$ in the $CH_3^{18}OH$–CO system, $D^{12}C^{16}O$ and $D^{13}C^{16}O$ in the CD_3OD–^{13}CO system, and $H^{12}C^{16}O$ and $H^{12}C^{18}O$ in the CH_3OH–$C^{18}O$ system are expected and can ultimately lead to the synthesis of four different isotopomers of glycolaldehyde. For instance, in the case of irradiated $CH_3^{18}OH$–CO ice, four different isotopomers of glycolaldehyde, $H^{18}OCH_2CH^{18}O$, $H^{18}OCH_2CHO$, $HOCH_2CH^{18}O$ and $HOCH_2CHO$, can be formulated. Here, the carbonyl group ($HC^{18}O$) in $H^{18}OCH_2CH^{18}O$ and $HOCH_2CH^{18}O$ can be formed *via* the decomposition of $CH_3^{18}OH$. Furthermore, the carbonyl unit (HCO), in $H^{18}OCH_2CHO$ and $HOCH_2CHO$ can be synthesized *via* the hydrogenation of the carbon monoxide (CO). Note that, despite the existence of four different isotopomers, the ν_{14} (CO stretching of the HCO unit) fundamental band will only indicate two glycolaldehyde isotopomers, as in the $CH_3^{18}OH$–$C^{18}O$ (1715 cm^{-1}) and CH_3OH–CO (1743 cm^{-1}) ices. Unfortunately, the isotopic combination is attributed to the alcohol vibrational mode, and is consequently masked by the

(C), (D), (E) and (G), the bands marked as (1′), (2′), (3′) and (4′) are assigned to ν_{14} of glycolaldehyde, ν_4 of formaldehyde, ν_{14} of methyl formate, and $2\nu_6$ bands of glycolaldehyde and their isotopomers, respectively. The dotted lines in (B) and (E) correspond to $2\nu_8$ of CD_3OD as in the pristine ice.

Table 1 Deconvoluted peak positions of the carbonyl absorption bands observed in the processed CH_3OH, CD_3OD, $^{13}CH_3OH$, and $CH_3^{18}OH$ ices

Band	CH_3OH (cm^{-1})	CD_3OD (cm^{-1})	$\Delta\nu$ (cm^{-1})	$^{13}CH_3OH$ (cm^{-1})	$\Delta\nu$ (cm^{-1})	$CH_3^{18}OH$ (cm^{-1})	$\Delta\nu$ (cm^{-1})	Assignment
1	1743	1711	−32	1703	−40	1713	−30	ν_{14} ($HOCH_2CHO$)
2	1726	1676	−50	1687	−39	1693	−33	ν_4 (H_2CO)
3	1714	1664	−50	1676	−38	1682	−32	ν_{14} ($HCOOCH_3$)
4	1697	1647	−50	1659	−38	1666	−31	$2\nu_6$ ($HOCH_2CHO$)

Table 2 Deconvoluted peak positions of the carbonyl absorption bands observed in the processed isotopically pure ices $CH_3OH–CO$, $CD_3OD–CO$, and $CH_3^{18}OH–C^{18}O$

Band	$CH_3OH–CO$ (cm^{-1})	$CD_3OD–CO$ (cm^{-1})	$\Delta\nu$ (cm^{-1})	$CH_3^{18}OH–C^{18}O$ (cm^{-1})	$\Delta\nu$ (cm^{-1})	Assignment
1	1743	1714	−29	1715	−28	ν_{14} ($HOCH_2CHO$)
2	1726	1686	−40	1693	−33	ν_4 (H_2CO)
3	1714	1668	−46	1680	−34	ν_{14} ($HCOOCH_3$)
4	1697	1651	−46	1662	−35	$2\nu_6$ ($HOCH_2CHO$)

broad parent methanol feature. Indeed, deconvolution of the carbonyl absorption in the 1800–1600 cm^{-1} region of irradiated $CH_3^{18}OH–CO$ ices (Fig. 3b) resulted in identified bands at 1708 cm^{-1} (band 1) and 1743 cm^{-1} (band 1′), which are in close agreement with the expected frequencies based on the isotopically labeled pure ices as mentioned above. Similar to the irradiated $CH_3^{18}OH–CO$ ices, deconvolution of the mixed isotopic ices of $^{13}CH_3OH–CO$, $CD_3OD–^{13}CO$ and $CH_3OH–C^{18}O$ reveal two distinct carbonyl bands as shown in Fig. 3b, with the assignments also listed in Table 3. Here, in the isotopically labeled $^{13}CH_3OH–CO$ irradiated ices, the ν_{14} fundamental of glycolaldehyde isotopomers with $H^{13}CO$ and HCO functional groups are expected at 1703 cm^{-1} and 1743 cm^{-1}, based on the observed band positions in the irradiated $^{13}CH_3OH$ ices and $CH_3OH–CO$ ices. As expected, deconvolution elucidates a band at 1743 cm^{-1} (band 1′) corresponding to the carbon monoxide hydrogenation products. Unfortunately however, the expected band at 1703 cm^{-1} could not be identified due to a strong transition of the ν_{14} fundamental of methyl formate ($HCOO^{13}CH_3$) at 1710 cm^{-1} (band 3′). In $CD_3OD–^{13}CO$ ices, the deconvolution of the relevant 1800–1600 cm^{-1} region required the inclusion of the $2\nu_8$ overtone band of CD_3OD (1666 cm^{-1}, ESI Table S3†). Consequently, any prediction may result in an ambiguous interpretation. However, the fully deuterated isotopomers of glycolaldehyde with the DCO carbonyl functional group (Fig. 4) can be attributed to the band observed at 1709 cm^{-1} (band 1), in agreement with irradiated $CD_3OD–CO$ ices (1714 cm^{-1}). Finally, in $CH_3OH–C^{18}O$ ices, the ν_{14} fundamental of the glycolaldehyde isotopomers with HCO and $HC^{18}O$ carbonyl functional groups are expected at 1715 cm^{-1} and 1743 cm^{-1}, based on the observed band positions in the irradiated $CH_3^{18}OH–C^{18}O$ and $CH_3OH–CO$ ices. Certainly, deconvolution reveals two bands centered at 1707 cm^{-1} (band 1′) and 1747 cm^{-1} (band 1), which are in close agreement with our expectations based on the irradiated isotopically pure ices as mentioned above.

Table 3 Deconvoluted peak positions of the carbonyl absorption bands observed in the processed isotopically mixed ices: $^{13}CH_3OH-CO$, $CH_3^{18}OH-CO$, $CD_3OD-^{13}CO$, and $CH_3^{18}OH-CO$

Bands	$^{13}CH_3OH-CO$ (cm^{-1})	$\Delta\nu$ (cm^{-1})	$CH_3^{18}OH-CO$ (cm^{-1})	$\Delta\nu$ (cm^{-1})	$CD_3OD-^{13}CO$ (cm^{-1})	$\Delta\nu$ (cm^{-1})	$CH_3OH-C^{18}O$ (cm^{-1})	$\Delta\nu$ (cm^{-1})	Assignment
1	—	—	1708	−35	1709	−34	1747	3	ν_{14} (HOCH$_2$CHO)
1′	1743	0	1743	0	—	—	1707	−36	
2	1692	−34	1692	−34	1678	−48	1726	0	ν_4 (H$_2$CO)
2′	1724	−2	1724	−2	1655	−71	1695	−31	
3	1678	−36	1680	−34	1666	−48	1716	2	ν_{14} (HCOOCH$_3$)
3′	1710	−4	1714	0	1647	−67	1684	−30	
4	1664	−33	—	—	—	—	—	—	$2\nu_6$ (HOCH$_2$CHO) (?)
4′	—	—	1654	−43	1633	−64	1670	−27	

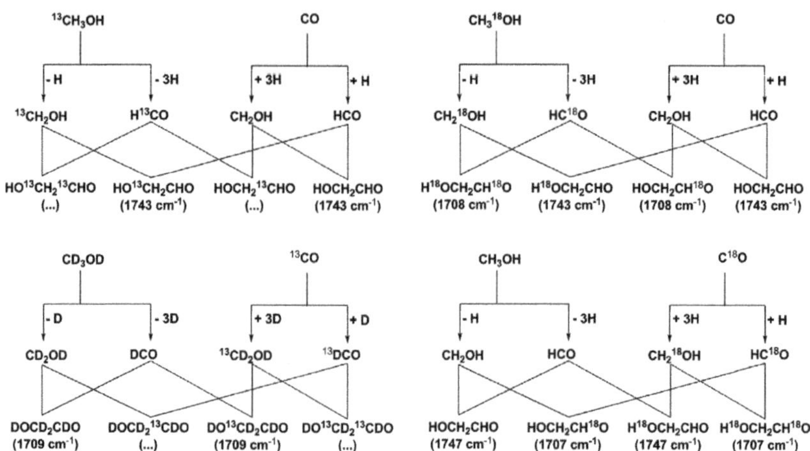

Fig. 4 Schematic representation of the formation of glycolaldehyde isotopomers in mixed isotopic ices of $^{13}CH_3OH$–CO, $CH_3^{18}OH$–CO, CD_3OH–^{13}CO and CH_3OH–$C^{18}O$. Here, two different carbonyl functional units ($H^{13}C^{16}O$ and $H^{12}C^{16}O$ in $^{13}CH_3OH$–CO; $H^{12}C^{18}O$ and $H^{12}C^{16}O$ in $CH_3^{18}OH$–CO; $D^{12}C^{16}O$ and $D^{13}C^{16}O$ in CD_3OD–^{13}CO; $H^{12}C^{16}O$ and $H^{12}C^{18}O$ in CH_3OH–$C^{18}O$) are expected to be observed in each system. The observed peak positions are shown in parentheses.

Fig. 3b and Tables 1–3 also show the identification of ν_4 of the formaldehyde isotopologues, and ν_{14} of the methyl formate isotopologues. In the case of isotopically pure ices of CH_3OH–CO, CD_3OD–CO and $CH_3^{18}OH$–$C^{18}O$, formaldehyde was confirmed via band positions at 1726 cm^{-1}, 1686 cm^{-1}, 1693 cm^{-1}, respectively, and methyl formate bands were observed at 1714 cm^{-1}, 1668 cm^{-1} and 1680 cm^{-1}. Both frequency shifts are in excellent agreement with the observed band positions in the irradiated methanol (isotopically pure) ices (Table 1). Similar to the detection of two different isotopomers of glycolaldehyde in the mixed isotopic ices of $^{13}CH_3OH$–CO, $CH_3^{18}OH$–CO, CD_3OD–^{13}CO and CH_3OH–$C^{18}O$, two isotopomers of formaldehyde and methyl formate were also observed in the carbonyl stretching regions in each system as expected. Here, H_2CO and $H_2^{13}CO$ were confirmed via bands at 1724 cm^{-1} (band 2′) and 1692 cm^{-1} (band 2) following irradiation of $^{13}CH_3OH$–CO ice, in agreement with the observed band positions of H_2CO (1726 cm^{-1}) and $H_2^{13}CO$ (1687 cm^{-1}) in irradiated CH_3OH and $^{13}CH_3OH$ ices. Further, after the irradiation of $CH_3^{18}OH$–CO and CH_3OH–$C^{18}O$ ices, $H_2C^{18}O$ was observed at 1692 (band 2) and 1695 cm^{-1} (band 2′), respectively, with H_2CO confirmed at 1724 cm^{-1} (band 2′) and 1726 cm^{-1} (band 2), respectively. These peak positions agree with the observed positions of formaldehyde in irradiated pure $CH_3^{18}OH$ (1693 cm^{-1}) and CH_3OH (1726 cm^{-1}) ices. In irradiated CD_3OD–^{13}CO ices, D_2CO and $D_2^{13}CO$ were observed at 1678 cm^{-1} (band 2) and 1655 cm^{-1} (band 2′), respectively, matching well with the band at 1676 cm^{-1} in irradiated CD_3OD ice, and the literature value for the isotope shift of $D_2^{13}CO$ (71 cm^{-1}) is close to the observed value of 76 cm^{-1}.[68] Similarly, two isotopomers of methyl formate were also observed in the isotopically mixed ices of methanol–carbon monoxide. Here, HCO and H^{13}CO of methyl formate in the irradiated $^{13}CH_3OH$–CO ice were detected at 1710 cm^{-1} (band 3′) and 1678 cm^{-1} (band 3), which are close to the observed values of methyl formate in CH_3OH (1714 cm^{-1})

and $^{13}CH_3OH$ (1676 cm^{-1}) ices. In irradiated $CH_3{}^{18}OH$–CO and CH_3OH–C^{18}O ices, $HC^{18}O$ units of methyl formate isotopomers were observed at 1680 cm^{-1} (band 3) and 1684 cm^{-1}(band 3′), respectively and the HCO carbonyl unit was observed at 1714 cm^{-1} (band 3′) and 1716 cm^{-1} (band 3), respectively. These peak positions agree with the observed positions of methyl formate in irradiated pure $CH_3{}^{18}OH$ (1682 cm^{-1}) and CH_3OH (1714 cm^{-1}) ices. In irradiated CD_3OD–^{13}CO ices, DCO and $D^{13}CO$ carbonyl units of methyl formate were observed at 1666 cm^{-1} (band 3) (1664 cm^{-1} in irradiated CD_3OD ice) and 1647 cm^{-1} (band 3′), respectively. The isotope shift observed in the case of the $D^{13}CO$ (67 cm^{-1}) carbonyl unit in methyl formate is reasonable based on the observed shift in the case of the formaldehyde isotopomer (the $\Delta\nu$ of $D_2{}^{13}CO$ is 71 cm^{-1}), as discussed earlier.

3.1.3 Quantitative analysis of glycolaldehyde formation. In the case of methanol ices, considering an infrared absorption coefficient of 1.3×10^{-17} cm molecule^{-1} for the ν_8 band of methanol,[40] the initial column density of $1.8 \pm 0.2 \times 10^{17}$ molecules cm^{-2} decreased to $1.2 \pm 0.2 \times 10^{17}$ molecules cm^{-2} after irradiation, with a decrease of $6.0 \pm 0.6 \times 10^{16}$ molecules of methanol (Table 4). During the irradiation by 5 keV electrons with a beam current of 30 nA, a total of 6.7×10^{14} electrons penetrated the sample. From here, we can estimate that 90 ± 9 molecules of methanol were destroyed by each impinging electron, which is equivalent to $1.9 \pm 0.2 \times 10^{-2}$ molecules eV^{-1}. During the same period of time, $3.1 \pm 0.3 \times 10^{14}$ molecules of glycolaldehyde were produced, with a production rate of $1.0 \pm 0.1 \times 10^{-4}$ molecules eV^{-1}. The glycolaldehyde abundance is thus calculated to be $0.17 \pm 0.02\%$ with respect to the initial methanol column density. Furthermore, in the case of methanol–carbon monoxide ices, $4.1 \pm 0.4 \times 10^{16}$ molecules of methanol out of $1.1 \pm 0.1 \times 10^{17}$ molecules cm^{-2}, and $5.4 \pm 0.5 \times 10^{16}$ molecules of carbon monoxide out of $1.8 \pm 0.2 \times 10^{17}$ molecules cm^{-2} were destroyed during the irradiation. This leads to the rates of destruction of methanol and carbon monoxide of $1.3 \pm 0.2 \times 10^{-2}$ and $1.8 \pm 0.2 \times 10^{-2}$ molecules eV^{-1}, respectively. Additionally, $1.6 \pm 0.1 \times 10^{15}$ molecules of glycolaldehyde were produced immediately following irradiation, with an estimated yield of $5.2 \pm 0.5 \times 10^{-4}$ molecules eV^{-1}. The glycolaldehyde abundance is thus calculated to be $1.4 \pm 0.1\%$ and $0.89 \pm 0.09\%$ with respect to the initial column densities of the methanol and carbon monoxide, respectively. Note that this production rate (molecules eV^{-1}) is almost five times greater than the rate of glycolaldehyde formation in pure methanol ices, suggesting the presence of a more favorable reaction pathway, most probably the hydrogenation of carbon monoxide to the formyl radical, as observed in previous studies.[41]

3.2 ReTOF mass spectroscopy

In addition to the detection of glycolaldehyde using FTIR spectroscopy, we employed complementary, highly sensitive single photoionization ReTOF mass spectrometry coupled with temperature programmed desorption (TPD) studies to identify glycolaldehyde in the irradiated methanol ices and mixed methanol–carbon monoxide ices based on the mass-to-charge ratios, the sublimation temperatures and the corresponding shifts in the mass-to-charge ratios upon isotope substitution. Fig. 5A and 5B show the complete ReTOF mass spectra as a function of temperature during the warm-up phase after the irradiation of the methanol and methanol–carbon monoxide ices, respectively. The spectra display the intensity as total counts of the ionized products subliming into the gas phase at well-defined temperatures.

In the case of the irradiated methanol ices, the molecular formulae of the products can be assigned by tracing the sublimation of the irradiated isotopically labeled ices (D4-methanol (CD_3OD), ^{13}C-methanol ($^{13}CH_3OH$) and ^{18}O-methanol ($CH_3{}^{18}OH$)) *via* a shift by 1 amu for each hydrogen (replacing H by D) and carbon (replacing ^{12}C by ^{13}C) atom, and 2 amu for each oxygen (replacing ^{16}O by ^{18}O) atom. Further, the isotopically labeled counterparts can be easily identified in all four ices by plotting the sublimation temperatures *versus* the ion counts and verifying that these species have identical sublimation profiles.[61,65] In the case of binary ices of methanol–carbon monoxide, in addition to the natural isotopic ices of CH_3OH–CO, selected isotopically labeled ices CD_3OD–CO, $^{13}CH_3OH$–CO, $CH_3{}^{18}OH$–CO, CD_3OD–^{13}CO, $CH_3{}^{18}OH$–$C^{18}O$ and CH_3OH–$C^{18}O$ were also exposed to ionizing radiation, and the radiation induced products were analyzed following TPD utilizing ReTOFMS. Here, the molecular formulae of the products were assigned based on the same methodology as discussed above. Further, the present set of isotopic ices helped us to assign whether the C/H/O bearing molecules followed the incorporation of the carbon monoxide unit. As an example, from the results from FTIR spectroscopy (Section 3.1), the carbonyl unit (HCO) of the glycolaldehyde ($HOCH_2CHO$) product can be formed from both methanol and carbon monoxide. Here, selected mixed isotopically labeled binary ices of $^{13}CH_3OH$–CO, $CH_3{}^{18}OH$–CO, CD_3OD–^{13}CO and CH_3OH–$C^{18}O$ were studied to assign the molecular formula along with their formation routes.

3.2.1 Glycolaldehyde ($HOCH_2CHO$). Specific attention is paid to the detection of glycolaldehyde ($HOCH_2CHO$) in the interstellar analogous ices of methanol and binary ices of methanol–carbon monoxide. In these irradiated CH_3OH ices and CH_3OH–CO ices, the molecular ion peak of glycolaldehyde ($C_2H_4O_2{}^+$) is expected at $m/z = 60$ amu. It should be mentioned that with the molecular formula $C_2H_4O_2$, four possible isomers exists with following ionization energies (IE): glycolaldehyde ($HOCH_2CHO$; IE = 10.20 eV), ethene-1,2-diol (HOCH= CHOH; IE = 9.62 eV), methyl formate ($HCOOCH_3$; IE = 10.84 eV) and acetic acid (CH_3COOH; IE = 10.63 eV). Among these isomers, methyl formate (10.84 eV) and acetic acid (10.63 eV) have ionization energies higher than the energy of the photoionization source (10.49 eV) and therefore do not contribute to the ion signal at $m/z = 60$ amu. However, the other two isomers, glycolaldehyde and ethene-1,2-diol, have ionization energies of 10.20 eV and 9.62 eV, respectively, and therefore can be ionized.

3.2.2 Glycolaldehyde in irradiated methanol ices. First, we would like to discuss the identification of glycolaldehyde in methanol ices. It should be mentioned here that, in ices of methanol isotopologues, molecular ion signals of $C_2H_4O_2$ and its isotopomers are expected at $m/z = 60$ amu ($C_2H_4O_2{}^+$; CH_3OH ice), $m/z = 64$ amu ($C_2D_4O_2{}^+$; CD_3OD ice), $m/z = 62$ amu ($^{13}C_2H_4O_2{}^+$; $^{13}CH_3OH$ ice), and at $m/z = 64$ amu ($C_2H_4{}^{18}O_2{}^+$; $CH_3{}^{18}OH$ ice). Table 5 shows feasible formation routes of $C_2H_4O_2$ isomers, glycolaldehyde and ethene-1,2-diol, through the recombination of radical reactants (HCO, CH_2OH, and CHOH) which can be formed upon radiolysis of methanol. Here, a glycolaldehyde molecule is formulated *via* the radical–radical combination of one HCO with a CH_2OH radical, and ethene-1,2-diol is formally formed *via* dimerization of CHOH or a tautomerization mechanism of glycolaldehyde.

Table 4 Fractional abundances of glycolaldehyde in methanol and methanol–carbon monoxide ices after the irradiation at 5.5 K, and during the warm up phase at 110 K (maximum value) and 150 K (at the time to sublimation)

Ice	Irradiation		Warm up phase			
	5.5 K	$f(CH_3OH)$	110 K	$f(CH_3OH)$	150 K	$f(CH_3OH)$
CH_3OH	$3.1 \pm 0.3 \times 10^{14}$	$1.7 \pm 0.2 \times 10^{-3}$	$8.0 \pm 0.8 \times 10^{14}$	$4.4 \pm 0.4 \times 10^{-3}$	$3.2 \pm 0.3 \times 10^{14}$	$1.8 \pm 0.2 \times 10^{-3}$
$CH_3OH–CO$	$1.6 \pm 0.1 \times 10^{15}$	$1.4 \pm 0.1 \times 10^{-2}$	$2.1 \pm 0.2 \times 10^{15}$	$1.9 \pm 0.2 \times 10^{-2}$	$6.3 \pm 0.6 \times 10^{14}$	$5.7 \pm 0.6 \times 10^{-3}$

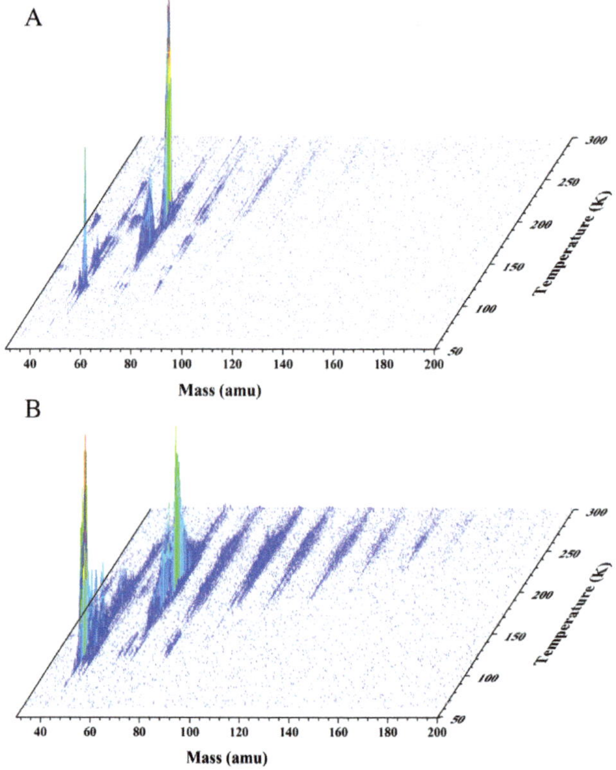

Fig. 5   (A) ReTOF mass spectra as a function of the temperature of the newly formed products subliming into the gas phase from the irradiated methanol (CH$_3$OH) ices. (B) ReTOF mass spectra as a function of the temperature of the newly formed products subliming into the gas phase from the irradiated methanol–carbon monoxide (CH$_3$OH–CO) ices.

Fig. 6 depicts the sublimation profile of integrated ion counts at $m/z = 60$ amu ($C_2H_4O_2{}^+$). In principle, the ion signal at $m/z = 60$ amu can also originate from the ionization of C_3H_8O ($m/z = 60$ amu) isomers such as 1-propanol (10.22 eV), 2-propanol (10.17 eV), and methoxyethane (9.72 eV); these molecules have ionization energies lower than the 10.49 eV ionization photon. The distinct peak at 120 K is assigned to C_3H_8O isomers, as observed in previously irradiated CH$_4$–CO ices.[65] Furthermore, the pronounced peak at 200 K displays a similar onset to the sublimation profile of $m/z = 62$ amu ($C_2H_6O_2{}^+$), as shown in Fig. 6. The similarity in the onsets of the sublimation profiles of $m/z = 60$ amu ($C_2H_4O_2{}^+$) and $m/z = 62$ amu ($C_2H_6O_2{}^+$) implies that the ion signal at $m/z = 60$ amu ($C_2H_4O_2{}^+$) could be due to the photofragmentation of the parent $C_2H_6O_2$ molecules. In an effort to explore this possibility, the TPD ReTOF mass spectra of pure ethylene glycol (HOCH$_2$CH$_2$OH; IE = 10.16 eV) and a 10% mixture of ethylene glycol in methanol and methanol–carbon monoxide ices were recorded as a function of temperature under identical experimental parameters. Here, ethylene glycol did not exhibit any fragmentation to $m/z = 60$ amu, indicating that co-sublimation of $C_2H_4O_2$ ($m/z = 60$ amu) along with ethylene glycol ($C_2H_6O_2$; $m/z = 62$ amu) is responsible for the signal detected at 200 K.

To verify the identification of $C_2H_4O_2$ isomers, we probed the sublimation of the isotopically substituted counterparts in the irradiated CD_3OD, $^{13}CH_3OH$, and $CH_3{}^{18}OH$ ices, and explored the expected isotope shifts of glycolaldehyde (Fig. 7). Here, we observed a signal at $m/z = 64$ amu ($C_2D_4O_2{}^+$) in CD_3OD ice, at $m/z = 62$ amu ($^{13}C_2H_4O_2{}^+$) in $^{13}CH_3OH$ ice, and at $m/z = 64$ amu ($C_2H_4{}^{18}O_2$) in $CH_3{}^{18}OH$ ices, and their temperature-dependent sublimation profiles are shown in Fig. 7. Here, the ReTOF sublimation profiles at $m/z = 60$ amu ($C_2H_4O_2{}^+$) in the CH_3OH system and at $m/z = 64$ amu ($C_2D_4O_2{}^+$) in the CD_3OD system show five distinct peaks (i–v), whereas the TPD ReTOF sublimation profiles at $m/z = 62$ amu ($^{13}C_2H_4O_2{}^+$) in the $^{13}CH_3OH$ system and at $m/z = 64$ amu ($C_2H_4{}^{18}O_2{}^+$) in the $CH_3{}^{18}OH$ system show four peaks (ii–v); the assignments of these peaks are listed in Table 6. As discussed earlier, the distinct peak (i) at 120 K in CH_3OH and CD_3OD ices can be assigned to the C_3H_8O ($m/z = 60$ amu) and C_3D_6O ($m/z = 64$ amu) isomers, respectively, as observed in a previous report of irradiated CH_4–CO ices.[65] The remaining peaks, labeled as (ii), (iii), (iv), and (v) in Fig. 7, are observed in all the isotopologous ices; the similar sublimation profiles and positions of the peaks in these systems support the formation of $C_2H_4O_2$ isomers. Here, peak (ii), which begins sublimation at 150 ± 2 K with a maximum at 166 ± 2 K, is assigned to glycolaldehyde ($HOCH_2CHO$). To support this assignment, it is important to recall that the formation of glycolaldehyde was also confirmed *via* the detection of a peak at 1743 cm^{-1} (ν_{14}) together with its isotopically labeled counterparts. Furthermore, the ν_{14} integrated band area displays a strong decline at 150 K, which correlates well with the increase in the ReTOF ion signal of the $C_2H_4O_2$ isotopologues at 150 ± 2 K (ESI, Fig. S11†). As we have discussed earlier, the glycolaldehyde absorption band becomes untraceable beyond 175 K, and any sublimation of $m/z = 60$ amu above 175 K cannot be explained by the sublimation of glycolaldehyde, but likely originates from the sublimation of the second isomer

Table 5 Mass-to-charge ratios of $C_2H_4O_2$ isomers along with the radical–radical combination routes in irradiated methanol ices

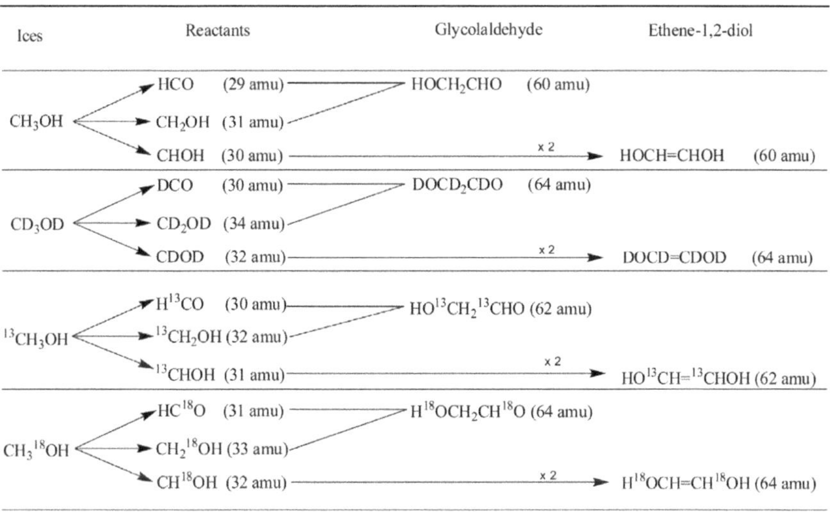

ethene-1,2-diol. Correspondingly, we assign peaks (iii) and (iv) at 200 K and 210 K to the sublimation of ethene-1,2-diol (HOCH=CHOH). Recall that the peak at 200 K is due to the co-sublimation of $C_2H_4O_2$ along with ethylene glycol as discussed previously. Note that ethene-1,2-diol, having two OH functional groups, can interact more strongly with ethylene glycol *via* hydrogen bonding, and co-sublimation is therefore reasonable. Furthermore, sublimation of ethene-1,2-diol peaking at 210 K (iv) is also reasonable, as ethene-1,2-diol has a higher polarity compared to glycolaldehyde (HOCH$_2$CHO). In addition to this, evidence of infrared absorption in the OH stretching region (see Fig. 8) after the complete sublimation of ethylene glycol at around 210 K also suggests the presence of ethene-1,2-diol in the ices, which peaks at 210 K. Finally, the distinct peak at around 234 K (v) can be assigned to fragmentation from glycerol ($C_3H_8O_3$), based on a recent study on the photoionization of glycerol ($C_3H_8O_3$) which identified a prominent photofragment at $m/z = 60$ amu ($C_2H_4O_2$) with an appearance energy of 9.9 eV.[69] The identification of glycerol in these experiments will be presented in a follow up article, as we are focusing solely on glycolaldehyde here.

3.2.3 Glycolaldehyde in irradiated methanol–carbon monoxide ices. Here we would like to discuss the detection of glycolaldehyde (HOCH$_2$CHO) during the TPD studies in irradiated methanol–carbon monoxide isotopologue ices. Radical reactants HCO, CH$_2$OH, and CHOH can be formed once again *via* radiolysis of CH$_3$OH, or specifically in this system *via* the stepwise hydrogenation of carbon monoxide molecules as shown in Table 7. Note that, in the case of isotopically pure ices, namely CH$_3$OH–CO, CD$_3$OD–CO, and CH$_3{}^{18}$OH–C^{18}O, where only one isotope of each atom is present, $C_2H_4O_2$ isotopologues can only be observed at m/z = 60 amu ($C_2H_4O_2{}^+$), 64 amu ($C_2D_4O_2{}^+$), and 64 amu ($C_2H_4{}^{18}O_2{}^+$), respectively. However, in the case of isotopically mixed ices such as ^{13}CH$_3$OH–CO, CH$_3{}^{18}$OH–CO, CD$_3$OD–^{13}CO and CH$_3$OH–C^{18}O, two different isotopes of either carbon (^{12}C

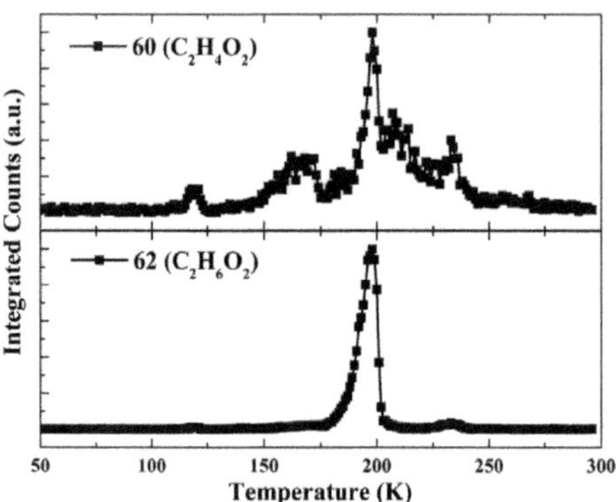

Fig. 6 Top: sublimation profile of integrated counts at m/z = 60 amu ($C_2H_4O_2{}^+$) in the CH$_3$OH system. Bottom: sublimation profile of m/z = 62 amu ($C_2H_6O_2{}^+$). The pronounced peak at 200 K of m/z = 60 amu is a result of the co-sublimation of $C_2H_4O_2$ with $C_2H_6O_2$. The signal at m/z = 62 amu ($C_2H_6O_2{}^+$) is due to ethylene glycol (HOCH$_2$CH$_2$OH; 10.16 eV). The distinct peak at 120 K corresponds to C_3H_8O (m/z = 60 amu; 1-propanol (10.22 eV), 2-propanol (10.17 eV)), as observed in previous results of irradiated CH$_4$–CO ices.[65] Further, the peak at around 234 K is assigned to the fragmentation from higher molecular mass $C_3H_8O_3$ (glycerol), which shows a prominent fragmentation at m/z = 60 amu ($C_2H_6O_2{}^+$, appearance energy 9.9 eV).[69]

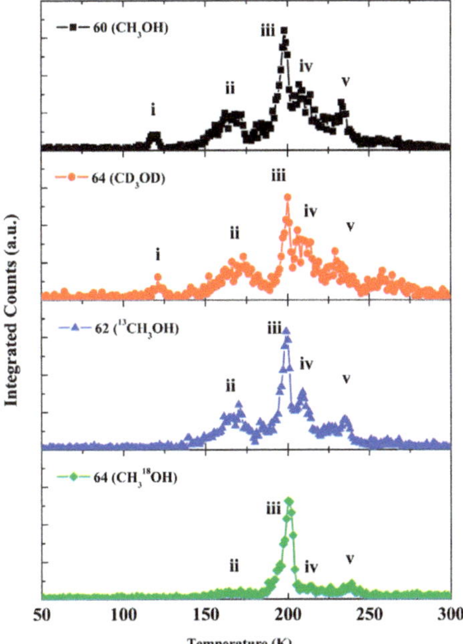

Fig. 7 Sublimation profiles of ion counts at $m/z = 60$ amu ($C_2H_4O_2^+$), 64 amu ($C_2D_4O_2^+$), 62 amu ($^{13}C_2H_4O_2^+$), and 64 amu ($C_2H_4{}^{18}O_2^+$) in CH_3OH, CD_3OD, $^{13}CH_3OH$, and $CH_3{}^{18}OH$ systems, respectively. In the case of CD_3OD ices, the peak at 120 K (i) corresponds to C_3D_6O ($m/z = 64$ amu; acetone (9.7 eV), propanal (10.0 eV)), as observed in a previous report of irradiated CD_4–CO ices.[65] The peak at 200 K is a result of the co-sublimation of $C_2H_4O_2^+$ ($m/z = 60$ amu) with ethylene glycol ($C_2H_6O_2$). Detailed assignments of the peaks (i–v) are listed in Table 6.

and ^{13}C) or oxygen (^{16}O and ^{18}O) are present together; hence two different isotopic sets of radical reactants are possible, as identified *via in situ* FTIR spectroscopy of the processed ices (Section 3.1.2). Here, radical–radical recombination leads to the observation of $C_2H_4O_2$ isotopologues at three different masses are expected to be observed in each system; at $m/z = 60$ amu ($C_2H_4O_2^+$), 61 amu ($C^{13}CH_4O_2^+$) and 62 amu ($^{13}C_2H_4O_2^+$) in the $^{13}CH_3OH$–CO system, at $m/z = 60$ amu ($C_2H_4O_2^+$), 62 amu ($C_2H_4O^{18}O^+$) and 64 amu ($C_2H_4{}^{18}O_2^+$) in the $CH_3{}^{18}OH$–CO system, at $m/z = 66$ amu ($^{13}C_2D_4O_2^+$), 65 amu ($C^{13}CD_4O_2^+$) and 64 amu ($C_2D_4O_2^+$) in the CD_3OD–^{13}CO system, and finally at $m/z = 64$ amu ($C_2H_4{}^{18}O_2^+$), 62 amu ($C_2H_4O^{18}O^+$) and 60 amu ($C_2H_4O_2^+$) in the CH_3OH–$C^{18}O$ system. Note that, in all the above cases (Table 7), both glycolaldehyde and ethene-1,2-diol cannot be separated based on isotopic mass shifts; however, we can contemplate their detection in irradiated methanol–carbon monoxide ices.

Fig. 9 depicts the sublimation profile of integrated ion counts at $m/z = 60$ ($C_2H_4O_2^+$) in CH_3OH–CO ices. The small peak at 120 K is assigned to C_3H_8O isomers, similar to the result for methanol ices. Here, the sublimation profile of $C_2H_6O_2$ is also presented in Fig. 9, and the peak at 196 ± 2 K can be attributed to the co-sublimation of $C_2H_4O_2$ isomers along with ethylene glycol ($C_2H_6O_2$), as discussed in Section 3.2.2. Note that, in isotopically pure ices of CH_3OH–CO, CD_3OD–CO and $CH_3{}^{18}OH$–$C^{18}O$, $C_2H_4O_2$ isotopologues are observed only at $m/z =$

Table 6 Assignment of the ion peaks observed in the sublimation profiles at $m/z = 60$, 64, 62, and 64 in irradiated CH_3OH, CD_3OD, $^{13}CH_3OH$, and $CH_3^{18}OH$ ices, respectively

Ice	Ion signal (m/z)	Peak i (120 K)	Peak ii (166 K)	Peak iii (200 K)	Peak iv (210 K)	Peak v (234 K)
CH_3OH	60	C_3H_8O	$HOCH_2CHO$	$HOCH{=}CHOH$	$HOCH{=}CHOH$	Fragmentation ($C_2H_4O_2$) of $C_3H_8O_3$
CD_3OD	64	C_3D_6O	$DOCD_2CDO$	$DOCD{=}CDOD$	$DOCD{=}CDOD$	Fragmentation ($C_2D_4O_2$) of $C_3D_8O_3$
$^{13}CH_3OH$	62	—	$HO^{13}CH_2{}^{13}CHO$	$HO^{13}CH{=}{}^{13}CHOH$	$HO^{13}CH{=}{}^{13}CHOH$	Fragmentation ($^{13}C_2H_4O_2$) of $^{13}C_3H_8O_3$
$CH_3^{18}OH$	64	—	$H^{18}OCH_2CH^{18}O$	$H^{18}OCH{=}CH^{18}OH$	$H^{18}OCH{=}CH^{18}OH$	Fragmentation ($C_2H_4^{18}O_2$) of $C_3H_8^{18}O_3$

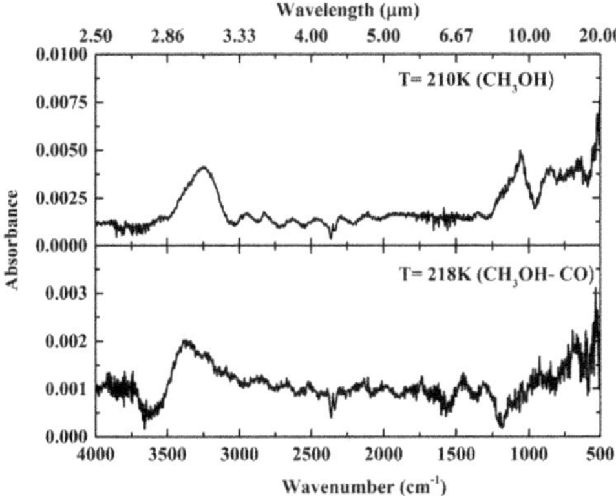

Fig. 8 Infrared absorption spectra of the irradiated methanol (CH_3OH) and methanol–carbon monoxide ($CH_3OH–CO$) ices at 210 K and 218 K. The broad peak from 3500 to 3050 cm^{-1} is assigned to the OH stretching vibrations of alcohols present in the ices. The temperatures of 210 K and 218 K in the CH_3OH and $CH_3OH–CO$ systems correspond to the sublimation of ethene-1,2-diol with ethylene glycol already being sublimed in the gas phase.

60 amu ($C_2H_4O_2^+$), 64 amu ($C_2D_4O_2^+$) and 64 amu ($C_2H_4{}^{18}O_2^+$), respectively, as derived in Table 7 and shown in Fig. 10. In the case of the $CD_3OD–CO$ system, the peak at 125 K can be assigned to C_3D_6O ($m/z = 64$ amu) isomers. Furthermore, the sublimation profiles of $C_2H_4O_2$ isotopologues depict two distinct peaks at 196 ± 2 K and 218 ± 2 K. The maximum at 196 ± 2 K is assigned to the sublimation of glycolaldehyde. To support these assignments, we recall that the formation of glycolaldehyde was also confirmed *via* the detection of peaks at 1743 cm^{-1} (ν_{14}) and 1062 cm^{-1} (ν_7) in the post-irradiated ices of methanol–carbon monoxide, together with the expected frequency shifts of the isotopically labeled counterparts. The integrated infrared absorption band area of glycolaldehyde at 1743 cm^{-1} (ν_{14}) shows a slow decline in column density at 150 K, followed by a rapid decline at 180 ± 2 K (ESI, Fig. S11†). Note that in ESI Fig. S12,† the sublimation profiles derived from TPD ReTOF spectroscopy correlate well with the slow (150 K) and then fast (178 ± 2 K) sublimation of glycolaldehyde observed in the FTIR data. However, the peak at 218 ± 2 K cannot be explained by the sublimation of glycolaldehyde, as no trace of glycolaldehyde can be observed in the infrared spectra after 195 K. Here, the peak at 218 ± 2 K is likely to be due to the sublimation of ethene-1,2-diol. As discussed earlier, ethene-1,2-diol is expected to sublimate at a higher temperature due to its higher polarity, and correspondingly stronger intermolecular interactions. Furthermore, the infrared spectrum at 218 K (Fig. 8) shows evidence of absorption in the OH stretching region, which suggests the presence of ethene-1,2-diol, and thus is attributed to the peak at 218 K, as shown in Fig. 9.

4. Discussion

We would like to discuss now the formation mechanisms of glycolaldehyde. The formation routes are based on the empirical evidence presented above, *i.e.* the observation of distinct isotopomers of glycolaldehyde following irradiation within

Table 7 Mass-to-charge ratios of the $C_2H_4O_2$ isomers, along with the radical–radical combination routes in irradiated methanol–carbon monoxide ices

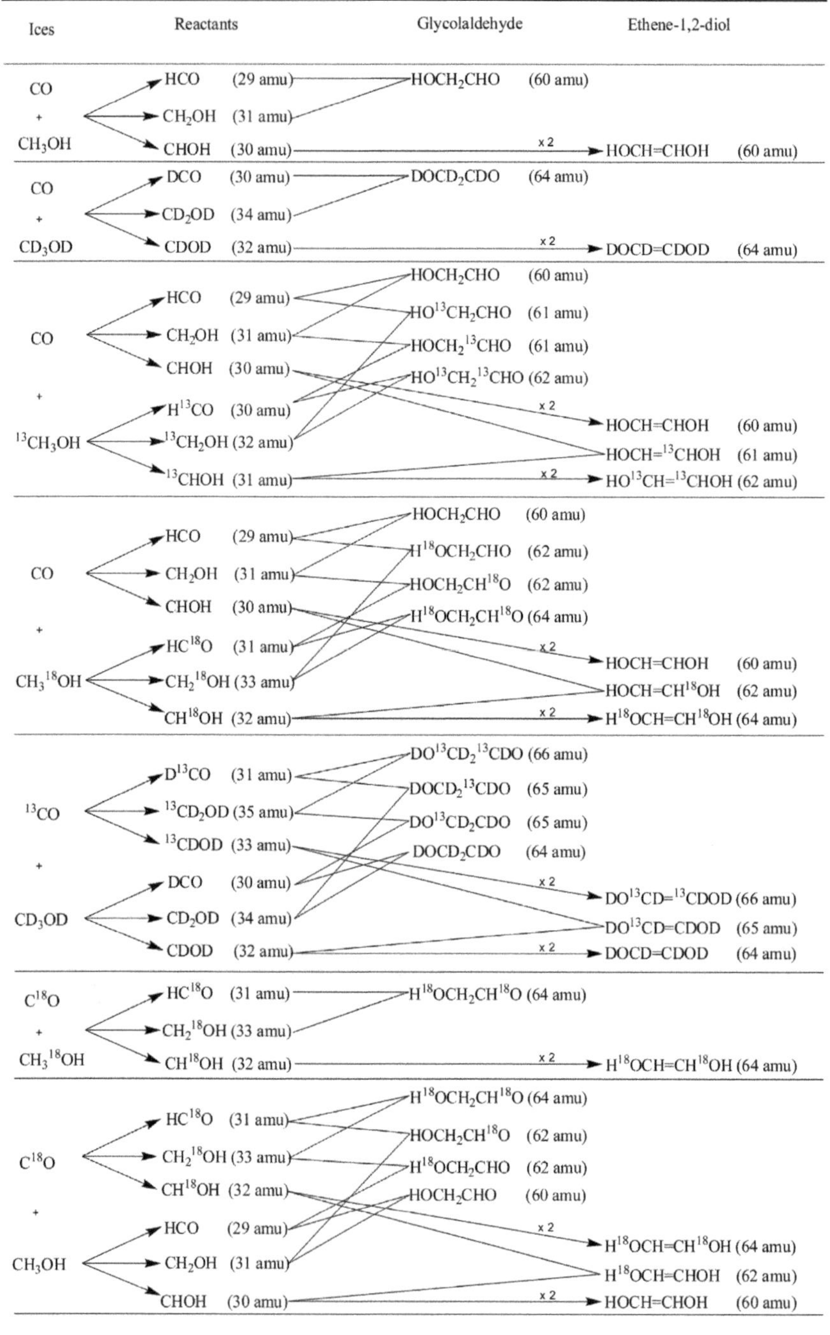

isotopically mixed ices, and in conjunction with previously identified routes guided by the kinetic modeling and numerical fitting of the temporal evolution of newly formed products in irradiated ices of methanol[40] and methanol–carbon monoxide ices[41] (Fig. 11). In the case of methanol ices, irradiation induced decomposition of methanol follows three major competing reaction pathways: (1) unimolecular decomposition to a hydroxymethyl (CH_2OH) radical and a suprathermal hydrogen atom, (2) unimolecular decomposition to a methoxy (CH_3O) radical and a suprathermal hydrogen atom, and (3) decomposition to formaldehyde (H_2CO) and molecular hydrogen (or two hydrogen atoms).

$$CH_3OH \rightarrow CH_2OH + H \tag{1}$$

$$CH_3OH \rightarrow CH_3O + H \tag{2}$$

$$CH_3OH \rightarrow H_2CO + H_2 \tag{3}$$

Among these, reaction (3) is the most dominating pathway, having the rate constant $4.4 \times 10^{-4}\ s^{-1}$, which is about six and four times faster than reaction (1) ($k = 6.95 \times 10^{-5}\ s^{-1}$) and reaction (2) ($k = 1.04 \times 10^{-4}\ s^{-1}$),[40] respectively. Furthermore, both the hydroxymethyl radical and the methoxy radical were found to undergo subsequent unimolecular decomposition to produce formaldehyde and a hydrogen atom, as in reaction (4) and reaction (5), respectively.

$$CH_2OH \rightarrow H_2CO + H \tag{4}$$

$$CH_3O \rightarrow H_2CO + H \tag{5}$$

Recall that, during the irradiation of all isotopologue ices, infrared absorptions due to the formaldehyde and hydroxymethyl radicals are indeed observed with corresponding frequency shifts. The kinetic fitting also suggests that formaldehyde is decomposed to form the formyl (HCO) radical and a hydrogen atom (reaction (6)).

$$H_2CO \rightarrow HCO + H \tag{6}$$

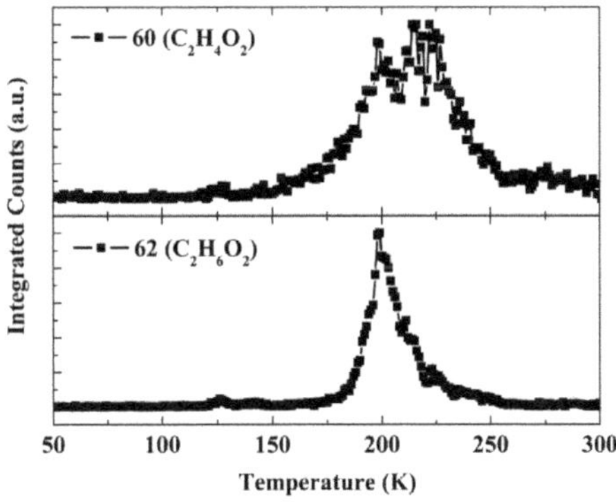

Fig. 9 Top: sublimation profile of ion counts at $m/z = 60$ amu ($C_2H_4O_2^+$) in the CH_3OH–CO system. Bottom: sublimation profile of $m/z = 62$ amu ($C_2H_6O_2^+$), suggesting that the first peak at 198 K in $m/z = 60$ amu is a result of co-sublimation of $C_2H_4O_2$ with ethylene glycol ($C_2H_6O_2$; $HOCH_2CH_2OH$). The peak at 120 K is due to C_3H_8O ($m/z = 60$ amu; 1-propanol (10.22 eV), 2-propanol (10.17 eV)), as observed previously for irradiated CH_4–CO ices.[65]

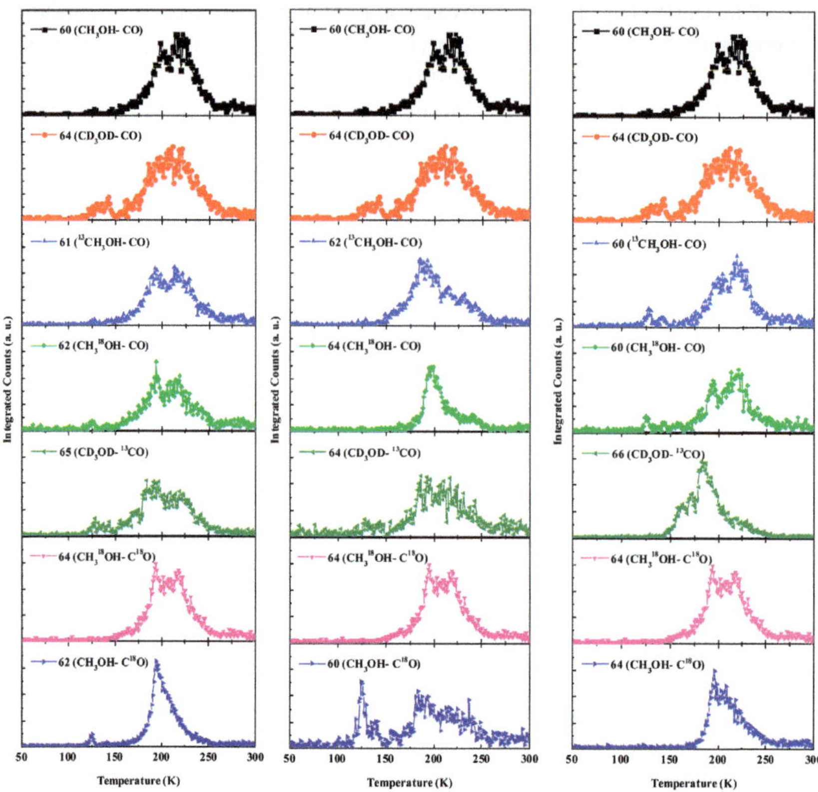

Fig. 10 Sublimation profiles of $C_2H_4O_2$ isotopomers in methanol–carbon monoxide ices. The graphs in the left panel depict the sublimation profile of $C_2H_4O_2$ isotopologues which can be formally formed from one methanol plus one carbon monoxide building block; the graphs in the central panel depict the sublimation profiles of $C_2H_4O_2$ isotopologues which can only be synthesized from two methanol building blocks. The graphs in the right panel show the sublimation profiles of $C_2H_4O_2$ isotopologues which can be formed from two carbon monoxide building blocks. The slight peak at 120 K in the sublimation profiles of $m/z = 60$ amu ($CH_3OH–CO$) and 62 amu ($CH_3^{18}OH–CO$ and $CH_3OH–C^{18}O$) are due to C_3H_8O and $C_3H_8^{18}O$, respectively. Furthermore, the slight peak at 125 K in the sublimation profiles of $m/z = 64$ amu ($CD_3OD–CO$), 60 amu ($^{13}CH_3OH–CO$), 60 amu ($CH_3^{18}OH–CO$), 65 amu ($CD_3OD–^{13}CO$) and 60 amu ($CH_3OH–C^{18}O$) are due to C_3D_6O, $C^{13}C_2H_6O$, $C_3H_6^{18}O$, $C_2^{13}CD_6O$, and $C_3H_6^{18}O$, respectively, as observed previously for irradiated $CH_4–CO$ ices.[65]

Finally, the barrierless recombination of the formyl radical with the hydroxymethyl radical can lead to the formation of glycolaldehyde (reaction (7)).

$$CH_2OH + HCO \rightarrow HOCH_2CHO \tag{7}$$

It should be mentioned here that radiolysis of CH_3OH to produce a CH_2OH radical and atomic hydrogen requires 4.03 eV molecule^{-1}, and the minimum energy required to produce one HCO radical is 8.06 eV molecule^{-1}.[40] As such, the energy required to produce one molecule of glycolaldehyde is 12.09 eV molecule^{-1}. Therefore, non-equilibrium chemistry is crucial to drive the formation of glycolaldehyde in low temperature ices.

In the presence of carbon monoxide, suprathermal atomic hydrogen produced during the decomposition of methanol can overcome the barrier of addition (4.03 eV molecule^{-1}) to the carbon monoxide molecule, leading to the formation of a formyl radical *via* reaction (8), which may be followed by reaction (7) leading to the formation of glycolaldehyde as well.

$$CO + H \rightarrow HCO \tag{8}$$

Note that in binary ices of methanol and carbon monoxide, the formyl radical can be produced *via* at least two different reaction mechanisms, either *via* radiolysis of methanol (reactions (1)–(6)), or *via* hydrogenation of carbon monoxide (reaction (8)). The reaction mechanisms discussed above suggest that, in the mixed isotopic ices, $^{13}CH_3OH$–CO, $CH_3{}^{18}OH$–CO, CD_3OD–^{13}CO and OH–C^{18}O, two different glycolaldehyde isotopomers can be formed with the CH_2OH unit from methanol building blocks and the HCO unit either from carbon monoxide building blocks or from methanol building blocks. The deconvolution of the carbonyl absorption in the FTIR spectra of these ices indeed revealed the production of glycolaldehyde isotopologues; thereby supporting the mechanism proposed by Bennett *et al.*[40,41]

In addition to this, successive hydrogenation of carbon monoxide could led to the formation of a CH_2OH radical unit which can recombine with the HCO radical unit, formed *via* the above mentioned reaction pathways (6) and (8), to form glycolaldehyde as well (see Fig. 11). In this regard, the hydrogenation of the formyl radical (HCO) formed *via* reaction (8) to form formaldehyde (H_2CO), and a subsequent hydrogenation leading to the formation of the CH_2OH radical *via* reaction pathways (9) and (10) is expected, as shown in Fig. 4 and 11.

$$HCO + H \rightarrow H_2CO \tag{9}$$

$$H_2CO + H \rightarrow CH_2OH \tag{10}$$

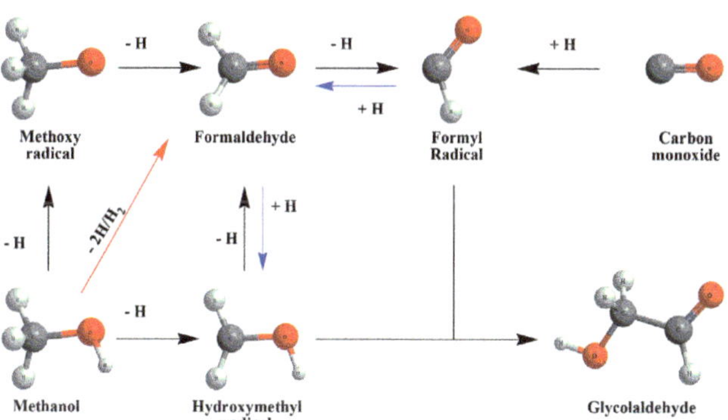

Fig. 11 Reaction scheme for the formation of glycolaldehyde (HOCH$_2$CHO) in irradiated methanol and in methanol–carbon monoxide mixed ices extracted from the kinetic fittings of the column density of the products. The red arrow indicates the dominant pathway among other pathways for the formation of formaldehyde.[40,41] The blue arrows indicate additional proposed reaction pathways based on the current experimental results.

Reaction pathway (9) suggests the formation of formaldehyde *via* hydrogenation of carbon monoxide. Recall that, in the mixed isotopic ices of methanol–carbon monoxide, $^{13}CH_3OH–CO$, $CH_3{}^{18}OH–CO$, $CD_3OD–^{13}CO$ and $CH_3OH–C^{18}O$, two isotopomers of formaldehyde were observed (Table 3) in each system. As an example, in $CH_3{}^{18}OH–CO$ ices, formaldehyde bands were identified at 1692 cm^{-1} and 1724 cm^{-1}, and these band positions were close to the formaldehyde bands observed in $CH_3{}^{18}OH$ ices ($H_2C^{18}O$; 1693 cm^{-1}) and CH_3OH ices (H_2CO; 1726 cm^{-1}). Consequently, these observations confirmed the detection of $H_2C^{18}O$ as well as H_2CO at 1692 cm^{-1} and 1724 cm^{-1}, respectively, in $CH_3{}^{18}OH–CO$ ices. The above observation provides clear evidence that reaction pathway (9) occurs. Furthermore, the hydrogenation of formaldehyde to form the CH_2OH radical, followed by recombination with the HCO radical unit (reaction (7)) can result in the formation of glycolaldehyde as well. Note that, in the mixed isotopic ices of methanol–carbon monoxide, two different isotopomers of CH_2OH radicals (*via* reaction pathways (1) and (10)) and two different isotopomers of HCO radicals (reaction pathways (6) and (8)) could lead to the formation of four isotopomers of glycolaldehyde molecules, as shown in Fig. 4. However, within these four isotopomers, only two isotopic carbonyl units (HCO) are present, which results in the detection of two different infrared absorptions of the glycolaldehyde carbonyl stretching vibration in the mixed isotopic ices (Table 3, Fig. 3b). As an example, in the $CH_3{}^{18}OH–CO$ system, CH_2OH, formed *via* the successive hydrogenation of carbon monoxide ((8)–(10)), can recombine with $HC^{18}O$ (*via* reactions (1)–(6)) or HCO (reaction (8)), resulting in $HOCH_2CH^{18}O$ and $HOCH_2CHO$ (see Fig. 4). These two isotopomers have identical carbonyl units of $H^{18}OCH_2CH^{18}O$ and $H^{18}OCH_2CHO$ (reactions (1)–(8)); consequently, the two carbonyl stretching frequencies detected at 1708 cm^{-1} ($HC^{18}O$) and 1743 cm^{-1} (HCO) correspond to the respective isotopomers of glycolaldehyde. Based on the *in situ* FTIR and TPD ReTOF mass spectroscopic evidence presented above, the overall reaction pathways ultimately resulting in the formation of glycolaldehyde are presented in Fig. 11.

Also note the increase of glycolaldehyde abundance during the warm up phase of the irradiated ices. Here, the increase of glycolaldehyde was associated with a simultaneous decrease in the amount of formyl radicals. In the case of irradiated methanol ices, the amount of glycolaldehyde increased by almost the exact same amount by which formyl radicals decreased, about 5×10^{14} molecules. Similarly, a corresponding increase in the amount of glycolaldehyde with the decrease in the amount of formyl radicals was observed following the warm up of the irradiated mixed methanol and carbon monoxide ices. Here, the amount of glycolaldehyde increases by a similar order of magnitude (2×10^{14} molecules); however, the amount of formyl radical decreased by almost an order of magnitude more (1.6×10^{15} molecules), accounting for only 12% of the apparent increase in glycolaldehyde, suggesting that an additional chemical route is involved. These observations imply additional thermal chemistry of the trapped radicals diffusing through the methanol matrix following reaction (7), ultimately yielding glycolaldehyde.

5.   Conclusions

The present experimental approach specifically focused on the detection of glycolaldehyde in irradiated methanol ices and methanol–carbon monoxide binary ices along with their isotopically labeled counterparts. Here we utilized two complementary detection techniques, infrared spectroscopy and single photo-ionization ReTOF mass spectrometry, in order to analyze the endogenous synthesized products formed *via* radiation induced chemical processing. The on line and *in situ* infrared spectroscopy identified the formation of the astro-biologically important molecule glycolaldehyde, based on the agreement of the observed infrared band positions and the associated isotopic shifts with the literature data. In the case of mixed isotopic ices of methanol–carbon monoxide ($^{13}CH_3OH$–CO, $CH_3^{18}OH$–CO, CD_3OD–^{13}CO and CH_3OH–$C^{18}O$) where two different isotopes of carbon or oxygen are present together, deconvolution of the broad carbonyl absorption features identified at least two isotopomers of glyco-laldehyde (Fig. 3B and Table 3). This confirms the presence of at least two reaction pathways in methanol–carbon monoxide ices for the formation of the carbonyl functional group (HCO) of glycolaldehyde, *via* the decomposition of methanol and the hydrogenation of carbon monoxide molecules. Accordingly, during the TPD studies of irradiated methanol and methanol–carbon monoxide ices and their isotopologues, using single photoionization ReTOF mass spectrometry we confirmed the detection of glycolaldehyde based on the identical sublimation profiles of the corresponding shifted masses. Furthermore, the agreement between the sublimation of glycolaldehyde obtained from ReTOF mass spec-troscopy and FTIR spectroscopy supports this detection. In the mixed isotopic ices ($^{13}CH_3OH$–CO, $CH_3^{18}OH$–CO, CD_3OD–^{13}CO and CH_3OH–$C^{18}O$) we were able to detect glycolaldehyde at three different isotopic masses, suggesting that at least three competing reaction mechanisms are involved in the formation of glyco-laldehyde in the irradiated ices at 5 K: (i) formation of glycolaldehyde *via* a hydrogenation mechanism of carbon monoxide, (ii) formation of glycolaldehyde *via* the reaction of one methanol unit with carbon monoxide unit, and (iii) *via* the decomposition of a methanol molecule followed by recombination of HCO and CH_2OH radicals.

Upon warming up of the irradiated samples, the column density of the gly-colaldehyde shows an increase, implying the presence of additional thermal chemistry, most probably *via* the diffusion and ensuing reaction of trapped radicals. Further, the fractional abundance of the glycolaldehyde in the irradiated methanol ices was estimated (see the ESI for relevant details†) within the range 4.5 ± 0.5 to $1.8 \pm 0.2 \times 10^{-8}$ with respect to molecular hydrogen; in methanol–carbon monoxide ices the fractional abundance of glycolaldehyde is from $1.9 \pm 0.2 \times 10^{-7}$ to $5.7 \pm 0.6 \times 10^{-8}$. These estimated glycolaldehyde abundances are close to the abundances of those calculated based on our observations, with doses relevant to the typical lifetime of interstellar ices prior to the star-formation induced warm up phase. Therefore, our laboratory simulation demonstrates that radiation exposure of bulk ices containing methanol and methanol–carbon monoxide, relevant to the actual physical environment of an ice covered grain mantle, will ultimately lead to the synthesis of glycolaldehyde.

Acknowledgements

The authors would like to thank the W. M. Keck Foundation, the University of Hawaii, and the NASA Exobiology Program MNX13AH62G.

References

1 J. M. Hollis, F. J. Lovas and P. R. Jewell, *Astrophys. J.*, 2000, **540**, L107–L110.

2 L. E. Snyder, *Proc. Natl. Acad. Sci. U. S. A.*, 2006, **103**, 12243–12248.

3 J. M. Hollis, S. N. Vogel, L. E. Snyder, P. R. Jewell and F. J. Lovas, *Astrophys. J.*, 2001, **554**, L81–L85.

4 D. T. Halfen, A. J. Apponi, N. Woolf, R. Polt and L. M. Ziurys, *Astrophys. J.*, 2006, **639**, 237–245.

5 M. T. Beltrán, R. Cesaroni, R. Neri, C. Codella, R. S. Furuya, L. Testi and L. Olmi, *Astron. Astrophys.*, 2005, **435**, 901–925.

6 M. T. Beltrán, C. Codella, S. Viti, R. Neri and R. Cesaroni, *Astrophys. J.*, 2009, **690**, L93.

7 M. A. Requena-Torres, J. Martín-Pintado, S. Martín and M. R. Morris, *Astrophys. J.*, 2008, **672**, 352.

8 J. K. Jørgensen, C. Favre, S. E. Bisschop, T. L. Bourke, E. F. van Dishoeck and M. Schmalzl, *Astrophys. J.*, 2012, **757**, L4.

9 J. L. Bada, *Earth Planet. Sci. Lett.*, 2004, **226**, 1–15.

10 D. P. Bartel and P. J. Unrau, *Trends Biochem. Sci.*, 1999, **24**, M9–M13.

11 A. Lazcano and S. L. Miller, *Cell*, 1996, **85**, 793–798.

12 O. Leslie E, *Crit. Rev. Biochem. Mol. Biol.*, 2004, **39**, 99–123.

13 L. E. Orgel, *Trends Biochem. Sci.*, 1998, **23**, 491–495.

14 F. H. C. Crick, *J. Mol. Biol.*, 1968, **38**, 367–379.

15 L. E. Orgel, *J. Mol. Biol.*, 1968, **38**, 381–393.

16 L. E. Orgel, *Cold Spring Harbor Symp. Quant. Biol.*, 1987, **52**, 9–16.

17 C. R. Woese, *The genetic code: The molecular basis for genetic expression*, Harper & Row, New York, 1967.

18 S. A. Benner, H.-J. Kim and M. A. Carrigan, *Acc. Chem. Res.*, 2012, **45**, 2025–2034.

19 M. Yarus, *Origins Life Evol. Biosphere*, 2013, **43**, 19–30.

20 M. Levy and S. L. Miller, *Proc. Natl. Acad. Sci. U. S. A.*, 1998, **95**, 7933–7938.

21 K. Ruiz-Mirazo, C. Briones and A. de la Escosura, *Chem. Rev.*, 2014, **114**, 285–366.

22 R. Breslow, M. Levine and Z.-L. Cheng, *Origins Life Evol. Biosphere*, 2010, **40**, 11–26.

23 M. A. Boutlerow, *Comptes rendus hebdomadaires des séances de l'Académie des Sciences.*, 1861, **53**, 145.

24 R. Shapiro, *Origins Life Evol. Biosphere*, 1988, **18**, 71–85.

25 D. Müller, S. Pitsch, A. Kittaka, E. Wagner, C. E. Wintner, A. Eschenmoser and G. Ohlofjgewidmet, *Helv. Chim. Acta*, 1990, **73**, 1410–1468.

26 S. A. Benner, H.-J. Kim, M.-J. Kim and A. Ricardo, *Cold Spring Harbor Perspectives in Biology*, 2010, 2.

27 M. W. Powner, B. Gerland and J. D. Sutherland, *Nature*, 2009, **459**, 239–242.

28 C. Harman, J. Kasting and E. Wolf, *Origins Life Evol. Biosphere*, 2013, **43**, 77–98.

29 M. E. Jacox and D. E. Milligan, *J. Mol. Spectrosc.*, 1973, **47**, 148–162.

30 K. Toriyama and M. Iwasaki, *J. Am. Chem. Soc.*, 1979, **101**, 2516–2523.

31 E. P. Kalyazin and G. V. Kovalev, *Khim. Vys. Energ.*, 1978, **12**, 371–373.

32 G. A. Baratta, A. C. Castorina, G. Leto, M. E. Palumbo, F. Spinella and G. Strazzulla, *Planet. Space Sci.*, 1994, **42**, 759–766.

33 M. H. Moore, R. F. Ferrante and J. A. Nuth III, *Planet. Space Sci.*, 1996, **44**, 927–935.

34 M. E. Palumbo, A. C. Castorina and G. Strazzulla, *Astron. Astrophys.*, 1999, **342**, 551–562.

35 G. Strazzulla, M. Arena, G. A. Baratta, C. A. Castorina, G. Celi, G. Leto, M. E. Palumbo and F. Spinella, *Adv. Space Res.*, 1995, **16**, 61–71.

36 G. Strazzulla, A. C. Castorina and M. E. Palumbo, *Planet. Space Sci.*, 1995, **43**, 1247–1251.

37 R. L. Hudson and M. H. Moore, *Icarus*, 2000, **145**, 661–663.

38 R. L. Hudson and M. H. Moore, *Radiat. Phys. Chem.*, 1995, **45**, 779–789.

39 P. A. Gerakines, W. A. Schutte and P. Ehrenfreund, *Astron. Astrophys.*, 1996, **312**, 289–305.

40 C. J. Bennett, S.-H. Chen, B.-J. Sun, A. H. H. Chang and R. I. Kaiser, *Astrophys. J.*, 2007, **660**, 1588.

41 C. J. Bennett and R. I. Kaiser, *Astrophys. J.*, 2007, **661**, 899–909.

42 P. Modica and M. E. Palumbo, *Astron. Astrophys.*, 2010, **519**, A22.

43 P. Modica, M. E. Palumbo and G. Strazzulla, *Planet. Space Sci.*, 2012, **73**, 425–429.

44 Y. J. Chen, A. Ciaravella, G. M. M. Caro, C. Cecchi-Pestellini, A. Jimenez-Escobar, K. J. Juang and T. S. Yih, *Astrophys. J.*, 2013, **778**, 162.

45 A. L. F. de Barros, A. Domaracka, D. P. P. Andrade, P. Boduch, H. Rothard and E. F. da Silveira, *Mon. Not. R. Astron. Soc.*, 2011, **418**, 1363–1374.

46 A. Ciaravella, G. M. Caro, A. J. Escobar, C. Cecchi-Pestellini, S. Giarrusso, M. Barbera and A. Collura, *Astrophys. J.*, 2010, **722**, L45.

47 O. S. Heavens, *Optical Properties of Thin Solid Films*, Butterworths Scientific Publications, London, 1955.

48 P. Winsemius, F. F. v. Kampen, H. P. Lengkeek and C. G. v. Went, *J. Phys. F: Met. Phys.*, 1976, **6**, 1583.

49 A. M. Goodman, *Appl. Opt.*, 1978, **17**, 2779–2787.

50 M. S. Westley, G. A. Baratta and R. A. Baragiola, *J. Chem. Phys.*, 1998, **108**, 3321–3326.

51 D. M. Hudgins, S. A. Sandford, L. J. Allamandola and A. G. G. M. Tielens, *Astrophys. J. Suppl.*, 1993, **86**, 713–870.

52 C. J. Bennett, C. Jamieson, A. M. Mebel and R. I. Kaiser, *Phys. Chem. Chem. Phys.*, 2004, **6**, 735–746.

53 A. Wada, N. Mochizuki and K. Hiraoka, *Astrophys. J.*, 2006, **644**, 300.

54 R. L. Hudson and M. H. Moore, *J. Geophys. Res.*, 2001, **106**, 33275–33284.

55 R. Brunetto, G. Caniglia, G. A. Baratta and M. E. Palumbo, *Astrophys. J.*, 2008, **686**, 1480.

56 M. Garozzo, D. Fulvio, Z. Kanuchova, M. E. Palumbo and G. Strazzulla, *Astron. Astrophys.*, 2010, **509**, A67.

57 P. Hovington, D. Drouin and R. Gauvin, *Scanning*, 1997, **19**, 1–14.

58 R. Luna, M. Á. Satorre, M. Domingo, C. Millán and C. Santonja, *Icarus*, 2012, **221**, 186–191.

59 B. D. Teolis, M. J. Loeffler, U. Raut, M. Famá and R. A. Baragiola, *Icarus*, 2007, **190**, 274–279.

60 K. I. Öberg, R. T. Garrod, E. F. van Dishoeck and H. Linnartz, *Astron. Astrophys.*, 2009, **504**, 891–913.

61 B. M. Jones and R. I. Kaiser, *J. Phys. Chem. Lett.*, 2013, **4**, 1965–1971.

62 A. Aspiala, J. Murto and P. Sten, *Chem. Phys.*, 1986, **106**, 399–412.

63 J. Ceponkus, W. Chin, M. Chevalier, M. Broquier, A. Limongi and C. Crépin, *J. Chem. Phys.*, 2010, **133**, 094502.

64 K. Cai and J. Wang, *J. Phys. Chem. B*, 2009, **113**, 1681–1692.

65 R. I. Kaiser, S. Maity and B. M. Jones, *Phys. Chem. Chem. Phys.*, 2014, **16**, 3399–3424.

66 J. Lohilahti, T. A. Kainu and V.-M. Horneman, *J. Mol. Spectrosc.*, 2005, **233**, 275–279.

67 T.-L. Tso and E. K. C. Lee, *J. Phys. Chem.*, 1984, **88**, 5475–5482.

68 J. Lohilahti, T. A. Kainu and V.-M. Horneman, *J. Mol. Spectrosc.*, 2005, **233**, 275–279.

69 F. Bell, Q. N. Ruan, A. Golan, P. R. Horn, M. Ahmed, S. R. Leone and M. Head-Gordon, *J. Am. Chem. Soc.*, 2013, **135**, 14229–14239.

PAPER

Application of a diffusion–desorption rate equation model in astrochemistry

Jiao He and Gianfranco Vidali*

Received 27th November 2013, Accepted 17th January 2014

DOI: 10.1039/c3fd00113j

Desorption and diffusion are two of the most important processes on interstellar grain surfaces; knowledge of them is critical for the understanding of chemical reaction networks in the interstellar medium (ISM). However, a lack of information on desorption and diffusion is preventing further progress in astrochemistry. To obtain desorption energy distributions of molecules from the surfaces of ISM-related materials, one usually carries out adsorption–desorption temperature programmed desorption (TPD) experiments, and uses rate equation models to extract desorption energy distributions. However, the often-used rate equation models fail to adequately take into account diffusion processes and thus are only valid in situations where adsorption is strongly localized. As adsorption–desorption experiments show that adsorbate molecules tend to occupy deep adsorption sites before occupying shallow ones, a diffusion process must be involved. Thus, it is necessary to include a diffusion term in the model that takes into account the morphology of the surface as obtained from analyses of TPD experiments. We take the experimental data of CO desorption from the MgO(100) surface and of D_2 desorption from amorphous solid water ice as examples to show how a diffusion–desorption rate equation model explains the redistribution of adsorbate molecules among different adsorption sites. We extract distributions of desorption energies and diffusion energy barriers from TPD profiles. These examples are contrasted with a system where adsorption is strongly localized – HD from an amorphous silicate surface. Suggestions for experimental investigations are provided.

1   Introduction

The kinetics of interstellar related species on dust grain surfaces are known to play an important role in astrochemistry. Astrochemical modeling shows that the abundance of molecules such as H_2, H_2O and CO_2 can't be explained by gas-phase reactions alone, and surface reactions must be involved.[1-3] Desorption and diffusion are the two most important processes on surfaces, and these determine the rates of surface reactions. In recent years there has been a considerable body of work on the surface kinetics of ISM (interstellar medium)-related species. Katz et al.[4] used rate equations to extract desorption and diffusion energy barriers for

Syracuse University, 201 Physics Bldg., Syracuse, NY, 13244-1130, USA. E-mail: gvidali@syr.edu

atomic hydrogen from the data on formation of molecular hydrogen of Pirronello *et al.*[5,6] on polycrystalline olivine and amorphous carbon surfaces. Cazaux and Tielens[7] came up with a similar model that included the possibility of diffusion *via* tunneling and the possibility of an atom dropping into a chemisorption site. Iqbal *et al.*[8] studied the formation of molecular hydrogen over a wider range of surface temperatures using a kinetic Monte Carlo simulation in order to simulate the formation of molecular hydrogen in regions such as PDRs (photodissociation regions), where grains are at a temperature higher than in diffuse and dense clouds. Amiaud *et al.*[9,10] conducted D_2 adsorption–desorption experiments on both porous and non-porous amorphous water ices, then used a rate equation model and a direct inversion method to obtain desorption energy barriers. Noble *et al.*[11] did experiments on the desorption of CO, O_2 and CO_2 from non-porous water ice, crystalline water ice and silicate surfaces, and used a Polanyi–Wigner equation to find the parameters governing desorption behaviors. He *et al.*[12] used multiple desorption energy levels to simulate the adsorption–desorption experiments of D_2 on both single crystalline and amorphous silicate surfaces, and obtained semi-continuous desorption energy distributions. Desorption energies for a variety of ices, including ice mixtures, are summarized by Burke and Brown.[13] All of these investigations assumed either non-hopping between different desorption sites or adlayer equilibration.[14] Thus only the desorption energy can be obtained.

The diffusion of radicals is an obviously important step in the formation of molecules in or on ices. Because of its abundance and low mass, the diffusion of H atoms is the most important process. H diffusion in CO ice leads to the formation of H_2CO and CH_3OH,[15-18] H diffusion in N_2 leads to ammonia formation,[19] and H diffusion in O_2 ices leads to the formation of water.[20,21] In general, the formation of molecules on surfaces requires diffusion to take place, such as in the case of the formation of water *via* the deposition of H, O and O_2 on water ice[22,23] and on bare amorphous silicates.[24,25] Similarly, the formation of molecules due to energetic particles or radiation involves the migration of super thermal radicals.[26] Except in the investigation of the formation of D_2 by the bombardment of CD_4 ice with 5 keV electrons, where rate equations were used to obtain the diffusion energy barrier for deuterium atoms,[27] the analyses of these and other reactions don't contain estimates of diffusion energy barriers or diffusion coefficients.

The diffusion of larger molecules can also take place in or on ices and on bare surfaces. Mispelaer *et al.*[28] studied the diffusion of CO, HNCO, H_2CO and NH_3 in amorphous ice using infrared spectroscopy, while He *et al.* investigated the formation of ozone *via* the diffusion of molecular and atomic oxygen on a bare amorphous silicate surface.[29] Zubkov *et al.*[30] studied the diffusion of nitrogen in amorphous solid water, and Roser *et al.* studied the formation of CO_2 *via* the migration of O in a water-ice capped CO ice.[31] Because diffusion is more difficult to study than desorption, both experimentally and theoretically, much less information is available about diffusion on interstellar grain surfaces than desorption. It is often assumed that the energy barrier for thermal diffusion, E_{diff}, is a fraction of the desorption energy, E_{des}; $E_{\text{diff}} = \alpha E_{\text{des}}$, where α is taken to be a constant with a value typically around 0.3 for weakly adsorbed systems on well ordered surfaces.[32] However, results of simulations of data of molecular hydrogen formation on disordered or amorphous surfaces of silicates and water ice yield a larger value of α (0.7–0.8).[4,33,34] Since these simulations can give only a lower bound for the hydrogen desorption energy (E_{des}), the value of α could be lower. Simulations of ISM chemistry that include surface reactions have used such a wide range of values as well.[8,35,36]

In any case, the relation $E_{diff} = \alpha E_{des}$ is an oversimplification. Since the diffusion energy barrier is in the exponential term of the Polanyi–Wigner equation, a slight change in α could affect the diffusion rate by orders of magnitude, resulting in unrealistic reaction rates. Surface diffusion occurs both by quantum tunneling and by thermally activated hopping. Quantum tunneling is efficient for very light adsorbed species, such as atomic[17] and molecular hydrogen,[37] but the efficiency decreases dramatically as the mass of the adsorbed species increases. Furthermore, quantum tunneling depends on both the energy barrier and the separation to the site the particle is tunneling into. It typically decreases dramatically for disordered or amorphous surfaces. Hama *et al.*[38] studied the diffusion of hydrogen atoms on the surface of amorphous ice, using a combination of photon-stimulated desorption and REMPI (Resonance Enhanced MultiPhoton Ionization) and found that hydrogen becomes trapped at deep sites, confirming the indirect evidence obtained in the study of H_2 formation on silicates and amorphous carbon[4,6,39] and on amorphous silicates.[34] For ad-species other than hydrogen, thermal diffusion is the dominant diffusion mechanism, and this will be the focus of this paper.

The remainder of this paper is organized as follows. In the next section we summarize the most general form of a rate equation model and discuss its limitations, then in Section 3 we present the diffusion–desorption rate equation model, and in Section 4 we apply it to the systems CO on MgO(100),[40] D_2 on non-porous water ice,[10] and HD on amorphous silicate surface, followed by a discussion of the limitations and suggestions for experimental investigations. In Section 5 we summarize the paper.

2   Rate equation model

Temperature programmed desorption (TPD)[41,42] has been used widely as a tool to study gas–surface interactions in astrochemistry, such as adsorption–desorption and surface diffusion.[9,10,12,13,40,43–45] A typical TPD experiment consists of two stages, the exposure stage in which the substrate surface is exposed to adsorbate particles, and the warm-up stage in which the substrate surface is warmed up to desorb adsorbate material. A mass spectrometer is placed in the vacuum chamber to monitor particles desorbing from the surface. Interpretation of TPD spectra is generally carried out using the Polanyi–Wigner rate equation; assuming first order desorption, the desorption rate can be written as:

$$R(t) \propto -\frac{d\theta(t)}{dt} = \nu\theta(t)\exp\left(-\frac{E_{des}}{k_B T(t)}\right) \tag{1}$$

where $R(t)$ is the desorption rate, ν is the desorption pre-exponential factor that depends on the substrate and adsorbate – hereafter we use the standard value of 10^{12} s^{-1}, $\theta(t)$ is the coverage, defined as the percentage of 1 ML, *i.e.*, the number of adsorbate particles divided by the number of adsorption sites on the surface, E_{des} is the desorption energy, k_B is the Boltzmann constant, and $T(t)$ is the temperature of the surface. This is a simple model assuming all surface sites are identical. In reality, however, a surface consists of different adsorption sites, and to better describe the desorption behavior a continuous desorption energy distribution $f(E_{des})$ is required. We modify the rate equation as:

$$\int f(E_{des})\mathrm{d}E_{des} = 1 \tag{2}$$

$$\frac{\mathrm{d}\theta(E_{des}, t)}{\mathrm{d}t} = \mathrm{flux}(t)(1 - \theta(E_{des}, t)) - \nu\theta(E_{des}, t)\exp\left(-\frac{E_{des}}{k_B T(t)}\right) \tag{3}$$

$$R(t) = \int \left(\mathrm{flux}(t)\theta(E_{des}, t) + \nu\theta(E_{des}, t)\exp\left(-\frac{E_{des}}{k_B T(t)}\right)\right) f(E_{des})\,\mathrm{d}E_{des} \tag{4}$$

Eqn (2) gives the desorption energy distribution, eqn (3) gives the coverage as a function of time for different sites, with $(1 - \theta(E_{des}, t))$ as a rejection term to avoid multiple occupancy of the same site, and eqn (4) describes the desorption rate plus reflection of the incoming flux, which is proportional to the measured mass spectra signal. The integration in eqn (2) and eqn (4) is over the whole desorption energy spectrum.

This set of equations forms the basis for the interpretation of most adsorption–desorption experiments. The more energy levels used in the modeling, the smoother the energy distribution becomes. In the work of He et al.,[12] more than 50 energy levels were used to fit the TPD spectra; a semi-continuous desorption energy distribution was obtained for both amorphous and single crystalline silicates. To obtain a continuous desorption energy distribution, some groups used a direct inversion method based on first order desorption to extract information from TPD spectra;[46] a more detailed analysis can be found in Barrie et al.[47] All these analyses are based on first order desorption without diffusion. We show below that diffusion is necessary to explain the experimental data.

In some systems, however, it is found that the desorption energy is coverage dependent; as coverage increases from 0 to 1 ML, the TPD peak shifts to lower temperatures, as shown in Dohnálek et al.[40] and Amiaud et al.[10] (referred to as Dohnálek 2001 and Amiaud 2007 hereafter). There are two different explanations for the temperature shift, adsorbate lateral interactions and hopping between different adsorption sites. For the former, lattice-gas modeling has been used to simulate the effect of lateral interactions on TPD shapes,[14,48,49] and fitting with TPD traces can be obtained. However, in the interstellar medium the coverage of ad-species on grain surfaces is typically very small (much smaller than 1 ML). In low coverage experiments that find applications in astrochemistry, the morphology and hopping between different adsorption sites of the surface contribute more to the desorption than lateral interactions. Therefore, in this paper we focus on the effects of hopping and ignore lateral interactions. In the exposure stage, particles tend to occupy deeper adsorption sites before they occupy the shallower sites. The intrinsic physical process underlying this phenomenon is surface diffusion, i.e., when adsorbate particles are in the shallow sites, the diffusion energy barrier is also low. Particles escape from the shallow adsorption site and move on the surface until they reach deeper sites where the diffusion energy barrier is high enough that they are trapped.

In an analysis of CO desorption from MgO TPD experimental data, Dohnálek 2001 used an inversion method to extract a continuous coverage dependent desorption energy distribution, and the extracted spectrum can reproduce the experimental traces very well. It should be noted that the direct inversion method is only applicable to the equilibrium diffusion state, in which the mobility of particles on a surface is so fast that particles are already in an equilibrium state before desorption begins. By using the direct inversion method, Dohnálek 2001 and Amiaud 2007 implicitly assume that the mobility of adsorbate particles is fast enough, *i.e.*, the diffusion rate is high enough, that the adsorbate is equilibrated. The effect of limited mobility of ad-species was discussed by Šurda *et al.*[50] but, to best of our knowledge, this work has not been used in astrochemistry. In the next section we introduce the diffusion–desorption rate equation that was inspired by the work of Šurda *et al.*[50]

3   The diffusion–desorption rate equation model

An adsorption–desorption experiment involves both desorption and diffusion processes. Hereby, the desorption rate and diffusion rate both depend on the substrate surface temperature. Below, we assume that both obey an Arrhenius-like expression. The desorption rate for sites with desorption energy E_{des} can be expressed as $\nu\theta(E_{\text{des}}, t)\exp\left(-\dfrac{E_{\text{des}}}{k_{\text{B}}T(t)}\right)$; similarly, the diffusion rate is

$\nu_{\text{diff}}\theta(E_{\text{des}}, t)\exp\left(-\dfrac{E_{\text{diff}}}{k_{\text{B}}T(t)}\right)$, where E_{diff} is the diffusion energy barrier for those

sites with desorption energy E_{des}. There is a direct positive relationship between E_{des} and E_{diff}, *i.e.*, sites with higher desorption energy also have higher diffusion energy barriers. A distinction is made between the two pre-factors ν_{diff} and ν, since they correspond to different physical processes and could have different values. Diffusion consists of two subprocesses: 1) particles hop out of adsorption sites and go into a transition state; 2) particles in the transition state go back to the adsorption sites. There is a redistribution in the second subprocess depending on the availability of adsorption sites. If a particle can go on top of another in the same adsorption site, particles in transition states will be redistributed evenly among all adsorption sites. The redistribution obeys the same distribution as the desorption energy. Otherwise, if a particle cannot go on top of another, the particle in the transition state will be redistributed evenly among all *empty* sites; hereafter we assume the latter case.

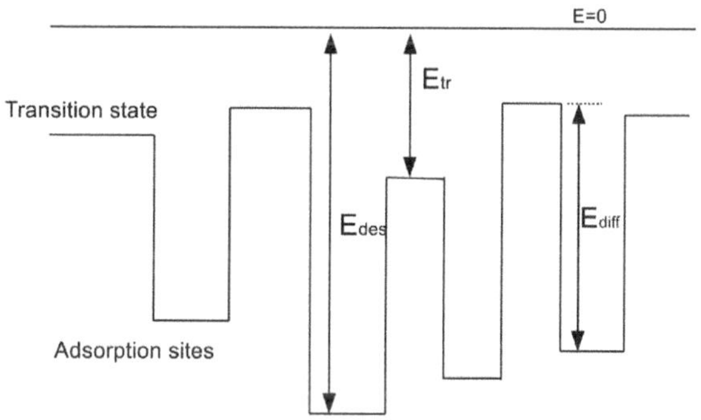

Fig. 1   Diagram of surface adsorption sites and transition states.

The desorption energy is a continuous distribution. Let us assume it can be represented by a single Gaussian distribution. As is shown in Fig. 1, the diffusion energy barrier E_{diff} is the difference between the desorption energy E_{des} and the energy at the transition state E_{tr}, $E_{\text{diff}} = E_{\text{des}} - E_{\text{tr}}$. Supposing that E_{tr} also follows a Gaussian distribution, then we express these two distributions as:

$$f(E_{\text{des}}) = \frac{1}{\sigma_{\text{des}}\sqrt{2\pi}}\exp\left(-\frac{(E_{\text{des}} - \overline{E_{\text{des}}})^2}{2\sigma_{\text{des}}^2}\right)$$

$$f(E_{\text{tr}}) = \frac{1}{\sigma_{\text{tr}}\sqrt{2\pi}}\exp\left(-\frac{(E_{\text{tr}} - \overline{E_{\text{tr}}})^2}{2\sigma_{\text{tr}}^2}\right)$$

Assuming that these two distributions are independent, then E_{diff} is also a Gaussian distribution, with $\overline{E_{\text{diff}}} = \overline{E_{\text{des}}} - \overline{E_{\text{tr}}}$ and $\sigma_{\text{diff}} = \sqrt{\sigma_{\text{des}}^2 + \sigma_{\text{tr}}^2}$. If we further assume that $\overline{E_{\text{diff}}}$ is not far from $\overline{E_{\text{des}}}$ – it has been suggested that $E_{\text{diff}} \sim 0.5$–$0.7\ E_{\text{des}}$[4,34,51] – then σ_{tr}^2 should be much smaller than σ_{des}^2, and we assume $\sigma_{\text{diff}} = \sqrt{\sigma_{\text{des}}^2 + \sigma_{\text{tr}}^2} \approx \sigma_{\text{des}}$, which means that E_{diff} can be approximated by $E_{\text{diff}} = E_{\text{des}} - \overline{E_{\text{tr}}} = E_{\text{des}} - \Delta E$. If E_{tr} and E_{des} are dependent, perhaps the approximation $E_{\text{diff}} = \alpha E_{\text{des}}$[52] is more appropriate. A more general expression would be $E_{\text{diff}} = \alpha(E_{\text{des}} - \Delta E)$.

Now eqn (3) should be modified as follows:

$$\frac{\mathrm{d}\theta(E_{\text{des}}, t)}{\mathrm{d}t} =$$

$$\underbrace{flux(t)(1 - \theta(E_{\text{des}}, t))}_{\text{term 1}} - \underbrace{v\theta(E_{\text{des}}, t)\exp\left(-\frac{E_{\text{des}}}{k_{\text{B}}T(t)}\right)}_{\text{term 2}}$$

$$\underbrace{- v_{\text{diff}}\theta(E_{\text{des}}, t)\exp\left(-\frac{\alpha(E_{\text{des}} - \Delta E)}{k_{\text{B}}T(t)}\right)}_{\text{term 3}} \tag{5}$$

$$\underbrace{+ \frac{1 - \theta(E_{\text{des}}, t)}{1 - \Theta(t)}\int v_{\text{diff}}\theta(E_{\text{des}}', t)\exp\left(-\frac{\alpha(E_{\text{des}}' - \Delta E)}{k_{\text{B}}T(t)}\right)f(E_{\text{des}}')\mathrm{d}E_{\text{des}}'}_{\text{term 4}}$$

$$\Theta(t) = \int\theta(E_{\text{des}}, t)f(E_{\text{des}})\mathrm{d}E_{\text{des}} \tag{6}$$

Eqn (2) and eqn (4) are unchanged. In eqn (5), term 1 is the flux term, assuming particles cannot go on top of each other; term 2 is the desorption term; term 3 is the diffusion term that describes particles that go into transition states; term 4 is the redistribution term from the transition states, assuming particles can only go to empty sites; and the factor $(1 - \theta(E_{\text{des}},t))/(1 - \Theta(t))$ accounts for the redistribution of particles in transition states among empty adsorption sites. In eqn (6), $\Theta(t)$ is the overall coverage. Note that integration of terms 3 and 4 over the whole energy spectrum gives zero; this means that in the diffusion processes the total particle number is conserved, which should hold true for the model to be correct, since diffusion alone won't change the total number of particles on the surface. The diffusion term is no longer a first order term, in fact it behaves like a second order term. Thus the modified rate equation model is a candidate to interpret coverage dependent desorptions. A comparison with the complete model in Li *et al.*[53] (referred to as Li 2010 hereafter) is worth mentioning here. It can be shown that in the low coverage limit, this model is equivalent to eqn (5) in Li 2010; however, as the coverage approaches 1 ML, the models differ. In Li 2010 it is assumed that at a coverage close to 1 ML the diffusion rate is close to 0; however in this paper we assume that the probability that particles jump out from adsorption sites to transition sites is independent of coverage, which leads to the exchange of particles among adsorption sites and a faster redistribution. The calculated diffusion rate in this work is therefore faster than that in Li 2010.

4   Simulations and discussion

4.1   CO on MgO(100)

Below we illustrate how to utilize the rate equation model to extract the adsorption–diffusion energy parameters. The experimental data are taken from Dohnálek 2001, where a detailed description of the experiment can be found.[40] A highly ordered MgO(100) surface kept at 22 K was exposed to different doses of CO gas in an ultra-high vacuum. The MgO(100) film was grown epitaxially on a Mo(100) substrate at 600 K by the evaporation of Mg metal in an O_2 atmosphere. The dose of CO was calculated by integration of TPD traces to be $\theta = (0.09, 0.16, 0.23, 0.32, 0.42, 0.48, 0.58, 0.71, 0.90, 1.00, 1.14, 1.27)$ ML. Below we only discuss the submonolayer range, omitting the last two coverage values. After exposure the thin film surface was subjected to a linear heating ramp of $\beta = dT/dt = 0.6$ K s^{-1} to desorb the CO molecules. The TPD traces are shown in Fig. 2. We begin with the direct inversion of eqn (1); the coverage dependent desorption energy can be calculated for each TPD trace as follows:

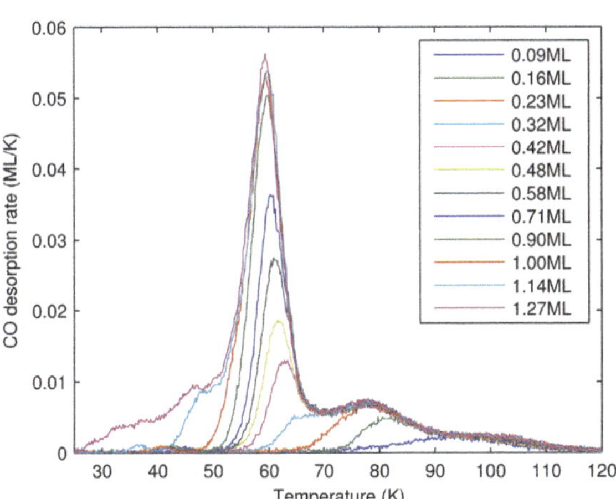

Fig. 2   CO TPD spectra for monolayer and submonolayer coverages of CO/MgO(100) ($\theta = 0.09, 0.16, 0.23, 0.32, 0.42, 0.48, 0.58, 0.71, 0.90, 1.00, 1.14, 1.27$). Replotted from Dohnálek 2001.[40]

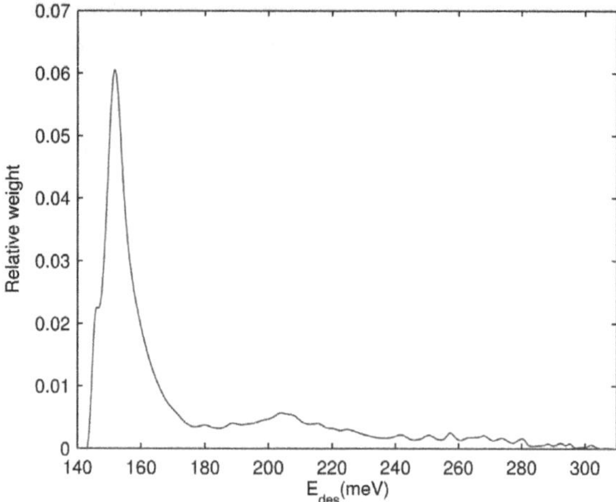

Fig. 3   Desorption energy barrier distribution for CO on MgO(100), obtained using data from Dohnálek 2001.[40]

$$E(\theta) = -k_{\mathrm{B}}T\ln\left(-\frac{\beta}{\nu\theta}\frac{\mathrm{d}\theta}{\mathrm{d}T}\right) \tag{7}$$

The resulting desorption energy distribution $f(E_{\mathrm{des}})$ is shown in Fig. 3.

In Dohnálek 2001, the TPD traces for different initial coverages are inverted and different pre-exponential factors are tried until convergence is achieved among the inverted distributions from different initial coverages. They found $\nu = 10^{15}$ s^{-1}. In our opinion this value should be revised. Nordholm $et\ al.$[54] found that abnormally large pre-exponential factors might come from dispersion in the desorption energy. Thus, to determine the pre-exponential factor using convergence might be incorrect. ν is a fundamental factor that depends on both the substrate and the adsorbate, and it is better to determine its value by $ab\ initio$ calculations. We are not aware of such calculations for the present system. Therefore, we prefer to use the widely accepted value $\nu = 10^{12}$ s^{-1}. For ν_{diff} there is even less information available, and we use the same value as ν for simplicity. To do the inversion, we pick the trace with the highest submonolayer coverage from the TPD data.

In eqn (5), if one lets $\nu_{\mathrm{diff}} = 0$ then the diffusion–desorption rate equation is a rate equation for non-hopping ad-species. The simulated TPD traces are shown in Fig. 4. Next, we set ν and the flux equal to 0, $i.e.$, we assume no incoming flux and no desorption, and we focus on the diffusion process. We start from an initial coverage of 0.3 ML for all desorption sites with $\Delta E = 30$ meV and $\nu_{\mathrm{diff}} = 10^{12}$ s^{-1}, and then monitor the redistribution process during warm-up. Fig. 5 shows the fractional occupation distribution at temperatures 22 K, 40 K, 45 K, 50 K, 55 K, 60 K, 65 K, 70 K and 75 K. As shown in the figure, at 22 K and 40 K, the surface has a uniform coverage of 0.3 ML. When the surface temperature rises to 45 K, molecules in shallow sites ($E_{\mathrm{des}} < 150$ meV) become active and begin to diffuse, distributing themselves evenly among deeper sites that have not been occupied yet. As the temperature rises further, deeper sites come into play and molecules begin to diffuse. After the temperature reaches the point at which the deepest

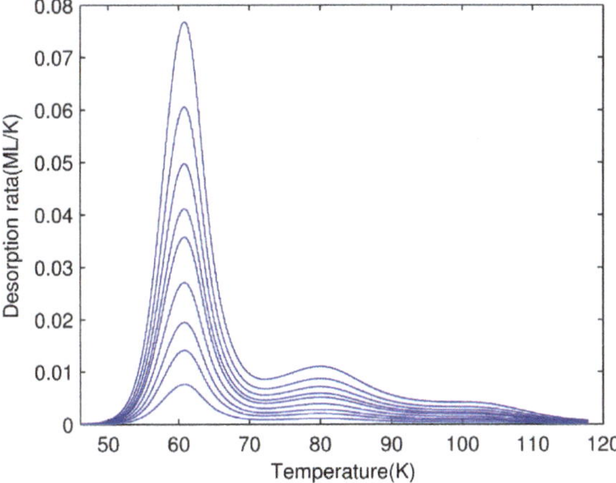

Fig. 4 Simulation of TPD spectra for different initial coverages using the desorption energy distribution obtained from direct inversion of the CO on MgO(100) TPD spectra of Dohnálek 2001,[40] see Fig. 3.

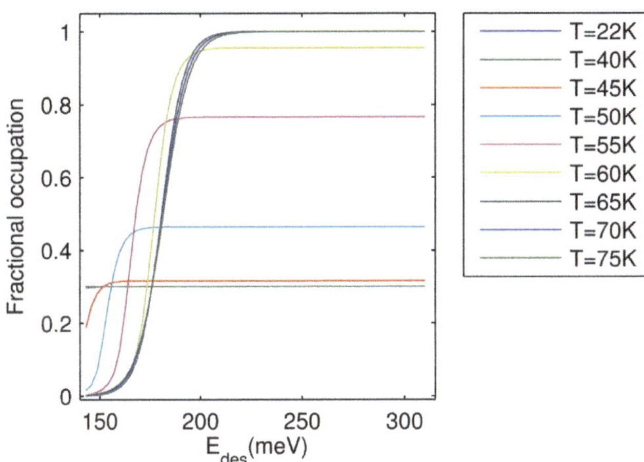

Fig. 5 Fractional occupation of adsorption sites at different temperatures: 22 K, 40 K, 45 K, 50 K, 55 K, 60 K, 65 K, 70 K and 75 K. No desorption is assumed and only the diffusion process is simulated. The initial coverage and ΔE are 0.3 ML and 30 meV, respectively. As the temperature increases, molecules move from shallow adsorption sites to deeper ones.

sites are fully occupied (about 65 K in this case), the distribution doesn't change any more as we increase the temperature further. This means that when the diffusion rate is high enough, the coverage distribution doesn't change as the diffusion rate is further increased. In the fast diffusion case one cannot get the absolute value of the diffusion energy barrier and only a lower bound can be obtained, or equivalently, a lower bound for ΔE. What one sees in the TPD is that molecules occupy deeper sites before they go to shallower sites, and the TPD trailing edges for different initial coverages should converge. This justifies the use of the direct inversion method, since the direct inversion method assumes that molecules automatically occupy the deepest sites available on the surface. On the other hand, if the diffusion rate is limited, ad-species start to desorb and leave the surface before an equilibrium state is achieved.[50] If one plots the weighted overall desorption rate and diffusion rate as a function of temperature, the two traces might overlap with each other. If the overlapping is small, *i.e.* the desorption starts after most molecules have begun to diffuse, the assumption of fast diffusion holds true. However, if the two traces overlap significantly, an equilibrium state cannot be obtained before desorption begins, and the direct inversion method is less applicable. In this latter case, the diffusion–desorption rate equation model should be used.

Now we show how to obtain the E_{diff} distribution, or equivalently, the value (or range) of ΔE. We try different values of ΔE in eqn (5) and simulate TPD for different initial values of the coverage until we find the value that best fits the experimental data. The results are shown in Fig. 6. It can be seen from the simulation results that when ΔE is small, the TPD traces are similar to first order desorption with high localization. As ΔE increases, the gaps between trailing edges decrease. When ΔE is greater than 30 meV there are almost no gaps, and the shapes of the TPD traces do not change as the ΔE value is increased further. Thus we only get a lower boundary for ΔE in the case of fast diffusion. This is consistent with our previous argument.

4.2   D$_2$ on amorphous water ice

H_2/D_2 plays an important role in interstellar chemistry. Here we examine D_2 adsorption–desorption on non-porous amorphous water ice as a second example of how to obtain desorption and diffusion energy barriers. The experiments were reported in Amiaud *et al.*[10] The water ice films were grown on a copper surface using a microchannel array doser. The morphology was found to be non-porous amorphous by N_2 adsorption–desorption TPD experiments. The sample was kept

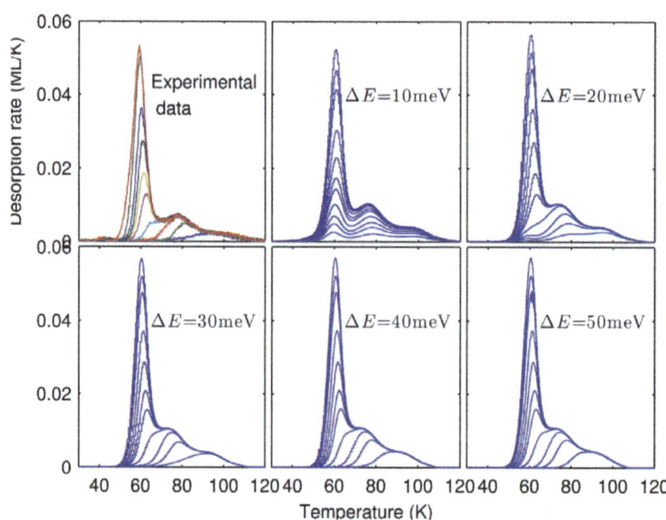

Fig. 6   Simulation of Dohnálek 2001[40] experimental data using different ΔE values, based on the diffusion–desorption rate equation model. The experimental data are shown in the first panel.

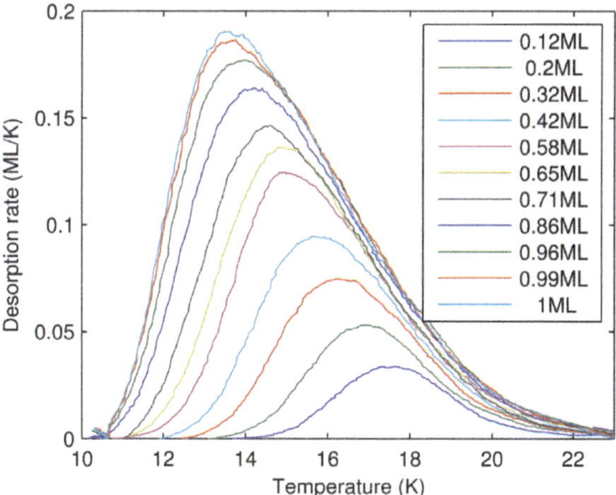

Fig. 7   TPD spectra for D_2 adsorption–desorption from non-porous amorphous water ice; the surface temperature during exposure to D_2 was 10 K. Data from Amiaud et al.[10]

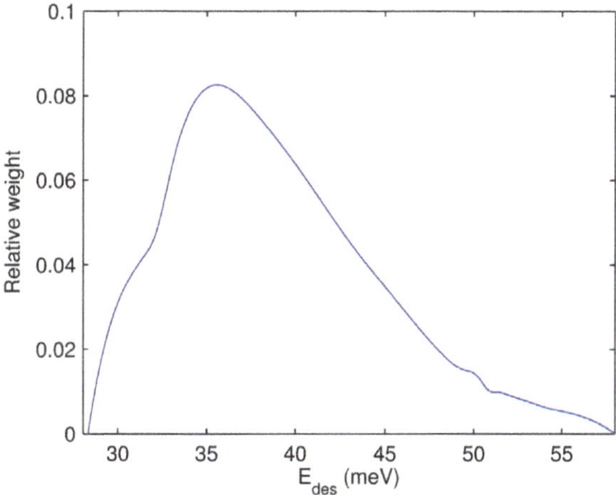

Fig. 8   Desorption energy barrier distribution for D_2 obtained from the model using the non-porous amorphous water ice data of Amiaud et al.[10]

at 10 K and exposed to various doses of D_2. After exposure, TPD was carried out with a heating rate of 10 K min^{-1}. The TPD spectra are shown in Fig. 7. Following the same direct inversion procedure as discussed above, we obtain a desorption energy barrier distribution E_{des} as shown in Fig. 8. From the thus-obtained E_{des} distribution, we simulate TPD spectra for different ΔE values. The results are shown in Fig. 9. Next, we compare the simulation results and the experimental TPD profiles to find out the best ΔE value. Note that in the trailing edge of the TPD profiles there are gaps between the traces. As is clear from our previous discussion and from Fig. 6, if the diffusion rate is fast enough there shouldn't be gaps in the trailing edges. Thus, the diffusion rate is not fast enough, and this appears to be desorption of a non-equilibrated film. A best fit gives $\Delta E = 2$ meV. If the gaps in the trailing edge are due to the limited pumping speed in the vacuum chamber,

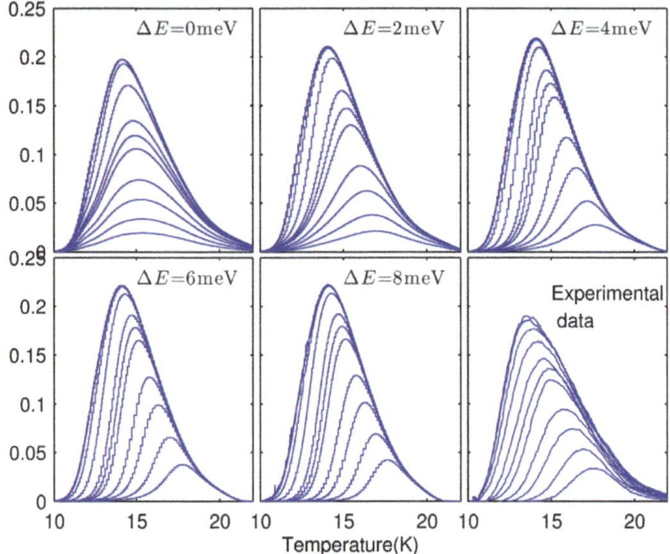

Fig. 9   Simulation of the adsorption–desorption TPD spectra of Amiaud et al.[10] using different ΔE values; the original experimental data are also shown for comparison.

the experimental TPD traces should be corrected for it. This may reduce the gaps in the trailing edges and make the experimental TPD peaks sharper. If after pumping speed corrections there are no gaps at all, then the best fit is $\Delta E \geq 6$ meV. In this case we would have a lower limit of ΔE. The pumping speed corrections are not presently available for this system.

4.3   HD on amorphous silicate

Here we analyze the HD desorption from an amorphous silicate surface. The sample was prepared by Dr. Brucato (then at the Osservatorio Astronomico di Capodimonte, Naples) by laser ablation of targets in an oxygen atmosphere. The experimental set-up and measuring methods are similar to the ones described by Perets et al.[34] In brief, an amorphous silicate ($FeMgSiO_4$) sample was kept at 10.5 K and exposed to a HD beam flux for 30 s, 1 min, 2 min, 4 min and 8 min. Coverage calibration is unavailable, but it is known that the coverages are well below 1 ML. After exposure, the surface was warmed up by cutting off the liquid helium flow. A reproducible heating curve is achieved for different TPD runs. The TPD traces are shown in Fig. 10. Applying the direct inversion method to the trace for 4 min exposure, which is a typical result, we obtain the desorption energy distribution shown in Fig. 11. The TPD spectra are typical of first order desorption without hopping; this indicates that the diffusion energy barriers are similar to or even higher than the desorption energies, so that diffusion doesn't take place before desorption.

4.4   Limitations and suggestions for experimental investigations

As we have seen in the first two examples, although we are able to obtain good fits for the adsorption–desorption TPD profiles, we can get only a lower limit of ΔE.

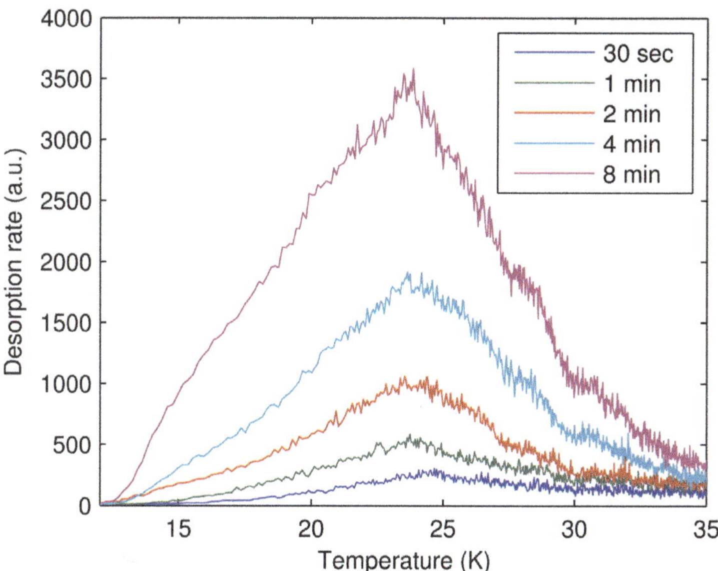

Fig. 10 HD desorption from an amorphous silicate (FeMgSiO$_4$) surface. The exposure doses were 30 s, 1 min, 2 min, 4 min and 8 min. The surface temperature during exposure with HD was 10.5 K.

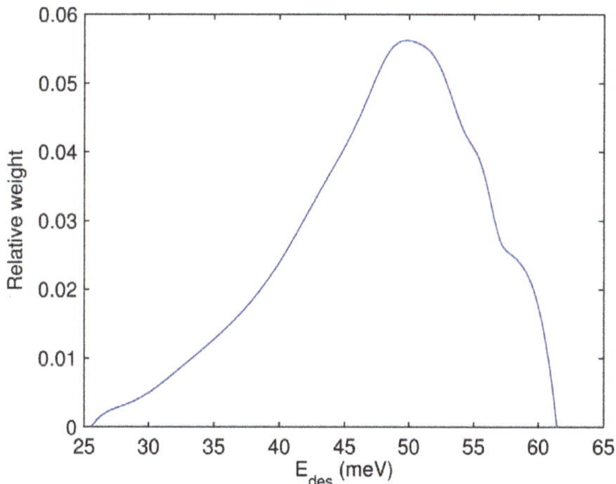

Fig. 11 HD desorption energy distribution obtained from direct inversion of the TPD spectra in Fig. 10.

Furthermore, we considered only thermally activated diffusion, since this is what several experiments suggest. Based on our analyses, we suggest steps in future investigations that could facilitate the extraction of important parameters. In experimental investigations, efforts should be made to push the film to a state far from equilibrium, to separate the trailing edges. This has been done with flash desorption in many surface science studies,[42] but it poses technical challenges for surfaces kept at liquid helium temperature. It would be advantageous to study films at very low coverages, since the differences in TPD traces with different ΔE are most significant when the coverages are low, as can be seen in Fig. 6. Furthermore, special attention needs to be paid to the pumping speed, background noise and desorption from parts other than the sample, since these affect the TPD traces, especially the tailing edges.

5   Conclusions

In this paper we discussed the limitations of the often-used rate equation model when simulating adsorption–desorption TPD experimental profiles, and introduced a diffusion–desorption rate equation model to take into account the diffusion processes. Desorption and diffusion energy barrier distributions were obtained for CO on MgO(100) and D_2 on non-porous amorphous water ice. We then presented a system, HD desorption from an amorphous silicate, in which there is hardly any evidence of diffusion. These distributions, and the information obtained from them, can be used in models of the chemical evolution of interstellar medium environments. For example, Herbst et al.[55] and Iqbal et al.[8] have taken into consideration the morphology of the surface in the formation of molecular hydrogen on surfaces in their kinetic Monte Carlo models. At that time, little or no information was available on the diffusion energy barriers and on distribution of adsorption of energy sites besides a little more than rule of thumb. Now we have shown that using existing TPD data it is possible to tease out important information on the energetics of adsorption–desorption and diffusion. Furthermore we have made suggestions for future experimental investigations so that analyses of the data will yield more detailed information on these important processes that are at the heart of molecule formation on interstellar dust grains.

Acknowledgements

We thank Z. Dohnálek and J.-L. Lemaire for providing data. We are grateful for partial financial support from NSF, Astronomy & Astrophysics Division (grant no. 0908108 and no. 1311958) and NASA (grant no. NNX12AF38G).

References

1 D. A. Williams, *J. Phys. Conf. Ser.*, 2005, **6**, 1–17.
2 R. T. Garrod, S. L. W. Weaver and E. Herbst, *Astrophys. J.*, 2008, **682**, 283–302.
3 G. Vidali, *J. Low Temp. Phys.*, 2013, **170**, 1–30.
4 N. Katz, I. Furman, O. Biham, V. Pirronello and G. Vidali, *Astrophys. J.*, 1999, **522**, 305.
5 V. Pirronello, O. Biham, C. Liu, L. O. Shen and G. Vidali, *Astrophys. J. Lett.*, 1997, **483**, L131.
6 V. Pirronello, C. Liu, L. Shen and G. Vidali, *Astrophys. J.*, 1997, **475**, L69–L72.
7 S. Cazaux and A. G. G. M. Tielens, *Astrophys. J.*, 2004, **604**, 222–237.
8 W. Iqbal, K. Acharyya and E. Herbst, *Astrophys. J.*, 2012, **751**, 58.
9 L. Amiaud, J.-H. Fillion, S. Baouche, F. Dulieu, A. Momeni and J. L. Lemaire, *J. Chem. Phys.*, 2006, **124**, 94702.
10 L. Amiaud, F. Dulieu, J.-H. Fillion, A. Momeni and J. L. Lemaire, *J. Chem. Phys.*, 2007, **127**, 144709.

11 J. A. Noble, E. Congiu, F. Dulieu and H. J. Fraser, *Mon. Not. R. Astron. Soc.*, 2012, **421**, 768–779.

12 J. He, P. Frank and G. Vidali, *Phys. Chem. Chem. Phys.*, 2011, **13**, 15803–9.

13 D. J. Burke and W. A. Brown, *Phys. Chem. Chem. Phys.*, 2010, **12**, 5947–69.

14 S. Sundaresan and K. Kaza, *Surf. Sci.*, 1985, **160**, 103–121.

15 N. Watanabe and A. Kouchi, *Astrophys. J.*, 2002, **571**, L173–L176.

16 H. Hidaka, N. Watanabe, T. Shiraki, A. Nagaoka and A. Kouchi, *Astrophys. J.*, 2004, **614**, 1124–1131.

17 N. Watanabe and A. Kouchi, *Prog. Surf. Sci.*, 2008, **83**, 439–489.

18 G. W. Fuchs, H. M. Cuppen, S. Ioppolo, C. Romanzin, S. E. Bisschop, S. Andersson, E. F. van Dishoeck and H. Linnartz, *Astron. Astrophys.*, 2009, **505**, 629–639.

19 H. Hidaka, M. Watanabe, A. Kouchi and N. Watanabe, *Phys. Chem. Chem. Phys.*, 2011, **13**, 15798–802.

20 S. Ioppolo, H. M. Cuppen, C. Romanzin, E. F. van Dishoeck and H. Linnartz, *Astrophys. J.*, 2008, **686**, 1474–1479.

21 N. Miyauchi, H. Hidaka, T. Chigai, A. Nagaoka, N. Watanabe and A. Kouchi, *Chem. Phys. Lett.*, 2008, **456**, 27–30.

22 Y. Oba, N. Miyauchi, H. Hidaka, T. Chigai, N. Watanabe and A. Kouchi, *Astrophys. J.*, 2009, **701**, 464–470.

23 F. Dulieu, L. Amiaud, E. Congiu, J.-H. Fillion, E. Matar, A. Momeni, V. Pirronello and J. L. Lemaire, *Astron. Astrophys.*, 2010, **512**, A30.

24 D. Jing, J. He, J. Brucato, A. De Sio, L. Tozzetti and G. Vidali, *Astrophys. J.*, 2011, **741**, L9.

25 D. Jing, J. He, J. J. R. Brucato, G. Vidali, L. Tozzetti and A. De Sio, *Astrophys. J.*, 2012, **756**, 98.

26 R. Kaiser, G. Eich, A. Gabrysch and K. Roessler, *Astron. Astrophys.*, 1999, **346**, 340–344.

27 J. He, K. Gao and G. Vidali, *Astrophys. J.*, 2010, **721**, 1656–1662.

28 F. Mispelaer, P. Theulé, H. Aouididi, J. Noble, F. Duvernay, G. Danger, P. Roubin, O. Morata, T. Hasegawa and T. Chiavassa, *Astron. Astrophys.*, 2013, **555**, A13.

29 J. He, D. Jing and G. Vidali, *Phys. Chem. Chem. Phys.*, 2014, **16**, 3493–3500.

30 T. Zubkov, R. S. Smith, T. R. Engstrom and B. D. Kay, *J. Chem. Phys.*, 2007, **127**, 184707.

31 J. E. Roser, G. Vidali, G. Manicò and V. Pirronello, *Astrophys. J.*, 2001, **555**, L61–L64.

32 G. Antczak and G. Ehrlich, *Surf. Sci.*, 2005, **589**, 52–66.

33 H. B. Perets, O. Biham, G. Manico, V. Pirronello, J. Roser, S. Swords and G. Vidali, *Astrophys. J.*, 2005, **627**, 850–860.

34 H. B. Perets, A. Lederhendler, O. Biham, G. Vidali, L. Li, S. Swords, E. Congiu, J. Roser, G. Manicó, J. R. Brucato and V. Pirronello, *Astrophys. J.*, 2007, **661**, L163–L166.

35 T. Stantcheva, V. I. Shematovich and E. Herbst, *Astron. Astrophys.*, 2002, **391**, 1069–1080.

36 J. Le Bourlot, F. Le Petit, C. Pinto, E. Roueff and F. Roy, *Astron. Astrophys.*, 2012, **541**, A76.

37 Y. Oba, N. Watanabe, T. Hama, K. Kuwahata, H. Hidaka and A. Kouchi, *Astrophys. J.*, 2012, **749**, 67.

38 T. Hama, K. Kuwahata, N. Watanabe, A. Kouchi, Y. Kimura, T. Chigai and V. Pirronello, *Astrophys. J.*, 2012, **757**, 185.

39 V. Pirronello, C. Liu, J. E. Roser and G. Vidali, *Astron. Astrophys.*, 1999, **344**, 681–686.

40 Z. Dohnálek, G. A. Kimmel, S. A. Joyce, P. Ayotte, R. S. Smith and B. D. Kay, *J. Phys. Chem. B*, 2001, **105**, 3747–3751.

41 J. T. Yates, *Methods of Experimental Physics: Solid State Physics: Surfaces*, Academic Press, 1985, vol. 22, p. 425.

42 K. W. Kolasinki, *Surface science: foundations of catalysis and nanoscience*, Wiley, Hoboken, NJ, 2008.

43 M. P. Collings, M. A. Anderson, R. Chen, J. W. Dever, S. Viti, D. A. Williams and M. R. S. McCoustra, *Mon. Not. R. Astron. Soc.*, 2004, **354**, 1133–1140.

44 S. E. Bisschop, H. J. Fraser, K. I. Öberg, E. F. van Dishoeck and S. Schlemmer, *Astron. Astrophys.*, 2006, **449**, 1297–1309.

45 L. Hornekaer, A. Baurichter, V. V. Petrunin, A. C. Luntz, B. D. Kay and A. Al-Halabi, *J. Chem. Phys.*, 2005, **122**, 124701.

46 G. Vidali and L. Li, *J. Phys.: Condens. Matter*, 2010, **22**, 304012.

47 P. J. Barrie, *Phys. Chem. Chem. Phys.*, 2008, **10**, 1688–96.

48 Y. Tovbin, *Prog. Surf. Sci.*, 1990, **34**, 1–235.

49 B. Meng and W. Weinberg, *Surf. Sci.*, 1997, **374**, 443–453.

50 A. Šurda and I. Karasova, *Surf. Sci.*, 1981, **109**, 605–620.

51 U. A. Yıldız, K. Acharyya, P. F. Goldsmith, E. F. van Dishoeck, G. Melnick, R. Snell, R. Liseau, J.-H. Chen, L. Pagani, E. Bergin, P. Caselli, E. Herbst, L. E. Kristensen, R. Visser, D. C. Lis and M. Gerin, *Astron. Astrophys.*, 2013, **558**, A58.

52 L. Bruch, R. Diehl and J. Venables, *Rev. Mod. Phys.*, 2007, **79**, 1381–1454.

53 G. Vidali, L. Li, H. Zhao, Y. Frank, I. Lohmar, H. B. Perets and O. Biham, *J. Phys. Chem. A*, 2010, **114**, 10575–83.

54 S. Nordholm, *Chem. Phys.*, 1985, **98**, 367–379.

55 E. Herbst and H. M. Cuppen, *Proc. Natl. Acad. Sci. U. S. A.*, 2006, **103**, 12257–62.

Faraday Discussions

PAPER

Wavelength resolved UV photodesorption and photochemistry of CO_2 ice

J.-H. Fillion,[*a] E. C. Fayolle,[b] X. Michaut,[a] M. Doronin,[a] L. Philippe,[a] J. Rakovsky,[a] C. Romanzin,[c] N. Champion,[d] K. I. Öberg,[e] H. Linnartz[b] and M. Bertin[a]

Received 11th December 2013, Accepted 6th January 2014

DOI: 10.1039/c3fd00129f

Over the last four years we have illustrated the potential of a novel wavelength-dependent approach in determining molecular processes at work in the photodesorption of interstellar ice analogs. This method, utilizing the unique beam characteristics of the vacuum UV beamline DESIRS at the French synchrotron facility SOLEIL, has revealed an efficient *indirect* desorption mechanism that scales with the electronic excitations in molecular solids. This process, known as DIET – desorption induced by electronic transition – occurs efficiently in ices composed of very volatile species (CO, N_2), for which photochemical processes can be neglected. In the present study, we investigate the photodesorption energy dependence of pure and pre-irradiated CO_2 ices at 10–40 K and between 7 and 14 eV. The photodesorption from pure CO_2 is limited to photon energies above 10.5 eV and is clearly initiated by CO_2 excitation and by the contribution of dissociative and recombination channels. The photodesorption from "pre-irradiated" ices is shown to present an efficient additional desorption pathway below 10 eV, dominating the desorption depending on the UV-processing history of the ice film. This effect is identified as an *indirect* DIET process mediated by photoproduced CO, observed for the first time in the case of less volatile species. The results presented here pinpoint the importance of the interconnection between photodesorption and photochemical processes in interstellar ices driven by UV photons having different energies.

Introduction

In recent years it has become clear that the non-thermal desorption of molecules from ice-covered dust grains plays an important role in determining the balance

[a]*Laboratoire de Physique Moléculaire pour l'Atmosphère et l'Astrophysique (LPMAA), CNRS UMR 7092, UPMC Univ. Paris 6, F-75252 Paris, France*
[b]*Sackler Laboratory for Astrophysics, Leiden Observatory, Leiden University, P. O. Box 9513, NL-2300 RA Leiden, The Netherlands*
[c]*Laboratoire de Chimie Physique (LCP), CNRS UMR 8000, Univ. Paris Sud, F-91400 Orsay, France*
[d]*LERMA, UMR 8112 du CNRS et Observatoire de Paris, 4 place J. Jansen, 92195 Meudon Cedex, France*
[e]*Harvard-Smithsonian Center for Astrophysics, 60 Garden Street, Cambridge, MA 02138, USA*

between gas phase and solid state species in the interstellar medium (ISM). Non-thermal desorption, for example, explains the astronomically observed gas phase abundances of species at temperatures well below their accretion value. Among the non-thermal desorption pathways, photon-stimulated desorption (PSD) appears to be a dominant mechanism, particularly in protoplanetary disks, but this process also is considered to be relevant in other regions where ices are exposed to UV photons.[1-4] Photodesorption rates, for instance, are needed to predict the position of snowlines in protoplanetary environments.[1,5] Quantifying and constraining photodesorption processes in sufficient detail is therefore a prerequisite for incorporation into astrochemical networks. Equally important is the thorough understanding of the underlying molecular processes.

The focus of this paper is carbon dioxide; CO_2 is one of the most abundant molecules present in interstellar ice after H_2O and CO, with abundances relative to solid H_2O varying from $\sim$15 to 40% along many lines of sight.[6-8] In the gas phase, typical abundances relative to H_2 of about $\sim 10^{-7}$ have been reported in warm and cold gas towards low and high mass protostars, where it acts as an important tracer of the chemical and physical history.[9,10] The observed solid state abundances in the ISM are 100-fold higher than in the gas phase and cannot be reproduced by gas phase chemical models, which is the reason why it is generally accepted that CO_2 forms in the solid state. In protostellar environments, CO_2 can be found in H_2O-rich or CO-rich ices, and it has also been observed as pure ice most likely originating from thermal processing and segregation of ice mixture components.[11,12] CO_2 has been observed in the upper layers of protoplanetary disks in the gas phase and in the condensed phase,[13] the latter resulting most likely from recondensation of upper layers and from incorporation of protostellar ice material.

The photodesorption of CO_2 ice has been the subject of several experimental studies over the last five years. Öberg et $al.$ investigated thin ice photodesorption by combining infrared reflection spectroscopy (RAIRS) and mass spectrometry (MS) to monitor ice depletion with irradiation time, and explored the influence of various physical parameters on the desorption efficiency.[14] They found a major effect following the photodissociation of CO_2 into CO and O and subsequent recombination to form energetic CO_2, resulting in thickness and temperature dependent photodesorption rates. Yuan and Yates[15,16] investigated isotopic effects in the photodesorption and photochemical decomposition of $^{12}CO_2/^{13}CO_2$ ice at 75 K, and concluded that these originate from differences in electronic energy transfer efficiencies from excited CO_2 molecules to the ice matrix. Combining quartz micro-balance techniques and mass spectrometry, Bahr and Baragiola[17] reported very large desorption yields for CO, O_2 and O from highly photoprocessed CO_2 ices at 40–60 K. Very recently, high desorption rates following radiation damage at high photon fluence combined with thermally activated desorption have been observed for crystalline CO_2 ice (75 K).[18] Besides these approaches based on discharge lamps, pump-probe laser-based experiments were performed by Kinugawa et $al.$[19] at 7.9 eV, yielding a state selective detection of $O(^3P)$ and CO ($v = 0, 1$) photofragments in the gas phase, arising from the photodissociation of CO_2 trapped in amorphous CO_2:H_2O samples at 90 K. Wavelength dependent photodesorption studies of CO_2 have not been reported to date.

UV-induced photo-processes in interstellar ice analogs have been simulated using different light sources. The most commonly applied source is the microwave discharge H_2 flow lamp, which provides high UV photon fluxes (typically 10^{13}–10^{15} photons cm^{-2} s^{-1}). Such "broad-band" photon-sources are believed to be well-suited to simulate UV interstellar fields, because they provide atomic hydrogen Lyman alpha emission at 10.2 eV (121.6 nm), which is quasi-ubiquitous in space, together with (depending on operating conditions) continuous emission at lower energies between 7 and 9 eV.[20] For decades such lamps have been used to study photochemical processes in generally thick (thousands of monolayers) ice mixtures, applying infrared and more recently also optical spectroscopy (*e.g.* ref. 21). Hydrogen discharge lamps were also the first systems widely used to derive absolute photodesoption rates of various and generally thinner (a few ML up to a few tens of ML thick) interstellar ice analogs, such as solid H_2O, CO, or CO_2.[14,22–25] The method applied in these studies is that the time dependent decrease in absolute surface density is recorded using infrared spectroscopy, eventually supported by simultaneous mass recording of the ejected species. With a known UV photon flux it is then possible to calculate the desorption rate.

In recent years the development of a new undulator-based VUV beamline (DESIRS) on the 2.75 GeV storage ring SOLEIL (Saint-Aubin, France) has provided an optimal flux-to-resolution solution, presenting the opportunity to investigate photo-processes with a continuously tunable monochromatic source combining (among other properties) high spectral purity and high photon fluxes.[26] In parallel to this development, the "Surface Processes & ICES" (SPICES, UPMC, Paris) set-up has been designed with the aim of coupling it with the DESIRS beamline, in order to initiate wavelength-dependent vacuum UV photodesorption studies. The approach relies on photon-stimulated desorption (PSD) experiments, in which atomic/molecular species are ejected from an ice film into the gas phase upon monochromatic vacuum UV (VUV) irradiation. The calibrated detection through mass spectroscopy as a function of the incident photon energy provides a unique wavelength-dependent signature of the ice photodesorption, reflected by a so-called PSD spectrum. The investigation of the VUV photodesorption of CO ice films at 18 K in the 7–14 eV energy range has benchmarked a series of new experimental studies supported by this approach.[27] Firstly, this has provided the absolute wavelength-dependent photodesorption rates needed in astrochemical models to interpret different radiation fields. Secondly, it has proven to be extremely powerful for revealing the underlying molecular processes. Our PSD spectra of pure ices composed of light volatiles species such as CO and N_2 have provided the first experimental proof of a desorption induced by electronic transition (DIET) mechanism, which dominates most of the desorption events taking place between 7 and 14 eV.[27,28] As an illustration, Fig. 1a displays the PSD spectrum recorded for VUV irradiated pure CO and N_2 ice samples. The spectra clearly reveal two vibrationally-resolved electronic transitions, identified as CO $A^1\Pi(v') - X^1\sum(v'' = 0)$ and N_2 $b\,^1\Pi_u(v') - X\,^1\sum_g(v'' = 0)$, which reflect the absorption properties of solid CO (below 10 eV) and solid N_2 (above 12 eV), respectively. Systematic investigations of these typical fingerprints in (isotopically labeled) ices layered samples led to the conclusion that DIET takes place in the near-surface region of the ice, involving at most the upper three molecular layers of the ice film.[29] By further investigating N_2:CO-mixed (and -layered) samples, the desorption from a mixed (or layered) film was found to be radically different than

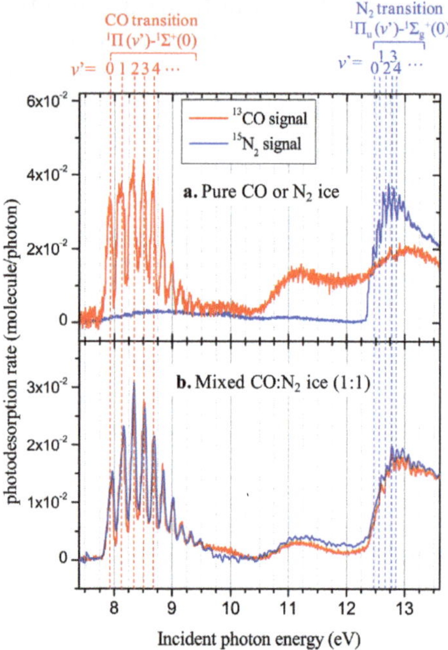

Fig. 1 Photon-stimulated desorption (PSD) spectra of (a) ^{13}CO from a pure ^{13}CO ice and N$_2$ from a pure ^{15}N$_2$ ice, and of (b) ^{15}N$_2$ and ^{13}CO from a condensed (1 : 1) mixture of ^{13}CO and ^{15}N$_2$. The spectra have been acquired from 30 ML-thick ices, deposited on HOPG, and kept at 15 K. The electronic transitions in condensed CO (in red) and condensed N$_2$ (in blue), responsible for the desorption are indicated. The spectra in the lower panel show evidence that the photodesorption rates of both CO and N$_2$ from the mixture are identical, and are the combination of the photo-excitation of each pure component.[30]

that observed for a pure ice sample. In particular, Bertin *et al.*[30] showed that the efficient desorption of one specific chemical component in a binary ice can be observed for wavelengths where the corresponding pure ice does not desorb upon VUV radiation.[30] This is linked to the concept of indirect desorption illustrated in Fig. 1b, which compares the CO and N$_2$ desorption from a N$_2$:CO (1 : 1) ice mixture. It is worth noting that the PSD spectra of the two species from the ice mixture are identical in Fig. 1b. The N$_2$ desorption below 10 eV is due to CO $A^1\Pi$ – $X^1\Sigma$ excitation, and in the same way the CO desorption above 12 eV is a direct consequence of N$_2$ $b^1\Pi_u$ – $X^1\Sigma_g$ excitation. These PSD spectra illustrate the indirect nature of the DIET process, since the prompt relaxation of the excited molecular states can be transferred to neighboring molecules and up to weakly bound surface species independent of their chemical nature.[30] Such an *indirect adsorbate-mediated* process has also been suggested to play a role in the desorption of water from layered ices of water and benzene irradiated by UV laser pulses.[31] The indirect nature of the desorption process reinforces the critical importance of the photon energy distribution of the light source in order to understand how the chemical composition and internal organization of the ice surface affect the photodesorption rates from a complex icy sample.

Since most stable molecules possess electronic states accessible in VUV, it is of particular interest to investigate possible indirect DIET desorption mechanisms for heavier and more strongly bound species. Moreover, previous studies have focused on molecules that are very stable upon VUV excitation, *i.e.* species that do not efficiently photodissociate. Especially for polyatomic molecules, energetic radiation can cause fragmentation and it is not fully clear how this affects or competes with other surface processes, such as photodesorption. This is the topic of the present study, in which VUV irradiation of CO_2 ice films is used to investigate the energy dependence of the photodesorption and its interconnection with photochemistry. The wavelength dependent approach, as described above, is used to investigate the primary processes at play to explain the desorption of the parent molecule and its photoproducts. We will show here that pure, as-grown CO_2 ices and photoprocessed ices behave differently. For a pure CO_2 ice sample, most of the intact CO_2 and photoproduced CO molecules are ejected in the gas phase for incident photon energies higher than 10.5 eV. For photoprocessed ices, the presence of CO as a photochemical product can explain the enhancement of CO_2 desorption below 10 eV due to an indirect desorption effect.

Experimental section

The photodesorption studies were performed in the SPICES set-up of UPMC (Paris, France). This device is an ultrahigh vacuum system with a base pressure of $\sim 1 \times 10^{-10}$ Torr. Molecular ices were grown either onto a polycrystalline gold or a highly oriented pyrolitic graphite (HOPG) substrate, mounted on the cold tip of a closed-cycle helium cryostat. Its temperature can be controlled in the 10–300 K range with 0.5 K precision, using a resistive heating system.

$^{13}CO_2$ (Eurisotop, 99.18% ^{13}C) ice samples were prepared *in situ* and introduced at low flow speed into the chamber *via* a dosing tube, positioned $\sim$2 mm in front of the substrate. This geometry allows the condensation of $^{13}CO_2$ ices onto the cold surface without a substantial increase in the background pressure of the set-up. The thickness of the resulting ice is expressed in monolayers (ML), which corresponds to the surface density of a compact molecular layer on a flat surface (1 ML $\approx 10^{15}$ molecules cm^{-2}). This thickness can be controlled by varying the exposure time during the ice growth. The experimental conditions allow for satisfactory ice thickness reproducibility, with a precision better than 1 ML.

The ices were irradiated across the 7 to 14 eV range using the monochromatized output of the undulator-based beamline DESIRS of the SOLEIL synchrotron facility.[26] The SPICES set-up was directly connected to the beamline end *via* a differential pumping stage, thus preventing any cut-off of high energy photons by a coupling window. The photon energy can be fixed or scanned over a wide range. A narrow bandwidth of typically 40 meV was selected by the 6.65 m normal incidence monochromator that is implemented on the beamline. A rare gas filter in the beamline suppressed the higher harmonics of the undulator that can be transmitted in higher diffraction orders of the grating. Calibrated photodiodes installed on the beamline were used to precisely determine the photon flux as a function of the photon energy. In our configuration, the applied energy-dependent photon flux varied from 0.8 to 2.6×10^{13} photons s^{-1} cm^{-2} over the 7–14 eV range. This monochromatized mode for the beamline was used for either irradiating the ices at a fixed energy and at a given dose, or scanning the incident photon energy to record energy-resolved photodesorption rates.

Energy-resolved photodesorption spectra were recorded as follows. The incident photon energy was scanned from 7 to 14 eV with 25 meV steps. Simultaneously, photodesorbed species were detected in the gas phase using a Quadrupole Mass Spectrometer (QMS). The gas phase mass signal was thus monitored as a function of the incident energy, resulting in a PSD spectrum. The mass signal can be converted into an absolute number of desorbing molecules. This calibration has been done for CO ices by correlating the depletion of the condensed molecules, probed by infrared spectroscopy, with the mass signal of ejected molecules during irradiation.[27] In the case of $^{13}CO_2$, this procedure needs special attention, as chemistry can occur during the irradiation; consequently, correlating the condensed $^{13}CO_2$ infrared signature to gas phase ejected $^{13}CO_2$ is not straightforward. Instead, the $^{13}CO_2$ gas phase signal has been calibrated using the pre-existing CO calibration, corrected by different electron-impact ionization cross sections in the QMS, as taken from Freund *et al.*[32] In addition, we have verified that the apparatus function of our QMS for the two masses, 28 amu (CO) and 45 amu ($^{13}CO_2$), is the same to within a few percent. In this way, using the measured energy-dependent photon flux as a function of the photon energy, we can derive the absolute photodesorption rate, in desorbed molecules per incident photon, for a given photon energy. It is important to note that this data treatment procedure does not significantly alter the shape of the spectra, since all the structures in the calibrated PSD spectra are already clearly observed in the raw mass signal data. More details on the data quantification procedure can be found in Fayolle *et al.*[28]

The $^{13}CO_2$ ices were also probed after deposition, and during or after irradiation, by means of RAIRS. In our case, this method mainly allows for the identification of the condensed phase composition, and in particular for the observation of photochemistry during irradiation at fixed energy. Fine quantification of the condensed phase composition was performed using the temperature programmed desorption (TPD) method; the sample was warmed up at a constant heating rate (3 K min^{-1}), and the thermally desorbed species were measured as a function of the temperature by mass spectrometry. Unlike RAIRS, this method is sensitive to very low amounts of molecules (less than 0.1 ML), and can differentiate species as a function of their binding energy to the ice, *i.e.* whether they are bulk or surface-located. Each irradiated sample was probed using TPD in order to quantify the eventual photoproducts formed upon UV irradiation. Obviously, a TPD experiment destroys the ice and is only performed after the photoprocessing has fully taken place.

In both PSD and TPD, the gas phase mass signal of desorbing ^{13}CO was corrected from the fragmentation of intact $^{13}CO_2$ in the QMS. The ^{13}CO signal originating from $^{13}CO_2$ fragmentation was systematically subtracted from the overall ^{13}CO signal, in order to probe only the desorbing ^{13}CO molecules.

Results

Fig. 2a shows the photodesorption rates measured between 7 and 14 eV from pure $^{13}CO_2$ ice at 10 K and 40 K. The quadrupole mass spectrometer can simultaneously detect $^{13}CO_2$ and ^{13}CO species (in the gas phase). As explained in the experimental section, absolute photodesorption rates are obtained (i) by calibrating the gas phase mass signal to the absolute amount of ejected molecules,

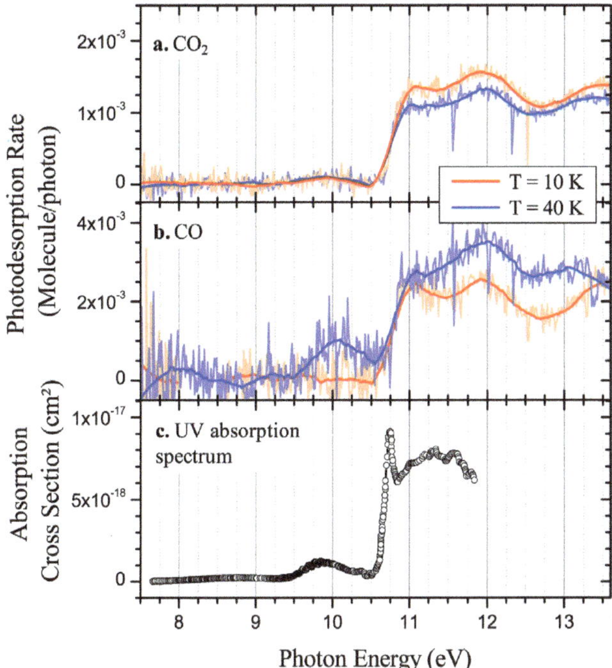

Fig. 2 PSD spectra of $^{13}CO_2$ (a) and ^{13}CO (b) from 10 ML-thick $^{13}CO_2$ ices. The ices were deposited on a polycrystalline Au substrate and kept at 10 K and 40 K. Panel (c) shows, for comparison, the absorption spectra of solid CO_2, as measured by Monahan and Walker.[33] The spectrum has been calibrated in the absorption cross section using the data from Mason *et al.*[34] and Cruz-Diaz *et al.*[35]

and (ii) by taking into account the variations in measured photon flux with photon energy. The PSD spectra reveal a very strong energy dependence of the $^{13}CO_2$ and ^{13}CO photodesorption rates. Below about 10.5 eV, the photodesorption rate for both species is essentially zero, but a sudden increase is found around this value, with a more or less continuous and efficient desorption at higher energies. Very similar and small variations are observed for the $^{13}CO_2$ and ^{13}CO species above 11 eV, with photodesorption rates varying between 1×10^{-3} to 3×10^{-3} molecules photon^{-1}.

The PSD spectra are compared to the absorption spectrum of solid $^{12}CO_2$ at 53 K from Monahan and Walker,[33] as shown in Fig. 2c, in which the absorption is measured in natural logarithmic absorbance ($\ln(I_0/I)$, in which I_0 is the incident intensity and I is the transmitted intensity, for different ice thicknesses. There is no absorption spectrum available for $^{13}CO_2$, but only small differences between the two isotopologues are expected. These values have been normalized for a given thickness, and absolute cross sections were calibrated using the quantitative data of Mason *et al.*[34] and Cruz-Diaz *et al.*[35] in the 9–10.5 eV energy range. In the lower energy range, $E < 10.5$ eV, the absorption spectrum is associated with valence state excitations and comprises a lower lying diffuse absorption band centered around 8.6 eV, overlapping with a second more intense absorption band centered around 9.8 eV with an irregular vibrational structure, as reported in detail by Mason *et al.*[34] These transitions are commonly assigned to the optically forbidden (but vibronically allowed) transitions $^1\Delta_u - {}^1\sum_g^+$ and $^1\Pi_g - {}^1\sum_g^+$ for the 8.6 eV and 9.8 eV band, respectively.[33,36] Recent high-level *ab initio* calculations of isolated CO_2 molecules indicate that this energy region is very complex due to multiple degeneracies of six valence states linked through a network of couplings due to the Renner–Teller effect, and conical intersections either at linearity in the Franck–Condon region or at bent geometries.[37] According to the gas-phase spectrum interpretation, the intensity distribution can be explained by non-vertical transitions originating from Herzberg–Teller effects in the transition dipole moments. The first band involves electronic transitions towards two bent electronic components, $2^1A'/^1A''$ ($1\,^1\Delta_u$ at linearity), and an additional weak transition towards the $2^1A''$ ($^1\sum_u^-$ at linearity) excited state. At higher energy, the second band is assigned to a transition to the linear $3^1A'/3^1A''$ ($1\,^1\Pi_g$) states, and is more intense due to "intensity borrowing", most probably from the allowed $1\,^1\sum_u^+$ Rydberg states lying at higher energy.[38] The absorption oscillator strength of this second band at 9.8 eV is even more enhanced in the solid phase.[33] A very weak photodesorption signal is visible around 10 eV in the PSD spectra of $^{13}CO_2$ and ^{13}CO (Fig. 2a and b).

The onset around 10.5 eV observed in the photodesorption rates correlates with the enhancement by almost one order of magnitude of the absorption cross sections in solid CO_2 above 10.5 eV. In the higher energy domain ($E > 10.5$ eV), the strong absorption is associated with dipole allowed transitions from the ground state towards $(np\sigma_u)^1\sum_u^+$ and $(np\pi_u)^1\Pi_u$ Rydberg states (with $n \geq 3$), and possible contributions of other series (s, d and f Rydberg series) converging to the first ionization potential, as observed in the gas phase.[39,40] The intense absorption peak observed at 10.7 eV (Fig. 2c) has been associated with the first dipole allowed electronic transition $(3p\sigma_u)^1\sum_u^+ - {}^1\sum_g^+$.[36] Apart from this first peak, which is not seen in any of the photodesorption spectra, the photodesorption efficiency follows the photoabsorption continuum above 11 eV, although the comparison cannot be extended beyond 11.75 eV, since absorption spectra of solid CO_2 are lacking above this value. The broad band observed at 11.6 eV in the CO_2 photodesorption spectrum may be associated with the excitation of the $(3p\pi_u)^1\Pi_u$ state, as suggested by electron energy loss spectroscopy.[41] Finally, another band observed at 13 eV matches the ionization energy of solid CO_2.[42]

In the gas phase, CO_2 is known to photodissociate upon energetic VUV irradiation. Experiments show that excitation of the diffuse absorption bands ($^1\Delta_u$ and $^1\Pi_g$) leads to atomic oxygen production with quantum yields close to unity. Consequently, the photodissociation of CO_2 into $O(^1D) + CO(^1\Sigma^+)$ and $O(^3P) + CO(^1\Sigma^+)$ dominates the photolysis below 10.5 eV.[43–45] Between 10.5 eV and 14.5 eV, new photodissociation channels open, providing additional pathways to produce efficiently excited atoms $O(^1S)$ and excited CO ($a^3\Pi$) molecules.[46,47] The formation of excited $O(^1S)$ atoms resulting from reactive processes following UV excitation of CO ($CO^* + O \rightarrow CO_2^* \rightarrow CO + O^*$) have been identified in Ar cryogenic matrices experiments.[48,49] In the solid phase, however, the photodecomposition efficiency is expected to be strongly reduced with respect to the gas phase. This is due either to possible reactive collisions between O and CO fragments on the surface or within the ice matrix, forming CO_2, or to the electronic quenching of the electronic excited states producing vibrationally hot CO_2 molecules.[16] In addition, the reaction between excited and non-excited photofragments such as CO ($a^3\Pi$) + CO ($X^1\Sigma$) $\rightarrow CO_2 + C$, cannot be excluded and may also decrease the CO_2 photolysis efficiency compared to the gas phase.[50] Nevertheless, excitation of CO_2 is expected to lead to partial photofragmentation into CO and O on a very short time scale (10^{-15} s). As shown in Fig. 2, the shape of the PSD spectra of CO and CO_2 recorded in the low temperature (10 K) experiments are very similar, showing a steep onset at 10.5 eV. It should be noted that the CO photodesorption rates are higher than those of CO_2, indicating that, despite their common origin, the desorption processes follow different underlying mechanisms. Photodesorption rates that are higher for CO than for CO_2 have also been reported by others.[14,17,18] This behavior is consistent with CO desorption arising from direct CO_2 photodissociation, kicking fragments into the gas phase. Oxygen atoms are indeed also detected simultaneously by mass spectrometry and give similar PSD spectra. No quantitative analysis of the oxygen desorption rates for the atoms has been performed. This is not possible as the fractionation of several heavier O-bearing species in the QMS contributes strongly to the overall signal at 16 amu.

Fig. 2a and b compare the CO_2 and CO PSD spectra obtained at 10 and 40 K. The 40 K CO_2 PSD spectrum does not show any desorption around 10 eV. A weak CO_2 desorption might be present at 10 K, but the desorption rates ($\sim 9 \times 10^{-5}$ molecules photon^{-1}) are close to the noise level and are highly uncertain. By contrast, a temperature effect is observed in the 40 K PSD spectrum of CO, which desorbs at around 10 eV, in correlation with the 9.8 eV absorption band of CO_2 (Fig. 2c). A temperature effect is even more clearly observed between 11 and 14 eV. As the temperature increases from 10 to 40 K, the CO_2 desorption rates decrease quasi-uniformly by 10% (Fig. 2a). Interestingly, an opposite behavior is simultaneously observed for CO; here the desorption rates are higher at 40 K (Fig. 2b). This effect can be explained by differences in the binding energies of the two species. As will be discussed in more detail in the next section, the adsorption energy of CO on CO_2 ice is about 90 meV, which corresponds to an onset for CO desorption at ~ 25 K, peaking at 40 K. Therefore, at 40 K, we expect that the CO desorbs just after its formation in the condensed phase. On the contrary, CO_2 is more strongly bound to the CO_2-rich surface, with a binding energy of 230 meV,[51] and its thermal desorption is negligible on the experimental time scale. The enhancement of the CO desorption at 40 K can thus be explained by the contribution of thermally activated desorption of CO molecules from the top ice layers. Contribution from the CO molecules in the underlying surface layers is also possible. The observation that the desorption of CO molecules is accompanied by a lowering of the CO_2 photodesorption rates is fully consistent with the picture in which the photoproduced CO molecules – lost upon thermal desorption – are in fact those contributing to the desorption of CO_2. This mechanism will be explained in detail in the discussion section.

In order to further investigate the possible role of CO in the photodesorption process, we irradiated a $^{13}CO_2$ ice film at higher photon fluences but only for a few selected energies. Typically, 10 ML thick $^{13}CO_2$ ice films at 10 K were pre-irradiated with a 7×10^{16} photon dose. This irradiation mode is well-suited to separately excite different regions of the CO_2 ice absorption spectrum. Irradiation with photons at 7.2 eV, 8.7 eV, 9.8 eV, and 11 eV was performed with a narrow 40 meV bandwidth resulting from the monochromator. The pre-irradiation at 7.2 eV can be used as a blank experiment, since the CO_2 ice film does not absorb at this energy. After each pre-irradiation sequence, the sample was again investigated in a similar way to that described above by recording the PSD spectrum. The results are shown in Fig 3.

After pre-irradiation at 8.7 eV, 9.8 eV, and more markedly at 11 eV, the CO desorption rates increase and a strong desorption pattern with a new structure appears in the 7–10 eV energy range (Fig. 3b). This structure originates from the CO $A^1\Pi(v')$ – $X^1\Sigma(v''=0)$ electronic transition, as shown in Fig. 1, that has been studied in much detail in previous works.[27,29,52] This feature is a clear signature of an ongoing DIET process, indicating the presence of high concentrations of CO in the upper layers of the ice film. Moreover, this was confirmed by complementary TPD measurements. Fig. 4 displays the ^{13}CO desorption peak corresponding to thermal desorption from the ice film between 30 and 75 K. A comparison can be made with the thermal desorption of 1 ML of CO deposited on CO_2 ice, *i.e.* recorded without photon irradiation. As expected, no significant CO signal, *i.e.*, clearly discernible from the background noise, was detected from the ice film after irradiation at 7.2 eV, for which the CO_2 ice is transparent. For all other energies, the TPD spectra can be integrated to estimate the CO ice concentration within the ice film. From the comparison with the desorption of 1 ML CO from CO_2, the CO surface concentration (*i.e.* in the top layers of the ice that contribute to the desorption process) was determined to amount to roughly 4%, 20% and 45% after pre-irradiation at 8.7 eV, 9.8 eV and 11 eV, respectively. The CO ice concentration increases with increasing photon energy due to the

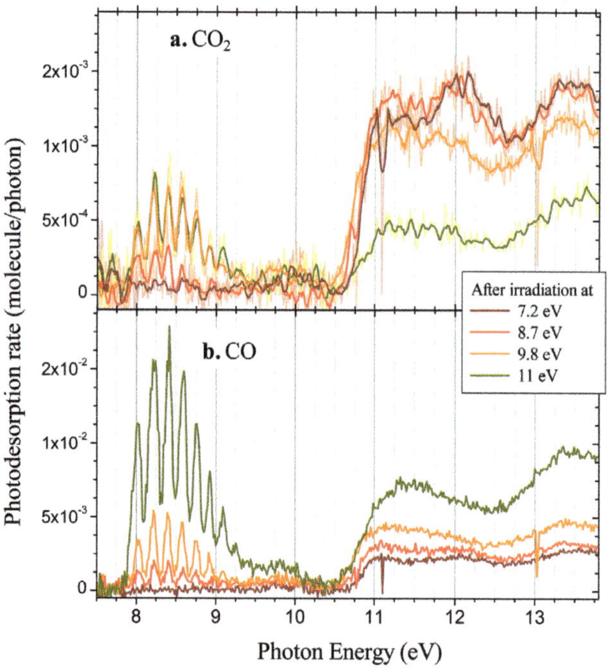

Fig. 3 PSD spectra of $^{13}CO_2$ (a) and ^{13}CO (b) from 10 ML-thick $^{13}CO_2$ ices, pre-irradiated at fixed energies (7.2, 8.7, 9.8 and 11 eV). The ices were deposited on a polycrystalline Au substrate and kept at 10 K. The pre-irradiation was realized with a fluence of 7×10^{16} photons for each energy. The structure appearing in the low energy part of the spectra is associated with the excitation of CO molecules formed in the ice during the pre-irradiation. The prevalence of this structure at higher pre-irradiation energies is linked to the greater quantity of CO formed in the solid.

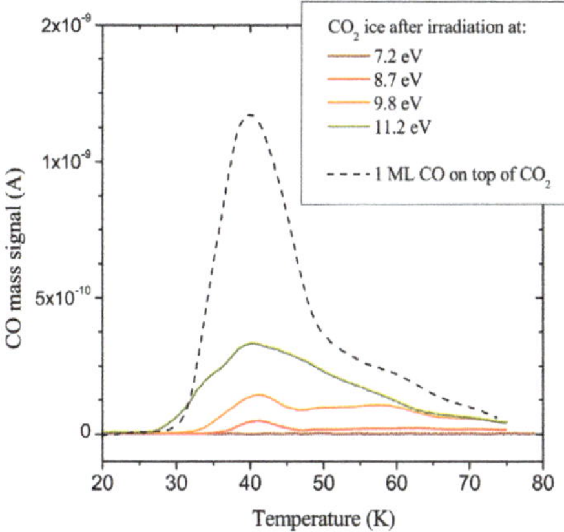

Fig. 4 Temperature programmed desorption (TPD) curves of ^{13}CO from 10 ML-thick ^{13}CO ices, deposited on polycrystalline Au kept at 10 K, and pre-irradiated with a fluence of 7×10^{16} photons at 7.2, 8.7, 9.8 and 11 eV. The TPD curve of 1 ML of ^{13}CO deposited at 10 K on 10 ML of ^{13}CO$_2$ is presented for comparison (dashed line). The TPD was carried out with a temperature ramp of 3 K min^{-1}. The shapes of the TPD curves for the pre-irradiated ices are very similar to that of surface CO on top of CO$_2$, which allows the conclusion that these molecules most probably originate from the top layers of the pre-irradiated ices. The integrals of the TPD signals give the amount of CO formed during the pre-irradiation located on the (sub)surface of the ice.

increasing values of the CO$_2$ absorption cross sections for these four selected energies (Fig. 2c).

The CO$_2$ PSD spectra after pre-irradiation are shown in Fig. 3a. The CO$_2$ desorption is significantly affected by the presence of CO on the surface of the ice film. At first, new desorption structures are observed between 8–9 eV, with relatively high CO$_2$ photodesorption rates. This effect is particularly noticeable since in this energy region only an extremely weak CO$_2$ desorption could be observed from fresh CO$_2$ films. The CO$_2$ PSD spectra present the same typical CO $A^1\Pi(v') - X^1\Sigma(v'' = 0)$ vibronic progression as observed for CO. It is therefore assigned to an *indirect* DIET process, driven by CO electronic excitation. From this observation, it is clear that the CO* electronic relaxation transfers some energy to the CO$_2$ molecules. It is worth noting that this effect appears after pre-irradiation at 8.7 eV, corresponding to only 4% concentration of CO. Significant photodesorption rates, reaching 7×10^{-4} molecules photon^{-1}, are observed for an ice film containing about 20% CO (9.8 eV pre-irradiation). At high CO coverage (11 eV pre-irradiation), a decrease in the CO$_2$ photodesorption rate above 10.5 eV is observed. In this case, the shape of the PSD spectrum is similar to that observed for a pure CO sample, indicating that the desorption is mainly driven by CO molecules. As the energy of the pre-irradiation increases, the CO concentration also increases, and the PSD spectra of CO$_2$ and CO are in between that of a desorption process dominated by CO$_2$ excitation and a desorption process dominated by CO excitation. This series of experiments thus simulates how the desorption from an ice sample evolves when continuously irradiated at high photon fluxes.

Discussion

A. Specificities of the energy-resolved and low fluence experimental method

Before discussing the involved desorption mechanisms and photodesorption rates, important specificities of the present experimental conditions need to be pointed out. In this work the direct detection of CO_2 (and CO) molecules in the gas phase is realized by mass spectrometry. This is different from most previous experimental studies, in which the quantitative determination of the CO_2 photodesorption was based on a depletion measurement of CO_2 from the solid phase, either by following solid CO_2 infrared fingerprint spectra during irradiation,[14–16] or by probing the mass loss using a quartz microbalance,[17] but not *via* the gas phase detection of CO_2. In the present study, the analysis of the partial gas phase pressure is assumed to reflect the desorption kinetics from the ice surface, which is generally the case in ultrahigh-vacuum chambers with high pumping speed.[18] In addition, gas phase detection is found to have a much higher sensitivity to surface desorption processes as compared to IR spectroscopy (see also ref. 18). This allows the detailed exploration of desorption mechanisms at low photon fluence that are not accessible by IR spectroscopy. Indeed, the total photon fluence used in PSD never exceeds 5×10^{15} photons cm^{-2} over a complete energy scan, which is about three orders of magnitude lower than in previously reported experiments using H_2 microwave discharge lamps.[14,17,18,25] In this setting, very limited chemistry is induced, as attested by the extremely faint CO IR stretching band observed by RAIRS spectroscopy, together with the non-detection of other chemical species such as CO_3 and O_3, which have been reported in other studies.[14,17] TPD experiments following a PSD spectrum show that the conversion of CO_2 into CO is barely detectable, and doesn't exceed ~0.6% after a PSD scan performed on a fresh CO_2 sample at 10 K. Moreover, it should be pointed out that very little material is removed from the solid phase, since less than 0.04 ML is estimated to be desorbed during a PSD experiment. From these observations, we can conclude that, under the conditions used here, the influence of secondary chemical processes is not favored. For the same reason, the enhancement of desorption rates associated with ice re-structuring can be completely neglected. The data we present here thus focus on the basic molecular processes associated with the initial fast photodesorption of CO_2 and CO species from the outermost surface.

B. Photodesorption mechanisms

Several mechanisms are at play simultaneously in the photodesorption of CO_2 and CO from CO_2 ices. Our experiments, performed on as-grown ices and on pre-irradiated ices, allow us to identify two regimes: (i) photodesorption from as-grown CO_2 samples, consisting of primary processes triggered by the photon absorption of solid CO_2 molecules, and (ii) photodesorption from pre-irradiated films, reflecting *indirect* processes triggered by the photon absorption of solid CO molecules that have been formed and accumulated in the CO_2 ice during pre-irradiation. Fig. 5 gives a summary of the different mechanisms to be considered, which are discussed systematically below.

B1. Desorption from as-grown CO_2 samples. The initial step in the photodesorption is the photon absorption of CO_2 molecules, as illustrated in Fig. 2, where the energy-dependence of both CO and CO_2 desorption mimics the absorption spectrum of the CO_2 ice. Subsequently, the absorption can lead to dissociation into CO and O. If formed with enough kinetic energy, some of the photofragments can desorb directly after their formation, leading to a CO gas phase signal.

In the case where the photofragments remain trapped within the ice, their recombination–reforming of CO_2, which is exothermic by ~ 5 eV,[53] can also transfer energy to the ice, causing CO_2 to desorb. Indeed, this recombination reaction has a low barrier (40–60 meV) and is known to be active at low temperature ice surfaces.[54] It is interesting to note that H_2O photodesorption is expected to share common properties with CO_2 because of the well-known dissociative character of the first electronic states of H_2O under vacuum UV, producing OH and O photofragments. Indeed, theoretical simulations of H_2O photodesorption in the first absorption band reveal that the exothermic chemical recombination of O atoms and OH fragments contributes to the desorption rates.[55] For CO_2 ice, we estimate the role of this dissociation–recombination process in the overall photodesorption of CO_2. The production of CO_2 by chemical recombination consumes CO molecules. The anti-correlation observed between CO and CO_2 desorption rates when comparing PSD spectra recorded at 10 and 40 K supports this mechanism (Fig. 2). As already mentioned above, this behavior is most probably explained by the dramatic lowering of the surface residence time of CO. At 40 K, the induced thermal CO desorption inhibits chemical recombination, resulting in lower CO_2 desorption rates together with higher CO desorption rates. Assuming that the chemical recombination is completely avoided at 40 K, this effect would contribute to 10% of the total CO_2 photodesorption efficiency.

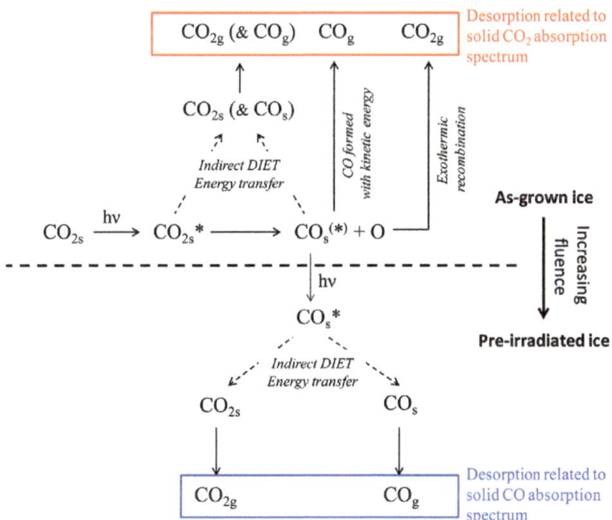

Fig. 5 Scheme of the proposed mechanisms involved in the photodesorption of CO and CO_2 from a CO_2 ice. The primary mechanisms, *i.e.* involving only the absorption of a photon by the CO_2 molecules, are presented in the top part of the figure, whereas the mechanisms requiring a subsequent absorption by the photo-produced CO molecules are shown in the lower part of the scheme.

Consequently, mechanisms other than dissociation–recombination must be at play in the overall CO and CO_2 desorption. This implies that reactive processes have a minor contribution to the overall photodesorption, in agreement with recent conclusions reached by Yuan and Yates.[18]

Theoretical work on water predicts its total photodesorption rate to be largely driven by a "kick-out" mechanism, where fast H atoms produced by photodissociation deeper within the ice travel to the surface and transfer their momentum to a surface molecule.[55] In the case of CO_2 ice, a similar role may be played by O atoms that will carry most of the energy. Taking into account that part of the photon energy is needed for dissociation into specific electronic states, the kinetic energy that is shared between O and CO fragments then increases with the incident photon energy, varying from zero (at the dissociation threshold) up to a few eV above the threshold. Thus, a kick-out mechanism would be observed in our PSD spectra as a continuous increase with increasing photon energy. Such a contribution is not clearly observed.

Beside mechanisms initiated by photodissociation, the DIET process, as previously discussed for N_2 and CO, is associated with the fast relaxation of electronically excited states revealed by strong resonances in the PSD spectra (Fig. 1). It is not possible to extract similar spectral resonances within the continuous desorption above 11 eV for CO_2, which is certainly due to the high density of dissociative states. In this case, due to the lengthening of the C–O bond in the upper repulsive states and subsequent quenching, it has been proposed that highly vibrationally excited molecules are produced within the ice matrix, originating from the Menzel–Gomer–Redhead effect.[56] This energy can be either locally transferred to the surrounding molecules leading to desorption, or dissipated within the ice matrix, depending on the strength of vibrational couplings between the ice components.[15,16] In the case where desorption is induced by energetic VUV photons however, we conclude that excited photofragments also contribute to the desorption process. Excess energy released upon relaxation of CO*, for example formed in its $a^3\Pi$ state with energies between 11 and 14 eV, could be transferred to surrounding CO_2 (or other CO molecules depending on the amount formed), initiating desorption.

Finally, when accurately comparing the PSD spectra to the regular CO_2 absorption spectrum, it should be noted that the sharp peak observed at 10.7 eV in the absorption spectrum is not seen in the PSD spectra. CO_2 dissociation into CO and O fragments therefore is expected to contribute to a direct "CO-from-CO_2" desorption. Because any corresponding CO desorption peak is observed at 10.7 eV, we conclude that this resonance does not induce dissociation. This is certainly linked with the excitonic, and therefore delocalized, character of this resonance, for which the energy is diluted in all the degrees of freedom within the matrix.[36] This energy dissipation may also explain the lack of photodesorption associated with this state.

B2. Desorption from pre-irradiated samples. Our previous investigations have highlighted the importance of indirect desorption in explaining the photodesorption of weakly bound species from a binary ice sample.[29,30] Indirect desorption also plays a role in the photodesorption of CO_2 ice, despite the fact that CO_2 molecules are more strongly bound to the surface. The presence of an indirect desorption mechanism is clearly evidenced by the experiments performed after CO_2 pre-irradiation (Fig. 3), which illustrate that some CO_2 is converted into CO.

In order to interpret these experiments, a comparison with the desorption results from $CO:N_2$ films as previously analyzed is useful (Fig. 1). In the case of binary films, the desorption can be described by a combination (50% each) of pure N_2 and CO desorption rates. Moreover, the desorption is dominated by CO excitation at low energies ($E < 10$ eV), because N_2 absorption is extremely weak below 12 eV. A very similar situation applies to "pre-irradiated" CO_2 samples (Fig. 3), for which low and high energy components can also be attributed to different contributions. In the low energy range, the desorption is dominated by CO excitation because of the low absorption cross section of CO_2. In the higher energy range ($E > 10.5$ eV) the desorption results from a combination of CO_2 and CO excitation. However, a major difference in the $CO:N_2$ case is that the separate weights are not equivalent. This is revealed by considering the ratio between the CO and CO_2 desorption rates. On a pure CO_2 sample, the CO desorption rate exceeds that of CO_2 by a factor of about 2 (Fig. 2). This can be understood by the contribution of CO-from-CO_2 desorption, related to the photodissociation of surface CO_2 molecules. By contrast, the ratio between CO and CO_2 desorption rates increases with increasing CO concentration (Fig. 3). The CO desorption rate exceeds that of CO_2 by one order of magnitude when about half of the CO_2 molecules have been converted into CO (from the experiments after 11 eV pre-irradiation). This huge effect highlights the fact that indirect desorption processes have very different yields for CO and CO_2 molecules. At first sight, this is not very surprising since the CO_2 physisorption energy is about twice that of CO. However, considering the large energy released into the ice film as compared to the binding energies, this suggests differences in the energy transfer efficiency between CO–CO and CO–CO_2 media. This again reinforces the crucial role played by the molecular environment in mediating the energy transfer towards the surface and into a desorption channel, as pointed out previously by Bertin *et al.*[29] and Yuan *et al.*[16]

The weight of the CO excitation-related desorption compared to the CO_2 excitation-related desorption in the total desorption spectra increases with the concentration of photoproduced CO on the surface. As stated previously, this leads to a notable increase in the CO photodesorption, especially in the low energy range, of about an order of magnitude. In addition, it also impacts the CO_2 photodesorption behavior. With increasing CO formation, CO_2 desorption becomes gradually dominated by the CO excitation profile (as can be seen in Fig. 1 for pure CO ice). Interestingly, this has the effect of activating the CO_2 photodesorption in the low energy range (below 10 eV), but also decreases the photodesorption efficiency in the high energy range, since CO-induced desorption is less efficient in this part of the spectrum (Fig. 1).

C.   Photodesorption rates

This investigation reveals for the first time a very strong energy dependence of the UV photodesorption rates from CO_2 ice films at very low temperature. The total photodesorption rate from pure CO_2 ices samples between 7 and 14 eV is dominated by energetic photons with an energy above 10.5 eV (Fig. 2). As noted above, this behavior mainly originates from the large difference in the absorption cross section of CO_2 ice below and above 10.5 eV. The maximum desorption rate is 1.6 $\times\ 10^{-3}$ molecules photon^{-1}, which is much lower than the desorption rates derived by the same method for pure CO or N_2 ices, which peak at 4×10^{-2}

molecules photon^{-1} (Fig. 1). However, a significant enhancement of the CO_2 desorption rate is observed in ices prepared by photolysis due to the additional contribution of photons below 10 eV, resulting in desorption rates as high as 8 $\times$ 10^{-4} molecules photon^{-1}. Such desorption rates are not far from the rates determined by Öberg *et al.*[14] using a hydrogen discharge source, which preferentially emits radiation at around 10.2 eV (Ly-α). It is worth mentioning that our investigation shows that neither CO_2 nor CO significantly photodesorb upon Ly-α excitation. The lamp used is not monochromatic, and although 10.2 eV photons might dominate in experiments using H_2 discharge lamps, "residual" molecular H_2 emission at lower energies can contribute as well, depending on the experimental operating conditions. As mentioned at the beginning of this discussion, a significant quantity of CO molecules is formed at the very beginning of the irradiation, due to the higher fluxes and fluences used in broad band discharge lamp experiments. We thus attribute the CO_2 desorption rates determined previously with hydrogen discharge lamps to indirect desorption induced by the photo-produced CO electronic excitation in the 8–10 eV range. We point out that, even though 10.2 eV photons do not directly induce strong desorption, their role in the surface chemistry should not be neglected in general, as the photoformation of new species can indirectly contribute to the desorption efficiency. It has been shown here that in particular the photons with an energy above 10.5 eV can efficiently convert CO_2 into CO, thus resulting in the enhancement of CO_2 photodesorption by lower energy photons. Hence, it is the cumulative effect of high and low energy photons that leads to effective desorption. Interestingly, a similar interplay between photochemistry and photodesorption at various energies has also been demonstrated in a very recent study by Chen *et al.*[20] on CO photodesorption rates under various operating conditions of an H_2 discharge lamp.

Finally, the relatively moderate desorption rates discussed here should be distinguished from the higher rates ($\sim 10^{-2}$ molecules photon^{-1}) derived for much higher fluences from CO_2 ices at 40–60 K, and for which thermal activation and ice restructuring enhance the desorption. This effect has been extensively described recently and does not contribute to the desorption under these experimental conditions.[17,18]

In the diffuse interstellar medium or at the edge of molecular clouds, highly energetic photons ($E > 10.5$ eV) may contribute to direct CO_2 (and CO) desorption. Deeper inside the molecular clouds, however, the highly energetic component of the spectrum is progressively shielded by gas phase H_2 and CO, which should strongly limit their contribution. However, molecular H_2 emission induced by cosmic rays penetrating into dense clouds produces photons *in situ* (at a lower flux), which are not very efficient in inducing CO_2 desorption. One might conclude that the UV photodesorption of CO_2 is not very efficient in most dense environments compared to other species. However, CO_2 is generally observed in CO_2:H_2O or CO:CO_2 phases.[12] Assuming that the CO:CO_2 component is formed over the H_2O rich layers, the high efficiency of photodesorption driven by CO should therefore contribute to the gas phase enrichment of CO and CO_2 in cold environments. The experiments described above indeed lead to the conclusion that low energy photons ($E < 10.5$ eV) can actually induce the desorption of CO_2 due to indirect desorption, with rates peaking close to 10^{-3} molecules photon^{-1}, despite the fact that pure CO_2 ice is not efficiently photodesorbed in this energy range. In order to further elucidate the relevance of such a process, UV desorption efficiencies for different concentrations of $CO:CO_2$ binary mixtures will be needed, which are scheduled for upcoming DESIRS runs using the wavelength-dependent approach described here.

Summary and conclusion

The energy dependence of absolute desorption rates from CO_2 ices have been obtained for the first time using tunable synchrotron radiation between 7–14 eV. The high sensitivity of mass spectrometry enables the quantitative determination of the desorption rates from the direct gas phase detection of CO and CO_2 molecules ejected from the ice. The chosen experimental conditions used in this experiment allow for a systematic comparison of the photodesorption spectra of pure and pre-irradiated CO_2 ices. Hence, the PSD spectra provide a "snapshot" of the different desorption mechanisms at play depending on the UV-processing history of the ice film. In the case of pure CO_2 samples at 10 K, CO_2 and correlated CO desorption are found to proceed mostly above 11 eV, with desorption rates ranging between 1–3×10^{-3} molecules photon^{-1}, whereas no desorption or weak desorption ($<10^{-4}$ molecules photon^{-1}) is observed below 10.5 eV. Chemical recombination, contributing 10% of the total desorption rate at low temperature (10 K), has been identified on the basis of a significant decrease in CO_2 desorption rates at 40 K confirmed by the enhancement of the CO desorption rate, preventing any recombination at the surface of the ice film.

In the case of pre-irradiated samples, the preliminary photolysis of the ice sample dramatically modifies the desorption behavior. A new photodesorption route, due to indirect desorption through CO excitation and fast relaxation is evidenced between 8 and 10 eV, for a concentration of only a few percent of CO within the upper layers of the ice film. This effect is an *indirect* DIET mechanism mediated by high CO absorption cross sections in this energy range. This leads to CO_2 desorption rates up to 10^{-3} molecules photon^{-1}. Although the efficiency of the energy transfer is much lower than that previously observed for the light and weakly bonded species CO and N_2, this mechanism also applies to CO_2. This process may explain the desorption efficiency at higher energies as well.

In conclusion, ice photolysis and photodesorption yields are closely interconnected because the desorption efficiency is very sensitive to the ice composition. Highly energetic VUV photons (10.5–14 eV) are efficient at photodesorbing CO_2 and/or converting CO_2 into CO. Low energetic photons (8–9 eV) are efficient at photodesorbing CO_2 *via* an *indirect* DIET mechanism mediated by CO. This illustrates the complementary role of low and high energy photons in the overall photodesorption efficiency. This interplay is believed to play a major role when induced by polychromatic photon sources, as typically encountered in the interstellar medium, and explains the desorption rates obtained in previous laboratory studies based on broad band discharge lamps, which we attribute mainly to the photoproduction of CO in CO_2 ice.

Acknowledgements

We acknowledge SOLEIL for providing of synchrotron radiation under project 20120834, and we thank Gustavo Garcia for assistance on the beamline DESIRS. Financial support from the French CNRS national program *Physique et Chimie du*

Milieu Interstellaire, the UPMC platform for Astrophysics *"ASTROLAB"*, the Dutch program Nederlandse Onderzoekschool voor Astronomie, and the Hubert Curien Partnership Van Gogh (25055YK) is gratefully acknowledged. The Leiden group acknowledges NWO support through a VICI grant. The authors are grateful to Guillermo Muñoz-Caro and Gustavo Adolfo Cruz-Diaz for sending VUV absorption cross sections prior to publication. J.-H. F. thanks Dr. M. Gudipati for fruitful discussions.

References

1 K. Willacy and W. D. Langer, *Astrophys. J.*, 2000, **544**, 903.

2 C. Dominik, C. Ceccarelli, D. Hollenbach and M. Kaufman, *Astrophys. J. Lett.*, 2005, **635**, L85.

3 D. Hollenbach, M. J. Kaufman, E. A. Bergin and G. J. Melnick, *Astrophys. J.*, 2009, **690**, 1497–1521.

4 V. Guzmán, J. Pety, J. R. Goicoechea, M. Gerin and E. Roueff, *Astron. Astrophys.*, 2011, **534**, A49.

5 A. Oka, A. K. Inoue, T. Nakamoto and M. Honda, *Astrophys. J.*, 2012, **747**, 138.

6 E. Dartois, *Space Sci. Rev.*, 2005, **119**, 293–310.

7 K. M. Pontoppidan, A. C. A. Boogert, H. J. Fraser, E. F. van Dishoeck, G. A. Blake, F. Lahuis, K. I. Öberg, N. J. Evans II and C. Salyk, *Astrophys. J.*, 2008, **678**, 1005.

8 D. C. B. Whittet, P. A. Gerakines, A. G. G. M. Tielens, A. J. Adamson, A. C. A. Boogert, J. E. Chiar, T. de Graauw, P. Ehrenfreund, T. Prusti, W. A. Schutte, B. Vandenbussche and E. F. van Dishoeck, *Astrophys. J. Lett.*, 1998, **498**, L159.

9 A. M. S. Boonman, E. F. van Dishoeck, F. Lahuis, S. D. Doty, C. M. Wright and D. Rosenthal, *Astron. Astrophys.*, 2003, **399**, 1047–1061.

10 N. Sakai, T. Sakai, Y. Aikawa and S. Yamamoto, *Astrophys. J. Lett.*, 2008, **675**, L89.

11 P. Ehrenfreund, A. C. A. Boogert, P. A. Gerakines, D. J. Jansen, W. A. Schutte, A. G. G. M. Tielens and E. F. van Dishoeck, *Astron. Astrophys.*, 1996, **315**, L341–L344.

12 K. M. Pontoppidan, A. C. A. Boogert, H. J. Fraser, E. F. van Dishoeck, G. A. Blake, F. Lahuis, K. I. Öberg, N. J. Evans II and C. Salyk, *Astrophys. J.*, 2008, **678**, 1005–1031.

13 K. M. Pontoppidan, C. P. Dullemond, E. F. van Dishoeck, G. A. Blake, A. C. Boogert, N. J. Evans II, J. E. Kessler-Silacci and F. Lahuis, *Astrophys. J.*, 2005, **622**, 463.

14 K. I. Öberg, E. F. van Dishoeck and H. Linnartz, *Astron. Astrophys.*, 2009, **496**, 281–293.

15 C. Yuan and J. T. Yates Jr, *J. Chem. Phys.*, 2013, **138**, 154302.

16 C. Yuan and J. T. Yates Jr, *J. Chem. Phys.*, 2013, **138**, 154303.

17 D. A. Bahr and R. A. Baragiola, *Astrophys. J.*, 2012, **761**, 36.

18 C. Yuan and J. T. Yates Jr, *Astrophys. J.*, in press, **2014**.

19 T. Kinugawa, A. Yabushita, M. Kawasaki, T. Hama and N. Watanabe, *Phys. Chem. Chem. Phys.*, 2011, **13**, 15785.

20 Y.-J. Chen, A. Chuang, K.-J., G. M. Muñoz Caro, C. Nuevo, M., C.-C. Chu, T.-S. Yih, W.-H. Ip and C.-Y. R. Wu, *ApJ*, 2014, **781**, 15.

21 G. M. Munoz Caro and W. A. Schutte, *Astron. Astrophys.*, 2003, **412**, 121–132.
22 M. S. Westley, R. A. Baragiola, R. E. Johnson and G. A. Baratta, *Planet. Space Sci.*, 1995, **43**, 1311–1315.
23 K. I. Öberg, G. W. Fuchs, Z. Awad, H. J. Fraser, S. Schlemmer, E. F. van Dishoeck and H. Linnartz, *Astrophys. J. Lett.*, 2007, **662**, L23–L26.
24 K. I. Öberg, H. Linnartz, R. Visser and E. F. van Dishoeck, *Astrophys. J.*, 2009, **693**, 1209–1218.
25 G. M. Muñoz Caro, A. Jiménez-Escobar, J. Á. Martín-Gago, C. Rogero, C. Atienza, S. Puertas, J. M. Sobrado and J. Torres-Redondo, *Astron. Astrophys.*, 2010, **522**, A108.
26 L. Nahon, N. de Oliveira, G. A. Garcia, J.-F. Gil, B. Pilette, O. Marcouillé, B. Lagarde and F. Polack, *J. Synchrotron Radiat.*, 2012, **19**, 508–520.
27 E. C. Fayolle, M. Bertin, C. Romanzin, X. Michaut, K. I. Öberg, H. Linnartz and J.-H. Fillion, *Astrophys. J.*, 2011, **739**, L36.
28 E. C. Fayolle, M. Bertin, C. Romanzin, H. A. M Poderoso, L. Philippe, X. Michaut, P. Jeseck, H. Linnartz, K. I. Öberg and J.-H. Fillion, *Astron. Astrophys.*, 2013, **556**, A122.
29 M. Bertin, E. C. Fayolle, C. Romanzin, K. I. Öberg, X. Michaut, A. Moudens, L. Philippe, P. Jeseck, H. Linnartz and J.-H. Fillion, *Phys. Chem. Chem. Phys.*, 2012, **14**, 9929.
30 M. Bertin, E. C. Fayolle, C. Romanzin, H. A. M. Poderoso, X. Michaut, L. Philippe, P. Jeseck, K. I. Öberg, H. Linnartz and J.-H. Fillion, *Astrophys. J.*, 2013, **779**, 120.
31 J. D. Thrower, M. P. Collings, M. R. S. McCoustra, D. J. Burke, W. A. Brown, A. Dawes, P. D. Holtom, P. Kendall, N. J. Mason, F. Jamme, H. J. Fraser, I. P. Clark and A. W. Parker, *J. Vac. Sci. Technol., A*, 2008, **26**, 919.
32 R. S. Freund, R. C. Wetzel and R. J. Shul, *Phys. Rev. A: At., Mol., Opt. Phys.*, 1990, **41**, 5861.
33 K. M. Monahan and W. C. Walker, *J. Chem. Phys.*, 1974, **61**, 3886.
34 N. J. Mason, A. Dawes, P. D. Holtom, R. J. Mukerji, M. P. Davis, B. Sivaraman, R. I. Kaiser, S. V. Hoffmann and D. A. Shaw, *Faraday Discuss.*, 2006, **133**, 311.
35 G. A. Cruz-Diaz, G. M. Munoz Caro, Y.-J. Chen and T.-S. Yih, *Astron. Astrophys.*, 2014, **562**, A119.
36 K. M. Monahan and W. C. Walker, *J. Chem. Phys.*, 1975, **63**, 1676–1681.
37 S. Y. Grebenshchikov, *J. Chem. Phys.*, 2013, **138**, 224106.
38 S. Y. Grebenshchikov, *J. Chem. Phys.*, 2013, **138**, 224107.
39 W. F. Chan, G. Cooper and C. E. Brion, *Chem. Phys.*, 1993, **178**, 401–413.
40 C. Cossart-Magos, M. Jungen and F. Launay, *Mol. Phys.*, 1987, **61**, 1077–1117.
41 M. C. Deschamps, M. Michaud and L. Sanche, *J. Chem. Phys.*, 2003, **119**, 9628.
42 J.-H. Fock, H.-J. Lau and E.-E. Koch, *Chem. Phys.*, 1984, **83**, 377–389.
43 T. G. Slanger and G. Black, *J. Chem. Phys.*, 1978, **68**, 1844.
44 J. A. Schmidt, M. S. Johnson and R. Schinke, *Proc. Natl. Acad. Sci. U. S. A.*, 2013, **110**, 17691–17696.
45 Y. Pan, H. Gao, L. Yang, J. Zhou, C. Y. Ng and W. M. Jackson, *J. Chem. Phys.*, 2011, **135**, 071101.
46 G. M. Lawrence, *J. Chem. Phys.*, 1972, **57**, 5616.
47 G. M. Lawrence, *J. Chem. Phys.*, 1972, **56**, 3435.
48 M. S. Gudipati, *J. Phys. Chem. A*, 1997, **101**, 2003–2009.

49 R. Wagner, F. Schouren and M. S. Gudipati, *J. Phys. Chem. A*, 2000, **104**, 3593–3602.

50 M. J. Loeffler, G. A. Baratta, M. E. Palumbo, G. Strazzulla and R. A. Baragiola, *Astron. Astrophys.*, 2004, **435**, 587.

51 S. A. Sandford and L. J. Allamandola, *Astrophys. J.*, 1990, **355**, 357–372.

52 H.-C. Lu, H.-K. Chen, B.-M. Cheng, Y.-P. Kuo and J. F. Ogilvie, *J. Phys. B: At., Mol. Opt. Phys.*, 2005, **38**, 3693–3704.

53 *CRC Handbook of Chemistry and Physics*, ed. William M. Haynes, CRC Press, 2012.

54 M. Minissale, E. Congiu, G. Manicò, V. Pirronello and F. Dulieu, *Astron. Astrophys.*, 2013, **559**, A49.

55 S. Andersson and E. F. van Dishoeck, *Astron. Astrophys.*, 2008, **491**, 907–916.

56 D. Menzel and R. Gomer, *J. Chem. Phys.*, 1964, **41**, 3311–3328.

PAPER

Free-electron laser induced processes in thin molecular ice

Björn Siemer, Sebastian Roling, Robert Frigge, Tim Hoger, Rolf Mitzner†
and Helmut Zacharias*

Received 2nd December 2013, Accepted 17th December 2013

DOI: 10.1039/c3fd00116d

Intermolecular reactions in and on icy films on silicate and carbonaceous grains constitute a major route for the formation of new molecular constituents in interstellar molecular clouds. In more diffuse regions and in protoplanetary discs, energetic radiation can trigger reaction routes far from thermal equilibrium. As an analog of interstellar ice-covered dust grains, highly-oriented pyrolytic graphite (HOPG) covered with D_2O, NO, and H atoms is irradiated by ultrashort XUV pulses and the desorbing ionic and neutral products are analysed. The yields of several products show a nonlinear intensity dependence and thus enable the elucidation of reaction dynamics by two-pulse correlated desorption.

1   Introduction

Molecular ice layers on grain particles are active environments for the trapping, formation, and release of new and even complex molecular species into the interstellar medium.[1,2] In dense molecular clouds at low temperatures ($T_s \sim 10$ to 50 K), physisorbed atoms and molecules may undergo exothermic reactions to form new species. In particular the formation of water and of molecular hydrogen is believed to take place under such conditions.[3-8] Inner cores of large molecular clouds in the interstellar medium (ISM) are largely opaque to radiation in the visible, ultraviolet (UV), and vacuum ultraviolet (VUV) spectral range, and only partially transparent to IR radiation.[9] Although only a low level of VUV and extreme ultraviolet (XUV) radiation is present[10] due to the interaction of cosmic rays with the constituents of the clouds, its impact on the desorption of simple and complex molecules is relevant.

At the outer rims of such clouds and particularly in the photodissociation regions, the temperatures rise and the densities are considerably lower.[1] In these regions more energetic photons can penetrate into the clouds and reach the surface of ice-covered grains with temperatures of up to about 100 K. Such

---

Physikalisches Institut, Westfälische Wilhelms-Universität, 48149 Münster, Germany. E-mail: hzach@uni-muenster.de

† present address: Helmholtz-Zentrum Berlin für Materialien und Energie, Berlin.

conditions also prevail in the outer regions of approximately solar-mass proto-planetary discs at distances of about 3 to 50 AU from the central object.[11,12] Typical grains consist of silicate and carbonaceous cores, and show sizes in the range from a few nm for polycyclic aromatic hydrocarbons (PAHs) to 1000 nm for amorphous silicates.[13-15] With graphite and graphene-like surfaces the (photo)chemistry on some of these grains can be simulated in laboratory experiments.

Figure 1 shows schematically the typical processes that might occur on such grains when they interact with UV/VUV radiation. These basic processes have been described in various review articles.[1,2,4,8,12] Besides the direct photo-desorption of molecules adsorbed on the icy surface, more involved processes can be triggered. Photodissociation of film constituents generates energetic frag-ments which may directly leave or induce, *via* collisions, other molecules to leave the grain particle. Furthermore, reactions between dissociation products and constituents of the ice layer or even of the grain core may occur. The high kinetic energy of the primary dissociation products and the solid state density of the film enable reaction cascades very different from equilibrium gas phase chemistry.

The interaction of VUV radiation with condensed molecular films has been studied in recent years extensively by various groups. Early studies involved broad band VUV radiation from hydrogen and helium discharge lamps,[16,17] while recently tunable radiation from frequency doubled lasers[18] and synchrotron light sources[19] has been employed. Besides simple monomolecular ices like H_2O, CO and N_2,[20] ice films of mixed constituents address issues of radical reactions in these films.[21] A prototypical radical is nitric oxide which has been observed in the gas phase of clouds[22] and which, according to simulations, builds up to signifi-cant concentrations in ices.[23] It also is the main precursor for nitrogen and oxygen chemical networks.[24,25] Such reactions become more important after the incor-poration of more complex molecules, like benzene,[18] small polycyclic aromatic hydrocarbons,[26] and C_{60} fullerenes[27] as dopants into water ice.

Important aspects of VUV and XUV induced processes in condensed molecular films are further addressed by low-energy electron beam irradiation of these systems. Such electrons are also liberated by XUV irradiation of the samples. The

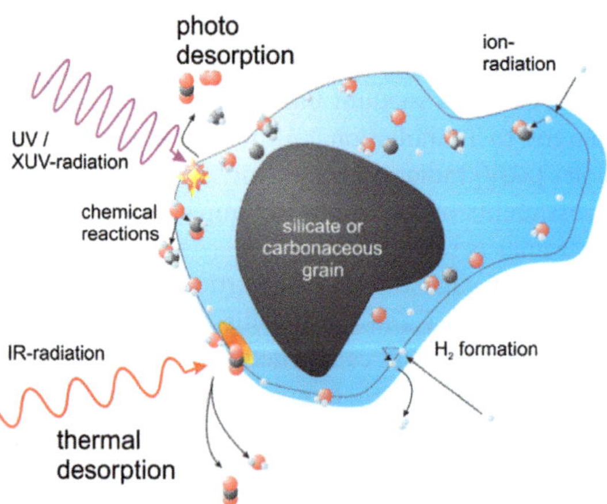

Fig. 1   Illustration of reactive processes on icy grains.

primary ions, either generated directly by the photons or by secondary photo-electrons, may initiate very effective ion-molecule reactions such as protonation of photoproducts and original constituents. Electron stimulated desorption of various hydrogen and oxygen containing positive ions[28] and aromatic molecules[29] from water ice has been reported, as well as the interaction with more complex ices.[30]

In this paper we report on preliminary studies of the interaction of XUV photons with graphite surfaces covered by hydrogen atoms, nitric oxide, or water ice. We make use of the ultrashort duration of XUV pulses from a free-electron laser, of the order of 30 fs, to shed light onto reaction pathways triggered by this radiation. Although the photon energy of these pulses is certainly higher than that from external irradiation sources of molecular clouds due to the atomic hydrogen ionization limit, the basic dynamic features induced by this radiation in the ice films are expected to be similar to the action of VUV photons. Moreover, internal radiation created by cosmic ray collisions produce XUV radiation, even in dense clouds.

2   Experimental

The experiments were performed at beamlines 1 and 3 of the free-electron laser at Hamburg (FLASH), which provides pulsed radiation in the XUV spectral range with fundamental photon energies from about $h\nu = 25$ to 300 eV.[31] In the present experiments, FLASH operated at photon energies of $h\nu = 38.15$ eV ($\lambda = 32.5$ nm) and 57.1 eV ($\lambda = 21.7$ nm) in a 5 Hz single bunch mode. The spectral energy width averaged to about $h\Delta\nu \sim 0.8$ eV, with pulse energies up to about (13.5 ± 3.5) μJ. The average pulse duration was measured to *ca.* $t_{\mathrm{L}} \sim 30$ fs.[32]

For time-resolved experiments, a wave front dividing beam splitter and pulse split delay unit (SDU) was employed.[33,34] This SDU provides two jitter-free replica pulses for XUV pump–XUV probe experiments (see Fig. 2a). The incoming FLASH beam is split into two parts by the first mirror which is partially moved into the beam path. Each partial beam propagates on a separated beam path (orange and green) and both are recombined again into the original direction with the final mirrors of their respective beam paths. Following recombination, the soft x-ray beams are focused by a toroidal mirror to ellipsoidal spot sizes of 220 μm × 175 μm and 235 μm × 125 μm for the variable and fixed beam paths, respectively. The spatial overlap of the two foci is monitored by an x-ray CCD camera in the equivalent focal position *via* a reflecting quartz plate which is brought into the beam path. In this way the focal diameter has been measured and the stability of the spatial overlap ensured over the whole delay range. This set-up further facil-itated a measurement of the different intensities in the two focal spots. The $\hat{p}$- polarized XUV pulses strike the highly oriented pyrolytic graphite (HOPG) surface at an angle of incidence of $\theta = 67.5°$ relative to the surface normal. At these photon energies of 38.15 eV and 57.1 eV and incident angle, about 42 % and 32 % of the incident radiation is reflected from graphite and about 12 % and 5 % from thick ice films, respectively.[35,36]

Figure 2b shows schematically the set-up for the detection of desorbing species. The HOPG sample exhibited a mosaic angle of 0.4° and a diameter of 10 mm. It was mounted on a liquid nitrogen containing reservoir attached to a holder on an X-Y-Z stage in an ultra-high vacuum (UHV) chamber. The UHV

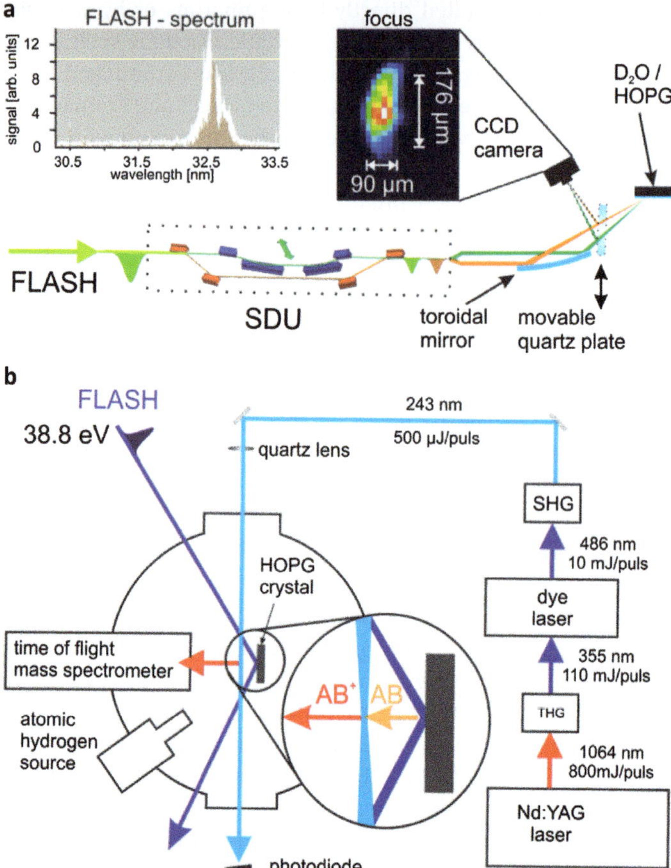

Fig. 2   (a) Schematic diagram of the optical layout of the experimental set-up. The spectrum shows in dark colour a single shot spectrum and in white the spectral envelop over 1000 pulses. (b) Experimental set-up for the detection of desorption products.

chamber was evacuated by an oil-free pumping system to a base pressure of $< 7 \cdot 10^{-10}$ mbar. The HOPG sample was prepared by cleaving with adhesive tape and long-time annealing at 800 K by electron bombardment from the rear side. After sample preparation the temperature of the graphite crystal was held at $T_s = 95$ K, as monitored with a K-type thermocouple attached to the surface.

Molecules were dosed onto the HOPG surface through a tube at normal incidence at a constant chamber pressure for several minutes or *via* atomic and molecular beams. Hydrogen and deuterium atoms were supplied by a heated tungsten capillary, using a design which avoids ionic species. Hydrogen only sticks as chemisorbed atomic hydrogen after surmounting a considerable adsorption barrier of $E_b \sim 200$ meV,[37,38] an energy which can be achieved when heating the tube to temperatures above 2000 °C. NO molecules were dosed by a doubly differentially pumped pulsed thermal molecular beam. Nitric oxide condenses as dimers in a self-terminating monolayer. At this surface temperature water condenses as microporous amorphous ice. For water ice, coverages of 80 to 200 ML have been prepared and investigated. At XUV photon energies of 38 eV and 57 eV the penetration depth of the radiation into ice amounts to about 9 nm and 15.7 nm, respectively.[35,36] Therefore, at an incident angle of $\theta = 67.5°$ the first 11 and 18 ML, respectively, are irradiated using a value of $1.15 \cdot 10^{15}$ molecule cm^{-2} for a monolayer (ML) of water on graphite.

Cations desorbing directly after irradiation by the XUV pulse were measured by a Wiley-McLaren type time-of-flight mass spectrometer (ToF) mounted along the surface normal direction. Then the sample holder served as the first repelling electrode. Desorbing neutral molecules were first ionized by tuneable laser radiation (see Fig. 2b) and then detected by the same ToF spectrometer. The output of the microchannel plates was directly monitored on a digital oscilloscope, gated and forwarded to a computer. For neutral hydrogen atoms a (2 + 1) REMPI scheme *via* the two-photon excitation 2s $^2S_{1/2}$ ← 1s $^2S_{1/2}$ at $\lambda \sim 243$ nm was chosen, and for NO a (1 + 1) REMPI scheme *via* the A $^2\Sigma^+$ ← X $^2\Pi$ γ-band transitions around $\lambda \sim 226$ nm. This UV probe laser radiation was provided by a frequency doubled, tuneable dye laser with nanosecond pulse duration, which was electronically synchronized to the FEL pulses.

For each delay time setting between the desorbing XUV and the detecting tuneable dye laser pulses, the time-of-flight mass spectra were summed over 1000 single shots. For each shot the signal and the relative FLASH pulse energy as given by the gas monitor detector was stored. Normalization and averaging were performed in the data analysis.

3 Results

Water ice

XUV photon induced reactions in undoped D_2O water ice multilayers with thicknesses of up to about 200 ML are observed. At the photon energies employed, 38.15 eV and 57.1 eV, the exciting radiation is absorbed in the top most ice layers and does not reach the substrate. At respective penetration depths of 9.0 and 15.7 nm into the ice and an angle of incidence of $\theta = 67.5°$, only the top most 11 and 18 ML of the ice, respectively, are irradiated. After illuminating the surface with soft x-ray radiation at h$\nu = 38.15$ eV ($\lambda = 32.5$ nm) with pulse energies up to (13.5 ± 3.5) μJ, the directly desorbing cations were detected with a time-of-flight mass spectrometer. When reducing the acceleration voltage in the spectrometer, an asymmetric line shape for the detected ions arises and thus the kinetic energy distribution of these ions can be derived.

A typical mass spectrum is shown in Fig. 3. Besides the expected masses of D_2O^+, OD^+, O^+, and D^+ resulting from fragmentation and ionization, products of reactions within the ice layers are also observed, in particular D_3O^+, $D_2O_2^+$, and O_2^+. In addition, water cluster ions $(D_2O)^+_n$ and deuteronated clusters $(D_2O)_nD^+$ which are formed by dissociative deuteronation are observed at higher masses.[39] The yield of these reactively formed products exhibits a nonlinear dependence on the FEL intensity which is exemplarily shown in Fig. 4 for O_2^+ ions. The signals can be fitted by a power law, $Y \sim I^n$, where $n = 3$, even for unfocussed FEL pulses. In an oversimplified picture, such a power dependence then suggests that three XUV photons are involved in the formation of the product species. Strong pulse-to-pulse fluctuations of the FEL output power causes a broad scatter of single shot data, which leaves some variations in the exponent n.

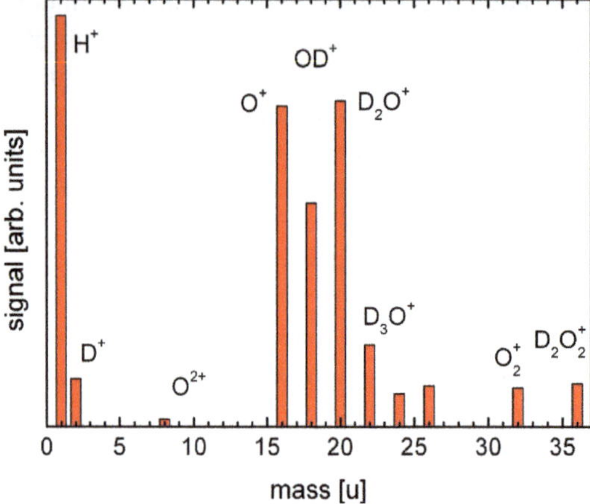

Fig. 3   Time-of-flight mass spectrum of species desorbed form ice/graphite by XUV FEL pulses at hν = 38.15 eV.

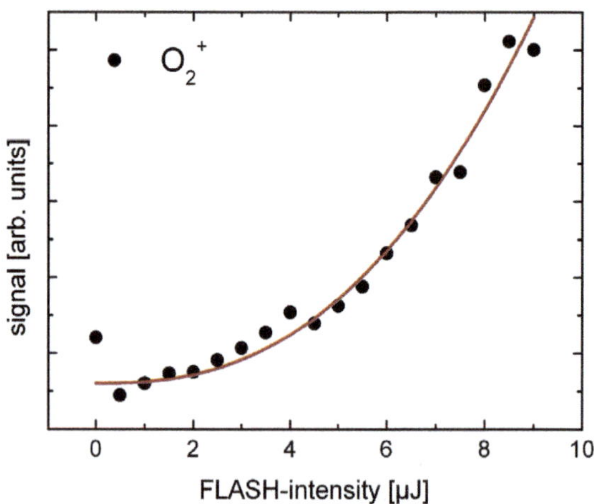

Fig. 4   Dependence of the O_2^+ ion signal on the pulse energy of the desorbing FEL pulses at hν = 57.1 eV. The other oxygen containing ions observed (Fig. 5) show similar intensity dependencies.

Such a nonlinear intensity dependence allows the investigation of the temporal development of the signal *via* an XUV pump–XUV probe set-up, thus partially elucidating the reaction dynamics. For the present experiment the SDU was employed with an asymmetric intensity splitting of the FEL pulse. In the overlapping foci of both beams this resulted in an averaged intensity of $(6.5 \pm 1.2) \cdot 10^{10}$ W/cm^2 and $(1.1 \pm 0.1) \cdot 10^{11}$ W/cm^2 for the variable and fixed beam, respectively. Thus, an asymmetric yield is expected as a function of temporal delay between both pulses. For each delay, Δt, the spatial overlap of both beams was controlled by measuring the beam profile by reflecting the beam off a quartz plate onto a CCD camera positioned at the equivalent focal position before entering the experimental chamber (see Fig. 2a).

Figure 5 shows the yield of various product ions as a function of delay times $\Delta t = -1$ and $+5$ ps between both XUV pulses. The data points are averages of several runs. Besides the expected D_2O^+ ions, several fragment ions - D^+, O^+, and OD^+ - and reaction products - $D_2O_2^+$, O_2^+, and D_3O^+ (see Fig. 6) - are also observed. The temporal structure of the two-pulse correlation curves are very similar for all ions shown. Fitting the longer main signal by a Gaussian curve yields a maximum at $\Delta t = 2.2$ ps and a signal width of about 2.7 ps (FWHM). In addition, for the oxygen containing products, a second maximum is observed for $\Delta t = 0.6$ ps. Despite the fact that this relative maximum is supported by just one data point, it was observed in every run. The very limited beam time at the FEL prohibited a finer temporal grid. It is worth noting that the signal of nearly all ions, including the D^+ ion, approach zero for longer temporal delays. The D_2O^+ ion signal, however, shows a large background signal of more than 50% of the peak signal which does not change with delay. A large part of this signal is thus generated by a single photon process, as expected.

From the observation that the maximum of the signals do not occur when both pulses overlap in time, it may be concluded that not only pure photonic processes are involved in forming these cations. Instead, collisions between fragments

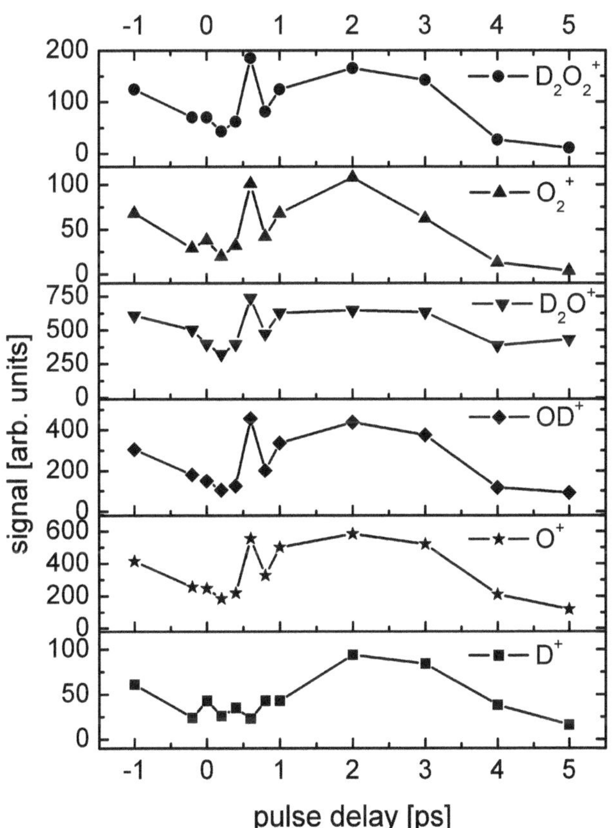

Fig. 5 Two-pulse correlated XUV desorption of various cation fragments from D_2O ice on graphite. The photon energy of the desorbing laser was $h\nu = 38.15$ eV.

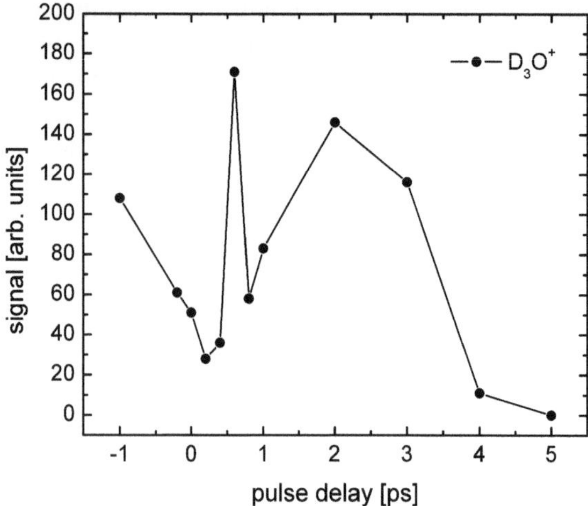

Fig. 6   Two-pulse correlated XUV desorption of D_3O^+ from ice/graphite. The photon energy of the desorbing laser was $h\nu = 38.15$ eV.

within the ice layer appear to be important. The first step for the generation of these products certainly involves the photodissociation of a water molecule:

$$D_2O + h\nu\ (38.15\ \text{eV}) \rightarrow D_2O^+ + e^-\ (E_{kin} = 25.55\ \text{eV})^{[40]} \tag{1}$$

$$\rightarrow OD + D\ (E_{exc} = 33.05\ \text{eV})^{[43]} \tag{1a}$$

$$\rightarrow OD + D^+ + e^-\ (E_{exc} = 19.45\ \text{eV})^{[41]} \tag{1a'}$$

$$\rightarrow OD^+ + D + e^-\ (E_{exc} = 19.96\ \text{eV}),^{[41]} \tag{1a''}$$

where E_{kin} and E_{exc} denote the kinetic energy of the liberated electron and the excess energy available for distribution over internal and translational degrees of freedom, respectively.

Since all products detected in Figs. 5 and 6, except D_2O^+, show a yield close to zero at large pulse separations in the two-pulse correlation, these products are not generated in a single photon event. The hydroxyl radicals produced in the beam path of the FEL beam penetrating the ice layer may then collide with each other and form hydrogen peroxide:

$$OD + OD + M \rightarrow D_2O_2 + M, \tag{2}$$

where also substrate phonons can act as the third collision partner to carry away the excess energy of the reaction.

The temporal behaviour of the two-pulse correlation signals suggests that hydrogen peroxide serves as a reservoir molecule for the other products. A second photon then either ionizes hydrogen peroxide, produces the other fragments, or strips off both hydrogen atoms:

$$D_2O_2 + h\nu \rightarrow D_2O_2^+ + e^-\ (E_{kin} = 27.5\ \text{eV})^{[42]} \tag{3a}$$

$$\rightarrow \text{OD} + \text{OD} \ (E_{\text{exc}} = 35.92 \text{ eV})[43] \tag{3b}$$

$$\rightarrow \text{OD}^+ + \text{OD} + \text{e}^- \ (E_{\text{exc}} = 22.8 \text{ eV})[42] \tag{3b'}$$

$$\rightarrow \text{DO}_2 + \text{D} \ (E_{\text{exc}} = 34.32 \text{ eV})[43] \tag{3c}$$

$$\rightarrow \text{DO}_2{}^+ + \text{D} + \text{e}^- \ (E_{\text{exc}} = 22.79 \text{ eV})[42] \tag{3c'}$$

$$\rightarrow \text{O}_2 + \text{D} + \text{D} \ (E_{\text{exc}} = 32.22 \text{ eV})[44] \tag{3d}$$

$$\rightarrow \text{O}_2 + \text{D}^+ + \text{D} + \text{e}^- \ (E_{\text{exc}} = 19.62 \text{ eV})[42] \tag{3d'}$$

$$\rightarrow \text{O}_2{}^+ + \text{D} + \text{D} + \text{e}^- \ (E_{\text{exc}} = 20.15 \text{ eV})[45,46] \tag{3d''}$$

In the gas phase the three product channels show similar reaction probabilities in the VUV.[43] This second step may be followed by the absorption of a third photon which then yields the corresponding atomic and molecular ions, if they are not already produced in the previous step.

$$\text{OD} + \text{h}\nu \rightarrow \text{OD}^+ + \text{e}^- \ (E_{\text{kin}} = 25.13 \text{ eV})[47] \tag{4a}$$

$$\text{D} + \text{h}\nu \rightarrow \text{D}^+ + \text{e}^- \ (E_{\text{kin}} = 24.55 \text{ eV})[48] \tag{4b}$$

$$\text{O}_2 + \text{h}\nu \rightarrow \text{O}_2{}^+ + \text{e}^- \ (E_{\text{kin}} = 26.08 \text{ eV})[45] \tag{4c}$$

The nonlinear yield dependence, which is equal or larger than $n = 3$, suggests that this last step indeed occurs.

In this reaction system, process (2) which describes the formation of hydrogen peroxide, is certainly the step which governs the temporal evolution of the signal. Since the excess energy release in processes (1a) through (1a'') is high, with a maximum kinetic energy in the OD radicals of up to $E_{\text{kin}} = 3.3$ eV, neighbouring hydroxyl radicals can even react within the temporal overlap of both FEL pulses at $\Delta t = 0$ ps. This part then contributes to the small signal at time zero. However, it is evident that it requires some more time to accumulate an optimal concentration of hydrogen peroxide. The two-pulse correlation results (Fig. 5) suggest that this occurs at about $\Delta t = 2.2$ ps for hydroxyl radicals, presumably produced in the bulk ice. Surface hydroxyl may react on a faster time scale as is suggested by the peak observed at $\Delta t = 0.6$ ps. This mechanism requires at least two XUV photons from the FEL beam, in accordance with the recorded power dependencies of $n \geq 3$. Hydrogen peroxide is thus an essential intermediate precursor for the formation of oxygen containing ions. This conclusion is supported by earlier experiments where ice layers have been irradiated by electron beams with kinetic energies in the same region as the photon energies employed here.[49,50]

Other oxygen containing cations observed, as shown in Fig. 5, may have additional channels which contribute to the signal height. In particular, D_2O^+ shows a large background of about one half of the peak intensity, when both pulses are separated in time. This can easily be understood by the direct and dissociative ionization of the ice constituents. It is interesting to note that for D^+, the fast peak at $\Delta t = 0.6$ ps is missing. A direct D^+ ion abstraction seems not to take place. Moreover, since we observe a significant two-pulse correlation signal, which can only be observed for a nonlinear intensity dependence, a multiphoton process must also be invoked for this product channel.

The hydronium cation, D_3O^+, which is also observed to show a pronounced temporal dependence in two-pulse correlated desorption (see Fig. 6), is easily formed by deuteronation reactions. For this product, a fast correlation peak after *ca.* $\Delta t = 0.6$ ps is also observed, followed by a major peak at about $\Delta t \sim 2.2$ ps. This is again interpreted as a distinction between surface and bulk reactions. The relative integrated yield of the surface contributions is about 0.09.

In this case either a process *via* the D^+ ion

$$D^+ + D_2O + M \rightarrow D_3O^+ + M \tag{5a}$$

or a reaction like

$$D_2O^+ + D_2O \rightarrow D_3O^+ + OD \tag{5b}$$

is conceivable. The D^+ ions observed (Fig. 5) display only the broad peak at $\Delta t = 2.2$ ps. The D_3O^+ ion products (Fig. 6) exhibit both peaks in the signal at $\Delta t = 0.6$ ps and at 2.2 ps. At longer pulse delays the D_3O^+ signal approaches zero. This last observation argues against reaction (5b), because the generation of the water cation requires only a single XUV photon. This is an efficient process also for widely separated pulses, as is witnessed by the large background in the temporal dependence of the D_2O^+ signal (Fig. 5). The non-appearance of the fast peak at $\Delta t = 0.6$ ps in the D^+ ion signal (Fig. 5) remains to be explained. It is conceivable that these early D^+ ions react with other products or neutral film constituents, see reaction (5a), before they were able to leave the surface. Certainly, additional experiments are required and in particular the kinetic energy release into the different channels need to be measured before a definite conclusion can be drawn.

It is important to note that photoinduced processes within the bulk ice lead not only to the formation of new species, in this case hydrogen peroxide, but to the release of these species and dissociation products from the grain surface into the gas phase. Collisions occurring between photodissociation products and ice constituents are central to the formation of new species. Furthermore, even without reactions, such energetic collisions will lead to the desorption of neutral species from the ice layer, as has been observed for vibrational resonant IR-induced desorption from molecular ices, like N_2O,[51,52] CD_3F,[53] and methane.[54]

Kinetic energies of directly desorbing ions

As is evident from Fig. 5, the XUV pulses also induce the desorption of hydrogen ions from ice layers. In this case, the temporal behaviour of the two-pulse correlated desorption signal suggests that D^+ ions liberated in the bulk predominantly contribute to the signal observed. These hydrogen ions show a high kinetic energy as shown in Fig. 7, where both the velocity and kinetic energy distributions of H^+ are plotted. The velocity distribution in Fig. 7a can be fitted by two Maxwellian distributions with respective temperatures of $T = 7000$ K and 19000 K and relative integrated yields of 45% and 55%. From this analysis, an average kinetic energy of $E_{kin} = 1.8$ eV for the H^+ ions is obtained. When the

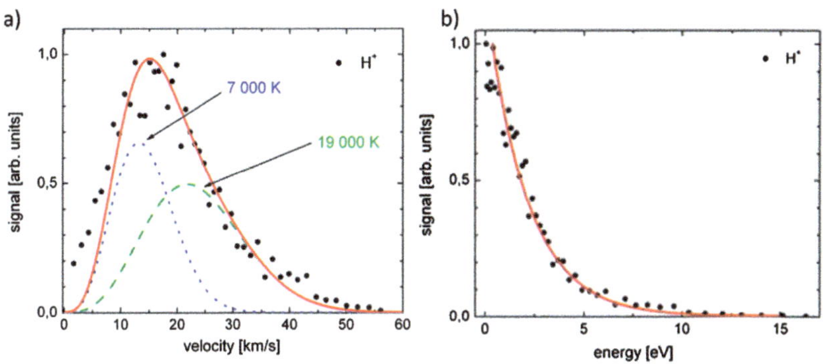

Fig. 7   (a) Velocity distribution of H^+ after XUV desorption from ice/graphite. Two Maxwellian velocity distribution with $T = 7000$ K and 19000 K approximate most of the distribution. Some deviation at low velocities can be noticed. (b) Kinetic energy distribution of H^+ from ice/graphite. The red curve is a fit to a single sided exponential yielding a kinetic temperature of 16245 K. Photon energy: 38.15 eV.

measured distribution is transformed into the energy picture (Fig. 7b) a single-sided exponential distribution, $W(E_{\text{kin}})$, is obtained

$$W(E_{\text{kin}}) \sim a \; 1/\sqrt{T}\exp\{- E_{\text{kin}}/kT\}. \tag{6}$$

Such distributions are often found in plasma physics to determine kinetic temperatures. In this case a best fit yields a kinetic temperature of $T_{\text{kin}} = 16245$ K, corresponding to an energy of about $E_{\text{kin}} = 2.1$ eV. For O^+ and O_2^+ ions, high kinetic energies of 0.84 eV and 0.39 eV, respectively, have also been observed.[55]

Neutral hydrogen

Different from other groups which investigate the reaction dynamics of physisorbed hydrogen on ice,[4] graphite,[56] and silicate surfaces[7,57] we concentrate here on chemisorbed hydrogen atoms on a graphite substrate. Such conditions are reached in warmer parts of the ISM where molecular gases have already been desorbed from the grain particles. Such grain temperatures prevail, *e.g.*, at the outer rims of molecular clouds and of protoplanetary discs.[12]

The high photon energy of the FEL radiation enables the direct desorption of ionized and neutral hydrogen species. For this experiment on HOPG (0001) we kept, after cleaning, the surface temperature at room temperature to prohibit the adsorption of water from the UHV background gas. Hydrogen atoms were dosed onto the surface from a heated tungsten capillary. In this way the adsorption barrier of *ca.* 200 meV can be overcome. In this experiment, a hydrogen coverage of about 0.25 ML is adsorbed on the surface.

Neutral H atoms were desorbed by $h\nu = 38.8$ eV radiation using p-polarized FEL pulses with an average pulse energy of 13 µJ, incident under an angle of $\theta = 67.5°$ relative to the surface normal. Slight focusing resulted in an illuminated area of about 0.19 mm², an applied fluence of 6.9 mJ/cm² and a fluence absorbed by the substrate of about 3.9 mJ/cm². By changing the temporal delay between the probe and desorption laser pulses, an arrival time spectrum of the H atoms in the probe volume is recorded.[58] After transformation into the velocity space, rather

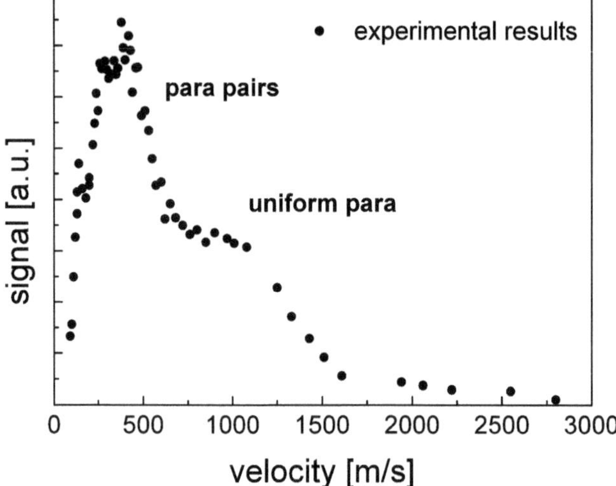

Fig. 8 Velocity distribution of neutral H atoms desorbing from chemisorbed H on graphite in a uniform para adsorbate configuration. This results in a coverage of 0.25 ML H on graphite. Average kinetic energies of $<E_{kin}> = 0.8$ meV and 5.3 meV are deduced for the para pairs and uniform para configurations, respectively.

slow velocities for the desorbing hydrogen atoms are observed (see Fig. 8). A theoretical analysis of the desorption process reveals that multiple electron scattering on the H adsorbate causes the desorption, thus a typical DIMET (desorption induced by multiple electronic transitions) process is observed.[59] These hot electrons are generated by the absorption of the high energy XUV photon by the graphite substrate, producing a dense cloud of hot electrons in the vicinity of the surface with a rather short lifetime of about half a picosecond. Multiple electron scattering leads to a successive population of higher vibrational states of the C–H vibration, and, in a final step, the desorption out of one of these vibrational states. A peak at a velocity of $v = 425$ ms^{-1}, and two shoulders at 255 ms^{-1} and 136 ms^{-1} are apparent. These very slow H atoms show an average kinetic energy of only $E_{kin} = 0.8$ meV. A second maximum at a velocity of $v = 1000$ ms^{-1} yields an average kinetic energy of $E_{kin} = 5.3$ meV. The relative integrated intensities of the very slow and the faster peak are about the same.

The theoretical analysis shows good agreement with the experimental observations when the very slow peak is caused by desorption of H atoms out of a hydrogen surface dimer pair which occupies the para position on the underlying graphene lattice.[38,60,61] The faster peak can be reconciled when one assumes that the desorption occurs out of a uniform para configuration of adsorbed H atoms.[58] In this configuration all H atoms are only adsorbed in a para position relative to each other on the graphene lattice. Thereby additional energy is gained by the adsorbate system, as depicted in Fig. 9. There, the adsorption energies gained by both atoms when adsorbing a second hydrogen atom in the ortho-, para-, and uniform para-position relative to the first one is shown. Besides a barrier-free adsorption, as for a para pair, the energy gain is even higher by about $\Delta E \sim 0.6$ eV in the uniform para configuration. Thus, instead of both hydrogen atoms exhibiting a binding energy of about $E_b \sim 0.8$ eV each as for the adsorption of

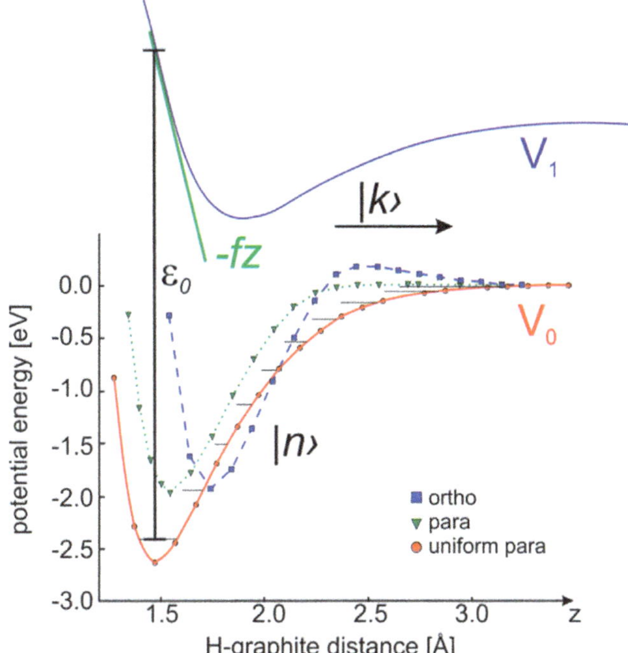

Fig. 9   Potential energy diagram for H atoms chemisorbed on graphene for the ortho-, para-, and uniform para configurations.

single isolated H atoms,[61] the uniform para pair show together an adsorption energy of $E_b \sim 2.6$ eV, thus 1.0 eV larger than the sum of two single H atoms on graphene.[62] This theoretical result is also in good agreement with the experimental observation.

Nitric oxide

The investigation of the XUV light induced desorption process of nitric oxide from graphite is driven by the desire to learn more about the fundamental microscopic desorption dynamics of small molecules. NO can be detected with very high sensitivity and with internal state selection by resonantly enhanced (1 + 1) REMPI. It is the population distributions over the internal states of a product molecule which allow a deeper understanding of the microscopic dynamics. Figure 10 shows a rotational state resolved REMPI spectrum of the A $^2\Sigma^+ \leftarrow$ X $^2\Pi$ (v″ = 0, 1) γ-bands around $\lambda = 226$ nm after excitation of the nitric oxide covered graphite surface by XUV light at h$\nu = 57.1$ eV. It is evident that besides molecules in the vibrational ground state, also vibrationally excited molecules are detected. An analysis of the rotational state distribution yields a non-Boltzmann distribution with an average rotational energy in v″ = 0 of $<E_{rot}> = 311$ cm^{-1} or a rotational temperature of about $<E_{rot}>/k = 448$ K.[63] For the vibrationally excited state the same distribution of the population in the rotational states is observed. The high intensity of the lines belonging to the γ(1-1) band suggest already a considerable vibrational excitation. For these two vibrational states a vibrational

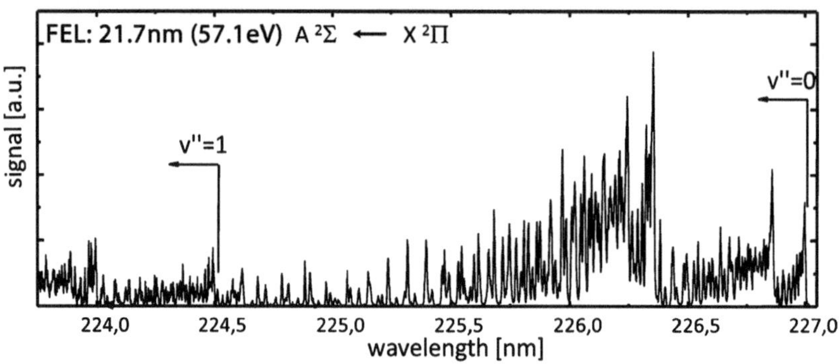

Fig. 10   Spectrum of the NO γ(0-0) and γ(1-1) bands around $\lambda = 227$ nm and 224.5 nm, respectively, after desorption from $(NO)_2$/graphite by XUV pulses of hν = 57.1 eV.

temperature of about $T_{vib} \sim 1580$ K is found. Higher vibrational states could not be measured due to the limited beam time at the FEL facility. The translational energy of neutral NO molecules was obtained by changing the temporal delay between the FEL pulse (at hν = 38 eV) and the detection laser pulse. A distribution is observed which can be fitted by the combination of two Maxwellian velocity distributions. They show the temperature parameters of $T_{kin} = 480$ K and $T_{kin} = 1717$ K, and relative yields of 35% and 65%, respectively. On average, a kinetic energy of $E_{kin} = 170$ meV is released in the desorption of neutral NO molecules. At this photon energy the desorption cross section was determined to about $\sigma = (1.1 \pm 0.4) \cdot 10^{-17}$ cm^2.[63] The desorption yield of NO from this $(NO)_2$/graphite system exhibits a slightly nonlinear dependence on the XUV pulse intensity.[63] Exponents between $n = 1.4$ and 3 have been observed in different runs. Such a dependence would then enable the application of two-pulse correlated excitation of the system in order to learn more about the temporal behaviour of the excited system.

At surface temperatures below $T_s \sim 100$ K nitric oxide condenses as cis-dimer $(NO)_2$ on graphite.[64] The release of NO molecules thus has to be preceded by the dissociation of the NO dimer. Direct dissociative ionisation cross sections are about one order of magnitude lower than the desorption cross section measured in this study.[65]

Therefore, a direct photoexcitation of the adsorbate is unlikely, and an indirect excitation mechanism is suggested. The XUV photons are predominantly absorbed by the graphite substrate, thereby creating a cloud of hot electrons which interacts with the adsorbate. This interaction leads to a temporarily excited $(NO)_2{}^-$ dimer anion, where one NO molecule lies flat on the surface. For the dimer anion the trans geometry is, however, energetically more favourable than the cis geometry.[66] Therefore, the temporary formation of a dimer anion leads to both a strong rotational excitation of the leaving NO fragment due to this transition from a cis to a trans conformation of the dimer while in the excited state, as well as to a high vibrational excitation due to the transient negative ion, as in the case of metal substrates. Both conclusions are supported by the experimental observations.

Conclusion

Femtosecond pulsed XUV radiation from free-electron laser sources is a valuable tool to induce reactive processes in molecular ices and elucidate the dynamics of such systems. Preliminary results of such reactions in water ice leading to cationic products have been presented. The high photon energy induces the formation of new species, in this case D_2O_2, which may act as a reservoir for further reactions. The observation of extremely slow H atoms from the chemisorbed state on graphite is really surprising and may have an influence on the understanding of the dynamics in accretion discs. The extension to other systems, such as doped or molecular ices where the neutral constituents themselves can be detected with internal state selection, will shed light onto the microscopic desorption dynamics. The study of the $(NO)_2$ thin film presented provides a first glimpse of the wealth of information which can be obtained.

However, many interesting questions remain at present unanswered, including the velocity distribution of most of the oxygen containing cation products observed, and the relative roles of direct photoprocesses and processes induced by fast electrons liberated by the ionization of ice constituents. Also, the detection of neutral desorbing products, *e.g.* D atoms or OD radicals would contribute to a deeper understanding. Future experiments on VUV and XUV FEL facilities will allow us to address these questions and to contribute to a deeper microscopic understanding of reaction and desorption processes in complex ices.

Acknowledgements

The authors thank the FLASH team at HASYLAB for providing the XUV pulses, D. Gerlich, T. Olsen and J. Thrower for helpful discussions, and S. Eppenhoff and F. Wahlert for extensive technical support. The Bundesministerium für Bildung und Forschung provided partial financial support *via* grant nos. 05 KS4PMC/8 and 05 KS7PM1 within the research program FSP 301 "FLASH" which is gratefully acknowledged. The European Union provided partial support via the FP7-PEOPLE-ITN-2008 programme "LASSIE" no. 238258.

References

1 A. G. G. M. Tielens, *Rev. Mod. Phys.*, 2013, **85**, 1021.

2 E. Herbst and E. F. van Dishoeck, *Annu. Rev. Astron. Astrophys.*, 2009, **47**, 427.

3 R. J. Gould and E. E. Salpeter, *Astrophys. J.*, 1963, **138**, 393.

4 D. A. Williams and E. Herbst, *Surf. Sci.*, 2002, **500**, 823.

5 L. Hornekaer, A. Baurichter, V. Petrunin, D. Field and A. C. Luntz, *Science*, 2003, **302**, 1943.

6 S. C. Creighan, J. S. A. Perry and S. D. Price, *J. Chem. Phys.*, 2006, **124**, 114701.

7 J. L. Lemaire, G. Vidali, S. Baouche, M. Chehrouri, H. Chaabouni and H. Mokrane, *Astrophys. J.*, 2010, **725**, L156.

8 G. Vidali, *J. Low Temp. Phys.*, 2013, **170**, 1.

9 J. S. Mathis, P. G. Mezger and N. Panagia, *Astron. Astrophys.*, 1983, **128**, 212.

10 S. S. Prasad and S. P. Tarafdar, *Astrophys. J.*, 1983, **267**, 603.

11 J. Najita, E. A. Bergin and J. N. Ullom, *Astrophys. J.*, 2001, **561**, 880.

12 T. Henning and D. Semenow, *Chem. Rev.*, 2013, **113**, 9016.

13 J. S. Mathis, W. Rumpl and K. H. Nordsiek, *Astrophys. J.*, 1977, **217**, 425.

14 B. T. Draine and H. M. Lee, *Astrophys. J.*, 1984, **285**, 89.

15 D. Hollenbach, M. J. Kaufman, E. A. Bergin and G. J. Melnick, *Astrophys. J.*, 2009, **690**, 1497.

16 M. S. Westley, R. A. Baragiola, R. E. Johnson and G. A. Baratta, *Nature*, 1995, **373**, 405.

17 S. R. Baggott, K. W. Kolasinski, L. M. A. Perdigao, D. Riedel, Q. Guo and R. E. Palmer, *J. Chem. Phys.*, 2002, **117**, 6667.

18 J. D. Thrower, D. J. Burke, M. P. Collings, A. Dawes, P. D. Holton, F. Jamme, P. Kendall, W. A. Brown, I. P. Clark, H. J. Fraser, M. R. S. McCoustra, N. J. Mason and A. W. Parker, *Astrophys. J.*, 2008, **673**, 1233.

19 E. C. Fayolle, M. Bertin, C. Romanzin, X. Michaut, K. I. Öberg, H. Linnartz and J.-H. Fillion, *Astrophys. J.*, 2011, **739**, L36.

20 K. I. Öberg, E. F. van Dishoeck and H. Linnartz, *Astron. Astrophys.*, 2009, **496**, 281.

21 K. I. Öberg, E. F. van Dishoeck, H. Linnartz and S. Andersson, *Astrophys. J.*, 2010, **718**, 832.

22 D. McGonagle, L. M. Ziurys, W. M. Irvine and Y. C. Minh, *Astrophys. J.*, 1990, **359**, 121.

23 S. B. Charnley, S. D. Rogers and P. Ehrenfreund, *Astron. Astrophys.*, 2001, **378**, 1024.

24 D. T. Halfen, A. J. Apponi and L. M. Ziurys, *Astrophys. J.*, 2001, **561**, 244.

25 E. Congiu, H. Chaabouni, C. Laffon, P. Parent, S. Baouche and F. Dulieu, *J. Chem. Phys.*, 2012, **137**, 054713.

26 J. Bouwman, H. M. Cuppen, M. Steglich, L. J. Allamandola and H. Linnartz, *Astron. Astrophys.*, 2011, **529**, A46.

27 S. H. Cuylle, H. Linnartz and J. D. Thrower, *Chem. Phys. Lett.*, 2012, **550**, 79.

28 T. M. Orlando, A. B. Alexandrov and J. Herring, *J. Phys. Chem. B*, 2003, **107**, 9370.

29 J. D. Thrower, M. P. Collings, F. J. M. Rutten and M. R. S. McCoustra, *Chem. Phys. Lett.*, 2011, **505**, 106.

30 A. G. M. Abdulgalil, D. Marchione, J. D. Thrower, M. P. Collings, M. R. S. McCoustra, F. Islam, M. E. Palumbo, E. Congiu and F. Dulieu, *Philos. Trans. R. Soc. London, Ser. A*, 2013, **371**, 20110586.

31 K. Tiedtke, *et al.*, *New J. Phys.*, 2009, **11**, 023029.

32 R. Mitzner, A. A. Sorokin, B. Siemer, S. Roling, M. Rutkowski, H. Zacharias, M. Neeb, T. Noll, F. Siewert, W. Eberhardt, M. Richter, P. Juranic, K. Tiedtke and J. Feldhaus, *Phys. Rev. A: At., Mol., Opt. Phys.*, 2009, **80**, 025401.

33 R. Mitzner, B. Siemer, M. Neeb, T. Noll, F. Siewert, S. Roling, M. Rutkowski, A. A. Sorokin, M. Richter, P. Juranic, K. Tiedtke, J. Feldhaus, W. Eberhardt and H. Zacharias, *Opt. Express*, 2008, **16**, 19909.

34 M. Wöstmann, R. Mitzner, T. Noll, S. Roling, B. Siemer, F. Siewert, S. Eppenhoff, F. Wahlert and H. Zacharias, *J. Phys. B: At., Mol. Opt. Phys.*, 2013, **46**, 164005.

35 B. L. Henke, E. M. Gullikson and E. M. Davis, *At. Data Nucl. Data Tables*, 1993, **54**, 18.

36 J. H. Hubell, W. J. Feigele, E. A. Briggs, R. T. Brown, D. T. Cromer and R. J. Howerton, *J. Phys. Chem. Ref. Data*, 1977, **4**, 471.

37 L. Jeloica and V. Sidis, *Chem. Phys. Lett.*, 1999, **300**, 157.

38 X. Sha and B. Jackson, *Surf. Sci.*, 2002, **496**, 318.

39 V. Hermann, B. D. Kay and A. W. Castleman Jr., *Chem. Phys.*, 1982, **72**, 185; F. Dong, S. Heinbuch, J. J. Rocca and E. R. Bernstein, *J. Chem. Phys.*, 2006, **124**, 224319.

40 J. E. Reutt, L. S. Wang, Y. T. Lee and D. A. Shirley, *J. Chem. Phys.*, 1986, **85**, 6928.

41 J. H. D. Eland, *Adv. Mass Spectrom.*, 1974, **6**, 917.

42 F. S. Ashmore and A. R. Burgess, *J. Chem. Soc., Faraday Trans. 2*, 1997, **73**, 1247; S. N. Foner and R. L. Hudson, *J. Chem. Phys.*, 1962, **36**, 2676.

43 R. Atkinson, D. L. Baulch, R. A. Cox, J. N. Crowley, R. F. Hampson, R. G. Hynes, M. E. Jenkin, M. J. Rossi and J. Troe, *Atmos. Chem. Phys.*, 2004, **4**, 1461.

44 L. J. Stief and V. J. De Carlo, *J. Chem. Phys.*, 1969, **50**, 3.

45 R. G. Tonkyn, J. W. Winniczek and M. G. White, *Chem. Phys. Lett.*, 1989, **164**, 137.

46 *NIST – JANAF Thermochemical Tables, 4th edition*, M. W. Chase, ed., 1998, Monograph 9, ISBN 1-56396-831-2.

47 R. T. Wiedmann, R. G. Tonkyn, M. G. White, K. Wang and V. McKoy, *J. Chem. Phys.*, 1992, **97**, 768.

48 R. L. Kelly, *J. Phys. Chem. Ref. Data*, 1987, 16 (suppl. 1), 1.

49 T. M. Orlando and M. T. Sieger, *Surf. Sci.*, 2003, **528**, 1.

50 X. N. Pan, A. D. Bass, J. P. Jay-Gerin and L. Sanche, *Icarus*, 2004, **172**, 521.

51 B. Redlich, H. Zacharias, G. Meijer and G. von Helden, *Surf. Sci.*, 2002, **502–503**, 325.

52 B. Redlich, L. van der Meer, H. Zacharias, G. Meijer and G. von Helden, *Nucl. Instrum. Methods Phys. Res., Sect. A*, 2003, **507**, 556.

53 B. Redlich, H. Zacharias, G. Meijer and G. von Helden, *Phys. Chem. Chem. Phys.*, 2002, **4**, 3448.

54 B. Redlich, H. Zacharias, G. Meijer and G. von Helden, *J. Chem. Phys.*, 2006, **124**, 044704.

55 B. Siemer, T. Hoger, M. Rutkowski, R. Treusch and H. Zacharias, *J. Phys.: Condens. Matter*, 2010, **22**, 084013.

56 E. R. Latimer, F. Islam and S. D. Price, *Chem. Phys. Lett.*, 2008, **455**, 174.

57 L. Gavilan, J. L. Lemaire, G. Vidali, T. Sabri and C. Jäger, *Astrophys. J.*, 2014, **781**, 79.

58 B. Siemer, T. Olsen, T. Hoger, M. Rutkowski, C. Thewes, S. Düsterer, J. Schiøtz and H. Zacharias, *Chem. Phys. Lett.*, 2010, **500**, 291.

59 J. A. Misewich, T. F. Heinz and D. M. Newns, *Phys. Rev. Lett.*, 1992, **68**, 3737.

60 N. Rougeau, D. Teillet-Billy and V. Sidis, *Chem. Phys. Lett.*, 2006, **431**, 135.

61 L. Hornekaer, E. Rauls, W. Xu, Z. Sljivancanin, R. Otero, I. Stensgaard, E. Laegsgaard, B. Hammer and F. Besenbacher, *Phys. Rev. Lett.*, 2006, **97**, 186102.

62 R. Frigge, T. Hoger, B. Siemer, H. Witte, M. Silies, H. Zacharias, T. Olsen and J. Schiøtz, *Phys. Rev. Lett.*, 2010, **104**, 256102.

63 B. Siemer, T. Hoger, M. Rutkowski, S. Düsterer and H. Zacharias, *J. Phys. Chem. A*, 2011, **115**, 7356.

64 J. P. Coulomb, J. Suzanne, M. Bienfait, M. Matecki, A. Thomy, B. Croset and C. Marti, *J. Phys.*, 1980, **41**, 1155.

65 J. A. R. Samson, T. Masuoka and P. N. Pareek, *J. Chem. Phys.*, 1985, **83**, 11.

66 B. Urban, A. Strobel and V. E. Bondybey, *J. Chem. Phys.*, 1999, **111**, 8939.

DISCUSSIONS

General discussion

DOI: 10.1039/c4fd90002b

Dr Cuppen opened the discussion of the paper by Yasuhiro Oba: In our 2010 PCCP paper[1] we used a similar approach to make H_2O_2 ice by co-deposition of O_2 and H, and subsequently heating the sample to remove all O_2. Upon hydrogenation of this sample we also found that the 2850 cm^{-1} band of H_2O_2 behaves differently than in our previous samples. It does not appear to grow linearly with the amount of H_2O_2 in the sample, but it appears to be highly sensitive to the local environment. I am glad to see that you have similar findings. Could you please comment on what you believe is the origin of this behaviour?

1 H. M. Cuppen, S. Ioppolo, C. Romanzin and H. Linnartz, *Phys. Chem. Chem. Phys.*, 2010, **12**, 12077.

Dr Oba responded: We agree with your suggestion that the 2850 cm^{-1} band may be sensitive to the local environment. Unfortunately, however, we cannot fully explain this unexpected behavior of the 2850 cm^{-1} band of H_2O_2. In fact, a similar result was obtained in our previous study,[1] in which we studied H_2O formation by the codeposition of cold OH with H_2 onto a substrate at 10 K, but the unexpected behavior was not discussed in that paper. As briefly noted in the present paper, we think that this behavior could be related to the number of hydrogen bonds between H_2O_2 and H_2O. In our high-purity H_2O_2 solid, H_2O abundance is very small (less than 5% of H_2O_2); therefore, the number of hydrogen bonds between H_2O_2 and H_2O should be small.[2] When H atoms were exposed to H_2O_2, H_2O was formed as shown in Fig. 2 in the paper, which would lead to an increase in the number of hydrogen bonds between unreacted H_2O_2 and the formed H_2O. Moreover, if we compare the FT-IR spectrum of the pure H_2O_2 with that of H_2O_2 on a-D_2O, we found that the intensity of the 2850 cm^{-1} band for H_2O_2 on a-D_2O was twice as large as that for pure H_2O_2, although the intensity of other bands at ~3300 and ~1400 cm^{-1} were quite similar each other (Fig. 1 and 8 in the paper). This behavior is reasonable if (i) the intensity of the ~2850 cm^{-1} band is correlated with the number of hydrogen bonds between H_2O_2 and H_2O (D_2O), and (ii) H_2O_2 forms lots of hydrogen bonds with surrounding D_2O molecules. Further experimental and theoretical studies are required to confirm our hypothesis.

1 Y. Oba, N. Watanabe, T. Hama, K. Kuwahata, H. Hidaka and A. Kouchi, *Astrophys. J.*, 2012, **749**, 67.
2 P. G. Sennikov, S. K. Ignatov and O. Schrems, *ChemPhysChem*, 2005, **6**, 392.

Dr Cuppen opened the discussion of the paper by Emanuele Congiu: Looking at Fig. 2 in your paper, I do not understand the underlying physics. It appears that the ozone production is temperature dependent if deposited at different temperatures. However, if one heats the sample, no further increase is observed. What happens to the oxygen atoms and molecules that are still present on the surface? There should be more reactive material present at the end of the deposition. Not everything can react instantaneously, or it should occur through a hot atom mechanism, where the species are hot from the deposition or from the large exothermicity of the reaction.

Different processes appear to occur during the deposition period and the heating period. I noticed in your recent *J. Chem. Phys.* paper[1] that you use two types of diffusion rates, k_{td} and k_x, which are active in the two different periods. A consistent model should be able to describe both phases simultaneously. Can you explain this? Related to this model, why did you not include the diffusion of molecular oxygen? The binding energy of molecular oxygen is much lower than that of atomic oxygen and it is therefore likely that it will diffuse as well.

1 M. Minissale, E. Congiu and F. Dulieu, *J. Chem. Phys.*, 2014, **140**, 074705.

Dr Congiu answered: In Fig. 2, we chose to display only the ozone yield as a function of surface temperature to show that more ozone is formed at higher temperatures because of the enhanced diffusion of oxygen atoms. It should be noted, however, that ozone is not the only product, as molecular oxygen can be formed as well. The $O_3 : O_2$ ratio increases with temperature and there is always a good correspondence between the number of O atoms on the surface and the product yields (see Fig. 11 of our recent paper[1]). It is true that not everything can react instantaneously upon irradiation, but if the diffusion of O atoms proceeds quickly, all the chemistry is completed within a millisecond timescale.

The use of two diffusion rates, k_x during deposition and k_{td} during the heating phase, was chosen to model the diffusion rates at a fixed temperature and during the TPD separately. The rate k_x comprises the quantum tunnelling diffusion term and the classical thermal hopping term, while k_{td} only accounts for thermal diffusion, as most authors assume during the heating phase. Our model is, however, perfectly consistent because both phases could be described simultaneously if it was necessary. This is not the case because the two phases do not occur concurrently. The diffusion of molecular oxygen was not included in the model to avoid adding more complexity to the system, although we do not believe that adding the O_2 mobility would change our interpretation of the experimental results between 6 and 20 K. It is true, however, that the significant increase in diffusion rate at temperatures greater than 20 K is very likely due to the onset of efficient thermal diffusion of O_2 molecules on the surface, which suggests in a way that O atoms diffuse at $T_s < 20$ K, following a quantum diffusion trend (also, because the reverse could not occur, namely O_2 diffusing quantum mechanically, could it?).

1 M. Minissale, E. Congiu and F. Dulieu, *J. Chem. Phys.*, 2014, **140**, 074705.

Professor Dulieu added: The chemical desorption of O_2 is dependent on the substrate. We could not detect any chemical desorption on water substrates. On the contrary, on graphite it could be up to 80% at very low coverage, and it decreases with the presence of other adsorbates and therefore with increasing coverage. It is thus possible to almost cancel the chemical desorption, and at least to reduce it by more than one order of magnitude. The chemical desorption is very dependent on substrate–adsorbate interaction, and can vary between <5% and 80% for the same reaction. This work is detailed in a paper that has been submitted to *J. Chem. Phys.*

Professor Kaiser opened the discussion of the paper by Stephen Price: First, a general comment. All authors and speakers should use scientifically correct terminologies. There is no "evaporation" from ices in the interstellar medium. By definition, "evaporation" requires a liquid phase. There is clearly no "liquid" phase on interstellar grains. The correct vocabulary is "sublimation". Second, a question directed to Professor Price. In your experiments, you start with oxygen atoms in their triplet ground states, and the reaction products are clearly closed shell singlets. So the reaction requires intersystem crossing (ISC) to be efficient. Based on your experimental data and global rate constant, do you have any idea how fast ISC is in these experiments, and how this compares to reactions in the gas phase?

Professor Price responded: Indeed, I agree, "evaporation" is not appropriate terminology, and I hope we have not used it! A chemical pedant could perhaps argue that "desorption" is perhaps an even better term than "sublimation" – in that "sublimation" carries with it, to a chemist, the implication of a system where the gas-phase and adsorbed species are in thermodynamic equilibrium (have equal chemical potential), which is not necessarily the case in the ISM. Regarding your question, we can see spectroscopically (with REMPI) that the vast majority of the O atoms incident on our surface are in the ^{3}P state. As you say, this means that the reaction must involve ISC, if we are to produce singlet state products. It would be interesting to try and consider the rate of this ISC process on the surface. I think we would need some theoretical input to try and extract this rate from our rate constants. However, one could speculate that interaction with the O_2 molecules (which are in a triplet state) on the surface might enhance the rate of ISC in comparison with that in the gas-phase.

Professor Meuwly continued the discussion of the paper by Emanuele Congiu: This comment – and the following question – concerns the proposal that quantum mechanical tunnelling is the explanation for the diffusion behavior of oxygen atoms on surfaces at low temperatures.

In formulating my questions I would like to mention a few insights from ligand diffusion and rebinding in biological systems, mainly carbon monoxide (CO) in myoglobin.[1–3,6] In very early work on this problem it was found that at low temperatures the rebinding rate of CO to the heme-iron becomes temperature-independent.[1] This was interpreted as quantum tunnelling becoming operative at sufficiently low temperatures. This proposal was later confirmed by investigating the isotope effect of the rebinding kinetics.[2] Hence, simple deviation from the expected "Arrhenius behavior" does not seem to be sufficient to infer that tunnelling is operative.

Furthermore, it may be relevant to mention that physical diffusion in the ice systems under consideration takes place on rough surfaces. It has been shown that diffusion on a rough surface (with sufficiently small surface roughness) amounts to rescaling the diffusion coefficient.[4,5] Hence the following questions:

1. Would it be possible to repeat the experiments with oxygen-18 and demonstrate the mass effect predicted in ref. 2?
2. What is the role of zero-point vibrational energy in influencing the diffusion process?
3. How would the explicit consideration of a barrier distribution (as in ref. 3) affect the analysis?

Finally, a comment on semantics: it would be safer to discuss diffusion in the context of a "diffusion coefficient", rather than "diffusion constants".

1 N. Alberding *et al.*, *Science*, 1976, **192**, 1002.
2 J. O. Alben *et al.*, *Phys. Rev. Lett.*, 1980, **44**, 1157.
3 P. J. Steinbach *et al.*, *Biochem.*, 1991, **30**, 3988.
4 R. Zwanzig, *Proc. Natl. Acad. Sci. U. S. A.*, 1988, **85**, 2029.
5 P. Banushkina and M. Meuwly, *J. Chem. Phys.*, 2007, **127**, 135101.
6 P. Banushkina and M. Meuwly, *J. Phys. Chem. B*, 2005, **109**, 16911.

Dr Congiu answered: 1. Yes, it would be possible to repeat the experiments using ^{18}O and that was something that we actually planned to do in the future. After the great interest raised by our article and the different interpretations proposed by others, however, this has become a priority and has to be done as soon as possible. In fact, only a possible isotope effect would confirm our conclusions that tunnelling is at play and that the classical Arrhenius process (high temperature-dependence, low mass-dependence) can be discarded.

2. In our simplified description where a rectangular barrier was used, the zero-point vibrational energy cannot play an important role. On the other hand, in a more accurate treatment that involves a more realistic barrier shape, the zero-point energy would almost certainly affect the diffusion process and would probably reduce the width of the barrier.

3. A more accurate description of the barrier distribution on the surface would be more realistic, since we are sure that the diffusion processes occur on rough surfaces characterized by sites of diverse energy depth. In light of our experimental data, however, we believe that such an improvement of the model would not change the physical interpretation of the diffusion process of oxygen atoms.

Dr Hama remarked: I am concerned that you only tried to fit the experimental data with an Arrhenius-law form with a single diffusion barrier. This is true only for diffusion of a particle on an ideal single crystalline surface, but there are multiple (or continuously distributed) potential sites with different depths for adsorption and diffusion on an amorphous surface. You should take this into account before introducing a quantum tunnelling effect for the diffusion of oxygen atoms.

Although our understanding of the thermal diffusion of adsorbed atoms or molecules on an amorphous surface is still rather qualitative, a basic concept has been developed both experimentally and theoretically.[1–5] In brief, at low

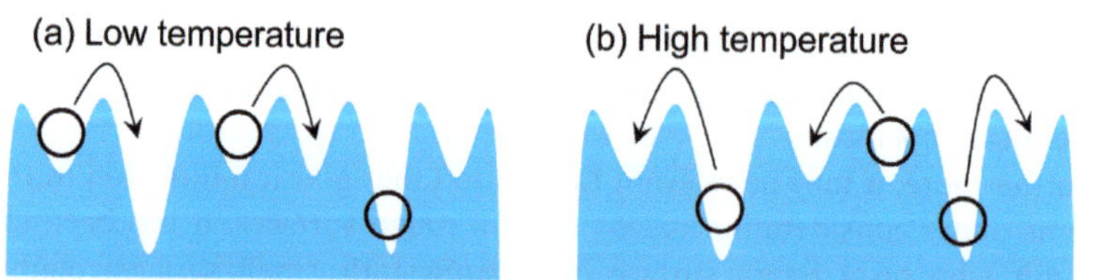

Fig. 1   Schematic illustration of the diffusion of an oxygen atom on an amorphous surface at (a) low temperature and (b) high temperature (see text for details).

temperatures adsorbates can diffuse only through low barriers on an amorphous surface, while atoms in deep sites surrounded by high barriers cannot move and thus be observable in your experiment. As a result, the observed diffusion rate is limited by the lower barriers (Fig. 1(a)). After moving between shallow sites through low barrier pathways, atoms finally encounter deep sites and are trapped there. The trapped atoms can diffuse from the deep sites over the high barriers only at higher temperatures (Fig. 1(b)), and macroscopically observed diffusion is limited by the higher activation barriers compared with the low-temperature case. This picture can explain the results shown in Fig 5 of your paper. In other words, the Arrhenius curve for the thermal diffusion constant tends to be steeper as the temperature increases, while the temperature dependence becomes weaker at very low temperatures (Fig. 2).

Below I briefly show a result of the rate equation model developed by Vidali and He (paper 23, ref. 5) for thermal diffusion on an amorphous surface. This simulation demonstrates how continuous desorption/diffusion energy distributions affect the diffusion rate of atomic oxygen. For simplicity, O_2 diffusion leading to O atom consumption is neglected. In this simulation, the desorption energy has a normal distribution centered at 600 K. Fig. 3 shows the desorption energy distribution used in the simulation. Here, the desorption energy (E_{des}) and the diffusion energy barrier (E_{diff}) are assumed to be related by: $E_{diff} = \alpha E_{des}$, and α = 0.6. The standard deviation of the normal distribution is varied so that the effect of spatial distribution on the effective O diffusion rate can be obtained as a function of surface temperature. We assume the coverage of O on the surface is 0.3 ML, and a pre-exponential factor $\nu = 10^{12}$ s^{-1}. Since these simulations are not intended to accurately reproduce the diffusion of O atoms on an amorphous

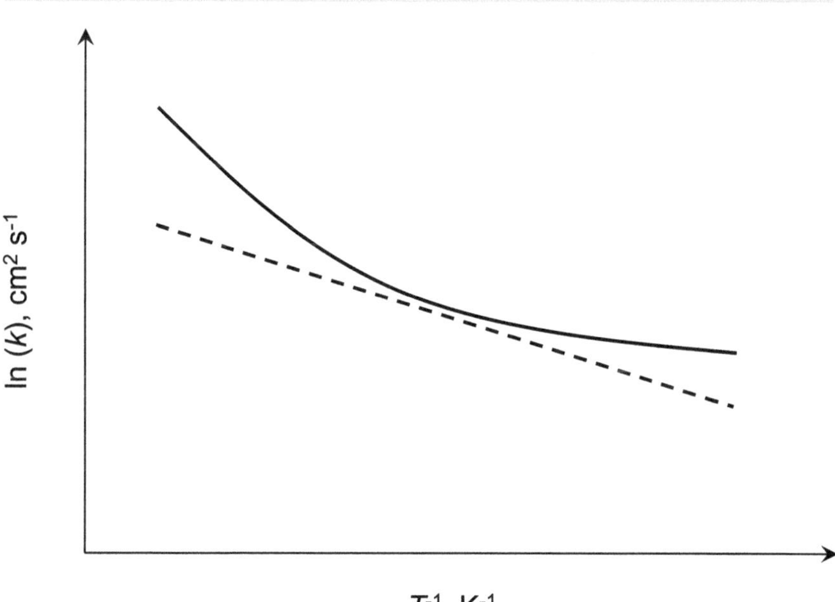

Fig. 2 Expected behavior of Arrhenius curves of diffusion coefficient k as a function of temperature (T) for an ideal single crystalline surface without kinks, steps, or terraces (dashed line), and for an amorphous material with a broad barrier distribution (solid line).

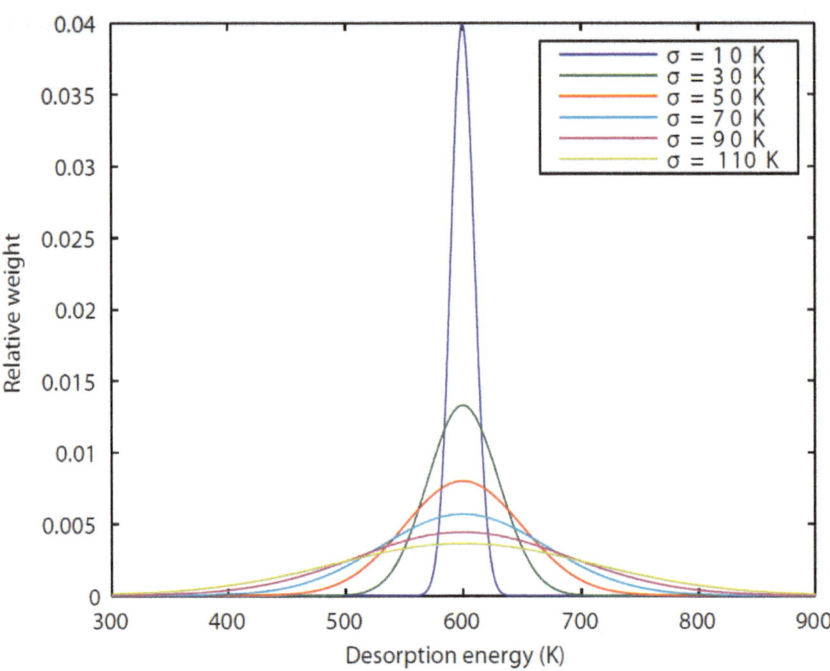

Fig. 3   Normal desorption energy distributions centered at 600 K with different standard deviations.

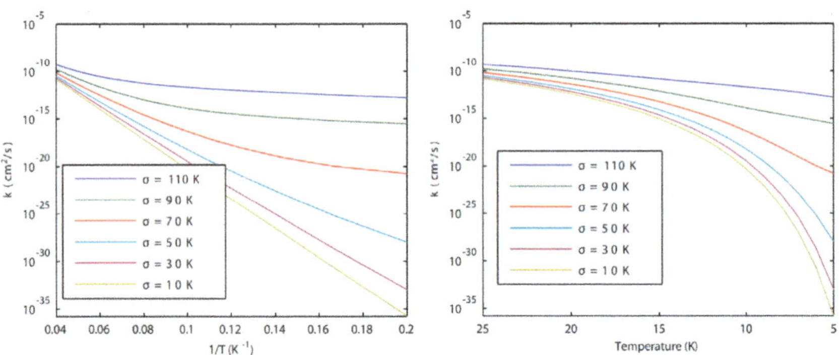

Fig. 4   The O diffusion coefficient (k) as a function of surface temperature (T) for different desorption/diffusion energy distributions. Left panel: log k vs. T^{-1}. Right panel: log k vs. T. The desorption energy distributions are normal distributions centered at 600 K, but with standard deviations 10 K, 30 K, 50 K, 70 K and 90 K. $E_{\mathrm{diff}} = 0.6 E_{\mathrm{des}}$ is assumed.

surface, the parameters are somewhat arbitrary. The effective diffusion rate is calculated by averaging the diffusion rate over all occupied sites. Fig. 4 shows the corresponding diffusion rate as a function of temperature. It can be seen that as the distribution becomes broader, the diffusion rate becomes further away from that predicted by the Arrhenius law using a single energy value. In particular, at low surface temperature, the diffusion rate is enhanced. This result is consistent with the experimental plots shown in Fig. 6 in your paper. This result stems from the redistribution of adsorbed oxygen atoms following diffusion. At very low surface temperature, the distribution of adsorbed oxygen atoms between different adsorption sites is more or less even. At higher temperature, the very shallow sites become empty following diffusion. When the temperature is high enough, only the deep sites are occupied. Thus, the effective diffusion energy barrier at lower temperature should be lower than the effective diffusion energy barrier at higher temperature (Fig. 2 and 4). Although we do not exclude the possibility of quantum tunnelling diffusion of oxygen atoms, your experimental results can be explained by thermal diffusion with a barrier distribution. As long as an alternative explanation is possible, it is risky to conclude the occurrence of quantum tunnelling from the present experimental results. Since diffusion by quantum tunnelling is of great importance, not only in the astrochemistry of ice and dust, but also in other fields of surface science, careful analysis and discussion are required.

1 D. Llera-Hurlburt, A. S. Dalton and E. G. Seebauer, *Surf. Sci.*, 2002, **504**, 244.
2 N. Watanabe, Y. Kimura, A. Kouchi, T. Chigai, T. Hama and V. Pirronello, *Astrophys. J.*, 2010, **714**, L233.
3 T. Hama, K. Kuwahata, N. Watanabe, A. Kouchi, Y. Kimura, T. Chigai and V. Pirronello, *Astrophys. J.*, 2012, **757**, 185.
4 T. Hama and N. Watanabe, *Chem. Rev.*, 2013, **113**, 8783.
5 G. Vidali and J. He, *Faraday Discuss.*, 2014, **168**, DOI: 10.1039/c3fd00113j.

Dr Congiu replied: Thank you very much for your detailed critique concerning our interpretation of the mechanisms of O diffusion. We perfectly agree on the fact that a disordered surface has a complex distribution of adsorption and diffusion barrier energies, and for each surface there exists a unique distribution that can be quite different, even in substrates of the same composition and morphology. In relation to your results obtained using the rate equation model by Vidali and He, we think that 600 K is a rather low temperature for centering a distribution of energies for O desorption, which is more likely to be near 1000 K, and the value of α usually considered is ~0.3. We have actually tried to implement this kind of complication in our model, by adding other free parameters to account for the energy distribution of physisorption sites, and we were able to obtain a good fit for the data due to the increased number of parameters, although this comes at the expense of the uniqueness of the solution. Because we also managed to reproduce the experimental values of the diffusion coefficients using only two free parameters by assuming quantum tunnelling diffusion of O atoms in the temperature range 6–20 K on all five different substrates, we considered it reasonable to accept and retain the most simple solution of all. Another reason why we discarded the classical solution, as we explain in the paper, is that it did not convince us from a physical point of view. In fact, we stated that an Arrhenius law form in which only one E_{diff} exists and is independent of the temperature, or where a fixed distribution of $E_{\text{diff}}(T)$ is given, is not able to fit the data; we explain this point in detail in the following. In Fig. 5 we present an attempt to model the distribution of diffusion energies on amorphous water as a function of surface temperature, calculated assuming an Arrhenius behaviour and according to the experimental values of diffusion coefficients k. The colour scale represents the fraction of adsorption sites with energy $E_{\text{diff}}(T)$ populating 1 ML. From Fig. 5 it is clear that the contour lines are not horizontal, as they should be in a fixed distribution of diffusion energies characterizing a

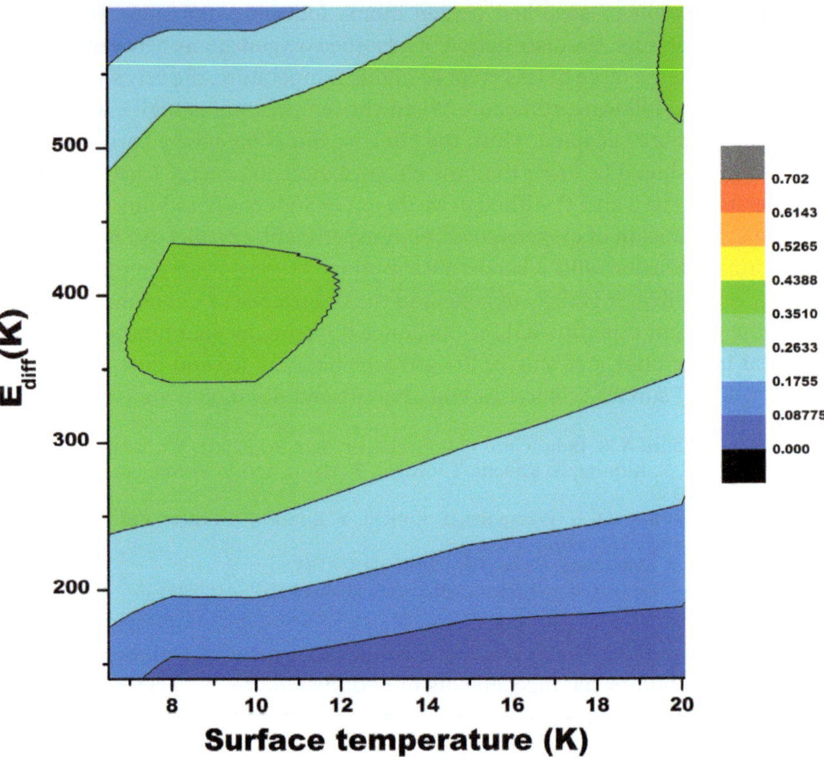

Fig. 5   Distribution of O-atom diffusion energies on amorphous water ice as a function of surface temperature calculated according to the experimental values of the diffusion coefficients and assuming an Arrhenius (thermal) behaviour. The colour scale represents the fraction of adsorption sites ($1 = 10^{15}$ sites cm^{-2} = 1 ML) with $E_{diff}(T)$ at temperature T.

given substrate. This implies that the distribution of diffusion energies cannot be constant in the narrow (6–20 K) temperature interval investigated, but its centre has to shift to higher energies with temperature to reproduce the observed trend in diffusion coefficients. We believe that a change in the distribution of barrier energies within such a narrow temperature range is not physically plausible, although not impossible, in the case of "stiff" surfaces such as silicate and graphite, nor in the case of the various morphologies of water ice at such low temperatures. Finally, we understand the criticism of our interpretation of diffusion *via* quantum tunnelling because of the many implications it has in astrochemistry and surface science. If we then don't change our point of view is not that we ignore the remarks made by others, but because we remain convinced that the views expressed are, relative to our present state of knowledge, substantially correct.

Professor Dulieu continued the discussion of the paper by Yasuhiro Oba: I would like to point out a very interesting fact presented in your study. I am happy to see here that you confirm that the water could undergo chemical desorption,[1] and especially that you noticed that the water (D$_2$O) substrate is able to retain a larger fraction of the molecules formed on its surface. This phenomenon is due to the excess energy of the exothermic reaction, and one could imagine that this excess energy could trigger other reactions, like abstraction or proton exchange with the underlying substrate. In your experiments you showed that H_2O_2 can be grown on a D_2O substrate, without any visible production of HDO. This seems to show that proton exchange with the substrate is inefficient. We already know that this kind of exchange happens upon desorption (see ref. 2, and references therein), so I think that this specific result is very meaningful. It allows us to validate the current scenario of water formation. The H : D ratio found in water is an indication of the H : D ratio once the water has been formed. The later deuteration in denser parts of the cloud where the D fraction increases does not affect the solid water already formed, and molecules that are formed later, like CH_3OH, can undergo high deuteration (especially on the methyl group). If the water substrate could integrate D atoms by proton exchange during the formation of other molecules, it would be difficult to understand the variety of deuteration observed.

1 F. Dulieu, E. Congiu, J. Noble, S. Baouche, H. Chaabouni, A. Moudens, M. Minissale and S. Cazaux, *Sci. Rep.*, 2013, **3**, 1338.
2 F. Dulieu, L. Amiaud, E. Congiu, J.-H. Fillion, E. Matar, A. Momeni, V. Pironello and J. L. Lemaire, *Astron. Astrophys.*, 2010, **512**, A30.

Professor Dulieu continued the discussion of the paper by Stephen Price: Your estimate of the high value of the binding energy of O atoms corroborates our own findings. We have also studied the CO + O reaction[1] and noticed a similar shape of the temperature dependency of the products. We have interpreted this behaviour as a competition between ozone formation and CO_2 formation on the surface. I was wondering if the minimum value you found of around 30 K, which anti-correlates with the ozone formation, could also give a reasonable explanation of your data. This, however, should not change the broad picture you described, and especially the efficiency of the process and the rather high barrier to desorption of atomic oxygen.

1 M. Minissale, E. Congiu, G. Manicò, V. Pironello and F. Dulieu, *Astron. Astrophys.*, 2013, **559**, A49.

Professor Price answered: I think this is an excellent and perceptive observation. Quite possibly, the minimum in our product yield around 30 K results from a reduction in the available number of O atoms on the surface due to their reaction with O_2 to form O_3. We will extend our very simple model to include this reaction and see if that allows an improved fit with the experimental data. It would be good to collaborate on improving the modelling of our experimental data.

Mr Minissale addressed Professor Price: In your paper you claim that the desorption energy of O atoms is around 13–15 kJ mol^{-1}, which means a desorption temperature of 50–60 K for O atoms. I totally agree with these results, but I am not sure if I have understood your model correctly (*i.e.* the lines in Fig. 2 and 4 in your paper). The Langmuir–Hinshelwood mechanism should be dependent on surface density of O atoms; a surface temperature higher than 50–60 K should decrease the residence time of O atoms on the surface and, as a consequence, O atom surface density. Why is the Langmuir–Hinshelwood (LH) mechanism more efficient than the Eley–Rideal (ER) mechanism at high surface temperatures? Moreover, why does ER mechanism efficiency change quickly as a function of surface temperature? A plausible scenario could consider three phases: a first phase, before O_2 desorption (<30 K), where O atoms are consumed to form O_3 rather than to oxidise propyne; a second one, between 30 and 55 K, where O atoms can oxidise propyne more readily due to their increased mobility (mainly *via* LH); finally a third phase, over 55 K, where the yield of C_3H_4O drops to due the lack of O atoms on the surface and the inefficiency of the LH mechanism. Could you comment on this?

Professor Price replied: The interpretation of Fig. 2 and 4 in the paper was complicated by an error in the caption (corrected in the final version published here). In these figures the dashed line actually shows the yield from the LH mechanism and the dotted line that from the "ER" mechanism. The output from the model then is exactly in accord with your expectations. Specifically, as the surface temperature rises from 30 K the rate of the LH process rises faster than that of the ER process, as the surface concentration of the reactant O atoms is markedly larger than the O atom fluence. Thus, up to a surface temperature of 50 K, the large surface concentration of O atoms means the LH rate is larger than the ER rate. Above a surface temperature of 50 K the O atoms have a very short residence time on the surface so the surface concentration of O atoms drops dramatically. Thus the LH rate drops sharply. Since the "ER" rate (where we only consider the reaction of a gas-phase O atom with a surface bound propyne molecule) does not depend on the surface concentration of O atoms, the "ER" rate continues to rise until desorption of the propyne molecule becomes significant at surface temperatures of about 90–100 K. You are absolutely correct that our model would be improved by including the reaction of O with O_2. The removal of O atoms by this reaction could nicely account for the drop in the product yield at a surface temperature of about 30 K. We are working on such a development of our model. In addition we will try to more realistically incorporate the effect of surface temperature on the rate of the ER process (as performed by your group[1]) using an effective temperature.

Our model pictures the direct process as more of a "hot atom" event, where the temperature of the surface has more influence on the rate of the reaction than in a pure ER process. I'm sure our model can be improved and refined and we will discuss this further with your group and others. In conclusion though, it is important from the astrochemical perspective not to lose sight of the fact that we all observe significant reactivity of O atoms with molecules on surfaces at low temperatures.

1 M. Minissale, E. Congiu, G. Manicò, V. Pirronello and F. Dulieu, *Astron. Astrophys.*, 2013, **559**, A49.

Dr Ioppolo addressed Dr Congiu: In the results section of your paper, you showed that there is no appreciable difference between the efficiency of O atom mobility on three types of water ice (porous, non-porous, and crystalline ice). The procedure used here to obtain such ices is described in detail in the Experimental section. In this work, amorphous porous water ice is deposited at 10 K and then annealed to 90 K, amorphous non-porous ice is deposited at 110 K, and crystalline water is deposited at 120 K and then annealed to 145 K. I understand that porous water ice had to be annealed to 90 K to stabilize the surface morphology before several temperature cycles were performed during these experiments. However, I would expect the porous water ice to become less porous when annealed at 90 K, and therefore to become to some extent similar to the non-porous water sample. Moreover, I have some concerns regarding the reason why the crystalline ice is not directly deposited at 145 K. In paper 21,[1] THz spectra of water ice deposited at different temperatures and in some cases annealed to higher temperatures showed that the degree of crystallinity of the water samples depended on the thermal history of the ices (see Fig. 5–7 of paper 21). Therefore, annealed water ice is likely to be less crystalline than a water ice directly deposited at temperatures above 145 K. All the aforementioned arguments make me think that the reason why the authors were not able to see differences in the efficiency of the O atom mobility on different water ices was partially caused by the structure of their water samples (*i.e.*, the porous water ice was not fully porous and the crystalline water ice was not completely crystalline). Can the authors comment on this?

1 S. Ioppolo, B. A. McGuire, M. A. Allodi and G. A. Blake, *Faraday Discuss.*, 2014, **168**, DOI: 10.1039/c3fd00154g.

Dr Congiu answered: The procedure we used to grow crystalline water ice is a standard procedure that we usually adopt in our laboratory. We determined, *via* isothermal desorptions at 145 K, that all the amorphous phase of the ice sample had desorbed, and this was also confirmed by subsequent RAIR spectra in which the amorphous ice feature disappeared, replaced by the characteristic crystalline ice band. We also determined that this is a suitable method to grow a smoother surface since we avoid the deposition of water molecules due to background residual H_2O in the gas phase following the ice growth. In addition, previous works by our group have demonstrated that we have a very good control of the morphology of the very surface *via* D_2 TPDs,[1–3] which we also adopted to probe the water ice samples used in some experiments concerning O atom mobility.

Our experimental data clearly show that there is a systematic trend showing an unexpected correlation between ordered surfaces and efficient diffusion of O atoms. In fact, the most ordered surface, oxidised graphite, turns out to be the surface where the mobility is the lowest, while porous water ice, the most disordered surface of all, exhibits the highest mobility. We believe that this is due to preferential pathways along which atoms overcome only the narrowest barriers, which is a more realistic scenario the more disordered the surface becomes. That being said, although it may be true that the bulk structure of our ice samples was not perfectly determined, we think that this only implies small quantitative differences in the derived O diffusion coefficients, while no qualitative changes are to be expected, nor is a different physical interpretation of the diffusive mechanism.

1 J. H. Fillion, L. Amiaud, E. Congiu, F. Dulieu, A. Momeni and J. L. Lemaire, *Phys. Chem. Chem. Phys.*, 2009, **11**, 4396.
2 M. Accolla, E. Congiu, F. Dulieu, G. Manicò, H. Chaabouni, E. Matar, H. Mokrane, J. L. Lemaire and V. Pirronello, *Phys. Chem. Chem. Phys.*, 2011, **13**, 8037.
3 M. Accolla, E. Congiu, G. Manicò, F. Dulieu, H. Chaabouni, J. L. Lemaire, and V. Pirronello, *Mon. Not. R. Astron. Soc.*, 2013, **429**, 3200.

Professor McCoustra communicated: The issue of crystallisation of amorphous ice in a moderately fast annealing cycle was discussed. The problem is that this process does not produce the same type of crystalline ice as is produced by deposition at high temperature and cooling. Some years ago, we demonstrated this in a comparative TEM–RAIRS study.[1] While rapid annealing produces a crystalline-like RAIR spectrum, only very slow (hours as opposed to minutes) annealing in water vapour at a pressure designed to balance the rate of loss by sublimation was able to produce a TEM consistent with a largely crystalline material.

1. P. Jenniskens, S. Banham, D. F. Blake and M. R. S. McCoustra, *J. Chem. Phys.*, 1997, **107**, 1232.

Dr Ioppolo asked Yasuhiro Oba: In your paper, you studied the formation of water ice upon hydrogenation of 1 ML of hydrogen peroxide previously formed on different substrates (Al and a-D_2O). In this case, because of the low coverage of hydrogen peroxide, the final water yield is clearly substrate dependent. Did the authors consider investigating the hydrogenation of peroxide in the multilayer regime, where the substrate effect is negligible?

Moreover, the reaction H_2O_2 + H leads to the formation of H_2O + OH. The authors discussed in the text the formation of water, as it only comes from the aforementioned reaction. However, in their experiments water ice can also form through the OH + H, OH + OH, and OH + H_2 channels. How do the authors distinguish the water formed through the hydrogenation of peroxide from the water formed through the other channels?

Dr Oba responded: We did actually perform some experiments similar to the one reported here, where H atoms were exposed to ~10 ML of H_2O_2 produced by the same method. However, the final water yield was not very different from that observed in the experiment which used 1 ML of H_2O_2 as an initial reactant. This is probably because penetration of H atoms into H_2O_2 solid is highly restricted, as suggested previously.[1,2] In addition, when we plotted the variations in the column density of H_2O_2 with time, as in Fig. 4 in the paper, for example, fitting to rate equations was not successful. This could be partly due to the strong interactions between H_2O_2 molecules. Anyway, to avoid undesirable effects on the reaction kinetics, we used a thin (1 ML) sample for the initial reactant. As for the formation pathway of H_2O, unfortunately we cannot distinguish between the reaction channels in the present study. I suppose H_2O is formed by multiple reactions.

1 N. Miyauchi, H. Hidaka, T. Chigai, A. Nagaoka, N. Watanabe and A. Kouchi, *Chem. Phys. Lett.*, 2008, **456**, 27.
2 S. Ioppolo, H. M. Cuppen, C. Romanzin, E. F. van Dishoeck and H. Linnartz, *Phys. Chem. Chem. Phys.*, 2010, **12**, 12065.

Dr Zins asked Dr Oba: I would like to ask you some further questions about the mechanism involved in the hydrogenation of hydrogen peroxide. Some time ago, we also investigated the reaction between non-deuterated hydrogen peroxide and atomic deuterium. In the solid phase, when we exposed an ice of hydrogen peroxide to atomic deuterium at 10 K, we observed, as you did, the formation of a water ice and the decrease of bands due to hydrogen peroxide (as shown in Fig. 6A). To further investigate the occurrence of tunnelling processes, we used the matrix isolation technique. Reagents are trapped in an inert medium, and they can further react in the dark by means of a tunnelling process. A sample

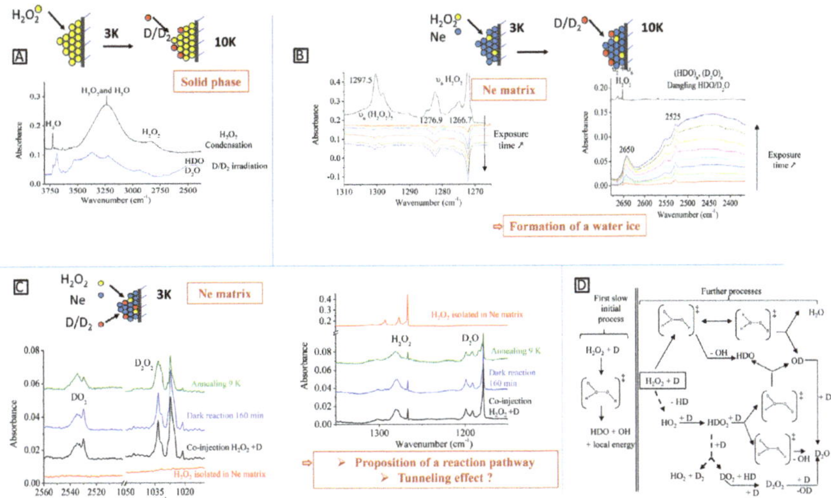

Fig. 6   IR spectra obtained during experiments in the solid phase or in a Ne matrix. A: Hydrogen peroxide ice was exposed to atomic deuterium. The spectrum of the hydrogen peroxide ice before the irradiation is shown in black, and irradiation led to the formation of a water ice, with a simultaneous decrease of the peaks due to H_2O_2 (blue: difference spectrum). All the spectra were recorded at 10 K. B: Hydrogen peroxide trapped in a Ne matrix (spectrum shown in black, top) was exposed to atomic deuterium. The irradiation leads to the formation of a water ice, with a simultaneous decrease of the peaks due to H_2O_2 (below: difference spectra, after 5 to 40 minutes of irradiation). All the spectra were recorded at 10K. C: Co-injection of H_2O_2/Ne and D/D_2. When the sample is maintained at 3 K with neither radiation nor reagents, no change is observed in the spectrum, whereas annealing up to 9 K leads to further reaction. The formation of HDO (not shown), D_2O, DO_2, HO_2 and D_2O_2 is observed. D: Based on these experimental results, a reaction pathway including successive reactions that may occur under our experimental conditions may be proposed.[1]

containing H_2O_2 isolated in a Ne matrix was generated and further exposed to atomic deuterium (Fig. 6B). Here we observed the decrease of peaks due to hydrogen peroxide during the exposition – these spectra correspond to difference spectra obtained after 5 to 40 minutes of exposure to atomic deuterium. At the same time, the formation of a water ice is observed, even after 5 minutes. I would like to underline the fact that ice formation is not expected in matrix experiments, and this observation proves that an exothermic process takes place. This result was not really a surprise, since theoretical calculations have shown that the reaction leading to water formation from hydrogen peroxide is indeed strongly exothermic. But we still wanted to have some proof that a tunnelling process takes place.

This is the reason why we carried out another series of experiments, in which the two reagents were co-injected and trapped at 3 K in a Ne matrix. From this initial sample, the formation of HDO, D_2O and DO_2 was already observed, even without any dark reaction or annealing (Fig. 6C). So, since several intermediates were observed under such experimental conditions, we thought about an alternative reaction pathway that may lead to water formation (Fig. 6D). We wondered whether the difference in the reactivity you observed between atomic H and D may not be due to differences in (i) the diffusion of these two atoms in an ice, (ii) the dissociation rates of D_2 and H_2 in the microwave-driven atomic source, and (iii) the sticking coefficients of D and H as a function of the temperature. Moreover, since the hydrogenation process of hydrogen peroxide is very much exothermic, do you think that this exothermicity will have an effect in your experiments?

1 E. L. Zins and L. Krim, *RSC Adv.*, 2014, **4**, 22172.

Dr Oba replied: Although I am not sure of your experimental conditions in detail, if the reaction scheme you propose in your Fig. 6D were correct, I think D atoms would need to diffuse rapidly in the Ne matrix at 3 K and multiple tunnelling reactions would have to take place during a $D/H_2O_2/Ne$ co-deposition experiment. Therefore, if I obtained the same result in the same experiment as yours, I would first suspect the possibility of O_2 contamination in the sample. If this is the case, O_2 reacts with D to yield DO_2 and D_2O_2.

There could be a very small difference between the diffusion rates of H and D, the dissociation rates of H_2 and D_2 in the atomic source, and the sticking coefficients of H and D atoms on surfaces. However, even so, their differences should not be so large as to explain the large differences in the rate of each reaction in the present experiments. The difference in the tunnelling mass can explain well the differences in the rates of the reactions.

I do not think the exothermicity of reactions has an effect on the reactions in the present experiments.

Dr Garrod continued the discussion of the paper by Stephen Price: You mentioned that the energy involved in a hot-atom/Eley–Rideal-mediated surface reaction is dependent on the thermal energy of the incident particle. However, as the particle approaches the surface, it will be accelerated into the surface potential, so that the amount of energy available for diffusion or for barrier-mediated reactions may be significantly greater than just the incident particle's thermal energy. These considerations make the analysis of surface experiments into barrier-mediated reactions difficult to parameterize, when dealing with conditions of high surface coverage of one of the reaction partners. Unless and until experimental conditions are achieved in which reaction partners are widely spaced on the surface – ensuring that the reactants are thermalized, and that Langmuir–Hinshelwood is the dominant reaction process – then the importance of activation energy barriers and thermal diffusion rates on surface processes will be difficult to disentangle from the effects of having suprathermal reaction partners.

Professor Price answered: I totally agree that to provide data of real *quantitative* value to astrochemical modellers we need to extend our experiments to lower (sub-monolayer) coverages. This will ensure that the LH process is the dominant process and we are not influenced by "suprathermal" reactions – this is a major focus of our current experimental effort and requires an increase in our detection efficiency. The initial objective of this current series of experiments was to see if O atoms are reactive under a set of pseudo-interstellar conditions – and indeed they are – as other papers in this session also show. Experiments at lower coverages are indeed our next objective. The point regarding the effect of the surface potential on the approach of the reactant to the surface is perhaps particularly pertinent in our current coverage regime.

Dr Garrod asked Emanuele Congiu: You suggest that the reaction proceeds as a result of diffusion *via* tunnelling, but an Eley–Rideal or hot-atom process is also likely to show a similarly small dependence on surface temperature as would be observed in the case of tunnelling. Therefore, those also seem to me to be plausible explanations for your results.

Under the assumption that tunnelling by atomic oxygen is indeed occurring, you derive tunnelling parameters that include a diffusion barrier width of 0.7 Å. This seems very narrow; it is smaller than typical chemical bonds within molecules. What does such a narrow barrier mean? Is it physically meaningful? Doesn't the requirement for such a narrow barrier suggest that tunnelling is not a plausible explanation for your results?

Dr Congiu replied: We did not discard an Eley–Rideal (ER) or hot-atom (HA) process *a priori*. These two mechanisms were actually implemented in our model so we could explore several values of the different parameters. We discussed this issue extensively in our previous study.[1] Briefly, the HA mechanism was studied from a theoretical point of view (in ref. 2 and 3, and references therein), but experimental works have also been reported (ref. 4 and 5, and references therein). In all these works only metallic surfaces were studied, using atoms with energies greater than 0.5 eV or very light atoms (H or D); under the above conditions there is a low energy transfer with the surface and therefore a larger probability that HA occurs. Conversely, our experiments were carried out under almost opposite conditions: non-metallic surfaces, heavy atoms ($mass_O/mass_H = 16$) with an energy < 0.03 eV. Furthermore, at least in the case of water ice, the molecular substrate is a very soft substrate known to have very good energetic exchange with the impinging atoms, as was shown by studying the sticking coefficients of light species.[6] These considerations suggest that, even though nobody can exclude that the HA process takes place on the surface, its effects may not be important in ozone formation, especially at low coverages, as is the case in our experiments. Another issue – still unresolved – is the temperature surface dependence of the HA process. Some experimental and theoretical works (ref. 7 and references therein) show that a temperature dependence exists, but the range of temperatures investigated is very broad (more than 300 K). In our case the range of surface temperature is very small (about 25 K) and it is fair to imagine that the energy transfer between the adsorbed particles and the surface is quite constant within this range of temperatures. On the other hand, one of the most solid experimental facts of our work is the dependence on temperature of the O_3 yield. The branching ratios in the kinetics of the reaction vary by two orders of magnitude in the 6–25 K temperature range. Is it then wise to explain and justify these results with a mechanism about which very little is known, like the hot-atom process?

The lack of theoretical and experimental studies on HA mechanisms under our experimental conditions (non-metallic surfaces, low energy and heavy atoms) together with our modelling results – concerning either a coverage or temperature dependence – support our conclusions that HA cannot explain the experimental results involving O atoms, and corroborate the hypothesis that the diffusion mechanism is dominant at low temperatures and in low-coverage regimes.

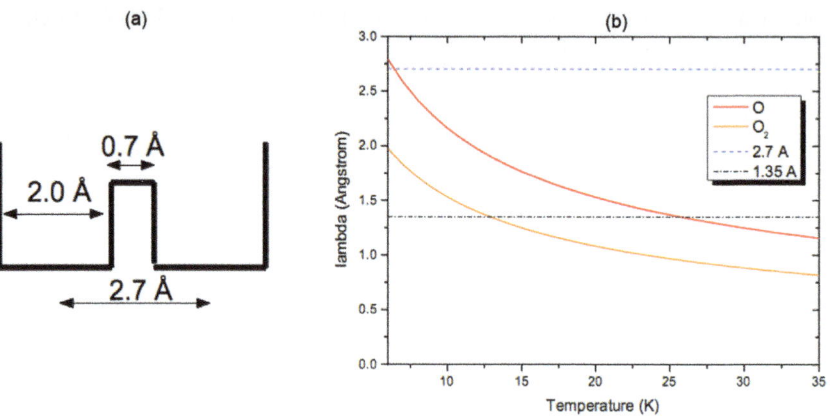

Fig. 7   (a) The rectangular potential between two adsorption sites used in the model shown in paper 7. (b) Variation of the de Broglie wavelength of O and O_2 with temperature. Because the de Broglie wavelength of O atoms at $T < 16$ K is larger than the mean adsorption site width, quantum tunnelling is thought to become the dominant mechanism for diffusion.

As far as the diffusion barrier width is concerned, the value of 0.7 Å that we derive can be considered an effective barrier width between physisorption sites used by O atoms to diffuse on preferential pathways. It is probably not wise to give it a physical meaning since we only assumed a rectangular potential, which is certainly not the case. We can, however, make some physical considerations about the simplified system used in our model. The mean distance between two adsorption sites on crystalline water ice is about 2.7 Å (see Fig. 7a), and the distribution varies between 2.1 and 3.2 Å (a range of values adapted from S. Andersson, private communication). Even if the usually accepted value is 3 Å (the mean distance between two water molecules), taking into account the disordered nature of an amorphous substrate, distances between adjacent adsorption sites lie in the 2.4–3.6 Å range. In our simplified rectangular potential scheme, the derived barrier width of 0.7 Å would then imply that adsorption sites have a width between 1.7 and 2.9 Å. If we consider the de Broglie wavelength (dBw), we find that, for O atoms, it varies between 2.8 Å at 6 K and 1.5 Å at 20 K. Therefore, two atoms sitting on adjacent sites could easily interact at very low temperature. In Fig. 7b we display the variation of the de Broglie wavelength with temperature for O (and O_2), to show in what range of temperatures the de Broglie wavelength of O atoms may be used to rule out quantum mechanical effects. An optimistic point of view would consider twice the length of the dBw and compare it to the distance between two adsorption sites. We then find that the dBw is larger than the mean distance between two adsorption sites for $T < 25$ K, which is exactly the range of temperatures of interest in our experiments. Another way of looking at it is that the de Broglie wavelength of O atoms can be larger than the size of one adsorption site at $T < 16$ K, which implies that quantum tunnelling diffusion becomes dominant.

1 M. Minissale, E. Congiu and F. Dulieu, *J. Chem. Phys.*, 2014, **140**, 074705.
2 E. Molinari and M. Tomellini, *Chem. Phys.*, 2002, **277**, 373.
3 R. Martinazzo, S. Assoni, G. Marinoni and G. F. Tantardini, *J. Chem. Phys.*, 2004, **120**, 8761.
4 C. Wei and G. L. Haller, *J. Chem. Phys.*, 1996, **105**, 810.
5 A. Dinger, C. Lutterloh, and J. Küppers, *J. Chem. Phys.*, 2001, **114**, 5338.

6 H. Chaabouni, M. Minissale, G. Manicò, E. Congiu, J. A. Noble, S. Baouche, M. Accolla, J. L. Lemaire, V. Pirronello and F. Dulieu, *J. Chem. Phys.*, 2012, **137**, 234706.
7 E. Quintas-Sánchez, C. Crespos, P. Larrégaray, J.-C. Rayez, L. Martin-Gondre and J. Rubayo-Soneira, *J. Chem. Phys.*, 2013, **138**, 024706.

Dr Cuppen addressed Professor Dulieu, Dr Congiu and Dr Garrod: A width of 1 Å is usually considered for a reaction, but since this is a non-reactive process, surely we would expect the barrier to be larger as opposed to smaller? Moreover, within the square potential approximation, if one triples the width of the barrier to obtain a reasonable value, one has to divide the barrier height by a factor of nine to obtain a similar rate. This would mean that even on an oxidised highly oriented pyrolytic graphite surface the barrier would be below 100 K, which can very easily be surpassed thermally.

Professor Dulieu responded: Our analysis of the data and our model, and also many other models, considers that diffusion occurs on a two dimensional grid, which is already a non-realistic approach. On water substrates, the mean distance between two adsorption sites can be accurately calculated, and it is in the 2.4–2.7 Å range. The square shape of quantum wells is also unrealistic, as we have always stressed. Nevertheless, a square shape is a convenient and very simple way to gain a basic picture of what could be the complex reality. Furthermore, the de Broglie wavelength of O atoms at 10 K is 2.16 Å. Again, nobody expects that the de Broglie wavelength is not affected by the properties of the adsorption potential wells, but it gives us an idea of what could be the wave-packet expansion of O atoms.

Now, we can imagine that one atom is almost as large as the distance between two adsorption sites. Taking into account that the square shape is not a realistic description, why should a diffusion barrier that is slightly narrower than expected (0.7 Å instead of 1 Å) be non-physical? Only detailed quantum calculations can correctly address this question.

Our data seems to be better fitted by high values of diffusion barriers (we found in an separate method between 600 and 900 K for the diffusion of O atoms on graphite, in the thermal hopping regime (a surface temperature of 50 K)). This high diffusion barrier cannot be crossed at low temperatures (<10 K) in the time scale of our experiments (a typical ms regime, observed at the 10–1000 s scale) unless another efficient mechanism occurs, such as quantum tunnelling promoted by a narrow diffusion barrier. It is possible that a looser definition of the positions of the atoms and the adsorption wells is the origin of the narrowness of the barrier, obtained with the help of the simple square barrier model.

Dr Congiu responded: On a given rough surface there is a distribution of barrier heights and barrier widths, so the 1 Å width usually considered is seen as a mean value. That being said, we believe that the diffusion of species on the surface is likely to occur through preferential pathways along which atoms overcome only the narrowest barriers. Based on the modelling of our experimental values, we found an effective barrier width of 0.7 Å because O atoms can probably only diffuse between the sites separated by barrier widths of the low-value end of the barrier width distribution.

Professor Watanabe continued the discussion of the paper by Stephen Price: I would like to ask why the ER process is so sensitive to the surface temperature in your model, as shown in Fig. 2 and 4 in your paper? The ER process should normally show very little dependence on the surface temperature, especially for barrierless or low barrier reactions. In fact, you derived a very low barrier of as low as 160 K. I understand that the LH process competes with the ER process in a certain temperature range. However, I feel that the temperature dependence of the ER process in your model is too strong. Can you explain what causes the significant drop-off of the ER process at around 60 K? I think that the O atom exposure of predeposited propyne solid instead of co-deposition would be useful to evaluate the contribution of the ER process. Have you tried this experiment?

Professor Price replied: In our model, the rate of the direct (ER, or perhaps pseudo-ER would be a better terminology) process is influenced by the surface temperature as we allow the energy of the surface bound species (at the surface temperature) to contribute to determining the rate of the direct reaction. We considered such a model as crudely representative of a "hot-atom style" ER process, where the reactant from the gas-phase is partially thermalized by the encounter with the surface. The effect of the surface temperature on the rate of the ER (or hot atom) process could perhaps be better incorporated (as done by others[1]) by determining an effective temperature of the reactant–adsorbate encounter using the reduced mass of the system. It seems to me, since we routinely discuss the desorption of adsorbates by local heating of the surface (*e.g.* by a chemical reaction) that there is coupling of the thermal bath of the surface to the adsorbate and hence to the atom–adsorbate activated complex. Thus one would certainly expect the surface temperature to influence the rate of the direct reaction. Perhaps our simple model overestimates that influence; a more sophisticated model will be explored. Also, as explained in answer to Mr Minissale's question, the interpretation of Fig. 2 and 4 in the paper is not helped by an error in the caption (corrected in the final version published here). In these figures, the dashed line shows the yield from the LH mechanism and the dotted line shows that from the "ER" mechanism. As the surface temperature rises from 30 K, the rate of the LH process rises faster than that of the ER process, as the surface concentration of the reactant O atoms is markedly larger than the O atom fluence. Thus, up to a surface temperature of 50 K, the large surface concentration of O atoms means that the LH rate is larger than the ER rate. Above a surface temperature of 50 K, the O atoms have a very short residence time on the surface, so the surface concentration of O atoms drops dramatically. Thus the LH rate drops sharply. Since the "ER" rate (where we only consider the reaction of a gas-phase O atom with a surface bound propyne molecule) does not depend on the surface concentration of O atoms, the "ER" rate continues to rise until desorption of the propyne molecule becomes significant at surface temperatures of about 90–100 K. The idea of exposing a predeposited propyne film to the O atom beam is an excellent one and we will try such an experiment.

1 M. Minissale, E. Congiu, G. Manicò, V. Pirronello and F. Dulieu, *Astron. Astrophys.*, 2013, 559, A49.

Professor Watanabe addressed Emanuele Congiu: It is not easy to accept your conclusion of the tunnelling diffusion of oxygen atoms on H_2O ice. As Tetsuya Hama pointed out in this Discussion, your observed temperature dependence of the O atom diffusion rate can be explained by thermal hopping without tunnelling diffusion if you include the barrier distribution. Furthermore, there have been several reports in which the diffusion of hydrogen atoms on amorphous ice was found to be dominated by thermal hopping[1] at around 10 K under similar experimental conditions to yours, *e.g.* low coverages. A basic question is why oxygen atoms, having a 16-times heavier mass than hydrogen atoms, can move by tunnelling diffusion at 10 K, while hydrogen diffusion is dominated by thermal hopping? You have tried to explain this by referring to Senevirathne *et al.* in ref. 48 in your paper, which shows that tunnelling diffusion will be enhanced at around 6–13 K. However, Senevirathne *et al.* did not use a simple rectangular potential, while you adopted the rectangular potential to evaluate the O atom tunneling diffusion. Of course, potentials for the diffusion are not the same between H and O atoms, but the potential shapes should be fairly similar to each other because both originate from van der Waals interactions. It is not appropriate to use quite different potential shapes to explain the difference in the appearance temperatures for tunnelling diffusion between H and O atoms. Instead, if we use the rectangular potential for H atom diffusion on ice as you did for O atom diffusion, the tunnelling diffusion must dominate the H atom diffusion process as well, and should be observed at temperatures significantly above 20 K (which is not the case). For example, assuming for H atoms the rectangular potential with a width of 1 Å and a depth of 255 K, which was determined for thermal hopping by Hama *et al.* in ref. 1 (it should be noted that this rectangular potential was not assumed in Hama *et al.*), the thermal hopping and tunnelling rates are approximately 3×10^6 and 1×10^9 s^{-1} at 20 K, respectively. The tunnelling rate becomes much faster. That is, if we adopt rectangular potentials for both H and O atoms, the H atom diffusion should show the tunnelling feature at much higher temperatures than O atoms. When using the rectangular potential both for H and O atoms, the appearance temperature of tunnelling diffusion for H atoms should be higher than that for O atoms. I do not reject the possibility of tunnelling diffusion, especially for H atoms, but it is not necessary to invoke tunnelling to explain your experimental results. I feel that your conclusion about tunnelling diffusion lacks sufficient evidence.

Furthermore, since the potential of adsorption sites on amorphous surfaces is non-periodic, unlike for single-crystalline surfaces, tunnelling diffusion over long distances should be highly suppressed, even for hydrogen, whereas it may be enhanced over a short range. Therefore, the diffusion distance is important when identifying the diffusion mechanism. Information about the diffusion distance and the distribution of activation barriers should be highlighted more clearly in your model to appropriately evaluate the conclusions. In addition, it is desirable that a more realistic potential is used when considering energy-level matching for tunnelling.

1 T. Hama *et al.*, *Astrophys. J.*, 2012, **757**, 185.

Dr Congiu answered: In our experiments we have observed that O atoms diffuse quickly on five different surface analogues of astrophysical interest, and this is very robust experimental evidence. Concerning the physical interpretation of this process, criticism of our views is acceptable since there is no definitive proof that O atoms dart around the surface *via* quantum tunnelling. However, following some considerations involved in a purely thermal diffusion process – that we have summarized in the response to Dr Hama's question – we are still convinced that a quantum tunnelling diffusion remains the most plausible explanation. To come back to your initial question concerning the difference in the masses and diffusion processes of O and H, we believe that these two particles are not suitable to be compared in this way. In the case of H,[1–3] the water ice morphology affects the diffusion of H atoms considerably. H diffusion turns out to be several orders of magnitude faster in the case of poly-crystalline water ice than that on porous amorphous ice. This is not the case for O atoms. O diffusion has been measured and lies within the same order of magnitude, regardless of the surface morphology of the water substrate. Classical diffusion barriers of H on amorphous water ice lie in the 200–350 K range, and this makes thermal diffusion competitive with quantum tunnelling effects. This is why, at 10 K, classical diffusion prevails over quantum tunnelling as far as H atoms are concerned. On the other hand, it would be surprising if thermal diffusion barriers of O atoms were as low as those of H atoms, or even significantly lower (*e.g.*, 175 K), as proposed by Mr Hama earlier. There exist many qualitative arguments and also some calculations on specific surfaces (like graphite) suggesting that the binding energies of O and H atoms differ by a factor of between 2 and 4 (as well as, by a rough extrapolation, a difference between their diffusion coefficients, although this has not been studied yet). For this reason, we believe that the distribution of the diffusion energies of O atoms should reasonably lie in the 400–1400 K range, which is why quantum tunnelling may predominate at very low temperatures. We consider our derived value of E_{diff} of 500 K to be a reasonable value for the lower end of the diffusion barrier distribution.

The purpose of our work and the following papers was not to solve the entire problem of O diffusion on five different surfaces, but simply to share our first insight into this interesting and open question.

Our experimental work will hopefully be a first step in pursuing new molecular dynamic calculations, or quantum chemistry investigations of ice models. Only detailed calculations of this kind will allow us to begin a discussion about what temperature marks the borderline between quantum tunnelling diffusion and classical motion. Before such time-consuming and sophisticated calculations come to solid conclusions, one cannot ignore the fact that the most efficient and simplest model able to completely fit our data (in terms of both temperature and coverage dependencies), surprising though may it seem, is the quantum tunnelling description.

1 L. Amiaud, F. Dulieu, J.-H. Fillion, A. Momeni and J. L. Lemaire, *J. Chem. Phys.*, 2007, **127**, 144709.
2 E. Matar, E. Congiu, F. Dulieu, A. Momeni and J. L. Lemaire, *Astron. Astrophys.*, 2008, **492**, L17.
3 B. Senevirathne, S. Andersson, F. Dulieu and G. Nyman, *Phys. Chem. Chem. Phys.*, submitted manuscript available at: https://dl.dropboxusercontent.com/u/9601281/Senevirathne_etal2014.pdf).

Professor Herbst continued the discussion of the paper by Stephen Price: Modelers need to know the absolute rate coefficients of surface reactions, but also the mechanisms so as to understand how to transfer the results to interstellar processes. As an example, the rate coefficients determined for the O + ethylene surface reaction to form ethylene oxide[1] have barriers that are lower than the diffusion barriers estimated for O atoms, so cannot be transformed easily for diffusive reactions on interstellar ices. Can the diffusive mechanism be studied using sub-monolayer amounts of reactants in the case where the product signal is rather strong?

1 A. Occhiogrosso, A. Vasyunin, E. Herbst, S. Viti, M. D. Ward, S. D. Price and W. A. Brown, *Astron. Astrophys.*, 2014, **564**, A123.

Professor Price replied: Yes, clearly trying to extend our investigations to sub-monolayer coverages is important. Such investigations are a current experimental objective and this work involves trying to improve our detection efficiency to get usable product signals at lower reactant doses. We are also currently trying to "dilute" our reactants in an inert matrix to try and increase the importance of the diffusion step in the reactivity we observe.

Professor Heard said: There will be competition between the oxygen atom reacting with propyne *via* the title reaction and the reaction with molecular oxygen to form ozone (in a manner involving O atom diffusion on the surface as described in other papers presented at this Discussion). The form of the surface temperature dependence of the yield of C_3H_4O measured could be influenced by the temperature dependence of the loss of O atoms by reaction with O_2 to form O_3. In order to avoid the complication of the reaction of O atoms with O_2 on the surface, it would be advantageous to generate O atoms in an incident beam in the absence of O_2. In the atmospheric and chemical dynamics surface communities, O atoms have been generated, for example, by photolysis of a suitable precursor present in an inert carrier before reaction on a surface held at higher tempera-tures than in this study. Examples include NO_2 photolysis at 355 nm to generate O(3P) atoms to study surface reactions of O atoms with the liquid hydrocarbon squalane.[1] Potential complications of laser-generated O atoms are the enhanced kinetic energy of the O atom as it strikes the surface, and the possibility of generating electronically excited O(1D) atoms, which are likely to be much more reactive at the surface, but in principle the careful choice of wavelength and laser pulse energies, and the avoidance of two-photon dissociation, would enable laser-generated O(3P) atoms to be generated in the absence of O_2.

1 S. P. K. Köhler, M. Allan, H. Kelso, D. A. Henderson and K. G. McKendrick, *J. Chem. Phys.* 2005, **122**, 024712.

Professor Price replied: The idea of generating the O atoms photolytically above the surface is an interesting one, and one we will explore. Clearly, as the question states, this removes the complication of the $O + O_2$ reaction on the surface. The experimental issues in implementing such a strategy will be (as partly indicated in the question): (1) avoiding photon-induced effects at the surface; (2) generating a sufficient flux of O atoms, and (3) understanding (and minimizing) the influence of un-dissociated precursor molecules (*e.g.* NO_2) on the surface chemistry. The latter is a problem as the precursor will freeze out on the cold surface. This is a very helpful experimental suggestion which we will pursue.

Professor Schram addressed Professor Price: In how far can the presence of negatively charged particles arriving at the surface from the O atom source be

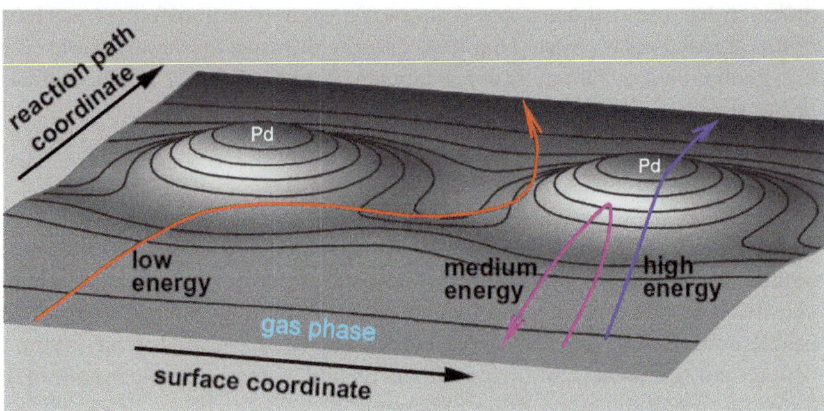

Fig. 8   A sketch of dissociative adsorption pathways for molecular hydrogen on Pd(100) which exhibits a barrier distribution within the surface unit cell. Different reaction pathways all aiming at the same point within the surface unit cell are shown, differing only by their kinetic energy. Detailed balance arguments hold for the association reaction, after Lischka and Groß.[1]

completely avoided? Or, to put it another way, could the presence of (negative) charge at the surface be important in promoting the mobility and the chemistry at the surface?

Professor Price answered: In optimizing the O atom source we checked carefully that no charged species were impinging upon the surface from the source. The gas path from the source to the surface is quite circuitous, and so we rationalized this observation by considering that this long path length from source to surface in a flow of neutral gas allowed any ions to be neutralized. Regarding the role of charge at the surface, I think this would be interesting to investigate, and I know my colleague at UCL (Dr Daren Caruana) has been interested in redox reactivity on model astrochemical surfaces and plans to investigate the effect of surface charge on the reactivity.[1]

1 D. J. Caruana and K. B. Holt, *Phys. Chem. Chem. Phys.*, 2010, **12**, 3072–3079.

Professor Zacharias continued the discussion of the paper by Emanuele Congiu: It is known that the recombination of two hydrogen atoms, even on well-prepared and smooth metal surfaces, is governed by a distribution of various barriers with heights which depend on the local position within the surface unit cell as well as on the spatial orientation of the two atoms with respect to the surface. For a hydrogen molecule approaching in the most favorable orientation onto a Pd(100) surface, there exists at the position on top of a Pd atom a barrier against dissociation about 150 meV in height, while at a bridge position between two Pd atoms the barrier essentially vanishes.[1] Moreover, if slow molecules approach the surface they can be steered by the potential from an unfavorable position into a path where the barrier is easier to overcome.[2,3] Fig. 8 illustrates this behavior. All three reaction pathways depicted are aimed at the same position within the surface unit cell, only with different kinetic energies. Detailed balance arguments then hold for the reverse reaction of associative desorption. For an adequate description of the associative reaction of a diatomic molecule on a surface, a six-dimensional potential is required without even taking the degrees of freedom of the substrate into account. Such potentials can be checked by comparing sticking probabilities with detailed experiments where the internal rovibrational, translational and orientational degrees of freedom of the desorbing molecule are measured, as has been done, for example, for hydrogen desorption from Pd(100),[4] and from Cu(100) and Cu(111).[5,6] The derivation of a single diffusion or desorption barrier for a complex reaction like the formation of ozone will therefore be an objectionable oversimplification, even in the case of smooth metal surfaces. In addition, the studied molecular ice surfaces comprise a multitude of different adsorption sites, all with different binding energies, and each with various barriers against diffusion or associative desorption.

1 M. Lischka and A. Groß, in *Recent Developments in Vacuum Science and Technology*, ed. J. Dabrowski, Research Signpost, Trivandrum, 2003, p. 111.
2 A. Groß, S. Wilke and M. Scheffler, *Phys. Rev. Lett.*, 1995, **75**, 2718.
3 G. J. Kroes, A. Gross, E. J. Baerends, M. Scheffler and D. A. McCormack, *Acc. Chem. Res.*, 2002, **35**, 193.
4 D. Wetzig, M. Rutkowski, H. Zacharias and A. Gross, *Phys. Rev. B*, 2001, **63**, 205412.
5 H. A. Michelsen and D. J. Auerbach, *J. Chem. Phys.*, 1991, **94**, 7502; C. T. Rettner, H. A. Michelsen and D. J. Auerbach, *J. Chem. Phys.*, 1995, **102**, 4625.
6 L. Sementa, M. Wijzenbroek, B. J. van Kolck, M. F. Somers, A. Al-Halabi, H. F. Busnengo, R. A. Olsen, G. J. Kroes, M. Rutkowski, C. Thewes, N. F. Kleimeier and H. Zacharias, *J. Chem. Phys.*, 2013, **138**, 044708.

Dr Congiu responded: You are perfectly right when you say that even a smooth metal surface exhibits a distribution of adsorption sites with various energy depths, let alone a rough surface like those studied in our work. We are well aware that a complex distribution of barrier heights and widths exists, and that is why, though not explicitly stated, we consider the values we provide as effective values of the barrier height and width through which oxygen atoms diffuse. In fact, we believe that the diffusion process of such heavy atoms can proceed only along preferential pathways as a result of the position and spatial orientation of the substrate atoms with respect to the surface. The values of height and width of our simplified rectangular barrier thus give a mean of the separation between the adsorption sites (and their mean depth) used to tunnel through the surface. As you clearly describe in the case of H_2 on Pd(100), the same mechanism might be at play on the surfaces we studied, where low-barrier pathways exist along bridge positions, used by O atoms to diffuse faster on water ice than on amorphous silicate and graphite.

Professor McCoustra communicated: The discussion of the mechanism of diffusion of O atoms on surfaces and in ices has been entertainingly robust. But it is clear that the robustness of the argument perhaps alarms our astronomical and astrophysical colleagues. They are still left with the question: what values should I use in my models? Could I suggest a solution to the senior colleagues engaged in this discussion, Professors Dulieu, Watanabe and Vidali (and perhaps others)? Some years ago our modelling colleagues had a similar issue over TPD data. Different groups looking at the same system got different results. They didn't know which to believe! Our solution was a paper co-authored by my group and that of Professor Brown,[1] that collated all the current data on one system and systematically demonstrated that the variation in our results was far outweighed by the environmental and other uncertainties on the astrophysical models. In essence, we undertook the validation process mentioned by others in this session in relation to databases of rate coefficients for atmospheric chemistry on the TPD of methanol.

1 S. D. Green, A. S. Bolina, R. Chen, M. P. Collings, W. A. Brown and M. R. S. McCoustra, *Mon. Not. R. Astron. Soc.*, 2009, **398**, 357.

Professor Vidali communicated in reply: Certainly, the understanding of the processes of gas–surface interaction in astrochemistry is not at the same level as what is achieved in typical surface science studies on single crystal surfaces. While at the beginning of the study of molecule formation on dust grain analogs, as for example in H_2 formation on silicates,[1] the main emphasis was on formation efficiency, it soon became clear that information on other processes could be obtained from kinetics.[2,3] In order to understand what happens in actual ISM environments, it was necessary to learn about sticking, diffusion, and the partition of the energy released in the reaction between the molecule and the surface, to name a few. These processes are hard to measure under the best of circumstances. For example, the surface science literature on single crystal, low-index surfaces of TiO_2 is very large, and controversy still rages about water adsorption/desorption. Thus, it is not surprising that in more difficult conditions, such as at low surface temperature and on poorly characterized surfaces, there are problems with interpretation. I suggest that it is important to sometimes move to experiments done in well controlled conditions with well characterized surfaces, even if the experiments themselves don't have an immediate astrochemical application. For example, in one study, we compared the adsorption/desorption of D_2 and the formation of molecular hydrogen on a single crystal forsterite surface and on an amorphous silicate surface.[4] As this Discussion has made it clear, the determination of the mobility of oxygen is particularly important in astrochemistry, but eventually a combination of experiments[5,6] and theoretical analysis[7] should sort the issue out. In a way, if this is frustrating to modelers, it is good for science, because it will spur more work by more groups, hopefully to find breakthrough techniques.

1 V. Pirronello *et al.*, *Astrophys. J.*, 1997, **475**, L69.
2 V. Pirronello *et al.*, *Astrophys. J.*, 1997, **483**, L131.
3 M. P. Collings *et al.*, *Mon. Not. R. Astron. Soc.*, 2004, **354**, 1133.
4 J. He *et al.*, *Phys. Chem. Chem. Phys.*, 2011, **13**, 15803.
5 J. He *et al.*, *Phys. Chem. Chem. Phys.*, 2014, **16**, 3493.
6 E. Congiu *et al.*, *Faraday Discuss.*, 2014, **168**, DOI: 10.1039/c4fd00002a.
7 J. He and G. Vidali, *Faraday Discuss.*, 2014, **168**, DOI: 10.1039/c3fd00113j.

Dr McGregor continued the discussion of the paper by Stephen Price by communicating: The paper appears to assume that the double addition of oxygen atoms is to the initial reactant or to an oxirene intermediate. Is it possible that this is a sequential reaction oxidising acrolein (2-propenal) to acrylic acid (which has the correct chemical formula)? In catalysis, under "industrial" as opposed to astrochemical conditions, it is well-established that acrylic acid can be formed through a 2-step oxidation process with propylene (so, a different starting molecule) being oxidised to acrolein, followed by acrolein oxidation to acrylic acid. The second step takes place at a lower temperature than the first.

Professor Price communicated in reply: Our model assumes the reaction is sequential, adding a second oxygen atom to the C_3H_4O product resulting from the first O atom addition to propyne (see eqn 10 and 11 in the paper). Our interpretation of the drop in the relative yield of the double-addition product at surface temperatures above 40 K is that, above this surface temperature, the single-addition product (maybe an oxirene) isomerizes to a less reactive species. Such an interpretation conforms to, at least our, chemical intuition. Your question invites us to consider that the oxirene may isomerize to 2-propenal before the second O atom addition.

As you say, we see that the second addition has a lower barrier than the first – indeed, it appears to be barrierless. Whether the "reactive" single-addition product is the oxirene or 2-propenal is not determined in our experiment, so these mechanistic "deductions" are to some greater or lesser extent speculation. We do however know that the product from the addition of O atoms to ethene is the epoxide, leading to us postulating the oxirene as the initial product from the reaction of O with propyne. Of course, our speculation may be incorrect and the second oxygen atom addition may occur after isomerisation. IR spectroscopy of the species on the surface may prove valuable in elucidating the intermediates in this reaction.

Professor Linnartz addressed Yasuhiro Oba by communicating: The method used in your paper to prepare a rather pure H_2O_2 surface is very elegant, and nicely described. The concept, however, is not new, and was first applied by Cuppen *et al.*[1] Moreover, in Ioppolo *et al.*,[2] the influence of the penetration depth of H atoms was discussed, and this is an effect which, for a pure H_2O_2 surface, may impact on the final yields. A clear isotopic (D) effect is observed that may be more typical for the experiment than for space, and I think one should be careful when using the $HDO : H_2O$ ratios found here for estimations of $D : H$ ratios in molecular clouds.

1 H. M. Cuppen, S. Ioppolo and H. Linnartz, *Phys. Chem. Chem. Phys.*, 2010, **12**, 12077.
2 S. Ioppolo *et al.*, *Phys. Chem. Chem. Phys.*, 2010, **12**, 12065.

Miss Lamberts communicated: In your 2012 paper dealing with the reaction $H_2 + OH$,[1] you point out explicitly which expression you used for determining the effective mass in the case of a bimolecular (linear) atom transfer reaction. The reaction that you are studying in the current paper is of a different type, *i.e.* an atom reacting with a molecule, and seems not to proceed through a linear transition state. Can you comment on the calculation of the reduced mass mentioned in Table 1 in the current paper?

1 Y. Oba, N. Watanabe, T. Hama, K. Kuwahata, H. Hidaka and A. Kouchi, *Astrophys. J.*, 2012, **749**, 67.

Dr Oba communicated in reply: This is a good question. In the case of a bimolecular atom transfer reaction like $H_2 + OH \rightarrow H_2O + H$, the tunnelling mass is represented as the center of the mass transferred from the initial state to the final state on the potential energy surface.[1] In the case of a simple atom-addition reaction, like $CO + H \rightarrow HCO$, the tunnelling mass in the reaction coordinate is simply defined by the reduced mass.[1] So, in the case of $H_2O_2 + H \rightarrow H_2O + OH$, for example, the proposed first step of the reaction is the addition of a H atom to one of the O atoms in H_2O_2, followed by the dissociation of the O–O bond in H_3O_2.[2] Koussa $et\ al.$ proposed that there is a large barrier on the entrance channel (H addition to H_2O_2), but not on the exit channel (O–O bond dissociation). If this is the case, the overall reaction can be regarded as a kind of simple atom-addition reaction, which means that the tunnelling mass can be represented by the reduced mass. In fact, the rate of each reaction is well correlated with the reduced mass (see Table 1 in the paper), while if we calculate the effective mass of each reaction, we did not find any correlation between the effective mass and the obtained reaction rate.

1 H. Hidaka, M. Watanabe, A. Kouchi and N. Watanabe, $Astrophys.\ J.$, 2009, **702**, 291–300.
2 H. Koussa, M. Bahri, N. Jaïdane and Z. B. Lakhdar, $J.\ Mol.\ Struct.:\ THEOCHEM$, 2006, **770**, 149–156.

Dr Hama opened the discussion of the paper by Markus Meuwly by communicating: On page 15 of your paper, you calculated solvation free energies of the atomic oxygen in amorphous ice, and suggested that atomic oxygen thermodynamically prefers to be localized "inside" amorphous ice and is available for chemical reaction. Unlike liquid water, water molecules in amorphous ice are rigid, and the reorientation of water molecules occurs less easily at low temperatures. May I ask how atomic oxygen is solvated in such a situation? For example, can oxygen atoms diffuse into the bulk amorphous water ice? If so, did the reorientation of the water molecules occur during the bulk diffusion, or were the water molecules still rigid? Otherwise, did oxygen atoms simply adsorb on the ice surface?

Professor Meuwly communicated in reply: The motivation to determine the solvation free energy for oxygen atoms in amorphous solid water (ASW) was related to the question of whether solvation is thermodynamically possible in ASW. A favourable solvation free energy does not make any statement about the kinetics or the pathway of atomic oxygen solvation.

The calculated barriers suggest that diffusion of oxygen atoms into bulk ASW should be possible on the nanosecond time scale. To quantitatively answer your question, an explicit simulation would be required.

The reorientational dynamics of the water molecules making up the ASW was not explicitly analyzed in our simulations.

Dr Hama communicated: Although there are only a small number of previous studies on the interactions between oxygen atoms and water ice, Murray and Plane[1] reported two types of adsorption sites for O atoms on crystalline ice, by using a combination of the flow-tube experiment, quantum calculations, and kinetic modeling. They showed that an oxygen atom adsorbs weakly ($E_{ad} = 11$ kJ mol^{-1}) on a crystalline ice surface interacting with a single dangling hydrogen atom, but can also adsorb much more strongly at a small number of disordered sites ($E_{ad} > 35$–40 kJ mol^{-1}). These values are larger than the calculated solvation free energies at low temperature (50 K) shown in Table 1 in your paper. Although I may misunderstand the relationship between the solvation free energy and adsorption energy, can you comment on this difference?

1 B. J. Murray and J. M. C. Plane, *Phys. Chem. Chem. Phys.*, 2003, **5**, 4129.

Professor Meuwly communicated in reply: The binding energy of an O atom on a perfectly crystalline ice surface computed by quantum calculations in the work of Murray and Plane is ~2.6 kcal mol^{-1} (or 11 kJ mol^{-1}). This value could be considered as the binding energy at 0 K. The free energy of adsorption would include not only the binding energy but also other terms, such as entropic contributions from the vibration of the O atom relative to the surface. Therefore, the free energy of adsorption is not the same as binding energy, although they are related.

The computed solvation free energy of atomic oxygen in amorphous ice at 50 K in the present manuscript is in the range of -1.7 to -1.4 kcal mol^{-1}, which is close to the binding energy computed by Murray and Plane.

Considering the differences between the work of Murray and Plane and the present work, *i.e.*, (1) different temperature, (2) different system (crystalline surface *vs.* bulk ASW), (3) different quantity (binding energy *vs.* free energy), it is expected that the two quantities differ.

Dr Hama communicated: In relation to my previous question, since the adsorption energies of H, C, N and O atoms on water ice are still poorly known,[1] they are also still intriguing issues in addition to the diffusion barrier heights of these atoms. Moreover, although modeling studies of interstellar chemistry often assume that the diffusion energy barrier (E_{diff}) is a fraction of the desorption energy (E_{des}), *i.e.*, $E_{\mathrm{diff}} = \alpha E_{\mathrm{des}}$, where α is typically 0.3–0.6, there is no direct evidence so far to support a relationship of $E_{\mathrm{diff}} = \alpha E_{\mathrm{des}}$. Can you calculate E_{des}, and find any correlation between E_{diff} and E_{des}?

1 T. Hama and N. Watanabe, *Chem. Rev.*, 2013, **113**, 8783.

Professor Meuwly communicated in reply: To rigorously address this point, simulations of atomic oxygen on ASW surfaces would need to be carried out. This is certainly possible, along similar lines to those shown in our manuscript for bulk diffusion. The desorption energy can be determined from free energy or umbrella sampling simulations, as done in our contribution. The diffusion energy barrier would probably be most conveniently determined from umbrella sampling simulations between neighboring surface cavities. From such simulations it will be possible to answer the question of whether or not a linear relationship between E_{diff} and E_{des} is meaningful. An important point to remember, however, is that in disordered environments we will always experience distributed barriers.

Professor Vidali continued the discussion of the paper by Stephen Price by communicating: In your paper, you report an estimate of the oxygen adsorption energy on HOPG of 14.0 ± 1.2 kJ mol^{-1}; recently we studied the oxygen mobility on an amorphous silicate and found that the desorption energy is 1764 ± 232 K,[1] which corresponds to 14.6 ± 1.9 kJ mol^{-1}. In that paper we also discuss an estimate of the diffusion energy barrier for O on silicates. These new values of oxygen atom adsorption energy are about twice the value that is typically used in codes of simulation of the chemical evolution of the ISM. Melnick *et al.* searched for O_2 in the Orion bar and concluded that the discrepancy between the model prediction and the observations could be reconciled if the adsorption energy for atomic oxygen was about twice the previously assumed value,[2] thus placing it very close to the value that both your group and mine obtained.

1 J. He *et al.*, *Phys. Chem. Chem. Phys.*, 2014, **16**, 3493.
2 G. J. Melnick *et al.*, *Astrophys. J.*, 2012, **752**, 26.

Professor Price communicated in reply: This agreement is, of course, satisfying. I think the discussion in this session, and others, should reinforce to experimentalists the value for modellers of reliable and reproducible determinations of desorption energies.

Professor Rawlings opened the discussion of the paper by Christopher Arumainayagam: As Eric Herbst has already suggested, chemical modellers like to have simple, reduced quantities to incorporate into their codes. Many of your results, concerning the importance of electron-induced reactions, may be very significant. However, the information that you have presented is in the form of cross-sections and yields, *etc.*

There is a danger that if these results are not presented in a form that is "accessible" to modellers, then they may be ignored by the astrophysical community.

What is required is rate-coefficients – that is to say, integrations of those yields across the nominal relevant electron energy distributions. Is anyone actually doing this and generating something that can be incorporated directly into models?

Professor Arumainayagam responded: For low-energy (<20 eV) electron-induced condensed phase reactions, determining rate constants that can be incorporated into astrochemical models is quite challenging. For example, calculating the ionizing radiation induced energy-dependent electron flux produced within cosmic ices requires Monte Carlo simulations.

Professor Herbst commented: To utilize the experimental results of chemistry occurring on analogs of interstellar ices *via* bombardment by energetic electrons at a variety of energy ranges is not easy. Firstly, for detailed chemical simulations, we need to have rate coefficients or cross sections as functions of electron energies along with penetration depths. Secondly, we have to know something about the flux of electrons as a function of energy striking dust particles. This problem is more difficult than that for photons, since for photons continuous extinction is mainly caused by the interaction with dust particles, while for non-thermal electrons, we have to worry about continuous processes in the gas such as the secondary ionization of neutral species, mainly H_2. In a submitted paper on the abundance of the NH_2 anion in space, we did estimate the energetic electron flux, and how much of the anion can be produced by dissociative attachment.[1]

1 C. M. Persson, M. Hajigholi, A. O. H. Olofsson, J. H. Black, E. Herbst, H. S. P. Muller, J. Cernicharo, E. S. Wirstrom, M. Olberg, A. Hjalmarson, D. C. Lis, H. M. Cuppen and K. M. Menten, *Astron. Astrophys.*, 2014, in press.

Professor Arumainayagam replied: Low-energy (<20 eV) electrons result from the interaction of cosmic rays with gaseous molecular hydrogen present in dark, dense molecular clouds. These low-energy electrons will interact with only the top few surface layers of cosmic ices. Professor Herbst has done an excellent job in describing the challenges faced in quantifying the contributions of these low-energy electrons.

There is another source of low-energy electrons. As mentioned in my response to the previous question, cosmic rays and other ionizing radiation incident upon cosmic ices will generate low-energy (<20 eV) electrons within the bulk as they traverse the ices. Quantifying the contributions from these low-energy electrons requires knowledge of (1) the flux of ionizing radiation incident on cosmic ices, (2) the energy-dependent electron flux produced within the ices by the ionizing radiation traversing the ices, and (3) the energy-dependent cross section for electron-induced molecular fragmentation. Requirement (2) involves Monte Carlo track simulations, similar to those performed by Pimblott and LaVerne.[1]

1 S. M. Pimblott and J. A. LaVerne, *Radiat. Phys. Chem.*, 2007, **76**, 1244.

Dr Woods opened the discussion of the paper by Nigel Mason: To what extent do energetic processes, such as UV/electron irradiation, contribute to the formation of complex organic molecules (COMs) in/on grain-surface ices in space? It appears that methanol, for example, can be formed by the hydrogenation of CO without the need for energetic processing, and several radicals are produced along the way. Is energetic processing (just) a laboratory technique used to "speed up" a mechanism which can take thousands of years in space, or does it significantly contribute to grain-surface complex molecule formation?

Professor Kamp added: As an astronomer, I am worried about incoming photon and electron fluxes. The fluxes produced in the lab cannot be as low as in space. Can you have one photon exciting a molecule on the surface and the next one facilitating a reaction of the excited molecule? How important is this with regards to the enormous spacing of events we expect in space?

Professor Mason answered: These are core questions for the field of astro-chemistry (as well as planetary science and astrobiology). The field of astro-chemistry has had several 'phases' in its history. The first was the discovery of molecules in the ISM (which was unexpected, since at such temperatures and pressures chemistry was felt to be impractical). Then came the recognition that ion molecule chemistry with its low (no) activation barriers may lead to molecular synthesis, but when this chemistry failed to explain H_2 formation, the role of the dust and surface chemistry was introduced. Having accepted that surface chemistry plays a role, the role of hydrogenation and neutral surface reactions have been probed and, as you say, they provide possible pathways to COMs. However, laboratory studies are equally clear that UV/cosmic ray bombardment is an equally effective route to the formation of COMs in an ice mixture. As Professor Linnartz commented, it would be useful to identify molecular species that are formed by one route and not the other, so that we can use these as 'fingerprints' for specific synthetic pathways. Unfortunately to date we have not identified such species. Perhaps nature is not as simple as we would like and in reality all processes play a role, and in different parts of the ISM and in different objects (YSOs, diffuse *vs.* dense regions, shocks) molecular synthesis proceeds by different routes. This is why your models are so important; if we can give you cross sections/rate constants for the different processes, you might be able to determine the relative importance of each process and test the model *vs.* measured concentrations in different regions/objects.

Your question about laboratory experiments and time is also important. Laboratory studies can mimic physical conditions in the ISM, but cannot replicate the time taken. This fact is important, for example, is the morphology of the ice dependent on deposition rate? Again, models can help here.

Finally, the role of electrons. Electron induced chemistry is becoming recognised as fundamental to many phenomena, *e.g.* the radiation damage of DNA and cellular material, and may be exploited by industry, *e.g.* in new generation lithography. I wouldn't claim that electron induced chemistry is the major process in surface chemistry in the ISM *etc.*, but studies show that primary radiation (cosmic rays) liberates tens of thousands of electrons on impact with a surface, and it is these electrons that really drive the chemistry, not the cosmic ray itself. UV photons can of course liberate electrons too (the photoelectric effect), and perhaps ISM dust grains are negatively charged. So, we need to do the experiments and put data in the models to answer your question.

Professor McCoustra commented: In relation to the previous comment on the comparison of laboratory irradiation of ices by photons or charged particles with irradiation in the astrophysical environment, as experimentalists we routinely measure the fluence dependence of the effects induced by irradiation. Those processes that clearly involve single photon or single particle excitations can with significant confidence be scaled directly from laboratory fluences to their astro-physical equivalent.

Professor Mason answered: I agree it is useful (and possibly should be required) to present such experimental data as a function of fluence (rather than, as is often customary, time) so that experiments can be compared. Nevertheless, it is usual to perform experiments to maximise signal/count rates and collect data "quickly". There are few experiments that radiate ice samples for extended periods (hours or even days) with low fluences. Experiments by Field *et al.* in Aarhus reveal that if you irradiate condensed films with very low fluences you see new phenomena (spontelectrics).[1] These effects are masked at high fluences (currents) due to induced charging of the surface. It would therefore be inter-esting to conduct such astrochemically relevant experiments under such low fluence/long time irradiation conditions to test whether there are any changes in the products formed or relative contributions from each product species.

1 D. Field *et al.*, *Int. Rev. Phys. Chem.*, 2013, **32**, 345.

Dr Meijer opened the discussion of the paper by Liv Hornekær: In your paper (and presentation) we do not see a dependence of the rate of hydrogenation on the total number of hydrogen atoms. I would have expected to see such a dependence, if anything, because the number of free carbons will decrease as more carbon atoms are hydrogenated. Can you please give some more detail on this?

Dr Hornekær answered: We had also expected that the addition cross-section would scale with the number of available addition sites, and hence would decrease with increased hydrogenation. However, this is not what we observe. A possible explanation is that the H atoms are very mobile on the polycyclic aromatic hydrocarbon (PAH) molecule, and hence are able to scan the molecule and find the free adsorption sites. Furthermore, barriers to addition are often lower in the vicinity of already adsorbed H atoms, which might mean that these sites would be the most active sites for addition anyway. A similar effect is observed on graphite, where H atoms are extremely mobile and are able to scan the surface and adsorb onto specific sites in the vicinity of already adsorbed H atoms.

Dr Meijer addressed Nigel Mason: You have shown in your paper an efficient way to make molecules that are relevant for astrobiology. However, such molecules (particularly on meteorites) have been shown to have a certain enantiomeric excess. I cannot see how you can introduce chirality using your methodology. Can you please comment on this?

Professor Mason responded: The question of how the molecular enantiomeric excess is created in meteorites and how this may lead to the selectivity in life (amino acids and sugars) remains one of the "great unknowns". Indeed, several people suggest that if we could resolve this question it would reveal how prebiotic chemistry developed and created the biomolecules that are central to life (as we know it), *e.g.* DNA. There are many suggestions, including the role of polarised light or polarised leptons (the Vester–Ulbricht hypothesis), but experiments have yet to recreate large enantiomeric excesses.[1]

I should note that it is possible that, rather than finding routes that produce one enantiomer in excess over another, one could explore how one enantiomer is preferentially destroyed over the other.

1 F. Cataldo *et al.*, *J. Phys.: Conf. Ser.*, 2005, **6**, 139.

Professor McCoustra commented: In relation to the previous question, concerning the origin of the homochirality of life, you do not require a synthetic or destructive process to produce homochirality to end up with a homochiral biochemistry. You just need to generate an initial small enantiomeric excess, and that will be amplified; this even happens stochastically from achiral solutions during crystallisation.[1] The questions then is, how do we generate that initial small enantiomeric excess? Of course, circularly polarised UV and VUV radiation can enhance the destruction rates of specific enantiomers,[2] but the comment that ion and electron irradiation are synthetically achiral is incorrect. Under certain conditions, electrons can be chiral. Two types of chiral electrons exist: spin polarised electrons generated by either scattering of electrons from a magnetised ferromagnetic material (perhaps charged graphite or iron nanoparticles?), or through ionisation by circularly polarised light from transition metals (iron nanoparticles?) and helical electrons which have gained handed cyclotronic motion as they pass through an axial magnetic field. Such electrons have a demonstrated handedness in their interaction with organic molecules.[3]

1 D. K. Kondepudi, R. J. Kaufman and N. Singh, *Science*, 1990, **250**(4983), 975.

2 C. Meinert, S. V. Hoffmann, P. Cassam-Chenaï, A. C. Evans, C. Giri, L. Nahon and U. J. Meierhenrich, *Angew. Chem., Int. Ed.*, 2014, **53**, 210.
3 R. A. Rosenberg, M. Abu Haija and P. J. Ryan, *Phys. Rev. Lett.*, 2008, **101**, 178301.

Professor Mason answered: I agree with these comments. However it is probable that the assembly of larger biomolecules may depend on the environment in which they form, and, at least in the case of evolution of life on Earth, we should not forget role of geology with, for example, ideas of a "clay world" in which chirality arises from the geological scaffold on which they assemble.[1]

1 D. G. Fraser *et al.*, *Phys. Chem. Chem. Phys.*, 2011, **13**, 825.

Dr Jones opened the discussion of the paper by Belén Maté: The hydrogenated amorphous carbon (HAC) samples would appear to be optically thick to the 7.6 eV UV photons used in your experiment. At this energy an optical depth of unity occurs at a depth into the HAC of about 30 nm, independent of the atomic hydrogen fraction (as shown in Fig. 15 of Jones[1]). Thus, most (~90%) of the sample is not UV photo-processed and that is why you appear to observe little evolution under the effects of UV irradiation.

1 A. P. Jones, *Astron. Astrophys.*, 2012, **540**, A2.

Dr Maté replied: We agree with this comment. The penetration depth of the 7.6 eV UV photons in our HAC samples is very low, and this is the reason for the small relative variations of the 3.4 μm band intensity. We are also grateful to Dr Jones for drawing our attention to his article on the optical properties of HAC that provides an estimate of the penetration depth. We would like to take this opportunity to provide some quantitative figures extracted from our experiments that could be relevant to the astrophysical community. We have estimated the number of C–H bonds destroyed per photon. To evaluate this number from our experiment (Fig. 4 in the paper), we have used a value of 1.08×10^{-17} cm per group for the integrated infrared absorption coefficient for the antisymmetric stretching modes of CH_3 and CH_2 groups, averaged from literature values for the absorption coefficients of individual CH_3 and CH_2 groups.[1] We estimate that 5×10^{17} CH bonds cm^{-2} were destroyed with a dose of 3.5×10^{19} photons cm^{-2} (2.6×10^{20} eV cm^{-2}). This means that 69 photons are needed to destroy a C–H bond.

1 E. Dartois *et al.*, *Astron. Astrophys.*, 2004, **423**, 549.

Dr Jones continued the discussion of the paper by Liv Hornekær: Firstly, a comment: super-hydrogenated PAHs are not PAHs because they are not aromatic, but they are polycyclic aliphatic hydrocarbons (PALs). For mixed aromatic/ aliphatic nanoparticles or clusters, we coined the term "arophatic".[1]

Secondly, a question: could the HOPG substrate (under the coronene layer) in the experiments being playing a significant role? For instance, is it possible that the incident H and D atoms are physisorbed onto the HOPG and coronene, and that they only react chemically during the TPD warm-up analysis ($T = 290$–1200 K), *i.e.*, how do you know that the H and D atoms are chemisorbed (on coronene) at the deposition temperature (290 K)?

Maybe I missed this somewhere, but what was the energy/temperature of the incident H and D atoms in the experiments? Again, maybe I missed or misunderstood something, but could this PAH "aliphatisation" processes still operate at typical interstellar grain temperatures, *i.e.*, at $T \sim 20$–30 K, rather than the 290 K substrate temperature used in the experiments?

Have you tried the same experiments on other types of (less adsorptive) substrate and at lower substrate temperatures?

1 E. R. Micelotta *et al.*, *Astrophys. J.*, 2012, **761**, 35.

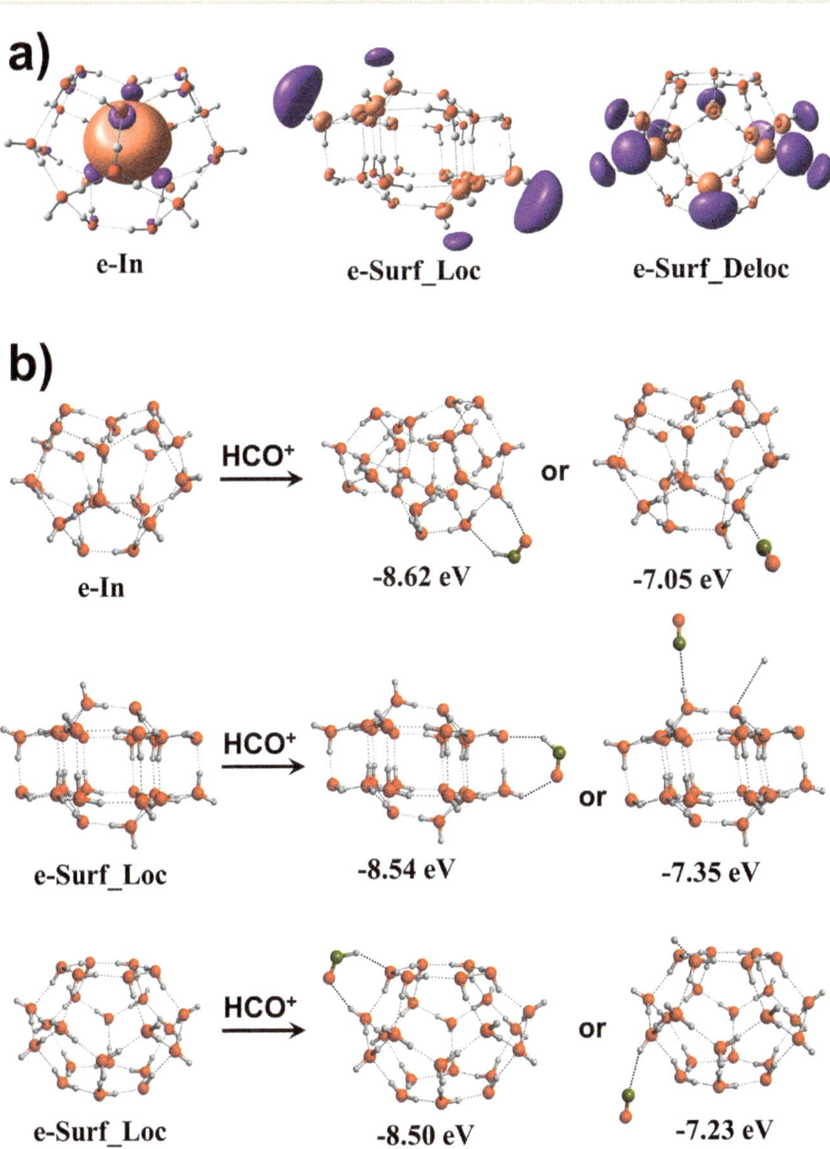

Fig. 9   (a) Singly occupied molecular orbitals (SOMOs) representing the location of the extra electron in negatively charged 24-molecule water clusters. (b) Results of DFT studies on the interaction of HCO$^+$ with these three negatively charged 24-molecule water clusters.

Dr Hornekær replied: In the present discussion paper, 2000 K deuterium atoms were dosed onto a coronene monolayer on a 290 K HOPG substrate. However, measurements taken by us using 500 K D atom beams show similar behaviour with very high degrees of super-hydrogenation of coronene. Furthermore, infrared spectroscopy measurements by Vito Mennella using 300 K H atom beams on a coronene thin film also show super-hydrogenation and abstraction reactions.[1] Also, experiments on a coronene monolayer on an HOPG substrate at 150 K show similar behaviour. It would be surprising if this changes at lower temperatures, as long as one stays above the surface temperature where physisorbed states begin to have significant lifetimes. Hence, I expect that these results are generalizable to PDR conditions. At 290 K, physisorbed H and D atoms on the graphite surface or coronene molecule would have extremely short (ps) lifetimes, and hence would have desorbed from the surface before the heat-up phase under the TPD which takes place several minutes after the original dose. We do, however, see that it is possible to transfer chemisorbed H atoms from the HOPG substrate onto the coronene molecules.[2] However, in this case only fairly low degrees of super-hydrogenation are observed for monolayer coronene coverage. Futhermore, for 500 K H atom beams it is not possible to hydrogenate the HOPG surface, and very high degrees of super-hydrogenation of coronene on the surface are still observed. Finally, in the 300 K H atom beam IR spectroscopy experiments done by Vito Menella, there is no heating step before the measurement and a film of coronene molecules was used, making substrate and heating effects irrelevant. We have not performed similar experiments with PAHs on other substrates – both ice and silicates would be interesting. However, I note that for H atom temperatures of 500 K or lower, chemisorption of hydrogen on graphite is inactive, making this surface pretty much the "least adsorptive" to hydrogen that one could imagine.

1 V. Mennella *et al.*, *Astrophys. J.*, 2012, **745**, L2.
2 J. D. Thrower *et al.*, *Phys. Chem. Chem. Phys.*, 2014, **16**, 3381.

Professor Kamp asked: Previous experiments on coronene cations in the gas phase to study H_2 formation indicate that the first H addition has no barrier, the second one has a barrier and the fourth one has a smaller barrier again.[1] How do you explain your results in comparison to these findings, where you claim that only the first few H additions show a barrier and the bulk of the addition happens barrier free?

1 L. Boschman, G. Reitsma, S. Cazaux, T. Schlathölter, R. Hoekstra, M. Spaans and O. González-Magaña, *Astrophys. J., Lett.*, 2012, **761**, L33.

Dr Hornekær responded: According to our DFT calculations,[1] barriers exist for the addition of the first and third H atom, while the remaining additions are barrierless. It is not surprising that we see the opposite "odd–even" behaviour as observed for the cations, since this presumably reflects the hydrogenation degree needed to form closed shell molecules, with increased stability and binding energy, in the two cases. The calculations indicate that the vanishing subsequent barriers are related to both electronic effects and a geometric deformation of the molecule following the initial hydrogenation steps. Our measurements and kinetic simulations seem to confirm this picture by giving initial low addition cross-sections, which then increase for intermediate hydrogenation degrees.

1 E. Rauls and L. Hornekær, *Astrophys. J.*, 2008, **679**, 531.

Dr Rimola commented: I would like to show you some of our preliminary results related to low-energy electron-induced chemistry, which complement the results presented here by the authors of papers 12, 13 and 14. In particular I would like to show some quantum mechanical results based on DFT methods that demonstrate that electron charge effects can also play a significant role in the chemistry occurring in dense clouds. As is well-known, dust grains can capture free electrons from interstellar gas, the density of which is estimated to be about 10^4–10^5 e m^{-3}.[1,2] A large fraction of these electrons are low-energy secondary electrons produced from the ionization of H_2 caused by the collision of cosmic rays and the incidence of cosmic rays traversing the ices. In the case that these low-energy electrons are mopped up by interstellar water ices, the most immediate consequence is the transfer of the excess energy of the electrons to the ice, resulting in the excitation and the homolytic breakage of water ice molecules. However, we wondered about the fate of this electron once it has released its excess energy. One reasonable possibility is that it remains within the bulk of the water ice in such a way that it becomes a solvated electron, and the water ice particle acts as a negatively charged one. The problem of the solvated (or wet) electron has been a subject of great interest in theoretical chemistry due to its fundamental character, and it has been deeply studied. The water cluster made up of 24 molecules and containing an extra electron is a paradigmatic case of the solvated electron, because it covers different possibilities of where the electron is localized. The electron can be found either inside the cluster cavity, at the surface of the cluster in a localized fashion, or fully delocalized on the surface. These three situations are shown in Fig. 9(a), which depicts the singly occupied molecular orbital (SOMO) representing where the extra electron is. In view of that, we addressed the question of whether this solvated electron can play any role in ice-based astrochemical processes. As a first step, we studied the interaction of HCO^+ with these three negatively charged 24-molecule water clusters, since this interaction leads to the formation of the formyl radical (HCO) through an electron transfer from the negatively charged cluster to HCO^+, *i.e.*, $(H_2O)_{24}^- + HCO^+ \rightarrow (H_2O)_{24} + HCO$. It is worth mentioning that HCO is an important intermediate species in the formation of H_2CO *via* H addition to HCO. However, the interaction of HCO^+ with the negatively charged 24-water cluster can also drive the formation of CO due to H^+ transfer from HCO^+ to the water cluster, *i.e.*, $(H_2O)_{24}^- + HCO^+ \rightarrow (H_3O)(H_2O)_{23} + CO$. Accordingly, we have calculated the reaction energy of both processes. Calculations have been carried out using the BHLYP hybrid density functional employing the 6-31(1+,3+)G* basis set, *i.e.*, the standard Pople 6-31G* basis set with one set of *sp* functions on each O atom and three diffuse *s* functions on each hydrogen, as this was shown by Herbert and Head-Gordon to be accurate enough for geometry optimizations of solvated-electron clusters.[3,4] The results obtained are presented in Fig. 9(b). As one can see, for the three water clusters, the formation of HCO is significantly more favorable than the formation of CO (by about 1.2–1.5 eV). This is an important result, as it points out that negatively charged dust particles can serve as reservoirs of electrons, which can in turn participate in electron transfer processes, driving an unexplored electron-induced chemistry.

The work currently in progress is designed to confirm that the formation of HCO is indeed more favorable than the formation of CO by using other water clusters with an excess of electrons that are well-described in the literature (for instance, consisting of 14 or 20 water molecules), and to improve the accuracy of the calculated reaction energies through single-point energy calculations on the BHLYP-optimized geometries using highly-correlated post-Hartree–Fock methods with extended basis sets.

1 D. A. Williams and T. W. Hartquist, *The Cosmic-Chemical Bond*, RSC Publishing, 2013.
2 A. Evans, *The Dusty Universe*, Ellis Horwood, New York, 1993.
3 J. M. Herbert and M. Head-Gordon, *J. Phys. Chem. A*, 2005, **109**, 5217.
4 J. M. Herbert and M. Head-Gordon, *Phys. Chem. Chem. Phys.*, 2006, **8**, 68.

Dr Fukushima continued the discussion of the paper by Nigel Mason: I would like to comment on the electronic state analysis of anions. The state of an anion is influenced by surrounding charged particles, and careful analysis is required for anions. The difficulty encountered is as follows: in solids and molecules, O^{2-} is stable, whereas in a vacuum, O^{2-} is not observed (although O^- is observed). The conventional discrete variational (DV) method of linear combination of atomic orbitals (LCAO) in density functional theory (DFT) involves the addition of a well potential for generating basis atomic orbitals of electrons on an anion. The anion well potential depth has a large value of -1 Eh, and the anion well radius is 3.5 a.u. (neither of the values are determined theoretically). The conventional DFT method predicts a metallic non-magnetic state for the insulating antiferromagnetic state, contradicting experimental results.

To resolve this contradiction, I developed the surrounding or solid Coulomb-potential-induced well for basis set (SIWB) method for theoretically determining the well values.[1–3] The anion well potential depth is equivalent to the Coulomb potential originating from surrounding nuclei and electron clouds and is shallow, having a value, for example, of -0.34 Eh. The anion well radius is the radius at which the electron charge densities of the anion and cation are equal. The calculated wave function is confirmed by the finite element method (FEM) analysis for a small molecule. The SIWB method reproduces insulating antiferromagnetism.

1 K. Fukushima, *Int. J. Quantum Chem.*, 2012, **112**, 44.
2 K. Fukushima, *J. Comput. Chem., Jpn.*, 2013, **12**, 95.
3 K. Fukushima and H. Sato, *J. Comput. Chem., Jpn.*, 2013, **12**, 197.

Professor Mason replied: I would comment that the formation and role of anions on dust grain surfaces is largely unknown. Most experiments that have explored anion formation on surfaces have been conducted in thin films (or monolayers) on well characterised metallic surfaces. There are few experiments exploring their role and presence in mixed ices and in thicker ice films. At the Open University we are developing an experiment to conduct such experiments.

Dr Ellinger addressed Belén Maté: We also considered the survival of amino acids in the ice under (cosmic) irradiation, as discussed in a previous report[1] and presented in poster P41. Mixed ices were obtained by co-injection, in which the glycine : water ratio could be monitored. Irradiations at 30 K were realized at the SOLEIL synchrotron (TEMPO beam line) coupled with quantum theory simulations. In-flight near edge X-ray absorption fine structure (NEXAFS) spectral analysis at the O1s and N1s K-edges showed that glycine has a limited photo-resistance. However, about 30% of the initial load survives, as in the electron bombardment experiments. The same major fragmentation products CO_2 and CH_3NH_2 were identified, which confirms that this finding is a robust fact. In addition, we also identified secondary fragmentations showing the imine ($H_2C{=}NH$) and nitrile HCN groups. Reversible reactions leading back to glycine are necessary to account for the equilibrium observed in the end. They suggest that the fragments are mostly confined in a cage structure within the ice. Similar results have been obtained for alanine, the next member of the amino acid series.

1 A. Pernet *et al.*, *Astron. Astrophys.*, 2013, **552**, A100.

Dr Maté responded: This is an interesting comment which also leads to experiments and theoretical modelling that we would be eventually willing to perform.

Professor Arumainayagam commented: I agree with what Nigel Mason said. Electron irradiation of ice films causes chemical synthesis as well as chemical destruction. For example, during the low-energy electron irradiation of condensed methanol, at low incident electron fluences (doses), the yields of methoxymethanol and ethylene glycol demonstrate a quadratic dependence on irradiation time (fluence). At higher electron fluences, product yields decrease with increasing fluence, indicating electron-induced decomposition of the products.

Miss Drozdovskaya asked Christopher Arumainayagam: Concerning electron-induced chemistry, in your experiments do you see any dependence on the amount of ice that you have, *i.e.* the number of monolayers present? In the physicochemical models that I work on,[1] we see that the UV flux is very important for methanol chemistry. If I were to include electron-induced chemistry in my models, would it be safe to treat the ice as a whole (*i.e.* a two-phase chemical model) or would I have to treat the surface ice layers separately (*i.e.* a three-phase chemical model or perhaps even a multilayer system)?

1 M. Drozdovskaya *et al.*, submitted.

Professor Arumainayagam replied: In a previous publication,[1] we showed that ethylene glycol is not produced when very thin films of methanol (one monolayer or two monolayers) are irradiated by 55 eV electrons. We attributed this lack of electron-induced reactivity to quenching by the metal surface. This observation, I believe, is not relevant to your studies of cosmic ices. In our current paper, we did most of our experiments with 20 monolayer thick methanol thin films. When we did experiments with five monolayer thick methanol films, we did not observe the formation of two species (glycolic acid and glycerol) that are likely not nascent primary radiolysis products of methanol. When considering interstellar ices, we need to consider three sources of low-energy electrons. The first source is external and is produced by the interaction of cosmic rays with gaseous molecular hydrogen present in the dark, dense molecular clouds. Given that most of these electrons will have energies less than 20 eV, only the top few (surface) layers of cosmic ices will be processed. The second source of low-energy electrons is internal and is produced during the inelastic collisions that the cosmic ray experiences as it traverses through the ices. According to calculations reported in a recent publication,[2] the penetration depth of heavy ions in ices is 10^1–10^4 µm, while that of light ions is 10^1–10^6 µm. Given that interstellar ices are typically less than 1 µm thick, it is clear that the low-energy electrons from this second source will process the bulk ice. The interactions of the cosmic rays with dust grains within the ice mantles could be a third source of low-energy electrons that can interact with interstellar ices. All of the above would also be broadly applicable to gamma rays (from sources such as galactic centers or gamma ray bursts) inter-acting with interstellar ices. Even UV photons with energies above ~10 eV can produce low-energy electrons below the surface of interstellar ice. The bottom line is that low-energy (<20 eV) electron processing of interstellar ices happens both on surfaces as well as in the bulk. It is important to note that the interaction between high-energy radiation (*e.g.*, γ-rays, X-rays, cosmic rays) and matter produces copious numbers (~4 × 10^4 electrons per MeV of energy deposited) of non-thermal secondary low-energy electrons.[3]

1 T. D. Harris, D. H. Lee, M. Q. Blumberg and C. R. Arumainayagam, *J. Phys. Chem.*, 1995, **99**, 9530.
2 A. L. F. de Barros, E. F. da Silveira, S. Pilling, A. Domaracka, H. Rothard and P. Boduch, *Mon. Not. R. Astron. Soc.*, 2014, **438**, 2026.
3 I. G. Kaplan and A. M. Miterev, *Adv. Chem. Phys.*, 1987, **68**, 255.

Professor Mason added: Most experiments exploring astrochemical ices use "thick" ices, that is ices that are many monolayers thick such that the primary radiation incident upon them does not penetrate into the substrate on which the ice is grown. If the primary beam does penetrate into the substrate, secondary electrons may be liberated from the substrate and react with the lower layers of the ice, whereas in the "thick ice" experiments, chemical processing is localised in the upper layers. In order to determine the depth to which the primary particles penetrate we use SRIM and TRIM models. Thus, to answer your question (at least partially), in most experimental cases we expect the chemical processing to occur in the upper layers and we ignore the substrate. This may not be appropriate for the dust grain model, but is a better model for planetary ices (*e.g.*, irradiation of Jovian icy moons by the Jovian magnetosphere).

Professor Linnartz continued the discussion of the paper by Belén Maté: In your experiments you have chosen to use a commercial D_2 lamp. Such a device is very well specified, both in terms of flux and spectral energy distribution (specifically around 160 nm). Most groups use microwave driven H_2 lamps. These are harder to characterize,[1] but have the advantage that they also emit at more energetic Ly-α wavelengths (121.6 nm) – in addition to 160 nm – and consequently these are considered to be better simulators of radiation fields in dark interstellar clouds. I was wondering whether your choice to work with a D_2 lamp may affect your conclusion on the stability of glycine under UV irradiation, both in the laboratory – as wavelength dependent chemical effects have been found, for example in the contribution of Fillion *et al.*[2] – and in the ISM, where the influence of Ly-α may not be negligible.

1 Y.-J. Chen, K.-J. Chuang, G. M. Muñoz Caro, M. Nuevo, C.-C. Chu, T.-S. Yih, W.-H. Ip and C.-Y. R. Wu, *Astrophys. J.*, 2014, **781**, 15.
2 J.-H. Fillion, E. C. Fayolle, X. Michaut, M. Doronin, L. Philippe, J. Rakovsky, C. Romanzin, N. Champion, K. I. Öberg, H. Linnartz and M. Bertin, *Faraday Discuss.*, 2014, **168**, DOI: 10.1039/c3fd00129f.

Dr Maté replied: It is certainly true that the shorter wavelength of the Ly-α emission of the H_2 lamp with respect to the D_2 lamp emission may have different effects on the chemistry taking place in the glycine layers. Several studies on the photostability of glycine under UV radiation, using H_2 or D_2 lamps, have been conducted and were commented on in our work; however it is difficult to conclude anything about the wavelength dependence of the processes due to the different characteristics of the samples. It would be very interesting to investigate wavelength dependent chemical effects on this amino acid.

The proper way to investigate the effects of UV radiation in the ISM would be the consideration of the whole UV component in the interstellar radiation field,[1] which is not a trivial matter. In this respect, investigations based on the radiation of a D_2 lamp (or of any other lamp) will provide only a partial view. We still think that the results are useful and can serve as a guide for future work.

1 J. S. Mathis *et al.*, *Astron. Astrophys.*, 1983, **128**, 212.

Professor Zacharias asked: What is the influence of the penetration depth of the VUV radiation or of the electrons on the measured IR signal? Usually the penetration depth of VUV photons is of the order of 15 to 20 nm, and that of electrons only of the order of 0.5 to about 5 nm, depending on the electron energy.[1] In the present case the films studied show thicknesses up to several hundreds of nm. The IR spectra therefore might measure predominantly molecular layers which are not modified by the radiation.

1 M. P. Seah and W. A. Dench, *Surf. Interf. Anal.*, 1979, **1**, 2.

Dr Maté replied: The HAC materials investigated in this work were optically thick for UV photons and electrons and for that reason the relative variation of the IR band intensity at 3.4 μm is only a few percent. Most of the molecular layers measured have not been affected by the radiation. However, our IR spectra are of high enough quality to allow us to make a quantitative comparison of the effects of the UV photons and the electrons on the HAC sample, which was the main objective of the work. From the CASINO code it was estimated that the penetration depth of 2 keV electrons in HAC is around 100 nm. For the penetration depth of the 7.6 eV UV photons, see the previous comment from Dr Jones.

Professor McCoustra commented: A previous question was asked on whether modellers should consider a two region model in developing grain chemistry models for the formation of complex organic molecules. We have recently published a collaborative paper with colleagues from Catania and Paris comparing the 200 keV proton and 200 eV electron irradiation of acetonitrile (CH_3CN)[1] in which we introduced the concept of the *selvedge*; a region physically and chemically distinct from the bulk which has surface-like properties and reflects the depth of penetration of the incident radiation and the depth of escape of species whose desorption was promoted by the irradiation. The cryogenic proton irradiation experiments resulted in the rapid growth of chemical complexity in the thick films employed in that work. Note however, that the films were not so thick as to stop the proton beam entirely within the film, and hence first order kinetics can be used to analyse the decay of CH_3CN with respect to the ion dose. In stark contrast, during the 200 eV electron irradiation of an ultrathin film (a few layers thick, indeed thin enough to observe exponential decay of the column density reflecting a first order desorption process), all that is observed is desorption; there is no evidence of chemical reactions and the accumulation of reaction products. This may be in part explained by the volatility of the products of the electron irradiation of solid CH_3CN at our somewhat elevated substrate temperature of *ca.* 100 K. Our measurements indicate that both processes occur at similar rates. This can be reconciled if we accept the interpretation that electron-promoted desorption (or indeed any energetically promoted desorption process) is a selvedge process, while the selvedge acts as a cap on the bulk to retain energetically produced reaction products. Realistically therefore, this division should be incorporated into models of the synthesis of complex organic molecules.

Of course, porous and open grain morphologies would add an additional complication, though the basic concept of bulk ice *versus* the selvedge of the ice remains a logical direction to pursue in models.

1 A. Abdulgalil, D. Marchione, M. P. Collings, M. R. S. McCoustra, F. Islam, M. E. Palumbo, E. Congiu and F. Dulieu, *Philos. Trans. R. Soc., A*, 2013, **371**, 20110586.

Professor Dulieu continued the discussion of the paper by Liv Hornekær: The study you present here was conducted at a high temperature for the atoms and at room temperature for the surface. (i) Can you comment on the effect of the temperatures you use? Do you expect to see any change by lowering the surface temperature? (ii) When one H atom is added to a PAH the carbonaceous skeleton is bent and changes shape. I suppose that when abstraction occurs the skeleton relaxes, and in doing so creates mechanical waves, another way of saying that the high internal energy of the PAH is released. I was wondering if this internal energy release could trigger rearrangements in the position of H atoms (secondary abstractions?), and finally if the addition–abstraction cycles could eventually lead to the most stable configurations of the hydrogenated PAH?

Dr Hornekær responded: (i) In the present discussion paper, 2000 K deuterium atoms were dosed onto a coronene monolayer on a 290 K HOPG substrate. However, measurements taken by us using 500 K D atom beams show similar behaviour with very high degrees of super-hydrogenation of coronene. Furthermore, infrared spectroscopy measurements by Vito Mennella using 300 K H atom beams on a coronene thin film also show super-hydrogenation and abstraction reactions.[1] Also, experiments on a coronene monolayer on an HOPG substrate at 150 K show similar behaviour. It would be surprising if this changes at lower temperatures, as long as one stays above the surface temperature where physisorbed states begin to have significant lifetimes.

(ii) We are not able to directly detect such reactions as you propose. However, as mentioned in the discussion paper we do see a loss of coronene molecules from the surface during the H (D) atom exposure. This could be related to the release of energy during H atom abstraction. Furthermore, in STM we do see that the coronene monolayer becomes disordered following hydrogenation, which again indicates that energy is released leading to a re-arrangement of molecules on the surface. With regard to H atom rearrangements and secondary abstraction reactions on the coronene molecule, the only thing we can say is that the binding energies that we observe between the super-hydrogenated coronene molecules and the graphite substrate are smaller than what one would expect based on density functional theory calculations for perhydrocoronene, which has the most stable hydrogenation configuration. They rather match the binding energies found for less stable hydrogenation structures.

1 V. Mennella *et al.*, *Astrophys. J.*, 2012, **745**, L2.

Professor Price asked Nigel Mason: Are the cross sections for ion pair formation following electron impact excitation well known? For photo-excitation the cross sections for ion pair formation are markedly smaller than for ionization; is the same true for excitation by electrons? The data in Fig. 11 in your paper indicate that the ion signals from ion pair formation are comparable to dissociative electron attachment (DEA); is this representative?

Professor Mason responded: Electron induced ion pair formation is perhaps the least studied electron collision process, despite it being known for nearly one hundred years. There are very few quantitative experimental studies revealing branching ratios, and fewer still revealing cross sections. Indeed, there are still very few measurements of absolute cross sections for DEA, a much more widely studied process. Nevertheless, cross sections are low for DEA and ion pair formation (between 10 and 18 or 19 cm^2). Despite these low cross sections, these processes are important in many phenomena where anions play a key role in local chemistry (*e.g.* in many technological plasmas) and they play a key role in Titan's ionosphere, where we recently showed that ion pair formation may account for 20–40% of the product ions.[1] The reason for the anion yields in hydrocarbons often being higher than that of the DEA cross section is that, in hydrocarbons, DEA cross sections are (very) low since neither C or H are electronegative and the temporary negative anion formed as part of the DEA process decays by auto-detachment rather than dissociative electron attachment.

1 E. Szymanksa *et al.*, *Phys. Chem. Chem. Phys.*, 2014, **16**, 3425.

Dr Jäger addressed Liv Hornekær: Le Page *et al.* have modeled the state of hydrogenation of PAHs and claimed that the rates for hydrogenation and dehydrogenation of PAHs are size-dependent.[1] You have used coronene as a model substance for your hydrogenation and H_2 abstraction studies. I wonder how much the size and shape of the PAHs influence the hydrogenation and H_2 abstraction of PAHs? Is the cross section for hydrogenation the same for smaller PAHs, such as naphthalene, as it is for much larger PAHs (hexabenzocoronene and even larger ones)? And how much does the shape of the PAHs affect the hydrogenation and abstraction process? Do PAHs with bay or armchair structures in the periphery show the same hydrogenation cross section as coronene?

1 V. Le Page, T. P. Snow and V. M. Bierbaum, *Astrophys. J.*, 2003, **584**, 316.

Dr Hornekær answered: Calculations show that hydrogenation starts from the edges and then proceeds inwards. Furthermore, the data presented here indicate that hydrogen atoms are very mobile on the PAH molecule and can scan the molecule to find the most preferential addition site. Taken together, these results may suggest that the addition cross section will scale with the geometric area of the PAH, within some upper limit. Eventually, as the PAH approaches graphene, the addition cross section may decrease due to a lack of access to edge sites. Regarding the type of edge site, calculations do show variations in binding energy of different types of site,[1] which will presumably also be reflected in addition barriers and thus cross sections. I expect that abstraction cross sections will only scale by the number of super-hydrogenated sites on the molecule, independent of PAH size and shape.

I believe that the origin of the size dependence in the hydrogenation degree found in the model of Le Page *et al.* was mainly related to UV-induced desorption of hydrogen from the PAHs, which increases for smaller PAHs due to the smaller number of internal degrees of freedom.

1 J. A. Rasmussen, *J. Phys. Chem. A*, 2013, **117**, 4279.

Dr McGregor enquired: The studies reported here use single-phase ices (*e.g.* water only or methanol only). Is this representative of the astrochemical environment? Small amounts of impurities in both liquid and solid-phase materials can have a significant impact on structure (both bulk and surface), and hence change the interaction of adsorbates with ices. How important is this in translating from the lab to "real" environments?

Professor Mason responded: You are quite right that current experiments have tended to use pure, clean ices of known composition, whereas in space the ices are likely to be "dirty" ice, made up of a very inhomogeneous mixture of ice compounds which will strongly influence the chemistry. In defence of the experiments, we are still exploring mechanisms which require well-characterised parameters. Earlier experiments with larger ice mixtures were performed and revealed complex molecular formation (*e.g.* amino acids),[1,2] but such "cooking" experiments could not reveal much information about the routes by which such complex molecules were formed. You are correct that in the future we need to do more "realistic" experiments, but interpretation of the results needs these "simpler", well-characterised experiments to be performed.

1 M. P. Bernstein *et al.*, *Nature*, 2002, **416**, 401.
2 G. M. Muñoz Caro *et al.*, *Nature*, 2002, **416**, 403.

Professor Watanabe addressed Liv Hornekær: I am interested in the influence of steps created by the substrate and edge of PAH molecules on hydrogenation. I wonder if hydrogenation starts at the edge of a PAH molecule rather than the midplane of it. Do you have any suggestion of which part of a PAH is first

Table 1 Electronic excitations induced by photons and charged particle scattering are important progenitors of physics (desorption) and chemistry, except at the highest of cosmic ray energies, where sputtering (nuclear–nuclear collisions) is most important

Progenitor	Mediator	
Photon (UV)	Exciton (1)	
Photon (X-ray)	Exciton (1–100s)	
CR (MeV)	Exciton (100–10 000s)	Sputtering
CR (GeV/TeV)	Exciton (100–10 000s)	Sputtering

hydrogenated? Your sample has a coverage of unity, where PAH molecules compose a well-packed flat PAH surface, in which the edges of PAHs are obscured. You may be able to gain information about this by using coverages of less than unity, or PAH solids where many steps (the edges of the PAHs) are exposed.

Dr Hornekaer responded: According to DFT calculations,[1] hydrogenation of larger PAHs like coronene most likely starts at the edges as you suggest, since the barriers to addition are smaller, and then proceeds inwards, since edge hydrogenation leads to reduced barriers for the hydrogenation of neighbouring inner sites. This is in accordance with IR spectroscopy measurements showing the formation of C–HD groups upon deuteration of thick coronene films. I do not expect that this will be affected by the fact that we use a full monolayer, since coronene molecules do not pack that closely on the surface. Hence, I would expect that for low H atom energies we would see a similar behaviour, while for very high H atom energies (2000 K), we could potentially lose some of this selectivity since it would be possible to hydrogenate centre sites directly. Hopefully we will soon be able to identify addition sites directly *via* low temperature scanning tunnelling microscopy of super-hydrogenated coronene with sub-molecular resolution.

1 E. Rauls and L. Hornekær, *Astrophys. J.*, 2008, **679**, 531.

Professor Watanabe added: We have performed experiments on H atom addition to solid benzene at around 10 K, which was presented by Hama *et al.* in the poster session. Our experiment may be relevant to your experiment as benzene is the elementary unit of PAHs. We found that benzene can be fully hydrogenated immediately to form cyclohexane, and few intermediates were observed. These results indicate that the first hydrogenation reaction is rate-limited, and thus the cross section of cyclohexane formation from benzene can be represented effectively by that of the first hydrogenation. This seems to be consistent with your findings that the cross section for addition does not scale with the number of sites available for addition on the molecule, but rather has a fairly constant value over a large range of super-hydrogenation levels.

Professor Fillion remarked: Our recent experiments on photodesorption have revealed a strong wavelength dependence on the photodesorption rates. We don't know yet whether the photochemistry is also sensitive to the incident photon wavelength. However, in order to be able to compare experiments performed on different set-ups, it seems to me to be crucial that in each experiment, the emission spectrum and photon flux of the source are characterized.[1] This message is addressed to groups that are using a broad-band discharge lamp: please characterize and publish the spectra of your lamps.

1 Y.-J. Chen *et al.*, *Astrophys. J.*, 2014, **781**, 15.

Dr Heays continued the discussion of the paper by Nigel Mason: It seems difficult to uniquely translate even simple laboratory electron/ice experiments into fundamental chemical properties suitable for combination into realistic models of astrophysical ices. Would it be more efficient to combine an approximately realistic electron energy distribution with an interstellar ice-analogue and measure the results directly?

Professor Mason replied: I think you might be proposing that we should irradiate ice-analogues with an electron "beam" with a characteristic energy distribution. There is no simple method for doing this and it is hard to determine what electron energy distribution to use. Furthermore it would still be necessary to know which part of the electron distribution leads to which results.

Professor Linnartz addressed Professor Mason and Professor Arumainayagam: So far, three solid state concepts for molecule formation have been discussed: atom (especially hydrogen) addition reactions, UV irradiation, and electron bombardment. I think that it is generally accepted that (sequential) H-addition is the most efficient way to make molecules like H_2O (*via* $O/O_2/O_3$ + H), NH_3 (*via* N + H), CH_4 (*via* C + H), CH_3OH (*via* CO + H), NH_2OH (*via* NO + H), CO_2 or HCOOH (*via* CO/O_2 + H), and models explaining the observed abundances in space are largely in agreement with this process. The method of formation of complex species, let's say with 8–10 atoms, is ambiguous. The message here is that electron bombarded ices have the same potential to form complex species as UV irradiated ones.[1] The relative importance of both methods depends, amongst other things, on the interstellar fluxes of UV photons compared to that of free electrons, and one of the points to note is that the (relative) numbers describing the underlying elementary processes are missing. Therefore, it would be important to know whether there are clear chemical differences between UV photon or electron induced chemistry, in the ice (specific molecules may have formed upon electron attachment), or in the gas phase upon desorption. So, would there be a way to astronomically discriminate between the two pathways?

1 K. I. Öberg, R. T. Garrod, E. F. van Dishoeck and H. Linnartz, *Astron. Astrophys.*, 2009, **504**, 891.

Professor Arumainayagam answered: Hudson and Moore have demonstrated that the azide radical is produced when molecular nitrogen ices were irradiated by 0.8 MeV protons but not when irradiated by far-UV photons.[1] If methoxy methanol can be conclusively shown not to result from condensed phase methanol photolysis, methoxy methanol could potentially serve as a chemical tracer for the importance of low-energy electrons in the synthesis of complex organics in space.

To determine the relative astrochemical importance of photons and low-energy electrons, these types of experiments should be done in the same UHV chamber using similar protocols and conditions.

1 R. L. Hudson and M. H. Moore, *Astrophys. J.*, 2002, **568**, 1095–1099.

Professor Mason replied: This is a very good and timely comment. We do indeed need to quantify the relative importance of electron/photon (and ion) induced chemistry in astrophysics. Clearly these may vary across astrophysical objects (dense clouds, protostars, shocks *etc.* are all different), however, we must first establish if cross sections differ for electron/photon induced chemistry at the same energy. Measuring such cross sections is not easy (due to, for example, different penetration depths), and to date a standard method for defining cross sections on or in such ices has not been developed.

The suggestion of identifying a route that can distinguish electron from photon induced processes and hence allow them to be distinguished in observations is a good one, but, to date, I cannot think of such a route, though current experiments may reveal one (*e.g.* methoxy methanol).

Professor McCoustra and **Professor Fillion** communicated: We would like to make a general comment about photon- and charged particle-induced physics and chemistry in condensed materials. With photons, the progenitor of the physics and chemistry is clear. It is the photon (actually the electronic excitation or exciton induced by that photon) that mediates the physics or chemistry. But is it? Above the ionisation threshold, we might form excitonic states on resonance with an electronic transition in the material, but we are more likely to generate secondary electrons which, in losing energy by dipole scattering, will generate more electronic and vibrational excitations in the solid.

With cosmic rays, the situation is even more complex. Dipole scattering of the cosmic ray itself will induce electronic excitations at multiple centres on its trajectory; some of these will release secondary electrons, while others may directly induce physics or chemistry. The secondary electrons also undergo dipole scattering, losing energy in quanta equivalent to the electronic and vibrational transitions in the material, and hence producing multiple electronic and vibrational excitations for each electron. Again, these excitations will induce physics or chemistry. The impact of this is summarised in Table 1, and suggests that the enhanced yield of excitations from charged particle processes may overcome the relatively low flux of charged particles in defining the efficacy of charged particle-induced processes. This also hints that overall the non-thermal physics and chemistry observed may, to a large extent, be independent of the nature of the incident particle inducing the excitation. Although, we might expect that electronic excitations (excitons) are likely to be the most significant promoter of non-thermal physics (desorption) and chemistry (chemical transformations) in icy solids, especially in H_2O-rich ices.

Faraday Discussions

PAPER

Concluding remarks: astrochemistry of dust, ice and gas

Eric Herbst*

Received 9th May 2014, Accepted 5th June 2014

DOI: 10.1039/c4fd00104d

In this closing article, we first introduce the topics of dust and ice chemistry and their role in astrochemistry. We then discuss the invited contributions and discussions concerning these topics, dividing the papers into groupings by subject: (i) astronomical sources, (ii) basic properties of dust, (iii) processes on bare grains, (iv) processes on and in ice mantles, and (v) complex organic molecules. A sample of poster contributions is included in the text, when they complement the discussion. The article ends with some suggestions for future research.

1   Introduction

Although the title of this Faraday Discussion suggests that dust, ice and gas are three important environments for chemical reactions in the interstellar medium, the written contributions heavily emphasize the dust and ice environments. Readers more interested in the current state of research in gas-phase interstellar chemistry can look at a number of review articles.[1-4] Dust chemistry occurs on tiny particles of silicates and amorphous carbonaceous material of initial size range 5–250 nm, until coagulation starts to occur during stellar and planetary formation. The dust particles comprise roughly 1% of the mass of an interstellar cloud, and the dust-to-gas ratio by number is approximately 10^{-12}.[5]

The new emphasis on dust chemistry has been long in coming; for much of its existence as a research field, astrochemistry has been dominated by simulations in which the gas played the major role. But this domination began to ebb as it became reasonably clear that gas-phase reactions cannot explain the synthesis of the dominant species that constitute the ice mantles surrounding cold interstellar grains, as well as the partially saturated gas-phase organic molecules found in warm regions known as "hot cores".[6] In the last decade, laboratory studies of chemistry occurring on analogues of both bare grain surfaces and on and in ice mantles have grown in number and stature, although we are still at an early stage of understanding these processes in great detail.

Departments of Chemistry and Astronomy, University of Virginia, Charlottesville, VA 22904, USA. E-mail: eh2ef@virginia.edu; Fax: +1-434-924-3710; Tel: +1-434-243-0535

As brought out during the discussion sections, even the chemistry occurring on bare grain surfaces presents difficulties, and the chemistry on and in ice is even more complex. One problem is distinctly related to astrochemistry: the conditions in interstellar clouds are unlike those in the laboratory, and extrapolations between laboratory and interstellar conditions for surface processes can be much more complex than in the gas-phase. In addition, unlike the situation in the gas-phase, there are three elementary mechanisms by which surface processes are thought to occur, although these pertain most exactly to systems in which adsorbed species constitute less than one full monolayer, as occurs most frequently for chemisorption. With physisorption, on the other hand, multiple monolayers can be built up, as occurs, of course, to produce the ices on cold interstellar grains. In this case, chemistry can occur on the outer surface of the ices, which often cannot be regarded as a simple monolayer, and even inside the ices by a variety of mechanisms.

The three basic mechanisms[7] include the Langmuir–Hinshelwood process, where reactive species diffuse thermally on a surface until meeting each other; the Eley–Rideal process, in which a gas-phase species lands atop or near an adsorbed species and reacts with it; and the hot-atom mechanism, in which a non-thermal species moves a large distance along the surface to react with another species. The non-thermal species can achieve its condition by accreting from a gas higher in temperature than the grain, or as a product of an exothermic chemical reaction occurring on the grain, as mentioned by Dr Cuppen and discussed in the paper by Lamberts *et al.* (DOI: 10.1039/c3fd00136a).

Perhaps the simplest chemical process on interstellar dust grains is the formation of molecular hydrogen from two hydrogen atoms, weakly bound on a bare silicate or carbonaceous grain, as occurs in diffuse interstellar clouds. In such clouds, the gas temperature is 50–100 K, the granular temperature ranges from 15–20 K, and the gas density is 10–100 cm^{-3},† consisting mainly of atomic and molecular hydrogen. The Langmuir–Hinshelwood mechanism occurs within a temperature range governed by the flux of hydrogen atoms hitting and sticking to the grain surface and the mobility of these atoms afterwards.[8] The lower temperature limit is defined by the ability of H atoms to move appreciably, while the upper temperature limit is defined by whether the two H atoms find one another before desorbing back into the gas. The range of temperatures for which the formation of H_2 is efficient can be very narrow.[8,9] Although there is strong evidence from the laboratory studies of Vidali and co-workers that interstellar H_2 is indeed formed by such a mechanism,[8] it is not clear from detailed theoretical studies that the process is sufficiently efficient to produce a significant fraction of the interstellar H_2.[10] Alternatives include a chemisorption-based Eley–Rideal process involving carbonaceous grains. Even the hot-atom mechanism has been invoked to explain some of the data measured for H_2 reactions.[11] The removal of H_2 from the surface following formation can occur *via* simple thermal desorption but perhaps more rapidly by ejection into the gas, which requires only a small fraction of the exothermicity of the H + H reaction, which is over 4.5 eV. The efficiency of this ejection mechanism, sometimes called reactive desorption, is

---

† The units used throughout the paper are those in common use in astronomy, which are mainly cgs units.

unclear for the case of hydrogen formation, although a high efficiency has been measured for several other surface reactions.[12]

Although the chemistry of the formation of molecular hydrogen dominates the bare surface chemistry in diffuse interstellar clouds, other reactions are also possible. The detection of gaseous NH in diffuse clouds has often been ascribed to the recombination of N and H atoms on a grain surface followed by desorption, and more recent suggestions of granular formation following the observation of NH_2 and NH_3 with the Herschel Space Observatory in spiral arm diffuse clouds have also been advanced.[13]

In addition to the surface chemistry that occurs in diffuse clouds, two additional environments have been discussed, both of which occur during the evolutionary stages of star and planetary formation. The first of these occurs on cold dust grains in so-called pre-stellar cold cores found in dense interstellar clouds, which are dense agglomerations formed by gravitational contraction from more diffuse sources. These cores have a temperature of 10 K, appropriate to both the gas and the dust, and a gas density of 10^4 cm^{-3}, which consists overwhelmingly of molecular hydrogen. Here, some molecular hydrogen is still being produced *via* the surface H + H process, but the unique processes occurring on these cold grains are reactions involving atomic hydrogen and other atoms and small molecules that combine, probably *via* a Langmuir–Hinshelwood-type diffusive mechanism, to form icy mantles, dominated by water ice, which can grow to over 100 monolayers of material. The major reactants are atoms and diatoms because they are typically bound less strongly to the bare and ice surfaces and so can diffuse more appreciably at the low temperature of 10 K. A diagram of the various reactions leading to the formation of water ice on grains is shown in a recent review by van Dishoeck *et al.*,[14] and was discussed at the meeting. Other major species formed by surface/ice chemistry and detected *via* infrared absorption include CO_2, H_2CO, CH_3OH, NH_3, CH_4, and OCN^-, the anion presumably attached to a suitable cation. Possible synthetic routes to more complex molecules can be found in a diagram by S. Charnley in the review article by Herbst and van Dishoeck.[6] According to current simulations, CO ice is formed *via* direct accretion from the gas, where it is quite abundant. The line shapes of infrared absorption spectra indicate the ices to be composed of both polar inner layers, dominated by water, and non-polar outer layers, in which a significant amount of CO exists, while CO_2 is found in both layers. The large mantles that form indicate that non-thermal desorption mechanisms, which are needed to desorb molecules that bind more strongly than H_2 and He, are not exceedingly efficient in this environment. Simulations of the nature of the ice indicate it to be rather amorphous with significant porosity at higher densities, as discussed at the meeting. Measurements of ice porosity were discussed in poster P14 by Bossa *et al.*

The second environment to occur sequentially is one that is warming to a temperature of perhaps 300 K, as the initially cold material of a pre-stellar cold core collapses towards a star-in-formation, labelled a protostar or young stellar object (YSO). As the temperature increases from 10 K to 100 K, the surface chemistry changes, as atomic hydrogen can no longer reside on grains long enough to be a major reactant, and the larger molecules formed at 10 K begin to diffuse. With some exceptions, however, these larger moving species are non-reactive, and must be activated by some mechanism, most likely to form reactive radicals. One suggestion[15–17] is that photons, produced *via* the interaction of cosmic rays and hydrogen molecules, can indeed produce radicals by the photodissociation of species throughout the ice mantles. The radicals then recombine to form so-called complex organic molecules, consisting of six or more atoms, which can undergo non-thermal desorption at lower temperatures but eventually desorb thermally and fully into the gas at temperatures exceeding 100 K. Detailed predictions of abundances using this model are rendered difficult by the physical complexity and dynamics of hot-core sources. Nevertheless, as discussed by Öberg *et al.* (DOI: 10.1039/c3fd00146f), analysis of interferometric observations buttressed by single-dish measurements can yield information on the abundances of these complex species as functions of temperature. These observations have been compared with current models supplied by Garrod. Studies of thermal desorption, however, with temperature programmed desorption (TPD) show that even this process on mixed ices is complex (as discussed in poster P08 by Martin-Doménech and Muñoz Caro) and must be treated in some detail in simulations.[18]

As will be discussed in more detail later in this closing paper, there are other forms of energy that can be used to activate granular molecules to become reactive. Perhaps the best studied are interactions between non-thermal electrons and molecules on and in ices.

2 Papers

In the following section, we group papers according to their subject content, and discuss some important aspects of the individual papers and related posters.

Astronomical source types

The paper by Guzmán *et al.* (DOI: 10.1039/c3fd00114h) concerns the famous photon-dominated region (PDR) known as the Horsehead Nebula, which is located in the larger Orion Nebula. The authors announce that, for the first time, complex organic molecules (COMs) have been detected in this region, in particular $HCOOH$, CH_2CO, CH_3CHO, CH_3CCH, and CH_3CN. These molecules can be classified as COMs using the definition advanced by Öberg *et al.* (DOI: 10.1039/c3fd00146f) in their paper presented here, where COMs are defined as "hydrogen-rich organics with three or more heavy atoms." In the older definition of six or more atoms, only acetaldehyde, acetonitrile, and methyl acetylene would be counted as complex. Nevertheless, the discovery of these species is of great interest, since COMs are now being found in a multitude of different sources, suggesting that more than one mechanism is needed to explain their formation. In other sources, however, the major COMs include dimethyl ether (CH_3OCH_3) and methyl formate ($HCOOCH_3$), and these have not yet been detected in PDRs. Interestingly, of the two portions of the PDR studied, the abundances are found to be enhanced in the warm PDR nearer to the edge and richer in UV photons than in the dense cold core. According to the authors, this difference suggests that photodesorption of COMs produced on grains is an efficient mechanism for releasing them into the gas, where they are observed. There is also the possibility that the enhanced photon field is instrumental in the formation of surface radicals, which can then react to form the COMs on grain mantles.[15-17]

Öberg *et al.* (DOI: 10.1039/c3fd00146f) are also interested in the topic of complex organic molecules, but in the more standard environment of hot cores, or more generally, massive young stellar objects (MYSOs). These authors studied three MYSOs using the Submillimeter Array to achieve high spatial resolution, along with single-dish data, and were able to determine how molecular abundances and ratios depend on the temperature, which itself is dependent on the proximity to the forming star. For example, acetaldehyde (CH_3CHO) is typically found in "lukewarm" regions, whereas acetonitrile (CH_3CN) is found in warmer regions. Their results are consistent with theoretical models of grain photolysis in warming regions,[16] in which different processes are likely to occur at different temperatures.

Three papers on the subject of the chemistry that occurs in protoplanetary disks were presented. The paper by Pontoppidan and Blevins (DOI: 10.1039/c3fd00141e) concerns molecular observations and modelling (with radiative transfer) of the inner, planet-forming portion of the protoplanetary disk RNO 90, near to the budding star, which is much warmer than dense clouds and protostellar envelopes. The authors show that the molecular inventory and abundances do not resemble those of the colder regions. The case of CO_2 is particularly strong; this molecule has a particularly low gaseous abundance in the inner disk compared with its high concentration in dense clouds and protostellar envelopes, although here the carbon dioxide is mainly found in ice mantles.

Bergin *et al.* (DOI: 10.1039/c4fd00003j) have looked at the origins of carbon in terrestrial worlds such as our own. It is clear that CO is lower in abundance in meteorites, comets, and the disk TW Hya than in the interstellar medium. The explanation for TW Hya is found in gas-phase reactions starting from the destruction of gaseous CO by reaction with He^+, which, given enough time, can lead to the conversion of CO into more complex organic material under certain conditions; the material can then be incorporated into comets and planets. This type of process can also occur in cold dense clouds, but only after non-physically long periods of time.

No COMs have yet been found in the gas of protoplanetary disks, although these molecules are found in many other sources. Two major reasons for this are that the disks are small, and that their high densities lead to much of the material accreting onto ices or remaining there from an earlier evolutionary stage. The chemistry of these species in disks has been studied by Walsh *et al.* (DOI: 10.1039/c3fd00135k) along an inward accretion flow with a gas-grain chemical model. Two situations were considered: in one the disk is isolated, but in the other, the disk is located near a massive star, the radiation from which affects the disk. In addition, two different initial conditions were used: simple ices and ices containing COMs. The different initial conditions and the proximity of a massive star can affect the balance between the chemistry occurring in the disk and that resulting simply from the initial conditions. Under some conditions, reasonable abundances of gas-phase COMs are predicted, leading to the hope that observations with ALMA will soon result in their detection.

Basic properties of dust

Jones *et al.* (DOI: 10.1039/c3fd00128h) have considered the cycling of carbon in dust particles in the interstellar medium, and deduced that this cycling is rapid in astronomical terms, especially for small hydrocarbon grains. In other words, carbonaceous dust cannot survive indefinitely in the interstellar medium (ISM), and its re-formation must be considered under low density conditions. In particular, the stage from diffuse cloud to outer portions of dense clouds is looked at carefully. In diffuse clouds, the major form of carbon in the outer layers of the dust is likely to be aromatic in nature, whereas the core is more likely to contain aliphatic-rich material. In the outer layers of dense clouds, the amount of aliphatic carbon becomes more dominant, whereas in the inner regions of dense clouds, accreting carbon will form aromatic species once again. Despite the nature of the carbon, it is clear that this model posits layers of carbonaceous material below the ice that forms in the cold core stage. Discussion has brought out the possibility that the carbon can play a role in the formation of complex organic molecules during star formation.

The condensation of tiny dust particles into much larger solid objects such as comets, planetesimals, meteors, and eventually planets is a complex process, with many poorly understood facets. Rouillé *et al.* (DOI: 10.1039/c4fd00010b) have reported experiments to understand aspects of condensation. In their experiments, refractory species accrete onto cold substrates; the process does not mimic condensation from the gas, but represents the growth of dust particles from smaller ones, perhaps those that survived a shock wave or other catastrophic event.

Processes on bare grains

Although much grain chemistry in the ISM occurs on ices, it is still useful to understand reactions on bare grain surfaces, especially if these surfaces are rough or amorphous, *i.e.*, they have a variety of different types of binding sites or a continuous distribution of such sites, so that normal rate equations are inadequate. The use of microscopic kinetic Monte Carlo methods solves the problem of diffusion with different types of sites quite easily, and can also be used with distributions of binding energies,[19] but is very computer intensive and difficult to use in simulations where a large number of different molecules are undergoing reactions.[20] So, the use of a rate equation treatment that treats diffusion properly in the presence of different types of binding sites is a useful advance, and this is what has been accomplished by He and Vidali (DOI: 10.1039/c3fd00113j), whose paper contains their method. These authors are also able to obtain distributions of desorption energies and diffusion barriers from TPD profiles with their approach.

Despite its role as the most basic surface reaction in the universe, even the production of H_2 on bare grain surfaces in diffuse interstellar conditions is still not fully understood, as briefly discussed above. On grains within the normal size range, current microscopic Monte Carlo calculations involving physisorption of atomic hydrogen on flat silicate surfaces indicate that the efficiency of the process is probably insufficient to account for the abundance of molecular hydrogen in diffuse clouds. The basic problem is the estimated temperature of the dust particles (20 K), which is higher than the temperature range of optimum efficiency for silicate grains.[8] The efficiency can be improved with carbonaceous grains.[10] One method of broadening the temperature range of high efficiency for silicate grains is to consider rough or amorphous surfaces,[8,9] in which sites of different energy are used, as in the He and Vidali paper discussed above. However, it is unclear whether the effect of roughness is strong enough to fully account for H_2 formation. Another possibility is to invoke chemisorption, which can be efficient only if the barrier to production of the chemisorption bond is sufficiently low.[10] The utility of chemisorption at dust temperatures relevant to diffuse clouds has been debated for some time now with no clear answer, because the barrier is still not well determined and the role of tunnelling under this barrier is poorly constrained. The formation of molecular hydrogen on crystalline models of forsterite has been studied theoretically by Rimola *et al.* (poster P03), who find both physisorption and chemisorption pathways.

The group of Hornekaer has studied the role of chemisorption for the production of H_2 first on graphite surfaces and subsequently on polycyclic aromatic hydrocarbons (PAHs), where the mechanism appears to be at least partially an Eley–Rideal one, in which H atoms stick to the PAH to make it supersaturated, and subsequent H atoms abstract the H atoms from the PAH into the gas. The paper from this group (Skov *et al.*, DOI: 10.1039/c3fd00151b) contains a discussion of experiments with hot deuterium atoms and the PAH coronene; the results are interpreted in terms of warm interstellar regions such as photon-dominated regions, where the hindering effect of barriers is less constraining. There appears to be little to no barrier for the abstraction step. These interesting results are not directly applicable to diffuse clouds, where H_2 is first produced. The role of PAHs as catalysts for H_2 production was also discussed in poster P05 by Mennella, with similar conclusions to those of Skov *et al.*

Congiu *et al.* (DOI: 10.1039/c4fd00002a) have looked carefully at the diffusive motion of another atom – atomic oxygen – in the laboratory, and found that this atom moves unexpectedly quickly on a variety of surfaces, and so can be an important reactant. The mechanism for the rapid diffusion is thought to be tunnelling, although this suggestion met with some disagreement in the discussion following the introduction of the paper.

Processes on and in ice mantles

Physical processes. The diffusion of atomic oxygen on water ice was found to be even faster than on silicate by Congiu *et al.* (DOI: 10.1039/c4fd00002a), suggesting that the mechanism does not depend entirely upon the structure of the surface, and might indeed be tunnelling. Molecular dynamics simulations have been reported by Lee and Meuwly (DOI: 10.1039/c3fd00160a) on the diffusion of atomic oxygen in amorphous ice at temperatures between 50 K and 100 K. These authors found that atomic oxygen is mobile "to a certain degree", without the need for tunnelling.

The process of photodesorption from ice mantles has been studied in the laboratory for a number of years, especially in Leiden.[20,21,22] Fillion *et al.* (DOI: 10.1039/c3fd00129f) reported the latest laboratory work on the subject regarding CO_2 ice, utilizing the vacuum UV beamline DESIRS at SOLEIL. Their work reveals a complex situation in which both direct photodesorption and photodesorption *via* photodissociation play a role. On CO_2 ice, photodissociation leads to CO, which can then undergo the indirect process of photodesorption, which occurs *via* its excited $a^1\Pi$ state. The direct desorption of CO_2 ice occurs at higher photon energies than the indirect process. One other possible distinction between indirect photodesorption initiated by photodissociation and photodesorption without photodissociation would be the depth dependence of the process originating in the ice mantle.

Chemical processes. Cuppen *et al.* have been exploring the use of microscopic Monte Carlo techniques both for physical and chemical processes involving interstellar grains and mantles for some time now.[9,19,23,24] For their paper here, the group presents a new facet of this approach: a departure from the Markovian, or stochastic, approximation (Lamberts *et al.*, DOI: 10.1039/c3fd00136a). In particular, they look at the non-equilibrium chemical process in which the products of a chemical reaction retain some of the exothermicity of the reaction long enough to move with considerable momentum. This process has an analogue in gas-phase reactions. In the interstellar medium, gaseous reaction products will most likely lose their excess rotational and vibrational energy *via* radiative and collisional processes and, by the time of their next reaction, these modes will be at local thermodynamic equilibrium. On the other hand, translational energy takes longer to relax, and collisional relaxation may well lose out to chemical reaction in the disposal of energy. Other forms of energy that relax very slowly are nuclear spin (*ortho–para* forms) and atomic fine structure. In current gas-phase chemical simulations, the most important molecules with abundances affected by non-thermal considerations are ammonia and the deuterated isotopomers of H_3^+. For ammonia, the reaction between N^+ and H_2 leading to the products $NH^+ + H$ has a small barrier, which, at low temperatures, can be surmounted by excess fine-structure energy of N^+ or excess (non-thermal) energy of the $J = 1$ *ortho* rotational state of H_2.[25,26]

Lamberts *et al.* consider the granular case of $H + HO_2$ leading to two OH radicals, which can then recombine to form hydrogen peroxide, written as HOOH or H_2O_2. The initial reaction is exothermic by 1.5 eV, or 1800 K, so that if a significant fraction of this energy remains in the translation of the OH products for some time, the dynamics of the reaction to form H_2O_2 will certainly change due to the competing effects of directionality and energy dissipation. Lamberts *et al.* constrained the dynamics with experimental results, and used the resulting parameters to constrain the larger problem of the interstellar formation of water ice.

Two other papers presented here are concerned with HOOH, perhaps the simplest internal rotor in the interstellar gas, and how it possibly affects the rather complex synthesis of water ice at low temperatures in the ISM.[27] A recent review of this synthesis is contained in a review article on water in the ISM by van Dishoeck *et al.*[14] Oba *et al.* (DOI: 10.1039/c3fd00112a) studied the reaction between H atoms and H_2O_2 and their deuterated counterparts at 10–30 K using hydrogen peroxide ice. These authors found that, despite an activation energy of 2000 K to form the products H_2O and OH, the reaction occurs appreciably. They also found that the reaction with deuterated reactants is significantly slower, suggesting a tunnelling mechanism. Yet, the reaction rate coefficients decrease with increasing temperature, whereas a simple tunnelling mechanism should show little if any change. Moreover, it is not obvious whether the detailed mechanism involves diffusion or not. The hydrogenation of H_2O_2 with deuterium atoms in a neon matrix was studied by Zins *et al.* (poster P02), and was found to occur slowly.

Parise *et al.* (DOI: 10.1039/c3fd00115f) discussed the observation of H_2O_2 in space, in particular the fact that in the gas phase it is found only in one source – ρ

Oph A. This seems peculiar because H_2O_2 in ice is a key intermediate in the formation of water ice. But it should be noted that although the ice mantle is dominated by water, there is little water in the gas in cold regions. The gas-grain chemical simulations done by these authors indicate that hydrogen peroxide production is very sensitive to temperature, with the ideal temperature for its observation in the gas to be in the 20 K–30 K range. The analysis suggests that much of the gas in ρ Oph A lies in this temperature range, and that this situation is unusual.

What can happen to deuterated water ice bombarded by XUV photons has been studied by Siemer *et al.* (DOI: 10.1039/c3fd00116d) using the free electron laser at Hamburg (FLASH). Photo-ionisation is one possibility, and it can be followed by dissociative recombination, a well-known gas-phase process, to form OD and D, as well as the higher energy channels $OD + D^+ + e$ and $OD^+ + D + e$. The OD radicals can recombine to form DOOD, as also discussed by Lamberts *et al.*, which can then suffer photodissociation and photoionisation.

Complex organic molecules (COMs)

The production of COMs is thought to occur mainly on and in icy grain mantles, although gas-phase processes can also be of importance.[6,15] Although a variety of grain mechanisms have been discussed in the literature, the currently dominant one involves photolysis of warming grains leading to radicals, which recombine to form more complex species.[6,15–17] An interesting variant on this idea was presented here by Rawlings *et al.* (DOI: 10.1039/c3fd00155e), who amplify their earlier theory in which gas-phase three-body association reactions occur in the transitory dense gas surrounding an interstellar dust particle that has just undergone an explosive event, catalysed by an ice mantle explosion caused by the catastrophic recombination of trapped hydrogen atoms and other radicals in the ice. Recent kinetic Monte Carlo calculations of ice mantle chemistry indicate that the radical abundances are higher than previously thought, lending some support to this theory.

With the exception of photon-dominated regions, the recent detection of complex organic molecules in a variety of new sources – cold cores, protostellar envelopes, infrared dark clouds, and circumnuclear disks of active galaxies – is not discussed in detail in any of the invited papers. But this diversity of sources shows that it is likely that more than one mechanism for the formation of complex organic molecules is at play. Several posters reported observations of COMs or discussions of the chemistry leading to complex organic molecules. Jaber *et al.* (poster P19) discussed a census of COMs towards the solar type protostar IRAS 16293-2422. Their observations provide evidence that at least some of the COM spectra arise from cold regions. Fayolle *et al.* (poster P57) have observed high abundances of COMs with respect to methanol in massive young stellar objects that do not possess strong hot core features. A new model for the formation of COMs under low temperature conditions was discussed in poster P68 by Vasyunin *et al.* Their network contains enhanced rates for reactive desorption and radiative association reactions in the gas not previously studied. Observations of complex organics and deuteration in the inner regions of low-mass protostars have been reported by Taquet *et al.* (poster P29). Furuya and Aikawa (poster P37) discuss the role of turbulence in the chemistry of COMs.

The molecule glycolaldehyde has been singled out by several authors. Woods *et al.* (poster P67) discussed theoretical approaches to its formation, while Brown *et al.* (poster P60) reported experimental studies on its thermal and UV processing. Maity *et al.* (DOI: 10.1039/c3fd00121k) discuss the formation of glycolaldehyde in ices exposed to ionizing radiation.

Although COMs usually refer to oxygen-containing and nitrogen-containing molecules, it is also important to point out that the syntheses of species as complex as PAHs can also occur on solid surfaces. The poster by Tian *et al.* (P71) reported that PAHs can be synthesized on silicates and silicon carbide by reactions starting from acetylene. The formation of PAHs in so-called planetary nebulae (PNe) is linked to small hydrocarbon precursors by Guzman-Ramirez *et al.* (poster P01). The gas-phase chemistry of PAH cations and nitrogen-containing analogs was discussed in poster P58 by Demarais *et al.*

One problem with the confirmation of granular hypotheses is that it is hard to detect complex molecules on ice mantles because the infrared spectroscopy needed consists of broad bands and is not very sensitive, even with satellite observations. Measurements are normally made on complex molecules after their post-synthesis desorption into the gas. A new approach has been discussed by Ioppolo *et al.* (DOI: 10.1039/c3fd00154g), in which THz spectra of solid objects have been studied in the laboratory, and the preliminary results do indicate that such spectra might be able to distinguish chemically some aspects of the chemistry that occurs on and in ice matrices.

A variety of processes are thought to occur on ices. At low temperatures, organic molecules can possibly be produced by single atoms diffusing rapidly on 10 K grains and adding chemically to smaller species to produce larger ones.[6] Methanol is formed in this manner, and some larger species might be formed also. Duflot *et al.* (poster P06) recalculated the barrier to the initial reaction in the formation of methanol ice: the hydrogenation of CO by reaction with atomic hydrogen. Subsequent reactions involving H atoms and HCO, H_2CO, and CH_3O (CH_2OH) lead to methanol. Kobayashi *et al.* (poster P17) find evidence for saturation of hydrocarbons on grains by reactions with atomic hydrogen.

The importance of adsorbed O atoms to form water ice in cold sources is well-known:

$$O(ads) + H(ads) \rightarrow OH(ads); OH(ads) + H(ads) \rightarrow H_2O(ads)$$

but atomic oxygen is reactive with species other than hydrogen atoms on ices, including organic molecules. The formation of CO_2 on ices has been a weak link in chemical simulations of ices, although a number of possible simulations, both thermal and requiring activation, have been suggested. Minissale *et al.* (poster P59) reported experimental evidence for the formation of CO_2 *via* the reaction of O and H_2CO. Following earlier low-temperature experiments between O atoms and ethylene to form ethylene oxide, the UCL group (Kimber *et al.*, DOI: 10.1039/c3fd00130j) report a study of the reaction between O atoms and propyne (C_3H_4) to form C_3H_4O. In both cases, the kinetic model for reaction involves a diffusive (Langmuir–Hinshelwood) channel and an Eley–Rideal channel. The kinetic parameters obtained for $O + C_3H_4$ include a small reaction barrier of 160 K for both processes, which, in the diffusive case, probably consists of some sort of chemical barrier plus a diffusion barrier (presumably for the more mobile O atoms), and desorption energies of 2500 K for propyne and 1700 K for O atoms. For the diffusive case, the amount of energy left for the diffusion barrier is unphysically small. In fact, a newly measured value for the diffusive barrier of O on amorphous silicates is 1785 K, which is considerably greater than the standard assumption of a fraction of the desorption energy.[28] Even if this value is somewhat large, it is likely that the process measured by this group does not involve diffusion. This hypothesis can possibly be checked by using sub-monolayer quantities of reactants to see if any role for diffusion can be found. In addition, the unusual temperature dependence might suggest a tunnelling mechanism. In an earlier study of the O + C_2H_4 reaction, the reaction barrier reported (190 K) for the Langmuir–Hinshelwood mechanism is once again lower than the diffusive value for O.

Although the surface reactions between O and organic molecules do not require anything but thermal energy, the syntheses of most organic molecules on grains are expected to be non-thermal in nature, and to require activation of some kind. Photolysis has been studied most extensively; as discussed above, photons, produced indirectly *via* cosmic ray bombardment, activate precursor species on grains such as methanol into radicals and the radicals recombine into larger species. Some laboratory evidence for this mechanism exists.[29]

In this Discussion however, more emphasis has been placed on electron bombardment upon surfaces to produce complex organic molecules in ice mantles. Although effective cross sections for such processes can be high, it is not easy to simulate what is actually happening on grains and ice mantles, despite a number of interesting experiments. Problems include an incomplete knowledge of the flux of non-thermal electrons interacting with the ice mantle of a grain. Energetic electrons can arise from interactions in the gas and in the ice mantle. Non-thermal electrons produced in the gas, however, may relax appreciably before striking grains.[30] For visible and UV photons, the major cause of extinction is interaction with dust grains, since most gas-phase processes are discrete in character to a significant extent. For electrons, on the other hand, inelastic processes in the gas tend to be continuous, so that the electrons that strike dust particles have been significantly relaxed by collisions with gas-phase species such as H, He, and H_2.[30]

For low-energy non-thermal electrons, formation in diffuse regions is mainly by the photoelectric effect between photons and grains. For significantly higher energy electrons, cosmic rays, with energies in the MeV–GeV range, are involved in a variety of ways. Although protons represent most of the cosmic ray flux, electrons are also present at the 1% level, presumably with energies reduced by about a factor of 2000 by the mass factor. Cosmic rays can also produce secondary electrons, both in the gas and as they traverse through grains. The gas-phase secondary electrons initially have perhaps 30 eV of energy. As discussed by Mason *et al.* (DOI: 10.1039/c4fd00004h), cosmic rays interacting with ice-covered dust particles can produce up to 10^4 secondary electrons, with a mean kinetic energy under 20 eV. Simulations indicate that cosmic ray bombardment of grains leads to both thermal and ionization effects, with heavy cosmic rays playing an important role.[31] The thermal effect is utilized as a desorption mechanism in interstellar gas-grain chemical simulations.[32]

So, experiments utilizing a variety of energy ranges for electrons are useful. Of course, it must be mentioned that both photon-irradiation and electron-irradiation experiments performed in the laboratory cannot be compared directly with inter-stellar processes, because the laboratory work utilizes much higher fluxes over much shorter periods of time. So, simulations must be constructed for the laboratory work and, if successful, perhaps carried over into the different physical conditions of the ISM. These simulations will not be facile.

There are a number of papers in this Discussion on electron-induced chemistry, with an introductory work by Mason *et al.*, who give a simple example of the formation of ozone on O_2 ice. Irradiation of the ice by electrons with energy of 10 eV or more dissociates the oxygen to produce atoms, which then recombine with molecular oxygen to form ozone, with the bulk ice acting as the third body. Mason *et al.* also discuss the production of anions on ices *via* dissociative attachment leading to some interesting chemistry.

The interstellar synthesis of pre-biotic molecules *via* low-energy-induced chemistry in ices is discussed in the paper by Boamah *et al.* (DOI: 10.1039/ c3fd00158j), who use condensed methanol as the target. With 7 eV and 20 eV electrons under high-vacuum conditions, these authors produced a number of organic molecules also detected in the interstellar medium and thought to be produced *via* photolysis. The molecules include formaldehyde, dimethyl ether, methyl formate, acetaldehyde, glycolaldehyde, acetic acid, ethanol, and ethylene glycol.

Experiments using keV electrons are reported in papers by Maté *et al.* (DOI: 10.1039/c3fd00132f) and Maity *et al.* (DOI: 10.1039/c3fd00121k) Maté *et al.* are interested in the stability of carbonaceous dust analogues and glycine against both photon and electron bombardment. The authors find that, for glycine, 2 keV electrons are less efficient than both UV photons and MeV protons for its destruction, and caution investigators who use keV electrons to simulate cosmic ray effects, given the low penetration depth of the electrons. Maity *et al.*, on the other hand, are interested in the synthesis of glycolaldehyde from methanol ice and methanol–carbon monoxide ices exposed to 5 keV electron irradiation. Evidence is found for three mechanisms involving radical–radical recombination (similar to the photolysis mechanism suggested in hot cores).

One concludes from a perusal of the papers on electron-induced chemistry that initial chemical simulations will have to make some drastic simplifying assumptions. It would be useful to find signatures to distinguish among thermal reactions, photolysis, and electron-induced processes. For example, the isomeric abundances of complex organic molecules may well betray how the molecules are formed.

The role of gas-phase processes in the chemistry of interstellar COMs and smaller species has not been fully represented in this Faraday Discussion, but some posters have reported exciting new results in this area. de Ruette *et al.* (poster P21) have studied the reaction between C and H_3^+, one of the primary reactions in the initial syntheses of organic molecules in cold clouds. Novotny *et al.* (poster P22) have measured dissociative recombination rate coefficients for the ions HCl^+, D_2Cl^+, HF^+, D_2F^+, and CF^+. Caravan *et al.* (poster P29) have measured enhanced rates of gas phase neutral–neutral reactions involving OH at very low temperatures due to a potential surface with a well in the entrance channel and a subsequent barrier that can be tunnelled under.[33] Harada *et al.* (poster P36) have observed and analysed *via* a gas-grain simulation the chemistry that occurs in the circumnuclear disk at the center of our galaxy. Hincelin *et al.*

(poster P31) have assembled a huge new gas-grain chemical network for deuterated species that takes into account the ortho, para, and meta spin states of simple hydrogen/deuterium species. Quantum chemical calculations have been performed by Zicler *et al.* (poster P12) to understand why HeH$^+$ has not been observed, while Ellinger *et al.* (poster P41) have studied the survival of glycine and alanine both experimentally and theoretically under irradiation. The gas-phase chemistry of carbon chains in dark cloud models has been reviewed by Loison *et al.* in poster P43. The low temperature reactivity of water and of methanol in the gas-phase has been studied by Hickson and Loison (poster P45).

3   Towards the future: what remains to be done?

It is always difficult to predict the future. Both Nigel Mason and I quoted Donald Rumsfeld, who divided knowledge into "known knowns, known unknowns, and unknown unknowns." A prediction of the future is normally a list of known unknowns. But the path of science is in reality more complex. Nevertheless, here is my list of known unknowns, divided into topics within two areas of research: observations/simulations, and laboratory and theoretical treatments of individual processes.

Topics in observations/simulations

(i) To determine the extent of homogeneity in different types of sources.

(ii) To understand the chemistry at high spatial resolution and high sensitivity.

(iii) To understand the chemistry in external galaxies to the extent obtained from single-dish observations of our own galaxy.

(iv) To obtain an increased understanding of the synthesis and distribution of complex organic molecules.

(v) To increase the use of molecular dynamics and kinetic Monte Carlo methods in simulations if needed.

Molecular observations are undergoing a revolutionary transformation because of the construction of a number of new and more powerful telescopes for the observations of molecules throughout much of the known universe. The telescopes are of two types: (i) "space" telescopes, designed to look at frequencies higher than those observable from the ground, so that rotational transitions of light hydrides or low frequency vibrations such as torsional motions can be explored, and (ii) interferometers, which consist of arrays of radio telescopes at excellent ground-based sites acting in a coherent manner. Interferometers have the capability of high sensitivity given their large combined surface area, but also have ultra-high spatial resolution, which enables them to observe the details of small nearby objects in molecular lines such as protoplanetary disks, as well as to see structure in distant galaxies. The space telescopes include the Herschel Space Telescope, which operated in solar orbit at frequencies up to 1910 GHz, until recently exhausting its supply of liquid helium, and SOFIA (Stratospheric Observatory for Infrared Astronomy), which operates from an airplane above most of the atmosphere. These telescopes have enabled astronomers to detect a number of new and exotic molecules and to probe environments not fully probed before. The interferometers include the mother-of-all-such instruments, the so-called ALMA (Atacama Large Millimeter/Submillimeter Array), which is undergoing construction in a high desert site in the Andes. When complete, it will consist of 60 dishes and probe the sky with ultra-high spatial resolution and unprecedented sensitivity.

Keeping up with the advances from such telescopes will keep molecular astronomers quite occupied. Observations from ALMA made even before its completion show that sources treated as nearly homogeneous can turn out to be dramatically inhomogeneous in molecular intensities.[34] The inhomogeneity can be caused by differences in physical conditions, which cause excitation differences, such as large variations in density over small distances, or can be caused by differences in chemistry, which can themselves stem from density and temperature differences as well as diverse time scales. In other words, chemical simulations will have to be much more detailed in their treatment of physical conditions on small distance scales, with three-dimensional hydrodynamic calculations[35] or semi-analytic methods[36] to understand dynamics combined with the chemistry. Such considerations lie behind my inclusion of research areas (i), (ii), and (iii) above. This viewpoint is not universal, with some investigators still preferring simplicity over what might be regarded as excess detail. But, how can simple homogeneous time-independent models obtain correct molecular abundances of different molecules if the molecules are not even totally located in the same regions of space, and if the physical conditions are different, either statically or dynamically? Even the well-studied "homogeneous" cold core source TMC-1 is certainly not homogeneous on scales already studied. The study of so-called "data cubes," consisting of spatial resolution across the sky and Doppler resolution in the direction towards the observer, seems to be the path many will follow.

Topics (iv) and (v) above concern the chemistry of simulations, which still must be improved in several areas, including the pathways leading to complex molecules, especially in new types of sources such as the atmospheres of extra-solar planets, and improvements in how surface chemistry, especially the chemistry occurring on and in ice mantles, is treated. As computers become more powerful, it should be possible to replace approximate methods, such as the use of rate equations, with more exact treatments, such as microscopic stochastic approaches, including the new "off-lattice" approaches mentioned here.[37]

Topics in laboratory and theoretical treatments of individual processes

(i) To obtain absolute rate coefficients for processes on surfaces and in ice mantles with understanding of the mechanism and reasonable agreement among different research groups.

(ii) To simulate laboratory rate experiments to allow subsequent extrapolation to rates under interstellar conditions.

(iii) To understand the differences between processes on surfaces and processes inside ices.

(iv) To include electron-induced processes on grain mantles in simulations.

(v) To improve understanding of poorly known gas-phase reactions, particularly those involving radiative emission.

Astronomical problems and problems in kinetic treatments are not the only difficulties faced when attempting to simulate the chemistry of regions alien to the earth. Some of the processes thought to be of importance in interstellar clouds, for example, cannot easily be studied in the laboratory and require quantum chemical treatments. Others can be studied in the laboratory, but not at temperature regions much lower or higher than room temperature, leading to the problem of whether long temperature extrapolations of experimental work are accurate. An example of the problem in extrapolations to low temperature (topic (ii)) has been highlighted recently with the discovery that the so-called Arrhenius rate law, which typically governs the temperature dependence of rate coefficients for gas-phase reactions with activation energy barriers, can be quite wrong at low temperatures where quantum mechanical tunnelling under barriers can increase the rate coefficient by many orders of magnitude.[33] In collaboration with the Leeds group, we are currently beginning an investigation of how important this effect will be in cold interstellar clouds. Similar laboratory research is also being undertaken in Rennes and Ciudad Real. At present, only the reaction

$$NH_3^+ + H_2 \rightarrow NH_4^+ + H,$$

which occurs with a high rate coefficient at low temperatures because of an entrance channel potential minimum and tunnelling through a small barrier, is included in standard networks to the best of our knowledge.[38]

Perhaps the most difficult problem facing modellers at present is to utilize the fascinating results being obtained on chemical reactions on and in ice mantles, as we have heard during this Discussion. This difficulty underlies topics (i)–(iii). Such utilization requires first that the laboratory scientists understand the mechanism for the reaction being studied in the laboratory. If a process occurring *via* an Eley–Rideal mechanism is instead labelled diffusive in nature, astrochemists will certainly not utilize these results correctly when they incorporate the rate coefficient into their networks. Similarly, if a process thought to be occurring on a flat source is really occurring in pores in the ice, the interstellar transformation will not be successful. In addition, laboratory scientists must understand differences that arise among measurements in different laboratories and must be able to obtain absolute rate coefficients based on well-determined parameters such as diffusion barriers and desorption energies. Finally, a collaboration between laboratory scientists and astrochemists is needed to transfer the laboratory results to interstellar conditions, where physical conditions are likely to be significantly different. In our estimation, the best manner in which to achieve this transfer is to first simulate the results of the laboratory experiments using stochastic treatments if necessary and, in so doing, achieve a sufficiently detailed understanding of the process or processes occurring so that the rate coefficient can be extrapolated to very different conditions. The closest we have come to success in this area appears to be our understanding of the synthesis of water ice.

There are some processes that we understand sufficiently poorly that it is currently exceedingly difficult to incorporate them into chemical simulations. Of these, the most important in the area of ice chemistry concerns electron-induced processes (topic (iv)). A variety of experiments discussed here show that electron bombardment of ices can indeed lead to new chemical pathways both for simple and complex molecule formation, but our detailed understanding of what exactly happens chemically as electrons penetrate ices is less than what is needed. Perhaps it might be possible to incorporate chemical processes and the details of Monte Carlo calculations that can follow the penetration into the ice mantles of cosmic rays and secondary electrons of various energies.

Finally, although much of the emphasis here has been on dust chemistry, there are some critical gas-phase processes that are very poorly understood. Topic (v) concerns our understanding of exotic chemical processes that are based on the emission of radiation. Known as radiative association in the atomic/molecular community and radiative capture in the nuclear physics community, the recombination of two heavy species to form one compound product can occur at low densities only by the emission of radiation. If we consider two molecules A and B, radiative association can be modelled as

$$A + B \leftrightarrow AB^* \rightarrow AB + h\nu$$

where AB^* is referred to as a complex, and sits atop a potential minimum. Depending upon the competition between radiative stabilisation, typically *via* vibrational relaxation, and re-dissociation of the complex, the rate coefficient can vary over 10 orders of magnitude. Very few experimental studies of radiative association have been undertaken, because the process requires very low densities, such as those pertaining to ion traps, or three-body association will dominate, where collision with a third body stabilizes the complex.[39] Moreover, to the best of our knowledge, no experiment has resulted in unambiguous measurement of any emitted radiation. In addition, detailed quantum treatments have not been applied except in rare instances. Simple statistical theories are customarily used; these do show a strong inverse temperature dependence to the rate coefficients, and very strong direct dependencies on the bond energy of the complex and its size, essentially because these parameters determine the complex lifetime.[40] According to a number of studies, radiative association is likely to help in the synthesis of complex organic molecules under cold conditions.[41] An analogous process, radiative attachment:

$$A + e^- \leftrightarrow A^{-*} \rightarrow A^- + h\nu$$

is thought to be a major process leading to the formation of negative molecular ions in cold sources.[42] Yet, there have been only a few studies of this process, and, to the best of our knowledge, no recent ones. Again, simple statistical approximations are used, and, when included in simulations of dense clouds and the AGB star IRC + 10216, seem to be adequate to reproduce the abundances of the larger anions, *e.g.* C_6H^-.[43]

In summary, the future promises to be a time of rapidly increasing knowledge of molecules throughout the universe, which will aid both our understanding of chemistry under exotic conditions, as well as our understanding of the physical conditions of the sources through both molecular spectroscopy and chemical simulations. When my research in this field started in 1972, there were probably fewer practitioners throughout the world than there were in the auditorium where our Discussions were held. The growth since then has been phenomenal. To maintain this growth in astrochemistry, further research in molecular observations, simulations, experiments in both spectroscopy and kinetics, and quantum chemical theory will be needed. As we have seen in this Discussion, astrochemistry is already a vibrant subject in which all of these modes of research are being employed.

Acknowledgements

E. H. wishes to acknowledge the support of the National Science Foundation for his astrochemistry programme. He also acknowledges support from the NASA Exobiology and Evolutionary Biology programme through a subcontract from Rensselaer Polytechnic Institute.

References

1 A. G. G. M. Tielens, *Rev. Mod. Phys.*, 2013, **85**, 1021–1081.

2 E. Herbst, *Phys. Chem. Chem. Phys.*, 2014, **16**, 3344–3359.

3 V. Wakelam, I. W. M. Smith, E. Herbst, J. Troe, W. Geppert, H. Linnartz, K. Oberg, E. Roueff, M. Agundez, P. Pernot, H. M. Cuppen, J. C. Loison and D. Talbi, *Space Sci. Rev.*, 2010, **156**, 13–72.

4 E. Herbst and T. J. Millar, in *Low Temperatures and Cold Molecules*, ed. I. W. M. Smith, Imperial College Press, London, 2008, pp. 1–54.

5 D. C. B. Whittet, *Dust in the Galactic Environment*, Inst. Of Physics, Bristol, 2003.

6 E. Herbst and E. F. van Dishoeck, *Annu. Rev. Astron. Astrophys.*, 2009, **47**, 427–480.

7 K. K. Kolasinski, *Surface Science: Foundations of Catalysis and Nanoscience*, John Wiley & Sons, Ltd, 2012.

8 (*a*) N. Katz, I. Furman, O. Biham, V. Pirronello and G. Vidali, *Astrophys. J.*, 1999, **522**, 305–312; (*b*) For experiments on amorphous silicate, see H. B. Perets, A. Lederhendler, O. Biham, G. Vidali, L. Li, S. Swords, E. Congiu, J. Roser, G. Manicó, J. R. Brucato and V. Pirronello, *Astrophys. J.*, 2007, **661**, L163–L166.

9 H. M. Cuppen and E. Herbst, *Mon. Not. R. Astron. Soc.*, 2005, **361**, 565–576.

10 W. Iqbal, K. Acharyya and E. Herbst, *Astrophys. J.*, 2014, **784**, 139.

11 J. L. Lemaire, G. Vidali, S. Baouche, M. Chehrouri, H. Chaabouni and H. Mokrane, *Astrophys. J.*, 2010, **725**, L156–L160.

12 F. Dulieu, E. Congi, J. Noble, S. Baouche, H. Chaabouni, A. Moudens, M. Minissale and S. Cazaux, *Sci. Rep.*, 2013, **3**, 1338.

13 C. M. Persson, J. H. Black, J. Cernicharo, J. R. Goicoechea, G. E. Hassel, E. Herbst, M. Gerin, M. de Luca, T. A. Bell, A. Coutens, E. Falgarone, P. F. Goldsmith, H. Gupta, M. Kazmierczak, D. C. Lis, B. Mookerjea, D. A. Neufeld, J. Pearson, T. G. Phillips, P. Sonnentrucker, C. Vastel, S. Yu, F. Boulanger, E. Dartois, P. Encrenax, T. R. Geballe, T. Giesen, B. Godard, C. Gry, P. Hennebelle, P. Hily-Blant, C. Joblin, R. Kolos, J. Kerlowski, J. Martin-Pintado, K. Menten, R. Monje, M. Perault, R. Plume, M. Salez, S. Schlemmer, M. Schmidt, D. Tevssier, I. Peron, P. Cais, P. Gaufre, A. Cros, L. Ravera, P. Morris, S. Lord and P. Planesas, *Astron. Astrophys.*, 2010, **521**, L45.

14 E. F. van Dishoeck, E. Herbst and D. A. Neufeld, *Chem. Rev.*, 2013, **113**, 9043–9085.

15 R. T. Garrod and E. Herbst, *Astron. Astrophys.*, 2006, **457**, 927–936.

16 R. T. Garrod, S. L. Widicus Weaver and E. Herbst, *Astrophys. J.*, 2008, **682**, 283–302.

17 J. C. Laas, R. T. Garrod, E. Herbst and S. L. Widicus Weaver, *Astrophys. J.*, 2011, **728**, 71.

18 S. Viti, M. P. Collings, J. W. Dever, M. R. S. McCoustra and D. A. Williams, *Mon. Not. R. Astron. Soc.*, 2004, **354**, 1141–1145.

19 Q. Chang, H. M. Cuppen and E. Herbst, *Astron. Astrophys.*, 2005, **434**, 599–611.

20 Q. Chang and E. Herbst, *Astrophys. J.*, 2012, **759**, 147.

21 K. Öberg, E. F. van Dishoeck and H. Linnartz, *Astron. Astrophys.*, 2009, **496**, 281–293.

22 K. Öberg, H. Linnartz, R. T. Visser and E. F. van Dishoeck, *Astrophys. J.*, 2009, **693**, 1209–1218.

23 H. M. Cuppen, E. F. van Dishoeck, E. Herbst and A. G. G. M. Tielens, *Astron. Astrophys.*, 2009, **508**, 275–287.

24 H. M. Cuppen, L. J. Karssemeijer and T. Lamberts, *Chem. Rev.*, 2013, **113**, 8840–8871.

25 E. T. Galloway and E. Herbst, *Astron. Astrophys.*, 1989, **211**, 413–418.

26 R. Le Gal, P. Hily-Blant, A. Faure, G. Pineau des Forêts, C. Rist and S. Maret, *Astron. Astrophys.*, 2014, **562**, A83.

27 Q. Chang and E. Herbst, *Astrophys. J.*, 2014, **787**, 135.

28 (*a*) D. Jing, J. He, J. R. Brucato, G. Vidali, L. Tozzetti and A. De Sio, *Astrophys. J.*, 2012, **756**, 98; (*b*) See also erratum: D. Jing, J. He, J. R. Brucato, G. Vidali, L. Tozzetti and A. De Sio, *Astrophys. J.*, 2014, **780**, 113.

29 K. I. Öberg, R. T. Garrod, E. F. van Dishoeck and H. Linnartz, *Proc. 2010 NASA Laboraotry Astrophysics Workshop*, ed. D. R. Schultz, 2011, p. C44.

30 A. Dalgarno, M. Yan and W. Liu, *Astrophys. J. Suppl.*, 1999, **125**, 237–256.

31 E. M. Bringa and R. E. Johnson, *Astrophys. J.*, 2004, **603**, 159–164.

32 T. Hasegawa and E. Herbst, *Mon. Not. R. Astron. Soc.*, 1993, **261**, 83–102.

33 R. J. Shannon, R. L. Caravan, M. A. Blitz and D. E. Heard, *Phys. Chem. Chem. Phys.*, 2014, **16**, 3466.

34 N. Sakai, T. Sakai, T. Hirota, Y. Watanabe, C. Ceccarelli, C. Kahane, S. Bottinelli, E. Caux, K. Demyk, C. Vastel, A. S. Coutens, V. Taquet, N. Ohashi, S. Takakuwa, H.-W. Yen, Y. Aikawa and S. Yamamoto, *Nature*, 2014, **507**, 78–80.

35 U. Hincelin, V. Wakelam, B. Commerçon, F. Hersant and S. Guilloteau, *Astrophys. J.*, 2013, **775**, 44.

36 R. Visser, E. F. van Dishoeck, S. D. Doty and C. P. Dullemond, *Astron. Astrophys.*, 2009, **495**, 881–897.

37 L. J. Karssemeijer, S. Ioppolo, M. V. van Hemert, A. van der Avoird, M. A. Alklodi, G. A. Blake and H. M. Cuppen, *Astrophys. J.*, 2014, **781**, 16.

38 E. Herbst, D. J. DeFrees, D. Talbi, F. Pauzat and W. Koch, *J. Chem. Phys.*, 1991, **94**, 7842–7849.

39 D. Gerlich and S. Horning, *Chem. Rev.*, 1992, **92**, 1509–1540.

40 D. R. Bates and E. Herbst, 1988, in *Rate Coefficients in Astrochemistry*, ed. T. J. Millar and D. A. Williams, Kluwer Academic Publishers, pp. 17–40.

41 A. I. Vasyunin and E. Herbst, *Astrophys. J.*, 2013, **769**, 34.

42 T. J. Millar, C. Walsh, M. A. S. Cordiner, R. Ní Chiumin and E. Herbst, *Astrophys. J.*, 2007, **662**, L87–L90.

43 E. Herbst and Y. Osamura, *Astrophys. J.*, 2008, **679**, 1670–1679.

Poster titles

The mixed chemistry problem, **L. Guzman-Ramirez, A. A. Zijlstra, E. Lagadec, K. Gesicki, T. J. Millar and P. M. Woods**, *ESO, Chile*

Elementary processes of hydrogenation of hydrogen peroxide on of interstellar ice grains, **E.-L. Zins, S. Nourry and L. Krim**, *Université Pierre et Marie Curie, France*

Quantum mechanical insights into molecular hydrogen formation on interstellar dust grains, **A. Rimola, J. Navarro-Ruiz, M. Sodupe and P. Ugliengo**, *Universitat Autònoma de Barcelona, Spain*

Synthesis of molecules of astrophysical interest by irradiation of methanol ice by soft X-rays, **A. Ciaravella, C. Cecchi-Pestellini, Y.-J. Chen, G. M. Muñoz Caro and A. Jiménez-Escobar**, *INAF-OAPa, Italy*

H_2 formation on aromatic carbon grains surface, **V. Mennella**, *INAF-Osservatorio Astronomico di Capodimonte, Italy*

A complete theoretical study of the formation of H_2CO and H_3COH at the surface of interstellar grains, **D. Duflot, P. S. Peters, C. Toubin, C. Ceccarelli, C. Kahane and L. Wiesenfeld**, *Université Lille 1, France*

Photon energy effect on the formation of S-bearing products of VUV/EUV irradiation of H_2S-containing ice mixtures, **Y.-J. Chen, K.-J. Juang, M. Nuevo, G. M. Muñoz Caro, C.-C. Chu, T.-S. Yih, C.-Y. R. Wu and W.-H. Ip**, *National Central University, Chinese Taipei*

Thermal desorption of circumstellar and cometary ice analogs, **R. Martín-Doménech, G. M. Muñoz Caro and F. Goesmann**, *Centro de Astrobiología, Spain*

Free-electron laser induced processes in thin molecular ice films, **H. Zacharias, B. Siemer, S. Roling, R. Frigge, T. Hoger and R. Mitzner**, *Westfälische Wilhelms-Universität, Germany*

Fragmentation and reaction pathways of heterocyclic anions of prebiotic relevance, **C. Cole, N. Demarais, T. Snow and V. Bierbaum**, *University of Colorado at Boulder, USA*

Dynamics of dissociative electron–molecule interactions in condensed methanol, **M. Boamah, M. Boyer, A. Bass, C. Arumainayagam, K. Sullivan, M. Bazin and L. Sanche**, *Wellesley College, USA*

About the non-observation of HeH^+, **E. Zicler, O. Parisel and Y. Ellinger**, *Laboratoire de Chimie Theorique, France*

The nature of energy barriers of the HCN→CNH isomerization. The reaction force perspective, **S. Gutiérrez-Oliva, S. Díaz-Acosta and A. Toro-Labbé**, *Pontificia Universidad Católica de Chile, Chile*

Porosity measurements of interstellar ice mixtures, **J.-B. Bossa, K. Isokoski, D. Paardekooper, M. Bonnin, E. P. van der Linden, T. Triemstra, S. Cazaux, A. G. G. M. Tielens and H. Linnartz**, *Leiden University, Netherlands*

Heterogeneous catalysis in extraterrestrial environments: the role of coke, **J. McGregor**, *University of Sheffield, UK*

Electron loss of one-electron projectiles by N_2, O_2 and H_2O molecules, **G. M. Sigaud**, *Pontifícia Universidade Católica do Rio de Janeiro (PUC-Rio), Brazil*

Hydrogen addition reactions of C_2H_2 on cold grains: clue to the formation mechanism of cometary C_2H_6, **H. Kobayashi, N. Watanabe, H. Hidaka, T. Hama, Y. Watanabe and H. Kawakita**, *Kyoto Sangyo University, Japan*

Dynamics of CO in amorphous water ice environments, **L. Karssemeijer, S. Ioppolo, M. van Hemert, A. van der Avoird, M. Allodi, G. Blake and H. Cuppen**, *Radboud University Nijmegen, Netherlands*

The census of complex organic molecules in the solar type protostar IRAS16293-2422, **A. Al-Edhari, C. Ceccarelli, C. Kahane and E. Caux**, *IPAG CNRS, France*

Mass-analytical tool for reactions in interstellar ices (MATRIICES) **D. Paardekooper, J.-B. Bossa, K. Isokoski and H. Linnartz**, *Leiden University, Netherlands*

Merged beam studies for astrobiology, **N. de Ruette, K. A. Miller, A. P. O'Connor, J. Stützel, X. Urbain and D. W. Savin**, *Columbia Astrophysics Laboratory, USA*

Dissociative recombination measurements on halogen-hydride ions relevant for astrochemistry, **O. Novotny, A. Becker, H. Buhr, W. Geppert, M. Hamberg, C. Krantz, H. Kreckel, D. Schwalm, K. Spruck, J. Stützel, A. Wolf, B. Yang and D. W. Savin**, *Columbia Astrophysics Laboratory, USA*

The interaction between neutral PAHs and atomic hydrogen, **A. L. Skov, J. D. Thrower, P. A. Jensen, H. Lemaître, E. E. Petersen-Friis, B. Jørgensen and L. Hornekær**, *Aarhus University, Denmark*

Surface formation routes of HDOin interstellar ice analogues, **T. Lamberts, X. de Vries, G. Fedoseev, S. Ioppolo, H. Cuppen and H. Linnartz**, *Leiden University, Netherlands*

Diffusion-limited reactivity in interstellar ices, **P. Ghesquiere, J. Noble, P. Theulé, F. Mispelaer, F. Duvernay, G. Danger, T. Mineva, D. Talbi and T. Chiavassa**, *CNRS/UM2 LUPM, France*

Formation of superhydrogenated PAHs through interaction with atomic hydrogen and hydrogenated carbonaceous grains, **J. D. Thrower, E. E. Friis, A. L. Skov, B. Jørgensen, L. Nilsson, S. Baouche, R. Balog and L. Hornekær**, *Westfälische Wilhelms-Universität Münster, Germany*

Surface chemistry on interstellar grains: the role of oxygen, **G. Vidali and J. He**, *Syracuse University, USA*

[PAH] concentration dependent photochemistry in interstellar ices, **S. Cuylle, L. Allamandola and H. Linnartz**, *Leiden University, Netherlands*

Multilayer formation and evaporation of deuterated ices in prestellar and protostellar cores, **V. Taquet, S. Charnley and O. Sipila**, *NPP/NASA GSFC, USA*

Complex organics and deuteration in the inner regions of low-mass protostars, **V. Taquet, A. López-Sepulcre, C. Ceccarelli, R. Neri and S. Charnley**, *NPP/NASA GSFC, USA*

Modeling of the early phases of star formation: deuterium chemistry, **U. Hincelin, E. Herbst, K. Furuya, Y. Aikawa and B. Commerçon**, *University of Virginia, USA*

Chemical mechanisms operating at very low temperatures to enhance the rates of gas phase reactions relevant to astrochemical environments, **R. Caravan, R. Shannon, M. Blitz and D. Heard**, *University of Leeds, UK*

Methanol along the path from envelope to protoplanetary disk, **M. Drozdovskaya, C. Walsh, R. Visser, D. Harsono and E. van Dishoeck**, *Leiden Observatory, Netherlands*

Catalysis on warm phyllosilicate grains in the solar nebula, **M. D'Angelo, I. Kamp and P. Rudolf**, *University of Groningen, Netherlands*

Variations of the 3.3 μm feature in AKARI IRC spectra, **M. Hammonds, T. Mori, F. Usui and T. Onaka**, *University of Tokyo, Japan*

Molecules in the circumnuclear disk of the galactic center, **N. Harada, D. Riquelme, S. Viti, K. Menten, M. Requena-Torres, R. Guesten and S. Hochguertel**, *Max Planck Institute for Radio Astronomy, Germany*

Reprocessing of ices in turbulent protoplanetary disks, **K. Furuya and Y. Aikawa**, *Kobe University, Japan*

Long time-scale simulations of ice surfaces, **K. T. Wikfeldt, V. Ásgeirsson and H. Jónsson**, *University of Iceland, Iceland*

Cyclohexane formation following hydrogenation of amorphous solid benzene at 20 K, **T. Hama, H. Ueta, A. Kouchi and N. Watanabe**, *Hokkaido University, Japan*

Synthesis of cosmic dust analogues, **V. L. Frankland, A. James and J. M. C. Plane**, *University of Leeds, UK*

Survival of COM in ices under cosmic irradiation: examples of glycine and alanine, **Y. Ellinger, F. Pauzat, J. Pilmé, G. Marloie, A. Pernet, M. Lattelais, C. Laffon and P. Parent**, *UPMC University Paris 06, France*

The thermal reaction of acrylonitrile and oxygen radicals on analogues of interstellar dust grains, **H. Kimber and S. Pricen**, *University College London, UK*

The gas-phase chemistry of carbon chains in dark cloud chemical models, **J.-C. Loison, V. Wakelam, K. M. Hickson, A. Bergeat and R. Mereau**, *CNRS - Université de Bordeaux, France*

Solid state chemistry of nitrogen oxides: surface consumption of NO, **M. Minissale, G. Fedoseev, E. Congiu, S. Ioppolo, F. Dulieu and H. Linnartz**, *University of Cergy-Pontoise, France*

The low temperature reactivity of water and methanol in the gas-phase, **K. M. Hickson and J.-C. Loison**, *Institut des Sciences Moleculaires, France*

Theoretical study of CH_4 + H reaction on surface of methane ice, **G. Jumet and N. Vaeck**, *Université Libre de Bruxelles, Belgium*

Adsorption of simple and complex organic molecules on interstellar grains, **F. Pauzat, Y. Ellinger and M. Lattelais**, *UPMC, France*

Ionisation and dissociation of PAH cations in the FELICE-FTICR mass spectrometer, **A. Petrignani, J. R. Eyler, A. F. G. van der Meer, J. Oomens, B. Redlich, M. Vala and A. G. G. M. Tielens**, *Leiden Observatory, Netherlands*

The impact of chemical networks and ice adsorption energies on protoplanetary disk models and the interpretation of their observations, **I. Kamp, W.-F. Thi, P. Woitke and C. Rab**, *Kapteyn Astronomical Institute, Netherlands*

Effect of the initial elemental abundances on radiation thermo-chemical models of protoplanetary discs, **C. Rab, P. Woitke, I. Kamp, W.-F. Thi and M. Güdel**, *University of Vienna, Austria*

CO ice mixed with CH_3OH: the answer to the non-detection of the 2152 cm^{-1} band?, **E. Penteado, A. Boogert and H. Cuppen**, *Radboud University Nijmegen, Netherlands*

Herschel Planetary Nebula Survey (HerPlaNS): first detection of OH^+ in planetary nebulae, **I. Aleman, T. Ueta, D. Ladjal, K. M. Exter, J. Kastner, R. Montez, A. G. G. M. Tielens, Y.-H. Chu, H. Izumiura, I. McDonald, R. Sahai, N. Siódmiak, R. Szczerba and P. A. M. van Hoof**, *Leiden University, Netherlands*

Femtosecond laser-induced desorption of atomic and molecular hydrogen isotopes from graphite, **R. Frigge, J. D. Thrower, H. Witte, Henrik and H. Zacharias**, *University of Münster, Germany*

A curiously line-free massive core, **C. Cyganowski, C. Brogan, K. Öberg, T. Hunter and D. Graninger**, *University of St Andrews/CfA, UK*

Ammonia radiolysis: an interstellar source of nitrogen, **C. Arumainayagam, K. Sullivan, M. Boamah, K. Shulenberger, S. Chapman, K. Atkinson and M. Boyer**, *Wellesley College, USA*

Organic molecules in high mass protostars: finding an ice–gas connection, **E. Fayolle, K. Öberg, R. Garrod, E. van Dishoeck and S. Bisschop**, *Harvard-Smithsonian Center for Astrophysics, USA*

Gas-phase chemistry of carbonaceous cations relevant to the interstellar medium: PA(N)H$^+$ and CF$^+$, **N. Demarais, Z. Yang, T. Snow and V. Bierbaum,** *University of Colorado, Boulder, USA*

Experimental evidence of CO_2 formation *via* the H_2CO + O reaction, **F. Dulieu, M. Minissale, E. Congiu, S. Baouche, H. Chaabouni and A. Moudens,** *LERMA, Cergy Pontoise University and Paris Observatory, France*

Experimental studies of the thermal and ultra-violet processing of glycolaldehdye and its isomers in model interstellar ices, **W. Brown, D. Burke, F. Puletti, P. Woods, S. Viti and B. Slater,** *University of Sussex, UK*

Solid state chemistry of nitrogen oxides: surface consumption of NO_2, **S. Ioppolo, G. Fedoseev, M. Minissale, E. Congiu, F. Dulieu and H. Linnartz,** *California Institute of Technology, USA*

Unveiling the surface structure of amorphous solid water *via* selective infrared irradiation of OH stretching modes, **J. Noble, C. Martin, H. Fraser, P. Roubin and S. Coussan,** *Aix-Marseille Université, France*

High resolution spectroscopy of neutral acenes at 3 μm, **E. Maltseva, A. Candian, A. Petrignani, J. Oomens, A. G. G. M. Tielens and W. J. Buma,** *Leiden Observatory, Netherlands*

RAIRS investigation of dipole alignment in spontelectric nitrous oxide (N_2O) layers, **A. Rosu-Finsen, J. Lasne, M. P. Collings, M. R. S. McCoustra, D. Field, A. Cassidy, O. Plekan, J. D. Thrower, R. Balog and N. Jones,** *Heriot-Watt University, UK*

Hydrogen bonding and interaction of benzene with icy films, **D. Marchione, M. P. Collings and M. R. S. McCoustra,** *Heriot-Watt University, UK*

In situ MALDI mass spectrometric studies of interstellar ice analogs: understanding plume dynamics, **B. Henderson and M. Gudipati,** *NASA, USA*

A multidisciplinary approach to understanding the formation of glycoladehyde in space, **P. Woods, S. Viti, Z. Raza, B. Slater, D. Burke and W. Brown,** *Queen's University Belfast, UK*

Complex organic molecules in cold clouds: an extended model, **A. Vasyunin, P. Caselli, E. Herbst and S. Cazaux,** *University of Leeds, UK*

Nitrogen isotope fractionation by photodissociation, **A. Heays, R. Visser, R. Gredel, W. Ubachs, B. Lewis, S. Gibson and E. van Dishoeck,** *Leiden University/ Sterrewacht, Netherlands*

A non-equilibrium *ortho*-to-*para* ratio of H_2O in the Orion PDR, **F. van der Tak, Y. Choi, E. Bergin and R. Plume,** *SRON, Netherlands*

Catalytic conversion of acetylene to polycyclic aromatic hydrocarbons (PAHs) over particles of olivine, pyroxene, and forsterite-type silicates and silicon carbide, **A. Cheung, M. Tian, B. Liu, M. Hammonds, T. Zhao and P. Sarre,** *University of Hong Kong, China*

On the role of substitutional grain defects in the adsorption of interstellar hydrogen, **A. J. H. M. Meijer and B. J. Irving,** *University of Sheffield, UK*

Atomic oxygen diffusion relevant to water formation in amorphous interstellar ices, **M. Meuwly and M. W. Lee,** *University of Basel, Switzerland*

Discerning ionization and deuterium fractionation in DM Tau protoplanetary disk: comparison of observational and computational approaches, **R. Teague, D. Semenov, S. Guilloteau, T. Henning and the CID Team,** *Max Planck Institute for Astronomy, Germany*

Experimental setup for recognition of chiral metal clusters, **K. Lange, B. Visser, D. Neuwirth, J. Eckhard and U Heiz,** *Technische Universität München, Germany*

Reflectron time-of-flight mass spectroscopic detection of complex organic molecules in CH_3OH and CH_3OH–CO ices exposed to ionization radiation, **S. Maity, R. Kaiser and B. M. Jones,** *University of Hawai'i at Manoa, USA*

N_2 photodissociation and chemistry in AGB stars, **X. Li, T. J. Millar, C. Walsh, A. N. Heays and E. F. van Dishoeck,** *Leiden University, Netherlands*

The Skinner Prize for the best poster was jointly awarded to Miss Rebecca Caravan from the University of Leeds, UK, for her poster on chemical mechanisms operating at very low temperatures to enhance the rates of gas phase reactions relevant to astrochemical environments, and Mr Robert Frigge of the University of Münster, Germany, for his poster on femtosecond laser-induced desorption of atomic and molecular hydrogen isotopes from graphite.

List of participants

Mr Ali Al-Edhari, *IPAG CNRS, France*
Dr Isabel Aleman, *Leiden University, Netherlands*
Professor Chris Arumainayagam, *Wellesley College, United States*
Professor Edwin Bergin, *University of Michigan, United States*
Dr Will Bergius, *Royal Society of Chemistry, United Kingdom*
Miss Mavis Boamah, *Wellesley College, United States*
Dr Jean-Baptiste Bossa, *Leiden University, Netherlands*
Miss Catherine Brown, *Royal Society of Chemistry, United Kingdom*
Professor Wendy Brown, *University of Sussex, United Kingdom*
Dr Alessandra Candian, *Leiden Observatory, Netherlands*
Ms Rebecca Caravan, *University of Leeds, United Kingdom*
Dr Yu-Jung Chen, *Department of Physics, National Central University, Chinese Taipei*
Professor Allan Cheung, *University of Hong Kong, Hong Kong*
Ms Yunhee Choi, *Kapteyn Astronomical Institute & SRON, Netherlands*
Mr Ko-Ju Chuang, *Leiden Observatory, Netherlands*
Dr Angela Ciaravella, *INAF-OAPA, Italy*
Ms Callie Cole, *University of Colorado at Boulder, USA*
Dr Emanuele Congiu, *Université de Cergy-Pontoise, France*
Dr Herma Cuppen, *Radboud University Nijmegen, Netherlands*
Mr Steven Cuylle, *Leiden University, Netherlands*
Dr Claudia Cyganowski, *University of St Andrews/CfA, United Kingdom*
Miss Martina D'Angelo, *University of Groningen, Netherlands*
Dr Nathalie de Ruette, *Columbia Astrophysics Laboratory/Columbia University, United States*
Mr Nicholas Demarais, *University of Colorado, Boulder, United States*
Mr Mikhail Doronin, *Université Pierre et Marie Curie - LERMA2, France*
Miss Maria Drozdovskaya, *Leiden Observatory, Netherlands*
Dr Denis Duflot, *Université Lille 1, Sciences et Technologies, France*
Professor François Dulieu, *LERMA, Cergy Pontoise University and Paris Observatory, France*
Dr Rob Eagling, *Royal Society of Chemistry, United Kingdom*
Dr Yves Ellinger, *UPMC Univ. Paris 06, France*
Dr Edith Fayolle, *Harvard-Smithsonian Center for Astrophysics, United States*
Mr Gleb Fedoseev, *Leiden Observatory/Leiden University, Netherlands*
Ms Siyi Feng, *Max-Planck Institute for Astronomy, Germany*
Professor Jean-Hugues Fillion, *Université Pierre et Marie Curie, France*
Dr Victoria Frankand, *University of Leeds, United Kingdom*
Mr Robert Frigge, *University of Münster, Germany, Germany*
Dr Kimichika Fukushima, *Advanced Reactor System Engineering Department, Toshiba Nuclear Engineering Service Corporation, Japan*
Mr Kenji Furuya, *Kobe University, Department of Earth and Planetary Sciences, Japan*
Dr Robin Garrod, *Cornell University, USA*
Dr Wolf Geppert, *Stockholm University, Sweden*

Mrs Maryvonne Gerin, *LERMA, France*
Mr Pierre Ghesquiere, *CNRS/UM2 LUPM, France*
Miss Berare Göktürk, *Boğaziçi University, Turkey*
Professor Soledad Gutiérrez-Oliva, *Pontificia Universidad Católica de Chile, Chile*
Dr Viviana Guzmán Veloso, *Harvard-Smithsonian Center for Astrophysics, United States*
Dr Lizette Guzman-Ramirez, *ESO, Chile*
Dr Tetsuya Hama, *Institute of Low Temperature Science, Hokkaido University, Japan*
Dr Mark Hammonds, *University of Tokyo, Japan*
Dr Nanase Harada, *Max Planck Institute for Radio Astronomy, Germany*
Professor Dwayne Heard, *University of Leeds, United Kingdom*
Dr Alan Heays, *Leiden University/Sterrewacht, Netherlands*
Dr Bryana Henderson, *NASA, United States*
Professor Eric Herbst, *University of Virginia, United States*
Dr Kevin Hickson, *Institut des Sciences Moleculaires, France*
Dr Ugo Hincelin, *University of Virginia, USA*
Dr Liv Hornekær, *Aarhus University, Denmark*
Dr Sergio Ioppolo, *California Institute of Technology, USA*
Dr Cornelia Jäger, *Max Planck Institute for Astronomy, Heidelberg, Germany, Germany*
Dr Anthony Jones, *Institut d'Astrophysique Spatiale, France*
Mr Brant Jones, *University of Hawaii, USA*
Mr Guillaume Jumet, *Université Libre de Bruxelles, Belgium*
Professor R Kaiser, *University of Hawaii, USA*
Dr Mihkel Kama, *Leiden Observatory, Netherlands*
Professor Inga Kamp, *Kapteyn Astronomical Institute, Netherlands*
Mr Leendertjan Karssemeijer, *Theoretical Chemistry, Institute for Molecules and Materials, Radboud University Nijmegen, Netherlands*
Ms Helen Kimber, *UCL, United Kingdom*
Dr Hitomi Kobayashi, *Kyoto Sangyo University, Japan*
Dr Fred Lahuis, *SRON Netherlands Institute for Space Research, Netherlands*
Miss Thanja Lamberts, *Leiden University, Netherlands*
Miss Kathrin Lange, *Technical University Munich, Germany*
Mr Yuandi Li, *Royal Society of Chemistry, United Kingdom*
Mr Xiaohu Li, *Leiden Observatory, Netherlands*
Mr Niels Ligterink, *Leiden University, Netherlands*
Professor Harold Linnartz, *Leiden Observatory, Netherlands*
Dr Jean-Christophe Loison, *CNRS -Université de Bordeaux, France*
Dr Robert Loughnane, *UNAM-CRyA, Mexico*
Mr Demian Marchione, *Heriott-Watt University, UK*
Mr Rafael Martín Doménech, *Centro de Astrobiología, Spain*
Professor Nigel Mason, *The Open University, United Kingdom*
Dr Belén Maté, *Instituto de Estructura de la Materia, IEM-CSIC, Spain*
Dr June McCombie, *University of Nottingham, United Kingdom*
Professor Martin McCoustra, *Heriot-Watt University, Scotland*
Dr James McGregor, *University of Sheffield, United Kingdom*
Dr Anthonius Meijer, *University of Sheffield, United Kingdom*
Dr Vito Mennella, *INAF-Osservatorio Astronomico di Capodimonte, Italy*
Professor Markus Meuwly, *University of Basel, Switzerland*
Professor Tom Millar, *QUB, United Kingdom*
Mr Marco Minissale, *University of Cergy-Pontoise, France*
Professor Andrew Mount, *The University of Edinburgh, United Kingdom*
Dr Jennifer Noble, *Aix-Marseille Université, France*

Dr Yasuhiro Oba, *Institute of Low Temperature Science, Hokkaido University, Japan*
Dr Karin Öberg, *Harvard University, United States*
Mr Daniel Paardekooper, *Leiden University, Netherlands*
Dr Bérengère Parise, *Cardiff University, United Kingdom*
Dr Françoise Pauzat, *UPMC, France*
Mr Eduardo Penteado, *Institute for Molecules and Materials - Radboud University Nijmegen, Netherlands*
Dr Magnus Persson, *Leiden Observatory, Netherlands*
Dr Annemieke Petrignani, *Leiden Observatory, Netherlands*
Professor John Plane, *University of Leeds, United Kingdom*
Dr Klaus Pontoppidan, *Space Telescope Science Institute, United States*
Professor Stephen Price, *UCL, United Kingdom*
Mr Christian Rab, *University of Vienna - Department of Astrophysics, Austria*
Professor Jonathan Rawlings, *Department of Physics & Astronomy, UCL, UK*
Dr Albert Rimola, *Universitat Autònoma de Barcelona, Spain*
Mr Alexander Rosu-Finsen, *Heriot-Watt University, United Kingdom*
Mr Roy Scheidsbach, *Radboud University Nijmegen, Netherlands*
Professor Daniel Schram, *Eindhoven University of Technology, Netherlands*
Dr Dmitry Semenov, *Max Planck Institute for Astronomy, Germany*
Miss Katherine Shulenberger, *Wellesley College, United States*
Mr Anders Lind Skov, *Department of Physics and Astronomy, Aarhus University, Denmark*
Ms Lei Song, *Radboud University Nijmegen, Netherlands*
Miss Claire Springett, *Royal Society of Chemistry, United Kingdom*
Mr Tushar Suhasaria, *WWU Münster, Germany*
Miss Dahbia Talbi, *LUPM/CNRS UM2, France*
Dr Vianney Taquet, *NPP/NASA GSFC, United States*
Mr Richard Teague, *Max-Planck-Institut für Astronomie, Germany*
Dr Patrice Theule, *Aix-Marseille University, France*
Mrs Rachel Thompson, *Royal Society of Chemistry, United Kingdom*
Dr John Thrower, *Westfälische Wilhelms-Universität Münster, Germany*
Professor Alejandro Toro-Labbé, *Pontificia Universidad Católica de Chile, Chile*
Professor Floris van der Tak, *SRON, Netherlands*
Professor Ewine van Dishoeck, *Leiden Observatory, Netherlands*
Dr Anton Vasyunin, *University of Leeds, United Kingdom*
Professor Gianfranco Vidali, *Syracuse University, United States*
Dr Catherine Walsh, *Leiden Observatory, Netherlands*
Professor Naoki Watanabe, *Institute of Low Temperature Science, Hokkaido University, Japan*
Dr Kjartan Thor Wikfeldt, *University of Iceland, Iceland*
Dr Paul Woods, *Queen's University Belfast, United Kingdom*
Mr Helmut Zacharias, *WWU Münster Physikalisches Institut, Germany*
Dr Daniel Zaleski, *Newcastle University, United Kingdom*
Miss Eleonore Zicler, *Laboratoire de Chimie Theorique, France*
Dr Emilie-Laure Zins, *Université Pierre et Marie Curie, France*